VISCOUS FLUID FLOW

McGraw-Hill Series in Mechanical Engineering

Alciatore and Histand: *Introduction to Mechatronics and Measurement Systems*
Anderson: *Computational Fluid Dynamics: The Basics with Applications*
Anderson: *Fundamentals of Aerodynamics*
Anderson: *Introduction to Flight*
Anderson: *Modern Compressible Flow*
Barber: *Intermediate Mechanics of Materials*
Beer/Johnston: *Vector Mechanics for Engineers*
Beer/Johnston/DeWolf: *Mechanics of Materials*
Budynas: *Advanced Strength and Applied Stress Analysis*
Cengel and Boles: *Thermodynamics: An Engineering Approach*
Cengel and Turner: *Fundamentals of Thermal-Fluid Sciences*
Cengel: *Heat Transfer: A Practical Approach*
Cengel: *Introduction to Thermodynamics & Heat Transfer*
Crespo da Silva: *Intermediate Dynamics*
Dieter: *Engineering Design: A Materials & Processing Approach*
Dieter: *Mechanical Metallurgy*
Doebelin: *Measurement Systems: Application & Design*
Dunn: *Measurement & Data Analysis for Engineering & Science*
EDS, Inc.: *I-DEAS Student Guide*
Hamrock/Schmid/Jacobson: *Fundamentals of Machine Elements*
Heywood: *Internal Combustion Engine Fundamentals*
Holman: *Experimental Methods for Engineers*
Holman: *Heat Transfer*
Hutton: *Fundamentals of Finite Element Analysis*
Kays/Crawford/Weigand: *Convective Heat and Mass Transfer*
Meirovitch: *Fundamentals of Vibrations*
Norton: *Design of Machinery*
Palm: *System Dynamics*
Reddy: *An Introduction to Finite Element Method*
Schaffer et al.: *The Science and Design of Engineering Materials*
Schey: *Introduction to Manufacturing Processes*
Shames: *Mechanics of Fluids*
Shigley/Mischke/Budynas: *Mechanical Engineering Design*
Smith: *Foundations of Materials Science and Engineering*
Suryanarayana and Arici: *Design and Simulation of Thermal Systems*
Turns: *An Introduction to Combustion: Concepts and Applications*
Ugural: *Mechanical Design: An Integrated Approach*
Ullman: *The Mechanical Design Process*
White: *Fluid Mechanics*
White: *Viscous Fluid Flow*
Zeid: *Mastering CAD/CAM*

VISCOUS FLUID FLOW

Third Edition

Frank M. White
University of Rhode Island

 **Higher Education**

Boston   Burr Ridge, IL   Dubuque, IA   Madison, WI   New York
San Francisco   St. Louis   Bangkok   Bogotá   Caracas   Kuala Lumpur
Lisbon   London   Madrid   Mexico City   Milan   Montreal   New Delhi
Santiago   Seoul   Singapore   Sydney   Taipei   Toronto

Higher Education

VISCOUS FLUID FLOW, THIRD EDITION

Published by McGraw-Hill, a business unit of The McGraw-Hill Companies, Inc., 1221 Avenue of the Americas, New York, NY 10020. Copyright © 2006, 1991, 1974 by The McGraw-Hill Companies, Inc. All rights reserved. No part of this publication may be reproduced or distributed in any form or by any means, or stored in a database or retrieval system, without the prior written consent of The McGraw-Hill Companies, Inc., including, but not limited to, in any network or other electronic storage or transmission, or broadcast for distance learning.

Some ancillaries, including electronic and print components, may not be available to customers outside the United States.

This book is printed on acid-free paper.

4 5 6 7 8 9 0 DOC/DOC 1 0 9 8 7 6 5 4 3 2 1

ISBN: 978-0-07-240231-5
MHID: 0-07-240231-8

Senior Sponsoring Editor: *Suzanne Jeans*
Developmental Editor: *Amanda J. Green*
Senior Marketing Manager: *Mary K. Kittell*
Project Manager: *Peggy S. Lucas*
Senior Production Supervisor: *Sherry L. Kane*
Media Technology Producer: *Eric A. Weber*
Designer: *Laurie B. Janssen*
Cover Design: *Lisa Gravunder*
(USE) Cover Image: Cover photo: The wake of a cylinder at Re = 35, by C.-Y. Wen and C. Y. Lin. Originally published in the *Physics of Fluids,* vol. 13, March 2001, pp. 557—558.
Supplement Producer: *Brenda A. Ernzen*
Compositor: *International Typesetting and Composition*
Typeface: *10/12 Times Roman*
Printer: *R. R. Donnelley Crawfordsville, IN*

Library of Congress Cataloging-in-Publication Data

White, Frank M.
    Viscous fluid flow/Frank M. White.—3rd ed.
       p. cm.—(McGraw-Hill series in mechanical engineering)
    Includes bibliographical references and index.
    ISBN 0–07–240231–8 (alk paper)
    1. Viscous flow. I. Title. II. Series.

QA929.W48 2006
532'.0533   dc22   2004058182

www.mhhe.com

ABOUT THE AUTHOR

Frank M. White is Professor Emeritus of Mechanical and Ocean Engineering at the University of Rhode Island. He is a native of Augusta, Georgia, and went to undergraduate school at Georgia Tech, receiving a B.M.E. degree in 1954. Then he attended the Massachusetts Institute of Technology for an S.M. degree in 1956, returning to Georgia Tech to earn a Ph.D. degree in mechanical engineering in 1959. He began teaching aerospace engineering at Georgia Tech in 1957 and moved to the University of Rhode Island in 1964. He retired in January 1998.

At the University of Rhode Island, Frank became interested in oceanographic and coastal flow problems and in 1966 helped found the first Department of Ocean Engineering in the United States. His research interests have mainly been in viscous flow and convection heat transfer. Known primarily as a teacher and writer, he received the ASEE Westinghouse Teaching Excellence Award in addition to seven University of Rhode Island teaching awards. His modest research accomplishments include some 80 technical papers and reports, the ASME Lewis F. Moody Research Award in 1973, and the ASME Fluids Engineering Award in 1991. He is a Fellow of the ASME and for 12 years served as editor-in-chief of the ASME *Journal of Fluids Engineering*. He received a Distinguished Alumnus award from Georgia Tech in 1990 and was elected to the Academy of Distinguished Georgia Tech Alumni in 1994.

In addition to the present text, he has written three undergraduate textbooks: *Fluid Mechanics, Heat Transfer*, and *Heat and Mass Transfer*. He continues to serve on the ASME Publications Committee and has been a consulting editor of the McGraw-Hill Encyclopedia of Science and Technology since 1992. He lives with his wife, Jeanne, in Narragansett, Rhode Island.

My wife, Jeanne Faucher White, is the key to this book.
Without her love and encouragement,
I can't even get started.

CONTENTS

Preface   xiii
List of Symbols   xvii

1   Preliminary Concepts   1

1-1   Historical Outline 1
1-2   Some Examples of Viscous-Flow Phenomena 4
1-3   Properties of a Fluid 15
1-4   Boundary Conditions for Viscous-Flow Problems 45
       Summary 54
       Problems 55

2   Fundamental Equations of Compressible Viscous Flow   59

2-1   Introduction 59
2-2   Classification of The Fundamental Equations 59
2-3   Conservation of Mass: The Equation of Continuity 60
2-4   Conservation of Momentum: The Navier–Stokes Equations.......................62
2-5   The Energy Equation (First Law of Thermodynamics)................................69
2-6   Boundary Conditions for Viscous Heat-Conducting Flow...........................74
2-7   Orthogonal Coordinate Systems 75
2-8   Mathematical Character of The Basic Equations 77
2-9   Dimensionless Parameters in Viscous Flow 81
2-10   Vorticity Considerations in Incompressible Viscous Flow 84
2-11   Two-Dimensional Considerations: The Stream Function 86
2-12   Noninertial Coordinate Systems 88
2-13   Control-Volume Formulations 89
       Summary 92
       Problems 92

3 Solutions of The Newtonian Viscous-Flow Equations — 96

- 3-1 Introduction and Classification of Solutions 96
- 3-2 Couette Flows Due to Moving Surfaces 98
- 3-3 Poiseuille Flow through Ducts 106
- 3-4 Unsteady Duct Flows 125
- 3-5 Unsteady Flows with Moving Boundaries 129
- 3-6 Asymptotic Suction Flows 135
- 3-7 Wind-Driven Flows: The Ekman Drift 141
- 3-8 Similarity Solutions 144
- 3-9 Low Reynolds Number: Linearized Creeping Motion 165
- 3-10 Computational Fluid Dynamics 183
  - Summary 205
  - Problems 205

4 Laminar Boundary Layers — 215

- 4-1 Introduction 215
- 4-2 Laminar Boundary-Layer Equations 225
- 4-3 Similarity Solutions for Steady Two-Dimensional Flow 230
- 4-4 Free-Shear Flows 251
- 4-5 Other Analytic Two-Dimensional Solutions 257
- 4-6 Approximate Integral Methods 261
- 4-7 Digital-Computer Solutions 271
- 4-8 Thermal-Boundary-Layer Calculations 278
- 4-9 Flow in the Inlet of Ducts 287
- 4-10 Rotationally Symmetric Boundary Layers 290
- 4-11 Asymptotic Expansions and Triple-Deck Theory 300
- 4-12 Three-Dimensional Laminar Boundary Layers 307
- 4-13 Unsteady Boundary Layers: Separation Anxiety 318
- 4-14 Free-Convection Boundary Layers 321
  - Summary 328
  - Problems 328

5 The Stability of Laminar Flows — 337

- 5-1 Introduction: The Concept of Small-Disturbance Stability .. 337
- 5-2 Linearized Stability of Parallel Viscous Flows 344
- 5-3 Parametric Effects in the Linear Stability Theory 357
- 5-4 Transition to Turbulence 370
- 5-5 Engineering Prediction of Transition 378
  - Summary 394
  - Problems 394

6 Incompressible Turbulent Mean Flow — 398

- 6-1 Physical and Mathematical Description of Turbulence 398
- 6-2 The Reynolds Equations of Turbulent Motion 406
- 6-3 The Two-Dimensional Turbulent-Boundary-Layer Equations 411
- 6-4 Velocity Profiles: The Inner, Outer, and Overlap Layers 414
- 6-5 Turbulent Flow in Pipes and Channels 425
- 6-6 The Turbulent Boundary Layer on a Flat Plate 433
- 6-7 Turbulence Modeling 440
- 6-8 Analysis of Turbulent Boundary Layers with a Pressure Gradient 454
- 6-9 Free Turbulence: Jets, Wakes, and Mixing Layers 473
- 6-10 Turbulent Convective Heat Transfer 485
- Summary 498
- Problems 498

7 Compressible-Boundary-Layer Flow — 505

- 7-1 Introduction: The Compressible-Boundary-Layer Equations 505
- 7-2 Similarity Solutions for Compressible Laminar Flow 511
- 7-3 Solutions for Laminar Flat-Plate and Stagnation-Point Flow 514
- 7-4 Compressible Laminar Boundary Layers under Arbitrary Conditions 525
- 7-5 Special Topics in Compressible Laminar Flow 539
- 7-6 The Compressible-Turbulent-Boundary-Layer Equations 544
- 7-7 Wall and Wake Laws for Turbulent Compressible Flow 547
- 7.8 Compressible Turbulent Flow Past a Flat Plate 553
- 7-9 Compressible-Turbulent-Boundary-Layer Calculation with a Pressure Gradient 561
- Summary 566
- Problems 566

Appendices — 571

- A Transport Properties of Various Newtonian Fluids 571
- B Equations of Motion of Incompressible Newtonian Fluids in Cylindrical and Spherical Polar Coordinates 581
- C A Runge–Kutta Subroutine for N Simultaneous Differential Equations 585

Bibliography — 590

Index — xxx

PREFACE

OVERVIEW

The third edition of this book continues the goal of serving as a senior or first-year graduate textbook on viscous flow with engineering applications. Students should be expected to have knowledge of basic fluid mechanics, vector calculus, ordinary and partial differential equations, and elementary numerical analysis. The material can be selectively presented in a one-semester course or, with fuller coverage, in two quarters or even two semesters. At the author's institution, the text is used in a first-semester graduate course that has, as a prerequisite, a one-semester junior course in fluid mechanics.

The evolution of viscous-flow prediction continues toward CFD instead of physical insight and mathematical analysis. However, this book still exists to introduce viscous-flow *concepts,* not software. Dozens of new books and monographs on CFD are discussed and listed here for further specialized study. Since the second edition appeared in 1991, more than 10,000 new articles have been published on viscous flows. Clearly, the present book is an introductory textbook, not a comprehensive state-of-the-art treatment of the entire field. The goal is to make the book readable and informative and to introduce graduate students to the field.

New to this edition:

- Over 30 percent of the problems are new or revised.
- New material has been added to Chapters 1, 3, and 4 on microflows, slip in liquids, gas slip flow in tubes and channels, and a novel micropump.
- Section 2-9 on Dimensional Analysis is completely rewritten. Material has been added on Euler's equation and inviscid flow analysis and their relation to viscous flow.

- Chapter 5 begins with the classic Kelvin–Helmholtz wind-wave instability.
- Turbulence modeling in Chapter 6 is completely rewritten, expanded, and updated.
- Chapter 7 includes a detailed discussion of isentropic flow analysis to increase its relation to the understanding of viscous effects.
- References are completely updated.
- An Instructor and Student Resource Web Site is available to users of the text.

ORGANIZATION

The seven-chapter format of the book remains the same. Chapter 1 covers the basic properties of fluids and introductory concepts. New material has been added on microflows, slip in liquids, and an improved discussion of boundary conditions for flow.

Chapter 2 covers the basic equations of flow, with a bit of condensing of Secs. 2-8, 9 and 2-11, 12. Section 2-9 on Dimensional Analysis has been completely rewritten. Material has been added on Euler's equation and inviscid flow analysis and their relation to viscous flow.

Chapter 3 treats a variety of laminar-flow solutions, both analytical and numerical, of the Navier–Stokes equation. A number of new exact solutions are discussed and the Stokes paradox is illuminated a bit more. The creeping-flow discussion is updated. Two new interesting engineering CFD applications are given for liquid spheres and a novel micropump.

Chapter 4 has some obsolete material deleted, such as the Stratford separation criterion. A new section has been added on unsteady boundary layers, including acoustic streaming and the Goldstein/MRS separation criteria. Numerical solutions are covered, but the traditional integral methods remain.

Chapter 5 has dropped the beam-buckling instability analogy and now begins with the classic Kelvin–Helmholtz wind–wave instability. A great wind-shear photo by Brooks Martner has been added. The concept of pseudoresonance has been added. Section 5-4 on transition processes has been completely rewritten. New results of DNS transition prediction are now discussed.

Chapter 6 has been updated with many new references, but the basic outline of turbulent mean-flow prediction remains. Section 6-7 on modeling has been completely rewritten. The power-law overlap layer controversy is now included. DNS predictions are augmented.

Chapter 7 has two new photos of supersonic boundary-layer flow. Isentropic flow analysis is added to increase our understanding of viscous effects. There is a new discussion of Morkovin's hypothesis. Section 7-7 on compressible wall–wake laws has been rewritten.

The three Appendices are pretty much the same. More fluid property data have been added to App. A. Appendix C, a Runge–Kutta subroutine, is still useful and clarifies numerical integration. However, more and more readers are changing to spreadsheet calculations.

SUPPLEMENTS

The new Instructor and Student Resource Web Site, http://www.mhhe.com/white3e, will house general text information, the solutions to end-of-chapter problems (*under password-protection*), additional problems (*with password-protected solutions*), and helpful Web links.

ACKNOWLEDGMENTS

There are many people to thank. Much appreciated comments, suggestions, photos, charts, corrections, and encouragement were received from Leon van Dommelen of Florida State University; Gary Settles of Penn State University; Steven Schneider of Purdue University; Kyle Squires of Arizona State University; Chihyung Wen of Da-Yeh University, Taiwan; Brooks Martner of the NOAA Environmental Technology Laboratory; Jay Khodadadi of Auburn University; Philipp Epple of Friedrich-Alexander-Universität; Jürgen Thoenes of the University of Alabama at Huntsville; Luca d' Agostino of Università Degli Studi di Pisa; Raul Machado of the Royal Institute of Technology (KTH), Sweden; Gordon Holloway of the University of New Brunswick; Abdulaziz Almukbel of George Washington University; Dale Hart of Louisiana Tech University; Debendra K. Das of the University of Alaska Fairbanks; Alexander Smits of Princeton University; Hans Fernholz of Technische Universitaet Berlin; Peter Bernard of the University of Maryland; John Borg of Marquette University; Philip Drazin of the University of Bristol, UK; Ashok Rao of Rancho Santa Margarita, CA; Deborah Pence of Oregon State University; Joseph Katz of Johns Hopkins University; Pierre Dogan of the Colorado School of Mines; Philip Burgers of San Diego, CA; Beth Darchi of the American Society of Mechanical Engineers; and Norma Brennan of the American Institute of Aeronautics and Astronautics.

I have tried to incorporate almost all of the reviewer comments, criticisms, corrections, and improvements. The third edition has greatly benefited from the reviewers of the second edition text, as well as the reviewers of the third edition manuscript:

Malcolm J. Andrews, *Texas A&M University*
Mehdi Asheghi, *Carnegie Mellon University*
Robert Breidenthal, *University of Washington*
H. A. Hassan, *North Carolina State University*
Herman Krier, *University of Illinois, Urbana-Champaign*
Daniel Maynes, *Brigham Young University*
Suresh Menon, *Georgia Institute of Technology*
Meredith Metzger, *University of Utah*
Kamran Mohseni, *University of Colorado*
Ugo Piomelli, *University of Maryland*
Steven P. Schneider, *Purdue University*

Kendra Sharp, *Pennsylvania State University*
Marc K. Smith, *Georgia Institute of Technology*
Leon van Dommelen, FAMU-FSU
Steve Wereley, *Purdue University*

    The editors and staff at McGraw-Hill Higher Education, Amanda Green, Jonathan Plant, Peggy Lucas, Rory Stein, Mark Neitlich, and Linda Avenarius, were constantly helpful and informative. The University of Rhode Island continues to humor me, even in retirement.

Frank M. White
whitef@egr.uri.edu

LIST OF SYMBOLS

English Symbols

a	speed of sound; acceleration (Chap. 2), body radius (Chap. 4)
A	area; amplitude, Eq. (5-40); damping parameter, Eq. (6-90)
b	jet or wake width, Fig. 6-35
B	stagnation-point velocity gradient (Sec. 3-8.1); turbulent wall-law intercept constant, Eq. (6-38a)
ΔB	wall-law shift due to roughness, Eq. (6-60)
c, c_i, c_r	wave phase speeds (Chap. 5)
c_p, c_v	specific heats, Eq. (1-69)
C	Chapman–Rubesin parameter, Eq. (7-20)
C_i	species concentrations (Chap. 1)
D	diameter; drag force (Chap. 4); diffusion coefficient (Chap. 1)
D_h	duct hydraulic diameter, Eq. (3-55)
D_{ij}	turbulent transport or diffusion, Eq. (6-111)
e, E	internal energy
e_t	internal plus kinetic plus potential energy, Eq. (2-113)
f, F	force
f, F, g	similarity variables
g	acceleration of gravity
$G(Pr)$	heat-transfer parameter, Eqs. (3-172) and (4-78)
h	enthalpy; duct width; heat-transfer coefficient
h_i	metric coefficients, Eqs. (2-58) and (4-229)
h_0	stagnation enthalpy, $h + V^2/2$
H	shape factor, δ^*/θ; stagnation enthalpy, Eq. (7-3)
H_1	alternate shape factor, $(\delta - \delta^*)/\theta$
J	jet momentum, Eqs. (4-97), (4-206), and (6-144)
k	thermal conductivity; roughness height (Chaps. 5 and 6)

LIST OF SYMBOLS

K	bulk modulus, Eq. (1-84); duct pressure-drop parameter, Eq. (4-176); turbulence kinetic energy, Eq. (6-16); stagnation-point velocity gradient, Fig. 7-6
ℓ	mean-free path (Chap. 1); mixing length, Eq. (6-88)
L	characteristic length
L_{slip}	slip length of a liquid, Eq. (1-89)
m	mass; wedge-velocity exponent, Eq. (4-69)
$\dot{m}$	mass rate of flow
M	molecular weight; moment, Eq. (3-190)
n	normal to the wall; power-law exponent, Eq. (1-35)
p	pressure
$\hat{p}$	effective pressure, $p + \rho g z$
P	pressure gradient parameter, Eq. (3-42); duct perimeter
q	heat-transfer rate per unit area; turbulence level, Eq. (5-43)
Q	heat; volume flow rate, Eq. (3-35)
r	radial coordinate; recovery factor, Eq. (7-16)
r, θ, z	cylindrical polar coordinates, Eq. (2-63)
r, θ, λ	spherical polar coordinates, Eq. (2-65)
r_0	cylinder surface radius, Fig. 4-34
R	gas constant; body radius
s	entropy
S	Sutherland constant, Eq. (1-36); laminar shear parameter, Eq. (4-134); van Driest parameter, Eq. (7-130)
t	time
T	temperature; percent turbulence, Eq. (5-43)
T^*	wall heat-flux temperature, $q_w/(\rho c_p v^*)$; compressible-flow reference temperature, Eq. (7-42)
$\mathcal{T}$	surface tension coefficient
u, v, w	Cartesian velocity components
u_θ, u_r, u_z	cylindrical polar velocity components
u', v', w'	turbulent velocity fluctuations
Δu	wake velocity defect, Fig. 6-35c and Eq. (6-155)
U, W	freestream velocity components
v^*	wall-friction velocity, $(\tau_w/\rho_w)^{1/2}$
v_β	wake velocity, Eq. (6-137)
V	velocity; also U_e/U_0, Eq. (6-133)
$\mathcal{V}$	volume (Chap. 2)
w	rate of work done on an element, Eq. (2-36)
x, y, z	Cartesian coordinates
Z	gas compressibility factor, $p/(\rho RT)$

LIST OF SYMBOLS

Greek Symbols

α	thermal diffusivity, $k/\rho c_p$; wedge angle (Fig. 3-32); wave number, Eq. (5-12); angle of attack
α, β, γ	finite-difference mesh-size parameters, Eq. (4-146); also compressible wall-law parameters, Eqs. (7-111)
α^*, β, ζ	compressible finite-difference mesh-size parameters, Eq. (7-67)
β	thermal expansion coefficient, Eq. (1-86); Falkner–Skan parameter, Eq. (4-71); Clauser parameter, Eq. (6-42)
γ	specific-heat ratio, c_p/c_v; finite-difference parameter Eq. (4-163) intermittency, Fig. 6-5; compressibility parameter, Eq. (7-111)
δ, δ_u	velocity boundary-layer thickness
δ^*	displacement thickness, Eq. (4-4)
δ_c	conduction thickness, Eq. (4-156)
δ_h	enthalpy thickness, Eq. (4-22)
δ_T	temperature boundary-layer thickness
δ_3	dissipation thickness, Eq. (4-128)
δ_{ij}	Kronecker delta
Δ	defect thickness, Eq. (6-43)
ϵ	perturbation parameter (Sec. 4-11); turbulent dissipation [term V of Eq. (6-17)]
ϵ_{ij}	strain-rate tensor; Reynolds stress dissipation, Eq. (6-111)
κ	Kármán constant, ≈ 0.41
λ	second viscosity coefficient (Chap. 2); Darcy friction factor, Eq. (3-38); Thwaites' parameter, Eq. (4-132); $(2/C_f)^{1/2}$ (Chap. 6)
Λ	Kármán–Pohlhausen parameter, $\delta^2 (dp/dx)/\mu U$; pipe-friction factor, Eq. (6-54)
λ_n	Graetz function eigenvalues, Table 3-1
η	similarity variable; free-surface elevation (Chap. 5)
μ	viscosity
ν	kinematic viscosity μ/ρ
π	3.14159 ...
Π	Coles' wake parameter, Eq. (6-47)
ϕ	velocity potential (Chap. 2); latitude (Chap. 3); wave angle, Eq. (5-12); dimensionless disturbance, Eq. (5-23)
Φ	dissipation function, Eq. (2-46)
ψ	stream function
θ	polar coordinate angle; momentum thickness, Eq. (4-6)
Θ	dimensionless temperature ratio, Eq. (3-167) or (4-56)
ρ	density
σ	molecular collision diameter (Chap. 1); numerical mesh parameter, Eq. (3-247); turbulent jet growth parameter, Eq. (6-147)

τ	boundary-layer shear stress
τ_{ij}	stress tensor
χ	hypersonic interaction parameter, Eq. (7-86)
ξ	dimensionless pressure gradient, Eq. (6-36); similarity variable, Eq. (7-19)
ω	vorticity; angular velocity; frequency
Ω_v, Ω_D	molecular potential functions, Chap. 1
Ω	angular velocity
ζ	heat-transfer coefficient, Eq. (3-14); ratio δ_T/δ, Eq. (4-24)

Dimensionless Groups

Br	Brinkman number, $\mu V^2/k\Delta T$
Ca	cavitation number, $(p_\infty - p_{sat})/\rho V^2$
C_D	drag coefficient, $2\,(\text{drag})/\rho V^2 A$
C_f	skin-friction coefficient, $2\tau_w/\rho V^2$
C_h	Stanton number, $q_w/\rho V c_p \Delta T$
C_L	lift coefficient, $2\,(\text{lift})/\rho V^2 A$
C_p	pressure coefficient, $2(p - p_\infty)/\rho V^2$
Ec	Eckert number, $V^2/c_p \Delta T$
Eu	Euler number, $2\Delta p/\rho U^2$
Fr	Froude number, V^2/gL
Gr	Grashof number, $g\beta\,\Delta T L^3/\nu^2$
Gr^*	modified Grashof number, $GrNu = g\beta q_w L^4/k\nu^2$
Kn	Knudsen number, ℓ/L
L^*	Graetz number (Sec. 3-3.8), $L/(d_0\,Re_D\,Pr)$
Le	Lewis number, D/α
Ma	Mach number, V/a
Nu	Nusselt number, $q_w L/k\Delta T$
Pe	Peclet number, $RePr$
Po	Poiseuille number, $2\tau L/\mu V$
Pr	Prandtl number, $\mu c_p/k$
Pr_t	turbulent Prandtl number, $\mu_t c_p/k_t$
Ra	Rayleigh number, $GrPr = g\beta\,\Delta T L^3/\nu\alpha$
Re	Reynolds number, $\rho VL/\mu$
Ro	Rossby number, $V/\Omega L$
Sc	Schmidt number, ν/D
St	Strouhal number, fL/V
Ta	Taylor number, Eq. (5-34)
We	Weber number, $\rho V^2 L/\mathcal{T}$

Subscripts

aw	adiabatic wall
∞	far field
e	freestream, boundary-layer edge
0	initial or reference value
c, crit	critical, at the point of instability
m	mean
n	normal
rms	root mean square
sep	separation point
t	turbulent, tangential
tr	transition
r	recovery or adiabatic wall
w	wall
x	at position x

Superscripts

$-$	time-mean
$'$	differentiation; turbulent fluctuation
$*$	dimensionless variable (Chaps. 2, 3, and 4)
$+$	law-of-the-wall variable
$\wedge$	small-disturbance variable (Chap. 5)

CHAPTER 1

PRELIMINARY CONCEPTS

1-1 HISTORICAL OUTLINE

By stretching a point one could say that the study of viscous fluid flow reaches back into antiquity, for it was probably in prehistoric times that human weaponry developed from simple sticks and stones into streamlined, weighted spears and slim, pointed, fin-stabilized arrows. One can conclude that primitive man recognized and solved in part the problem of viscous resistance.

The exact solution for the problem of the viscous fluid at rest was correctly given by the Greek mathematician Archimedes (287–212 B.C.) as his two postulates of buoyancy. Subsequently, in order to derive expressions for the buoyant force on various-shaped bodies, Archimedes actually developed a version of the differential calculus. At about the same time, the Romans were building their magnificent water-supply systems and in so doing demonstrated some intuitive understanding of the effect of viscous resistance in long conduits. However, the Romans contributed little to a systematic solution of this problem, and in fact no significant progress on channel resistance was made until Chézy's work in 1768.

The period from the birth of Christ to the fifteenth century produced the same impact on viscous-flow analysis as it did on other fields of science, i.e., little if any. But the mountains of conjecture and superstition accumulated in these unscientific centuries certainly contained nuggets of fact which the great thinkers of the Renaissance finally mined. In 1500, the equation of conservation of mass for incompressible one-dimensional viscous flow was correctly deduced by Leonardo da Vinci, the Italian painter, sculptor, musician, philosopher, anatomist, botanist, geologist, architect, engineer, and scientist. Leonardo's notes also contain accurate sketches and descriptions of wave motion, hydraulic jumps, free jets, eddy formation behind bluff bodies (see Example 2 of Sec. 1-2), reduction of drag by streamlining, and the velocity distribution in a vortex.

The next notable achievement was by Evangelista Torricelli (1608–1647), who in 1644 published his theorem that the velocity of efflux of a (viscous) liquid from a hole in a tank is equal to the velocity which a liquid particle would attain in free fall from its surface. Torricelli termed his discovery "almost useless," but history has seen fit to disagree. From the point of view of this text, the efflux principle is unusually interesting, since it is one of the few flow phenomena for which viscous effects are often negligible.

The above achievements do not relate directly to viscous motion. That is, these early workers were probably studying a fluid they thought to be inviscid, or *perfect*; it happens that their results are also true for a viscous, or *real* fluid. The first to make a direct study of fluid friction was probably Edme Mariotte (1620–1684), who invented a balance system to measure the drag of a model held stationary in a moving stream, the first wind tunnel. Mariotte's text, "Traité du mouvement des eaux," was published in 1686, a year before the incomparable "Principia Mathematica" of Sir Isaac Newton.

In 1687, Newton published in his "Principia" the simple statement which delineates the viscous behavior of nearly all common fluids: "The resistance which arises from the lack of lubricity in the parts of a fluid—other things being equal—is proportional to the velocity by which the parts of the fluid are being separated from each other." Such fluids, water and air being prominent examples, are now called *newtonian* in his honor. With the law of linear viscosity thus proposed, Newton contributed the first viscous-flow analysis by deriving the correct velocity distribution about a rotating cylinder.

But the world was apparently not ready for viscous-flow theory. This was probably due to Newton himself, because of his more famous discovery, the differential calculus. Whereas those who preceded Newton were essentially limited to discussion of fluid-flow problems, those who followed him could use the calculus to attack such problems directly. It is natural that the first efforts were directed toward the idealized frictionless fluid. First to succeed was Daniel Bernoulli, who in 1738 demonstrated the proportionality between pressure gradient and acceleration in inviscid flow. Subsequently, the master of the calculus, Leonhard Euler, derived in 1755 the famous frictionless equation which now bears Bernoulli's name. Euler's magnificent derivation is essentially unchanged today in ideal-fluid theory, or *hydrodynamics*, as Bernoulli termed it. Paralleling Euler, Jean d'Alembert published in 1752 his famous paradox, showing that a body immersed in a frictionless flow would have zero drag. Shortly afterward, Lagrange (1736–1813), Laplace (1749–1827), and Gerstner (1756–1832) carried the new hydrodynamics to elegant heights of analysis.

But theoretical results such as the d'Alembert paradox were too much for practical engineers to bear, with the tragic result that fluid mechanics was rent into two parts: hydrodynamics, under whose banner mathematicians soared to new frictionless summits, and hydraulics, which abandoned theory entirely and relied on experimental measurements. This schism continued unhealed for 150 years, to the beginning of the twentieth century. Indeed, the separation of fluid mechanics theory from experiment is not extinct even today, as witness the divergent views of the subject now held among aeronautical, chemical, civil, and mechanical engineers.

After Euler and his colleagues, the next significant analytical advance was the addition of frictional-resistance terms to Euler's inviscid equations. This was done, with varying degrees of elegance, by Navier in 1827, Cauchy in 1828, Poisson in 1829, St. Venant in 1843, and Stokes in 1845. The first four wrote their equations in terms of an unknown molecular function, whereas Stokes was the first to use the first coefficient of viscosity μ. Today these equations, which are fundamental to the subject, are called the *Navier–Stokes relations*, and this text can do little to improve upon Stokes' analysis.

The Navier–Stokes equations, though fundamental and rigorous, are non-linear, nonunique, complex, and difficult to solve. To this day, only a relatively few particular solutions have been found, although mathematicians are now taking an interest in the general properties of these remarkable equations [Constantin and Foias (1988)]. Meanwhile, the widespread use of digital computers has given birth to many numerical models and published computations of viscous flows. Certain of these models can be implemented, for simple geometries, on a small personal computer and are described here in Chaps. 3, 4, and 6. Experimentation remains a strong component of viscous-flow research because even the largest supercomputers are incapable of resolving the fine details of a high-Reynolds-number flow.

For practical fluids engineering, the biggest breakthrough was the demonstration, by Ludwig Prandtl in 1904, of the existence of a thin *boundary layer* in fluid flow with small viscosity. Viscous effects are confined to this boundary layer, which may then be patched onto the outer inviscid flow, where so many powerful mathematical techniques obtain. Boundary-layer theory applies to many, but definitely not all, engineering flows. The concept makes it possible, as Leslie Howarth said, "to think intelligently about almost any problem in real fluid flow."

The second most important breakthrough, also accomplished at the turn of the twentieth century, was to put fluid-flow experimentation on a solid basis, using dimensional analysis. Leaders in this effort were Osborne Reynolds (1842–1912), Lord Rayleigh (1842–1919), and Ludwig Prandtl (1875–1953). Modern engineering studies—and textbooks—routinely place their results in dimensionless form, thus making them applicable to any newtonian fluid under the same flow conditions.

With thousands of researchers now active in fluid mechanics, present progress is incremental and substantial. Instrumentation has advanced greatly with the inventions of the hot-wire, the hot-film, the laser-Doppler velocimeter, and miniature pressure and temperature sensors. Visualization of flow—through bubbles, smoke, dye, oil-films, holography, and other methods—is now outstanding [see, e.g., Van Dyke (1982), Nakayama and Tanida (1996), Smits and Lim (2000), and Nakayama (1988)]. Computational fluid dynamics (CFD) has grown from a special topic to infiltrate the entire field: Many user-friendly CFD codes are now available so that ordinary engineers can attempt to model a realistic two- or three-dimensional viscous flow.

The literature in fluid mechanics is now out of control, too much to keep up with, at least for someone as dedicated as this writer. The first edition of this text had a figure to show the growth of viscous-flow papers during the twentieth century. Prandtl's 1904 breakthrough could be considered as "Paper 1," and the research output rose at a 7 percent annual rate to 70 papers per year in 1970. Well, that annual increase has continued to this day, so that thousands of papers are now being published each year. A dozen new fluids-oriented journals have been introduced, plus a half-dozen serials related to computational fluid dynamics. There are dozens of conferences and symposia every year devoted to fluids-oriented topics. Consider the following statistics:

- In 2002, the *Journal of Fluid Mechanics* printed 450 papers covering 9000 journal pages.
- In 2002, the *Physics of Fluids* printed 410 papers covering 4500 journal pages.
- In 2002, the *Journal of Fluids Engineering* printed 120 papers covering 1068 journal pages.
- In 2002, the Fluids Engineering Division of the American Society of Mechanical Engineers sponsored 650 papers.
- The *May 2003 Meetings Calendar of the Journal of Fluids Engineering* listed 38 fluids-related conferences for the coming year.
- The *2001 International Symposium on Turbulent Shear Flow Phenomena* presented 300 papers.

This situation is, of course, much the same in other pure and applied sciences. The net effect of the above incomplete statistics is that the present text can only be an introduction to some of the many topics of ongoing research into laminar and turbulent shear flows. Many specialized monographs will be cited throughout this text for further reading.

The historical details in this present section were abstracted from the excellent history of hydraulics by Rouse and Ince (1957).

1-2 SOME EXAMPLES OF VISCOUS-FLOW PHENOMENA

Before embarking upon the inevitable detailed studies of theoretical and experimental viscous flows, let us discuss four examples, chosen to illustrate both the strength and the limitations of the subject: (1) airfoil flow, (2) a cylinder in cross-flow, (3) pipe-entry flow, and (4) prolate spheroid flow. These examples remind one that a textbook tends to emphasize analytical power while deemphasizing practical difficulties. Viscous-flow theory does have limitations, especially in the high-Reynolds-number *turbulent flow* regime, where the flow undergoes random fluctuations and is only modeled on a semiempirical time-mean or statistical basis.

Although geometry and fluid buoyancy and compressibility will be important, in all viscous flows, the primary controlling parameter is the dimensionless *Reynolds number*.

$$Re_L = \rho U L/\mu \qquad (1\text{-}1)$$

where U is a velocity scale, L is a characteristic geometric size, and ρ and μ are the fluid density and viscosity, respectively. Fluid properties alone can cause dramatic differences in the Reynolds number and, consequently, the flow pattern. For example, if $U = 1$ m/s and $L = 1$ m at 20°C, $Re_L =$ 9E3, 7E4, and 1E6 for SAE-10 oil, air, and water, respectively. By adding in changes in size and speed, the Reynolds number can vary from a small fraction (falling dust particles) to 5E9 (a cruising supertanker). For a given geometry, as Re_L increases, the flow pattern changes from smooth or *laminar* through a transitional region into the fluctuating or *turbulent* regime.

Example 1 Flow past a thin airfoil. Consider flow past a thin airfoil at small angle of incidence, $\alpha < 5°$, as sketched in Fig. 1-1a. In practical applications, the *Reynolds number*, Re_L, is large. For example, if $L = 1$ m, $U = 100$ m/s, and $\nu \approx$ 1.5E–5 m²/s (air at 20°C and 1 atm), $Re_L =$ 6.7E6. In these circumstances the flow creates a thin boundary layer near the airfoil surface and a thin wake downstream. The measured surface pressure distribution on the foil can be predicted by inviscid-flow theory [e.g., White (2003), Sec. 8.7], and the wall shear stress can be computed with the boundary-layer theory of Chaps. 4 to 6. The sharp trailing edge establishes the flow pattern, for a viscous fluid cannot go around such a sharp edge but instead must leave smoothly and tangentially, as shown in Fig. 1-1a.

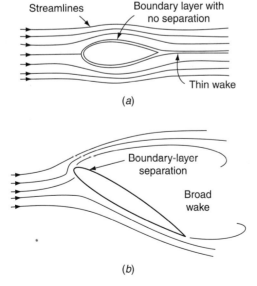

FIGURE 1-1
Flow past a thin airfoil: (*a*) low incidence angle, smooth flow, no separation; (*b*) high incidence angle, upper surface separates or "stalls," lift decreases.

According to inviscid theory, if F is the lift force per unit depth on a symmetric (two-dimensional) airfoil, the dimensionless lift coefficient C_L is given by

$$C_L = \frac{F}{\frac{1}{2}\rho U^2 L} \approx 2\pi \sin(\alpha) \qquad (1\text{-}2)$$

where L is the chord length of the airfoil.

At larger incidence angles (10–15°), boundary-layer separation, or *stall*, will occur on the upper or *suction* (low-pressure) surface, as shown in Fig. 1-1b. For thicker airfoils, separation generally occurs at the trailing edge. Thin airfoils may form a partial separation, or *bubble*, near the leading edge. In stalled flow, the upper surface pressure distribution deviates considerably from inviscid theory, resulting in a loss of lift and an increase in drag force.

When an airfoil flow separates, its lift coefficient levels off to a maximum and then decreases, sometimes gradually, sometimes quickly as a short leading-edge bubble suddenly lengthens. Figure 1-2 compares typical theoretical and experimental lift curves for a symmetric airfoil.

As long as the angle of attack is below stall, the lift can be predicted by inviscid theory and the friction by boundary-layer theory. The onset of stall can be predicted. In the stalled region, however, boundary-layer theory is not valid, and one must resort either to experimentation or—with increasing success as turbulence modeling improves—numerical simulation on a digital computer.

Figures 1-1 and 1-2 are for two-dimensional airfoils—of infinite span into the paper. Practical wings of course have tips and can have leading edges *swept* or nonorthogonal to the oncoming stream. The flow over them is three-dimensional. An example is the 45° swept wing shown in Fig. 1-3. The flow on the upper surface is visualized by streaks in an oil-film coating. At $\alpha = 12°$, Fig. 1-3a, there is a leading-edge separation bubble (the white strip), but the remaining flow is attached and moves toward the (low-pressure) tip region. At 15° incidence, Fig. 1-3b, the rear surface flow moves parallel to the trailing edge, toward a separated region at the tip. Finally, at $\alpha = 20°$, Fig. 1-3c, the wing is heavily stalled and the surface flow actually moves backward or upstream toward the leading edge. Inviscid-flow computation is common for three-dimensional flows, e.g., Anderson (2001), and viscous-flow simulation of such flows is now possible with the advent of supercomputers [Chung (2002)].

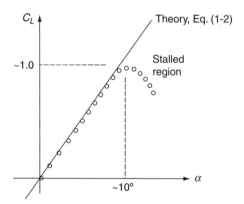

FIGURE 1-2
Typical comparison of theory and experiment for lift coefficient on a symmetric airfoil.

PRELIMINARY CONCEPTS 7

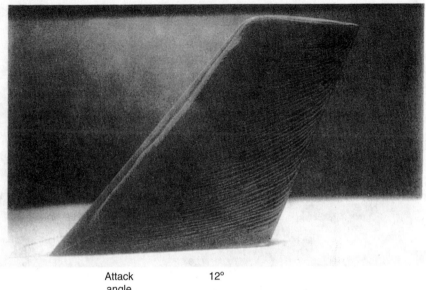

Attack angle 12°

(a)

15°

(b)

FIGURE 1-3
Oil-film visualization of suction-surface flow past a 45° swept wing at $Re_L = 2.6\text{E}5$: (a) $\alpha = 12°$, flow attached except for leading-edge bubble; (b) at 15°, surface flow moves toward a stalled tip region; (c) at 20°, the wing is almost entirely stalled and surface flow moves upstream. [*From Nakayama (1988), courtesy of Hitoshi Murai, Hiroshima Institute of Technology.*]

8 VISCOUS FLUID FLOW

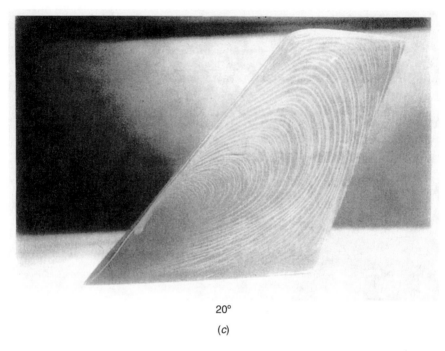

20°

(c)

FIGURE 1-3
Continued.

Example 2 Flow past a circular cylinder. A very common geometry in fluids engineering is crossflow of a stream at velocity U_∞ past a circular cylinder of radius R. For plane inviscid flow, the solution superimposes a uniform stream with a line doublet and is given in polar coordinates by [see, e.g., White (2003), p. 537]

$$v_r = U_\infty \left(1 - \frac{R^2}{r^2}\right) \cos\theta$$
$$v_\theta = -U_\infty \left(1 + \frac{R^2}{r^2}\right) \sin\theta \tag{1-3}$$

The streamlines of this flow can then be plotted as in Fig. 1-4. At the surface of the cylinder, $r = R$, we have $v_r = 0$ and $v_\theta = -2U_\infty \sin\theta$, the latter velocity being finite and thus violating the no-slip condition imposed by intermolecular forces between the fluid and the solid.

The pressure distribution at the cylinder surface can be found from Bernoulli's equation, $p + \tfrac{1}{2}\rho V^2 = \text{const}$, where ρ is the fluid density. The result is

$$p_s = p_\infty + \tfrac{1}{2}\rho U_\infty^2(1 - 4\sin^2\theta)$$

or

$$C_p = \frac{p_s - p_\infty}{\tfrac{1}{2}\rho U_\infty^2} = 1 - 4\sin^2\theta$$

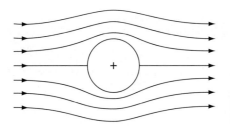

FIGURE 1-4
Perfect-fluid flow past a circular cylinder.

This distribution is shown as the dash-dot line in Fig. 1-5. Equations (1–3) illustrate a characteristic of inviscid flow without a free surface or "deadwater" region: There are no parameters such as Reynolds number and no dependence upon physical properties. Also, the symmetry of $C_p(\theta)$ in Fig. 1-5 indicates that the integrated surface-pressure force in the streamwise direction—the cylinder *drag*—is zero. This is an example of the d'Alembert paradox for inviscid flow past immersed bodies.

The experimental facts differ considerably from this inviscid symmetrical picture and depend strongly upon Reynolds number. Figure 1-5 shows measured C_p by Flachsbart (1932) for two Reynolds numbers. The pressure on the rear or lee side of the cylinder is everywhere less than the freestream pressure. Consequently, unlike the d'Alembert paradox, the real fluid causes a large pressure-drag force on the body.

Nor are the real streamlines symmetrical. Figure 1-6 shows the measured flow pattern in water moving past a cylinder at $Re_D = 170$. The flow breaks away or "separates" from the rear surface, forming a broad, pulsating wake. The pattern is visualized by releasing hydrogen bubbles at the left of the photograph, in *streaklines* parallel to the stream and *timelines* normal to the flow. Note that the wake consists of pairs of vortices shed alternately from the upper and lower part of the rear surface. They are called Kármán vortex streets, after a paper by Kármán (1911) explaining this alternation to be a stable configuration for vortex pairs. Beginning for $Re_D > 35$, the vortex streets occur in almost any bluff-body flow and persist over a wide range of Reynolds

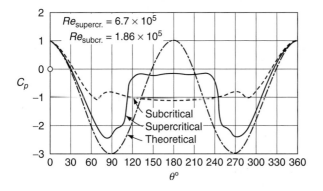

FIGURE 1-5
Comparison of perfect-fluid theory and an experiment for the pressure distribution on a cylinder. [*After Flachsbart (1932)*.]

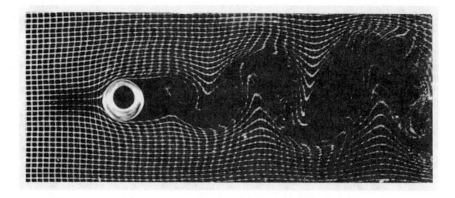

FIGURE 1-6
Timelines and streaklines for flow past a cylinder at $Re_D = 170$. [*From Nakayama (1988), courtesy of Y. Nakayama, Tokai University.*]

numbers, as shown in Fig. 1-7. As the Reynolds number increases, the wake becomes more complex—and turbulent—but the alternate shedding can still be detected at $Re = 10^7$.

As shown in Fig. 1-8, the dimensionless cylinder shedding frequency or *Strouhal number*, $St = fD/U \approx 0.2$ for Reynolds numbers from 100 to 10^5. Thus the shedding cycle takes place during the time that the freestream moves approximately five cylinder diameters. Vortex shedding is one of many viscous flows which, though posed with fixed and steady boundary conditions, evolve into unsteady motions because of flow instability. The pressure distributions in Fig. 1-5 are time averages for this reason.

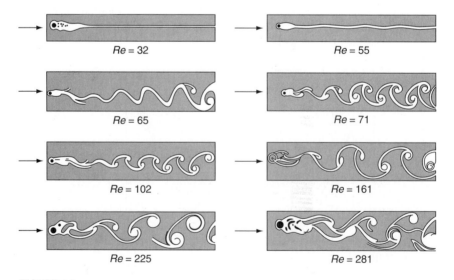

FIGURE 1-7
The effect of the Reynolds number on the flow past a cylinder. [*After Homann (1936).*]

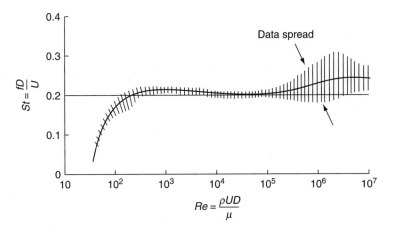

FIGURE 1-8
Measured Strouhal number for vortex shedding frequency behind a circular cylinder.

The drag coefficient on a cylinder, defined as $C_D = F/(\rho U^2 RL)$, is plotted in Fig. 1-9 over a wide range of Reynolds numbers. The solid curve, for a smooth wall, shows a sharp drop, called the *drag crisis*, at about $Re_D \approx 2E5$, which occurs when the boundary layer on the front surface becomes turbulent. If the surface is rough, the drag crisis occurs earlier, a fact exploited in sports by the deliberate dimpling of golf balls. As shown in Fig. 1-9, freestream turbulence also causes an early drag crisis.

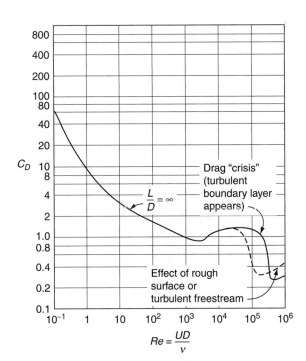

FIGURE 1-9
Drag coefficient of a circular cylinder.

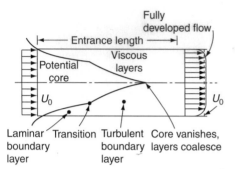

FIGURE 1-10
Flow in the entry region of a tube.

To summarize, real fluid flow past a bluff body such as a circular cylinder can differ markedly from the inviscid-flow prediction. Viscous forces which are extremely small (only a few percent of the total drag) actually control the flow by inducing separation and wake formation at the rear of the body. Boundary-layer theory can predict the onset of separation, but the surface pressure distribution changes so markedly from inviscid theory (Fig. 1-6) that Prandtl's matching scheme of Chap. 4 fails to be quantitative. For $Re \ll 1$, Stokes' creeping-flow theory can be used effectively (Sec. 3-9). For higher Reynolds numbers, both laminar and turbulent, numerical modeling on a digital computer is possible, e.g., Breuer (1998).

Example 3 Flow in a circular pipe. Consider now the flow illustrated in Fig. 1-10, where a steady viscous flow enters a tube from a reservoir. Wall friction causes a viscous layer, probably laminar, to begin at the inlet and grow in thickness downstream, possibly becoming turbulent further inside the tube. Unlike the *external* flows of Examples 1 and 2, this is an *internal* flow constrained by the solid walls, and inevitably the viscous layers must coalesce at some distance x_L, so that the tube is then completely filled with boundary layer. Slightly further downstream of the coalescence, the flow profile ceases to change with axial position and is said to be fully developed.

The developed flow in Fig. 1-10 ends up turbulent, which typically occurs for a Reynolds number $U_0 D/\nu > 2000$. At lower Re, both the developing and developed regions remain laminar. Figure 1-11 shows such a laminar experiment, using hydrogen bubbles in water flow. Note that the wall flow slows down and the central core accelerates. The bubble profiles change from near-slug flow at the inlet to near-parabolic downstream.

Pipe flow is common in engineering. The theory of constant-area duct flow, for both developing and developed laminar and turbulent conditions, is well formulated and satisfying. Analytical difficulties arise if the duct diameter is tapered. Tapered flow does not become fully developed and, if the area increases in the flow direction (subsonic diffuser), separation, backflow, and unsteadiness complicate the flow pattern. A sketch of flow in a "stalled" diffuser is shown in Fig. 1-12, after Kline et al. (1959). Diffusers can now be analyzed by numerical methods, using either boundary-layer theory, Johnston (1998), or computational fluid dynamics, Muggli (1997). If the duct area *decreases* in the flow direction (subsonic nozzle), there is no flow separation or unsteadiness, and the nozzle is readily analyzed by elementary boundary-layer theory.

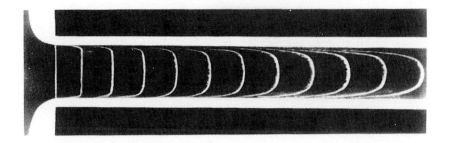

FIGURE 1-11
Hydrogen-bubble visualization of laminar flow in the entrance of a tube at $Re = 1600$. [*From Nakayama (1988), courtesy of Y. Nakayama, Tokai University.*]

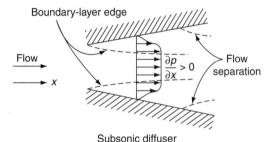

FIGURE 1-12
Flow separation in a diffuser.

Example 4 Flow past a prolate spheroid. Flows involving complicated three-dimensional effects, such as the cylinder flow of Example 2, are rightly termed *complex viscous flows* and cannot be analyzed by traditional boundary-layer methods. Complex flows are studied either experimentally or, increasingly, by computational fluid dynamics (CFD).

Figure 1-13 illustrates the computation of viscous flow past a 6:1 prolate spheroid (a slender football shape) at a high angle of attack of 20°. The angle of attack is defined as the angle between the oncoming flow and the central axis of the body. The Reynolds number is large, $\rho UL/\mu = 4.2E6$, and the flow near the body is turbulent. How might one analyze this flow with CFD? Laminar (smooth, nonfluctuating) flow can be computed accurately with a suitable fine mesh. However, the wide spectrum of random fluctuations of turbulence will yield to direct numerical simulation (DNS) only for low Reynolds numbers of order 10^4. Higher Reynolds numbers require *modeling* the small scale eddies with a time-averaging scheme (Chap. 6).

The prolate spheroid flow in Fig. 1-13 was computed using large-eddy simulation (LES) by Constantinescu et al. (2002). The authors directly simulated the large eddies but used a *turbulence model* for the fine-scale motions. The three-dimensional grid in Fig. 1-13a uses 2.6 million nodes, yet it still cannot resolve the important small fluctuations. Figure 1-13b shows the computed surface streamlines when the body is placed at a 20° angle to the freestream. The computations are in reasonable agreement

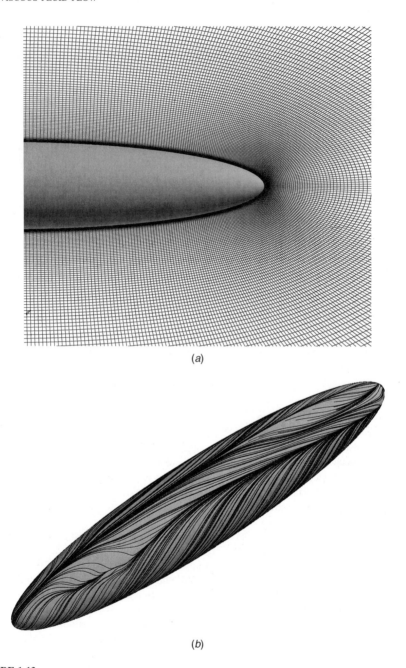

FIGURE 1-13
Computational fluid dynamics model of flow past a prolate spheroid: (*a*) meridian slice showing the million-node grid system; (*b*) surface pattern at Re_L = 4.2E6 and an angle of attack of 20°. [*After Constantinescu et al. (2002).*]

with the data of Chesnakas and Simpson (1997) for surface pressures and shear stresses, separation lines, and surface turning angles.

Prolate spheroid flow is a good example of fluids engineering research. Detailed *benchmark* experiments are followed by prediction methods that are gradually improved. In addition to the LES computations of Constantinescu et al. (2002), the same flow has been modeled by a multivortex simulation, Dimas et al. (1998), and by traditional turbulence modeling without large-eddy simulation, Paneras (1998).

The general subject of computational fluid dynamics is beyond the scope of the present text. We will briefly discuss CFD simulations in Chaps. 3, 4, and 6 but always refer the reader to advanced monographs for details of the subject.

1-3 PROPERTIES OF A FLUID

It is common in introductory physics to divide materials into the three classes of solids, liquids, and gases, noting their different behavior when placed in a container. This is a handy classification in thermodynamics, for example, because of the strong differences in state relations among the three. In fluid mechanics, however, there are only two classes of matter: fluids and nonfluids (solids). A solid can resist an applied shear force and remain at rest, while a fluid cannot. This distinction is not completely clear-cut. Consider a barrel full of pitch at room temperature. The pitch looks hard as a rock and will easily support a brick placed on its surface. But if the brick is left there for several days, one will have trouble retrieving the brick from the bottom of the barrel. Pitch, then, is usually classed as a fluid. Consider the metal aluminum. At room temperature, aluminum is solid to all appearances and will resist any applied shear stress below its strength limit. However, at 400°F, well below its 1200°F melting point, aluminum flows gently and continuously under applied stress and has a measurable viscosity. Nor is high temperature the criterion for fluid behavior in metals, since lead exhibits this gentle viscous creep at room temperature. Note also that mercury is a fluid and has the lowest viscosity relative to its own density (kinematic viscosity) of any common substance.

This text is primarily concerned, then, with easily recognizable fluids which flow readily under applied shear, although some slight attention will be paid to the borderline substances which partly flow and partly deform when sheared. All gases are true fluids, as are the common liquids, such as water, oil, gasoline, and alcohol. Some liquid substances which may not be true fluids are emulsions, colloids, high-polymer solutions, and slurries. The general study of flow and deformation of materials constitutes the subject of *rheology*, of which viscous flow is a special case [see, e.g., the texts by Reiner (1969), Bird et al. (2001), Hutton (1989), and Owens and Phillips (2002)].

Restricting ourselves to true fluids, we now define and illustrate their properties. These properties are of at least four classes:

1. *Kinematic* properties (linear velocity, angular velocity, vorticity, acceleration, and strain rate). Strictly speaking, these are properties of the flow field itself rather than of the fluid.
2. *Transport* properties (viscosity, thermal conductivity, mass diffusivity).

3. *Thermodynamic* properties (pressure, density, temperature, enthalpy, entropy, specific heat, Prandtl number, bulk modulus, coefficient of thermal expansion).
4. Other miscellaneous properties (surface tension, vapor pressure, eddy-diffusion coefficients, surface-accommodation coefficients).

Some items in class 4 are not true properties but depend upon flow conditions, surface conditions, and contaminants in the fluid.

The use of class 3 properties requires hedging. It is a matter of some concern that classical thermodynamics, strictly speaking, does *not* apply to this subject, since a viscous fluid in motion is technically not in equilibrium. Fortunately, deviations from local thermodynamic equilibrium are usually not significant except when flow residence times are short and the number of molecular particles few, e.g., hypersonic flow of a rarefied gas. The reason is that gases at normal pressures are quite dense in the statistical sense: A cube of sea-level air 1 μm on a side contains approximately 10^8 molecules. Such a gas, when subjected to a change of state—even a shock change—will rapidly smooth itself into local equilibrium because of the enormous number of molecular collisions occurring in a short distance. A liquid is even more dense, and thus we accept thermodynamic equilibrium as a good approximation in this text.[†]

1-3.1 The Kinematic Properties

In fluid mechanics, one's first concern is normally with the fluid velocity. In solid mechanics, on the other hand, one might instead follow particle displacements, since particles in a solid are bonded together in a relatively rigid manner.

Consider the rigid-body dynamics problem of a rocket trajectory. After solving for the paths of any three noncollinear particles on the rocket, one is finished, since all other particle paths can be inferred from these three. This scheme of following the trajectories of individual particles is called the *Lagrangian description* of motion and is very useful in solid mechanics.

But consider the fluid flow out of the nozzle of that rocket. Surely we cannot follow the millions of separate paths. Even the point of view is important, since an observer on the ground would perceive a complicated unsteady flow, while an observer fixed to the rocket would see a nearly steady flow of quite regular pattern. Thus it is generally useful in fluid mechanics (1) to choose the most convenient origin of coordinates, with luck making the flow appear steady, and (2) to study only the velocity *field* as a function of position and time, not trying to follow any specific particle paths. This scheme of describing the flow at every fixed point as a function of time is called the *Eulerian formulation* of motion. The Eulerian velocity vector field can be defined in the following Cartesian form:

$$\mathbf{V}(\mathbf{r}, t) = \mathbf{V}(x, y, z, t)$$
$$= \mathbf{i}u(x, y, z, t) + \mathbf{j}v(x, y, z, t) + \mathbf{k}w(x, y, z, t) \qquad (1\text{-}4)$$

[†]Note, however, that flows involving chemical or nuclear reactions require an extended concept of equilibrium. Such flows typically involve knowledge of reaction rates and are not treated here.

Complete knowledge of the scalar variables u, v, and w as functions of (x, y, z, t) is often the solution to a problem in fluid mechanics. Note that we have used the notation (u, v, w) to mean velocity components, not displacement components, as they would be in solid mechanics. Displacements are of so little use in fluid problems that they have no symbol reserved for themselves.

The Eulerian, or velocity-field system, is certainly the proper choice in fluid mechanics, but one definite conflict exists. The three fundamental laws of mechanics—conservation of mass, momentum, and energy—are formulated for particles (systems) of fixed identity, i.e., they are Lagrangian in nature. All three of these laws relate to the time rate of change of some property of a fixed particle. Let Q represent any property of the fluid. If dx, dy, dz, dt represent arbitrary changes in the four independent variables, the total differential change in Q is given by

$$dQ = \frac{\partial Q}{\partial x} dx + \frac{\partial Q}{\partial y} dy + \frac{\partial Q}{\partial z} dz + \frac{\partial Q}{\partial t} dt \tag{1-5}$$

Since we are deliberately following an infinitesimal particle of fixed identity, the spatial increments must be such that

$$dx = u\, dt \qquad dy = v\, dt \qquad dz = w\, dt \tag{1-6}$$

Substituting in Eq. (1-5), we find the proper expression for the time derivative of Q of a particular elemental particle:

$$\frac{dQ}{dt} = \frac{\partial Q}{\partial t} + u\frac{\partial Q}{\partial x} + v\frac{\partial Q}{\partial y} + w\frac{\partial Q}{\partial z} \tag{1-7}$$

The quantity dQ/dt is variously termed the *substantial derivative, particle derivative*, or *material derivative*—all names which try to invoke the feeling that we are following a fixed fluid particle. To strengthen this feeling, it is traditional to give this derivative the special symbol DQ/Dt, purely a mnemonic device, not intended to frighten readers. In Eq. (1-7), the last three terms are called the *convective derivative*, since they vanish if the velocity is zero or if Q has no spatial change. The term $\partial Q/\partial t$ is called the *local derivative*. Also note the neat vector form

$$\frac{DQ}{Dt} = \frac{\partial Q}{\partial t} + (\mathbf{V} \cdot \nabla)Q \tag{1-8}$$

where ∇ is the gradient operator

$$\mathbf{i}\frac{\partial}{\partial x} + \mathbf{j}\frac{\partial}{\partial y} + \mathbf{k}\frac{\partial}{\partial z}$$

1-3.2 Acceleration of a Fixed-Identity Particle

If Q is $\mathbf{V}$ itself, we obtain our first kinematic property, the particle-acceleration vector:

$$\frac{D\mathbf{V}}{Dt} = \frac{\partial \mathbf{V}}{\partial t} + (\mathbf{V} \cdot \nabla)\mathbf{V} = \mathbf{i}\frac{Du}{Dt} + \mathbf{j}\frac{Dv}{Dt} + \mathbf{k}\frac{Dw}{Dt} \tag{1-9}$$

Note that the acceleration is concerned with u, v, and w and 12 scalar derivatives, i.e., the local changes $\partial u/\partial t$, $\partial v/\partial t$, and $\partial w/\partial t$ and the nine spatial derivatives of the form $\partial u_i/\partial x_j$, where i and j denote the three coordinate directions. Henceforth, we shall not use i, j, and k to denote unit vectors but instead use them as Cartesian indexes.

The convective terms in D/Dt present unfortunate mathematical difficulty because they are non-linear products of variable terms. It follows that viscous flows with finite convective accelerations are non-linear in character and present such vexing analytical problems as failure of the superposition principle; nonunique solutions, even in steady laminar flow; and coupled oscillating motion in a continuous frequency spectrum, which is the chief feature of a high Reynolds number, or turbulent, flow. Note that these non-linear terms are accelerations, not viscous stresses. It is ironic that the main obstacle in viscous-flow analysis is an inviscid term; the viscous stresses themselves are linear if the viscosity and density are assumed constant.

In frictionless flow, non-linear convective accelerations still exist but do not misbehave, as can be seen with reference to the valuable vector identity

$$(\mathbf{V} \cdot \nabla)\mathbf{V} \equiv \nabla \frac{V^2}{2} - \mathbf{V} \times (\nabla \times \mathbf{V}) \tag{1-10}$$

As we shall see, the term $\nabla \times \mathbf{V}$ usually vanishes if the viscosity is zero (irrotationality), leaving the convective acceleration equal only to the familiar kinetic-energy term of Bernoulli's equation. Inviscid flow, then, is non-linear also, but the non-linearity is confined to the calculation of static pressure, not to the determination of the velocity field, which is linear.

If we agree from this brief discussion that viscous flow is mathematically formidable, we deduce that a very important problem is that of modeling a viscous-flow experiment. That is, when can a velocity distribution $\mathbf{V}_1$ measured in a flow about or through a model shape B_1 be scaled by (say) a simple multiplier to yield the velocity distribution $\mathbf{V}_2$ about or through a geometrically similar but larger (say) model shape B_2? This happy condition is called *similarity*, and the conditions for achieving it or the frustrations in not achieving it are discussed in Chap. 2.

1-3.3 Other Kinematic Properties

In fluid mechanics, as in solid mechanics, we are interested in the general motion, deformation, and rate of deformation of particles. Like its solid counterpart, a fluid element can undergo four different types of motion or deformation: (1) translation; (2) rotation; (3) extensional strain, or dilatation; and (4) shear strain. The four types are easy to separate geometrically, which is of course why they are so defined. The reader familiar with, say, the theory of elasticity for solids will find the following analysis of fluid kinematic properties almost identical to that in solid mechanics. Consider an initially square fluid element at time t and then again later at time $t + dt$, as illustrated in Fig. 1-14 for motion in the xy plane. We can see by inspection the four types of motion which acted on the element. There has been translation of the reference corner B to its new position B'. There has been a counterclockwise rotation of the diagonal BD to the new position $B'D'$. There has been

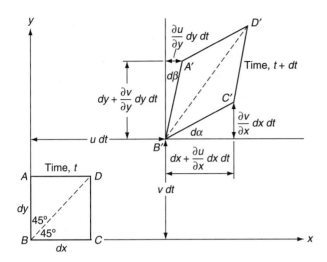

FIGURE 1-14
Distortion of a moving fluid element.

dilatation; the element looks a little bigger. There has been shear strain, i.e., the square has become rhombic.

Now put this discussion on a quantitative basis. Note in each case that the final result will be a *rate*, i.e., a time derivative.

The translation is defined by the displacements $u\,dt$ and $v\,dt$ of the point B. The *rate* of translation is u, v; in three-dimensional motion, the rate of translation is the velocity u, v, w.

The angular rotation of the element about the z axis is defined as the *average* counterclockwise rotation of the two sides BC and BA. As shown in Fig. 1-14, BC has rotated an amount $d\alpha$. Meanwhile, BA has rotated clockwise, thus its counterclockwise turn is $(-d\beta)$. The average rotation is thus

$$d\Omega_z = \tfrac{1}{2}(d\alpha - d\beta) \tag{1-11}$$

where the subscript z denotes rotation about an axis parallel to the z axis. Thus we perceived the counterclockwise rotation in Fig. 1-14 because $d\alpha$ was drawn larger than $d\beta$. Referring to Fig. 1-14, we find that both $d\alpha$ and $d\beta$ are directly related to velocity derivatives through the calculus limit:

$$d\alpha = \lim_{dt \to 0}\left(\tan^{-1}\frac{\frac{\partial v}{\partial x}\,dx\,dt}{dx + \frac{\partial u}{\partial x}\,dx\,dt}\right) = \frac{\partial v}{\partial x}\,dt$$

$$d\beta = \lim_{dt \to 0}\left(\tan^{-1}\frac{\frac{\partial u}{\partial y}\,dy\,dt}{dy + \frac{\partial v}{\partial y}\,dy\,dt}\right) = \frac{\partial u}{\partial y}\,dt \tag{1-12}$$

Substituting Eq. (1-12) into (1-11), we find that the *rate* of rotation (angular velocity) about the z axis is given by

$$\frac{d\Omega_z}{dt} = \frac{1}{2}\left(\frac{\partial v}{\partial x} - \frac{\partial u}{\partial y}\right) \tag{1-13}$$

In exactly similar fashion, the *rates* of rotation about the x and y axes are

$$\frac{d\Omega_x}{dt} = \frac{1}{2}\left(\frac{\partial w}{\partial y} - \frac{\partial v}{\partial z}\right) \qquad \frac{d\Omega_y}{dt} = \frac{1}{2}\left(\frac{\partial u}{\partial z} - \frac{\partial w}{\partial x}\right) \tag{1-14}$$

These are clearly the three components of the angular velocity vector $d\Omega/dt$. The three factors of one-half are irritating, and it is customary to work instead with a quantity $\boldsymbol{\omega}$ equal to twice the angular velocity:

$$\boldsymbol{\omega} = 2\frac{d\Omega}{dt} \tag{1-15}$$

The new quantity $\boldsymbol{\omega}$, of vital interest in fluid mechanics, is called the *vorticity* of the fluid. By inspecting Eqs. (1-13) to (1-15), we see that vorticity and velocity are related by the vector calculus:

$$\boldsymbol{\omega} = \operatorname{curl} \mathbf{V} = \nabla \times \mathbf{V} \tag{1-16}$$

and hence the divergence of vorticity vanishes identically:

$$\operatorname{div} \boldsymbol{\omega} = \nabla \cdot \boldsymbol{\omega} = \operatorname{div} \operatorname{curl} \mathbf{V} = 0 \tag{1-17}$$

Mathematically speaking, we say the vorticity vector is *solenoidal*. Note also that vorticity is intimately connected with convective acceleration through Eq. (1-10). If $\boldsymbol{\omega} = 0$, the flow is *irrotational*.

Now consider the two-dimensional shear strain, which is commonly defined as the average *decrease* of the angle between two lines which are initially perpendicular in the unstrained state. Taking AB and BC in Fig. 1-14 as our initial lines, the shear–strain increment is obviously $\frac{1}{2}(d\alpha + d\beta)$. The shear–strain *rate* is

$$\epsilon_{xy} = \frac{1}{2}\left(\frac{d\alpha}{dt} + \frac{d\beta}{dt}\right) = \frac{1}{2}\left(\frac{\partial v}{\partial x} + \frac{\partial u}{\partial y}\right) \tag{1-18}$$

Similarly, the other two components of shear–strain *rate* are

$$\epsilon_{yz} = \frac{1}{2}\left(\frac{\partial w}{\partial y} + \frac{\partial v}{\partial z}\right) \qquad \epsilon_{zx} = \frac{1}{2}\left(\frac{\partial u}{\partial z} + \frac{\partial w}{\partial x}\right) \tag{1-19}$$

By analogy with solid mechanics, the shear–strain rates are symmetric, that is, $\epsilon_{ij} = \epsilon_{ji}$.

The fourth and final particle motion is dilatation, or extensional strain. Again with reference to Fig. 1-14, the extensional strain in the x direction is defined as the fractional increase in length of the horizontal side of the element. This is given by

$$\epsilon_{xx}\, dt = \frac{(dx + \partial u/\partial x\, dx\, dt) - dx}{dx} = \frac{\partial u}{\partial x}\, dt \tag{1-20}$$

with exactly similar expressions for the other two strains. Thus the three extensional strain *rates* are

$$\epsilon_{xx} = \frac{\partial u}{\partial x} \qquad \epsilon_{yy} = \frac{\partial v}{\partial y} \qquad \epsilon_{zz} = \frac{\partial w}{\partial z} \tag{1-21}$$

Taken as a whole, the strain rates, both extensional and shear, constitute a symmetric second-order tensor, which may be visualized as the array

$$\epsilon_{ij} = \begin{pmatrix} \epsilon_{xx} & \epsilon_{xy} & \epsilon_{xz} \\ \epsilon_{yx} & \epsilon_{yy} & \epsilon_{yz} \\ \epsilon_{zx} & \epsilon_{zy} & \epsilon_{zz} \end{pmatrix} \tag{1-22}$$

Although the component magnitudes vary with a change of axes x, y, and z, the strain-rate tensor, like the stress tensor [Eq. (2-14)] and the strain tensor of elasticity, follows the transformation laws of symmetric tensors [see, e.g., Aris (1990), Spain (2003), or Talpaert (2003)]. In particular, there are three invariants that are independent of direction or choice of axes:

$$\begin{aligned} I_1 &= \epsilon_{xx} + \epsilon_{yy} + \epsilon_{zz} \\ I_2 &= \epsilon_{xx}\epsilon_{yy} + \epsilon_{yy}\epsilon_{zz} + \epsilon_{zz}\epsilon_{xx} - \epsilon_{xy}^2 - \epsilon_{yz}^2 - \epsilon_{zx}^2 \\ I_3 &= \begin{vmatrix} \epsilon_{xx} & \epsilon_{xy} & \epsilon_{xz} \\ \epsilon_{yx} & \epsilon_{yy} & \epsilon_{yz} \\ \epsilon_{zx} & \epsilon_{zy} & \epsilon_{zz} \end{vmatrix} \end{aligned} \tag{1-23}$$

A further property of symmetric tensors is that there exists one and only one set of axes for which the off-diagonal terms (the shear–strain rates in this case) vanish. These are the *principal axes*, for which the strain-rate tensor becomes

$$\begin{pmatrix} \epsilon_1 & 0 & 0 \\ 0 & \epsilon_2 & 0 \\ 0 & 0 & \epsilon_3 \end{pmatrix} \tag{1-24}$$

The quantities ϵ_1, ϵ_2, ϵ_3 are called the *principal strain rates*. For this special case, the three tensor invariants become

$$\begin{aligned} I_1 &= \epsilon_1 + \epsilon_2 + \epsilon_3 \\ I_2 &= \epsilon_1\epsilon_2 + \epsilon_2\epsilon_3 + \epsilon_3\epsilon_1 \\ I_3 &= \epsilon_1\epsilon_2\epsilon_3 \end{aligned} \tag{1-25}$$

If the invariants are known, Eqs. (1-25) are necessary and sufficient to solve for the principal strain rates, although there are actually much simpler ways to do this [see, e.g., Timoshenko and Goodier (1970)].

Finally, if we adopt the short notation $u_{i,j} = \partial u_i/\partial x_j$, where i and j are any two coordinate directions, note that any velocity derivative of this type can be split into two parts, one symmetric and one antisymmetric:

$$u_{i,j} = \tfrac{1}{2}(u_{i,j} + u_{j,i}) + \tfrac{1}{2}(u_{i,j} - u_{j,i}) \tag{1-26}$$

By comparing Eq. (1-26) with Eqs. (1-13), (1-14), (1-18), (1-19), and (1-21), we see that Eq. (1-26) can also be written as

$$\frac{\partial u_i}{\partial x_j} = \epsilon_{ij} + \frac{d\Omega_{ij}}{dt} \qquad (1\text{-}27)$$

That is, each velocity derivative can be resolved into a strain rate plus an angular velocity. The angular velocity does not distort the element, so that only the strain rate will cause a viscous stress, a fact which is exploited in Chap. 2.

To summarize, we have shown that all the kinematic properties of fluid flow—acceleration, translation, angular velocity, rate of dilatation, and shear–strain rate—are directly related to the fluid velocity vector $\mathbf{V} = (u, v, w)$. These relations are identical to the equivalent expressions from infinitesimal solid mechanics, if (u, v, w) are instead taken to be components of the *displacement* vector. This analogy between fluid and solid continuum mechanics is sometimes used to produce the equations of motion of a viscous linear fluid as a direct carryover from the equations of linear (Hookean) elasticity, with which students are often more familiar.

1-3.4 The Transport Properties of a Fluid

The three so-called transport properties are the coefficients of viscosity, thermal conductivity, and diffusion, so named because of the relation they bear to movement, or transport, of momentum, heat, and mass, respectively. Thus it is now popular in the field of chemical engineering to refer to a viscous-flow study as a problem in *momentum transport*, giving pause to the casual reader. The idea is not without merit. Each of the three coefficients relates a flux or transport to the gradient of a property. Viscosity relates momentum flux to velocity gradient, thermal conductivity relates heat flux to temperature gradient, and the diffusion coefficient relates mass transport to concentration gradient. Further, the mathematical properties of momentum, heat, and mass-flux problems are often similar and sometimes genuinely analogous. The perfect analogy, when it exists, is so striking that Bird et al. (2001) have devoted an entire textbook to an exposition of analogies among this trinity of problems. We should note now that the analogy fails in multidimensional problems because heat and mass flux are vectors while momentum flux (stress) is a tensor.

1-3.5 The Coefficient of Viscosity

The layperson knows from experience what *viscosity* means and associates it with the ability of a fluid to flow freely. Heavy oil takes a long time to flow out of a can. Light oil flows out quickly. The disastrous Boston molasses-tank explosion of 1919 [see Shank (1954)] caused one of the slowest floods in history. The idea of viscosity being proportional to time to flow has become accepted practice in the petroleum industry. Thus the motorist purchases oil with a viscosity labeled SAE 30. This means that 60 ml of this oil at a specified temperature takes 30 s to run out of a 1.76-cm hole in the bottom of a cup. This experiment is convenient and reproducible for very viscous

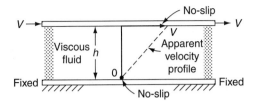

FIGURE 1-15
A fluid sheared between two plates.

liquids such as oil, but the time to flow is *not* viscosity, any more than the speed of sound is the time it takes an echo to return from a mountainside. It is an intriguing fact that the flow of a viscous liquid out of the bottom of a cup is a difficult problem for which no analytic solution exists at present.

A more fundamental approach to viscosity shows that it is the property of a fluid which relates applied stress to the resulting strain rate. The general relations are considered in Sec. 2-4. Here we consider a simple and widely used example of a fluid sheared between two plates, as in Fig. 1-15. This geometry is such that the shear stress τ_{xy} must be constant throughout the fluid. The motion is in the x direction only and varies with y, $u = u(y)$ only. Thus there is only a single finite strain rate in this flow:

$$\epsilon_{xy} = \frac{1}{2}\left(\frac{\partial u}{\partial y} + \frac{\partial v}{\partial x}\right) = \frac{1}{2}\frac{\partial u}{\partial y} = \frac{1}{2}\frac{du}{dy} \qquad (1\text{-}28)$$

If one performs this experiment, one finds that, for all the common fluids, the applied shear is a unique function of the strain rate:

$$\tau_{xy} = f(\epsilon_{xy}) \qquad (1\text{-}29)$$

Since, for a given motion V of the upper plate, τ_{xy} is constant, it follows that in these fluids ϵ_{xy}, and hence du/dy, is constant, so that the resulting velocity profile is linear across the plate, as sketched in Fig. 1-15. This is true regardless of the actual form of the functional relationship in Eq. (1-29). If the no-slip condition holds, the velocity profile varies from zero at the lower wall to V at the upper wall (Prob. 1-16 considers the case of a slip boundary condition). Repeated experiments with various values of τ_{xy} will establish the functional relationship Eq. (1-29). For simple fluids such as water, oils, or gases, the relationship is linear or *newtonian*:

$$\tau_{xy} \sim \epsilon_{xy}$$

or

$$\tau_{xy} = \mu \frac{V}{h} = 2\mu\epsilon_{xy} = \mu \frac{du}{dy} \qquad (1\text{-}30)$$

The quantity μ, called the *coefficient of viscosity of a newtonian fluid*, is what handbooks commonly quote when listing the viscosity of a fluid (see App. A). Actually, there is also a second coefficient, λ, related to bulk fluid expansions, but it is rarely encountered in practice (see Sec. 2-4). Equation (1-30) shows that the dimensions

24 VISCOUS FLUID FLOW

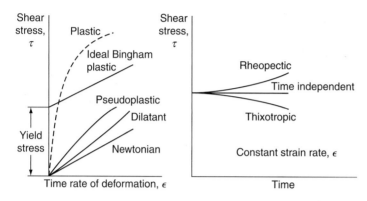

FIGURE 1-16
Viscous behavior of various materials.

of μ are stress-time: $N \cdot s/m^2$ [or $kg/(m \cdot s)$] in metric units and $lbf \cdot s/ft^2$ [or $slugs/(ft \cdot s)$] in English units. The conversion factor is

$$1 \text{ N} \cdot s/m^2 = 0.020886 \text{ lbf} \cdot s/ft^2 \quad (1\text{-}31)$$

The coefficient μ is a thermodynamic property and varies with temperature and pressure. Data for common fluids are given in App. A.

If the functional relationship in Eq. (1-29) is nonlinear, the fluid is said to be *nonnewtonian*. Some examples of nonnewtonian behavior are sketched in Fig. 1-16. Curves for true fluids, which cannot resist shear, must pass through the origin on a plot of τ vs. ϵ. Other substances, called *yielding fluids*, show a finite stress at zero strain rate and are really part fluid and part solid.

The curve labeled *pseudoplastic* in Fig. 1-16 is said to be shear-thinning, since its slope (local viscosity) decreases with increasing stress. If the thinning effect is very strong, the fluid may be termed *plastic*, as shown. The opposite case of a shear-thickening fluid is usually called a *dilatant* fluid, as shown.

Also illustrated in Fig. 1-16 is a material with a finite yield stress, followed by a linear curve at finite strain rate. This idealized material, part solid and part fluid, is called a *Bingham plastic* and is commonly used in analytic investigations of yielding materials under flow conditions. Yielding substances need not be linear but may show either dilatant or pseudoplastic behavior.

Still another complication of some nonnewtonian fluids is that their behavior may be time-dependent. If the strain rate is held constant, the shear stress may vary, and vice versa. If the shear stress decreases, the material is called *thixotropic*, while the opposite effect is termed *rheopectic*. Such curves are also sketched in Fig. 1-16.

A simple but often effective analytic approach to nonnewtonian behavior is the power-law approximation of Ostwald and de Waele:

$$\tau_{xy} \approx 2K\epsilon_{xy}^n \quad (1\text{-}31a)$$

where K and n are material parameters which in general vary with pressure and temperature (and composition in the case of mixtures). The exponent n delineates three cases on the left-hand side of Fig. 1-16.

$$\begin{aligned} n < 1 &\quad \text{pseudoplastic} \\ n = 1 &\quad \text{newtonian } (K = \mu) \\ n > 1 &\quad \text{dilatant} \end{aligned} \quad (1\text{-}31b)$$

Note the Eq. (1-31a) is unrealistic near the origin, where it would incorrectly predict the pseudoplastic to have an infinite slope and the dilatant a zero slope. Hence many other formulas have been proposed for nonnewtonian fluids; see, e.g., Bird et al. (2001) or Hutton et al. (1989). The reader is advised to become familiar with such flows although the present text is confined to the study of newtonian flow.

1-3.6 Viscosity as a Function of Temperature and Pressure

The coefficient of viscosity of a newtonian fluid is directly related to molecular interactions and thus may be considered a thermodynamic property in the macroscopic sense, varying with temperature and pressure. The theory of the transport properties of gases and liquids is still being developed, and a comprehensive review is given by Hirschfelder et al. (1954). Extensive data on properties of fluids are given by Reid et al. (1987).

No single functional relation $\mu(T, p)$ really describes any large class of fluids, but reasonable accuracy (± 20 percent) can be achieved by nondimensionalizing the data with respect to the critical point (T_c, p_c). This procedure is the so-called principle of corresponding states [Keenan (1941)], wherein the given property, here μ/μ_c, is found to be roughly a function of T/T_c and p/p_c, the reduced temperature and pressure. This principle is *not* justified on thermodynamic grounds but arises simply from dimensional analysis and experimental observation. Since changes occur very rapidly near the critical point, T_c and p_c are known only approximately, and it is essentially impossible to measure μ_c accurately. Appendix A contains a table of critical constants (T_c, p_c, μ_c, k_c) for common fluids, which should be regarded as best-fit values.

Figure 1-17 shows a recommended correlation of reduced viscosity μ/μ_c vs. reduced temperature T/T_c and reduced pressure p/p_c. As stated, the accuracy for any given fluid is about ± 20 percent. By examining this figure, we can make the following general statements:

1. The viscosity of liquids decreases rapidly with temperature.
2. The viscosity of low-pressure (dilute) gases increases with temperature.
3. The viscosity always increases with pressure.
4. Very poor accuracy obtains near the critical point.

Since p_c for most fluids is greater than 10 atm (App. A), typical gas flow problems are at low reduced pressure and approximate the low-density-limit curve of Fig. 1-17.

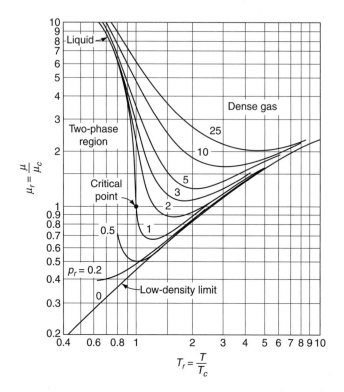

FIGURE 1-17
Reduced viscosity vs. reduced temperature for several values of reduced pressure $p_r = p/p_c$. [After Uyehara and Watson (1944).]

Thus it is common in aerodynamics, for example, to ignore the pressure dependence of gas viscosity and consider only temperature variations.

1-3.7 The Low-Density Limit

The kinetic theory of dilute gases is now highly refined and has been the subject of several excellent texts, e.g., Brush (1972), Present (1958), and Hirschfelder et al. (1954). In these theories, which date back to Maxwell, the macroscopic concept of viscosity is related to a statistical average of the momentum exchange occurring between molecules of the fluid. For dilute gases, the viscosity is found to be proportional to the density ρ, the mean free path ℓ, and the speed of sound a in the gas. The accepted approximation for nonpolar gases is a slight adjustment of Maxwell's original expression:

$$\mu \approx 0.67 \rho \ell \, a \qquad (1\text{-}32)$$

The product $\rho \ell$ is approximately constant for dilute gases, but more refined calculations by Chapman and Cowling (1970) show that it varies slightly with temperature due to the so-called collision integral Ω_v, computed from an approximate intermolecular force potential between the given molecules of effective collision diameter σ. Thus the generalized kinetic-theory formula for dilute-gas viscosity is of the form.

$$\mu = \frac{2.68\text{E}-6\sqrt{MT}}{\sigma^2 \Omega_v} \quad \text{[Chapman and Cowling (1970)]} \quad (1\text{-}33)$$

where σ = collision diameter, Å
M = molecular weight of gas
μ = viscosity, kg/(m · s)
T = absolute temperature, K

Ω_v is dimensionless and equal to unity for noninteracting molecules. Note that the mean free path can be computed by comparing Eqs. (1-32) and (1-33). In the general theory, Ω_v is found to depend upon the ratio T/T_ϵ, where T_ϵ is an effective temperature characteristic of the force potential chosen and the particular molecule. Most calculations are based either upon the Stockmayer potential or the Lennard–Jones potential function. Some calculated values of Ω_v for the Stockmayer potential function are given in Table 1-1. A curve-fit formula which fits these computed values to ±2 percent is

$$\Omega_v \approx 1.147 \left(\frac{T}{T_\epsilon}\right)^{-0.145} + \left(\frac{T}{T_\epsilon} + 0.5\right)^{-2.0} \quad (1\text{-}34)$$

The second term is negligible for T/T_ϵ greater than about 10. Values of the molecular parameters T_ϵ and σ are given in App. A for the common gases.

For routine calculations, a simpler formula than Eq. (1-33) may be desired. At higher temperatures, Eq. (1-34) indicates that $\Omega_v \sim T^{-0.145}$ and hence μ from Eq. (1-33) is proportional to $T^{0.645}$, which is approximately correct for all gases. Figure 1-17 also suggests the same power-law behavior. Thus a common approximation for the viscosity of dilute gases is the power law

$$\frac{\mu}{\mu_0} \approx \left(\frac{T}{T_0}\right)^n \quad (1\text{-}35)$$

TABLE 1-1
Collision integrals computed from the Stockmayer potential

$T^* = T/T_\epsilon$	Ω_v	Ω_v [Eq. (1-34)]
0.3	2.840	2.928
1.0	1.593	1.591
3.0	1.039	1.060
10.0	0.8244	0.8305
30.0	0.7010	0.7015
100.0	0.5887	0.5884
400.0	0.4811	0.4811

Source: Data from Hirschfelder et al. (1954).

TABLE 1-2
Power-law and Sutherland-law viscosity parameters for gases
[Eqs. (1-35) and (1-36)][†]

Gas	T_0, K	μ_0, N·s/m²	n	Error, %, temperature range, K		S, K	Temperature range for 2% error
Air	273	1.716E−5	0.666	±4	210–1900	111	170–1900
Argon	273	2.125E−5	0.72	±3	200–1500	144	120–1500
CO_2	273	1.370E−5	0.79	±5	209–1700	222	190–1700
CO	273	1.657E−5	0.71	±2	230–1500	136	130–1500
N_2	273	1.663E−5	0.67	±3	220–1500	107	100–1500
O_2	273	1.919E−5	0.69	±2	230–2000	139	190–2000
H_2	273	8.411E−6	0.68	±2	80–1100	97	220–1100
Steam	350	1.12E−5	1.15	±3	280–1500	1064	360–1500

Source: Data from Hilsenrath et al. (1955).
[†]No data given above maximum temperature listed. Formulas inaccurate below minimum temperature listed.

where n is of the order of 0.7 and T_0 and μ_0 are reference values. This formula was suggested by Maxwell also and later deduced on purely dimensional grounds by Rayleigh. Table 1-2 lists empirical values of n for various gases and the accuracy obtainable for a given temperature range.

A second widely used approximation resulted from a kinetic theory by Sutherland (1893) using an idealized intermolecular-force potential. The final formula is

$$\frac{\mu}{\mu_0} \approx \left(\frac{T}{T_0}\right)^{3/2} \frac{T_0 + S}{T + S} \qquad (1\text{-}36)$$

where S is an effective temperature, called the *Sutherland constant*, which is characteristic of the gas. The accuracy is slightly better than that of Eq. (1-35) for the same temperature range. Values of S for common gases are also given in Table 1-2. Less common gases are tabulated in App. A.

These dilute-gas formulas are strictly valid only for a single component substance; air qualifies only because its two principal components, oxygen and nitrogen, are nearly identical diatomic molecules. For mixtures of gases of markedly different species, the mixture viscosity varies strongly with species concentration. A good discussion of transport properties of gas mixtures is given in Bird et al. (2001).

1-3.8 The Coefficient of Thermal Conductivity

It is well established in thermodynamics that heat flow is the result of temperature variation, i.e., a temperature gradient. This can be formally expressed as a proportionality between heat flux and temperature gradient, i.e., Fourier's law:

$$\mathbf{q} = -k\nabla T \qquad (1\text{-}37)$$

where **q** is the vector rate of heat flow per unit area. The quantity k is our second transport coefficient, the thermal conductivity. Solid substances often show anisotropy, or directional sensitivity:

$$-\mathbf{q}_{sol} = \left(k_x \frac{\partial T}{\partial x}, k_y \frac{\partial T}{\partial y}, k_z \frac{\partial T}{\partial z} \right) \qquad (1\text{-}38)$$

Fortunately, a fluid is isotropic, i.e., has no directional characteristics, and thus k is a thermodynamic property and, like viscosity, varies with temperature and pressure. Also, like viscosity, the common fluids will correlate with their critical properties, as shown in Fig. 1-18. Note that the general remarks for Fig. 1-17 also apply to Fig. 1-18 but the actual numerical values are quite different. The low-density limit for k is quite practical for the problems in this text, and the kinetic theory of dilute gases applies once more.

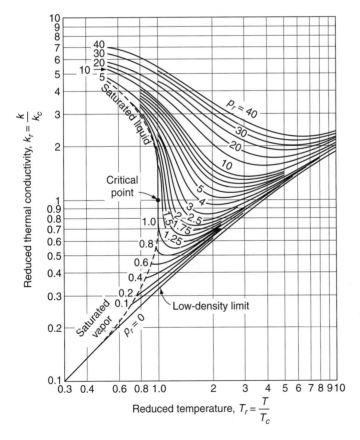

FIGURE 1-18
Reduced thermal conductivity vs. reduced temperature. [*After Owens and Thodos (1957).*]

By inspection of Eq. (1-37), we see that thermal conductivity k should have dimensions of heat per time per length per degree, the common engineering unit being Btu/(h)(ft)(°R). The metric unit is W/(m · K), and the conversion factor is

$$1 \text{ W/(m·K)} = 0.5778 \text{ Btu/(h)(ft)(°R)} \tag{1-39}$$

Also, k has the dimensions of viscosity times specific heat, so that the ratio of these is a fundamental parameter called the *Prandtl number*:

$$\text{Prandtl number} = Pr = \frac{\mu c_p}{k} \tag{1-40}$$

This parameter involves fluid properties only, rather than length and velocity scales of the flow. As we shall see, the Prandtl number is important in heat-transfer calculations but does not enter into friction computations.

The kinetic theory for conductivity of dilute gases is very similar to the viscosity analysis and leads to an expression similar to Eq. (1-33):

$$k \text{ W/(m·K)} = \frac{0.0833\sqrt{T}}{\sigma^2 \Omega_v \sqrt{M}} \tag{1-41}$$

where σ, Ω_v, and M have the same meanings and the same numerical values as in Eq. (1-33) and T is in kelvins. The collision integral Ω_v is again given by Table 1-1. If Eq. (1-41) is divided by Eq. (1-33), the result is

$$Pr = \frac{\mu c_p}{k} \approx \frac{4\gamma}{15\gamma - 15} \qquad \gamma = \frac{c_p}{c_v} \tag{1-42}$$

which relates the Prandtl number to the specific-heat ratio of the gas. The accuracy is only fair, however, and this relation was modified by Eucken (1913) and now serves as a purely empirical correlation formula:

$$Pr \approx \frac{4\gamma}{7.08\gamma - 1.80} = \begin{cases} 0.690 & \text{if } \gamma = 1.40 \text{ (diatomic gas)} \\ 0.667 & \text{if } \gamma = 1.67 \text{ (monatomic gas)} \end{cases} \tag{1-43}$$

This is seen to be a fair approximation to the Prandtl numbers of the gases plotted in Fig. 1-26.

For routine calculations with dilute gases, the power law and Sutherland formula, which correlated viscosity data, can also be used for thermal conductivity:

Power law:
$$\frac{k}{k_0} \approx \left(\frac{T}{T_0}\right)^n \tag{1-44a}$$

Sutherland:
$$\frac{k}{k_0} \approx \left(\frac{T}{T_0}\right)^{3/2} \frac{T_0 + S}{T + S} \tag{1-44b}$$

The accuracy is from 2 to 4 percent, depending upon the gas. Values of n, S, k_0, and T_0 are given for the common gases in Table 1-3, along with their accuracy when compared with the data compiled by White (1988).

TABLE 1-3
Power-law and Sutherland-law thermal-conductivity parameters for gases [Eqs. (1-44a) and (1-44b)]

Gas	T_0, K	k_0, W/m·K	n	Error, % temperature range, K	S, K	Temperature range for $\pm 2\%$ error, K
Air	273	0.0241	0.81	±3 210–2000	194	160–2000
Argon	273	0.0163	0.73	±4 210–1800	170	150–1800
CO_2	273	0.0146	1.30	±2 180–700	1800	180–700
CO	273	0.0232	0.82	±2 210–800	180	200–800
N_2	273	0.0242	0.74	±3 210–1200	150	200–1200
O_2	273	0.0244	0.84	±2 220–1200	240	200–1200
H_2	273	0.168	0.72	±2 200–1000	120	200–1000
Steam	300	0.0181	1.35	±2 300–900	2200	300–700

Source: Data from White (1988).

1-3.9 The Coefficient of Mass Diffusivity

The third and final transport coefficient is associated with the movement of mass due to molecular exchange. This process, called *diffusion*, constantly occurs in a fluid because of random molecular motion; the air in a given portion of a room continually loses some particles and gains others, so that the specific content of a given volume is always changing. We cannot notice the change on a macroscopic basis because the air mixture remains the same; such diffusion of nearly identical particles is called *self-diffusion*.

Diffusion becomes macroscopically evident when a variable mixture of two or more species is involved. If we stand on one side of a room while an amount of hydrogen sulfide gas is released at the other side, we soon notice the odor as the hydrogen sulfide diffuses to our side, replacing some of our air, which diffuses to the other side. The transfer of mass is equal and opposite, i.e., diffusion is an equal exchange of species, and the final state in this case would be a uniform mixture of air and hydrogen sulfide throughout the room. It is well to note that the rate of diffusion is *not* equal to the mean molecular speed $\bar{c}$ but is far slower. Many molecular collisions occur in a short distance, the mean free path; hence the hydrogen sulfide "front" advances slowly, a mean free path at a time, so to speak.

Mass diffusion occurs, then, whenever there is a gradient in the proportions of a mixture, i.e., a *concentration gradient*. There are two different definitions of concentration in frequent use: (1) the volume concentration $\rho_i = m_i/\text{volume} = $ mass of component i per unit volume and (2) the mass concentration $C_i = \rho_i/\rho = $ the mass of species i per unit mass of the mixture. The second definition or mass concentration will be more useful in this text if for no other reason than that it is a dimensionless fraction less than or equal to unity. By analogy with viscosity and thermal conductivity, we postulate that mass diffusion per unit area is proportional to concentration gradient. In vector form, we have

$$\frac{\dot{m}_i}{A} = -D\nabla(\rho_i) \tag{1-45}$$

where $\dot{m}_i$ is the mass flux of species i in the direction of decreasing concentration (which accounts for the minus sign). The quantity D is called the *coefficient of mass diffusivity* and has dimensions of (length)^2/time, usually either square feet per second or square centimeters per second. Equation (1-45) is called *Fick's law of diffusion*.

In terms of mass concentration $C_i = \rho_i/\rho$, Eq. (1-45) becomes

$$\frac{\dot{m}_i}{A} = -\rho_i D\nabla(\ln C_i) \tag{1-46}$$

where ln denotes the natural logarithm. Since mass flux per unit area equals density times velocity, we can write, by definition,

$$\frac{\dot{m}_i}{A} = \rho_i \mathbf{V}_i \tag{1-47}$$

where $\mathbf{V}_i$ is called the *diffusion velocity* of species i. In this simpler form, then, Fick's law can be written as

$$\mathbf{V}_i = -D\nabla(\ln C_i) \tag{1-48}$$

Here we account only for concentration gradients. Rigorous kinetic theory shows that species mass fluxes may also be caused by pressure gradients or by temperature gradients, with a general expression

$$\mathbf{V}_i = -D\nabla(\ln C_i) - D_p\nabla(\ln p) - D_T\nabla(\ln T) \tag{1-49}$$

as explained in Chapman and Cowling (1970), for example. The pressure-diffusion term is usually small because pressure gradients are smaller than concentration gradients in flow problems. The thermal-diffusion term is usually neglected because D_T is typically much smaller than D.

The diffusion coefficient D has the same dimensions as the kinematic viscosity $\nu = \mu/\rho$ or the thermal diffusivity $\alpha = k/\rho c_p$. Thus their ratios, like the Prandtl number considered earlier, are important dimensionless parameters in a fluid-diffusion problem:

$$\text{Schmidt number} = Sc = \frac{\mu}{\rho D} = \frac{\nu}{D}$$

$$\text{Lewis number} = Le = \frac{\rho c_p D}{k} = \frac{D}{\alpha} \tag{1-50}$$

The Schmidt number relates viscous diffusion to mass diffusion, and the Lewis number compares mass diffusion to thermal diffusion.

For dilute gases, kinetic theory yields the following expression for the binary-diffusion coefficient D_{12} between species 1 and 2 [Chapman and Cowling (1970)]:

$$D_{12} = \frac{0.001858 T^{3/2}[(M_1 + M_2)/M_1 M_2]^{1/2}}{p\sigma_{12}^2 \Omega_D} \quad (1\text{-}51)$$

where
- D_{12} = binary-diffusion coefficient, cm^2/s
- M_1, M_2 = molecular weight of two gases
- Ω_D = diffusion collision integral
- σ_{12} = effective collision diameter, Å
- T = absolute temperature, K
- p = pressure, atm

The collision integral is about 10 percent smaller than Ω_v from Table 1-1 and can be approximated closely by modifying Eq. (1-34):

$$\Omega_D \approx 1.0 T^{*-0.145} + (T^* + 0.5)^{-2.0} \quad (1\text{-}52)$$

where $T^* = T/T_{\epsilon_{12}}$. The effective temperatures and diameters are averages computed from the separate molecular properties of each species:

$$\begin{aligned}\sigma_{12} &= \tfrac{1}{2}(\sigma_1 + \sigma_2) \\ T_{\epsilon_{12}} &= (T_{\epsilon_1} T_{\epsilon_2})^{1/2}\end{aligned} \quad (1\text{-}53)$$

The molecular properties listed in App. A are valid for these diffusion calculations also. Note that, unlike viscosity and conductivity, diffusion coefficient D varies inversely with pressure in Eq. (1-51). Thus, at low pressures, product pD is a function only of temperature. For dense gases, there is considerable pressure dependence also, but very few data are available. Figure 1-19 is a tentative plot of this pressure effect for self-diffusion D_{11} (nearly identical molecules). Note that reduced

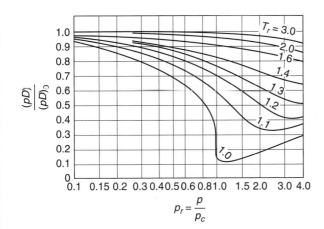

FIGURE 1-19
Tentative generalized chart for self-diffusivity of a dense gas. [*After Slattery and Bird (1958).*]

diffusivity D/D_c cannot be used. The plot involves instead the ratio of pD to its value $(pD)_0$ at the same temperature but very low pressures.

Since binary-diffusion data are scarce, we shall not attempt to present a power law or a Sutherland formula for estimating D_{12} of dilute gases. For further information on diffusion estimates, consult Reid et al. (1987).

1-3.10 Transport Properties of Dilute-Gas Mixtures

The previous discussion of viscosity has been confined essentially to fluids of a single species. For gases, the kinetic theory has been extended, for the low-density limit, to multicomponent mixtures. The details are given by Hirschfelder et al. (1954). For routine calculations, the semiempirical formula of Wilke (1950) is recommended:

$$\mu_{\text{mix}} \approx \sum_{i=1}^{n} \frac{x_i \mu_i}{\sum_{j=1}^{n} x_j \phi_{ij}} \qquad (1\text{-}54)$$

where

$$\phi_{ij} = \frac{[1 + (\mu_i/\mu_j)^{1/2}(M_j/M_i)^{1/4}]^2}{(8 + 8M_i/M_j)^{1/2}}$$

This formula is for a mixture of n gases, where M_i are the molecular weights and x_i are the mole fractions. In terms of the *mass* fractions $C_i = \rho_i/\rho$, we have

$$x_i = \frac{C_i/M_i}{\sum_{j=1}^{n}(C_j/M_j)} \qquad (1\text{-}55)$$

A similar formula is recommended by Wilke for the thermal conductivity k of a mixture of n gases except that μ_i are replaced by k_i.

1-3.11 Transport Properties of Liquids

The theoretical analysis of liquid transport properties is not nearly as well developed as that for gases. The difficulty is that liquid molecules are very closely packed compared to gases and thus dominated by large intermolecular forces. Momentum transfer by collisions—so dominant in gases—is small in liquids. The kinetic theory of liquids is summarized in the texts by Bird et al. (1977) and Reid et al. (1987). However, the theory is not yet quantitative, in the sense that a given liquid's properties cannot be predicted from other thermodynamic data.

If no data are available, we recommend the reduced temperature and pressure plots, Figs. 1-17 and 1-18, for ±20 percent accuracy. If data are available for calibration, however, they may be fit accurately to the empirical formula:

$$\text{Liquid:} \qquad \ln \frac{\mu}{\mu_0} \approx a + b\left(\frac{T_0}{T}\right) + c\left(\frac{T_0}{T}\right)^2 \qquad (1\text{-}56)$$

where (μ_0, T_0) are reference values and (a, b, c) are dimensionless curve-fit constants. For nonpolar liquids, $c \approx 0$, i.e., the plot is linear.

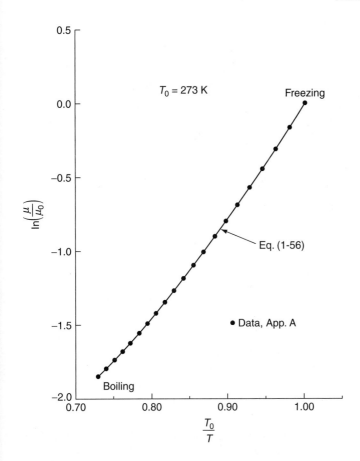

FIGURE 1-20
Empirical plot of the viscosity of water in the manner suggested by Eq. (1-56).

For example, App. A tabulates the viscosity of water at atmospheric pressure. If these data are plotted in the manner suggested by Eq. (1-56), the results are shown in Fig. 1-20. The plot is nearly linear, and the curve-fit values are

$$a = -2.10 \quad b = -4.45 \quad c = 6.55 \qquad (1\text{-}57)$$

corresponding to $T_0 = 273$ K and $\mu_0 = 0.00179$ kg/m·s. The accuracy of the curve fit is ± 1 percent when compared to the data of App. Table A.1.

For thermal conductivity of liquids, Eq. (1-56) may not be a good fit—see, e.g., data for water in App. Fig. A-5. Reid et al. (1987) recommend the simple linear fit

$$k_{\text{liq}} \quad \text{or} \quad D_{\text{liq}} \approx a + bT \qquad (1\text{-}58)$$

which will be accurate at least over a limited temperature range.

1-3.12 The Thermodynamic Properties

As stated before, by their very nature viscous flows are technically not in equilibrium, but at normal densities the deviations from equilibrium are negligible. Two exceptions to this are chemically reacting flows and very sudden state changes, as in a strong shock wave. Generally, though, it is quite reasonable to assume that a moving viscous fluid is a pure substance whose properties are related by ordinary equilibrium thermodynamics. The properties most important to this subject are pressure, density, temperature, entropy, enthalpy, and internal energy. Of these six, two may be regarded as the independent variables from which all others follow through experimental (or theoretical) equations of state. In Chap. 2, it will be convenient to consider p and T as primary variables, but technically this is a flawed choice.

The first law of thermodynamics can be written as

$$dE = dQ + dW \tag{1-59}$$

where dE = change in total energy of system
dQ = heat added to system
dW = work done on system

For a substance at rest with infinitesimal changes, we have

$$dW = -p\, d\,(\text{volume})$$
$$dQ = T\, dS \tag{1-60}$$

Substituting in Eq. (1-59) and expressing the result on a unit mass basis, we have

$$de = T\, ds + \frac{p}{\rho^2}\, d\rho \tag{1-61}$$

which is one form of the first and second laws combined for infinitesimal processes. Equation (1-61) implies that the single state relation

$$e = e(s, \rho) \tag{1-62}$$

is entirely sufficient to define a fluid thermodynamically. For, in terms of the calculus of two variables,

$$de = \frac{\partial e}{\partial s}\, ds + \frac{\partial e}{\partial \rho}\, d\rho \tag{1-63}$$

from which, by comparison with Eq. (1-61), the temperature and pressure can be calculated by

$$T = \left.\frac{\partial e}{\partial s}\right|_\rho \qquad p = \left.\rho^2 \frac{\partial e}{\partial \rho}\right|_s \tag{1-64}$$

after which the enthalpy can be calculated from its definition:

$$h = e + \frac{p}{\rho} \tag{1-65}$$

Thus a single chart of e vs. s for lines of constant ρ is sufficient to calculate all thermodynamic properties. Equation (1-62) is therefore called a *canonical* equation of state.

A second and more popular canonical equation of state is obtained by using Eq. (1-65) to eliminate de in Eq. (1-61), with the result

$$dh = T\,ds + \frac{1}{\rho}\,dp \tag{1-66}$$

Again using calculus, we find that from the canonical relation

$$h = h(s, p) \tag{1-67}$$

all other properties can be calculated:

$$T = \left.\frac{\partial h}{\partial s}\right|_\rho \qquad \frac{1}{\rho} = \left.\frac{\partial h}{\partial p}\right|_s$$
$$e = h - \frac{p}{\rho} \tag{1-68}$$

In this case, a chart of h vs. s for lines of constant p will define the substance completely. Such a plot is very popular and is called a *Mollier chart*, after Richard Mollier (1863–1935), a German engineering professor who first proposed it.

Figure 1-21 is a sketch of the Mollier chart for air as given by Little (1963) from many sources of data. Tabular properties for eight common gases are given by Hilsenrath et al. (1955). Figure 1-21 shows only the constant-pressure lines, which

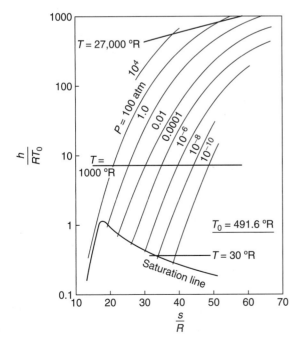

FIGURE 1-21
Mollier chart for equilibrium air. [*After Little (1963)*.]

are quite sufficient thermodynamically, as discussed previously. However, in preparing such a Mollier chart for publication, it is usual to include other lines (constant density, temperature, speed of sound). Data for water and steam are given by Parry et al. (2000).

1-3.13 Secondary Thermodynamic Properties

Still other properties are often used in flow analyses, particularly with idealized equations of state such as the perfect-gas law. Two of these are the specific heats (so-called) at constant pressure and constant volume:

$$c_p = \left.\frac{\partial h}{\partial T}\right|_p \qquad c_v = \left.\frac{\partial e}{\partial T}\right|_v \tag{1-69}$$

which of course are not heats at all but expressions of energy change. The ratio of specific heats

$$\gamma = \frac{c_p}{c_v} \tag{1-70}$$

is an important dimensionless parameter in high-speed (compressible) flow problems. This ratio lies between 1.0 and 1.7 for all fluids. For a liquid, which is nearly incompressible, $c_p \approx c_v$ and $\gamma \approx 1.0$. Figure 1-22 shows values of γ at atmospheric

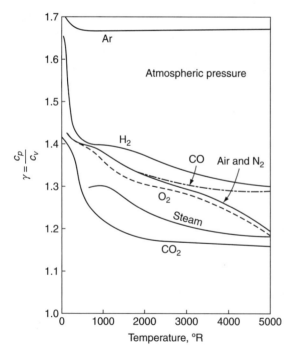

FIGURE 1-22
Specific-heat ratio for eight common gases. [*Data from Hilsenrath et al.* (*1955*).]

pressure for various gases as a function of absolute temperature. The pressure dependence of γ is very weak, and the figure shows that the approximation (γ = const) is accurate over fairly wide temperature variations.

Another minor but important thermodynamic property is the *speed of sound* a, defined as the rate of propagation of infinitesimal pressure pulses:

$$a^2 = \left.\frac{\partial p}{\partial \rho}\right|_s \tag{1-71}$$

[see, e.g., White (2003), Sec. 9.2]. The partial derivative in Eq. (1-71) is often rather clumsy to handle, in which case an alternate relation can be used:

$$a^2 = \gamma \left(\frac{\partial p}{\partial \rho}\right)_T \tag{1-72}$$

The student may prove as an exercise that Eqs. (1-71) and (1-72) are thermodynamically identical. Since viscous flows are definitely *not* isentropic in general, the speed of sound is not a natural inhabitant of viscous analyses but enters instead through the assumption of perfect-gas relations, as does the Mach number.

1-3.14 The Perfect Gas

All the common gases follow with reasonable accuracy, at least in some finite region, the so-called ideal or perfect-gas law:

$$p = \rho R T \tag{1-73}$$

where R is called the gas constant. Equation (1-73) has a solid theoretical basis in the kinetic theory of dilute gases, e.g., Brush (1972), and should not be regarded as an empirical formula. The gas constant R is the ratio of Boltzmann's constant to the mass of a single molecule:

$$R = \frac{K}{m} \tag{1-74}$$

Alternately, R may be written in terms of the molecular weight M of the gas:

$$R_{\text{gas}} = \frac{R_0}{M_{\text{gas}}} \tag{1-75}$$

where R_0 is a universal constant similar to Boltzmann's constant. In metric units,

$$R_0 = 8314 \text{ J/(kg} \cdot \text{mol} \cdot \text{K)} \tag{1-76}$$

which is too many significant figures, since no gas really fits the law that well. Equations (1-73) and (1-75) are also suitable for mixtures of gases if the equivalent molecular weight is properly defined in terms of the mass fractions $C_i = \rho_i/\rho$:

$$M_{\text{mix}} = \frac{1}{\sum(C_i/M_i)} \tag{1-77}$$

In terms of the mole fractions x_i (number of moles of species i per mole of mixture), we have

$$M_{\text{mix}} = \sum x_i M_i \qquad (1\text{-}78)$$

As a classic example, air at ordinary temperatures has mole fractions of approximately 78 percent nitrogen, 21 percent oxygen, and 1 percent argon. Then, from Eq. (1-78),

$$M_{\text{air}} = 0.78(28.016) + 0.21(32.000) + 0.01(39.944) = 28.97$$

from which

$$R_{\text{air}} = \frac{8314}{28.97} = 287 \text{ J/(kg} \cdot \text{K)}$$

These are the accepted values for room-temperature dry air.

According to Eq. (1-73), the so-called compressibility factor

$$Z = \frac{p}{\rho RT} \qquad (1\text{-}79)$$

should be unity for gases. Actually, Z varies from zero to 4.0 or greater, depending upon temperature and pressure. To good accuracy, Z is a function only of the reduced variables $p_r = p/p_{\text{crit}}$ and $T_r = T/T_{\text{crit}}$ referred to the critical point. This is illustrated in Figs. 1-23 and 1-24, which are representative of all gases. Examining these figures, we see that the perfect-gas law ($Z = 1$) is accurate to ± 10 percent in the range $1.8 \leq T_r \leq 15$ and $0 \leq p_r \leq 10$, which is the range of interest of the majority of viscous-flow problems. The higher temperatures should be viewed with

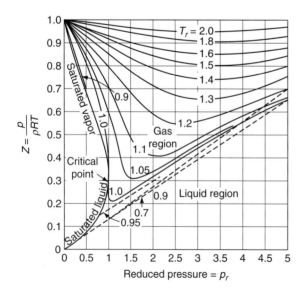

FIGURE 1-23
Compressibility factors for gases. [*After Weber (1939)*.]

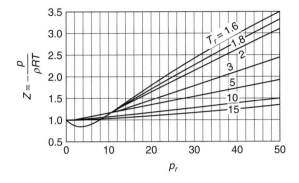

FIGURE 1-24
Compressibility at higher p_r and T_r. [*After Weber (1939)*.]

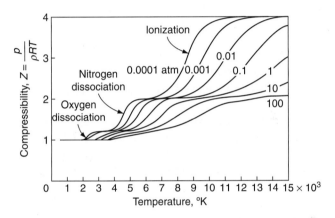

FIGURE 1-25
Calculated compressibility of air. [*After Hansen (1958)*.]

suspicion for multiatom molecules, which may dissociate into smaller particles and raise the value of Z markedly. In air, for example, dissociation happens twice: at 2000 K, where $O_2 \rightarrow 2O$, and at 4000 K, where $N_2 \rightarrow 2N$. Figure 1-25 shows the compressibility factor of air as calculated by Hansen (1958). Note the first and second plateaus corresponding to O_2 and then N_2 dissociation. The effect is particularly strong at low pressure.

1-3.15 Other Properties of the Perfect Gas

As a direct consequence of the perfect-gas law [Eq. (1-73)], many useful and simple relations arise. The specific heats are constrained to be functions of temperature only:

$$c_p = c_p(T) \text{ only}$$

and

$$c_v = c_v(T) \text{ only} \qquad (1\text{-}80)$$

42 VISCOUS FLUID FLOW

Hence, from Eqs. (1-69), we have

$$de = c_v\, dT$$
$$dh = c_p\, dT \quad (1\text{-}81)$$

Further, it follows that

$$\gamma = \gamma(T) \text{ only}$$

$$c_p(T) = c_v(T) + R = \frac{\gamma(T)R}{\gamma(T) - 1} \quad (1\text{-}82)$$

$$c_v(T) = \frac{R}{\gamma(T) - 1}$$

Finally, from Eq. (1-72), the speed of sound takes the simple form

$$a^2_{\text{perfect gas}} = \gamma RT \quad (1\text{-}83)$$

Because of reasonable accuracy and delightful simplicity, the perfect-gas relations are understandably popular in viscous-flow analysis. We make liberal use of the perfect-gas law in Chap. 7.

As shown in Fig. 1-22, the specific-heat ratio γ tends to decrease with temperature. Since all these gases are reasonably perfect, it follows from Eqs. (1-82)

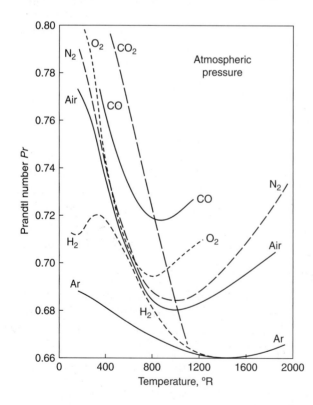

FIGURE 1-26
Prandtl number vs. temperature for seven gases. [*Data from Hilsenrath et al. (1955).*]

that c_p and c_v must increase with temperature. Since R is known from Eq. (1-75), Fig. 1-22 may be used to calculate c_p and c_v to good accuracy.

Finally, Fig. 1-26 shows the Prandtl number, $Pr = \mu c_p/k$, for various common gases. Note by comparison with Fig. 1-22 that Eucken's formula, Eq. (1-43), is only a fair approximation.

1-3.16 Bulk Modulus

In flow problems which involve sound-wave propagation, it is useful to have a thermodynamic property which expresses the change of density with increasing pressure. This property is the *bulk modulus K*:

$$K = \rho\left(\frac{\partial p}{\partial \rho}\right)_T \tag{1-84}$$

From Eq. (1-72), then, we see that the speed of sound can be written in terms of K:

$$a^2 = \frac{\gamma K}{\rho} \tag{1-85}$$

where γ may be taken as 1.0 for liquids and solids.[†] For a perfect gas, the reader may verify that $K = p$ itself. For liquids, however, K is fairly constant, varying slightly with pressure and temperature. For water, a good average value is $K = 2.2\text{E}9$ Pa. Taking $\gamma = 1.0$ and an average water density of 998 kg/m^3, we calculate the speed of sound in water to be approximately 1480 m/s.

1-3.17 Coefficient of Thermal Expansion

There is a subset of flow problems, called *natural convection*, where the flow pattern is due to buoyant forces caused by temperature differences. Such buoyant forces are proportional to the coefficient of thermal expansion β, defined as

$$\beta = -\frac{1}{\rho}\left(\frac{\partial \rho}{\partial T}\right)_p \tag{1-86}$$

For a perfect gas, the reader may show that $\beta = 1/T$. For a liquid, β is usually smaller than $1/T$ and may even be negative (the celebrated inversion of water near the freezing point). For imperfect gases, β can be considerably larger than $1/T$ near the saturation line, particularly at high pressures. This is illustrated in Fig. 1-27 for steam. We see that steam is well approximated by the perfect-gas result $\beta T = 1$ at low pressures and high temperatures.

[†]A solid has *two* sound speeds: $(K/\rho)^{1/2}$ is the dilatation, or longitudinal-wave, speed, and $(G/\rho)^{1/2}$ is the rotational, or shear-wave, speed. Fluids have only one sound speed.

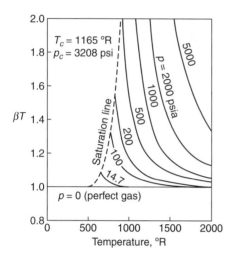

FIGURE 1-27
Thermal-expansion coefficient for steam.

The quantity β is also useful in estimating the dependence of enthalpy on pressure, from the thermodynamic relation

$$dh = c_p \, dT + (1 - \beta T)\frac{dp}{\rho} \tag{1-87}$$

where we remark that of course T must be absolute temperature. For the perfect gas, the second term vanishes, so that $h = h(T)$ only. Figure 1-27 shows that imperfect gases like steam also often fit this approximation $dh \approx c_p \, dT$.

Table 1-4 gives measured values of the bulk modulus K and the thermal-expansion coefficient β for water at saturation pressures.

TABLE 1-4
Bulk modulus K and expansion coefficient β for water at saturation conditions

T, K	p_{sat}, kPa	βT	K, MPa
273	0.61	−0.019	2,062
293	2.34	0.057	2,230
313	7.38	0.119	2,304
333	19.92	0.176	2,301
353	47.35	0.230	2,235
373	101.3	0.281	2,120
423	461	0.447	1,692
473	1,580	0.637	1,190
523	3,970	0.985	716
573	8,560	1.80	342
623	16,500	4.8	82
647[†]	22,090[†]	∞	0

[†]Critical point.

1-4 BOUNDARY CONDITIONS FOR VISCOUS-FLOW PROBLEMS

The equations of motion to be discussed in Chap. 2 will require mathematically tenable and physically realistic boundary conditions. It is of interest here to study the underlying physical mechanisms of the boundary approximations commonly used. For fluid flow, there are five types of boundaries considered:

1. A solid surface (which may be porous)
2. A free liquid surface
3. A liquid–vapor interface
4. A liquid–liquid interface
5. An inlet or exit section

Cases 2 and 3 are related in the sense that a free liquid surface is a special case of the liquid–vapor interface where the vapor causes a negligible interaction. Let us take up these cases in order.

1-4.1 Conditions at a Solid Surface: Microflow Slip in Liquids

Wall boundary conditions depend upon whether the fluid is a liquid or a gas. For *macroflows*, system dimensions are large compared to molecular spacing, so that both liquid and gas particles contacting the wall must essentially be in equilibrium with the solid. At a solid surface, the fluid will assume the velocity of the wall (the *no-slip* condition) and the temperature of the wall (the *no-temperature-jump* condition):

$$\mathbf{V}_{\text{fluid}} = \mathbf{V}_{\text{sol}}$$
$$T_{\text{fluid}} = T_{\text{sol}} \quad (1\text{-}88)$$

We shall use Eqs. (1-88) throughout this text for newtonian liquids. However, certain liquid/solid combinations are known to *slip* under small-scale *microflow* conditions. The subject is somewhat controversial and is reviewed in the monograph edited by Gad-el-Hak (2001). Some liquid microflows studies show slip and others do not.

One way to characterize slip in liquids is the *slip length* L_{slip} relating slip velocity to the local velocity gradient, a model first suggested by Navier himself:

$$u_{\text{wall}} = L_{\text{slip}} \left(\frac{\partial u}{\partial n}\right)_{\text{wall}} \quad (1\text{-}89)$$

The slip length depends upon the liquid, the geometry, and the shear rate. Tretheway and Meinhart (2002) test water flowing in a 30 μm width microchannel coated with hydrophobic octadecyltrichlorosilane (OTS). They measure $L_{\text{slip}} \approx 1$ μm and a wall-slip velocity equal to 10 percent of the centerline velocity. Choi et al. (2003), studying hydrophilic and hydrophobic coated 1–2 μm width microchannels, report $L_{slip} \approx$ 5–35 nm, increasing with shear rate. Lin and Schowalter (1989) measure both slip and stick-slip behavior when high molecular weight polybutadienes flow through a capillary. Brutin and Tadrist (2003) report an unexpected wall friction increase when water flows through fused silica microtubes. They hypothesize a fluid ionic coupling with the silica surface. Denn (2001) reviews the extrusion of polymer melts and reports slip-flow conditions. A recent conference, CECAM (2004), is entirely devoted to experimental measurements of microfluidic interfaces.

To the writer's knowledge, there is no fundamental theory yet that can predict the slip of liquids flowing past solid walls. This is not the case for gases.

1-4.2 Kinetic Theory for Slip Velocity in Gases

If the fluid is a gas, Eqs. (1-88) will fail when the mean free path ℓ is comparable to the length-scale L of the flow. The ratio of these two lengths is called the Knudsen number, Kn:

$$Kn = \frac{\ell}{L} \tag{1-90}$$

Thus, if $Kn \ll 1$, there is negligible slip. If $Kn = \mathcal{O}(0.1)$, there is slip. If $Kn = \mathcal{O}(1)$ or greater, the flow is molecular and slip is an inadequate concept. There are two ways for the Knudsen number to dominate: (1) if ℓ is very large, that is, a very rarefied gas, Cercignani (2000); and (2) if L is very small, as occurs in micro- and nanoflows, Karniadakis and Bestok (2001).

Consider gas molecules as they strike and reflect from a solid wall, as in Fig. 1-28. If the wall is perfectly smooth, the molecules would impinge and reflect at the same angle θ, similar to light rays from a mirror. This is *specular reflection*, where molecules conserve their tangential momentum and exert no wall shear. In this case, there is perfect slip flow at the wall.

On a molecular scale, however, even the most highly polished wall appears rough. It is more likely that the impinging molecules view the wall as rough and reflect at random angles uncorrelated with their entry angle. This is termed *diffuse*

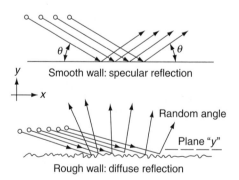

FIGURE 1-28
Specular and diffuse reflection of particles.

reflection and is shown in the lower half of Fig. 1-28. Equilibrium across a plane y near the wall requires that the loss of tangential momentum be balanced by a finite slip velocity u_w to transmit shear to the wall. As postulated by Navier in 1827, the slip velocity is proportional to the wall shear:

$$u_w \approx \alpha \tau_w$$

where α is a constant. Later, Maxwell in 1879 showed with kinetic theory that $\alpha \approx \ell/\mu$, or

$$u_w \approx \ell \left(\frac{\partial u}{\partial y}\right)_w \qquad (1\text{-}91)$$

where ℓ is the mean free path of the gas. See, for example, Chap. 8 of the text by Kennard (1938). One can approximate the fluid away from the wall as a newtonian continuum. If we assume that a fraction f of the molecules is reflected diffusely, the slip velocity becomes

$$u_w \approx \left(\frac{2-f}{f}\right)\ell\left(\frac{\partial u}{\partial y}\right)_w \qquad (1\text{-}92)$$

However, this correction seems superfluous, as a recent review by Sharipov and Seleznev (1998) indicates that the simpler Eq. (1-91) is quite accurate for single-component gases. Further, the recent theory of Sharipov and Kalempa (2003), based upon the Boltzmann equation, shows that Eq. (1-91) is only a few percent low for a wide variety of gas mixtures.

Is slip important in boundary-layer flows? When we substitute for ℓ from Eq. (1-32) and for $(\partial u/\partial y)$ from Eq. (1-91), viscosity cancels and the slip velocity becomes

$$u_w = \frac{3}{2}\frac{\tau_w}{\rho a} \qquad (1\text{-}93)$$

We can then compare the slip velocity to freestream velocity U by dividing and rearranging:

$$\frac{u_w}{U} = \frac{3}{4}\frac{U}{a}\frac{2\tau_w}{\rho U^2} = 0.75\, Ma\, C_f \qquad (1\text{-}94)$$

where $Ma = U/a$ is the Mach number of the freestream and C_f is called the *skin friction coefficient* of the flow. In turbulent flow, C_f is no larger than 0.005 and decreases with Mach number. In laminar flow, C_f increases as the Reynolds number decreases and can be quite large. Thus we conclude:

Boundary layers: Turbulent gas flow: no-slip

Laminar gas flow: possible slip at low Re_L and high Ma $\qquad (1\text{-}95)$

For further study of micro- and nanoflows and rarefied gases, see the monographs by Karniadakis and Bestok (2001) and Cercignani (2000). For most gas flow analyses in this text, we will use the standard no-slip condition, Eqs. (1-88).

For a new method of computer calculation of molecular flows, formulated by following and averaging thousands of individual particles, see the text by Bird (1994). This type of calculation is now called the direct simulation Monte Carlo method (DSMC).

1-4.3 Kinetic Theory for Wall-Temperature Jump

The temperature condition in Eq. (1-88) will also fail for a gas flow if the mean free path is large compared to flow dimensions. This effect is called *temperature jump* and is analogous to velocity slip. The kinetic-theory expression for the jump $T_{gas} - T_w$ in a gas flow was given by Smoluchowski (1898):

$$T_{gas} - T_w \approx \left(\frac{2}{\alpha} - 1\right)\frac{2\gamma}{\gamma + 1}\frac{\ell k}{\mu c_p}\left(\frac{dT}{dy}\right)_w \quad (1\text{-}96)$$

where α is the so-called thermal-accommodation coefficient, defined as the fraction of impinging molecules which becomes accommodated to the temperature of the wall. Clearly α is analogous to the diffuse-reflection coefficient f in Eq. (1-92). Experiments have shown that α is also fairly close to unity. With $\alpha = 1.0$, we can again substitute for ℓ from Eq. (1-32) and for dT/dy from Eq. (1-37) to obtain

$$T_{gas} - T_w \approx \frac{3\gamma}{\gamma + 1}\frac{\mu k}{\rho a \mu c_p}\frac{q_w}{k} \quad (1\text{-}97)$$

Eliminating μ and k, we divide by the temperature difference $T_r - T_w$ which controls the wall heat transfer (see Sec. 7-3) and rearrange:

$$\frac{T_{gas} - T_w}{T_r - T_w} \approx \frac{3\gamma}{\gamma + 1}\frac{U}{a}\frac{q_w}{\rho c_p U(T_r - T_w)} = \frac{3\gamma}{\gamma + 1}MaC_h \quad (1\text{-}98)$$

where C_h is the so-called Stanton number, or wall heat-transfer coefficient of the flow. It will be shown in Chap. 4 that in boundary-layer flows, the Stanton number is approximately one-half the friction coefficient, an effect which is called the *Reynolds analogy*:

$$C_h \approx \tfrac{1}{2}C_f \quad (1\text{-}99)$$

If finally we combine Eqs. (1-98) and (1-99) and take $\gamma = 1.4$ (air), we find that the temperature jump, expressed as a fraction of the driving temperature difference $T_r - T_w$, becomes

$$\frac{T_{gas} - T_w}{T_r - T_w} \approx 0.87 MaC_f \quad (1\text{-}100)$$

which is nearly identical to the slip relation, Eq. (1-94). Thus the same remarks as before apply here. Temperature jump in turbulent flow is entirely negligible, and the jump in laminar flow is also extremely small except at low Reynolds number and high Mach number. It is customary, then, to adopt the no-slip and no-temperature-jump conditions in routine analysis of viscous gas-flow problems:

$$\mathbf{V}_{gas} \approx \mathbf{V}_w$$
$$T_{gas} \approx T_w \tag{1-101}$$

In many cases, of course, the coordinate system is such that the wall is stationary, so that the velocity condition is simply $\mathbf{V}_{fluid} = 0$.

1-4.4 Conditions at a Permeable Wall

In the event that the wall is porous and can permit fluid to pass through, the no-slip condition is relaxed on the velocity component normal to the wall. The proper conditions are complicated by the type of porosity of the wall, but, in general, we assume

$$V_{tangential} = 0 \quad \text{no slip}$$
$$V_{normal} \neq 0 \quad \text{flow through wall} \tag{1-102}$$

The temperature condition is also complicated by a porous wall. For wall *suction*, where the fluid leaves the main flow and passes into the wall, it is sufficiently accurate to assume

$$T_{fluid} = T_w \quad \text{suction} \tag{1-103}$$

For *injection* through a porous wall into the main stream (sometimes called *transpiration*), the injected fluid may be, say, a coolant at a temperature considerably different from the wall, and one needs to consider an energy balance at the wall. A good approximation for coolant injection is to use the boundary condition proposed by Roberts [in Truitt (1960, Chap. 11)]:

Injection: $$k\frac{dT}{dy}\bigg|_w \approx \rho_w V_n c_p (T_w - T_{coolant}) \tag{1-104}$$

where $\rho_w V_n$ is the mass flow of coolant per unit area through the wall. The actual numerical value of V_n depends largely upon the pressure drop across the porous wall. In general, although a few simple porous-wall solutions will be given in this text, the flow characteristics of porous media are a self-contained specialized subject of fluid mechanics. For more information, consult such texts on porous media as Bear (2000) or Crolet (2000).

1-4.5 Conditions at a Free Liquid Surface

There are many flow problems where the liquid fluid ends, not at a solid wall, but at an open or free surface exposed to an atmosphere of either gas or vapor. We distinguish between two cases: (1) the ideal or classic free surface that exerts only a known pressure on the liquid boundary and (2) a more complicated case where the atmosphere exerts not only pressure but also shear, heat flux, and mass flux at the

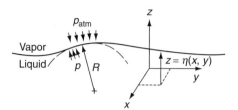

FIGURE 1-29
Conditions at an ideal free surface.

surface. This latter, more involved surface is more properly termed a liquid–vapor interface, the ocean surface being a splendid example.

The classic free surface is sketched in Fig. 1-29. Let us assume that the xy plane is more or less parallel to the free surface, so that the actual deflected shape of the surface can be denoted by $z = \eta(x, y)$.

The two required conditions for this surface are (1) the fluid particles at the surface must remain attached (kinematic condition) and (2) the liquid and the atmospheric pressure must balance except for surface-tension effects. The kinematic condition is specified mathematically by requiring the particle's upward velocity to equal the motion of the free surface:

$$w(x, y, \eta) = \frac{D\eta}{Dt} = \frac{\partial \eta}{\partial t} + u \frac{\partial \eta}{\partial x} + v \frac{\partial \eta}{\partial y} \qquad (1\text{-}105)$$

The pressure equilibrium is expressed by

$$p(x, y, \eta) = p_a - \mathcal{T}\left(\frac{1}{R_x} + \frac{1}{R_y}\right) \qquad (1\text{-}106)$$

where R_x and R_y are the radii of curvature of the surface and where $\mathcal{T}$ is the coefficient of surface tension of the interface. For a two-dimensional surface deflection $\eta = \eta(x)$ only, Eq. (1-106) becomes

$$p(x, \eta) = p_a - \frac{\mathcal{T} d^2\eta/dx^2}{[1 + (d\eta/dx)^2]^{3/2}} \qquad (1\text{-}107)$$

We see from this relation that, when the interface smiles (concave upward, positive curvature), $p < p_a$, while a frowning interface (concave downward) results in $p > p_a$. Equations (1-105) and (1-106) are complex non-linear conditions, but they can be evaluated by numerical computer modeling.

In the range 0 to 100°C, a clean air–water interface has a nearly linear variation of surface tension with temperature:

$$\mathcal{T}\left(\frac{N}{m}\right) \approx 0.076 - 0.00017T\,(°C) \qquad (1\text{-}108)$$

with accuracy of ±1 percent. Measured values of air–water surface tension are given in Table 1-5 for temperatures up to the critical point. These are ideal data for a very clean interface. Under field conditions, $\mathcal{T}$ can vary greatly due to the presence of surface contaminants or slicks.

TABLE 1-5
Surface-tension coefficient for an air–water interface

T, °C	Υ, N/m	T, °C	Υ, N/m
0	0.0757	200	0.0377
20	0.0727	220	0.0331
40	0.0696	240	0.0284
60	0.0662	260	0.0237
80	0.0627	280	0.0190
100	0.0589	300	0.0144
120	0.0550	320	0.0099
140	0.0509	340	0.0056
160	0.0466	360	0.0019
180	0.0422	374[†]	0.0[†]

[†]Critical point.

In large-scale problems, such as open-channel or river flow, the free surface deforms only slightly and surface-tension effects are negligible. Equations (1-105) and (1-106) then simplify to

$$w \approx \frac{\partial \eta}{\partial t} \qquad p \approx p_a \qquad (1\text{-}109)$$

These are obviously very attractive linearized conditions.

Note: The present discussion concerns the deformation of an interface by *uniform* surface tension. If the surface tension varies along the interface, due, for example, to a temperature gradient, a flow called *Marangoni convection* will be induced from the hot surface toward the cold surface. This condition is beyond the scope of the present text. For an example of Marangoni flow analysis, see the paper by Sasmal and Hochstein (1994) or the reviews by Ostrach (1982) and Davis (1987).

1-4.6 Conditions at a Liquid–Vapor or Liquid–Liquid Interface

The term *free surface* means that the gas lying over the liquid has no effect except to impose pressure on the interface. Heat transfer and shear effects are negligible. In a true liquid–vapor or liquid–liquid interface, the upper fluid is strongly coupled and exerts kinematic, stress, and energy constraints on the lower fluid. The motions of the two fluids are solved simultaneously and must match in certain ways at the interface.

Since an interface has vanishing mass, it cannot store momentum or thermal energy. Therefore the total velocity, shear stress, and temperature must be continuous across the interface. If "1" and "2" denote the upper and lower fluid, respectively, then, at a true fluid–fluid interface,

$$\mathbf{V}_1 = \mathbf{V}_2 \qquad \tau_1 = \tau_2 \qquad T_1 = T_2 \qquad (1\text{-}110)$$

The normal velocities match, $V_{n1} = V_{n2}$, and each can, if desired, be expressed in terms of the interface motion $\eta(x, y, t)$ as in Eq. (1-105). The tangential velocities also match, $V_{t1} = V_{t2}$, which is the equivalent of the no-slip condition, Eq. (1-101). The pressures satisfy Eq. (1-106) with $p_2 = p$ and $p_1 = p_a$ and do not match unless surface tension or interface curvature is negligible.

Although velocities and temperatures are continuous across the interface, their slopes generally do not match because of differing transport coefficients, that is,

$$\tau_1 = \mu_1 \frac{\partial V_{t1}}{\partial n} = \tau_2 = \mu_2 \frac{\partial V_{t2}}{\partial n}$$
$$q_1 = -k_1 \frac{\partial T_1}{\partial n} = q_2 = -k_2 \frac{\partial T_2}{\partial n} \tag{1-111}$$

where n is the coordinate normal to the interface. The slopes are not equal if $\mu_1 \neq \mu_2$ or $k_1 \neq k_2$. This situation is illustrated in Fig. 1-30 and is typical of momentum and energy interactions between dissimilar fluids.

Since k and μ for a vapor are usually much smaller than for a liquid, we can often approximate liquid conditions at the interface as

$$\left.\frac{\partial V_t}{\partial n}\right|_{liq} \approx 0 \qquad \left.\frac{\partial T}{\partial n}\right|_{liq} \approx 0 \tag{1-112}$$

Finally, if there is evaporation, condensation, or diffusion at the interface, the mass flows must also balance, $\dot{m}_1 = \dot{m}_2$. For diffusion,

$$D_1 \frac{\partial C_1}{\partial n} = D_2 \frac{\partial C_2}{\partial n} \tag{1-113}$$

This condition is of course analogous to Eqs. (1-111).

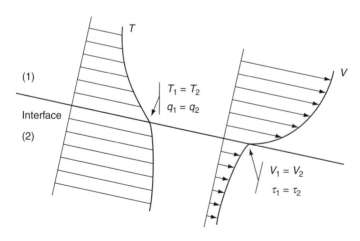

FIGURE 1-30
Conditions at an actual fluid interface. Velocities and temperatures match, but their slopes do not because of differing k and μ. Pressures also match except for the surface-tension effect of Eq. (1-106).

1-4.7 Conditions at a Deformable Fluid–Solid Interface

In most of the problems of this text, we shall take a solid boundary to be a rigid interface which merely imposes no-slip and no-temperature-jump conditions on the fluid. The dynamics and thermodynamics of the solid are neglected. However, there is a growing field of research in which the solid is coupled to the fluid through deformable and dynamic interactions. In this case, we must match the velocity, stress, temperature, and heat transfer across the interface, and the equations of motion of fluid and solid are to be solved simultaneously. Material-behavior relations analogous to Eqs. (1-111) must be used—perhaps the fluid is newtonian and the solid satisfies Hooke's law. For details, see the texts by Blevins (1977) or Au-Yang (2001).

1-4.8 Inlet and Exit Boundary Conditions

In many viscous-flow problems, it is convenient to limit the analysis to a finite region through which the flow passes. In the pipe-flow problem of Fig. 1-11, for example, we would like to limit our analysis to the short length of pipe shown.

Both finite and infinite regions require known property conditions at all boundaries. In pipe flow, Fig. 1-11, we specify no-slip and no-temperature-jump at the pipe walls. An alternate temperature condition would be known heat flux, $q_w = -k(\partial T/\partial n)_w$, which specifies the normal temperature gradient. In this latter case, the wall temperature would be found from the solution. At the pipe inlet, we would specify the distributions of $\mathbf{V}$, T, and p. Often the inlet pressure is assumed uniform as a simplification. At the pipe exit, we specify $\mathbf{V}$ and T. No exit condition is required upon p, which is then found from the solution. As we shall see, pressure occurs only as first derivatives in the equations of motion (Chap. 2); thus only one pressure condition is needed. We can disregard temperature conditions if the flow is assumed isothermal.

For the infinite-region example of flow about a cylinder (Fig. 1-6), we specify no-slip and no-temperature-jump at the cylinder walls. At the "inlet," or approach flow, we would specify the distributions of $\mathbf{V}$, T, and p. Far downstream, we specify $\mathbf{V}$ and T, which is very difficult, because the wake of the body will cause nonuniformity and possible unsteadiness in $\mathbf{V}$ and T. We would not know, in advance, the exact velocity and temperature far downstream, but there are several useful downstream approximations that lead to excellent CFD simulations of flow past immersed bodies. Pressure downstream will be calculated and need not be specified.

1-4.9 Résumé of Fluid-Flow Boundary Conditions

We have discussed the physical conditions which must hold at three different types of fluid-flow boundaries: (1) a fluid–solid interface, (2) a fluid–fluid interface, and (3) the inlet and exit of a flow. These various types of conditions are sketched in

54 VISCOUS FLUID FLOW

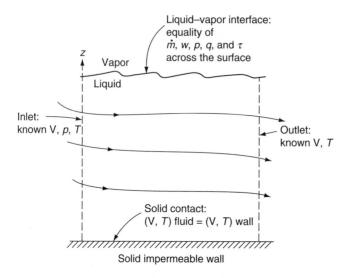

FIGURE 1-31
Various boundary conditions in fluid flow.

Fig. 1-31. The general results are that, for the majority of viscous-flow analyses, we must have the following information at the boundaries:

1. No slip or temperature jump at a fluid–solid interface:

$$\mathbf{V}_{\text{fluid}} = \mathbf{V}_{\text{sol}} \qquad T_{\text{fluid}} = T_{\text{sol}} \qquad (1\text{-}101a)$$

   Note that there is no condition on p unless the wall is permeable.

2. Equality of velocity, momentum flux, heat flux, and mass flux at a liquid–fluid interface (neglecting surface tension):

$$(\mathbf{V}, p, \tau, q, \dot{m})_{\text{fluid above}} = (\mathbf{V}, p, \tau, q, \dot{m})_{\text{liq below}} \qquad (1\text{-}114)$$

   If the fluid above is a gas with negligible interaction, the conditions on τ, q, and $\dot{m}$ may be ignored.

3. Known values of $\mathbf{V}$, p, and T at every point on an inlet section of the flow. Also, unless simplifying approximations are made, $\mathbf{V}$, and T must be known on any exit section.

SUMMARY

This chapter has reviewed the introductory concepts of fluid motion with which the reader should already be conversant. A brief history and some sample viscous-flow problems were outlined, followed by an extensive discussion of the different quantities which distinguish a fluid: (1) the kinematic properties, (2) the transport properties, and (3) the thermodynamic properties. The chapter closes with a detailed look at the various boundary conditions relevant to viscous-fluid flow.

PROBLEMS

1-1. A sphere 1.4 cm in diameter is placed in a freestream of 18 m/s at 20°C and 1 atm. Compute the diameter Reynolds number of the sphere if the fluid is (*a*) air, (*b*) water, (*c*) hydrogen.

1-2. A telephone wire 8 mm in diameter is subjected to a crossflow wind and begins to shed vortices. From Fig. 1-8, what wind velocity (in m/s) will cause the wire to "sing" at middle C (256 Hz)?

1-3. If the wire in Prob. 1-2 is subjected to a crossflow wind of 12 m/s, use Fig. 1-9 to estimate its drag force (in N/m).

1-4. For oil flow in a pipe far downstream of the entrance (Figs. 1-10 and 1-11), the axial velocity profile is a function of r only and is given by $u = (C/\mu)(R^2 - r^2)$, where C is a constant and R is the pipe radius. Suppose the pipe is 1 cm in diameter and $u_{max} = 30$ m/s. Compute the wall shear stress (in Pa) if $\mu = 0.3$ kg/(m·s).

1-5. A tornado may be simulated as a two-part circulating flow in cylindrical coordinates, with $v_r = v_z = 0$,

$$v_\theta = \omega r \quad \text{if } r \leq R \quad \text{and} \quad v_\theta = \omega R^2 / r \quad \text{if } r \geq R$$

Determine (*a*) the vorticity and (*b*) the strain rates in each part of the flow.

1-6. A plane unsteady viscous flow is given in polar coordinates by

$$v_r = 0 \qquad v_\theta = \frac{C}{r}\left[1 - \exp\left(-\frac{r^2}{4\nu t}\right)\right]$$

where C is a constant and ν is the kinematic viscosity. Compute the vorticity $\omega_z(r, t)$ and sketch an array of representative velocity and vorticity profiles for various times.

1-7. A two-dimensional unsteady flow has the velocity components:

$$u = \frac{x}{1+t} \qquad v = \frac{y}{1+2t}$$

Find the equation of the streamlines of this flow which pass through the point (x_0, y_0) at time $t = 0$.

1-8. Using Eq. (1-2) for inviscid flow past a cylinder, consider the flow along the streamline approaching the forward stagnation point $(r, \theta) = (R, \pi)$. Compute (*a*) the distribution of strain rates ϵ_{rr} and $\epsilon_{r\theta}$ along this streamline and (*b*) the time required for a particle to move from the point $(2R, \pi)$ to the stagnation point.

1-9. A commonly used equation of state for water is approximately independent of temperature:

$$\frac{p}{p_0} \approx (A + 1)\left(\frac{\rho}{\rho_0}\right)^n - A$$

where $A \approx 3000$, $n \approx 7$, $p_0 = 1$ atm, and $\rho_0 = 998$ kg/m³. From this formula, compute (*a*) the pressure (in atm) required to double the density of water, (*b*) the bulk modulus of water at 1 atm, and (*c*) the speed of sound in water at 1 atm.

1-10. As shown in Fig. P1-10, a 3 × 3-ft plate slides down a long 30° incline on which there is a film of oil 0.005 in. thick with viscosity $\mu = 0.0005$ slug/(ft·s). Assuming that the plate does not deform the oil film, estimate (a) the terminal sliding velocity (in ft/s) and (b) the time required for the plate to accelerate from rest to 99 percent of the terminal velocity.

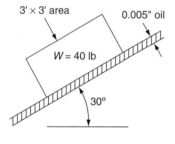

FIGURE P1-10

1-11. Estimate the viscosity of nitrogen at 86 MPa and 49°C and compare with the measured value of 45 μPa $\cdot$ s.

1-12. Estimate the thermal conductivity of helium at 420°C and 1 atm and compare with the measured value of 0.28 W/(m $\cdot$ K).

1-13. It is desired to form a gas mixture of 23% CO_2, 14% O_2, and 63% N_2 at 1 atm and 20°C. The constituent properties are as follows:

Constituent	Mole fraction	μ, Pa $\cdot$ s	k, W/(m $\cdot$ K)
CO_2	0.23	1.37E−5	0.0146
O_2	0.14	1.92E−5	0.0244
N_2	0.63	1.66E−5	0.0242

Estimate the viscosity and thermal conductivity of this mixture.

1-14. Some measured values for the viscosity of ammonia gas are as follows:

T, K	300	400	500	600	700	800
μ, Pa $\cdot$ s	1.03E−5	1.39E−5	1.76E−5	2.14E−5	2.51E−5	2.88E−5

Fit these data, in the least-square-error sense, to the power law, Eq. (1-35), and the Sutherland law, Eq. (1-36).

1-15. Analyze the flow between two plates of Fig. 1-15 by assuming the fluid is a de Waele power-law fluid as in Eq. (1-31a). Compute (a) the velocity profile $u(y)$ with the power n as a parameter and (b) the velocity at the midpoint $h/2$ for $n = 0.5, 1.0$, and 2.0.

Answer: (b) $u = V/2$ for all cases

1-16. Repeat the analysis of the velocity profile between two plates (Fig. 1-15) for a newtonian fluid but allow for a slip velocity $\delta u \approx \ell(du/dy)$ at both walls. Compute the shear stress at both walls. [$\tau_w = \mu V/(h + 2\ell)$ at both walls.]

1-17. By considering the equilibrium of forces on the element shown in Fig. P1-17, derive Eq. (1-106), which expresses the pressure jump across a curved surface due to surface tension.

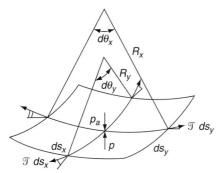

FIGURE P1-17

1-18. Two spherical bubbles of radii R_1 and R_2, respectively, containing air, coalesce into a single bubble of radius R_3. If the ambient air pressure is p_0 and the merging process is isothermal, derive a formula for relating R_3 to $(p_0, R_1, R_2, \mathcal{T})$.

1-19. In Prob. 1-1, if the temperature, sphere size, and velocity remain the same for airflow, at what air pressure will the Reynolds number Re_D be equal to 10,000?

1-20. A solid cylinder of mass m, radius R, and length L falls concentrically through a vertical tube of radius $R + \Delta R$, where $\Delta R \ll R$. The tube is filled with gas of viscosity μ and mean free path ℓ. Neglect fluid forces on the front and back faces of the cylinder and consider only shear stress in the annular region, assuming a linear velocity profile. Find an analytic expression for the terminal velocity of fall, V, of the cylinder (a) for no slip; (b) with slip, Eq. (1-91).

1-21. Oxygen at 20°C and approximately 1200 Pa absolute flows through a 35 μm diameter smooth capillary tube at an average velocity of 10 cm/s. Estimate the Knudsen number of the flow and whether slip flow will be important.

1-22. In Fig. P1-22 a disk rotates steadily inside a disk-shaped container filled with oil of viscosity μ. Assume linear velocity profiles with no slip and neglect stress on the outer edges of the disk. Find a formula for the torque M required to drive the disk.

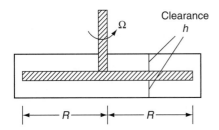

FIGURE P1-22

1-23. Show, from Eq. (1-86), that the coefficient of thermal expansion of a perfect gas is given by $\beta = 1/T$. Use this approximation to estimate β of ammonia gas (NH_3) at 20°C and 1 atm and compare with the accepted value from a data reference.

1-24. The *rotating-cylinder viscometer* in Fig. P1-24 shears the fluid in a narrow clearance Δr, as shown. Assuming a linear velocity distribution in the gaps, if the driving torque M is measured, find an expression for μ by (a) neglecting, and (b) including the bottom friction.

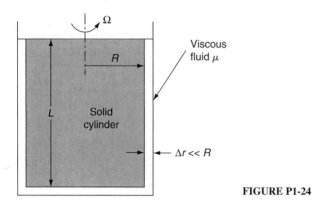

FIGURE P1-24

1-25. Consider 1 m³ of a fluid at 20°C and 1 atm. For an isothermal process, calculate the final density and the energy, in joules, required to compress the fluid until the pressure is 10 atm, for (*a*) air and (*b*) water. Discuss the difference in results.

1-26. Equal layers of two immiscible fluids are being sheared between a moving and a fixed plate, as in Fig. P1-26. Assuming linear velocity profiles, find an expression for the interface velocity U as a function of V, μ_1, and μ_2.

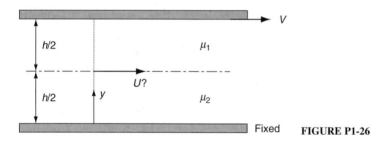

FIGURE P1-26

1-27. Use the inviscid-flow solution of flow past a cylinder, Eqs. (1-3), to (*a*) find the location and value of the maximum fluid acceleration along the cylinder surface. Is your result valid for gases and liquids? (*b*) Apply your formula for a_{max} to airflow at 10 m/s past a cylinder of diameter 1 cm and express your result as a ratio compared to the acceleration of gravity. Discuss what your result implies about the ability of fluids to withstand acceleration.

CHAPTER 2

FUNDAMENTAL EQUATIONS OF COMPRESSIBLE VISCOUS FLOW

2-1 INTRODUCTION

The equations of viscous flow have been known for more than 100 years. In their complete form, these equations are very difficult to solve, even on modern digital computers. In fact, at high Reynolds numbers (turbulent flow), the equations are, in effect, *impossible* to solve with present mathematical techniques because the boundary conditions become randomly time-dependent. Nevertheless, it is very instructive to derive and discuss these fundamental equations because they give many insights, yield several particular solutions, and can be examined for modeling laws. Also, these exact equations can then be simplified, using Prandtl's boundary-layer approximations. The resulting simpler system is very practical and yields many fruitful engineering solutions.

2-2 CLASSIFICATION OF THE FUNDAMENTAL EQUATIONS

The basic equations considered here are the three laws of conservation for physical systems:

1. Conservation of mass (continuity)
2. Conservation of momentum (Newton's second law)
3. Conservation of energy (first law of thermodynamics)

The three unknowns that must be obtained simultaneously from these three basic equations are the velocity **V**, the thermodynamic pressure p, and the absolute temperature T. We consider p and T to be the two required independent thermodynamic variables. However, the final forms of the conservation equations also contain four other thermodynamic variables: the density ρ, the enthalpy h (or the internal energy e), and the two transport properties μ and k. Using our tacit assumption of local thermodynamic equilibrium, the latter four properties are uniquely determined by the values of p and T. Thus the system is completed by assuming knowledge of four state relations

$$\rho = \rho(p, T) \qquad h = h(p, T)$$
$$\mu = \mu(p, T) \qquad k = k(p, T)$$
(2-1)

which can be in the form of tables or charts or semitheoretical formulas from kinetic theory. Many useful analyses simply assume that ρ, μ, and k are constant and that h is proportional to T ($h = c_p T$).

Finally, to specify a particular problem completely, we must have known conditions (of various types) for **V**, p, and T at *every* point of the boundary of the flow regime.

The preceding considerations apply to a fluid of assumed uniform, homogeneous composition, i.e., diffusion and chemical reactions are not considered. Multicomponent reacting fluids must consider at least two extra basic relations:

4. Conservation of species
5. Laws of chemical reaction

plus additional auxiliary relations such as knowledge of the diffusion coefficient $D = D(p, T)$, chemical-equilibrium constants, reaction rates, and heats of formation. This text does not consider reacting boundary-layer flows [see Kee et al. (2003)].

Finally, even more relations are necessary if one considers the flow to be influenced by electromagnetic effects. This is the subject of the field of *magnetohydrodynamics*. Such effects are not considered in the present text.

Let us now derive the three basic equations of a single-component fluid, bearing in mind that the results will also apply to uniform nonreacting mixtures, such as air or liquid solutions.[†]

2-3 CONSERVATION OF MASS: THE EQUATION OF CONTINUITY

As mentioned in the discussion of Eq. (1-5), all three of the conservation laws are Lagrangian in nature, i.e., they apply to fixed systems (particles). Thus, in the

---

[†]Note, however, that air flowing at very high temperatures will undergo spontaneous diffusion and chemical reactions, as will many other mixtures. Even single-component fluids, such as oxygen, will dissociate into atomic oxygen at high temperatures.

Eulerian system appropriate to fluid flow, our three laws must utilize the particle derivative

$$\frac{D}{Dt} = \frac{\partial}{\partial t} + (\mathbf{V} \cdot \nabla) \tag{1-8}$$

which is a formidable expression. In Lagrangian terms, the law of conservation of mass is surpassingly simple:

$$m = \rho \mathcal{V} = \text{const} \tag{2-2}$$

where $\mathcal{V}$ is the volume of a particle. In Eulerian terms, this is equivalent to

$$\frac{Dm}{Dt} = \frac{D}{Dt}(\rho \mathcal{V}) = 0 = \rho \frac{D\mathcal{V}}{Dt} + \mathcal{V}\frac{D\rho}{Dt} \tag{2-3}$$

We can relate $D\mathcal{V}/Dt$ to the fluid velocity by noticing that the total dilatation or normal-strain rate is equal to the rate of volume increase of a particle per unit volume:

$$\epsilon_{xx} + \epsilon_{yy} + \epsilon_{zz} = \frac{1}{\mathcal{V}}\frac{D\mathcal{V}}{Dt} \tag{2-4}$$

Further, we can substitute for the strain rates from our kinematic relations, Eqs. (1-21):

$$\epsilon_{xx} + \epsilon_{yy} + \epsilon_{zz} = \frac{\partial u}{\partial x} + \frac{\partial v}{\partial y} + \frac{\partial w}{\partial z}$$
$$= \text{div } \mathbf{V} = \nabla \cdot \mathbf{V} \tag{2-5}$$

Combining Eqs. (2-3) to (2-5) to eliminate $\mathcal{V}$, we obtain the equation of continuity for fluids in its most common general form:

$$\frac{D\rho}{Dt} + \rho \text{ div } \mathbf{V} = 0 \quad \text{or} \quad \frac{\partial \rho}{\partial t} + \text{div } \rho\mathbf{V} = 0 \tag{2-6}$$

If the density is constant (incompressible flow), Eq. (2-6) reduces to the simpler condition of velocity solenoidality:

$$\text{div } \mathbf{V} = 0 \tag{2-7}$$

which is equivalent to requiring particles of constant volume.[†]

2-3.1 A Useful Strategy: The Stream Function

If the continuity Eq. (2-6) reduces to only two nonzero terms, it can be satisfied identically, and therefore replaced, by the so-called *stream function* ψ. The idea dates back to the French mathematician J. L. Lagrange in 1755. The Cartesian

---

[†]But not necessarily constant *shape*, i.e., the element may undergo normal strains of opposite sign plus shear strains of any sign.

incompressible-flow stream function is treated in detail in Sec. 2-11. Here we illustrate with two-dimensional steady compressible flow in the *xy* plane, for which the continuity equation reduces to

$$\frac{\partial}{\partial x}(\rho u) + \frac{\partial}{\partial y}(\rho v) = 0 \tag{2-8}$$

If we now define the stream function ψ such that

$$\rho u = \frac{\partial \psi}{\partial y} \quad \text{and} \quad \rho v = -\frac{\partial \psi}{\partial x} \tag{2-9}$$

we see by direct substitution that Eq. (2-8) is satisfied identically, assuming of course that ψ is continuous to second-order derivatives. Thus the continuity equation can be discarded and the number of dependent variables reduced by one, the penalty being that the remaining velocity derivatives are increased by one order.

The stream function is not only useful but has physical significance:

$$d\psi = \frac{\partial \psi}{\partial x} dx + \frac{\partial \psi}{\partial y} dy = -\rho v\, dx + \rho u\, dy$$
$$= \rho \mathbf{V} \cdot d\mathbf{A} = d\dot{m} \tag{2-10}$$

which means that lines of constant ψ ($d\psi = 0$) are lines across which there is no mass flow ($d\dot{m} = 0$), that is, they are streamlines of the flow. Also, the difference between the values of ψ of any two streamlines is numerically equal to the mass flow between those streamlines.

2-4 CONSERVATION OF MOMENTUM: THE NAVIER–STOKES EQUATIONS

This relation, commonly known as Newton's second law, expresses a proportionality between applied force and the resulting acceleration of a particle of mass m:

$$\mathbf{F} = m\mathbf{a} \tag{2-11}$$

If the system is a fluid particle, it is convenient to divide Eq. (2-11) by the volume of the particle, so that we work with density instead of mass. It is also traditional to reverse the terms and place the acceleration on the left-hand side. Hence we write

$$\rho \frac{D\mathbf{V}}{Dt} = \mathbf{f} = \mathbf{f}_{\text{body}} + \mathbf{f}_{\text{surface}} \tag{2-12}$$

where $\mathbf{f}$ is the applied force per unit volume on the fluid particle. Note that in our chosen Eulerian system, the acceleration is the rather complicated particle derivative from Eq. (1-9). We have divided $\mathbf{f}$ into two types: surface forces and body forces.

The so-called body forces are those that apply to the entire mass of the fluid element. Such forces are usually due to external fields such as gravity or an applied

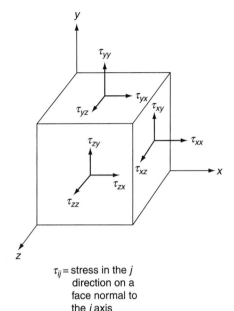

τ_{ij} = stress in the j direction on a face normal to the i axis

FIGURE 2-1
Notation for stresses.

electromagnetic potential. We ignore magnetohydrodynamic effects here and consider only the gravitational body force, which on our unit volume basis is

$$\mathbf{f}_{\text{body}} = \rho \mathbf{g} \tag{2-13}$$

where $\mathbf{g}$ is the vector acceleration of gravity.

The surface forces are those applied by external stresses on the sides of the element. The quantity stress τ_{ij} is a tensor, just as the strain rate ϵ_{ij} was in Sec. 1-3. The sign convention for stress components on a Cartesian element is shown in Fig. 2-1, where all stresses are positive. The stress tensor can be written as

$$\tau_{ij} = \begin{pmatrix} \tau_{xx} & \tau_{xy} & \tau_{xz} \\ \tau_{yx} & \tau_{yy} & \tau_{yz} \\ \tau_{zx} & \tau_{zy} & \tau_{zz} \end{pmatrix} \tag{2-14}$$

by analogy with ϵ_{ij} in Eq. (1-22). Like strain rate, τ_{ij} forms a symmetric tensor; that is, $\tau_{ij} = \tau_{ji}$. This symmetry is required to satisfy equilibrium of moments about the three axes of the element.[†]

The positions of the τ's in the array of Eq. (2-14) are not arbitrary; the rows correspond to applied force in each coordinate direction. Considering the

---

[†]Here we assume the absence of concentrated couple stresses.

front faces of the element in Fig. 2-1, the total force in each direction due to stress is

$$dF_x = \tau_{xx}\,dy\,dz + \tau_{yx}\,dx\,dz + \tau_{zx}\,dx\,dy$$
$$dF_y = \tau_{xy}\,dy\,dz + \tau_{yy}\,dx\,dz + \tau_{zy}\,dx\,dy \qquad (2\text{-}15)$$
$$dF_z = \tau_{xz}\,dy\,dz + \tau_{yz}\,dx\,dz + \tau_{zz}\,dx\,dy$$

preserving the positions in the array. For an element in equilibrium, these forces would be balanced by equal and opposite forces on the back faces of the element. If the element is accelerating, however, the front- and back-face stresses will be different by differential amounts. For example,

$$\tau_{xx,\,\text{front}} = \tau_{xx,\,\text{back}} + \frac{\partial \tau_{xx}}{\partial x}\,dx \qquad (2\text{-}16)$$

Hence the *net* force on the element in the x direction, for example, will be due to three derivative terms:

$$dF_{x,\,\text{net}} = \left(\frac{\partial \tau_{xx}}{\partial x}\,dx\right)dy\,dz + \left(\frac{\partial \tau_{yx}}{\partial y}\,dy\right)dx\,dz + \left(\frac{\partial \tau_{zx}}{\partial z}\,dz\right)dx\,dy$$

or, on a unit volume basis, dividing by $dx\,dy\,dz$, since $\tau_{ij} = \tau_{ji}$,

$$f_x = \frac{\partial \tau_{xx}}{\partial x} + \frac{\partial \tau_{xy}}{\partial y} + \frac{\partial \tau_{xz}}{\partial z} \qquad (2\text{-}17)$$

which we note is equivalent to taking the divergence of the vector $(\tau_{xx}, \tau_{xy}, \tau_{xz})$, the upper row of the stress tensor. Similarly, f_y and f_z are the divergences of the second and third row of τ_{ij}. Thus the total vector surface force is

$$\mathbf{f}_{\text{sur}} = \nabla \cdot \tau_{ij} = \frac{\partial \tau_{ij}}{\partial x_j} \qquad (2\text{-}18)$$

where the divergence of τ_{ij} is to be interpreted in the tensor sense, so that the result is a vector.

Newton's law, Eq. (2-12), now becomes

$$\rho\frac{D\mathbf{V}}{Dt} = \rho\mathbf{g} + \nabla \cdot \tau_{ij} \qquad (2\text{-}19)$$

and it remains only to express τ_{ij} in terms of the velocity $\mathbf{V}$. This is done by relating τ_{ij} to ϵ_{ij} through the assumption of some viscous deformation-rate law, e.g., the newtonian fluid.

2-4.1 The Fluid at Rest: Hydrostatics

From the definition of a fluid, Sec. 1-3, viscous stresses vanish if the fluid is at rest. The velocity and shear stresses are zero, and the normal stresses become equal to the hydrostatic pressure. Equation (2-19) reduces to the hydrostatic equation if $\mathbf{V} = 0$:

$$\tau_{xx} = \tau_{yy} = \tau_{zz} = -p$$
$$\tau_{ij} = 0 \quad \text{for } i \neq j \quad (2\text{-}20)$$
$$\nabla p = \rho \mathbf{g}$$

If we take the z coordinate as *up* and assume ρ and g are constant, the pressure varies linearly with z, $\delta p = -\rho g\, \delta z$. Pressure increases downward, proportional to the specific weight of the fluid. Recall that *hydrostatics* is treated extensively in undergraduate texts, e.g., White (2003). We must ensure here that our dynamic momentum equation reduces to Eq. (2-20) when $\mathbf{V} = 0$.

2-4.2 Deformation Law for a Newtonian Fluid

By analogy with Hookean elasticity, the simplest assumption for the variation of viscous stress with strain rate is a linear law. These considerations were first made by Stokes (1845), and, as far as we know, the resulting deformation law is satisfied by all gases and most common fluids. Stokes' three postulates are

1. The fluid is continuous, and its stress tensor τ_{ij} is at most a linear function of the strain rates ϵ_{ij}.
2. The fluid is isotropic, i.e., its properties are independent of direction, and therefore the deformation law is independent of the coordinate axes in which it is expressed.
3. When the strain rates are zero, the deformation law must reduce to the hydrostatic pressure condition, $\tau_{ij} = -p\delta_{ij}$, where δ_{ij} is the Kronecker delta function ($\delta_{ij} = 1$ if $i = j$ and $\delta_{ij} = 0$ if $i \neq j$).

Note that the isotropic condition 2 requires that the principal stress axes be identical with the principal strain-rate axes. This makes the principal axes a convenient place to begin the deformation-law derivation. Let x_1, y_1, and z_1 be the principal axes, where the shear stresses and shear strain rates vanish [see Eq. (1-24)]. With these axes, the deformation law could involve at most three linear coefficients, C_1, C_2, C_3. For example,

$$\tau_{11} = -p + C_1 \epsilon_{11} + C_2 \epsilon_{22} + C_3 \epsilon_{33} \quad (2\text{-}21)$$

The term $-p$ is added to satisfy the hydrostatic condition (condition 3 above). But the isotropic condition 2 requires that the crossflow effect of ϵ_{22} and ϵ_{33} be identical, i.e., that $C_2 = C_3$. Therefore, there are really only *two* independent linear coefficients in an anisotropic newtonian fluid. We can rewrite Eq. (2-21) in the simpler form

$$\tau_{11} = -p + K\epsilon_{11} + C_2(\epsilon_{11} + \epsilon_{22} + \epsilon_{33}) \quad (2\text{-}22)$$

where $K = C_1 - C_2$, for convenience. Note also that $\epsilon_{11} + \epsilon_{22} + \epsilon_{33}$ equals div $\mathbf{V}$ from Eq. (2-5).

Now let us transform Eq. (2-22) to some arbitrary axes x, y, z, where shear stresses are *not* zero, and thereby find an expression for the general deformation law. With respect to the principal axes x_1, y_1, z_1, let the x axis have direction cosines l_1, m_1, n_1, let the y axis have l_2, m_2, n_2, and let the z axis have l_3, m_3, n_3. Remember also that, for example, $l_1^2 + m_1^2 + n_1^2 = 1.0$ for any set of direction cosines. Then the transformation rule between a normal stress or strain rate in the new system and the principal stresses or strain rates is given by, for example,

$$\tau_{xx} = \tau_{11} l_1^2 + \tau_{22} m_1^2 + \tau_{33} n_1^2$$
$$\epsilon_{xx} = \epsilon_{11} l_1^2 + \epsilon_{22} m_1^2 + \epsilon_{33} n_1^2 \tag{2-23}$$

Similarly, the shear stresses (strain rates) are related to the principal stresses (strain rates) by the following transformation law:

$$\tau_{xy} = \tau_{11} l_1 l_2 + \tau_{22} m_1 m_2 + \tau_{33} n_1 n_2$$
$$\epsilon_{xy} = \epsilon_{11} l_1 l_2 + \epsilon_{22} m_1 m_2 + \epsilon_{33} n_1 n_2 \tag{2-24}$$

We can now eliminate τ_{11}, ϵ_{11}, τ_{22}, etc., from Eq. (2-23) by using the principal-axis deformation law, Eq. (2-22), and the fact that $l^2 + m^2 + n^2 = 1.0$. The result is

$$\tau_{xx} = -p + K\epsilon_{xx} + C_2 \operatorname{div} \mathbf{V} \tag{2-25}$$

with exactly similar expressions for τ_{yy} and τ_{zz}. Similarly, we can eliminate τ_{11}, ϵ_{11}, etc., from Eqs. (2-24) to give

$$\tau_{xy} = K\epsilon_{xy} \tag{2-26}$$

and exactly analogous expressions for τ_{xz} and τ_{yz}. Note that the direction cosines have all politely vanished. Equations (2-25) and (2-26) are, in effect, the desired general deformation law. By comparing Eq. (2-26) and Eq. (1-30) for shear flow between parallel plates, we see that the linear coefficient K is equal to twice the ordinary coefficient of viscosity, $K = 2\mu$. The coefficient C_2 is new and independent of μ and may be called the *second coefficient of viscosity*. In linear elasticity, C_2 is called Lamé's constant and is given the symbol λ, which we adopt here also. Since λ is associated *only* with volume expansion, it is customary to call it the *coefficient of bulk viscosity* λ.

Equations (2-25) and (2-26) can be combined, using the initial notation, and rewritten into a single general deformation law for a newtonian (linear) viscous fluid:

$$\tau_{ij} = -p\delta_{ij} + \mu\left(\frac{\partial u_i}{\partial x_j} + \frac{\partial u_j}{\partial x_i}\right) + \delta_{ij}\lambda \operatorname{div} \mathbf{V} \tag{2-27}$$

where we have written ϵ_{ij} in terms of the velocity gradients from Eq. (1-26). As mentioned, this deformation law was first given by Stokes (1845). In deriving it, we have assumed knowledge of the geometric rules for coordinate transformation of stresses and strain rates. The interested reader may wish to obtain further details from textbooks on continuum mechanics, e.g., Lai et al. (1995), Malvern (1997), or Talpaert (2003).

2-4.3 Thermodynamic Pressure versus Mechanical Pressure

Stokes (1845) pointed out an interesting consequence of Eq. (2-27). By analogy with the strain relation [Eqs. (1-25)], the sum of the three normal stresses $\tau_{xx} + \tau_{yy} + \tau_{zz}$ is a tensor invariant. Define the *mechanical pressure* $\bar{p}$ as the negative one-third of this sum, i.e., the average compression stress on the element. Then, by summing Eq. (2-27), we obtain

$$\bar{p} = -\tfrac{1}{3}(\tau_{xx} + \tau_{yy} + \tau_{zz}) = p - \left(\lambda + \tfrac{2}{3}\mu\right) \operatorname{div} \mathbf{V} \qquad (2\text{-}28)$$

Thus the mean pressure in a deforming viscous fluid is *not* equal to the thermodynamic property called pressure. This distinction is rarely important, since div $\mathbf{V}$ is usually very small in typical flow problems, but the exact meaning of Eq. (2-28) has been a controversial subject for more than a century. Stokes himself simply resolved the issue by an assumption:

Stokes' hypothesis (1845): $\lambda + \tfrac{2}{3}\mu = 0$

This simply assumes away the problem; it is essentially what we do in this book. However, the available experimental evidence from the measurement of sound-wave attenuation, as reviewed by Karim and Rosenhead (1952), indicates that λ for most liquids is actually positive, rather than $-2\mu/3$, and often is much larger than μ. The experiments themselves are a matter of some controversy [Truesdell (1954)].

A second type of assumption will also make $\bar{p}$ equal to p:

Incompressible flow: div $\mathbf{V} = 0$

Again this merely assumes away the problem. The bulk viscosity cannot affect a truly incompressible fluid, but in fact it *does* affect certain phenomena occurring in *nearly* incompressible fluids, e.g., sound absorption in liquids. Meanwhile, if div $\mathbf{V} \neq 0$, that is, compressible flow, we may still be able to avoid the problem if viscous normal stresses are negligible. This is the case in *boundary-layer flows* of compressible fluids, for which only the *first* coefficient of viscosity μ is important. However, the normal shock wave is a case where the coefficient λ cannot be neglected. The second case (only two cases are known to this author) is the previously mentioned problem of sound-wave absorption and attenuation.

It appears, then, that the second viscosity coefficient is still a controversial quantity. In fact, λ may not even be a thermodynamic property, since it is apparently frequency-dependent. Fortunately, the disputed term, λ div $\mathbf{V}$, is almost always so very small that it is entirely proper simply to ignore the effect of λ altogether. An interesting discussion of second viscosity is in Landau and Lifshitz (1959, Sec. 78). For further reading, see Panton (1996).

2-4.4 The Navier–Stokes Equations

The desired momentum equation for a general linear (newtonian) viscous fluid is now obtained by substituting the stress relations, Eq. (2-27), into Newton's law

[Eq. (2-19)]. The result is the famous equation of motion which bears the names of Navier (1823) and Stokes (1845). In scalar form, we obtain

$$\rho \frac{Du}{Dt} = \rho g_x - \frac{\partial p}{\partial x} + \frac{\partial}{\partial x}\left(2\mu \frac{\partial u}{\partial x} + \lambda \text{ div } \mathbf{V}\right) + \frac{\partial}{\partial y}\left[\mu\left(\frac{\partial u}{\partial y} + \frac{\partial v}{\partial x}\right)\right]$$

$$+ \frac{\partial}{\partial z}\left[\mu\left(\frac{\partial w}{\partial x} + \frac{\partial u}{\partial z}\right)\right]$$

$$\rho \frac{Dv}{Dt} = \rho g_y - \frac{\partial p}{\partial y} + \frac{\partial}{\partial x}\left[\mu\left(\frac{\partial v}{\partial x} + \frac{\partial u}{\partial y}\right)\right] + \frac{\partial}{\partial y}\left(2\mu \frac{\partial v}{\partial y} + \lambda \text{ div } \mathbf{V}\right)$$

$$+ \frac{\partial}{\partial z}\left[\mu\left(\frac{\partial v}{\partial z} + \frac{\partial w}{\partial y}\right)\right] \quad (2\text{-}29a)$$

$$\rho \frac{Dw}{Dt} = \rho g_z - \frac{\partial p}{\partial z} + \frac{\partial}{\partial x}\left[\mu\left(\frac{\partial w}{\partial x} + \frac{\partial u}{\partial z}\right)\right] + \frac{\partial}{\partial y}\left[\mu\left(\frac{\partial v}{\partial z} + \frac{\partial w}{\partial y}\right)\right]$$

$$+ \frac{\partial}{\partial z}\left(2\mu \frac{\partial w}{\partial z} + \lambda \text{ div } \mathbf{V}\right)$$

These are the Navier–Stokes equations, fundamental to the subject of viscous-fluid flow. Considerable economy is achieved by rewriting them as a single vector equation, using the indicial notation:

$$\rho \frac{D\mathbf{V}}{Dt} = \rho \mathbf{g} - \nabla p + \frac{\partial}{\partial x_j}\left[\mu\left(\frac{\partial v_i}{\partial x_j} + \frac{\partial v_j}{\partial x_i}\right) + \delta_{ij}\lambda \text{ div } \mathbf{V}\right] \quad (2\text{-}29b)$$

2-4.5 Incompressible Flow: Thermal Decoupling

If the fluid is assumed to be of constant density, div $\mathbf{V}$ vanishes due to the continuity Eq. (2-7) and the vexing coefficient λ disappears from Newton's law. Equations (2-29) are not greatly simplified, though, if the first viscosity μ is allowed to vary with temperature and pressure (and hence with position). If, however, we assume that μ is constant, many terms vanish, leaving us with a much simpler Navier–Stokes equation for *constant viscosity and density*:

$$\rho \frac{D\mathbf{V}}{Dt} = \rho \mathbf{g} - \nabla p + \mu \nabla^2 \mathbf{V} \quad (2\text{-}30)$$

Most of the problems and solutions in this text are for incompressible flow, Eqs. (2-7) and (2-30). Note that, if ρ and μ are constant, these equations are entirely *uncoupled* from temperature. One may solve continuity and momentum for velocity and pressure and then later, if one desires, solve for temperature from the *energy equation* of Sec. 2-5. This approximation often divides textbooks into "fluid mechanics" and, later, "heat transfer." However, the present text will maintain some heat-transfer discussions throughout.

2-4.6 Inviscid Flow: The Euler and Bernoulli Equations

If we further assume that viscous terms are negligible, Eq. (2-30) reduces to

$$\rho \frac{D\mathbf{V}}{Dt} \approx \rho \mathbf{g} - \nabla p \qquad (2\text{-}30a)$$

This is called *Euler's equation* (derived by Leonhard Euler in 1755) for inviscid flow. It is first order in **V** and p and thus rather simpler than the second-order Navier–Stokes Eq. (2-30). At a fixed wall, the no-slip condition must be *dropped*, and tangential velocity is allowed to slip. Much research has been reported for Euler's equation: analytical (Currie 1993), numerical (Tannehill et al. 1997), and mathematical theorems (Kreiss and Lorenz 1989). As covered in undergraduate texts, for example, White (2003), Euler's equation for steady, incompressible, frictionless flow may be integrated along a streamline between any points 1 and 2 to yield

$$\left(p + \tfrac{1}{2}\rho V^2 + \rho g z\right)_1 \approx \left(p + \tfrac{1}{2}\rho V^2 + \rho g z\right)_2 \qquad (2\text{-}30b)$$

where z is up, that is, opposite gravity. This is *Bernoulli's equation* for steady frictionless flow. Though approximate, since all fluids are viscous, it has many applications in aeronautics and hydrodynamics and serves as an outer boundary matching condition in boundary-layer theory (Chap. 4). The unsteady form of Bernoulli's equation will be given in Sec. 2-10.

2-5 THE ENERGY EQUATION (FIRST LAW OF THERMODYNAMICS)

The first law of thermodynamics for a system is a statement of the fact that the sum of the work and heat added to the system will result in an increase in the energy of the system:

$$dE_t = dQ + dW \qquad (2\text{-}31)$$

where Q = heat added
W = work done on system

The quantity E_t denotes the total energy of the system; hence, in a moving system, such as a flowing fluid particle, E_t will include not only internal energy but also kinetic and potential energy. Thus, for a fluid particle, the energy per unit volume is

$$E_t = \rho\left(e + \tfrac{1}{2}V^2 - \mathbf{g}\cdot\mathbf{r}\right) \qquad (2\text{-}32)$$

where e = internal energy per unit mass
r = displacement of particle

Just as in conservation of mass and momentum, the energy equation for a fluid is conveniently written as a time rate of change, following the particle. Thus

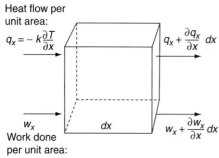

FIGURE 2-2
Heat and work exchange on the left and right sides of an element.

Eq. (2-31) becomes

$$\frac{DE_t}{Dt} = \frac{DQ}{Dt} + \frac{DW}{Dt} \tag{2-33}$$

From Eq. (2-32), we have

$$\frac{DE_t}{Dt} = \rho\left(\frac{De}{Dt} + V\frac{DV}{Dt} - \mathbf{g}\cdot\mathbf{V}\right) \tag{2-34}$$

It remains to express Q and W in terms of fluid properties.

It is assumed that the heat transfer Q to the element is given by Fourier's law. From Eq. (1-37), the vector heat flow per unit area is

$$\mathbf{q} = -k\nabla T \tag{1-37}$$

Referring to Fig. 2-2, the heat flow into the left face of the element is

$$q_x \, dy \, dz$$

while the heat flow out of the right face is

$$\left(q_x + \frac{\partial q_x}{\partial x}\,dx\right)dy\,dz$$

A similar situation holds for the upper and lower faces, involving q_y, and the front and rear faces, involving q_z. In each case, the net heat flow is out of the element. Hence the net heat transferred *to* the element is

$$-\left(\frac{\partial q_x}{\partial x} + \frac{\partial q_y}{\partial y} + \frac{\partial q_z}{\partial z}\right) dx\, dy\, dz$$

Dividing by the element volume $dx\, dy\, dz$, we have the desired expression for the heat-transfer term neglecting internal heat generation:

$$\frac{DQ}{Dt} = -\text{div}\,\mathbf{q} = +\text{div}(k\,\nabla T) \tag{2-35}$$

Referring again to Fig. 2-2, the rate of work done to the element per unit area on the left face is

$$w_x = -(u\tau_{xx} + v\tau_{xy} + w\tau_{xz})$$

and the rate of work done by the right-face stresses is

$$-w_x - \frac{\partial w_x}{\partial x} dx \quad \text{per unit area}$$

Again the other faces are similar. In just the same fashion as the heat transfer, then, the net rate of work done on the element is

$$\frac{DW}{Dt} = -\text{div } \mathbf{w} = \frac{\partial}{\partial x}(u\tau_{xx} + v\tau_{xy} + w\tau_{xz})$$

$$+ \frac{\partial}{\partial y}(u\tau_{yx} + v\tau_{yy} + w\tau_{yz})$$

$$+ \frac{\partial}{\partial z}(u\tau_{zx} + v\tau_{zy} + w\tau_{zz}) \tag{2-36}$$

Using the indicial notation, this becomes, very simply,

$$\frac{DW}{Dt} = \nabla \cdot (\mathbf{V} \cdot \tau_{ij}) \tag{2-37}$$

This expression can be decomposed in very convenient fashion into

$$\nabla \cdot (\mathbf{V} \cdot \tau_{ij}) = \mathbf{V} \cdot (\nabla \cdot \tau_{ij}) + \tau_{ij}\frac{\partial u_i}{\partial x_j} \tag{2-38}$$

the first term of which is directly related to the momentum equation

$$\nabla \cdot \tau_{ij} = \rho\left(\frac{D\mathbf{V}}{Dt} - \mathbf{g}\right) \tag{2-19}$$

Hence

$$\mathbf{V} \cdot (\nabla \cdot \tau_{ij}) = \rho\left(\mathbf{V}\frac{D\mathbf{V}}{Dt} - \mathbf{g} \cdot \mathbf{V}\right) \tag{2-39}$$

which are exactly the kinetic- and potential-energy terms in Eq. (2-34). Thus the kinetic and potential energy vanish when we substitute for E_t, Q, and W in Eq. (2-33) from Eqs. (2-34), (2-35), and (2-38):

$$\rho\frac{De}{Dt} = \text{div}(k\,\nabla T) + \tau_{ij}\frac{\partial u_i}{\partial x_j} \tag{2-40}$$

This is a widely used form of the first law of thermodynamics for fluid motion. If we split the stress tensor into pressure and viscous terms, using Eq. (2-27), we have

$$\tau_{ij}\frac{\partial u_i}{\partial x_j} = \tau'_{ij}\frac{\partial u_i}{\partial x_j} - p\,\text{div }\mathbf{V} \tag{2-41}$$

From the equation of continuity, Eq. (2-6), we have

$$p \text{ div } \mathbf{V} = -\frac{p}{\rho} \frac{D\rho}{Dt} = \rho \frac{D}{Dt}\left(\frac{p}{\rho}\right) - \frac{Dp}{Dt} \qquad (2\text{-}42)$$

Combining Eqs. (2-40) to (2-42) gives

$$\rho \frac{D}{Dt}\left(e + \frac{p}{\rho}\right) = \frac{Dp}{Dt} + \text{div}(k\,\nabla T) + \tau'_{ij}\frac{\partial u_i}{\partial x_j} \qquad (2\text{-}43)$$

The advantage of this rearrangement, if any, is that an old friend crops up, the fluid enthalpy:

$$h = e + \frac{p}{\rho} \qquad (2\text{-}44)$$

Generally, enthalpy will be a more useful function than internal energy, particularly in boundary-layer flows. Also, the pressure term Dp/Dt may often be neglected in Eq. (2-43), while the related term p div $\mathbf{V}$ in Eq. (2-40) will be nonnegligible for the same flow.

The last term in Eq. (2-43), involving viscous stresses, is customarily called the *dissipation function* Φ:

$$\Phi = \tau'_{ij}\frac{\partial u_i}{\partial x_j} \qquad (2\text{-}45)$$

The term Φ is always positive definite, in accordance with the second law of thermodynamics, since viscosity cannot add energy to the system. For a newtonian fluid, using τ' from Eq. (2-27), we obtain

$$\Phi = \mu\left[2\left(\frac{\partial u}{\partial x}\right)^2 + 2\left(\frac{\partial v}{\partial y}\right)^2 + 2\left(\frac{\partial w}{\partial z}\right)^2 + \left(\frac{\partial v}{\partial x} + \frac{\partial u}{\partial y}\right)^2 \right.$$
$$\left. + \left(\frac{\partial w}{\partial y} + \frac{\partial v}{\partial z}\right)^2 + \left(\frac{\partial u}{\partial z} + \frac{\partial w}{\partial x}\right)^2\right] + \lambda\left(\frac{\partial u}{\partial x} + \frac{\partial v}{\partial y} + \frac{\partial w}{\partial z}\right)^2 \qquad (2\text{-}46)$$

which is always positive since all the terms are quadratic. On the other hand, λ may actually be negative, e.g., Stokes' hypothesis, $\lambda = -\frac{2}{3}\mu$. It is an interesting exercise, using Eq. (2-46), to prove that the correct conditions in which Φ may not be negative are

$$\mu \geq 0 \qquad 3\lambda + 2\mu \geq 0 \qquad (2\text{-}47)$$

Using this short notation, the energy relation Eq. (2-43) takes the final form

$$\rho \frac{Dh}{Dt} = \frac{Dp}{Dt} + \text{div}(k\,\nabla T) + \Phi \qquad (2\text{-}48)$$

where Φ is given by Eq. (2-46).

2-5.1 The Incompressible-Flow Approximation

From the thermodynamic identity (1-87), we can rewrite Eq. (2-48) as

$$\rho c_p \frac{DT}{Dt} = \beta T \frac{Dp}{Dt} + \text{div}(k \nabla T) + \Phi \tag{2-49}$$

Now, if flow velocity scale U becomes smaller, while heat transfer remains important, the fluid kinetic energy U^2 will eventually become much smaller than the enthalpy change $c_p \Delta T$. Since both Dp/Dt and Φ are of order U^2, the limit of low-velocity or *incompressible* flow will be

$$\rho c_p \frac{DT}{Dt} \approx \text{div}(k \nabla T) \tag{2-50}$$

If we further assume constant thermal conductivity, we obtain the more familiar incompressible heat-convection equation:

$$\rho c_p \frac{DT}{Dt} \approx k \nabla^2 T \tag{2-51}$$

Note that the correct specific heat is c_p, not c_v, even in the incompressible-flow limit of near-zero Mach number. A very good discussion of this point is given in Panton (1996, Sec. 10.9).

2-5.2 Summary of the Basic Equations

Summarizing, the three basic laws of conservation of mass, momentum, and energy have been adapted for use in fluid motion. They are, respectively,

$$\frac{\partial \rho}{\partial t} + \text{div}\, \rho \mathbf{V} = 0 \tag{2-6}$$

$$\rho \frac{D\mathbf{V}}{Dt} = \rho \mathbf{g} + \nabla \cdot \tau'_{ij} - \nabla p \tag{2-19a}$$

$$\rho \frac{Dh}{Dt} = \frac{Dp}{Dt} + \text{div}(k \nabla T) + \tau'_{ij} \frac{\partial u_i}{\partial x_j} \tag{2-48a}$$

where, for a linear (newtonian) fluid, the viscous stresses are

$$\tau'_{ij} = \mu \left(\frac{\partial u_i}{\partial x_j} + \frac{\partial u_j}{\partial x_i} \right) + \delta_{ij} \lambda\, \text{div}\, \mathbf{V} \tag{2-27}$$

As mentioned in the beginning of this chapter, Eqs. (2-6), (2-19), and (2-48) involve seven variables, of which three are assumed to be primary: p, $\mathbf{V}$, and T (say). The remaining four variables are assumed known from auxiliary relations and data of the form

$$\begin{aligned} \rho &= \rho(p, T) & \mu &= \mu(p, T) \\ h &= h(p, T) & k &= k(p, T) \end{aligned} \tag{2-1}$$

Finally, we note that these relations are fairly general and involve only a few restrictive assumptions: (1) the fluid forms a (mathematical) continuum, (2) the particles are essentially in thermodynamic equilibrium, (3) the only effective body forces are due to gravity, (4) the heat conduction follows Fourier's law, and (5) there are no internal heat sources.

2-6 BOUNDARY CONDITIONS FOR VISCOUS HEAT-CONDUCTING FLOW

The various types of boundary conditions have been discussed in detail in Sec. 1-4 for the three different basic boundaries: (1) a fluid–solid interface, (2) a fluid–fluid interface, and (3) an inlet or exit section. A sketch was given in Fig. 1-31. In their full generality, the conditions in Sec. 1-4 can be quite complex and non-linear, necessitating digital-computer treatment. Thus the bulk of our analytical solutions in this text will be confined to simple but realistic approximations:

1. At a fluid–solid interface, there must be no slip

$$\mathbf{V}_{\text{fluid}} = \mathbf{V}_{\text{sol}} \tag{2-52}$$

and either no temperature jump (when the wall temperature is known)

$$T_{\text{fluid}} = T_{\text{sol}} \tag{2-53}$$

or equality of heat flux (when the solid heat flux is known)

$$\left(k\frac{\partial T}{\partial n}\right)_{\text{fluid}} = q \quad \text{from solid to fluid} \tag{2-54}$$

2. At the interface between a liquid and a gaseous atmosphere, there must be kinematic equivalence

$$V_{n,\,\text{liq}} \approx \frac{\partial \eta}{\partial t} \tag{2-55}$$

where η is the surface coordinate, and there must be equality of normal momentum flux

$$p_{\text{liq}} \approx p_{\text{atm}} \tag{2-56}$$

We must also have equality of tangential momentum flux and heat flux, which leads to the approximations

$$\left.\frac{\partial V}{\partial n}\right|_{\text{liq}} = \left.\frac{\partial T}{\partial n}\right|_{\text{liq}} \approx 0 \tag{2-57}$$

when the atmosphere has negligibly small transport coefficients. Also, in some problems, such as the analysis of confused, stormy seas, we need to know the moisture evaporation rate $\dot{m}$ at the interface.

3. At any inlet section of the flow, we need the three quantities $\mathbf{V}$, p, and T at every point on the boundary. At the exit section, we generally need to know $\mathbf{V}$ and T,

but not the pressure p, as you recall from the discussion of Fig. 1-31. Exit conditions are difficult because of the formation of wakes and other a priori unknown outflow behavior. One approximation is to let the streamwise flow gradients vanish far downstream of the flow field of interest.

Note: If the pressure can be eliminated, as, for example, with the vorticity/stream-function approach to be discussed in Sec. 2-11, then there is no need to impose a pressure boundary condition. The new computer-oriented *vortex methods*, explained in the monograph by Cottet and Koumoutsakos (1999), do not compute the pressure at all.

2-7 ORTHOGONAL COORDINATE SYSTEMS

Most of the previous discussion has been illustrated by Cartesian coordinates, and only lip service has been paid to other important orthogonal systems. The basic equations of motion, Eqs. (2-6), (2-19), and (2-48) are, of course, valid for any coordinate system when written in tensor form; the problem for non-Cartesian systems is to derive the correct formula for the gradient vector ∇ plus the related expressions for divergence and curl. A straightforward procedure is to use the metric *stretching factors* h_i to relate the new curvilinear coordinates to a Cartesian system, as discussed in standard mathematical references, e.g., Pipes (1958). Let the general curvilinear system (x_1, x_2, x_3) be related to a Cartesian system (x, y, z) so that the element of arc length ds is given by

$$(ds)^2 = (dx)^2 + (dy)^2 + (dz)^2$$
$$= (h_1\, dx_1)^2 + (h_2\, dx_2)^2 + (h_3\, dx_3)^2 \qquad (2\text{-}58)$$

which defines the factors h_i. Note that h_i in general will be functions of the new coordinates x_i. Then, in the new system, the components of the gradient of a scalar ϕ are

$$\frac{1}{h_1}\frac{\partial \phi}{\partial x_1} \qquad \frac{1}{h_2}\frac{\partial \phi}{\partial x_2} \qquad \frac{1}{h_3}\frac{\partial \phi}{\partial x_3} \qquad (2\text{-}59)$$

The divergence of any vector $\mathbf{A} = (A_1, A_2, A_3)$ is given by

$$\text{div}\, \mathbf{A} = \frac{1}{h_1 h_2 h_3}\left[\frac{\partial}{\partial x_1}(h_2 h_3 A_1) + \frac{\partial}{\partial x_2}(h_3 h_1 A_2) + \frac{\partial}{\partial x_3}(h_1 h_2 A_3)\right] \qquad (2\text{-}60)$$

and the components of the vector $\mathbf{B} = \text{curl}\, \mathbf{A}$ are given by

$$B_1 = \frac{1}{h_2 h_3}\left[\frac{\partial}{\partial x_2}(h_3 A_3) - \frac{\partial}{\partial x_3}(h_2 A_2)\right] \qquad (2\text{-}61)$$

with exactly similar expressions for B_2 and B_3. Equations (2-59) to (2-61) are sufficient to derive the equations of motion in the new coordinate system x_i. We illustrate with the two classic (and probably most important) systems, cylindrical and spherical.

76 VISCOUS FLUID FLOW

2-7.1 Cartesian Coordinates

Just to review our previous results, let $x_i = (x, y, z)$, from which $h_i = (1, 1, 1)$. If $\mathbf{V} = (u, v, w)$, the vector operations are

$$\nabla \phi = \left(\frac{\partial \phi}{\partial x}, \frac{\partial \phi}{\partial y}, \frac{\partial \phi}{\partial z} \right)$$

$$\text{div } \mathbf{V} = \frac{\partial u}{\partial x} + \frac{\partial v}{\partial y} + \frac{\partial w}{\partial z}$$

$$\text{curl } \mathbf{V} = \left(\frac{\partial w}{\partial y} - \frac{\partial v}{\partial z}, \frac{\partial u}{\partial z} - \frac{\partial w}{\partial x}, \frac{\partial v}{\partial x} - \frac{\partial u}{\partial y} \right) \quad (2\text{-}62)$$

$$\mathbf{V} \cdot \nabla = u \frac{\partial}{\partial x} + v \frac{\partial}{\partial y} + w \frac{\partial}{\partial z}$$

$$\nabla^2 \phi = \frac{\partial^2 \phi}{\partial x^2} + \frac{\partial^2 \phi}{\partial y^2} + \frac{\partial^2 \phi}{\partial z^2}$$

2-7.2 Cylindrical Polar Coordinates

These coordinates (r, θ, z) are related to the Cartesian system (x, y, z) by

$$x = r \cos \theta \qquad y = r \sin \theta \qquad z = z \quad (2\text{-}63)$$

from which we find that $h_1 = 1$, $h_2 = r$, and $h_3 = 1$. Let the cylindrical polar velocity components be (v_r, v_θ, v_z). Then the new vector relations are

$$\nabla \phi = \left(\frac{\partial \phi}{\partial r}, \frac{1}{r} \frac{\partial \phi}{\partial \theta}, \frac{\partial \phi}{\partial z} \right)$$

$$\nabla \cdot \mathbf{V} = \frac{1}{r} \frac{\partial}{\partial r}(r v_r) + \frac{1}{r} \frac{\partial v_\theta}{\partial \theta} + \frac{\partial v_z}{\partial z}$$

$$\mathbf{V} \cdot \nabla = v_r \frac{\partial}{\partial r} + \frac{v_\theta}{r} \frac{\partial}{\partial \theta} + v_z \frac{\partial}{\partial z} \quad (2\text{-}64)$$

$$\nabla^2 \phi = \frac{1}{r} \frac{\partial}{\partial r}\left(r \frac{\partial \phi}{\partial r} \right) + \frac{1}{r^2} \frac{\partial^2 \phi}{\partial \theta^2} + \frac{\partial^2 \phi}{\partial z^2}$$

2-7.3 Spherical Polar Coordinates

The spherical polar coordinates (r, θ, λ) are defined by the relations

$$x = r \sin \theta \cos \lambda \qquad y = r \sin \theta \sin \lambda \qquad z = r \cos \theta \quad (2\text{-}65)$$

from which we find that $h_1 = 1$, $h_2 = r$, and $h_3 = r \sin \theta$. Let the velocity components be $\mathbf{V} = (v_r, v_\theta, v_\lambda)$. Then the desired new vector relations are

$$\nabla \phi = \left(\frac{\partial \phi}{\partial r}, \frac{1}{r} \frac{\partial \phi}{\partial \theta}, \frac{1}{r \sin \theta} \frac{\partial \phi}{\partial \lambda} \right)$$

$$\text{div } \mathbf{V} = \frac{1}{r^2} \frac{\partial}{\partial r} (r^2 v_r) + \frac{1}{r \sin \theta} \frac{\partial}{\partial \theta} (v_\theta \sin \theta) + \frac{1}{r \sin \theta} \frac{\partial v_\lambda}{\partial \lambda} \qquad (2\text{-}66)$$

$$\mathbf{V} \cdot \nabla = v_r \frac{\partial}{\partial r} + \frac{v_\theta}{r} \frac{\partial}{\partial \theta} + \frac{v_\lambda}{r \sin \theta} \frac{\partial}{\partial \lambda}$$

$$\nabla^2 \phi = \frac{1}{r^2} \frac{\partial}{\partial r} \left(r^2 \frac{\partial \phi}{\partial r} \right) + \frac{1}{r^2 \sin \theta} \frac{\partial}{\partial \theta} \left(\sin \theta \frac{\partial \phi}{\partial \theta} \right) + \frac{1}{r^2 \sin^2 \theta} \frac{\partial^2 \phi}{\partial \lambda^2}$$

It is seen that the terms in non-Cartesian systems are somewhat more complicated in form but relatively straightforward. The complete equations of motion in cylindrical and spherical coordinates [the special-case forms of Eqs. (2-6), (2-19), and (2-48)] are given in App. B.

2-8 MATHEMATICAL CHARACTER OF THE BASIC EQUATIONS

The character of the basic relations, Eqs. (2-6), (2-19), and (2-48), is extremely complex. There are at least two factors that hinder our analysis: (1) the equations are coupled in the three variables $\mathbf{V}$, p, and T, and (2) each equation contains one or more non-linearities. Are these equations of the boundary-value type or are they wavelike in nature? The answer is that they contain mixtures of both boundary-value and wavelike characteristics.

The general theory concerning the character of partial differential equations has developed from the study of the interesting but rather specialized quasi-linear second-order equation studied in, for example, Kreyszig (1999):

$$A \frac{\partial^2 \phi}{\partial x^2} + B \frac{\partial^2 \phi}{\partial x \, \partial y} + C \frac{\partial^2 \phi}{\partial y^2} = D \qquad (2\text{-}67)$$

where the coefficients A, B, C, and D may be non-linear functions of x, y, ϕ, $\partial \phi / \partial x$, and $\partial \phi / \partial y$ but not of the second derivatives of ϕ (hence the term quasi-linear in the second derivatives). By developing an analytic continuation of these second derivatives, it is found that the character of Eq. (2-67) changes radically, depending upon the sign of the *discriminant* function $B^2 - 4AC$. In particular,

$$\text{If } B^2 - 4AC \begin{cases} < 0 & \text{the equation is } \textit{elliptical} \\ = 0 & \text{the equation is } \textit{parabolic} \\ > 0 & \text{the equation is } \textit{hyperbolic} \end{cases} \qquad (2\text{-}68)$$

The names elliptical, parabolic, and hyperbolic mean little by themselves, having arisen by analogy with the conic sections of analytic geometry. However, like the conic sections, these names denote a vastly different character for Eq. (2-67):

1. If the equation is *elliptical*, it can be solved only by specifying the boundary conditions on a *complete contour enclosing the region*; it is a boundary-value problem.
2. If the equation is *parabolic*, boundary conditions must be *closed in one direction* but remain *open at one end* of the other direction; it is a mixed initial- and boundary-value problem.
3. If the equation is *hyperbolic*, it can be solved in a given region by specifying conditions at only *one portion* of the boundary, the other boundaries remaining *open*; it is an initial-value problem.

Many of the partial differential equations of mathematical physics can be classified with reference to Eq. (2-67); for example,

Laplace equation: $\quad \dfrac{\partial^2 \phi}{\partial x^2} + \dfrac{\partial^2 \phi}{\partial y^2} = 0 \quad$ elliptical

Heat-conduction equation: $\quad \dfrac{\partial^2 \phi}{\partial x^2} - \dfrac{\partial \phi}{\partial y} = 0 \quad$ parabolic $\hspace{2em}$ (2-69)

Wave equation: $\quad \dfrac{\partial^2 \phi}{\partial x^2} - \dfrac{\partial^2 \phi}{\partial y^2} = 0 \quad$ hyperbolic

For further examples and discussion of these partial-differential-equation classifications, see Chap. 21 of the text by Kreyszig (1999). Certain special cases of fluid flow, such as two-dimensional inviscid flow, do fit the form of Eq. (2-67). But the full viscous-flow equations, (2-6), (2-19), and (2-48), are too complicated and display a variable mixture of elliptic, parabolic, and hyperbolic behavior. The monograph by Kreiss and Lorenz (1989) develops a variety of mathematical properties and theorems for the Navier–Stokes equations.

It is instructive now to consider the special case of *diffusion*.

2-8.1 Low-Speed Diffusion: The Prandtl Number

Consider a newtonian fluid at low speed with constant ρ, μ, and k. Let the streamwise changes be small, that is, neglect the convective derivatives compared to local changes. Then the momentum and energy equations reduce to

Momentum: $\hspace{4em} \rho \dfrac{\partial \mathbf{V}}{\partial t} \approx \rho \mathbf{g} - \nabla p + \mu \nabla^2 \mathbf{V} \hspace{3em}$ (2-70)

Energy: $\hspace{5em} \rho c_p \dfrac{\partial T}{\partial t} \approx k \nabla^2 T \hspace{6em}$ (2-71)

We recognize Eq. (2-71) as a multidimensional form of the heat-conduction equation (2-69). The temperature has elliptic boundary-value behavior in space (x, y, z)

and parabolic "marching" behavior in time. We only need an initial condition $T(x, y, z, 0)$ to get started and can "march" forward in time indefinitely if we know boundary values on T at all times.

The behavior of Eq. (2-70) is not so obvious. But we can eliminate **V** by taking the divergence of the entire equation, noting that div(**V**) = 0 for constant density. The result is

$$\nabla^2 p = 0 \qquad (2\text{-}72)$$

Thus, for these simplifications, the pressure shows elliptic behavior from Eq. (2-69). Finally, the pressure itself could be eliminated by taking the curl of Eq. (2-70), with the result

$$\frac{\partial \boldsymbol{\omega}}{\partial t} = \frac{\mu}{\rho} \nabla^2 \boldsymbol{\omega},$$

where

$$\boldsymbol{\omega} = \text{curl } \mathbf{V} = \text{fluid vorticity} \qquad (2\text{-}73)$$

This relation, like Eq. (2-71), is also the heat-conduction equation with diffusivity coefficients that are purely fluid properties:

$$\nu = \frac{\mu}{\rho} = \text{viscous } \textit{diffusivity}$$

$$\alpha = \frac{k}{\rho c_p} = \text{thermal } \textit{diffusivity} \qquad (2\text{-}74)$$

The coefficient ν is also called the *kinematic viscosity* because it has no mass units. The units of both coefficients are m^2/s, exactly the same as the *mass diffusivity D* from Eq. (1-45). Their ratios are important dimensionless fluid-property groups that give a measure of relative rates of diffusion:

Prandtl number: $\quad Pr = \dfrac{\nu}{\alpha} = \dfrac{\text{viscous diffusion rate}}{\text{thermal diffusion rate}}$

Schmidt number: $\quad Sc = \dfrac{\nu}{D} = \dfrac{\text{viscous diffusion rate}}{\text{mass diffusion rate}} \qquad (2\text{-}75)$

Lewis number: $\quad Le = \dfrac{D}{\alpha} = \dfrac{\text{mass diffusion rate}}{\text{thermal diffusion rate}}$

Note the identity $Pr = ScLe$. Since mass diffusion is not emphasized here, consider viscous and thermal effects. Table 2-1 gives values of the Prandtl number for various fluids at 20°C. It shows that liquid metals have small Prandtl number, gases are slightly less than unity, light liquids somewhat higher than unity, and oils have very large Pr. Thus there are dramatic differences among the fluids in their relative spreading of viscous and thermal effects.

Figure 2-3 shows low-speed viscous flow past a hot wall as an example. The sketches illustrate low Reynolds number or laminar-flow effects. High Reynolds number, or turbulent, flows agree qualitatively with Fig. 2-3, but the profile differences are not so broadly obvious.

TABLE 2-1
Prandtl number of various fluids at 20°C

Fluid	Prandtl number Pr	Fluid	Prandtl number Pr
Mercury	0.024	Benzene	7.4
Helium	0.70	Carbon tetrachloride	7.9
Air	0.72	Ethyl alcohol	16
Liquid ammonia	2.0	SAE 30 oil	3500
Freon-12	3.7	Castor oil	10,000
Methyl alcohol	6.8	Glycerin	12,000
Water	7.0		

A quantitative formula for the relative spreading is obtained by remembering that the two diffusion equations

$$\frac{\partial T}{\partial t} = \alpha \nabla^2 T \tag{2-76}$$

$$\frac{\partial \omega}{\partial t} = \nu \nabla^2 \omega \tag{2-77}$$

can be made independent of the constants α and ν by defining new spatial variables: $x_i/\sqrt{\alpha}$ for the T equation and $x_i/\sqrt{\nu}$ for the ω equation. The equations then have unit diffusion constants and hence both show unit spreading. For any given time, then, the thermal spreading distance L_T and the viscous spreading length L_V must be related by

$$\frac{L_T}{\sqrt{\alpha}} = \frac{L_V}{\sqrt{\nu}}$$

or

$$\frac{L_V}{L_T} \approx \sqrt{Pr} \tag{2-78}$$

Equation (2-78) is a good approximation for all laminar boundary-layer flows, even at high speeds.

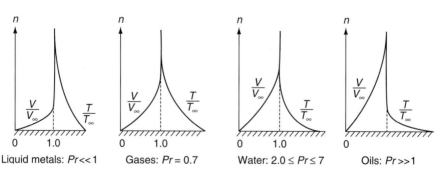

FIGURE 2-3
Prandtl number effects on viscous and thermal diffusion. (Hot wall shown for convenience.)

2-9 DIMENSIONLESS PARAMETERS IN VISCOUS FLOW

Since our basic equations of motion are extremely difficult to analyze in general, we should cast them in their most efficient form, thereby increasing the usefulness of whatever solutions we find. This is accomplished by nondimensionalizing the equations and boundary conditions.

For simplicity, assume constant c_p and c_v, approximately true for all gases, and also neglect the second coefficient of viscosity, λ, which is seldom needed. Then our four variables, p, ρ, $\mathbf{V}$, and T, will depend upon space and time and eight parameters that occur in the basic equations and boundary conditions:

$$\mathbf{V} \text{ or } p \text{ or } \rho \text{ or } T = f(x_i, t, \mu, k, g, c_p, T_w, q_w, \ell, \mathcal{T}) \qquad (2\text{-}79)$$

The eight parameters, μ through $\mathcal{T}$, are assumed to be known from data or thermodynamic state relations. Let g be constant. Now select *constant* reference properties appropriate to the flow:

1. Reference velocity U (the freestream velocity)
2. Reference length L (the writer suggests body length for external flows or duct diameter for internal flows)
3. Freestream properties p_0, ρ_0, T_0, μ_0, k_0

$(2\text{-}80)$

Steady viscous flows have no characteristic time of their own, so select particle residence time L/U as a reference time. (If the flow oscillates with frequency ω, one might select ω^{-1} as a reference time). Now define dimensionless variables and denote them by an asterisk.

$$x_i^* = \frac{x_i}{L} \quad t^* = \frac{tU}{L} \quad \mathbf{V}^* = \frac{\mathbf{V}}{U} \quad p^* = \frac{p - p_0}{\rho U^2} \quad \Phi^* = \frac{L^2}{\mu_0 U^2}\Phi$$

$$\rho^* = \frac{\rho}{\rho_0} \quad T^* = \frac{T - T_0}{T_w - T_0} \quad \mu^* = \frac{\mu}{\mu_0} \quad k^* = \frac{k}{k_0} \quad \nabla^* = L\nabla \qquad (2\text{-}81)$$

2-9.1 Nondimensionalizing the Basic Equations

Substitute the above variables from (2-81) into our basic equations (2-6), (2-19), and (2-48). Collect terms and divide out dimensional constants so that all remaining terms are dimensionless. The final results are our nondimensional equations of motion:

Continuity: $\dfrac{\partial \rho^*}{\partial t^*} + \nabla \cdot (\rho^* \mathbf{V}^*) = 0$

Navier–Stokes: $\rho^* \dfrac{D\mathbf{V}^*}{Dt^*} = \dfrac{1}{Fr}\rho^* - \nabla^* p^* + \dfrac{1}{Re}\nabla^* \cdot \left[\mu^*\left(\dfrac{\partial u_i^*}{\partial x_j^*} + \dfrac{\partial u_j^*}{\partial x_i^*}\right)\right]$ (2-82)

Energy: $\rho^* \dfrac{DT^*}{Dt^*} = Ec\dfrac{Dp^*}{Dt^*} + \dfrac{1}{RePr}\nabla^* \cdot (k^* \nabla^* T^*) + \dfrac{Ec}{Re}\Phi^*$

We see that continuity is devoid of parameters, while Navier–Stokes contains two:

Reynolds number: $$Re = \frac{\rho_0 U L}{\mu_0}$$

Froude number: $$Fr = \frac{U^2}{gL} \tag{2-83}$$

The Reynolds number is the most important dimensionless group in fluid mechanics. Almost all viscous-flow phenomena depend upon the Reynolds number. The Froude number is important only if there is a free surface in the flow.

The dimensionless energy equation also contains the Reynolds number, plus two additional parameters:

Prandtl number: $$Pr = \frac{\mu_0 c_p}{k_0}$$

Eckert number: $$Ec = \frac{U^2}{c_p T_0} \tag{2-84}$$

For a perfect gas, the more familiar *Mach* number, $Ma = U/a_0$, can replace the Eckert number:

$$Ec = \frac{U^2}{c_p T_0} = \frac{U^2}{[\gamma R/(\gamma - 1)]T_0} = (\gamma - 1)\frac{U^2}{\gamma R T_0}$$

$$= (\gamma - 1)\frac{U^2}{a_0^2} = (\gamma - 1) Ma^2 \tag{2-85}$$

The Prandtl number is always important in convective heat-transfer problems. In high-speed flows, Re, Pr, γ, and Ma are all important (Chap. 7), but at low speeds ($Ma < 0.3$), the Eckert/Mach number terms are often negligible and the energy equation reduces to Eq. (2-51). The product $RePr$ in the conduction term is often called the *Peclet* number, Pe.

An important point, sometimes neglected in elementary approaches, is that the variables ρ, μ, and k must be specified as thermodynamic functions for the particular fluid studied:

$$\rho^*, \mu^*, k^* = f(p^*, T^*)$$

Thus, if different fluids are being compared, Re, Pr, and Ec might not be sufficient if there are large temperature variations. As seen in Chap. 1, gases and liquids have quite different temperature-dependent ρ, μ, and k. Temperature corrections might be needed.

2-9.2 Momentum with Free Convection

In free or "natural" convection, there is no freestream U. The flow is driven by gravity acting on slight density changes. Pressure is nearly constant, and the density can

be modeled as
$$\rho \approx \rho_0[1 - \beta(T - T_0)]$$
where
$$\beta = -\frac{1}{\rho}\left(\frac{\partial \rho}{\partial T}\right)_p \tag{2-86}$$

The parameter U is replaced by the velocity grouping $[\mu/(\rho_0 L)]$. The reader is invited (as a problem exercise) to nondimensionalize the Navier–Stokes equation again and find a new parameter characteristic of free convection:

Grashof number: $\quad Gr = \dfrac{g\beta\rho_0^2 L^3(T_w - T_0)}{\mu_0^2} \tag{2-87}$

As we shall see in Sec. 4-13, Gr and Pr correlate heat-transfer results in free convection.

2-9.3 Nondimensionalizing the Boundary Conditions

Just analyzing the basic equations is not enough. The boundary conditions also contain important dimensionless parameters. In the freestream, where $V = U$ and $T = T_0$, the conditions simply become $V^* = 1$ and $T^* = 0$. If, then, we substitute the same variables from Eqs. (2-81) into the boundary conditions at a fixed wall in Sec. 2-6, we obtain

$$\mathbf{V}_w^* = 0 \quad T_w^* = 1 \quad \text{or} \quad \left(k^* \frac{\partial T^*}{\partial n^*}\right)_w = \frac{q_w L}{k_0(T_w - T_0)} = Nu \tag{2-88}$$

So the temperature conditions specify either a unit temperature ratio or a dimensionless wall heat transfer that is called the *Nusselt* number, Nu. The computed results at the wall will then be the opposite: either a Nusselt number or a (nonunit) temperature ratio.

If there is slip or temperature jump at the wall, Eqs. (1-91) and (1-96) become

$$u_w^* = Kn\left(\frac{\partial u^*}{\partial n^*}\right)_w$$

and

$$T_w^* = 1 + \frac{2\gamma}{Pr(\gamma + 1)} Kn\left(\frac{\partial T^*}{\partial n^*}\right)_w \tag{2-89}$$

Slip conditions introduce the Knudsen number, $Kn = \ell/L$, and the specific-heat ratio, $\gamma = c_p/c_v$.

Finally, at an interface where surface tension is important, Eq. (1-106) for interfacial pressure becomes

$$p_{\text{interface}}^* = Ca + \frac{1}{Fr}\eta^* - \frac{1}{We}\left(\frac{1}{R_x^*} + \frac{1}{R_y^*}\right) \tag{2-90}$$

This equation introduces our three final dimensionless groups:

Cavitation number: $$Ca = \frac{p_a - p_0}{\rho U^2}$$

Froude number: $$Fr = \frac{U^2}{gL} \qquad (2\text{-}91)$$

Weber number: $$We = \frac{\rho_0 U^2 L}{\mathcal{T}}$$

The Froude number is the most important of the three. It is fundamental to all free-surface flows and can never be neglected. The Weber number is important only if it is small, of $\mathcal{O}(10)$ or less, which is usually due to a small length scale L. The cavitation number is appropriate if the pressure p_0 is interpreted as the *vapor pressure* of the liquid, p_v. If $Ca \ll 1$, the liquid might vaporize (cavitate) when the local pressure drops below p_v. For further details, see the text by Brennen (1995).

To summarize, the following parameters are important for any particular flow:

1. All viscous flows: Reynolds number
2. Variable-temperature problems: Prandtl and Eckert (or Mach) numbers
3. Flow with free convection: Grashof and Prandtl numbers
4. Wall heat transfer: temperature ratio or Nusselt number
5. Slip flow: Knudsen number and specific-heat ratio
6. Free-surface conditions: Froude number (always); Weber number (sometimes), and cavitation number (sometimes)

It is interesting to note that, in spite of our care, we have missed two parameters which become important at higher Reynolds numbers, where the flow becomes turbulent (a mean flow plus a superimposed random unsteadiness). These two parameters, which occur in Chaps. 5 to 7, are the degree of surface roughness (a departure from geometric similarity) and the amount of turbulence (percentage of random fluctuation) in the reference velocity U (called freestream turbulence). Reference to these two effects was made in the discussion of the drag of a cylinder in Fig. 1-10. To the novice, these parameters express unexpected effects—timely reminders that fluid mechanics is a difficult subject, containing many hidden surprises to confound or delight the would-be analyst.

2-10 VORTICITY CONSIDERATIONS IN INCOMPRESSIBLE VISCOUS FLOW

As discussed in Chap. 1, the vorticity vector $\boldsymbol{\omega} = $ curl **V** is a measure of rotational effects, being equal to twice the local angular velocity of a fluid element. While vorticity is not a primary variable in flow analysis, the velocity itself being far more useful and significant, it is still extremely instructive to examine the impact of the vorticity vector on the Navier–Stokes Eq. (2-29). To keep the lesson from being lost in a maze of algebra, let us assume constant ρ and μ, so that Eq. (2-30) applies. Also let $\mathbf{g} = -g\mathbf{k}$, again for simplicity. Then the Navier–Stokes equation becomes

$$\rho \frac{D\mathbf{V}}{Dt} = -\nabla p - \rho g \mathbf{k} + \mu \nabla^2 \mathbf{V} \tag{2-92}$$

Introduce vorticity into both the acceleration and the viscous terms, using the two vector identities,

$$(\mathbf{V} \cdot \nabla)\mathbf{V} = \nabla \frac{V^2}{2} - \mathbf{V} \times \boldsymbol{\omega} \tag{1-10}$$

$$\nabla^2 \mathbf{V} = \nabla(\text{div } \mathbf{V}) - \text{curl } \boldsymbol{\omega} \tag{2-93}$$

and remember that div $\mathbf{V} = 0$ if ρ is constant. Equation (2-92) then becomes

$$\rho \frac{\partial \mathbf{V}}{\partial t} + \nabla\left(p + \tfrac{1}{2}\rho V^2 + \rho gz\right) = \rho \mathbf{V} \times \boldsymbol{\omega} - \mu \text{ curl } \boldsymbol{\omega} \tag{2-94}$$

Equation (2-94) is most illuminating. The left-hand side is the sum of the classic Euler terms of inviscid flow. The right-hand side vanishes if the vorticity is zero, regardless of the value of the viscosity. Thus, if $\boldsymbol{\omega} = 0$ identically, which is the classic assumption of irrotational flow, the velocity vector must by definition be a potential function, $\mathbf{V} = \nabla\phi$, where ϕ is the velocity potential, and Eq. (2-94) becomes the celebrated Bernoulli equation for unsteady incompressible flow:

$$\rho \frac{\partial \phi}{\partial t} + p + \tfrac{1}{2}\rho V^2 + \rho gz = \text{const} \tag{2-95}$$

It follows that Bernoulli's equation is valid even for *viscous* fluids if the flow is irrotational. Put another way, Eq. (2-94) shows that every potential-flow solution of classical hydrodynamics is in fact an exact solution of the full Navier–Stokes equations. The difficulty is, of course, that potential flows do not and cannot satisfy the no-slip condition at a solid wall, which requires both the normal and tangential velocities to vanish. This is because the assumption of irrotationality eliminates the second-order velocity derivatives from Eq. (2-92), leaving only first-order derivatives, so that only *one* velocity condition can be satisfied at a solid wall. In potential flow, then, we require only the normal velocity to vanish at a wall and put no restriction on the tangential velocity, which commonly slips at the wall. It would appear at this point that potential solutions are no aid in viscous analysis, but in fact at high Reynolds numbers, viscous flow past a solid body is primarily potential flow everywhere except close to the body, where the velocity drops off sharply through a thin viscous boundary layer to satisfy the no-slip condition. In many cases, the boundary layer is so thin that it does not really disturb the outer potential flow, which can be calculated by the methods of classical hydrodynamics described, for example, in the texts by Milne-Thomson (1968) or Robertson (1965). This was the case in Fig. 1-1a. It behooves the reader, therefore, to master potential-flow analysis as a natural introduction to the study of viscous flow.

We can ascertain when irrotational flow occurs by deriving an expression for the rate of change of vorticity. One form is obtained by taking the curl of Eq. (2-94), with the result

$$\frac{D\boldsymbol{\omega}}{Dt} = (\boldsymbol{\omega} \cdot \nabla)\mathbf{V} + \nu\nabla^2\boldsymbol{\omega} \tag{2-96}$$

which is the famous Helmholtz equation of hydrodynamics, usually written without the second term involving the kinematic viscosity. The first term on the right arises from the convective derivative and is called the *vortex-stretching term*; this term was zero in the special case of Eq. (2-77). The second term is obviously a viscous-diffusion term; if it is neglected (the ideal fluid), Eq. (2-96) leads to Helmholtz' theorem that the strength of a vortex remains constant and also to Lagrange's theorem that $\boldsymbol{\omega} = 0$ for all time if it is zero everywhere at $t = 0$. Still a third result of neglecting the viscous term is Kelvin's theorem that the circulation about any closed path moving with the fluid is a constant. All three theorems are valid and useful in the inviscid approximation, but all three fail when the viscosity is not zero because the second term in Eq. (2-96) is constrained by the no-slip condition not to vanish near solid walls. Thus, in all viscous flows, vorticity is generally present, being generated by relative motion near solid walls. If the Reynolds number is large, the vorticity is swept down stream and remains close to the wall. A boundary layer is formed, outside of which the vorticity may be taken as zero. If the flow is between two walls, as in duct flow, the two boundary layers will meet and fill the duct with vorticity, as in Fig. 1-11, so that the potential-flow model is generally not valid in duct flow.

2-11 TWO-DIMENSIONAL CONSIDERATIONS: THE STREAM FUNCTION

In Eq. (2-9) we noted that the stream function is an exact solution of the continuity equation when only two independent spatial variables are involved in the flow. For maximum benefit, let us examine the simple case of constant μ and ρ and two-dimensional flow in the xy plane, i.e., a flow with only the two velocity components $u(x, y, t)$ and $v(x, y, t)$. The equations of motion then are the continuity equation and the two components of the Navier–Stokes equations:

$$\frac{\partial u}{\partial x} + \frac{\partial v}{\partial y} = 0 \tag{2-97}$$

$$\frac{\partial u}{\partial t} + u\frac{\partial u}{\partial x} + v\frac{\partial u}{\partial y} = g_x - \frac{1}{\rho}\frac{\partial p}{\partial x} + \nu\left(\frac{\partial^2 u}{\partial x^2} + \frac{\partial^2 u}{\partial y^2}\right) \tag{2-98}$$

$$\frac{\partial v}{\partial t} + u\frac{\partial v}{\partial x} + v\frac{\partial v}{\partial y} = g_y - \frac{1}{\rho}\frac{\partial p}{\partial y} + \nu\left(\frac{\partial^2 v}{\partial x^2} + \frac{\partial^2 v}{\partial y^2}\right) \tag{2-99}$$

to be solved for the three dependent variables u, v, and p. Note that the momentum equation has been uncoupled from the energy equation by the assumption of constant μ and ρ, so that $T(x, y, t)$ can be solved later from Eq. (2-50) if desired.

Lest the reader think that the system, Eqs. (2-97) to (2-99), has been so oversimplified that it is not worth making a fuss over, the remark that this system has borne the brunt of much of the theoretical work on viscous flow to this date.

By analogy with Eq. (2-9), Eq. (2-97) yields immediately to a stream function $\psi(x, y, t)$ defined by

$$u = \frac{\partial \psi}{\partial y} \quad v = -\frac{\partial \psi}{\partial x} \tag{2-100}$$

where ψ now takes on the meaning of volume flow rather than mass flow, since the density is missing in the definition. Meanwhile, the pressure and gravity can be eliminated from Eqs. (2-98) and (2-99) by cross-differentiation, i.e., taking the curl of the two-dimensional vector momentum equaiton. The result is

$$\frac{\partial \omega_z}{\partial t} + u \frac{\partial \omega_z}{\partial x} + v \frac{\partial \omega_z}{\partial y} = \nu \left(\frac{\partial^2 \omega_z}{\partial x^2} + \frac{\partial^2 \omega_z}{\partial y^2} \right) \tag{2-101}$$

where

$$\omega_z = \left(\frac{\partial v}{\partial x} - \frac{\partial u}{\partial y} \right) \tag{1-13}$$

Since $\mathbf{V} = \mathbf{V}(x, y, t)$ only, the vorticity has only the single nonvanishing component ω_z. By comparison with the more general vorticity transport relation, Eq. (2-97), we see that the vortex-stretching term $(\boldsymbol{\omega} \cdot \nabla)\mathbf{V}$ has vanished identically because, for this special case of flow in the xy plane, the vorticity vector $\boldsymbol{\omega}$ is perpendicular to the gradient of $\mathbf{V}$. Combining Eqs. (2-100) and (1-13), we have the intriguing relation

$$\omega_z = -\nabla^2 \psi \tag{2-102}$$

so that Eq. (2-101) can be rewritten as a fourth-order partial differential equation with the stream function as the only variable:

$$\frac{\partial}{\partial t}(\nabla^2 \psi) + \frac{\partial \psi}{\partial y}\frac{\partial}{\partial x}(\nabla^2 \psi) - \frac{\partial \psi}{\partial x}\frac{\partial}{\partial y}(\nabla^2 \psi) = \nu \nabla^4 \psi \tag{2-103}$$

The boundary conditions would be in terms of the first derivatives of ψ, from Eq. (2-100). For example, in the flow of a uniform stream in the x direction past a solid body, the conditions would be

At infinity: $\quad \dfrac{\partial \psi}{\partial x} = 0 \quad \dfrac{\partial \psi}{\partial y} = U_\infty$

At the body (no slip): $\quad \dfrac{\partial \psi}{\partial x} = 0 = \dfrac{\partial \psi}{\partial y}$ (2-104)

This *stream-function vorticity* approach is an alternative to the direct or "primitive" variable approach of solving for (u, v, p) in Eqs. (2-97) to (2-99). Equations (2-103) and (2-104) are convenient to numerical analysis and were used in the first (hand-calculated) CFD solution, Thom (1933), known to this writer.

2-11.1 Creeping Flow: $Re \ll 1$

If the viscosity is very large, the Reynolds number UL/ν is very small, $Re \ll 1$, and the right-hand side of Eq. (2-103) dominates. We are left with a linear fourth-order equation:

$$Re \ll 1: \qquad \nabla^4 \psi = 0 \qquad (2\text{-}105)$$

This is the two-dimensional *biharmonic* equation for "creeping flow." Many solutions are known from elasticity, Timoshenko and Goodier (1970), which can be adapted to creeping flow. A wide variety of low Reynolds number flows are treated in the monographs by Langlois (1964) and by Happel and Brenner (1983). We shall briefly treat this subject in Sec. 3-9.

2-12 NONINERTIAL COORDINATE SYSTEMS

By far the majority of flow problems are attacked in a fixed, or inertial, coordinate system. Cases do arise, though, such as the flow over rotating turbine blades or the geophysical boundary layer on a rotating earth, where we may wish to use noninertial coordinates moving with the accelerating system. Then we must modify Newton's law, Eq. (2-11), which is valid only if **a** is the absolute acceleration of the particle relative to inertial coordinates.

Suppose that (X, Y, Z) are in an inertial frame and that our chosen coordinates (x, y, z) are translating and rotating relative to that frame. Let **R** and $\mathbf{\Omega}$ be the displacement and angular velocity vector of the (x, y, z) system relative to (X, Y, Z). Then, by straightforward vector calculus [see Greenwood (1988, pp. 49ff.)], we can relate the absolute acceleration **a** of a particle to its displacement **r** and velocity **V** relative to the moving system:

$$\mathbf{a} = \frac{d^2\mathbf{R}}{dt^2} + \frac{d\mathbf{\Omega}}{dt} \times \mathbf{r} + \mathbf{\Omega} \times (\mathbf{\Omega} \times \mathbf{r}) + \frac{d\mathbf{V}}{dt} + 2\mathbf{\Omega} \times \mathbf{V} \qquad (2\text{-}106)$$

Thus, if **V** is a noninertial velocity vector, the entire formidable right-hand side of Eq. (2-106) must replace the derivative $D\mathbf{V}/Dt$ in Eq. (2-12) or (2-19). However, we would be expected to know the functions $\mathbf{R}(t)$ and $\mathbf{\Omega}(t)$ which relate the two systems, so that in a sense we are merely adding known inhomogeneities to the problem.

2-12.1 Geophysical Flows: The Rossby Number

A classic example of a noninertial system is the use of earth-fixed coordinates to calculate large-scale (geophysical) motions, where the earth's rotation cannot be neglected. In this case, the first three terms of Eq. (2-106) can be neglected, for three different reasons: (1) the first term is the earth's acceleration relative to the fixed stars and is surely almost zero; (2) the second term vanishes because $d\mathbf{\Omega}/dt$ is nearly zero for the earth; and (3) the third term is nothing more than the earth's centripetal acceleration, which essentially accounts for the variation of the gravity vector **g** with latitude. Thus, for geophysical flows,

$$\mathbf{a} \approx \frac{d\mathbf{V}}{dt} + 2\mathbf{\Omega} \times \mathbf{V} \qquad (2\text{-}107)$$

the second term of which is called the *Coriolis acceleration*, named after the French mathematician G. Coriolis, who was the first, in 1835, to study this phenomenon. If, as before, we use U and L to nondimensionalize this expression, we obtain

$$\mathbf{a}^* \approx \frac{d\mathbf{V}^*}{dt^*} + \frac{1}{Ro}(2\mathbf{\Omega}^* \times \mathbf{V}^*) \qquad (2\text{-}108)$$

where $Ro = U/\Omega L$ is the dimensionless Rossby number. It follows that the Coriolis term can be neglected if Ro is large, which will be true if the motion scale L is small compared to the earth's radius. We shall study a geophysical viscous flow (Ekman flow) in Chap. 3.

2-13 CONTROL-VOLUME FORMULATIONS

The expressions so far in this chapter have all been differential relations: fluid flow in the small, so to speak. We shall also have use for integral relations to calculate the gross fluxes of mass, momentum, and energy passing through a finite region of the flow. This is the control-volume approach now widely used in fluid mechanics [see White (2003), Chap. 3]. We shall content ourselves here with a fixed control volume; the text by Hansen (1967) treats moving and deformable control volumes.

Let the closed solid curve in Fig. 2-4 represent a finite region (control volume) through which a fluid flow passes. At any instant t, the region is filled by an aggregate of fluid particles which we shall call the *system*, i.e., a quantity of known identity. We wish to calculate the rate of change dB/dt of any gross property B (mass, kinetic energy, enthalpy, etc.) of the system at that instant t. We do this by a limiting process: At time $t + \Delta t$, the system has passed on slightly to the right, as indicated by the dotted lines in Fig. 2-4 and suggested by the streamline arrows.

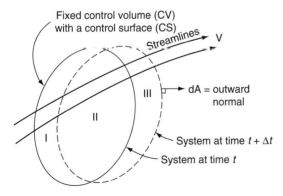

FIGURE 2-4
Sketch of a control volume.

The slight motion has outlined three regions, I, II, and III, as marked on the figure. The limit we are looking for is

$$\frac{dB}{dt}(\text{system at } t) = \lim_{\Delta t \to 0} \frac{(B_{II} + \Delta B_{III})_{t+\Delta t} - (\Delta B_{I} + B_{II})_{t}}{\Delta t}$$

$$= \lim_{\Delta t \to 0}\left[\frac{(B_{II})_{t+\Delta t} - (B_{II})_{t}}{\Delta t} + \frac{\Delta B_{III}}{\Delta t} - \frac{\Delta B_{I}}{\Delta t}\right] \quad (2\text{-}109)$$

As shown in undergraduate texts, such as White (2003), Sec. 3.2, the region II terms in the brackets become the rate of change of B within the control volume. The region III term is the outflow of B, and the region I term is the inflow of B into the control volume. By the device of defining the area vector $d\mathbf{A}$ as having the direction of the *outward* normal to the control surface (see Fig. 2-4), we can account for outflow and inflow with a single dot product. The final limit is

$$\frac{dB}{dt} = \frac{d}{dt}\int_{CV}\frac{dB}{dm}\rho\, d(vol) + \int_{CS}\frac{dB}{dm}\rho\mathbf{V}\cdot d\mathbf{A} \quad (2\text{-}110)$$

This general relation is often called the *Reynolds transport theorem*. We can apply this formulation to any property B. Here we select mass, momentum, and energy.

2-13.1 Conservation of Mass

The relevant property is $B = $ mass m, and $dm/dm = 1$. Equation (2-110) becomes

$$\frac{dm}{dt} = 0 = \frac{d}{dt}\int_{CV}\rho\, d(vol) + \int_{CS}\rho\mathbf{V}\cdot d\mathbf{A} \quad (2\text{-}111)$$

This relation must be true for any fixed control volume and will be very useful in boundary-layer theory, Chap. 4. As an exercise, one can show, using Gauss' theorem, that Eq. (2-111) is identical to the differential equation of continuity, Eq. (2-6).

2-13.2 Conservation of Linear Momentum

The relevant property is $B = $ linear momentum $m\mathbf{V}$, and $d(m\mathbf{V})/dm = \mathbf{V}$ itself. Equation (2-110) becomes equal to the total force $\mathbf{F}$ acting on the system passing through the control volume:

$$\mathbf{F} = \frac{d}{dt}(m\mathbf{V}) = \frac{d}{dt}\int_{CV}\mathbf{V}\rho\, d(vol) + \int_{CS}\mathbf{V}(\rho\mathbf{V}\cdot d\mathbf{A}) \quad (2\text{-}112)$$

It is emphasized that this relation holds only for an inertial control volume. The force $\mathbf{F}$ includes both surface forces on the control surface and body forces on the mass within. Equation (2-112) was first applied to viscous boundary layers by Kármán (1921).

2-13.3 Conservation of Energy

The relevant property is $B = $ total energy E, and we denote $dE/dm = e_t$. Recall the first law of thermodynamics, Eq. (1-59), $dE = dQ + dW$. Equation (2-110) becomes

$$\frac{dE}{dt} = \frac{dQ}{dt} + \frac{dW}{dt} = \frac{d}{dt}\int_{CV} e_t \rho \, d(vol) + \int_{CS} e_t \rho \mathbf{V} \cdot d\mathbf{A} \quad (2\text{-}113)$$

where

$$e_t = \frac{dE}{dm} = e + \tfrac{1}{2}V^2 - \mathbf{g} \cdot \mathbf{r}$$

The work term dW includes work on the boundaries of normal and shear stresses plus any shaft work added to the system:

$$W = W_{\text{normal stress}} + W_{\text{shear stress}} + W_{\text{shaft}} \quad (2\text{-}114)$$

Equation (2-113) may be applied to boundary layers to determine the heat transfer.

2-13.4 The Steady-Flow Energy Equation

An important special case of Eq. (2-113) occurs in steady flow when the control volume consists of fixed solid walls plus a simple one-dimensional inlet and outlet, as shown in Fig. 2-5. The work of boundary forces is zero at the fixed walls because of the no-slip condition. Further, if the inlet and outlet flows are approximately uniform and parallel, we may reasonably neglect viscous normal stresses there, so that the only important boundary work is due to pressure forces at inlet and outlet. The term d/dt vanishes for steady flow, and Eq. (2-113) reduces to

$$\frac{dQ}{dt} + \frac{dW_{\text{shaft}}}{dt} - \iint_{CS} p\mathbf{V} \cdot d\mathbf{A} = \iint_{CS} e_t \rho \mathbf{V} \cdot d\mathbf{A} \quad (2\text{-}115)$$

But the only contributions to $\mathbf{V} \cdot d\mathbf{A}$ are $-V_1 A_1$ and $+V_2 A_2$ at the inlet and outlet, respectively. Thus we can carry out the integration to yield two terms

$$\frac{dQ}{dt} + \frac{dW_s}{dt} = \left(\frac{p}{\rho} + e + \tfrac{1}{2}V^2 + gz\right)_2 \rho_2 A_2 V_2$$

$$- \left(\frac{p}{\rho} + e + \tfrac{1}{2}V^2 + gz\right)_1 \rho_1 A_1 V_1 \quad (2\text{-}116)$$

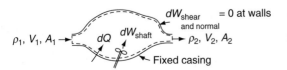

FIGURE 2-5
Sketch for the one-dimensional steady-flow energy equation.

where we have introduced $(-\mathbf{g} \cdot \mathbf{r}) = +gz$, implying as before the choice of the gravity direction $\mathbf{g} = -g\mathbf{k}$. For steady flow, $\rho_1 A_1 V_1 = \rho_2 A_2 V_2$ equals the mass flow rate dm/dt through the control volume. Also, the quantity $p/\rho + e$ is the fluid enthalpy h. We divide through by dm/dt to achieve the familiar form

$$\frac{dQ}{dm} + \frac{dW_s}{dm} = \left(h + \tfrac{1}{2}V^2 + gz\right)_2 - \left(h + \tfrac{1}{2}V^2 + gz\right)_1 \qquad (2\text{-}117)$$

which we recognize as the so-called steady-flow energy equation, stating that the change in total enthalpy $h + \tfrac{1}{2}V^2 + gz$ equals the sum of the heat and shaft-work addition per unit mass. The formula does not hold in boundary-layer theory unless we are careful to define our control volume to match Fig. 2-5 conditions.

SUMMARY

It is the role of this chapter to derive and discuss the basic equations of viscous fluid flow: conservations of mass, momentum, and energy plus the auxiliary state relations. The various boundary conditions and coordinate systems are discussed, and dimensionless parameters are derived and related to particular classes of flow. Stream-function and vorticity approaches are discussed, and the basic relations are also derived in integral form for finite control volumes. It is the purpose of the remainder of this text to attack these equations as best we can for particular classes of viscous flow.

PROBLEMS

2-1. By consideration of the cylindrical elemental control volume shown in Fig. P2-1, use the conservation of mass to derive the continuity equation in cylindrical coordinates [App. B, Eq. (B-3)].

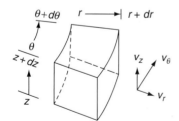

FIGURE P2-1

2-2. Simplify the equation of continuity in cylindrical coordinates (r, θ, z) to the case of steady compressible flow in polar coordinates $(\partial/\partial z = 0)$ and derive a stream function for this case.

2-3. Simplify the equation of continuity in cylindrical coordinates to the case of steady compressible flow in axisymmetric coordinates ($\partial/\partial\theta = 0$) and derive a stream function for this case.

Answer: $\partial\psi/\partial r = \rho r v_z$; $\partial\psi/\partial z = -\rho r v_r$

2-4. For steady incompressible flow with negligible viscosity, show that the Navier–Stokes relation [Eq. (2-30)] reduces to the condition that $p/\rho + |\mathbf{V}|^2/2 + gh$ is constant along a streamline of the flow, where h denotes the height of the fluid particle above a horizontal datum. This is the weaker form of the so-called Bernoulli relation.

2-5. Show that for incompressible steady flow with negligible viscosity and thermal conductivity, the energy Eq. (2-40) reduces to the condition that $e + p/\rho + |\mathbf{V}|^2/2 + gh$ is constant along a streamline of the flow. This is the stronger form of Bernoulli's relation.

2-6. Consider the proposed incompressible axisymmetric flow field $v_z = C(R^2 - r^2)$, $v_r = 0$ in the region $0 \leq z \leq L, 0 \leq r \leq R$, where C is a constant. (*a*) Determine if this is an exact solution of the Navier–Stokes equation. (*b*) What might it represent? (*c*) If an axisymmetric stream function $\psi(r, z)$ exists for this flow, find its form.

2-7. Investigate the two-dimensional stream function $\psi = Cxy$, with C a constant, to determine whether it can represent (*a*) a realistic incompressible frictionless flow and (*b*) a realistic incompressible viscous flow. Sketch some streamlines of the flow.

2-8. Investigate a proposed plane stream function for isothermal incompressible flow: $\psi = C(x^2 y - y^3/3)$, where C is a constant. Determine whether this flow is an exact solution for constant μ and, if so, find the pressure distribution $p(x, y)$ and plot a few representative streamlines.

2-9. Consider the following incompressible plane unsteady flow:

$$v_r = 0 \qquad v_\theta = \frac{C}{r}\left[1 - \exp\left(-\frac{r^2}{4t\nu}\right)\right]$$

where C and ν are constant and gravity is neglected. Is this an exact solution of the continuity and Navier–Stokes equations? If so, plot some velocity profiles for representative times. Is there any vorticity in the flow? If so, plot some vorticity profiles.

2-10. Using the expression for dissipation Φ from Eq. (2-46), prove the inequalities about μ and λ given by Eq. (2-47).

2-11. The differential equation for irrotational plane compressible gas flow [Shapiro (1954), Chap. 9] is

$$\frac{\partial^2 \phi}{\partial t^2} + \frac{\partial}{\partial t}(u^2 + v^2) + (u^2 - a^2)\frac{\partial^2 \phi}{\partial x^2} + (v^2 - a^2)\frac{\partial^2 \phi}{\partial y^2} + 2uv\frac{\partial^2 \phi}{\partial x \partial y} = 0$$

where ϕ is the velocity potential and a the (variable) speed of sound in the gas. In the spirit of Sec. 2-9.1, nondimensionalize this equation and define any parameters which appear.

2-12. In flow at moderate to high Reynolds numbers, pressure changes scale with $(\rho_0 U_0^2)$, as in Eq. (2-83). In very viscous (low Reynolds number) flow, pressure scales as

($\mu U_0/L$). Make this change in Eqs. (2-83) and repeat the nondimensionalization of the Navier–Stokes equation. Define any parameters which arise and show what happens if the Reynolds number is very small.

2-13. The equations of motion for free convection near a hot vertical plate for incompressible flow with constant properties are

$$\frac{\partial u}{\partial x} + \frac{\partial v}{\partial y} = 0$$

$$u\frac{\partial u}{\partial x} + v\frac{\partial u}{\partial y} = g\beta(T - T_1) + \nu\left(\frac{\partial^2 u}{\partial x^2} + \frac{\partial^2 u}{\partial y^2}\right)$$

$$\rho c_p\left(u\frac{\partial T}{\partial x} + v\frac{\partial T}{\partial y}\right) = k\left(\frac{\partial^2 T}{\partial x^2} + \frac{\partial^2 T}{\partial y^2}\right)$$

Introduce the dimensionless variables

$$u^* = \frac{uL}{\nu} \qquad v^* = \frac{vL}{\nu} \qquad x^* = \frac{x}{L} \qquad y^* = \frac{y}{L} \qquad T^* = \frac{T - T_1}{T_0 - T_1}$$

where L is the length of the plate. Use these variables to nondimensionalize the free convection equations and define any parameters which arise.

2-14. For laminar flow in the entrance to a pipe, as shown in Fig. P2-14, the entrance flow is uniform, $u = U_0$, and the flow downstream is parabolic in profile, $u(r) = C(r_0^2 - r^2)$. Using the integral relations of Sec. 2-13, show that the viscous drag exerted on the pipe walls between 0 and x is given by

$$\text{Drag} = \pi r_0^2\left(p_0 - p_x - \tfrac{1}{3}\rho U_0^2\right)$$

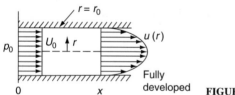

FIGURE P2-14

2-15. To illustrate "boundary-layer" behavior, i.e., the effect of the no-slip condition for large Reynolds numbers, Prandtl in a 1932 lecture proposed the following model (linear) differential equation:

$$\epsilon\frac{d^2 u}{dy^2} + \frac{du}{dy} + u = 0 \qquad \epsilon \ll 1$$

where ϵ mimics the small viscosity of the fluid. The boundary conditions are (1) $u(0) = 2$, and (2) u remains bounded as y becomes large. Solve this equation for these conditions and plot the profile $u(y)$ in the range $0 < y < 2$ for $\epsilon = 0, 0.01$, and 0.1. Comment on the results. Is the plot for $\epsilon = 0$ the same as that obtained by setting $\epsilon = 0$ in the original differential equation and then solving?

2-16. Consider the plane, incompressible, Cartesian stream function in the region $0 \leq y \leq \infty$:

$$\psi = ax + by + \frac{b}{c} e^{-cy}$$

where (a, b, c) are positive constants. (a) Determine if this is an exact solution to the continuity and Navier–Stokes equations if gravity and pressure gradient are neglected. (b) What are the dimensions of (a, b, c)? (c) If $y = 0$ represents a wall, does the no-slip condition hold there? (d) Is there any vorticity in the flow field? If so, what is its form?

2-17. Use Eq. (2-86) for density in the gravity term of the Navier–Stokes equation (2-29b), let $\rho \approx \rho_0$ in the acceleration term, and let μ = constant. For free convection, with U replaced by $\mu/(\rho_0 L)$, nondimensionalize Navier–Stokes and show that the Grashof number appears.

2-18. Flow through a well-designed contraction or nozzle is nearly frictionless, as shown, for example, in White (2003), Sec. 6.12. Suppose that water at 20°C flows through a horizontal nozzle at a weight flow of 50 N/s. If entrance and exit diameters are 8 cm and 3 cm, respectively, and the exit pressure is 1 atm, estimate the entrance pressure from Bernoulli's equation.

2-19. Show, using Gauss' theorem [Kreyszig (1999), Sec. 9.8] that the control-volume mass relation, Eq. (2-111), leads directly to the partial differential equation of continuity, Eq. (2-6).

2-20. In discussing incompressible flow with constant μ, Eq. (2-30), we cavalierly said, "many terms vanish" from Eq. (2-29a). Be less cavalier and show that the many viscous terms in Eqs. (2-29a) do indeed reduce to the single vector term $\mu \nabla^2 \mathbf{V}$, in three dimensions.

2-21. In deriving the basic equations of motion in this chapter, we skipped over the partial differential equation of *angular momentum*. Did we forget? Do some reading, perhaps in Lai et al. (1995) or Malvern (1997), and explain the significance of the angular momentum equation.

CHAPTER 3

SOLUTIONS OF THE NEWTONIAN VISCOUS-FLOW EQUATIONS

3-1 INTRODUCTION AND CLASSIFICATION OF SOLUTIONS

The equations of continuity, momentum, and energy, derived in Chap. 2, are a formidable system of non-linear partial differential equations. No general analytical method exists, and no general existence or uniqueness theorems exist. For example, no exact solution is known for the thin airfoil problem of Fig. 1-1*a*, and certainly not for the stalled airfoil of Fig. 1-1*b*. However, the boundary-layer techniques of Chaps. 4 and 6 will be a good approximate tool for many problems. Meanwhile, for *laminar* flows, where no fine-scale turbulent fluctuations occur, computational fluid dynamics (CFD) is quite accurate, if rather specific, for a variety of flows and geometries. See, for example, Orszag et al. (1991), Chung (2002), Lomax et al. (2001), or Tannehill et al. (1997). For turbulent flow (Chap. 6), except for *direct numerical simulation* (DNS) at low Reynolds numbers, we must resort either to empirical correlations or use approximate CFD models that are reliable only for a limited subset of flow patterns.

In accumulating exact solutions, then, we are poor but not destitute. Over the past 150 years, a considerable number[†] of exact but particular solutions have been

---

[†]There are about 80 different cases, depending upon how one counts them. A large number are given by Berker (1963). See also Rogers (1992) and Profilo et al. (1998).

found which satisfy the complete equations for some special geometry. This chapter will outline some of these particular solutions, many of which are very illuminating about viscous-flow phenomena.

As we might expect, almost all the known particular solutions are for incompressible newtonian flow with constant transport properties, for which the basic equations [(2-6), (2-19), and (2-48)] reduce to

Continuity: $$\text{div}\mathbf{V} = 0 \tag{3-1}$$

Momentum: $$\rho \frac{D\mathbf{V}}{Dt} = -\nabla \hat{p} + \mu \nabla^2 \mathbf{V} \tag{3-2}$$

Energy: $$\rho c_p \frac{DT}{Dt} = k\nabla^2 T + \Phi \tag{3-3}$$

where Φ denotes the dissipation function from Eq. (2-46) and $\hat{p}$ is the total hydrostatic pressure, i.e., it includes the gravity term for convenience:

$$\nabla \hat{p} = \nabla p - \rho \mathbf{g}$$

or $$\hat{p} = p + \rho g z \tag{3-2a}$$

where z is the vertical coordinate. The three unknowns in Eqs. (3-1) to (3-3) are $\mathbf{V}$, $\hat{p}$, and T. Note an important fact, mentioned earlier: Since we assume that μ is constant, Eqs. (3-1) and (3-2) are uncoupled from temperature and thus can be solved for $\mathbf{V}$ and $\hat{p}$ independently of T, after which T can be solved from Eq. (3-3). Note also that T itself is *not* independent of $\hat{p}$ or $\mathbf{V}$, since the velocity, which may be pressure-dependent, enters Eq. (3-3) through the terms Φ and $DT/Dt = \partial T/\partial t + \mathbf{V} \cdot \nabla T$. Because of this uncoupling of temperature in incompressible flows, many texts simply ignore the energy equation, but we consider T here to be very important, both practically and pedagogically.

Basically, there are two types of exact solutions of Eq. (3-2):

1. Linear solutions, where the convective acceleration $\mathbf{V} \cdot \nabla$ vanishes
2. Non-linear solutions, where $\mathbf{V} \cdot \nabla$ does not vanish

It is also possible to classify solutions by the type or geometry of flow involved:

1. Couette (wall-driven) steady flows
2. Poiseuille (pressure-driven) steady duct flows
3. Unsteady duct flows
4. Unsteady flows with moving boundaries
5. Duct flows with suction and injection
6. Wind-driven (Ekman) flows
7. Similarity solutions (rotating disk, stagnation flows, etc.)

These are the topics of the next seven sections of this chapter, after which we conclude with the creeping-flow approximation (Sec. 3-9) and digital computer solutions (Sec. 3-10).

Although a general solution is unattainable, particular exact solutions of Navier–Stokes are still being found. Here are some recent examples: a porous channel with moving walls, Dauenhauer and Majdalani (2003); oscillating flow in a rectangular duct, Tsangaris and Vlachakis (2003); chaotic flow in cylindrical tubes, Blyth et al. (2003); generalized Beltrami flows, Wang (1990); an unsteady stretching surface, Smith (1994); free shear layers, Varley and Seymour (1994); and two interacting vortices, Agullo and Verga (1997). For further study of laminar viscous flows, see the monographs by Constantinescu (1995), Ockendon and Ockendon (1995), and Papanastasiou et al. (1999).

3-2 COUETTE FLOWS DUE TO MOVING SURFACES

These flows are named in honor of M. Couette (1890), who performed experiments on the flow between a fixed and moving concentric cylinder. We consider several examples.

3-2.1 Steady Flow between a Fixed and a Moving Plate

In Fig. 3-1, two infinite plates are $2h$ apart, and the upper plate moves at speed U relative to the lower. The pressure $\hat{p}$ is assumed constant. The upper plate is held at temperature T_1 and the lower plate at T_0. These boundary conditions are independent of x or z ("infinite plates"); hence it follows that $u = u(y)$ and $T = T(y)$. Equations (3-1) to (3-3) reduce to

Continuity: $\quad \dfrac{\partial u}{\partial x} = 0$

Momentum: $\quad 0 = \mu \dfrac{d^2 u}{dy^2}$ $\quad\quad$ (3-4)

Energy: $\quad 0 = k \dfrac{d^2 T}{dy^2} + \mu \left(\dfrac{du}{dy}\right)^2$ $\quad\quad$ (3-5)

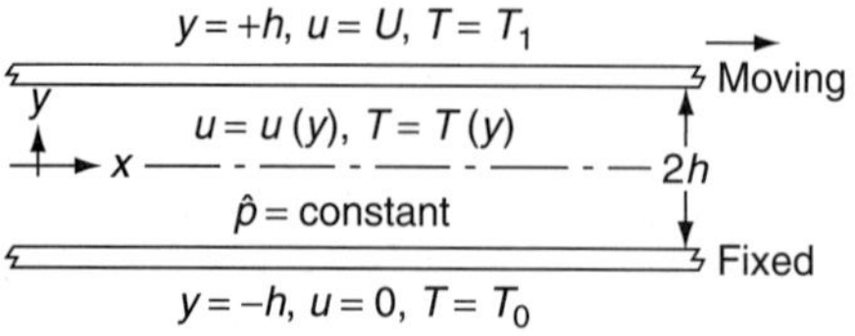

FIGURE 3-1
Couette flow between parallel plates.

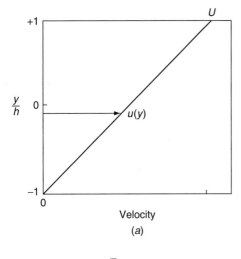

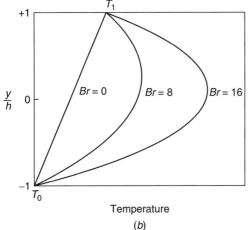

FIGURE 3-2
Couette flow between a fixed and a moving plate: (*a*) the velocity profile; (*b*) temperature profiles for various Brinkman numbers.

where continuity merely verifies our assumption that $u = u(y)$ only. Equation (3-4) can be integrated twice to obtain

$$u = C_1 y + C_2$$

The boundary conditions are no slip, $u(-h) = 0$ and $u(+h) = U$, whence $C_1 = U/2h$ and $C_2 = U/2$. Then the velocity distribution is

$$u = \frac{U}{2}\left(1 + \frac{y}{h}\right) \tag{3-6}$$

This is plotted in Fig. 3-2*a* and is seen to be a straight line connecting the no-slip condition at each plate. We term this plot a *velocity profile* and commonly sketch in line segments with arrows, as shown, to indicate motion. Since $\mu = $ const and we have neglected buoyancy, this profile is independent of the temperature distribution.

The shear stress at any point in the flow follows from the viscosity law:

$$\tau = \mu\left(\frac{\partial u}{\partial y} + \frac{\partial v}{\partial x}\right) = \mu\frac{du}{dy} = \frac{\mu U}{2h} = \text{const} \tag{3-7}$$

Thus for this simple flow, the shear stress is constant throughout the fluid, as is the strain rate—even a nonnewtonian fluid would maintain a linear velocity profile.

The dimensionless shear stress is usually defined in engineering flows as the *friction coefficient*,

$$C_f = \frac{\tau}{\frac{1}{2}\rho U^2} = \frac{\mu}{\rho U h} = \frac{1}{Re_h} \tag{3-8}$$

Thus the Reynolds number $Re_h = \rho U h/\mu$ arises naturally when the shear stress is nondimensionalized. Or perhaps *unnaturally* is more appropriate: Churchill (1988) points out that the Reynolds number is unsuitable for this nonaccelerating flow, since density does not play a part. He suggests that one should instead use the *Poiseuille number*,

$$Po = C_f Re_h = \frac{2h\tau}{\mu U} = 1 \tag{3-9}$$

Clearly a unit Poiseuille number is more convenient than a varying friction coefficient.

With $du/dy = U/2h$ known from Eq. (3-6), we may substitute into Eq. (3-5) and integrate twice to obtain the temperature distribution:

$$T = -\frac{\mu U^2}{4kh^2}\frac{y^2}{2} + C_3 y + C_4 \tag{3-10}$$

The no-temperature-jump conditions require $T(-h) = T_0$ and $T(+h) = T_1$, whence C_3 and C_4 can be evaluated. The final solution is

$$T = \left(\frac{T_1 + T_0}{2} + \frac{T_1 - T_0}{2}\frac{y}{h}\right) + \frac{\mu U^2}{8k}\left(1 - \frac{y^2}{h^2}\right) \tag{3-11}$$

The first term in parentheses represents the straight-line distribution which would arise due to pure conduction in the fluid. The second (parabolic) term is the temperature rise due to *viscous dissipation* in the fluid. The temperatures $T(y)$ from Eq. (3-11) are plotted in Fig. 3-2b. If T is nondimensionalized by $(T_1 - T_0)$, a dimensionless dissipation parameter arises naturally, the *Brinkman number*:

$$Br = \frac{\mu U^2}{k(T_1 - T_0)} = \frac{\mu c_p}{k}\frac{U^2}{c_p(T_1 - T_0)} = PrEc \tag{3-12}$$

From Fig. 3-2b, a Brinkman number of order unity or greater means that the temperature rise due to dissipation is significant. Qualitatively, it represents the ratio of dissipation effects to fluid conduction effects.

For low-speed flows, only the most viscous fluids (oils) have significant Brinkman numbers. For example, take $U = 10$ m/s and $(T_1 - T_0) = 10°C$ and compare the following numerical values for four fluids:

Fluid	μ(kg/m·s)	k(W/m·°C)	Brinkman number
Air	1.8E–5	0.26	0.0007
Water	1.0E–3	0.60	0.017
Mercury	1.54E–3	8.7	0.0018
SAE 30 oil	0.29	0.145	20.0

Thus, except for heavy oils, we commonly neglect dissipation effects in low-speed flow temperature analyses.

To complete this lengthy initial-case discussion, we compute the rate of heat transfer at the walls:

$$q_w = \left|k\frac{\partial T}{\partial y}\right|_{\pm h} = \frac{k}{2h}(T_1 - T_0) \pm \frac{\mu U^2}{4h} \qquad (3\text{-}13)$$

where the ($\pm$) refers to lower and upper surface, respectively. The first term on the right represents pure conduction through the fluid.

Since many convection analyses result in q_w being proportional to ΔT, it is customary to refer to their ratio as a *heat-transfer coefficient*:

$$\zeta = \frac{q_w}{T_1 - T_0} \qquad (3\text{-}14)$$

One then nondimensionalizes ζ either as the Stanton number from Eq. (1-98) or, alternately, as the *Nusselt number*:

$$Nu_L = \frac{\zeta L}{k} = C_h Re_L Pr \qquad (3\text{-}15)$$

where L is the characteristic length of the flow geometry. Here we select L as the plate separation distance ($2h$) and compute, from Eq. (3-13),

$$Nu_{2h} = \frac{2h\zeta}{k} = 1 \pm \frac{Br}{2} \qquad (3\text{-}16)$$

where again the (+) means the lower surface. If $Br > 2$, both the upper and lower surfaces must be cooled to maintain their temperatures. Note that, since $Nu \approx 1$ for pure conduction, the numerical value of Nu represents the ratio of convective heat transfer to conduction for the same value of ΔT.

3-2.2 Axially Moving Concentric Cylinders

Consider two long concentric cylinders with a viscous fluid between them, as in Fig. 3-3. Let either the inner ($r = r_0$) cylinder move axially at $u = U_0$ or the outer ($r = r_1$) cylinder move at $u = U_1$, as shown. Let the pressure and temperature be constant for this example. The no-slip condition will set the fluid into steady motion $u(r)$, and u_θ and u_r will be zero. The equations of motion in cylindrical coordinates

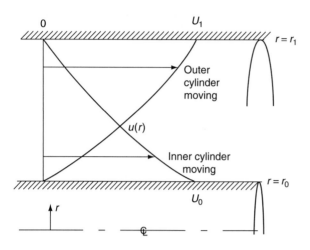

FIGURE 3-3
Axial annular Couette flow between concentric moving cylinders. The velocity distributions are plotted from Eqs. (3-18) and (3-19).

are given in App. B. Continuity is satisfied identically if $u = u(r)$, and the axial momentum equation reduces to

$$\nabla^2(u) = \frac{1}{r}\frac{\partial}{\partial r}\left(r\frac{\partial u}{\partial r}\right) = 0 \qquad (3\text{-}17)$$

which is, like the previous example, independent of fluid properties. The solution has the form

$$u = C_1 \ln(r) + C_2$$

If the inner cylinder moves, the no-slip conditions are $u(r_0) = U_0$ and $u(r_1) = 0$, whence $C_1 = U_0/\ln(r_0/r_1)$ and $C_2 = -C_1\ln(r_1)$. The solution is

$$u = U_0 \frac{\ln(r_1/r)}{\ln(r_1/r_0)}$$
$$\tau = \frac{-\mu U_0}{r\ln(r_1/r_0)} \qquad (3\text{-}18)$$

Similarly, if, instead, the outer cylinder is moving, the boundary conditions are $u(r_0) = 0$ and $u(r_1) = U_1$, and the solution is

$$u = U_1 \frac{\ln(r/r_0)}{\ln(r_1/r_0)}$$
$$\tau = \frac{\mu U_1}{r\ln(r_1/r_0)} \qquad (3\text{-}19)$$

These two velocity distributions are sketched in Fig. 3-3; they are exactly analogous to temperature distributions $T(r)$ in pure conduction through the fluid. Note the difference in curvature for the two cases: If the inner cylinder moves, $u(r)$ is concave, whereas it is convex if the outer cylinder moves.

The temperature distribution is given as a problem exercise, as is the question of what happens if both cylinders move at once.

3-2.3 Flow between Rotating Concentric Cylinders

Consider the steady flow maintained between two concentric cylinders by steady angular velocity of one or both cylinders. Let the inner cylinder have radius r_0, angular velocity ω_0, and temperature T_0, and the outer cylinder has r_1, ω_1, and T_1, respectively. The geometry is such that the only nonzero velocity component is u_θ and the variables u_θ, T, and p must be functions only of radius r. The equations of motion in polar coordinates [Eq. (2-64)] reduce to

Continuity: $\quad \dfrac{\partial u}{\partial \theta} = 0$

r momentum: $\quad \dfrac{dp}{dr} = \dfrac{\rho u_\theta^2}{r}$ (3-20)

θ momentum: $\quad \dfrac{d^2 u_\theta}{dr^2} + \dfrac{d}{dr}\left(\dfrac{u_\theta}{r}\right) = 0$

Energy: $\quad 0 = \dfrac{k}{r}\dfrac{d}{dr}\left(r\dfrac{dT}{dr}\right) + \mu\left(\dfrac{du_\theta}{dr} - \dfrac{u_\theta}{r}\right)^2$

with boundary conditions at each cylinder.

At $r = r_0$: $\quad u_\theta = r_0 \omega_0 \quad\quad T = T_0 \quad\quad p = p_0$ (3-21)

At $r = r_1$: $\quad u_\theta = r_1 \omega_1 \quad\quad T = T_1$

The solution to the θ-momentum equation has the form

$$u_\theta = C_1 r + \dfrac{C_2}{r}$$

i.e., the sum of a solid-body rotation and a "potential" vortex whose Laplacian is zero. We may find C_1 and C_2 from the boundary conditions Eq. (3-21) and write the solution as the sum of inner and outer driven flows:

$$u_\theta = r_0 \omega_0 \dfrac{r_1/r - r/r_1}{r_1/r_0 - r_0/r_1} + r_1 \omega_1 \dfrac{r/r_0 - r_0/r}{r_1/r_0 - r_0/r_1} \quad (3\text{-}22)$$

If this velocity distribution is substituted into the energy equation in Eq. (3-20), the temperature distribution may be found as follows:

$$\dfrac{T - T_0}{T_1 - T_0} = PrEc \dfrac{r_1^4}{(r_1^2 - r_0^2)^2}\left[\left(1 - \dfrac{r_0^2}{r^2}\right) - \left(1 - \dfrac{r_0^2}{r_1^2}\right)\dfrac{\ln(r/r_0)}{\ln(r_1/r_0)}\right] + \dfrac{\ln(r/r_0)}{\ln(r_1/r_0)}$$

(3-23)

where $PrEc = \mu r_0^2 \omega_0^2 / k(T_1 - T_0)$ is the Brinkman number expressing the temperature rise due to dissipation. If $PrEc = 0$, T reduces to the simple heat-conduction solution. The evaluation of the pressure $p(r)$ is left as an exercise.

Some special cases are of interest. In the limit as the inner cylinder vanishes $(r_0, \omega_0 = 0)$, we obtain from Eq. (3-22):

$$u_\theta = \omega_1 r = \text{const} \tag{3-24}$$

i.e., steady outer rotation of a tube filled with fluid induces solid-body rotation.

In the limit as the outer cylinder becomes very large and remains fixed $(\omega_1 = 0, r_1 \to \infty)$, we obtain

$$u_\theta = \frac{r_0^2 \omega_0}{r} \tag{3-25}$$

This is a *potential vortex*, driven by the rotating cylinder with the no-slip condition. As anyone who paddles a canoe knows, a vortex, if not maintained by a driving moment, will decay. Two examples of decaying vortex distributions are given as end-of-chapter Probs. 3-14 and 3-22.

Figure 3-4a shows velocity distributions for various (r_1/r_0), when only the inner cylinder is rotating. Figure 3-4b shows the temperature distributions for the single case $r_1/r_0 = 5$ from Fig. 3-4a.

Another interesting special case is that of very small clearance between the cylinders, $r_1 - r_0 \ll r_0$. Again let the outer cylinder remain stationary. Equation (3-22) becomes, in the limit,

Small clearance: $$\frac{u_\theta}{r_0 \omega_0} \approx 1 - \frac{r - r_0}{r_1 - r_0} \tag{3-26}$$

which is a linear Couette flow between effectively parallel plates, as in Fig. 3-1. In Fig. 3-4a, the approach to linearity is already being hinted at for $r_1/r_0 = 2$. The

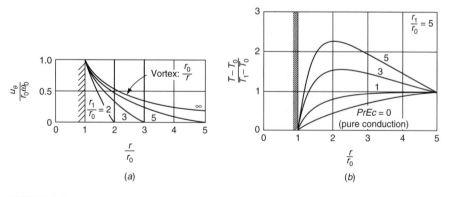

FIGURE 3-4
Velocity and temperature distributions between a stationary outer and a rotating inner cylinder, Eqs. (3-22) and (3-23): (a) velocity distribution; (b) temperature distribution for $r_1/r_0 = 5$.

velocity gradient is high, and hence a wide variety of shear stresses can be generated in a small-clearance apparatus.

Of interest in viscometry is the moment or torque exerted by the cylinders upon each other. This moment is independent of r and has the value (per unit depth of cylinder) of

$$M = 4\pi\mu \frac{r_1^2 r_0^2}{r_1^2 - r_0^2}(\omega_1 - \omega_0) \qquad (3\text{-}27)$$

By knowing the geometry and measuring M at either cylinder, one can calculate the viscosity of the fluid, as first suggested by Couette (1890). This is still a popular method in viscometry.

3-2.4 Stability of Couette Flows

All of the solutions in this section are exact steady-flow solutions of the Navier–Stokes equations. They are called *laminar* flows and have a smooth-streamline character. It is known that all laminar flows become *unstable* at a finite value of some critical parameter, usually the Reynolds number. A different type of flow then ensues, sometimes still laminar or, more often, an entirely new fluctuating flow regime called *turbulent*.

The laminar straight-line profile of flow between plates, Fig. 3-2a, is valid only up to $Re_h = Uh/\nu \approx 1500$. Above this value, the pattern changes into a randomly fluctuating flow whose time-mean S-shaped profile is shown in Fig. 3-5. This shape, sketched from data given by Reichardt (1956), varies slightly with the Reynolds number and increases the wall shear (and heat-transfer rate) by two orders of magnitude. We will discuss these effects in detail in Chaps. 5 and 6.

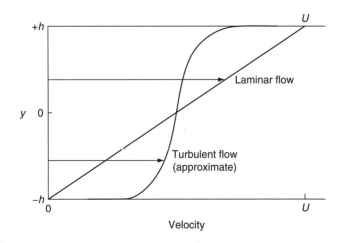

FIGURE 3-5
Instability of Couette flow between parallel plates: The straight-line laminar profile of Eq. (3-6) is valid only up to $Re_h = Uh/\nu = 1500$. [*Reichardt (1956)*.]

The axial annular flows of Fig. 3-3 are also unstable. Experiments by Polderman et al. (1986), with the inner cylinder moving, show fully turbulent flow at a clearance Reynolds number $Re_c = U_0(r_1 - r_0)/\nu \geq 7000$. The probable point of "transition" from laminar to turbulent flow, not reported, is about $Re_c \approx 2000$.

Although early work [Rayleigh (1916)] indicated that the rotating inner cylinder flows of Fig. 3-4a are unstable for all rotation rates, a classic paper by Taylor (1923) showed that the laminar profiles are valid until a critical rotation rate. For small clearance, $(r_1 - r_0) \ll r_0$, the critical value for instability is given by the *Taylor number*:

$$Ta = r_0(r_1 - r_0)^3 \frac{\omega_0^2}{\nu^2} \approx 1700$$

Above this value, the pattern changes into strikingly different three-dimensional laminar flow consisting of counterrotating pairs of toroidal vortices—see Fig. 5-22a.

All laminar flows are subject to instability. Therefore, all the exact solutions in this chapter are, in principle, valid only for a certain finite range of their governing parameter, e.g., the Reynolds number.

3-3 POISEUILLE FLOW THROUGH DUCTS

Whereas Couette flows are driven by moving walls, Poiseuille flows are generated by pressure gradients, with application primarily to ducts. They are named after J. L. M. Poiseuille (1840), as a French physician who experimented with low-speed flow in tubes.

Consider a straight duct of arbitrary but constant shape, as illustrated in Fig. 3-6. There will be an *entrance effect*, i.e., a thin initial shear layer and core acceleration as depicted in Fig. 1-10. The shear layers grow and meet, and the core disappears within a fairly short entrance length L_e. Shah and London (1978) show that, regardless of duct shape, the entrance length can be correlated for laminar flow

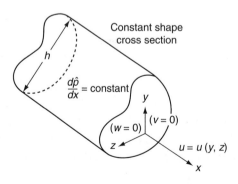

FIGURE 3-6
Fully developed duct flow.

in the form

$$\frac{L_e}{D_h} \approx C_1 + C_2 Re_{D_h} \tag{3-28}$$

where $C_1 \approx 0.5$, $C_2 \approx 0.05$, and D_h is a suitable "diameter" scale for the duct. Since transition to turbulence occurs at about $Re \approx 2000$, L_e is thus limited to, at most, 100 diameters.

For $x > L_e$, the velocity becomes purely axial and varies only with the lateral coordinates, that is, $v = w = 0$ and $u = u(y, z)$. The flow is then called *fully developed*, as depicted in Fig. 3-6. For fully developed flow, the continuity and momentum equations for incompressible flow, Eqs. (3-1) and (3-2) reduce to

Continuity: $\quad\quad \dfrac{\partial u}{\partial x} = 0$

Momentum: $\quad\quad 0 = -\dfrac{\partial \hat{p}}{\partial x} + \mu \left(\dfrac{\partial^2 u}{\partial y^2} + \dfrac{\partial^2 u}{\partial z^2} \right)$ (3-29)

$$0 = -\frac{\partial \hat{p}}{\partial y} = -\frac{\partial \hat{p}}{\partial z}$$

These indicate that the total pressure $\hat{p}$ is a function only of x for this fully developed flow. Further, since u does not vary with x, it follows from the x-momentum equation that the gradient $d\hat{p}/dx$ must only be a (negative) constant. Then the basic equation of fully developed duct flow is

$$\frac{\partial^2 u}{\partial y^2} + \frac{\partial^2 u}{\partial z^2} = \frac{1}{\mu} \frac{d\hat{p}}{dx} = \text{const} \tag{3-30}$$

subject only to the no-slip condition $u_w = 0$ everywhere on the duct surface. This is the classic Poisson equation and is equivalent to the torsional stress problem in elasticity [see, e.g., Timoshenko and Goodier (1970)] for a simply connected cross section.[†] Thus there are many known solutions for different duct shapes, especially well summarized by Berker (1963). Shah and London (1978) tabulate and chart both the laminar friction and heat-transfer characteristics of many different shapes. If new shapes arise, Eq. (3-30) is easily solved by complex-variable or numerical techniques.

Note that, like the Couette flow problems of the previous section, the acceleration terms vanish here, taking the density with them. These flows then are creeping flows in the sense that they are independent of density, even though the Reynolds number need not be small, as was required in Sec. 2-11. The Reynolds number is not even a required parameter (except to specify the stability limits),

---

[†]The duct problem is more restrictive for a multiply connected section than the torsion problem, which does not require that u vanish on the surfaces of the inner holes in the cross section.

but it is usually introduced artificially anyway. There is no characteristic velocity U and no axial length scale L either, since we are supposedly far from the entrance or exit. The proper scaling of Eq. (3-30) should include μ, $d\hat{p}/dx$, and some characteristic duct width h, as suggested in Fig. 3-6. Thus the dimensionless variables are

$$y^* = \frac{y}{h} \quad z^* = \frac{z}{h} \quad u^* = \frac{\mu u}{h^2(-d\hat{p}/dx)} \tag{3-31}$$

where the negative pressure gradient is needed to make u^* a positive quantity. In terms of these variables, Eq. (3-30) becomes

$$\nabla^{*2}(u^*) = -1 \tag{3-32}$$

subject to $u^* = 0$ at all points on the boundary of the duct cross section.

3-3.1 The Circular Pipe: Hagen–Poiseuille Flow

The circular pipe is perhaps our most celebrated viscous flow, first studied by Hagen (1839) and Poiseuille (1840). The single variable is $r^* = r/r_0$, where r_0 is the pipe radius. The Laplacian operator reduces to

$$\nabla^2 = \frac{1}{r}\frac{d}{dr}\left(r\frac{d}{dr}\right)$$

and the solution of Eq. (3-32) is

$$u^* = -\tfrac{1}{4}r^{*2} + C_1 \ln r^* + C_2 \tag{3-33}$$

Since the velocity cannot be infinite at the centerline, on physical grounds, we reject the logarithm term and set $C_1 = 0$. The no-slip condition is satisfied by setting $C_2 = +\tfrac{1}{4}$. The pipe-flow solution is thus

$$u = \frac{-d\hat{p}/dx}{4\mu}(r_0^2 - r^2) \tag{3-34}$$

Thus the velocity distribution in fully developed laminar pipe flow is a paraboloid of revolution about the centerline (the *Poiseuille paraboloid*). The total volume rate of flow Q is of interest, as defined for any duct by

$$Q = \int_{\text{section}} u \, dA$$

where the element of area is $2\pi r\, dr$ for this pipe case. The integration is simple and yields

$$Q_{\text{pipe}} = \frac{\pi r_0^4}{8\mu}\left(-\frac{d\hat{p}}{dx}\right) \tag{3-35}$$

The mean velocity is defined by $\bar{u} = Q/A$ and gives, in this case

$$\bar{u} = \frac{r_0^2(-d\hat{p}/dx)}{8\mu} = \frac{1}{2}u_{max} \qquad (3\text{-}36)$$

Finally, the wall shear stress is constant and is given by

$$\tau_w = \mu\left(-\frac{du}{dr}\right)_w = \frac{1}{2}r_0\left(-\frac{d\hat{p}}{dx}\right) = \frac{4\mu\bar{u}}{r_0} \qquad (3\text{-}37)$$

Even though τ_w is proportional to mean velocity (laminar flow), it is customary, anticipating turbulent flow, to nondimensionalize wall shear with the pipe *dynamic pressure*, $\rho\bar{u}^2/2$, by analogy with Eq. (3-8). Two different friction factor definitions are in common use in the literature:

$$\begin{aligned}\lambda &= \frac{8\tau_w}{\rho\bar{u}^2} = \text{Darcy friction factor} \\ C_f &= \frac{2\tau_w}{\rho\bar{u}^2} = \frac{1}{4}\lambda = \text{Fanning friction factor, or skin-friction coefficient}\end{aligned} \qquad (3\text{-}38)$$

By substituting into Eqs. (3-38), we obtain the classic relations

$$\begin{aligned}\lambda &= \frac{64}{Re_D} \\ C_f &= \frac{16}{Re_D}\end{aligned} \qquad (3\text{-}39)$$

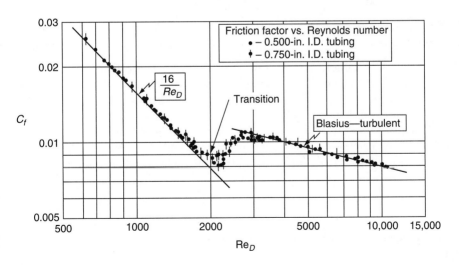

FIGURE 3-7
Comparison of theory and experiment for the friction factor of air flowing in small-bore tubes. [*After Senecal and Rothfus (1953)*.]

As with Couette flow, Eq. (3-9), the Poiseuille number makes a lot more sense in laminar tube flow, being a pure constant:

$$Po = C_f Re_D = \frac{2\tau_w D}{\mu \bar{u}} = 16$$

This classic laminar-flow solution is in good agreement with experiment, as shown in Fig. 3-7, from the data of Senecal and Rothfus (1953). The flow undergoes transition to turbulence at approximately $Re_D \approx 2000$, a value which can be raised somewhat by taking care to eliminate flow disturbances. Above $Re_D \approx 3000$, the pipe flow is fully turbulent. The curve labeled "Blasius" is a curve fit to turbulent-flow data, $C_f \approx 0.0791/Re_D^{1/4}$, by Prandtl's student H. Blasius (1913)—one of the first demonstrations of the power of dimensional analysis.

3-3.1.1 MICROFLOWS: TUBE FLOW OF GASES WITH SLIP. The classical Poiseuille flow, Eq. (3-34), is for no-slip conditions at the walls. If the Knudsen number, $Kn = \ell/D$, is not small, due to either small D or large ℓ, slip will occur at the walls, and the flow rate and velocity will increase for a given pressure gradient. In Eq. (3-33), C_1 will still be zero, but C_2 must satisfy the slip condition of Eq. (1-91): $u_w = \ell(\partial u/\partial r)$ at $r = r_0$. When the new value of C_2 is found, the velocity and flow rate become

$$u = \frac{-d\hat{p}/dx}{4\mu}\left(r_0^2 - r^2 + 2\ell r_0\right)$$

$$Q = \frac{\pi r_0^4}{8\mu}\left(-\frac{d\hat{p}}{dx}\right)\left(1 + 8\frac{\ell}{D}\right) \qquad (3\text{-}40)$$

Thus the flow rate is increased over the no-slip case by the fraction $8\ell/D = 8Kn$. If we require, for example, that the slip flow-rate effect be less than 2 percent, then Kn must be less than 0.0025. This is why we stated in Sec. 1-4.2 that $Kn = \mathcal{O}(0.1)$ is not small enough, slip would occur. The complete derivation of Eqs. (3-40) is given as an end-of-chapter problem.

Equations (3-40) are applicable to gases. For *liquid* slip flows, as a first approximation, one could replace the mean free path ℓ by the slip length L_{slip} defined in Eq. (1-89). For further reading on microchannel flows, see the monograph by Karniadakis and Bestok (2001).

3-3.2 Combined Couette–Poiseuille Flow Between Plates

Return now to our first Couette flow example, Fig. 3-1, and impose a constant pressure gradient $(d\hat{p}/dx)$ on the flow in addition to the moving upper wall. We now solve the differential equation

$$\mu \frac{d^2u}{dy^2} = \frac{d\hat{p}}{dx} = \text{const} \qquad (3\text{-}41)$$

subject to the no-slip condition $u(-h) = 0$. The solution is

$$\frac{u}{U} = \frac{1}{2}\left(1 + \frac{y}{h}\right) + P\left(1 - \frac{y^2}{h^2}\right) \quad P = \left(-\frac{dp}{dx}\right)\frac{h^2}{2\mu U} \qquad (3\text{-}42)$$

This relation is a superposition, possible because the non-linear convective acceleration is zero, of Couette wall-driven flow (the first term) and Poiseuille pressure-driven flow (the second term). It is plotted in Fig. 3-8 for various values of the dimensionless pressure gradient P. Of particular interest is the dashed line, $P = -\frac{1}{4}$, for which the shear stress $\mu(\partial u/\partial y)$ at the lower wall is zero. For $P < -\frac{1}{4}$ there is *backflow* at the lower wall, an indication of "flow separation" in unbounded shear layers (Chap. 4). In terms of the "thickness" $(2h)$ of the shear layer, we may write the separation criterion $P = -\frac{1}{4}$ in the form

$$\frac{dp}{dx}\frac{(2h)^2}{\mu U} = 2 \qquad (3\text{-}43)$$

This is identical in form to the laminar-boundary-layer separation estimates to be discussed in Chap. 4, except that the constant "2" is too small.

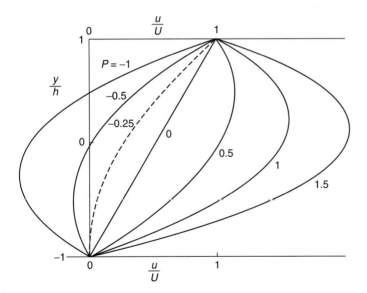

FIGURE 3-8
Combined Couette–Poiseuille flow between parallel plates, from Eq. (3-42). Backflow or "flow separation" occurs if $P < -\frac{1}{4}$.

If $U = 0$ (fixed walls), Eq. (3-42) reduces to pure Poiseuille flow between parallel plates:

$$u = u_{max}\left(1 - \frac{y^2}{h^2}\right)$$

$$u_{max} = \left(-\frac{dp}{dx}\right)\frac{h^2}{2\mu}$$

(3-44)

This is similar to laminar pipe flow, Eq. (3-34). The flow rate per unit depth is

$$Q = \int_{-h}^{+h} u\, dy = \frac{4}{3} h u_{max}$$

or

(3-45)

$$\bar{u} = \frac{Q}{2h} = \frac{2}{3} u_{max}$$

The wall shear stress is $\tau_w = 3\mu\bar{u}/h$ or, in dimensionless form,

$$C_f = \frac{6\mu}{\rho \bar{u} h}$$

or

(3-46)

$$Po = C_f Re_h = 6$$

Thus, except for differences in numerical constants, Poiseuille flow between plates is identical in form to laminar pipe flow.

3-3.3 Noncircular Ducts

Since Eq. (3-32) for fully developed duct flow is equivalent to a classic Dirichlet problem, it is not surprising that an enormous number of exact solutions are known

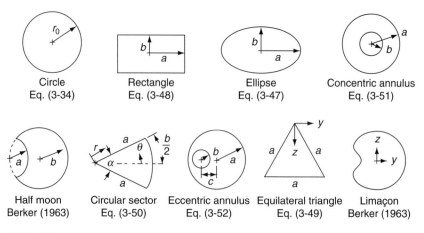

FIGURE 3-9
Some cross sections for which fully developed flow solutions are known; for still more, consult Berker (1963, pp. 67ff.) or Shah and London (1978).

for noncircular shapes, as reviewed by Berker (1963). Some of these shapes are shown in Fig. 3-9. Each solution is fascinating, but our mathematical ardor should be dampened somewhat by the practical fact that limaçon-shaped ducts, for example, are not commercially available at present. Nevertheless we list a few of these solutions because they lead to a valuable approximate principle, the *hydraulic radius*.

Elliptical section: $y^2/a^2 + z^2/b^2 \leq 1$:

$$u(y, z) = \frac{1}{2\mu}\left(-\frac{d\hat{p}}{dx}\right)\frac{a^2 b^2}{a^2 + b^2}\left(1 - \frac{y^2}{a^2} - \frac{z^2}{b^2}\right)$$

$$Q = \frac{\pi}{4\mu}\left(-\frac{d\hat{p}}{dx}\right)\frac{a^3 b^3}{a^2 + b^2}$$

(3-47)

Rectangular section: $-a \leq y \leq a, -b \leq z \leq b$:

$$u(y, z) = \frac{16a^2}{\mu\pi^3}\left(-\frac{d\hat{p}}{dx}\right)\sum_{i=1,3,5,\ldots}^{\infty}(-1)^{(i-1)/2}\left[1 - \frac{\cosh(i\pi z/2a)}{\cosh(i\pi b/2a)}\right]$$
$$\times \frac{\cos(i\pi y/2a)}{i^3}$$

(3-48)

$$Q = \frac{4ba^3}{3\mu}\left(-\frac{d\hat{p}}{dx}\right)\left[1 - \frac{192a}{\pi^5 b}\sum_{i=1,3,5,\ldots}^{\infty}\frac{\tanh(i\pi b/2a)}{i^5}\right]$$

Equilateral triangle of side a: coordinates in Fig. 3-9:

$$u(y, z) = \frac{-d\hat{p}/dx}{2\sqrt{3}\,a\mu}\left(z - \frac{1}{2}a\sqrt{3}\right)(3y^2 - z^2)$$

$$Q = \frac{a^4 \sqrt{3}}{320\mu}\left(-\frac{d\hat{p}}{dx}\right)$$

(3-49)

Circular sector: $-\frac{1}{2}\alpha \leq \theta \leq +\frac{1}{2}\alpha, 0 \leq r \leq a$:

$$u(r, \theta) = \frac{d\hat{p}/dx}{4\mu}\left[r^2\left(1 - \frac{\cos 2\theta}{\cos \alpha}\right) - \frac{16a^2 \alpha^2}{\pi^3}\right.$$
$$\left.\times \sum_{i=1,3,5,\ldots}^{\infty}(-1)^{(i+1)/2}\left(\frac{r}{a}\right)^i \frac{\cos(i\pi\theta/\alpha)}{i(i + 2\alpha/\pi)(i - 2\alpha/\pi)}\right]$$

$$Q = \frac{a^4}{4\mu}\left(-\frac{d\hat{p}}{dx}\right)$$

(3-50)

$$\times \left[\frac{\tan\alpha - \alpha}{4} - \frac{32\alpha^4}{\pi^5}\sum_{i=1,3,5,\ldots}^{\infty}\frac{1}{i^2(i + 2\alpha/\pi)^2(i - 2\alpha/\pi)}\right]$$

Concentric circular annulus: $b \leq r \leq a$:

$$u(r) = \frac{-d\hat{p}/dx}{4\mu}\left[a^2 - r^2 + (a^2 - b^2)\frac{\ln(a/r)}{\ln(b/a)}\right]$$

$$Q = \frac{\pi}{8\mu}\left(-\frac{d\hat{p}}{dx}\right)\left[a^4 - b^4 - \frac{(a^2 - b^2)^2}{\ln(a/b)}\right]$$
(3-51)

This is but a sample of the wealth of solutions available. The formula for a concentric annulus is important in viscometry, with a measured Q being used to calculate μ. To increase the pressure drop, the clearance $(a - b)$ is held small, in which case Eq. (3-51) for Q becomes the difference between two nearly equal numbers. However, if we expand the bracketed term [ ] in a series, the result is

$$(a^4 - b^4) - \frac{(a^2 - b^2)^2}{\ln(a/b)} = \frac{4}{3}b(a - b)^3 + \frac{2}{3}(a - b)^4 + \cdots + \mathcal{O}(a - b)^5$$

so that Q for small clearance is seen to be cubic in $(a - b)$.

The eccentric annulus in Fig. 3-9 has practical applications, for example, when a needle valve becomes misaligned. The solution was given by Piercy et al. (1933), using an elegant complex-variable method which transformed the geometry to a concentric annulus, for which the solution was already known, Eq. (3-51). We reproduce here only their expression for volume rate of flow:

$$Q = \frac{\pi}{8\mu}\left(-\frac{d\hat{p}}{dx}\right)\left[a^4 - b^4 - \frac{4c^2 M^2}{\beta - \alpha} - 8c^2 M^2 \sum_{n=1}^{\infty}\frac{ne^{-n(\beta+\alpha)}}{\sinh(n\beta - n\alpha)}\right]$$
(3-52)

where
$$M = (F^2 - a^2)^{1/2} \quad F = \frac{a^2 - b^2 + c^2}{2c}$$

$$\alpha = \frac{1}{2}\ln\frac{F + M}{F - M} \quad \beta = \frac{1}{2}\ln\frac{F - c + M}{F - c - M}$$

Flow rates computed from this formula are compared in Fig. 3-10 to the concentric result $Q_{c=0}$ from Eq. (3-51). It is seen that eccentricity substantially increases the flow rate, the maximum ratio of $Q/Q_{c=0}$ being 2.5 for a narrow annulus of maximum eccentricity. The curve for $b/a = 1$ can be derived from lubrication theory:

Narrow annulus:
$$\frac{Q}{Q_{c=0}} = 1 + \frac{3}{2}\left(\frac{c}{a - b}\right)^2$$
(3-53)

The reason for the increase in Q is that the fluid tends to bulge through the wider side. This is illustrated for one case in Fig. 3-11, where the wide side develops a set of closed high-velocity streamlines. This effect is well known to piping engineers, who have long noted the drastic leakage that occurs when a nearly closed valve binds to one side.

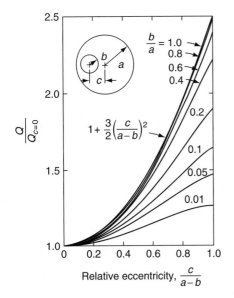

FIGURE 3-10
Volume flow through an eccentric annulus as a function of eccentricity, Eq. (3-52).

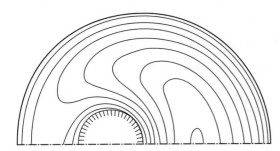

FIGURE 3-11
Constant-velocity lines for an eccentric annulus, $b/a = c/a = \frac{1}{4}$. [*After Piercy et al.* (*1933*).]

3-3.4 The Concept of Hydraulic Diameter

The definition of λ proposed in Eq. (3-39) fails for a noncircular duct since τ_w varies around the perimeter. For example, in the equilateral-triangle duct, Eq. (3-49), τ_w is zero in the corners and a maximum at the midpoints of the sides. The remedy, at least partially, is to define a mean wall shear stress

$$\bar{\tau}_w = \frac{1}{P}\int_0^P \tau_w \, ds$$

where ds = element of arc length
P = perimeter of section

It we isolate a slug of fluid passing through the duct as in Fig. 3-12 and note that there is no net momentum flux due to the fully developed flow, we can equate the

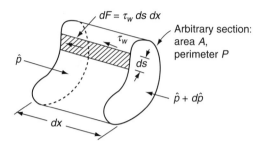

FIGURE 3-12
Force equilibrium in fully developed arbitrary duct flow.

net pressure and wall shear force on the fluid as follows:

$$dx \int_0^P \tau_w \, ds = -A \, d\hat{p}$$

or, from the definition of mean shear, we have

$$\bar{\tau}_w = \frac{A}{P}\left(-\frac{d\hat{p}}{dx}\right) \tag{3-54}$$

which is entirely analogous to Eq. (3-38) for a circular duct. The quantity A/P is a length and equals $r_0/2$ if the duct is circular. For a noncircular duct, then, we set $A/P = D_h/4$, where D_h is called the *hydraulic diameter* of the cross section:

$$D_h = \frac{4A}{P} = \frac{4 \times \text{area}}{\text{wetted perimeter}} \tag{3-55}$$

For a cross section with multiple surfaces, P must include all wetted walls. For example, for the concentric annulus in Fig. 3-9, we have

$$D_h(\text{annulus}) = \frac{4\pi(a^2 - b^2)}{2\pi a + 2\pi b} = 2(a - b) \tag{3-56}$$

or twice the clearance. By dimensional reasoning for laminar fully developed flow, we are guaranteed that the friction factor of a noncircular duct will vary inversely with a Reynolds number based on hydraulic diameter:

$$C_f = \frac{\lambda}{4} = \frac{\text{const}}{Re_{D_h}} \qquad Re_{D_h} = \frac{\rho \bar{u} D_h}{\mu} \tag{3-57}$$

In other words, the Poiseuille number $Po = C_f Re$ remains constant for a noncircular duct. However—and this is the critical flaw—the constant usually does *not* equal 16 as it did for a circular pipe.

Using our many exact solutions from Eqs. (3-47) to (3-51), we may compute the value of $Po = C_f Re_{D_h}$ and plot them versus section slenderness ratio b/a in Fig. 3-13. We see that some are higher and some are lower than the nominal circle value of 16, and the error can be as high as 50 percent. For laminar flow, then, one should use the exact values of Po from Fig. 3-13 or the exact formulas given here or in Shah and London (1978).

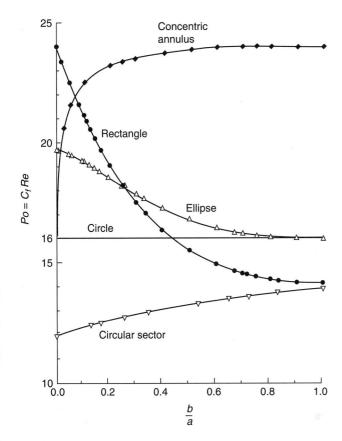

FIGURE 3-13
Comparison of Poiseuille numbers for various duct cross sections when the Reynolds number is scaled by the hydraulic diameter. [*Numerical data taken from Shah and London (1978).*]

We note for completeness that the equilateral triangle, Eq. (3-49), has a hydraulic diameter $D_h = a/\sqrt{3}$ and $C_f Re = 13.333$, or 17 percent low.

Especially vexing is the fact that the hydraulic diameter concept is insensitive to core eccentricity. The eccentric annulus in Fig. 3-10 has the same value $D_h = 2(a - b)$ regardless of the value of c, yet its flow rate may vary by over 100 percent depending upon c. Exact solutions are definitely needed.

The exact Poiseuille numbers in Fig. 3-13 are in very good agreement with the following laminar-flow experiments:

1. Rectangular and triangular ducts: Eckert and Irvine (1957)
2. Concentric annulus: Walker et al. (1957)

This is to be expected since, as far as we know, the Navier–Stokes equations are fundamentally correct for the common fluids and therefore should agree "exactly" with an accurate experiment.

Figure 3-13 has a surprising—and very relevant—use in an empirical method of determining the friction factor for *turbulent* flow in noncircular ducts for which the laminar solution is known—see Chap. 6.

3-3.5 Temperature Distribution in Fully Developed Duct Flow

The assumption of constant viscosity uncoupled the energy and momentum equations if natural convection was neglected, and we have solved for the velocity distribution in several ducts. With $u(y, z)$ known, Eq. (3-3) can be solved for T, which may be a function of (x, y, z), if the boundary conditions change with x. Equation (3-3) is linear in T, making it possible to examine some separate effects and add them together later, if desired. First let us look at the effect of viscous dissipation, assuming constant T_w. For the pipe case, T then equals $T(r)$ only, and Eq. (3-3) becomes

$$\frac{k}{r}\frac{d}{dr}\left(r\frac{dT}{dr}\right) = -\mu\left(\frac{du}{dr}\right)^2 = -\frac{16\mu \bar{u}^2 r^2}{r_0^4} \tag{3-58}$$

where we have introduced $u(r)$ from Eq. (3-34). Double integration leads to a logarithmic term which we discard to avoid a singularity at $r = 0$. Putting $T = T_w$ at $r = r_0$, we obtain

$$T = T_w + \frac{\mu \bar{u}^2}{k}\left(1 - \frac{r^4}{r_0^4}\right) \tag{3-59}$$

which is similar to Eq. (3-11) for flow between parallel plates. The maximum temperature rise $\mu \bar{u}^2/k$, at $\bar{u} = 100$ ft/s, is about 1°F for air and 3°F for water. Thus dissipation is usually neglected except for oils, where the viscosity is large, or in gas dynamics, where velocities are high (Chap. 7). The wall heat transfer is $q_w = k(dT/dr)$ at $r = r_0$, or $q_w = 4k(T_w - T_0)/r_0$, showing that the wall is being cooled to maintain constant T_w. The Nusselt number at the wall can be defined in the manner of Eq. (3-15), with L taken as the pipe diameter. Thus

$$Nu = \frac{q_w(2r_0)}{k(T_w - T_0)} = 8 \quad \text{for viscous dissipation} \tag{3-60}$$

This is a substantial number, but, as noted, the driving temperature difference $T_w - T_0$ is usually very small.

3-3.6 Asymptotic Uniform Heat-Flux Approximation

Before attacking the thermal-entrance problem of a sudden change in wall temperature, consider the conditions far downstream of such an entrance. The temperature varies with x, but the deviation $T_w - T$ is nearly independent of x, and q_w is nearly constant. Thus we specify

$$\frac{\partial}{\partial x}(T_w - T) = 0$$

$$\frac{q_w}{k} = \frac{\partial T}{\partial r} = \text{const} \tag{3-61}$$

Taken together, these require that the axial gradient be independent of r:

$$\frac{\partial T}{\partial x} = \frac{\partial T_w}{\partial x} = \text{const independent of } r \tag{3-62}$$

When we neglect (or separate out) dissipation, Eq. (3-3) becomes

$$\rho c_p u \frac{\partial T}{\partial x} = (\text{const}) \, u = \frac{k}{r}\frac{\partial}{\partial r}\left(r\frac{\partial T}{\partial r}\right) \tag{3-63}$$

which can be integrated twice; again we discard a logarithmic term. With $T = T_w$ at $r = r_0$, we obtain

$$T_w - T = \frac{c_p \bar{u} r_0^2}{8k}\frac{\partial T_w}{\partial x}\left(3 - \frac{4r^2}{r_0^2} + \frac{r^4}{r_0^4}\right) \tag{3-64}$$

from which we can calculate the Nusselt number. However, from a practical standpoint, Nu should not be based upon $T_w - T_0$ but upon a mean difference $T_w - T_m$, where T_m is the *cup-mixing temperature* of the fluid, computed by averaging T over the mass distribution of fluid in the pipe. Define

$$T_m = \frac{\int T \, dm}{\int dm} = \frac{\int_0^{r_0} T \rho u \, dA}{\int_0^{r_0} \rho u \, dA} \tag{3-65}$$

where $dA = 2\pi r \, dr$ for this case. For incompressible flow, the density cancels out. Using u from Eq. (3-34) for Poiseuille flow and T from Eq. (3-64), we obtain

$$T_w - T_m = \frac{11}{18}\left(\frac{3\rho c_p \bar{u} r_0^2}{8k}\frac{\partial T_w}{\partial x}\right) = \frac{11}{18}(T_w - T_0) \tag{3-66}$$

It is this temperature difference upon which engineers base the dimensionless wall heat-transfer, or Nusselt, number

$$Nu_m = \frac{q_w 2 r_0}{k(T_w - T_m)} = \frac{48}{11} = 4.36 \quad \text{asymptotic uniform heat flux} \tag{3-67}$$

A similar but algebraically more complicated solution for asymptotic constant wall temperature gives

$$Nu_m = 3.66 \tag{3-68}$$

as will be shown in the next section. These asymptotic values are more sensitive to duct cross section than to wall conditions. Calculations for other duct shapes (rectangular, triangular, etc.) are given by Shah and London (1978).

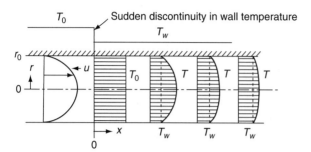

FIGURE 3-14
A thermal-entrance problem.

3-3.7 Thermal Entrance: The Graetz Problem

Now consider the problem of developing temperature profiles $T(x, r)$ in a pipe, due to a sudden change in wall temperature (Fig. 3-14). We neglect dissipation *and* axial heat conduction, for which Eq. (3-3) reduces to

$$u\frac{\partial T}{\partial x} \approx \frac{\alpha}{r}\frac{\partial}{\partial r}\left(r\frac{\partial T}{\partial r}\right) \quad (3\text{-}69)$$

where $\alpha = k/\rho c_p$ is the thermal diffusivity of the fluid. The velocity distribution $u(x, r)$ is assumed known in this equation and might be one of three types:

1. $u = \bar{u}$ = const, or slug flow: appropriate for low Prandtl number fluids (liquid metals), where T develops much faster than u
2. $u = 2\bar{u}(1 - r^2/r_0^2)$, or Poiseuille flow: appropriate for high Prandtl number fluids (oils) or when the thermal entrance is far downstream of the duct entrance
3. Developing u profiles (Sec. 4-9): suitable for any Prandtl number when the velocity and temperature entrance are at the same position

The thermal-entrance problem with a sudden change in wall temperature is illustrated in Fig. 3-14. The solution was given as an infinite series by Graetz (1883) for slug flow and another series in 1885 for Poiseuille flow. Further details and other geometries are given in the texts by Shah and London (1978) or by Kays and Crawford (1993).

With reference to Fig. 3-14, the proper boundary conditions on $T(x, r)$ for Eq. (3-69) are

$$T(0, r) = T_0; \quad T(r_0, x) = T_w \quad (3\text{-}70)$$

Let us outline Graetz' solution for Poiseuille flow (case 2). Define dimensionless variables:

$$T^* = \frac{T_w - T}{T_w - T_0} \quad r^* = \frac{r}{r_0} \quad x^* = \frac{x}{d_0 Re Pr} \quad (3\text{-}71)$$

where $Re = \bar{u}d_0/\nu$ is the diameter Reynolds number and Pr is the Prandtl number, or $RePr = \rho c_p \bar{u} d_0/k$ is the Peclet number, named after the French physicist

J. C. E. Peclet. The variables in Eq. (3-71) convert Eqs. (3-69) and (3-70) to

$$\frac{\partial T^*}{\partial x^*} = \frac{2}{r^*(1 - r^{*2})} \frac{\partial}{\partial r^*}\left(r^* \frac{\partial T^*}{\partial r^*}\right) \quad \begin{array}{l} T^*(r^*, 0) = 1 \\ T^*(1, x^*) = 0 \end{array} \quad (3\text{-}72)$$

Note that Eqs. (3-72) are entirely independent of parameters because of our choice of variables (T^*, r^*, x^*). It is clear that the variables are separable, so that a product solution will work. Let

$$T^*(r^*, x^*) = f(r^*) \, g(x^*) \quad (3\text{-}73)$$

be a particular solution and substitute into Eq. (3-72). We obtain

$$\frac{g'}{2g} = \frac{r^* f'' + f'}{r^*(1 - r^{*2})f} = -\lambda^2 = \text{const} \quad (3\text{-}74)$$

The g function clearly has the solution $g = C \exp(-2\lambda^2 x^*)$, whereas the equation for f is somewhat less familiar. To make the product solution $T = fg$ satisfy the condition $T^*(r^*, 0) = 1$ for all r^*, we take advantage of linearity and superimpose many such solutions, so that the proper formulation is, finally,

$$T^*(r^*, x^*) = \sum_{n=0}^{\infty} C_n f_n(r^*) \, e^{-2\lambda_n^2 x^*} \quad (3\text{-}75)$$

where the functions f_n are characteristic solutions of Eq. (3-74):

$$r^* f_n'' + f_n' + \lambda_n^2 r^*(1 - r^{*2}) f_n = 0 \quad (3\text{-}76)$$

We take $f_n(0) = 1$ for simplicity and force $f_n(1) = 0$ to satisfy the wall-temperature condition $T^*(1, x^*) = 0$ from Eq. (3-72). Then the solution is complete if the other (initial) condition from Eq. (3-72) is satisfied:

$$T^*(r^*, 0) = 1 = \sum_{n=0}^{\infty} C_n f_n(r^*) \quad (3\text{-}77)$$

Graetz showed that the eigenfunctions f_n are orthogonal over the interval 0 to 1 with respect to the weighting function $r^*(1 - r^{*2})$. Thus, if one multiplies Eq. (3-77) by $r^*(1 - r^{*2})f_m$ and integrates from 0 to 1, the constants may be found as follows:

$$C_n = \frac{\int_0^1 r^*(1 - r^{*2}) \, f_n \, dr^*}{\int_0^1 r^*(1 - r^{*2}) \, f_n^2 \, dr^*} \quad (3\text{-}78)$$

The first 10 values are tabulated in Table 3-1. We remark that Eq. (3-76), with its two conditions $f_n(0) = 1$ and $f_n(1) = 0$, is an eigenvalue problem and can be satisfied only for certain discrete values of λ_n, the *eigenvalues* of the Graetz functions f_n. Table 3-1 gives the first 10 eigenvalues and their associated constants. These are sufficient to calculate Nusselt numbers for almost any given wall conditions, even an arbitrary distribution of T_w^*.

TABLE 3-1
Important constants in the Graetz problem

n	λ_n	C_n	$-C_n f'_n(1)$
0	2.7043644	+1.46622	1.49758
1	6.679032	−0.802476	1.08848
2	10.67338	+0.587094	0.92576
3	14.67108	−0.474897	0.83036
4	18.66987	+0.404402	0.76474
5	22.67	−0.35535	0.71571
6	26.67	+0.31886	0.67798
7	30.67	−0.29049	0.64711
8	34.67	+0.26769	0.62119
9	38.67	−0.24890	0.59900

For large n, Sellars et al. (1956) give the following approximations:

$$\lambda_n \approx 4n + \tfrac{8}{3}$$

$$C_n \approx (-1)^n \frac{2\Gamma\left(\tfrac{2}{3}\right) 6^{2/3}}{\pi \lambda_n^{2/3}} = \frac{(-1)^n (2.8461)}{\lambda_n^{2/3}} \tag{3-79}$$

$$-C_n f'_n(1) \approx \frac{4\left(\tfrac{4}{3}\right)^{1/6} \Gamma\left(\tfrac{2}{3}\right)}{\pi \lambda_n^{1/3} \Gamma\left(\tfrac{4}{3}\right)} = \frac{2.0256}{\lambda_n^{1/3}}$$

These formulas have been used to compute the last five rows of Table 3-1 with an error of less than 0.15 percent.

To calculate the Nusselt number at the wall, we need the cup-mixing temperature, which we obtain by combining Eqs. (3-65) and (3-77):

$$T_m^* = 4 \int_0^1 T^*(1 - r^{*2}) r^* \, dr^* \tag{3-80}$$

Introducing T^* from Eq. (3-77), we obtain

$$Nu_x = \frac{2r_0 q_w}{k(T_w - T_m)} = \frac{\sum C_n f'_n(1) \exp(-2\lambda_n^2 x^*)}{2 \sum C_n \lambda_n^{-2} f'_n(1) \exp(-2\lambda_n^2 x^*)} \tag{3-81}$$

which converges well except at very small x^*. For large x^* (> 0.05), the first term of the series is dominant, giving the asymptotic result

$$Nu_x(x^* > 0.05) \approx \frac{\lambda_0^2}{2} = 3.66 \tag{3-82}$$

which is the thermally fully developed result already mentioned in Eq. (3-68). For small x^*, the results fit the approximation

$$Nu_x \approx 1.076 x^{*-1/3} - 1.064 \tag{3-83}$$

with an error no more than 1 percent for $x^* < 0.0004$.

3-3.8 Mean Nusselt Number

The total heat transferred to (or from) the wall is a useful quantity, since it equals the heat lost (or gained) by the fluid over the total length L of the tube. We define a mean wall heat flux per unit area

$$\bar{q}_w = \frac{1}{A_w}\int q_w \, dA_w = \frac{1}{L}\int_0^L q_w \, dx \qquad (3\text{-}84)$$

This mean heat flux must exactly balance the enthalpy change of the fluid between $x = 0$ and $x = L$:

$$\bar{q}_w A_w = \rho \bar{u} c_p (\pi r_0^2)[T_m(L) - T_0] \qquad (3\text{-}85)$$

so that $T_m(L)$ at the exit can be calculated from the mass flow rate in the tube. The dimensionless from $\bar{q}_w$ is called the mean Nusselt number:

$$Nu_m = \frac{2r_0 \bar{q}_w}{k \Delta T} \qquad (3\text{-}86)$$

and if T_w is variable, the choice of a suitable temperature difference ΔT is somewhat arbitrary. The reader may verify as an exercise that if one chooses the logarithmic mean-temperature difference,

$$\Delta T_{\ln} = \frac{[T_w(0) - T_m(0)] - [T_w(L) - T_m(L)]}{\ln\{[T_w(0) - T_m(0)]/[T_w(L) - T_m(L)]\}} \qquad (3\text{-}87)$$

then Nu_m is simply the average value of Nu_x in the tube between 0 and L:

$$Nu_m(\text{log mean}) = \frac{2r_0 \bar{q}_w}{k \Delta T_{\ln}} = \frac{1}{L}\int_0^L Nu_x \, dx \qquad (3\text{-}88)$$

For the present Graetz problem (large Prandtl number), we can evaluate Nu_m by using the differential form of Eq. (3-85):

$$q_w 2\pi r_0 \, dx = \rho \bar{u} \pi r_0^2 c_p \, dT_m$$

or $\qquad (3\text{-}89)$

$$4 Nu_x \, dx^* = -dT_m^*/T_m^*$$

We can integrate this from $x^* = 0$, $T_m^* = 1$ to $x^* = L^*$, $T_m^* = T_m^*(L)$, with a starkly simple result:

$$Nu_m = \frac{-1}{4L^*} \ln T_m^*(L) \qquad (3\text{-}90)$$

This is a general result for constant wall temperature. By carrying out the integration of Eq. (3-80), we find that

$$T_m^*(x^*) = \sum_{n=0}^{\infty} 4 C_n f'_n(1) \lambda_n^{-2} \exp(-2\lambda_n^2 x^*) \qquad (3\text{-}91)$$

for Poiseuille flow.

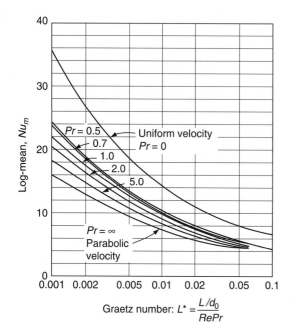

FIGURE 3-15
Finite-difference calculations for the log-mean Nusselt number in laminar pipe flow with developing velocity profiles. [After Goldberg (1958).]

Figure 3-15 shows the computations of Goldberg (1958) for a mean Nusselt number with velocity and temperature effects which are assumed to start at the same point, $x = 0$. The upper curve is for slug flow (type 1), valid for $Pr \ll 1$, that is, liquid metals. The lowest curve is for Poiseuille flow (type 2) for $Pr \gg 1$ (oils) and is computed from Eqs. (3-90) and (3-91). The intermediate curves are for type 3, where velocity and temperature develop together. Note that the abscissa is the *Graetz number*, $L^* = x^*(L)$, being based on diameter rather than r_0. The slug-flow curve ($Pr = 0$) is unique in that the parabolic velocity never develops and the limiting value of Nu_m is 5.78 at large L^* [Graetz (1883)].

All of the curves for finite Pr in Fig. 3-15 approach $Nu_m = 3.66$ at large L^*. We are reasonably close to this limit when $L^* \approx 0.05$. Thus we may define the *thermal-entrance length* in pipe flow as

$$L_{e,\text{thermal}} \approx 0.05 d_0 Re_D Pr \quad (3\text{-}92)$$

This is in the same sprit as the velocity-entrance-length formula proposed in Eq. (3-28).

The Graetz problem, especially for oil flow, $Pr \gg 1$, has many practical applications. For computation, one may curve-fit the lower curve in Fig. 3-15 to a type of formula proposed by Hausen (1943)

$$Nu_m(Pr \gg 1) \approx 3.66 + \frac{0.075/L^*}{1 + 0.05/L^{*2/3}} \quad (3\text{-}93)$$

where $L^* = (L/d_0)/(Re_D Pr)$. The error of this approximation is ± 5 percent.

Finally, we remark that the present analysis assumes constant fluid transport properties, whereas both liquids and gases have significant variations in μ and k with temperature. A first-order correction for this effect would be to evaluate fluid properties at the "film" temperature $(T_w + T_m)/2$. If the wall and fluid temperatures differ by more than 20°C, one should further correct the previous formulas for variable properties. Chapter 15 of the text by Kays and Crawford (1993) is a detailed discussion of these correction factors.

3-4 UNSTEADY DUCT FLOWS

Some interesting problems of unsteady duct flow can be worked out by retaining the fully developed flow assumption, and we give two examples here for circular pipe flow, with the same principle applying to other shapes. If we assume that the pipe axial velocity $u = u(r, t)$ only, with $v = w = 0$, then the continuity equation is identically satisfied and the momentum equation becomes

$$\rho \frac{\partial u}{\partial t} = -\frac{d\hat{p}}{dx} + \mu \left(\frac{\partial^2 u}{\partial r^2} + \frac{1}{r} \frac{\partial u}{\partial r} \right) \tag{3-94}$$

which is identical to the linear heat-conduction equation with a source term. The pressure gradient can vary only with time and thus represents a uniformly distributed heat source. Let us discuss two cases.

3-4.1 Starting Flow in a Circular Pipe

Suppose that the fluid in a long pipe is at rest at $t = 0$, at which time a sudden, uniform, and constant pressure gradient $d\hat{p}/dx$ is applied. An axial flow will commence which gradually approaches the steady Poiseuille flow

$$u = u_{\max}(1 - r^{*2})$$

where $r^* = r/r_0$. This problem was solved by Szymanski (1932). The boundary conditions are

Initial condition: $\quad u(r, 0) = 0$

No-slip condition: $\quad u(r_0, t) = 0$ \hfill (3-95)

The variables will separate in Eq. (3-94) if we subtract the steady Poiseuille flow and work with the *deviation* of u from the Poiseuille paraboloid. This removes the inhomogeneity $d\hat{p}/dx$, and the solution of Eq. (3-94) is then of the form $u = J_0(\lambda r^*) e^{-\lambda^2 t^*}$, where J_0 denotes the Bessel function of the first kind. The no-slip condition requires $J_0(\lambda_n)$ to vanish for each value of the separation constant, i.e., the λ_n are the roots of the Bessel function. These are tabulated in Table 3-2. Since the J_0 function is not a paraboloid, we must sum the functions $J_0(\lambda_n r^*)$ to obtain a negative paraboloid and thus satisfy the initial condition of the fluid at rest. The coefficients are obtained from the usual theory of orthogonal functions, and the final

TABLE 3-2
First ten roots of the Bessel function $J_0^\dagger$

n	λ_n	$J_1(\lambda_n)$
1	2.4048	0.5191
2	5.5201	−0.3403
3	8.6537	0.2715
4	11.7915	−0.2325
5	14.9309	0.2065
6	18.0711	−0.1877
7	21.2116	0.1733
8	24.3525	−0.1617
9	27.4935	0.1522
10	30.6346	−0.1442

† For $n > 10$:

$$\lambda_n = \frac{(4n-1)\pi}{4} \quad J_1(\lambda_n) \approx (-1)^{n+1}\left(\frac{2}{\pi\lambda_n}\right)^{1/2}$$

solution for this pipe starting-flow problem is

$$\frac{u}{u_{max}} = (1 - r^{*2}) - \sum_{n=1}^{\infty} \frac{8 J_0(\lambda_n r^*)}{\lambda_n^3 J_1(\lambda_n)} \exp\left(-\lambda_n^2 \frac{\nu t}{r_0^2}\right) \quad (3\text{-}96)$$

where $u_{max} = (-d\hat{p}/dx)r_0^2/4\mu$ from Eq. (3-37). Numerical results are plotted in Fig. 3-16 for various values of the dimensionless time $\nu t/r_0^2$. Two points are of particular interest: (1) initially a sort of boundary-layer effect occurs, where the central core of fluid accelerates uniformly (potential flow) but the wall region is retarded by friction; and (2) the flow is essentially the Poiseuille paraboloid when $t^* \approx 0.75$, giving us an estimate of how rapidly laminar tube flow responds to sudden change. Flows with small diameter and large viscosity will develop rapidly. Consider a 1 cm diameter tube. For air, $\nu = 1.5\text{E}{-}5 \text{ m}^2/\text{s}$, the value $t^* = 0.75$ translates to $t = 1.25$ s. For SAE 30 oil under the same conditions,

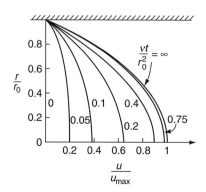

FIGURE 3-16
Instantaneous velocity profiles for starting flow in a pipe, Eq. (3-96). [*After Szymanski (1932)*.]

Poiseuille flow will develop in 0.06 s. Thus, in laminar small-bore pipe flows under varying pressure gradients, it is common to use a quasi-steady Poiseuille flow approximation.

Lefebvre and White (1989, 1991) have reproduced this flow experimentally for large tubes (D = 5 and 9 cm) and short times ($t^* < 0.01$). The measured velocity profiles are in good agreement with the Szymanski solution Eq. (3-96), and the laminar-flow state, stabilized by the acceleration, persists up to surprisingly high Reynolds numbers, $Re_D > 10^5$.

3-4.2 Pipe Flow Due to an Oscillating Pressure Gradient

As a second example, consider the solution of Eq. (3-94) if the pressure gradient varies sinusoidally with time:

$$\frac{d\hat{p}}{dx} = -\rho K e^{i\omega t}$$

where
$$e^{i\omega t} = \cos \omega t + i \sin \omega t \qquad (3\text{-}97)$$

$$i = \sqrt{-1}$$

We use the no-slip condition and look for a long-term steady oscillation, neglecting the transient, or start-up, of the flow. This is a classic problem in mathematical physics and leads to a Bessel function with imaginary arguments:

$$u = \frac{K}{i\omega} e^{i\omega t} \left[1 - \frac{J_0(r\sqrt{-i\omega/\nu})}{J_0(r_0\sqrt{-i\omega/\nu})} \right] \qquad (3\text{-}98)$$

which is neat but rather difficult to evaluate numerically. Equation (3-98) was first given by Sexl (1930), and further numerical calculations are given by Uchida (1956). Without bothering with exact calculations, we can show the general effect with the two overlapping series approximations

Small $z < 2$: $\qquad J_0(z) \approx 1 - \frac{z^2}{4} + \frac{z^4}{64} - \cdots$

$$\qquad (3\text{-}99)$$

Large $z > 2$: $\qquad J_0(z) \approx \sqrt{\frac{2}{\pi z}} \cos\left(z - \frac{\pi}{4}\right)$

Equation (3-98) shows that the proper dimensionless variables are

$$r^* = \frac{r}{r_0} \qquad \omega^* = \frac{\omega r_0^2}{\nu} \qquad u^* = \frac{u}{u_{\max}} \qquad (3\text{-}100)$$

where $u_{\max} = K r_0^2 / 4\nu$ is the centerline velocity for steady Poiseuille flow with a pressure gradient $-\rho K$. The quantity ω^* is sometimes called the *kinetic Reynolds number* and is a measure of viscous effects in oscillating flow. Similar to the static

Reynolds number, oscillating flows may become turbulent when ω^* exceeds approximately 2000.

By combining Eqs. (3-98) and (3-99), we obtain two series approximations for the velocity, which is the real part of the solution:

Small $\omega^* < 4$:

$$\frac{u}{u_{\max}} \approx (1 - r^{*2}) \cos \omega t + \frac{\omega^*}{16} (r^{*4} + 4r^{*2} - 5) \sin \omega t + \mathcal{O}(\omega^{*2}) \quad (3\text{-}101)$$

Large $\omega^* > 4$:

$$\frac{u}{u_{\max}} \approx \frac{4}{\omega^*} \left[\sin \omega t - \frac{e^{-B}}{\sqrt{r^*}} \sin (\omega t - B) \right] + \mathcal{O}(\omega^{*-2}) \quad (3\text{-}102)$$

where

$$B = (1 - r^*) \sqrt{\frac{\omega^*}{2}}$$

Remembering that dp/dx is proportional to $\cos \omega t$, we see that for very small ω^*, the velocity is nearly a quasi-static Poiseuille flow in phase with the slowly varying pressure gradient; the second term in Eq. (3-101) adds a lagging component which reduces the velocity at the centerline. At large ω^*, from Eq. (3-102), the flow approximately lags the pressure gradient by 90°, and again the centerline velocity is less than $u_{\max}$. However, near the wall, there is a region of high-velocity flow, as can be seen be averaging Eq. (3-102) over one cycle to obtain the mean square velocity $\overline{u^2}(r^*)$. We obtain

$$\frac{\overline{u^2}}{K^2/2\omega^2} = 1 - \frac{2}{\sqrt{r^*}} e^{-B} \cos B + \frac{e^{-2B}}{r^*} \quad (3\text{-}103)$$

This relation is plotted in Fig. 3-17 for two values of ω^*. There is an overshoot in mean velocity near the wall, which occurs when $\cos B + \sin B \approx e^{-B}$, or $B \approx 2.284$.

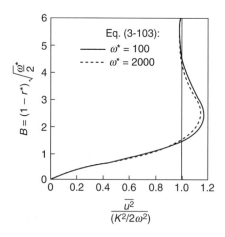

FIGURE 3-17
The near-wall velocity overshoot (Richardson's annular effect) due to an oscillating pressure gradient.

This effect is characteristic of oscillating duct flows and was first noted by Richardson and Tyler (1929) in a tube-flow experiment. The overshoot, now called *Richardson's annular effect*, was then verified theoretically through Eq. (3-102) by Sexl (1930). For a further discussion of these problems, the reader is referred to Uchida (1956) and the general review article by Rott (1964). A solution for oscillating flow in a rectangular duct is given by Tsangaris and Vlachakis (2003).

3-5 UNSTEADY FLOWS WITH MOVING BOUNDARIES

A variety of solutions are known for laminar flow with moving boundaries, some of which illustrate boundary-layer behavior. Again we make the parallel-flow assumption of $u = u(y, z, t)$, $v = 0$, $w = 0$, so that the momentum equation (3-2) becomes

$$\frac{\partial u}{\partial t} = -\frac{1}{\rho}\frac{d\hat{p}}{dx} + \nu\left(\frac{\partial^2 u}{\partial y^2} + \frac{\partial^2 u}{\partial z^2}\right) \quad (3\text{-}104)$$

The pressure gradient can only be a function of time for this flow and hence can be absorbed into the velocity by a change of variables. Define

$$u' = u + \int \frac{1}{\rho}\frac{d\hat{p}}{dx}\,dt$$

Then

$$\frac{\partial u'}{\partial t} = \nu\left(\frac{\partial^2 u'}{\partial y^2} + \frac{\partial^2 u'}{\partial z^2}\right) \quad (3\text{-}105)$$

Equation (3-105) is the homogeneous heat-conduction equation, for which a wealth of unsteady solutions are known [see, e.g., Carslaw and Jaeger (1959)]. One of these corresponds to the motion of a fluid above an infinite plane which is moved arbitrarily. Let the plane be at $y = 0$, so that $u = u(y, t)$. The boundary conditions are no slip at the plane and no initial motion of the fluid:

$$\begin{aligned} u(0, t) &= U(t) \quad \text{for } t > 0 \\ u(y, 0) &= 0 \quad \text{for } y > 0 \end{aligned} \quad (3\text{-}106)$$

No matter how complex $U(t)$ is, the solution can be obtained from superposition of the indicial, or step-function, solution, as shown by Watson (1958). We confine ourselves to two cases:

1. Sudden acceleration of the plane to constant U_0
2. Steady oscillation of the plane at $U \cos \omega t$

Case 1 is analogous to a conducting solid whose bottom plane is suddenly changed to a different temperature. The solution is well known to be the complementary

130 VISCOUS FLUID FLOW

error function or probability integral

$$\frac{u}{U_0} = 1 - \text{erf}\left(\frac{y}{2\sqrt{\nu t}}\right) = \text{erfc}\left(\frac{y}{2\sqrt{\nu t}}\right) \quad (3\text{-}107)$$

where

$$\text{erf}(\beta) = \frac{2}{\sqrt{\pi}} \int_0^\beta e^{-x^2} dx$$

Note that the independent variables y and t have been combined into a single dimensionless *similarity variable*, $\eta = y/[2\sqrt{\nu t}]$. Equation (3-105) becomes an ordinary differential equation for a variable $u'/U_0 = f(\eta)$. More discussion of similarity theory is given in Sec. 3-8.

Values of the complementary error function are tabulated in Table 3-3.

The solution can be reversed so that the fluid is moving at uniform speed U_0 and the plate is suddenly decelerated to zero velocity. In this case,

Suddenly stopped plate: $\quad u = U_0 \,\text{erf}\left(\dfrac{y}{2\sqrt{\nu t}}\right) \quad (3\text{-}108)$

These two complementary solutions are shown in Fig. 3-18. The suddenly started plate from Eq. (3-107) is shown in Fig. 3-18a; the suddenly stopped plate from Eq. (3-108) is shown in Fig. 3-18b. The units of y are arbitrary, and the curves (a) and (b) are of course mirror images.

In either (a) or (b) in Fig. 3-18, the plate's effect diffuses into the fluid at a rate proportional to the square root of the kinematic viscosity. It is customary to define the shear layer *thickness* as the point where the wall effect on the fluid has

TABLE 3-3
Numerical values of the complementary error function

β	erfc(β)	β	erfc(β)
0	1.0	1.1	0.11980
0.05	0.94363	1.2	0.08969
0.1	0.88754	1.3	0.06599
0.15	0.83200	1.4	0.04772
0.2	0.77730	1.5	0.03390
0.25	0.72367	1.6	0.02365
0.3	0.67137	1.7	0.01621
0.35	0.62062	1.8	0.01091
0.4	0.57161	1.9	0.00721
0.5	0.47950	2.0	0.00468
0.6	0.39615	2.5	0.000407
0.7	0.32220	3.0	0.0000221
0.8	0.25790	3.5	0.00000074
0.9	0.20309	4.0	0.00000001
1.0	0.15730	∞	0.0

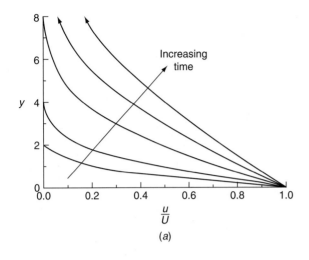

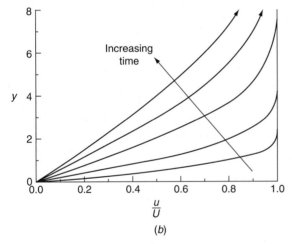

FIGURE 3-18
Stokes' first problem: (*a*) flow above a suddenly started plate; (*b*) streaming flow above a suddenly stopped plate.

dropped to 1 percent: in (*a*), where $u/U_0 = 0.01$; in (*b*), where $u/U_0 = 0.99$. These both correspond to $\text{erfc}(\beta) = 0.01$, or $\beta \approx 1.82$. Then the shear layer thickness in these flows is, approximately,

$$\delta \approx 3.64\sqrt{\nu t} \tag{3-109}$$

For example, for air at 20°C with $\nu = 1.5\text{E}{-}5 \text{ m}^2/\text{s}$, $\delta \approx 11$ cm after 1 min.

3-5.1 Fluid Oscillation above an Infinite Plate

Case 2, with an oscillating wall (or stream), is often called Stokes' second problem, after a celebrated paper by Stokes (1851). Consider first the oscillating wall, with $u(0, t) = U_0 \cos \omega t$, and the fluid in the far field at rest: $u(\infty, t) = 0$. The steadily

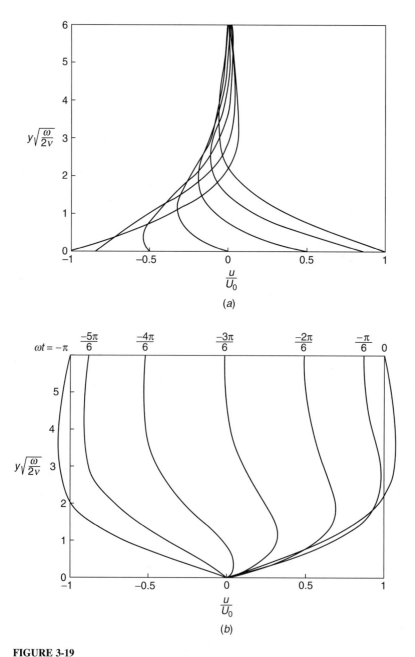

FIGURE 3-19
Stokes' second problem: (*a*) flow above an oscillating infinite plate, Eq. (3-111); (*b*) an oscillating stream above a fixed plate, Eq. (3-112). Velocity profiles shown for 30° increments over a half period.

oscillating solution to Eq. (3-105) must be of the form $u(y, t) = f(y)e^{i\omega t}$, where $i = \sqrt{-1}$. Substitution in Eq. (3-105) gives the ordinary differential equation

$$\frac{d^2 f}{dy^2} - \frac{i\omega}{\nu} f = 0 \tag{3-110}$$

and the solution is of the form $f = \exp[-y\sqrt{i\omega/\nu}]$. We may separate u into real and imaginary parts. If the wall is oscillating, the final result is

$$u_1 = U_0 \exp(-\eta) \cos(\omega t - \eta)$$
$$\eta = y\sqrt{\frac{\omega}{2\nu}} \tag{3-111}$$

If, instead, the stream is oscillating, $u(\infty, 0) = U_0 \cos \omega t$, and the wall is still, $u(0, t) = 0$, the solution is

$$u_2 = U_0 \cos \omega t - u_1 \tag{3-112}$$

where u_1 is given by Eq. (3-111).

The instantaneous velocity profiles for the oscillating wall or stream are shown in Fig. 3-19a and b, respectively. The curves differ in time by 30° during the half-cycle of sweep, from left to right, of the driving oscillation. In Fig. 3-19a, the waves created in the fluid by the moving wall lag behind in phase and damp out as y increases. The thickness δ of the oscillating layer can again be defined where $u/U_0 = 0.01$, that is, where $e^{-\eta} = 0.01$ or $\eta = 4.6$. Solving for y, we have

$$\delta \approx 6.5 \sqrt{\frac{\nu}{\omega}} \tag{3-113}$$

Again we have the characteristic laminar-flow dependence upon $\sqrt{\nu}$. For air at 20°C with a plate frequency of 1 Hz ($\omega = 2\pi$ rad/s), we compute $\delta \approx 1$ cm. The wall shear stress at the oscillating plate is given by

$$\tau_w = \mu \left(\frac{\partial u}{\partial y}\right)_\omega = U_0 \sqrt{\rho \omega \mu} \sin\left(\omega t - \frac{\pi}{4}\right) \tag{3-114}$$

and thus the maximum shear lags the maximum velocity by 135°.

The case of the fixed wall, Fig. 3-19b, is rather different. The low-momentum fluid near the wall actually leads in phase, and we see the "Richardson" type of velocity overshoot (Fig. 3-17) at both ends of the cycle.

3-5.2 Unsteady Flow between Two Infinite Plates

Recall the linear Couette flow between a fixed and a moving plate from Eq. (3-6) and Fig. 3-2a. Let us analyze the start-up of this flow from rest. Since the coordinate system of Fig. 3-2a is somewhat awkward, we will switch to the geometry shown in

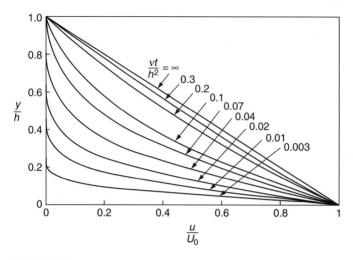

FIGURE 3-20
The development of plane Couette flow due to a suddenly accelerated lower wall, Eq. (3-119).

Fig. 3-20: The upper plate ($y = h$) is fixed; at time $t = 0$, the lower plate ($y = 0$) begins to move at uniform velocity U_0. In this system, the final steady flow is $u = U_0(1 - y/h)$. It is convenient to work with the *difference* between u and its final steady-flow value, i.e., define

$$u_1 = u - U_0\left(1 - \frac{y}{h}\right) \tag{3-115}$$

If we adopt the dimensionless variables $u_1^* = u_1/U_0$, $y^* = y/h$, and $t^* = \nu t/h^2$, the basic differential equation (3-105) becomes

$$\frac{\partial u_1^*}{\partial t^*} = \frac{\partial^2 u_1^*}{\partial y^{*2}} \tag{3-116}$$

subject to
$$u_1^*(0, t^*) = u_1^*(1, t^*) = 0$$
$$u_1^*(y^*, 0) = -1 + y^*$$

Equation (3-116) yields to a separation of variables $u_1^* = f(y^*)g(t^*)$, and substitution and separation give

$$\frac{g'}{g} = \frac{f''}{f} = -\lambda^2$$

or (3-117)

$$g = C\exp(-\lambda^2 t^*) \qquad f = A\sin(\lambda y^*) + B\cos(\lambda y^*)$$

where (A, B, C) are constants. To satisfy $u_1^*(0, t^*) = 0$, we must have $B = 0$. To satisfy $u_1^*(1, t^*) = 0$ with $A \neq 0$, it must be that $\lambda = n\pi$, where n is an integer.

Since no single sine wave will satisfy the initial condition, we form a Fourier sine series to require that $u_1^*(y^*, 0) = \Sigma A_n \sin(n\pi y^*) = -1 + y^*$. The standard Fourier orthogonality condition [Kreyszig (1999)] yields

$$A_n = \int_{-1}^{+1} (-1 + y^*) \sin(n\pi y^*) \, dy^* = -\frac{2}{n\pi} \qquad (3\text{-}118)$$

The final solution for Couette flow start-up is thus

$$\frac{u}{U_0} = \left(1 - \frac{y}{h}\right) - \frac{2}{\pi} \sum_{n=1}^{\infty} \frac{1}{n} \exp(-n^2 \pi^2 t^*) \sin \frac{n\pi y}{h} \qquad (3\text{-}119)$$

Velocity profiles plotted from this expression are shown in Fig. 3-20. The linear asymptote is approached at $t^* \approx 0.3$; for air at 20°C, this corresponds to $t = 2$ s if $h = 1$ cm.

In taking leave of these sections on steady and unsteady Couette and Poiseuille flows, we should note that other relatively straightforward one-coordinate solutions exist for the Navier–Stokes equations. An excellent summary is given in the review article by Berker (1963). Some of these solutions will be given as problem assignments:

1. Steady Couette flow where the moving wall suddenly stops
2. Unsteady Couette flow between a fixed and an oscillating plate
3. Radial outflow from a porous cylinder, Prob. 3-24.
4. Radial outflow between two circular plates, Probs. 3-23 and 3-36.
5. Combined Poiseuille and Couette flow in a tube or annulus, Prob. 3-11.
6. Gravity-driven thin fluid films, Probs. 3-15 to 3-18.
7. Decay of a line Oseen–Lamb vortex, Prob. 3-14.
8. The Taylor vortex profile, Prob. 3-22.

In the next three sections we will outline some more complex solutions, usually involving two- or three-dimensional flows.

3-6 ASYMPTOTIC SUCTION FLOWS

All the solutions discussed so far in this chapter have had vanishing convective acceleration, i.e., no non linear terms in the momentum equation. We now consider the case of simple but nonvanishing convection, i.e., flows with uniform suction (or injection) at the wall.

3-6.1 Uniform Suction on a Plane

Consider steady flow at velocity U (at $y = \infty$) past an infinite plane ($y = 0$). Let the plate be porous, and allow a normal velocity through the wall, so that $u = 0$ but

$v = v_w \neq 0$. The continuity equation for two-dimensional incompressible flow is satisfied if $u = u(y)$ and $v = v_w = $ const. The momentum equation then becomes

$$\rho v_w \frac{du}{dy} = \mu \frac{d^2u}{dy^2} \tag{3-120}$$

subject to $u = 0$ at $y = 0$. Since v_w is constant, the convective acceleration is linear, and the solution is readily obtained:

$$u = U(1 - e^{yv_w/\nu}) \tag{3-121}$$

Physically, v_w must be negative (wall suction), otherwise u would be unbounded at large y. As usual, define the boundary-layer thickness to be the point where $u = 0.99U$. This gives

$$\delta = -4.6 \frac{\nu}{v_w} \tag{3-122}$$

This thickness is constant, independent of y or U, because the convection toward the wall exactly balances the tendency of the shear layer to grow due to viscous diffusion. For air at 20°C, if $v_w = -1$ cm/s, $\delta \approx 7$ mm. For a plate with a leading edge ($x = 0$), a laminar shear layer would grow and approach this constant value at a distance estimated by Iglisch (1944) to be $x \approx 4\nu U/v_w^2$ (see Sec. 4.4). For air at 20°C with $U = 10$ m/s and $v_w = -1$ cm/s, this corresponds to a length $x \approx 6$ m.

3-6.2 Flow between Plates with Bottom Injection and Top Suction

Now consider the flow between two porous plates at $y = +h$ and $y = -h$, respectively, similar to Fig. 3-1. Let the main flow be generated by a constant pressure gradient $-d\hat{p}/dx$. Let the porous walls be such that a uniform vertical crossflow is generated:

$$v = v_w = \text{const} \tag{3-123}$$

Then the equation of continuity with $w = 0$ requires, as before, that $u = u(y)$ only. The momentum Eq. (3-2) reduces to

$$\rho v_w \frac{du}{dy} = -\frac{d\hat{p}}{dx} + \mu \frac{d^2u}{dy^2} \tag{3-124}$$

Since v_w is constant, the equation is linear. We retain the no-slip condition for the main flow:

$$u(+h) = u(-h) = 0 \tag{3-125}$$

Positive v_w corresponds to injection at the bottom plate and suction at the top plate (Fig. 3-21). The solution can be written in dimensionless form and contains the *wall*

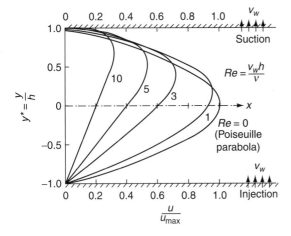

FIGURE 3-21
Velocity profiles for flow between parallel plates with equal and opposite porous walls, Eq. (3-126).

Reynolds number $Re = v_w h/\nu$ as a parameter:

$$\frac{u}{u_{max}} = \frac{2}{Re}\left(\frac{y}{h} - 1 + \frac{e^{Re} - e^{Re\, y/h}}{\sinh Re}\right) \quad (3\text{-}126)$$

where $u_{max} = h^2(-d\hat{p}/dx)/2\mu$ is the centerline velocity for imporous or Poiseuille flow, Eq. (3-42). For very small Re, the last term in the parentheses can be expanded in a power series, and the Poiseuille solution $1 - y^2/h^2$ is reclaimed. For very large Re, the same last term in the parentheses has the approximate value 2.0 except very near $y = +h$, so that

$$\frac{u}{u_{max}} \approx \frac{2(1 + y/h)}{Re}$$

a straight-line variation that suddenly drops off to zero at the upper wall. Some velocity profiles that illustrate these conditions are shown in Fig. 3-21, plotted from Eq. (3-126). Note that the average velocity is decreasing as Re increases, i.e., the friction factor increases as we apply more crossflow through the walls.

Similar analyses can be made with other geometries by imposing a known crossflow field, e.g., between rotating porous cylinders. The flow outside a single rotating cylinder with wall suction is given as a problem assignment; the character of this flow changes when the suction velocity v_w exceeds $2\nu/r_0$, where r_0 is the cylinder radius.

3-6.3 Non-Linear Effects: Flow in Porous Ducts

The previous examples in this section were linear because of the assumption of constant crossflow velocity. If we impose more realistic boundary conditions, e.g., suction at both walls, then the net mass flow will change with x, and at the very least we must have $u = u(x, y)$ and $v = v(y)$ to satisfy the continuity relation. This means that both of the axial convective-acceleration terms $u\,\partial u/\partial x$ and $v\,\partial u/\partial y$ will be products of variables and thus non-linear. The solutions may be vexed with both existence and nonuniqueness problems, and it is instructive to consider an example.

Let the duct be uniformly porous, i.e., the wall velocity v_w is a constant, independent of x. Then the average velocity $\bar{u}$ in the duct will vary linearly with x because of the mass flow through the walls. The two most studied geometries are the circular tube and channel flow between parallel plates. In practice, there must always be an entrance region, and, since the mean velocity continually varies, it is a controversial question as to whether a "fully developed" condition can be achieved.

For the present example, let us consider channel flow between uniformly porous parallel plates, with the geometry of Fig. 3-1. We assume, without proof, that we are far downstream of the entrance and that the boundary conditions

$$\text{At } y = +h: \quad u = 0 \quad v = +v_w$$
$$\text{At } y = -h: \quad u = 0 \quad v = -v_w \tag{3-127}$$

i.e., both walls have either equal suction ($v_w > 0$) or equal injection ($v_w < 0$). Let $\bar{u}(0)$ denote the average axial velocity at an initial section ($x = 0$). Then it is clear from a gross mass balance that $\bar{u}(x)$ will differ from $\bar{u}(0)$ by the amount $v_w x/h$. This observation led Berman (1953) to formulate the following relation for the stream function in the channel:

$$\psi(x, y) = (h\bar{u}(0) - v_w x)f(y^*) \tag{3-128}$$

where $y^* = y/h$ and f is a dimensionless function to be determined. The velocity components follow immediately from the definition of ψ:

$$u(x, y^*) = \frac{\partial \psi}{\partial y} = \left[\bar{u}(0) - \frac{v_w x}{h}\right] f'(y^*) = \bar{u}(x)f'(y^*)$$
$$v(x, y^*) = -\frac{\partial \psi}{\partial y} = v_w f(y^*) = v(y) \text{ only} \tag{3-129}$$

Thus the function f and its derivative f' represent the shape of the velocity profiles; these are independent of x, and the flow is thus termed *similar*; see Secs. 3-8 and 4-3 for further examples. The stream function must now be made to satisfy the momentum (Navier–Stokes) Eq. (3-2) for steady flow:

$$u\frac{\partial u}{\partial x} + v\frac{\partial u}{\partial y} = -\frac{1}{\rho}\frac{\partial p}{\partial x} + \nu\left(\frac{\partial^2 u}{\partial x^2} + \frac{\partial^2 u}{\partial y^2}\right)$$
$$u\frac{\partial v}{\partial x} + v\frac{\partial v}{\partial y} = -\frac{1}{\rho}\frac{\partial p}{\partial y} + \nu\left(\frac{\partial^2 v}{\partial x^2} + \frac{\partial^2 v}{\partial y^2}\right) \tag{3-130}$$

If we substitute u and v from (3-129) into (3-130), we find by cross-differentiation that

$$\frac{\partial^2 p}{\partial x\,\partial y} = 0 \tag{3-131}$$

showing that, unlike that for nonporous duct flow, the pressure gradient is *not* constant. When Eqs. (3-130) and (3-131) are combined, the result is a single fourth-order ordinary non-linear differential equation for $f(y^*)$:

$$f'''' + Re(f'f'' - ff''') = 0 \qquad (3\text{-}132)$$

where $Re = v_w h/\nu$ is the wall Reynolds number and primes denote differentiation with respect to y^*. The boundary conditions are converted from (3-127):

$$\begin{aligned} f'(1) &= 0 & f(1) &= 1 \\ f'(-1) &= 0 & f(-1) &= -1 \end{aligned} \qquad (3\text{-}133)$$

These show that $f(y^*)$ is antisymmetric about $y^* = 0$, so that at the centerline, $v = 0$ and $\partial u/\partial y = 0$, or $f(0) = f''(0) = 0$. Equation (3-132) has no known analytic closed-form solution, but it can be integrated once:

$$f''' + Re(f'^2 - ff'') = k(Re) = \text{const} \qquad (3\text{-}134)$$

Further progress requires some other technique, e.g.,

1. A perturbation solution: Berman (1953)
2. A numerical solution: Eckert et al. (1957)
3. A power-series solution: White (1959)

These are general techniques, and we shall have occasion later to use all three of them in analyzing problems for which the exact solution is not known. One should also look for special cases, and three of these have been found for Eq. (3-132) which satisfy conditions (3-133):

$Re = 0$ (Poiseuille flow):

$$f = \tfrac{3}{2}y^* - \tfrac{1}{2}y^{*3} \qquad f' = \tfrac{3}{2}(1 - y^{*2}) \qquad (3\text{-}135)$$

$Re = +\infty$ (infinite suction):

$$f = y^* \qquad f' = 1 \qquad \text{slug flow} \qquad (3\text{-}136)$$

$Re = -\infty$ (infinite injection):

$$f = \sin\frac{\pi y^*}{2} \qquad f' = \frac{\pi}{2}\cos\frac{\pi y^*}{2} \qquad (3\text{-}137)$$

For other Reynolds numbers, one could use, e.g., the Runge–Kutta program in App. C to solve Eq. (3-132) numerically. Two initial conditions are unknown, i.e., if one started at $y^* = -1$, $f''(-1)$ and $f'''(-1)$ would have to be guessed and the solution double-iterated until $f(1) = 1$ and $f'(1) = 0$. The most complete sets of profiles in the literature are given by White et al. (1958), and are shown in Fig. 3-22. The shapes change smoothly with Reynolds numbers over the full range $-\infty < Re < +\infty$ and show no odd or unstable behavior. Suction tends to draw the profile toward the wall, as it did in Fig. 3-21.

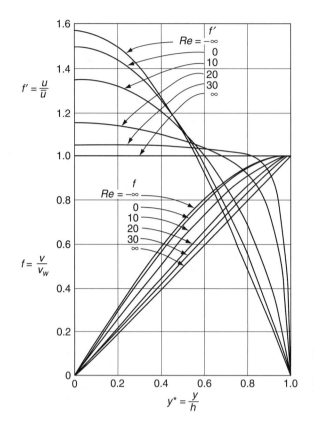

FIGURE 3.22
Similar velocity profiles for a symmetrically porous channel as a function of the wall Reynolds number $Re = v_w h/\nu$. [After White et al. (1958).]

The odd behavior shows up with uniformly porous *pipe*. White (1962a) gave an array of solutions for various values of $Re = v_w r_0/\nu$. The injection profiles ($Re < 0$) vary smoothly but are drawn *toward* the wall, which is physically unexpected. The suction solution ($Re > 0$) show the following pathological behavior:

1. Double solutions, one with backflow, for $0 < Re < 2.3$
2. No solutions whatever in the range $2.3 < Re < 9.1$
3. Multiple sets of double solutions for $Re > 9.1$

At the least, we learn that the Navier–Stokes equations, being non linear and complex, are subject to nonuniqueness and nonexistence difficulties. The previous results are so offbeat that we conclude that "fully developed" similar profiles probably do not occur in laminar porous pipe flow. An analysis by Weissberg (1959) for the special case $Re = 3$ (where no similar solutions have been found) shows that a Poiseuille parabolic entry flow does *not* develop into any kind of a similar profile downstream. One concludes that similar porous-pipe profiles, if they occur at all, are merely way stations along development of the flow into very complex and changing profiles.

3-7 WIND-DRIVEN FLOWS: THE EKMAN DRIFT

There are a number of problems where a liquid free surface is set into motion by a gas flowing over it, e.g., wind blowing over a pond. The shear stresses must match at the air–water interface. Let ($z = 0$) be the interface, with $z > 0$ upward. Then the free-surface condition is

$$\tau_0 = \left(\mu \frac{\partial u}{\partial z}\right)_{\text{air}} = \left(\mu \frac{\partial u}{\partial z}\right)_{\text{water}} \quad (3\text{-}138)$$

where we have assumed wind and water flow in the x direction. Since $\mu_{\text{air}} \ll \mu_{\text{water}}$, it follows that the gradient ($\partial u/\partial z$) in the water will be relatively small. Let us call this interface water-gradient $K = \tau_0/\mu_{\text{water}}$.

3-7.1 Start-Up of Wind-Driven Surface Water

Consider water at rest which is suddenly subjected to a wind stress τ_0. Assuming an unsteady parallel laminar flow $u(z, t)$ and neglecting Coriolis acceleration, we must solve the linear differential equation:

$$\frac{\partial u}{\partial t} = \nu \frac{\partial^2 u}{\partial z^2} \quad (3\text{-}139)$$

subject to the boundary conditions

$$\frac{\partial u}{\partial z}(z = 0) = K \qquad u(z \to -\infty) = 0$$

and the initial condition

$$u(z, 0) = 0 \qquad z < 0$$

This is similar to Stokes' first problem, Eq. (3-107), except that the surface condition involves ($\partial u/\partial z$) rather than u itself. The solution is

$$\frac{\partial u}{\partial z} = K\left[1 + \text{erf}\left(\frac{z}{2\sqrt{\nu t}}\right)\right]$$

or $(3\text{-}140)$

$$\frac{u}{K} = z\left[1 + \text{erf}\left(\frac{z}{2\sqrt{\nu t}}\right)\right] + 2\sqrt{\frac{\nu t}{\pi}} e^{-z^2/4\nu t}$$

This laminar-flow solution predicts that the water-interface velocity ($z = 0$) will grow rapidly, as $u_0 = 2K\sqrt{\nu t/\pi}$. We can try out some numerical values by using the estimate from the text by Roll (1965) of the wind stress for airflow over a water surface:

$$\tau_0 \approx 0.002\rho_{\text{air}}(V_{\text{wind}} - u_0)^2 \quad (3\text{-}141)$$

If we assume a brisk wind speed of 6 m/s (about 12 knots) and an air–water interface at 20°C, we may compute from Eqs. (3-140) and (3-141) that $u_0 \approx 0.6$ m/s after

1 min and 2.3 m/s after 1 h. These water velocities are greatly overestimated: In an actual flow of this type, the surface-water flow would rise in about 1 h to a nearly constant velocity of 0.2 m/s, or about 3 percent of the wind speed. The main reason for the discrepancy is that both the air and water flows, in such a large-scale (high Reynolds number) motion, are turbulent. In turbulent flow, we must replace the fluid's molecular velocity μ by a turbulent-mixing or *eddy* viscosity, μ_t, which is much larger and scaled with flow parameters. We will see how to model turbulent flows in Chap. 6 and can return to this problem more realistically.

3-7.1 Coriolis Effects: The Ekman Spiral

In practical ocean flows, the wind-driven water moves at such low velocities that the Coriolis acceleration [Eq. (2-137)] is not negligible. The resulting flow is steady but accelerating, i.e., the unbalanced momentum flux is in the direction of the Coriolis acceleration. The oceanographer F. Nansen (1902) found that drifting ice in the Arctic deviated from the wind direction by 20 to 40° to the right, which he correctly concluded was due to the earth's rotation. Nansen's student, V. W. Ekman, gave the following analytic solution (1905), which laid the foundation for modern dynamic oceanography. Let (x, y) be the plane of the horizontal, and let the z axis be directed upward. For convenience, let the applied surface wind stress τ_0 be in the y direction. Then, for steady flow with negligible pressure gradients and the acceleration given by Eq. (2-125), a realistic flow solution is possible with the horizontal velocities u and v as functions of z only. The momentum Eq. (3-2) becomes

$$\nu \frac{d^2}{dz^2}(u + iv) - 2i\omega \sin\phi\,(u + iv) = 0 \qquad (3\text{-}142)$$

Where $i = \sqrt{-1}$
ϕ = latitude angle
ω = earth rotation rate

This has an exponential solution if ϕ is assumed constant, i.e., localized wind drift at a given latitude. The boundary conditions are a y-directed shear at the surface and negligible velocity at great depth:

At $z = 0$: $\quad\quad \dfrac{du}{dz} = 0 \quad\quad \dfrac{dv}{dz} = K = \dfrac{\tau_0}{\mu_{\text{water}}}$

At $z = -\infty$: $\quad\quad u = v = 0$ $\qquad (3\text{-}143)$

The proper solution to Eq. (3-142) is of the form $u + iv = C_1 e^{\beta z}$, and C_1 and β can be found from Eq. (3-143). The results can then be decomposed

$$u = V_0 \exp\frac{\pi z}{D} \cos\left(\frac{\pi z}{D} + 45°\right)$$
$$v = V_0 \exp\frac{\pi z}{D} \sin\left(\frac{\pi z}{D} + 45°\right) \qquad (3\text{-}144)$$

where V_0 is the surface-water speed and D is a vertical decay distance, often called the *penetration depth* of the wind:

$$V_0 = \frac{\tau_0/\rho}{\sqrt{2\omega\nu\sin\phi}} \qquad D = \pi\sqrt{\frac{\nu}{\omega\sin\phi}} \qquad (3\text{-}145)$$

Equation (3-144) has several fascinating implications, particularly for engineers not used to Coriolis effects in fluid motion. First note that the surface-water velocities are $u = V_0 \cos 45°$ and $v = V_0 \sin 45°$, showing that the resultant surface-velocity vector is at a 45° angle to the right of the wind (in the Northern Hemisphere). As we move below the surface (negative z), the resultant current vector moves uniformly to the right and decreases exponentially in magnitude, forming a logarithmic spiral, as shown in Fig. 3-23, for equal-depth intervals. At $z = -3D/4$, the current is exactly opposite the wind and has a magnitude of only $V_0 e^{-3\pi/4} \approx 0.095 V_0$.

Another interesting feature is found by integrating the velocity distribution to find the net mass flux. For the x and y directions separately, this gives

$$\dot{m}_x = \int_0^{-\infty} u\,dz = \frac{\tau_0}{2\omega\sin\phi}$$

$$\dot{m}_y = \int_0^{-\infty} v\,dz = 0 \qquad (3\text{-}146)$$

Thus the net mass transport of water is perpendicular to the wind direction and is independent of ν if τ_0 is assumed known. We can explain this by noting that this is the direction of the Coriolis acceleration, which is the only dynamic imbalance on the system.

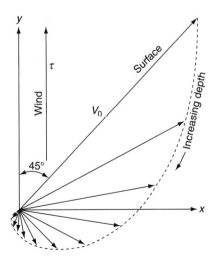

FIGURE 3-23
Wind-driven current vectors for depths of equal interval. The dashed line is a logarithmic spiral from Eq. (3-144). [*After Ekman (1905)*.]

As in the previous wind-start-up example, a few numbers will dispel the idea that ocean currents are laminar. Again take $V_{\text{wind}} = 6$ m/s over a 20°C air–water interface at a latitude of 41°N (Rhode Island). Equations (3-145) predict that $V_0 = 2.7$ m/s, which is far too high, and $D = 45$ cm, which is ridiculously low. The difficulty again is that real ocean flows are turbulent. If we scale the flow with an "eddy" viscosity as discussed in Chap. 6, we obtain the realistic results $V_0 \approx 2$ cm/s and $D \approx 100$ m. Thus an actual wind-driven Ekman current is a slow transport in a wide upper layer of the ocean.

For further details on ocean current, wind-driven or otherwise, consult a textbook on physical oceanography, such as Defant (1961) or Knauss (1978).

3-8 SIMILARITY SOLUTIONS

We complete our description of exact incompressible-flow solutions of the Navier–Stokes equations by discussing three similar flows. A *similarity solution* is one in which the number of independent variables is reduced by at least one, usually by a coordinate transformation. The idea is analogous to dimensional analysis. Instead of *parameters*, like the Reynolds number, the *coordinates* themselves are collapsed into dimensionless groups that scale the velocities.

We have seen two similarity solutions already, without much emphasis. In Stokes' suddenly moved wall problem, Eq. (3-107), the independent variables y and t were combined into the single similarity variable $\eta = y/[2\sqrt{(\nu t)}]$. The solution was then found analytically. In the porous-duct problem, Eq. (3-128), x was broken out into a linear expression, leaving a similarity variable $f(y^*)$, where $y^* = y/h$. In both cases, partial differential equations, perhaps requiring computer modeling, were reduced to ordinary differential equations.

Similarity solutions generally require a semi-infinite or infinite spatial and temporal extent. If Stokes' suddenly moved wall were only L long or if the wall stopped after a finite time t_0, the similarity variable idea would be compromised.

The general theory of similarity in physical problems has been examined in several textbooks: from a physical emphasis by Sedov (1959) and Hansen (1964), and from the mathematical viewpoint by Ames (1965), Bluman and Cole (1974), Dresner (1983), and Sachdev (2000). The benefits of a similarity analysis are signal: The three examples given here reduce a set of partial differential equations, which may yield only to the all-out computer assault of Sec. 3-10, into ordinary differential equations, which we can handle with an elementary numerical method such as Runge–Kutta integration. These mathematical gains are accompanied by a loss in generality: Similarity solutions are, without exception, limited to certain geometries and certain boundary conditions. For example, in the porous-channel problem, the solution of Eq. (3-132) is invalid if v_w varies with x, and the inlet velocity profile $u(0, y)$ must have the same shape as the curves of Fig. 3-22.

Let us now discuss three examples of laminar similarity solutions: (1) flow near a stagnation point, (2) flow near an infinite rotating disc, and (3) flow in a wedge-shaped region.

SOLUTIONS OF THE NEWTONIAN VISCOUS-FLOW EQUATIONS **145**

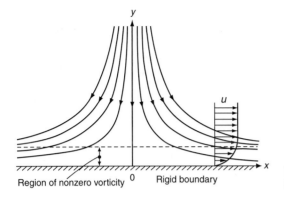

FIGURE 3-24
Stagnation-point flow.

3-8.1 Viscous Flow near a Stagnation Point

One of Prandtl's first students, Hiemenz (1911), discovered that stagnation-point flow can be analyzed exactly by the Navier–Stokes equations.† We piece together here several analyses in the literature: the two-dimensional velocity distribution by Hiemenz (1911) and the temperature distribution by Goldstein (1938), the axisymmetric velocity distribution by Homann (1936) and the temperature distribution by Sibulkin (1952).

The coordinate system is shown in Fig. 3-24. The origin is the stagnation point (where $u = v = 0$ in the frictionless solution), and y is the normal to the plane. For axisymmetric flow, x is to be interpreted as a radial coordinate. In order to satisfy the no-slip condition for $u(x, 0)$ along the wall, a viscous region must develop near the wall. It turns out that this shear layer has a constant thickness and has the effect of "displacing" the outer inviscid flow away from the wall. Let us first consider two-dimensional flow in detail and follow with a more cursory treatment of axisymmetric flow.

3-8.1.1 PLANE STAGNATION FLOW. For plane incompressible flow, the continuity Eq. (3-1) reduces to

$$\frac{\partial u}{\partial x} + \frac{\partial v}{\partial y} = 0$$

which can be satisfied by the plane stream function

$$u = \frac{\partial \psi}{\partial y} \qquad v = -\frac{\partial \psi}{\partial x} \qquad (3\text{-}147)$$

---

†Stagnation flow also happens to be an exact solution to the simpler boundary-layer equations. It is, in fact, one of the similar solutions to be discussed in Sec. 4-3.

Note that stream functions are valid for both viscous and potential flow. Once $\psi(x, y)$ is defined, it must satisfy the two-dimensional momentum relations, Eq. (3-2):

$$u\frac{\partial u}{\partial x} + v\frac{\partial u}{\partial y} = -\frac{1}{\rho}\frac{\partial p}{\partial x} + \nu\left(\frac{\partial^2 u}{\partial x^2} + \frac{\partial^2 u}{\partial y^2}\right)$$

$$u\frac{\partial v}{\partial x} + v\frac{\partial v}{\partial y} = -\frac{1}{\rho}\frac{\partial p}{\partial y} + \nu\left(\frac{\partial^2 v}{\partial x^2} + \frac{\partial^2 v}{\partial y^2}\right) \qquad (3\text{-}148)$$

Inviscid flow near a stagnation point of a body is described by the simple stream function

$$\psi = Bxy \qquad u = By \qquad v = -Bx$$

where B is a positive constant proportional to U_0/L, where U_0 is the stream velocity approaching the body and L a characteristic body length [see, e.g., White (2003, pp. 41–42)]. This flow slips at the wall, that is, $u \neq 0$ at $y = 0$. It must be modified to account for viscous effects.

Following Hiemenz (1911), we modify the stream function to vary with y so that the no-slip condition can be satisfied:

$$\psi_{\text{viscous}} = Bxf(y) \qquad u = \frac{\partial \psi}{\partial y} = Bx\frac{df}{dy} \qquad v = -\frac{\partial \psi}{\partial x} = -Bf$$

For no slip at the wall, we would require $f(0) = f'(0) = 0$. If we substitute this $\psi(x, y)$ into the y-momentum equation in Eq. (3-148), we find that the pressure gradient $\partial p/\partial y = fcn(y)$ only. Therefore, for this stream function,

$$\frac{\partial^2 p}{\partial x \, \partial y} = 0$$

By using this condition and the same stream function in the x-momentum relation of Eq. (3-148), we obtain the differential equation

$$f''' + \frac{B}{\nu}(ff'' - f'^2) = \text{const} = -\frac{B}{\nu} \qquad (3\text{-}149)$$

The constant is evaluated by noticing that the form of ψ is such that, as y becomes large, f''' and f'' must vanish and f' must approach unity.

Equation (3-149) immediately tells us that Hiemenz achieved *similarity* with this stream function. The coordinate x has disappeared, leaving only a single similarity variable, y, and an ordinary differential equation. However, this equation should be nondimensionalized to eliminate the dimensional constants B and ν. There is no body-length scale "L" for this flow. Rather, the proper length scale is $\sqrt{\nu/B}$, the velocity scale is $\sqrt{\nu B}$, and the appropriate dimensionless variables are

$$\eta = y\sqrt{\frac{B}{\nu}} \qquad \psi = xF(\eta)\sqrt{B\nu} \qquad (3\text{-}150)$$

from which

$$u = BxF'(\eta) \qquad \text{and} \qquad v = -F(\eta)\sqrt{B\nu}$$

where the prime denotes differentiation with respect to η. Substitution in Eq. (3-149) gives a differential equation free of parameters:

$$F''' + FF'' + 1 - F'^2 = 0 \qquad (3\text{-}151)$$

The boundary conditions are $u = v = 0$ at the wall ($\eta = 0$) and $u = Bx$ at a large distance from the wall:

$$F(0) = F'(0) = 0 \qquad F'(\infty) = 1 \qquad (3\text{-}152)$$

Equation (3-151) contains two non-linearities, and no analytic solution has ever been found. Hiemenz (1911) performed numerical calculations by hand. It is no trouble nowadays to solve Eq. (3-151) on a personal computer using, say, the Subroutine RUNGE in App. C. We would let $Y(1) = F''$, $Y(2) = F'$, and $Y(3) = F$, and the proper FORTRAN relations for Eq. (3-151) would be

$$\begin{aligned} F(1) &= Y(2) * Y(2) - 1 - Y(1) * Y(3) \\ F(2) &= Y(1) \\ F(3) &= Y(2) \end{aligned} \qquad (3\text{-}153)$$

As initial values at "X" = 0, we set $Y(2) = Y(3) = 0$ from the first two conditions of Eq. (3-152). The problem is to find the correct value of $Y(1) = F_0''$ which will cause $Y(2)$ to approach unity as η becomes very large. One must first address

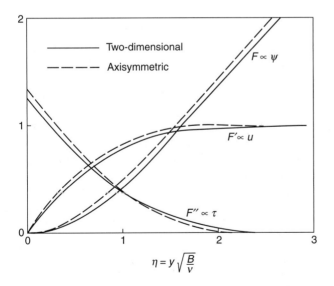

FIGURE 3-25
Numerical solutions of viscous stagnation flow for plane [Eq. (3-151)] and axisymmetric [Eq. (3-165)] conditions.

TABLE 3-4
Numerical solutions for stagnation flow

η	$F' = u/U$	
	Plane $F''(0) = 1.23259$ $\eta^* = 0.6479$	Axisymmetric $F''(0) = 1.31194$ $\eta^* = 0.5689$
0.1	0.11826	0.12619
0.2	0.22661	0.24239
0.3	0.32524	0.34863
0.4	0.41446	0.44499
0.5	0.49465	0.53160
0.6	0.56628	0.60871
0.7	0.62986	0.67663
0.8	0.68594	0.73577
0.9	0.73508	0.78666
1.0	0.77787	0.82987
1.1	0.81487	0.86608
1.2	0.84667	0.89598
1.3	0.87381	0.92032
1.4	0.89681	0.93983
1.5	0.91617	0.95522
1.6	0.93235	0.96718
1.7	0.94578	0.97631
1.8	0.95684	0.98316
1.9	0.96588	0.98822
2.0	0.97322	0.99190
2.2	0.98386	0.99635
2.4	0.99055	0.99847
2.6	0.99464	0.99940
2.8	0.99705	0.99979
3.0	0.99843	0.99993

the question, how large is "infinity"? One answer is, when F'' becomes very small, say, $<10^{-5}$. The following would be a typical asymptotic analysis: In Eq. (3-151), as η becomes large, $F \approx a + \eta$ and $(1 - F'^2)$ approaches zero. Therefore, at large η, a conservative view of Eq. (3-151) is

$$\frac{F'''}{F''} \approx -\eta$$

or

$$F'' \approx e^{-\eta^2/2}$$

With this estimate, we are reasonably confident that $F'' < 10^{-5}$ if $\eta > 4.8 =$ "infinity." Finally, to ensure numerical accuracy of the fourth-order-accurate Runge–Kutta method, we choose a step size $\Delta\eta = 0.03$ so that $(\Delta\eta)^4 \leq 10^{-6}$.

Using these preliminaries, the numerical solution of Eq. (3-151) is readily obtained by making an array of guesses of $F_0'' \approx 0$ to 1.5.

The complete solution for viscous stagnation flow is shown in Fig. 3-25 and tabulated in Table 3-4. Also shown—and to be discussed next—is the solution for axisymmetric stagnation flow. Represented in Fig. 3-25 are the stream function F, the velocity profile F', and the shear stress F''. The correct value of F_0'' turned out to be 1.2326.

3-8.1.2 HARBINGERS OF BOUNDARY-LAYER BEHAVIOR.

Stagnation flow, an exact solution of the Navier–Stokes equations, exhibits many characteristics of thin-shear-region or *boundary-layer* behavior. The no-slip condition creates a low-velocity region which merges smoothly with the outer inviscid flow along the wall. In boundary-layer theory, the outer flow is called the "freestream" velocity, $U(x)$:

Stagnation flow: $\quad\quad U = u(x, \infty) = Bx \quad\quad$ (3-154)

This type of accelerating freestream, whose pressure decreases in the flow direction, is called a *favorable* pressure gradient.

The thickness δ of this stagnation layer is defined as the point where $u/U = 0.99$, which occurs in Table 3-4 when $\eta \approx 2.4 = \delta\sqrt{B/\nu}$. Thus

$$\delta \approx 2.4\sqrt{\frac{\nu}{B}} \quad\quad (3\text{-}155)$$

The boundary-layer thickness is constant in this case because the thinning due to stream acceleration exactly balances the thickening due to viscous diffusion. In Sec. 4-3, we will see that if $U = Cx^m$, the boundary layer will grow with x if $m < 1$ and will become thinner if $m > 1$.

In typical engineering applications, the stagnation boundary layer is quite thin. For example, let air at 20°C approach a 10 cm diameter cylinder at a speed $U_0 = 10$ m/s. Then $B = 4U_0/D = 400$ s^{-1} (see Sec. 7.3), and Eq. (3-155) predicts that $\delta \approx 0.46$ mm, or only 0.5 percent of the body diameter.

Another boundary-layer effect is displacement of the outer stream by the shear layer, as hinted at in Fig. 3-24. We define the *displacement thickness* δ^* as the distance the outer inviscid flow is pushed away from the wall by the retarded viscous layer. In terms of the stream function F, we find that

$$\lim_{\eta \to \infty} F(\eta) = \eta - \eta^*$$

where $\quad\quad\quad\quad\quad\quad\quad\quad\quad\quad\quad\quad\quad\quad\quad\quad\quad\quad\quad\quad\quad\quad\quad\quad\quad\quad$ (3-156)

$$\eta^* = \delta^*\sqrt{\frac{B}{\nu}} = 0.6479$$

as listed in Table 3-4. In stagnation flow, then, $\delta^* \approx 0.27\delta$.

The pressure distribution also exhibits boundary-layer behavior. With u and v known from Eqs. (3-150), we can return to the momentum equations (3-148) and

integrate for the pressure $p(x, y)$. The result is
$$p(0, 0) - p(x, \eta) = \tfrac{1}{2}\rho(BxF')^2 + \tfrac{1}{2}B\mu F^2 + B\mu F'$$
or
$$p(0, 0) - p(x, \eta) = \tfrac{1}{2}\rho(u^2 + v^2) + B\mu F' \qquad (3\text{-}157)$$

Thus the pressure distribution is nearly the same as the frictionless Bernoulli equation except for the additional small term $B\mu F'$. It is instructive to calculate the pressure gradients in each direction:

$$\frac{\partial p}{\partial x} = -\rho B^2 x = -\rho U \frac{dU}{dx}$$
$$\frac{\partial p}{\partial y} = -\rho B \sqrt{B\nu}\,(FF + F'') = \mathcal{O}(\sqrt{\nu}) \qquad (3\text{-}158)$$

Thus the gradient parallel to the wall satisfies Bernoulli's equation, whereas the gradient normal to the wall is negligibly small if the fluid viscosity is small. As we shall see in Chap. 4, these are two of the fundamental assumptions of boundary-layer theory.

Finally, the wall shear is also, yes, a type of boundary-layer relation. With u and v known, we readily compute

$$\tau_w = \mu\left(\frac{\partial u}{\partial y} + \frac{\partial v}{\partial x}\right)_{y=0} = \mu BxF_0''\sqrt{\frac{B}{\nu}} = UF_0''\sqrt{\mu\rho B} \qquad (3\text{-}159)$$

In this flow, then, wall shear is proportional to freestream velocity. If we nondimensionalize in the manner of Eq. (3-39), we obtain

$$C_f = \frac{2\tau_w}{\rho U^2} = \frac{2F_0''}{\sqrt{Re_x}} \qquad Re_x = \frac{Ux}{\nu} \qquad (3\text{-}160)$$

This inverse variation of friction factor with the square root of the local Reynolds number is very common in laminar boundary layers.

3-8.1.3 AXISYMMETRIC STAGNATION FLOW. In plane flow, the stagnation "point" is really a *line*, i.e., there is no variation in the z direction. In axisymmetric flow, stagnation is a true point, we interpret x in Fig. 3-24 as the cylindrical radius coordinate r, and y is the axial coordinate. The cylindrical stream function is defined differently (see App. B) using axisymmetric continuity:

$$u = -\frac{1}{x}\frac{\partial \psi}{\partial y} \qquad v = \frac{1}{x}\frac{\partial \psi}{\partial x} \qquad (3\text{-}161)$$

Similarly, the x-(or radial) momentum equation is different:

$$u\frac{\partial u}{\partial x} + v\frac{\partial u}{\partial y} = -\frac{1}{\rho}\frac{\partial p}{\partial x} + \nu\left[\frac{1}{x}\frac{\partial}{\partial x}\left(x\frac{\partial u}{\partial x}\right) - \frac{u}{x^2} + \frac{\partial^2 u}{\partial y^2}\right] \qquad (3\text{-}162)$$

The stream function for inviscid flow toward an axisymmetric stagnation point is given by

$$\psi = -Bx^2 y$$

whence

$$u = Bx \quad \text{and} \quad v = -2By \tag{3-163}$$

Compare to the plane-flow case: Because of the circular geometry, with flow area increasing along the wall with x, an increase in u is balanced by twice as much of a decrease in v.

Following Hiemenz's analysis, Homann (1936) defined the appropriate dimensionless variables for axisymmetric flow:

$$\eta = y\sqrt{\frac{B}{\nu}} \quad \psi = -x^2 F(\eta)\sqrt{B\nu} \tag{3-164}$$

from which

$$u = BxF'(\eta)$$

and

$$v = -2F(\eta)\sqrt{B\nu}$$

Compare with Eqs. (3-150). Substitution in the y-momentum equation again yielded $\partial^2 p/\partial x\, \partial y = 0$, and then from Eq. (3-162), one obtains the basic axisymmetric differential equation:

$$F''' + 2FF'' + 1 - F'^2 = 0 \tag{3-165}$$

with, as before, $F(0) = F'(0) = 0$ and $F'(\infty) = 1$. Compare with Eq. (3-152). A numerical solution is found in the same manner as for plane flow, and the proper initial condition is $F_0'' = 1.31194$.

The axisymmetric solutions are also shown in Fig. 3-25 and tabulated in Table 3-4. They differ a little from plane flow, e.g., the displacement and boundary-layer thicknesses are slightly smaller and the wall shear stress slightly larger. Again there clearly is thin-shear-layer behavior.

3-8.1.4 STAGNATION-POINT TEMPERATURE DISTRIBUTIONS.

Once the velocities are known from the previous analysis, the temperatures can be found from the energy equation. A similarity solution exists if the wall and stream temperatures, T_w and T_∞, are constant—a realistic approximation in typical stagnation heat-transfer problems.

The plane-flow solution was given by Goldstein (1938). The two-dimensional energy, Eq. (3-3), may be written as

$$\rho c_p \left(u \frac{\partial T}{\partial x} + v \frac{\partial T}{\partial y} \right) = k \left(\frac{\partial^2 T}{\partial x^2} + \frac{\partial^2 T}{\partial y^2} \right) + \Phi \tag{3-166}$$

where Φ is the dissipation function. Following Goldstein (1938), we define a dimensionless temperature Θ which is zero at the wall and unity at ∞:

$$\Theta(\eta) = \frac{T - T_w}{T_\infty - T_w} \qquad \eta = y\sqrt{\frac{B}{\nu}} \qquad (3\text{-}167)$$

In other words, with constant T_w and T_∞, the fluid temperature $T = T(y)$ only. With u and v known from Eqs. (3-150), substitution in the energy Eq. (3-166) gives a second-order linear equation, with Φ neglected:

$$\frac{d^2\Theta}{d\eta^2} + PrF(\eta)\frac{d\Theta}{d\eta} = 0 \qquad (3\text{-}168)$$

subject to $\Theta(0) = 0$ and $\Theta(\infty) = 1$. The stream function $F(\eta)$ is known from Fig. 3-25, and the Prandtl number $Pr = \mu c_p/k$ is assumed constant.

We suppress the temptation to attack Eq. (3-168) by Subroutine RUNGE in App. C, for it is linear and has an exact solution which the reader should verify as an exercise:

$$\Theta(\eta) = \frac{\int_0^\eta d\eta \, \exp\left(-Pr\int_0^\eta F\,ds\right)}{\int_0^\infty d\eta \, \exp\left(-Pr\int_0^\eta F\,ds\right)} \qquad (3\text{-}169)$$

Figure 3-26 shows the temperature profiles from this relation for various Prandtl numbers. The region of large gradients in Θ may be termed the *thermal boundary layer* and, by analogy with the velocity shear layer thickness δ_u, its thickness δ_T is the point where $\Theta \approx 0.99$. A power law curve-fit to values computed from Fig. 3-26 gives the following:

$$\frac{\delta_u}{\delta_T} \approx Pr^{0.4} \qquad (3\text{-}170)$$

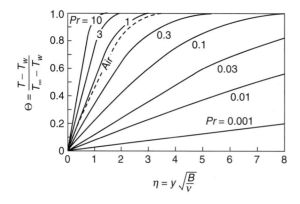

FIGURE 3-26
Stagnation-point temperature distributions for two-dimensional flow from Eq. (3-169).

Physically, the velocity boundary layer is thicker than the thermal boundary layer when $Pr > 1$, because viscous diffusion exceeds conduction effects. And the opposite is of course also true.

The heat transfer at the wall is computed from Fourier's law:

$$q_w = -k \left.\frac{\partial T}{\partial y}\right|_{y=0} = -k(T_\infty - T_w) G(Pr)\sqrt{\frac{B}{\nu}} \qquad (3\text{-}171)$$

where G^{-1} is the denominator in Eq. (3-169):

$$\frac{1}{G(Pr)} = \int_0^\infty d\eta \exp\left(-Pr \int_0^\eta F \, ds\right) \qquad (3\text{-}172)$$

We may compute G simply by adding two FORTRAN statements to our earlier analysis, Eqs. (3-153):

$$F(4) = Y(3)$$
$$F(5) = EXP(-PR * Y(4))$$

The output Y(5) from the subroutine equals $G(Pr)$. This effect of the Prandtl number on heat transfer is shown in Fig. 3-27. Some numerical values of G for both plane and axisymmetric flow follow:

Pr	G (plane flow)	G (axisymmetric)
0.01	0.076	0.106
0.1	0.220	0.301
1.0	0.570	0.762
10	1.339	1.752
100	2.986	3.870
1000	6.529	8.427

The axisymmetric values are about one-third higher. The log-log curves in Fig. 3-27 are nearly straight lines, so one can, if not too particular, fit them to a power law, at least near Prandtl number unity:

Plane flow: $\qquad\qquad G \approx 0.57 Pr^{0.4}$ $\qquad$ (3-173)

Axisymmetric flow: $\qquad G \approx 0.762 Pr^{0.4}$

For axisymmetric flow, the energy equation has the form (App. B)

$$\rho c_p \left(u \frac{\partial T}{\partial x} + v \frac{\partial T}{\partial y}\right) = k\left[\frac{1}{x}\frac{\partial}{\partial x}\left(x \frac{\partial T}{\partial x}\right) + \frac{\partial^2 T}{\partial y^2}\right] \qquad (3\text{-}174)$$

The solution was given by Sibulkin (1952), using exactly the same similarity variable $\Theta(\eta)$ from Eq. (3-167). However, with u and v now given by the axisymmetric

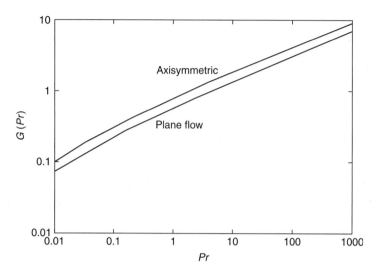

FIGURE 3-27
Variation of the heat-transfer parameter $G(Pr)$ for plane and axisymmetric stagnation flow.

forms Eq. (3-164), the equation for Θ becomes

$$\frac{d^2\Theta}{d\eta^2} + 2PrF\frac{d\Theta}{d\eta} = 0 \qquad (3\text{-}175)$$

The change is the factor "2" in the second term. The solutions for temperature distribution, Eq. (3-169), and heat-transfer parameter $G(Pr)$, Eq. (3-172), are valid for axisymmetric flow if the Prandtl number is replaced by "$2Pr$." Also, $F(\eta)$ must be taken from the axisymmetric stream function in Fig. 3-25.

The axisymmetric heat-transfer parameter $G(Pr)$ is also plotted in Fig. 3-27. Although Eq. (3-171) tells the whole story, namely, that q_w is constant independent of x, it is customary to nondimensionalize q_w as a local Nusselt number:

$$Nu_x = \frac{q_w x}{k(T_w - T_\infty)} = G(Pr)Re_x^{1/2} \qquad (3\text{-}176)$$

where $Re_x = Ux/\nu = Bx^2/\nu$ and $G(Pr)$ is given by Fig. 3-25 or the power-law approximations Eq. (3-173). As we shall see in Chap. 4, Eq. (3-176) has the typical form of laminar-boundary-layer heat-transfer relations, but in the present case, it is rather misleading because it contains a useless "x" on each side.

3-8.1.5 THE REYNOLDS ANALOGY. Osborne Reynolds (1874) postulated an approximation, now called the *Reynolds analogy*, for estimating heat transfer in shear layers. He found that, in pipe flow and "similar" boundary layers such as stagnation flow, the wall shear and heat-transfer rate are proportional:

$$\frac{|q_w|}{\tau_w} = \frac{|k(\partial T/\partial y)_w|}{\mu(\partial u/\partial y)_w} \approx \frac{k}{\mu}\left|\frac{dT}{du}\right|_w \qquad (3\text{-}177)$$

The analogy will not work unless u and T are similar in behavior, which is certainly true in stagnation flow from inspection of Figs. 3-25 and 3-26. Since T varies with Pr and u does not, the ratio above must vary with the Prandtl number. The appropriate way to nondimensionalize Eq. (3-177) is to compare the friction factor C_f to the Stanton number, $C_h = Nu/(RePr)$. Then, in general, the Reynolds analogy postulates that

$$\frac{C_h}{C_f} = fcn\left(Pr, \frac{x}{L}, \text{geometry}\right) \qquad (3\text{-}178)$$

We may rewrite our stagnation-flow heat-transfer results, Eqs. (3-173) and (3-176), in terms of the Stanton number:

$$0.1 < Pr < 10: \quad C_h\sqrt{Re_x} \approx \begin{cases} 0.570 Pr^{-0.6} & \text{(plane flow)} \\ 0.762 Pr^{-0.6} & \text{(axisymmetric)} \end{cases} \qquad (3\text{-}179)$$

Dividing this by the friction coefficient from Eq. (3-160), we obtain

$$\frac{C_h}{C_f} \approx \begin{cases} 0.23 Pr^{-0.6} & \text{(plane flow)} \\ 0.29 Pr^{-0.6} & \text{(axisymmetric)} \end{cases} \qquad (3\text{-}180)$$

Thus, if C_f is known, C_h follows immediately. Comparable relations hold for turbulent shear flow over simple geometries, notably the flat plate. The draw back to Eq. (3-180) is that the constants (0.23, 0.29) vary markedly with pressure gradient (Sec. 4-3). Also, the Reynolds analogy fails to be accurate for (1) varying wall temperature and (2) nonsimilar flows.

This concludes our detailed study of stagnation flow, an exact solution to the Navier–Stokes equations. The results demonstrate a variety of boundary-layer phenomena: a thin viscous layer, a displacement thickness, a thin thermal layer, pressure varying with Bernoulli's relation in the outer layer, very small normal velocity near the wall, very small normal pressure gradient, the Reynolds analogy, and power-law Reynolds and Prandtl number effects. The only important effect missing is shear layer *separation* (Fig. 1.11), which cannot occur in stagnation flow because the freestream velocity increases with x ("favorable" pressure gradient, Fig. 4-5). Our third example in this section, Jeffery–Hamel wedge flow, will illustrate flow separation.

3-8.2 The Flow near an Infinite Rotating Disk

Consider the steady flow which results if the infinite plane $z = 0$ rotates at constant angular velocity ω about the axis $r = 0$ beneath a newtonian viscous fluid which would otherwise be at rest. The viscous drag of the rotating surface would set up a swirling flow toward the disk, as illustrated in Fig. 3-28. All three velocity components v_r, v_θ, and v_z would be involved—a genuine three-dimensional motion—but because of radial symmetry they would be independent of θ, as would the pressure p. It is required then to solve for these four variables as functions of r and z from

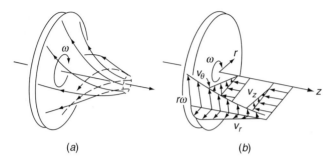

FIGURE 3-28
Laminar flow near a rotating disk: (*a*) streamlines; (*b*) velocity components.

the continuity equation and the momentum (Navier–Stokes) equations in the r, θ, and z directions:

$$\frac{1}{r}\frac{\partial}{\partial r}(rv_r) + \frac{\partial}{\partial z}(v_z) = 0$$

$$v_r\frac{\partial v_r}{\partial r} + v_z\frac{\partial v_r}{\partial z} - \frac{v_\theta^2}{r} = -\frac{1}{\rho}\frac{\partial p}{\partial r} + \nu\left(\frac{\partial^2 v_r}{\partial r^2} + \frac{1}{r}\frac{\partial v_r}{\partial r} + \frac{\partial^2 v_r}{\partial z^2} - \frac{v_r}{r^2}\right) \quad (3\text{-}181)$$

$$v_r\frac{\partial v_\theta}{\partial r} + v_z\frac{\partial v_\theta}{\partial z} + \frac{1}{r}v_r v_\theta = \nu\left(\frac{\partial^2 v_\theta}{\partial r^2} + \frac{1}{r}\frac{\partial v_\theta}{\partial r} + \frac{\partial^2 v_\theta}{\partial z^2} - \frac{v_\theta}{r^2}\right)$$

$$v_r\frac{\partial v_z}{\partial r} + v_z\frac{\partial v_z}{\partial z} = -\frac{1}{\rho}\frac{\partial p}{\partial z} + \nu\left(\frac{\partial^2 v_z}{\partial r^2} + \frac{1}{r}\frac{\partial v_z}{\partial r} + \frac{\partial^2 v_z}{\partial z^2}\right)$$

The boundary conditions are no slip at the wall and no viscous effect far from the wall (except an axial inflow):

At $z = 0$: $v_r = v_z = 0$ $v_\theta = r\omega$ $p = \text{const} = 0$

As $z \to \infty$: $v_r = v_\theta = \dfrac{\partial v_z}{\partial z} = 0$ (3-182)

This fully three-dimensional flow is often called "von Kármán's viscous pump" [Panton (1996)]. The rotating disk sets the near-wall fluid into circumferential motion v_θ. The flow would move in circular streamlines if the pressure increased radially to balance the inward centripetal acceleration. But, in fact, $p = p(z)$ only, and the radial imbalance causes an outward radial flow, $v_r > 0$ (Fig. 3-28*a*). This outward mass flow is balanced by an inward axial flow toward the disk, $v_z < 0$.

The solution to this problem was given in a remarkable paper by Kármán (1921), who considered not only the rotating disk but also all manner of laminar and turbulent shear flows. Kármán—whose autobiography (1964) is highly recommended by this writer—found a similarity solution by deducing that

v_r/r, v_θ/r, v_z, and p are all functions of z only. This reduces the problem to four coupled ordinary differential equations in the single variable z. Since the only parameters in the problem are ω and ν, it is easy to see that the correct dimensionless variable must be $z^* = z\sqrt{\omega/\nu}$. Following Kármán (1921), then, we propose the new dimensionless variables F, G, H, and P:

$$v_r = r\omega\, F(z^*)$$
$$v_\theta = r\omega\, G(z^*)$$
$$v_z = \sqrt{\omega\nu}\, H(z^*) \tag{3-183}$$
$$p = \rho\omega\nu\, P(z^*)$$

We can substitute these variables into Eqs. (3-181) and obtain the following set of non-linear ordinary coupled differential equations:

$$H' = -2F$$
$$F'' = -G^2 + F^2 + F'H$$
$$G'' = 2FG + HG' \tag{3-184}$$
$$P' = 2FH - 2F'$$

where primes denote differentiation with respect to z^*. The boundary conditions from (3-182) now become

$$F(0) = H(0) = P(0) = 0$$
$$G(0) = 1 \tag{3-185}$$
$$F(\infty) = G(\infty) = 0$$

Note that P is uncoupled: We can solve the first three of Eq. (3-184) for F, G, and H and then solve for P from the fourth equation.

Kármán used this flow to illustrate his celebrated momentum-integral relation (Sec. 4-5) derived in the same 1921 paper. Cochran (1934) improved the accuracy with matched inner and outer expansions. Rogers and Lance (1960) gave very accurate numerical solutions. The system Eq. (3-184) is easily solved by Subroutine RUNGE in App. C: Define six variables: $Y_1 = H$, $Y_2 = F'$, $Y_3 = F$, $Y_4 = G'$, $Y_5 = G$, and $Y_6 = P$. Equations (3-184) are then simulated on a digital computer by the six FORTRAN statements

$$F(1) = -2.\,*\,Y(3)$$
$$F(2) = -Y(5)\,*\,Y(5) + Y(3)\,*\,Y(3) + Y(2)\,*\,Y(1)$$
$$F(3) = Y(2)$$
$$F(4) = 2.\,*\,Y(3)\,*\,Y(5) + Y(1)\,*\,Y(4) \tag{3-186}$$
$$F(5) = Y(4)$$
$$F(6) = 2.\,*\,Y(3)\,*\,Y(1) - 2.\,*\,Y(2)$$

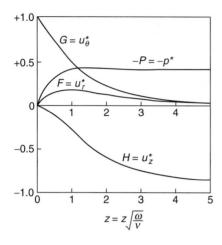

FIGURE 3-29
Numerical solutions of Eqs. (3-184) for the infinite rotating disk.

The reader should compare these statements with Eqs. (3-184). The known initial conditions from Eqs. (3-185) are $Y_1(0) = Y_3(0) = Y_6(0) = 0$ and $Y_5(0) = 1$. The unknown conditions are $F'_0 = Y_2(0)$ and $G'_0 = Y_4(0)$, which must be chosen to make Y_3 and Y_5 vanish at large η. An appropriate step size in $\Delta\eta \leq 0.1$ and "infinity" is reached at about $\eta \approx 10$. The correct initial conditions are found to be $F'_0 = 0.5102$ and $G'_0 = -0.6159$, and the complete numerical solutions are shown in Fig. 3-29 and Table 3-5.

We may define the thickness δ as the point where the circumferential velocity v_θ drops to 1 percent of its wall value, or $G \approx 0.01$, which occurs at about $\eta \approx 5.4$. Then the layer thickness is

$$\delta \approx 5.4\sqrt{\frac{\nu}{\omega}} \tag{3-187}$$

For a disk rotating at 1000 rpm (105 rad/s) in air at 20°C, this relation predicts a (laminar) shear layer thickness of approximately 2 mm.

The asymptotic value $H(\infty) = -0.8838$, which means that the disk draws fluid toward it at the rate

$$v_z(\infty) = -0.8838\sqrt{\omega\nu} \tag{3-188}$$

Thus the disk's pumping action increases with both viscosity and rotation rate. For the above numerical example (1000 rpm in air), this streaming velocity would be 3.5 cm/s toward the disk.

The circumferential wall shear stress on the disk is

$$\tau_{z\theta} = \mu \left.\frac{\partial u_\theta}{\partial z}\right|_{z=0} = \rho r G'_0 \sqrt{\nu\omega^3} \tag{3-189}$$

TABLE 3-5
Numerical solution for the rotating disk

z^*	F	F'	G	G'	H	$-P$
0.0	0.0	0.51023	1.0000	−0.61592	0.0	0.0
0.1	0.0462	0.4163	0.9386	−0.6112	−0.0048	0.0924
0.2	0.0836	0.3338	0.8780	−0.5987	−0.0179	0.1674
0.3	0.1133	0.2620	0.8190	−0.5803	−0.0377	0.2274
0.4	0.1364	0.1999	0.7621	−0.5577	−0.0628	0.2747
0.5	0.1536	0.1467	0.7075	−0.5321	−0.0919	0.3115
0.6	0.1660	0.1015	0.6557	−0.5047	−0.1239	0.3396
0.7	0.1742	0.0635	0.6067	−0.4763	−0.1580	0.3608
0.8	0.1789	0.0317	0.5605	−0.4476	−0.1933	0.3764
0.9	0.1807	0.0056	0.5171	−0.4191	−0.2293	0.3877
1.0	0.1801	−0.0157	0.4766	−0.3911	−0.2655	0.3955
1.2	0.1737	−0.0461	0.4037	−0.3381	−0.3364	0.4040
1.4	0.1625	−0.0640	0.3411	−0.2898	−0.4038	0.4066
1.6	0.1487	−0.0728	0.2875	−0.2470	−0.4661	0.4061
1.8	0.1338	−0.0754	0.2419	−0.2095	−0.5226	0.4042
2.0	0.1188	−0.0739	0.2034	−0.1771	−0.5732	0.4019
2.2	0.1044	−0.0698	0.1708	−0.1494	−0.6178	0.3997
2.4	0.0910	−0.0643	0.1433	−0.1258	−0.6568	0.3977
2.6	0.0788	−0.0580	0.1202	−0.1057	−0.6907	0.3961
2.8	0.0678	−0.0517	0.1008	−0.0888	−0.7200	0.3948
3.0	0.0581	−0.0455	0.0845	−0.0746	−0.7452	0.3938
3.2	0.0496	−0.0397	0.0709	−0.0625	−0.7666	0.3930
3.4	0.0422	−0.0344	0.0594	−0.0525	−0.7850	0.3924
3.6	0.0358	−0.0296	0.0498	−0.0440	−0.8005	0.3920
3.8	0.0303	−0.0254	0.0417	−0.0369	−0.8137	0.3917
4.0	0.0256	−0.0217	0.0349	−0.0309	−0.8249	0.3914
4.5	0.0167	−0.0144	0.0225	−0.0199	−0.8457	0.3911
5.0	0.0108	−0.0095	0.0144	−0.0128	−0.8594	0.3910
5.5	0.0070	−0.0062	0.0093	−0.0082	−0.8682	0.3910
6.0	0.0045	−0.0040	0.0059	−0.0053	−0.8739	0.3909
7.0	0.0018	−0.0017	0.0024	−0.0022	−0.8799	0.3909
8.0	0.0007	−0.0007	0.0010	−0.0009	−0.8824	0.3908
9.0	0.0002	−0.0003	0.0004	−0.0004	−0.8834	0.3908
10.0	0.0001	−0.0001	0.0001	−0.0001	−0.8838	0.3907
∞	0.0	−0.0	0.0	−0.0	−0.8838	0.3906

where $G'_0 = -0.6159$ from Table 3-5. We use this result with a radial strip integration to find the total torque required to turn a disk of radius r_0:

$$M = \int_0^{r_0} \tau_{z\theta} r (2\pi r\, dr) = \frac{\pi}{2} \rho r_0^4 G'_0 \sqrt{\nu \omega^3} \qquad (3\text{-}190)$$

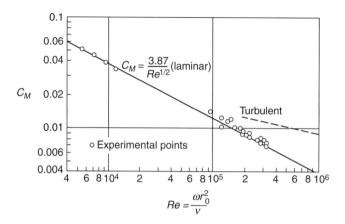

FIGURE 3-30
Theoretical and experimental torque coefficient for a rotating disk. [*Data from Theodorsen and Regier (1944).*]

Thus the required (laminar) moment increases as the fourth power of the disk radius. For our running example, 1000 rpm in air, if the disk radius is 10 cm, this torque is only 0.0005 N · m. A dimensionless torque coefficient can be defined for a disk wetted on both sides:

$$C_M = \frac{-2M}{\frac{1}{2}\rho\omega^2 r_0^5} \approx \frac{3.87}{\sqrt{Re}}$$

$$Re = \frac{\omega r_0^2}{\nu}$$

(3-191)

Again we have the characteristic inverse variation with the square root of the Reynolds number. Equation (3-191) is compared in Fig. 3-30 with the data of Theodorsen and Regier (1944). The agreement is good, but for $Re > 300{,}000$, the flow becomes turbulent and follows the dashed line.

The expected instability of laminar rotating-disk flow was clearly shown in the experiments of Kobayashi et al. (1980). They rotated a 40 cm diameter black aluminum disk in air at 1300 to 1900 rpm and visualized the flow with white titanium tetrachloride gas. A photograph of the flow is shown in Fig. 3-31. Instability is observed at $Re = \omega r^2/\nu \approx 8.8 \times 10^4$ and from 30 to 34 spiral laminar vortices from spreading out at an angle of 10 to 15°. Then, at $Re \approx 3.2 \times 10^5$, turbulence ensues.

Many other papers have been written about various flow problems associated with one or more circular disks. Of particular interest is rotating *flow* next to a fixed disk, first considered by Bödewadt (1940) and later by Rogers and Lance (1960). The picture is essentially reversed from Fig. 3-28. The rotating outer flow sets up a radial pressure gradient which, when acting on the low-velocity fluid near the disk, causes an *inward* radial flow, called "secondary" flow. This inward flow is balanced

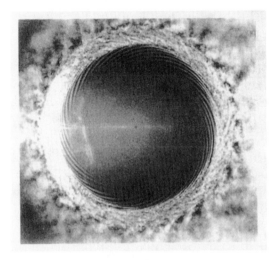

FIGURE 3-31
Flow pattern on a disk rotating in air at 1800 rpm. Laminar flow becomes unstable at $r = 8$ cm, laminar spiral vortices form at $r = 12$ cm, and transition to turbulence is at $r = 16$ cm. [*After Kobayashi et al.* (*1980*).]

by axial flow *away* from the disk. Secondary inward wall flow is familiar to anyone who stirs tea made with loose leaves.

3-8.3 Jeffery–Hamel Flow in a Wedge-Shaped Region

Our third and final example of an exact similarity solution is the radial flow caused by a line source or sink, first discussed by Jeffery (1915) and independently by Hamel (1917). Many subsequent analyses have been made for this flow, and we mention especially Rosenhead (1940) and Millsaps and Pohlhausen (1953).

As shown in Fig. 3-32, we consider the flow in polar coordinates (r, θ), generated by a source (or sink) at the origin and boundary by solid walls at $\theta = \pm\alpha$, as shown. Assume that the flow is purely radial, $u_\theta = 0$. Then, from the continuity equation in polar coordinates (App. B),

$$\frac{1}{r}\frac{\partial}{\partial r}(ru_r) = 0$$

or (3-192)

$$ru_r = fcn(\theta)$$

We expect that u_r will have a local maximum, $u_{\max}$, probably at $\theta = 0$. Then a convenient nondimensionalization for this problem is

$$\eta = \frac{\theta}{\alpha}$$

$$\frac{u_r}{u_{\max}} = f(\eta)$$

(3-193)

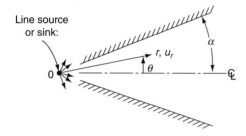

FIGURE 3-32
Geometry of the Jeffery–Hamel flow.

The momentum equations in polar coordinates, for $u_\theta = 0$, are

$$u_r \frac{\partial u_r}{\partial r} = -\frac{1}{\rho}\frac{\partial p}{\partial r} + \nu\left(\frac{\partial^2 u_r}{\partial r^2} + \frac{1}{r}\frac{\partial u_r}{\partial r} - \frac{u_r}{r^2} + \frac{1}{r^2}\frac{\partial^2 u_r}{\partial \theta^2}\right) \quad (3\text{-}194)$$

$$0 = -\frac{1}{\rho r}\frac{\partial p}{\partial \theta} + \frac{2\nu}{r^2}\frac{\partial u_r}{\partial \theta}$$

We can eliminate pressure by cross-differentiation and introduce the variables η and $f(\eta)$. The result is a third-order non-linear differential equation for f:

$$f''' + 2Re\,\alpha f f' + 4\alpha^2 f' = 0 \quad (3\text{-}195)$$

where $Re = u_{max} r\alpha/\nu$ is the characteristic Reynolds number of the flow. The boundary conditions are no slip at either wall and an assumed symmetric flow with a maximum at the centerline:

$$f(+1) = f(-1) = 0 \qquad f(0) = 1 \quad (3\text{-}196)$$

We could replace the second condition with the symmetry requirement $f'(0) = 0$ and confine the analysis to the upper half of the wedge region.

Since Eq. (3-195) is non-linear, we are tempted immediately to call upon Subroutine RUNGE, just as we did with stagnation flow and the rotating disk. In fact, however, an analytic solution is possible. We integrate once to obtain

$$f'' + Re\,\alpha f^2 + 4\alpha^2 f = \text{const}$$

Multiply this by f' and integrate again, using $f(0) = 1$ and $f'(0) = 0$:

$$f'^2 = (1-f)\left[\tfrac{2}{3}Re\,\alpha(f^2 + f) + 4\alpha^2 f + C\right]$$

which can be integrated again because the variables are separable:

$$\eta = \int_f^1 \frac{df}{\left\{(1-f)\left[\tfrac{2}{3}Re\,\alpha(f^2+f) + 4\alpha^2 f + C\right]\right\}^{1/2}} \quad (3\text{-}197)$$

where the boundary condition $f(0) = 1$ has been used again. As η approaches 1, the lower limit approaches zero, $f(1) = 0$, which specifies the constant C:

$$1 = \int_0^1 \frac{df}{\{(1-f)[\frac{2}{3}Re\,\alpha(f^2+f) + 4\alpha^2 f + C]\}^{1/2}} \tag{3-198}$$

Thus Eqs. (3-197) and (3-198) are the formal solution of the problem. The integral is an elliptical integral and can be evaluated from tables or by careful application of Subroutine RUNGE to avoid trouble as $(1-f) \to 0$.

Rosenhead (1940) made a very extensive analysis of this flow and cited the following observations:

1. For any given α and Re, there are an infinite number of possible solutions, both symmetric and asymmetric, corresponding to multiple regions of inflow and outflow.
2. For a given α and a specified number of inflow and outflow regions, there is a critical Re above which this specified solution in impossible.
3. For $\pi/2 < \alpha < \pi$, a solution with pure outflow is impossible, and pure-inflow solutions are limited in certain respects.
4. For $\alpha < \pi/2$, pure inflow is always possible and tends at large Re to have boundary-layer behavior, whereas pure outflow is limited to certain small Re whose approximate range is $Re < 10.31/\alpha$.

Thus once again we encounter nonuniqueness of the Navier–Stokes equations, although here it is regular and predictable.

3.8.3.1 SOLUTION FOR SMALL WEDGE ANGLE α. The most practical application of this flow is to large Re and small α. If both Re and α are small, we may neglect the second and third terms in Eq. (3-195) and integrate to

$$\alpha, Re\alpha \ll 1: \qquad f = \frac{u_r}{u_{max}} = 1 - \eta^2 \tag{3-199}$$

which is Poiseuille flow, Eq. (3-42), valid for either inflow or outflow.

Suppose instead that α is small but $Re\alpha$ is not. Then Eq. (3-198) reduces to

$$\left(\frac{2Re\,\alpha}{3}\right)^{1/2} = \int_0^1 \frac{df}{[(1-f)(f^2+f+K)]^{1/2}} \tag{3-200}$$

where $K = 3C/2Re\,\alpha$. Thus we can specify a range of values of K and carry out the integration to obtain $Re\alpha$, after which C follows from K. Values calculated in this manner are shown in Fig. 3-33. For negative Re (inflow), the values approach the straight line $C = -4\,Re\alpha/3$, which is a boundary-layer approximation first discovered by Pohlhausen (1921). For positive Re (outflow), C drops to zero at $Re\,\alpha \approx 10.31$; since $C = f'^2(1)$, this must be a point of zero wall shear stress, a separation point beyond which backflow will occur at the wall. After evaluating C

164 VISCOUS FLUID FLOW

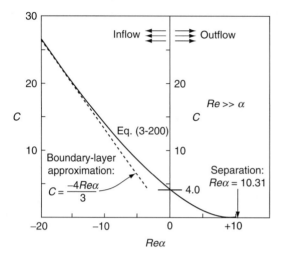

FIGURE 3-33
Values of the shear-stress constant C for wedge flow at large Reynolds numbers.

in this manner, Millsaps and Pohlhausen (1953) computed several velocity profiles from Eq. (3-197). Figure 3-34 shows an array of inflow ($Re < 0$) and outflow ($Re > 0$) profiles computed in the same manner. The inflow curves become flatter and more stable as Re increases and will be discussed again in Chap. 4.

The outflow profiles become S-shaped and have zero wall shear stress (the separation point) when $Re\,\alpha = 10.31$. If $Re\,\alpha > 10.31$, there is backflow at the wall. The difference in profile shape is the result of the change in sign of the

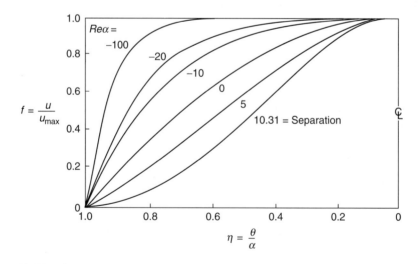

FIGURE 3-34
Velocity profiles for Jeffery–Hamel wedge flow at large $Re \gg \alpha$, from Eq. (3-197). [*Modified from the results of Millsaps and Pohlhausen (1953).*]

streamwise pressure gradient. For inflow, p decreases in the flow direction (favorable gradient) and there is no separation. For outflow, p increases downstream (adverse gradient), a point of inflection occurs in the profile, and separation is imminent.

A separation criterion can be found from the pressure gradient along the centerline, which from Eq. (3-194) is given by

$$\frac{\partial p}{\partial r}(r, 0) \approx \frac{\rho u_{max}^2}{r}$$

if $\alpha^2 \ll 1$. If we interpret $(r\alpha)$ as the thickness δ of the shear layer, the dimensionless form of the pressure gradient at separation would be

$$\frac{\delta^2}{\mu u_{max}} \frac{\partial p}{\partial r}(r, 0)_{separation} \approx (Re\ \alpha)_{separation} = 10.31 \qquad (3\text{-}201)$$

This is much more realistic than the Couette–Poiseuille flow estimate given earlier as Eq. (3-43). In laminar boundary layers, this quantity is called the Kármán–Pohlhausen parameter, $\Lambda = \delta^2(dp/dx)/\mu U$, which at the point of separation takes on values between $+8$ and $+12$.

With the velocity profiles given from Fig. 3-34, Millsaps and Pohlhausen (1953) also solved for temperature profiles, assuming dissipation only—no hot or cold walls. For inflow, there is a region of very strong dissipation temperature rise near the wall.

This completes our study of similar solutions to the Navier–Stokes equations. Other cases are given in the problem exercises.

3-9 LOW REYNOLDS NUMBER: LINEARIZED CREEPING MOTION

All the problems we have looked at so far in this chapter have been exact solutions, valid for an arbitrary Reynolds number, at least until instability sets in and turbulence ensues. While interesting, these exact solutions lack generality. In particular, there is no known solution for the very practical problem of viscous flow past an immersed body at an arbitrary Reynolds number. Other than direct experiment, immersed-body problems are presently solvable only by three approaches: (1) digital-computer simulations, as in Sec. 3-10 to follow; (2) boundary-layer viscous/inviscid patching schemes, as in Chap. 4; and (3) the creeping-flow approximation, to be discussed now.

The basic assumption of creeping flow, developed by Stokes (1851) in a seminal paper, is that density (inertia) terms are negligible in the momentum equation. In such a flow, with stream velocity U and body length L, pressure cannot scale with the "dynamic" or inertia term ρU^2 but rather must depend upon a "viscous" scale $\mu U/L$. If we nondimensionalize the Navier–Stokes Eq. (3-2) with the variables

$$x^* = \frac{x}{L} \qquad \mathbf{V}^* = \frac{\mathbf{V}}{U} \qquad t^* = \frac{tU}{L} \qquad p^* = \frac{p - p_\infty}{\mu U/L}$$

then we obtain the following dimensionless momentum equation:

$$Re \frac{D\mathbf{V}^*}{Dt^*} = -\nabla^* p^* + \nabla^{*2}\mathbf{V}^* \tag{3-202}$$

where $Re = \rho UL/\mu$. We can therefore neglect inertia (the left-hand side) if the Reynolds number is small. This is the creeping-flow or *Stokes flow* assumption:

$$Re \ll 1: \qquad \nabla p \approx \mu \nabla^2 \mathbf{V} \tag{3-203}$$

to be combined with the incompressible continuity relation,

$$\nabla \cdot \mathbf{V} = 0 \tag{3-204}$$

Note that inertia is also negligible if there is no convective acceleration, such as in fully developed duct flow, Sec. 3-3. In such a case, the creeping-flow approximation holds with no restriction on the Reynolds number.

By taking the curl and then the gradient of Eq. (3-203), we obtain two additional useful relations:

$$\begin{aligned} \nabla^2 \boldsymbol{\omega} &= 0 \\ \nabla^2 p &= 0 \end{aligned} \tag{3-205}$$

Thus both the vorticity and the pressure satisfy Laplace's equation in creeping flow.

In two-dimensional Stokes flow, $\omega = -\nabla^2 \psi$, where $\psi(x, y)$ is the stream function. The vorticity equation (3-205) may be rewritten as

Plane flow: $\qquad \nabla^4 \psi = 0 \tag{3-206}$

This is the *biharmonic* equation, much studied in plane elasticity problems. Typical boundary conditions for an immersed-body problem would be no slip at the body surface and uniform velocity and pressure in the freestream.

All of Eqs. (3-203) to (3-206) are linear partial differential equations and thus yield many solutions in closed form. Numerical methods are also very effective, including the boundary-element technique [Beer (2001) and Wrobel (2002)]. Whole textbooks are written about creeping flow, notably Happel and Brenner (1983). The drawback, of course, is that the condition $Re \ll 1$ is very restrictive and usually applies only to low-velocity, small-scale, highly viscous flows. At least four types of creeping flow arise:

1. *Fully developed duct flow.* We have given several examples here in Sec. 3-3 and many more can be found in Berker (1963), Ladyzhenskaya (1969), and Shah and London (1978).
2. *Flow about immersed bodies.* Stokes flow is the foundation of small-particle dynamics and is treated in a text by Happel and Brenner (1983), including multiple-body interaction effects.
3. *Flow in narrow but variable passages.* First formulated by Reynolds (1886) and now known as *lubrication theory*, these flows are covered in general Stokes flow

books such as Langlois (1964) and also in specialized texts such as Khonsari and Booser (2001), Szeri (1998), and Pirro et al. (2001).
4. *Flow through porous media*. This topic began with a famous treatise by Darcy (1856) and is now the subject of specialized texts such as Bear (2000), Ingham and Pop (2002), and Ehlers and Bluhm (2002). Civil engineers have long applied porous-media theory to groundwater movement [see, e.g., Charbeneau (1999)].

The fundamental principles of creeping, or small-inertia flows, are beautifully demonstrated in a 33-min color video, "Low Reynolds Number Flows," available from Encyclopedia Britannica Education Corp. The video was prepared from a color film produced 40 years ago under the sponsorship of the National Science Foundation. It is a historical document because its narrator is Sir Geoffrey Taylor, whose collected works contain scores of important contributions to our knowledge of fluid mechanics.

Finally, we should not fail to mention the interesting flow-visualization device constructed by Hele-Shaw (1898), wherein the streamlines of creeping motion in a narrow passage with an obstacle are shown to be identical to the potential (inviscid) flow about that same obstacle. As analogies go, this is certainly an ironic one.

3-9.1 Creeping Flow about Immersed Bodies: Stokes Paradox

This subject began with a solution for sphere motion given by Stokes (1851). Many solutions exist for three-dimensional bodies, but plane flows are fraught with paradox. As pointed out by Stokes himself, it is impossible to find a steady two-dimensional solution which satisfies both Eq. (3-206) and the no-slip and freestream boundary conditions. Three-dimensional flows do not have this problem, but they are slightly unrealistic in that the inertia terms are not strictly negligible in the far field of the body [Oseen (1910)].

We can illustrate the Stokes paradox with a dimensional argument: If inertia (density) is truly negligible, the force on the body must depend only upon freestream velocity U, fluid viscosity μ, and body-length scale L:

Two-dimensional flow:

$$F' = \text{force per unit length} = f(U, \mu, L) \quad (3\text{-}207a)$$

Three-dimensional flow:

$$F = \text{total drag force} = f(U, \mu, L) \quad (3\text{-}207b)$$

Dimensional analysis of these relations leads to the force laws

Two-dimensional: $$\frac{F'}{\mu U} = \text{const} \quad (3\text{-}208a)$$

Three-dimensional: $$\frac{F}{\mu U L} = \text{const} \quad (3\text{-}208b)$$

The second of these is quite realistic and verified by numerous experiments, but the first is physically in error, since it implies that the drag force is independent of the size L of the body, whether huge or tiny. It follows that there must *always* be a density effect in plane creeping motion:

Two-dimensional:

$$F' = f(\rho, U, \mu, L)$$

or (3-209)

$$\frac{F'}{\mu U} = f\left(\frac{\rho U L}{\mu}\right)$$

The Reynolds number is always important, then. Mathematically, the Stokes paradox means that a plane creeping solution will produce a logarithmic singularity at infinity unless inertia terms are accounted for. Physically, we can infer that a body of infinite depth produces such a profound disturbance in a viscous flow that inertia effects are always important, no matter how slow the freestream speed. Langlois (1964) explains these difficulties very well.

To this writer's surprise, it turns out that the Stokes paradox is not so robust for nonnewtonian fluids. With a simple analysis for power-law fluids in two-dimensional flow past a cylinder, Eq. (1-31a), Tanner (1993) shows that the paradox holds only for $n = 1$ (newtonian fluid) and for $n > 1$ (dilatant or shear-thickening fluid). Both these cases, $n \geq 1$, have density (inertia) effects no matter how large the effective viscosity. However, for $n < 1$ (pseudoplastic or shear-thinning fluid), Tanner shows that the Stokes paradox vanishes and the creeping-flow cylinder drag is independent of the density.

3-9.2 Stokes' Solution for an Immersed Sphere

Consider creeping motion of a stream of speed U about a solid sphere of radius a (Fig. 3-35). It is convenient to use spherical polar coordinates (r, θ), with $\theta = 0$ in the direction of U. The velocity components u_r and u_θ are related to the Stokes stream function ψ by the relations

$$u_r = \frac{1}{r^2 \sin \theta} \frac{\partial \psi}{\partial \theta} \qquad u_\theta = -\frac{1}{r \sin \theta} \frac{\partial \psi}{\partial r} \qquad (3\text{-}210)$$

These satisfy the continuity relation identically. The momentum equation then becomes

$$\left(\frac{\partial^2}{\partial r^2} + \frac{1}{r^2} \frac{\partial^2}{\partial \theta^2} - \frac{\cot \theta}{r^2} \frac{\partial}{\partial \theta}\right)^2 \psi = 0 \qquad (3\text{-}211)$$

Note that this is slightly more complicated than the biharmonic operator of

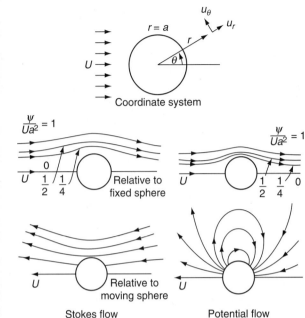

FIGURE 3-35
Comparison of creeping flow (*left*) and potential flow (*right*) past a sphere.

Eq. (3-206) because of the three-dimensional geometry. The boundary conditions are

At $r = a$:
$$\frac{\partial \psi}{\partial r} = \frac{\partial \psi}{\partial \theta} = 0 \qquad (3\text{-}212)$$

As $r \to \infty$:
$$\psi \to \tfrac{1}{2} U r^2 \sin^2 \theta + \text{const}$$

The problem appears formidable, but in fact it yields readily to a product solution $\psi(r, \theta) = f(r)g(\theta)$. With this broad hint, we substitute in Eq. (3-211), satisfy Eq. (3-212), and find the solution of Stokes (1851) for creeping motion past a sphere:

$$\psi = \frac{1}{4} U a^2 \sin^2 \theta \left(\frac{a}{r} - \frac{3r}{a} + \frac{2r^2}{a^2} \right) \qquad (3\text{-}213)$$

The velocity components follow immediately from Eq. (3-210):

$$u_r = U \cos \theta \left(1 + \frac{a^3}{2r^3} - \frac{3a}{2r} \right)$$
$$u_\theta = U \sin \theta \left(-1 + \frac{a^3}{4r^3} + \frac{3a}{4r} \right) \qquad (3\text{-}214)$$

Compared to the previous array of analyses in this chapter, this celebrated solution has several extraordinary properties:

1. The streamlines and velocities are entirely independent of the fluid viscosity. Upon reflection, we deduce that this is true of all creeping flows.
2. The streamlines possess perfect fore-and-aft symmetry: There is no wake of the type shown in Fig. 1-6. It is the role of the convective acceleration terms, here neglected, to provide the strong flow asymmetry typical of higher Reynolds number flows.
3. The local velocity is everywhere retarded from its freestream value: There is no faster region such as occurs in potential flow (Fig. 3-35) at the sphere shoulder (where $u_\theta = 1.5U$).
4. The effect of the sphere extends to enormous distances: At $r = 10a$, the velocities are still about 10 percent below their freestream values.

With u_r and u_θ known, the pressure is found by integrating the momentum relation, Eq. (3-203). The result is

$$p = p_\infty - \frac{3\mu a U}{2r^2} \cos \theta \qquad (3\text{-}215)$$

where p_∞ is the uniform freestream pressure. Thus the pressure deviation is proportional to μ and antisymmetric, being positive at the front and negative at the rear of the sphere. This creates a pressure drag on the sphere. There is also a surface shear stress which creates a drag force. The shear-stress distribution in the fluid is given by

$$\tau_{r\theta} = \mu\left(\frac{1}{r}\frac{\partial u_r}{\partial \theta} + \frac{\partial u_\theta}{\partial r} - \frac{u_\theta}{r}\right) = -\frac{\mu U \sin\theta}{r}\left(\frac{3a^3}{2r^3}\right) \qquad (3\text{-}216)$$

The total drag is found by integrating pressure and shear around the surface:

$$F = -\int_0^\pi \tau_{r\theta}\bigg|_{r=a} \sin\theta \, dA - \int_0^\pi p\bigg|_{r=a} \cos\theta \, dA$$

$$dA = 2\pi a^2 \sin\theta \, d\theta \qquad (3\text{-}217)$$

$$F = 4\pi\mu Ua + 2\pi\mu Ua = 6\pi\mu Ua$$

This is the famous sphere-drag formula of Stokes (1851), consisting of two-thirds viscous force and one-third pressure force. The formula is strictly valid only for $Re \ll 1$ but agrees with experiment up to about $Re = 1$.

The proper drag coefficient should obviously be $F/\mu Ua = 6\pi = $ const, but everyone uses the inertia type of definition $C_D = 2F/\rho U^2(\text{area})$, which is nearly constant at high Reynolds numbers. Thus the sphere drag is written as

$$C_D = \frac{2F}{\rho U^2 \pi a^2} = \frac{24}{Re}$$

where

$$Re = \frac{2a\rho U}{\mu} \tag{3-218}$$

As noted earlier, this introduces a Reynolds number effect where none exists. The formula underpredicts the actual drag when, for $Re > 1$, an unsymmetrical wake forms and, for $Re > 20$, the flow separates from the rear surface, causing markedly increased pressure drag.

The Stokes flow streamlines from Eq. (3-213) are compared in Fig. 3-35 with the streamlines for inviscid (potential) flow past a sphere, for which the classical solution [e.g., White (2003)] is

$$\psi = \frac{1}{2} U r^2 \sin^2 \theta \left(1 - \frac{a^3}{r^3}\right) \tag{3-219}$$

When we compare streamlines past a fixed sphere, the two are superficially similar, except that the Stokes streamlines are displaced further by the body. However, the difference is striking when we compare (Fig. 3-35) streamlines for a sphere moving through a fixed fluid, which we calculate by subtracting the freestream function $\frac{1}{2}Ur^2 \sin^2 \theta$ from Eq. (3-213). The sphere moving in potential flow shows circulating streamlines, indicating that it is merely pushing fluid out of the way. By contrast, the creeping sphere seems to drag the entire surrounding fluid with it, and recirculation is absent.

3-9.3 Other Three-Dimensional Body Shapes

In principle, a Stokes flow analysis is possible for any three-dimensional body shape, providing that one has the necessary analytical skill. A number of interesting shapes are discussed in the texts by Happel and Brenner (1983) and Clift et al. (1978). Of particular interest is the drag of a circular disk:

Disk normal to the freestream: $\quad F = 16\mu U a \tag{3-220a}$

Disk parallel to the freestream: $\quad F = \frac{32}{3}\mu U a \tag{3-220b}$

Here a is the radius of the disk. It is interesting that the values differ only by -15 and -43 percent, respectively, from the drag of a sphere, in spite of the vastly different geometry and orientation. Thus we would expect the Stokes sphere law to be accurate for roughly spherical bodies, such as grains of sand or dust particles, and the estimate $F = 6\pi\mu U a$ is often used in analysis of creeping motion of small particles.

Another shape of interest is the spheroid in Fig. 3-36a. The spheroid shown is *prolate*, $a > b$; it may also be *oblate*, $a < b$. The flow may either be tangential (U_t) or normal (U_n) to the axis of revolution. In either case, the drag force has the Stokes form

$$F = C\mu U b \qquad C = \text{const}$$

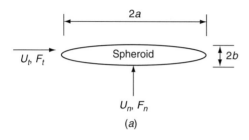

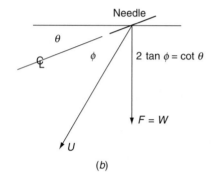

FIGURE 3-36
Forces on a body in creeping flow may be superimposed from tangential and normal components of the velocity vector: (*a*) ellipsoid geometry; (*b*) a nonhorizontal needle falls at an angle.

The exact solutions are rather lengthy and are given by Happel and Brenner (1983). Clift et al. (1978) give the following curve-fit formulas:

$$C_t \approx 6\pi \frac{4 + a/b}{5}$$

$$C_n \approx 6\pi \frac{3 + 2a/b}{5} \tag{3-221}$$

valid to ±10 percent error in the range $0 < a/b < 5$. When $a \gg b$, the spheroid resembles a rod or needle, and the following asymptotic formulas apply:

$a \gg b$:
$$C_t \approx \frac{4\pi a/b}{\ln(2a/b) - 0.5}$$

$$C_n \approx 2C_t \tag{3-222}$$

The drag of a needle is twice as large for flow normal to its axis compared to flow along the axis.

The linearity of creeping motion means that normal and tangential flows may be superimposed without interaction. Suppose that the flow approaches the spheroid in Fig. 3-36a with velocity U at an angle α to its axis. Then one may break the velocity into components $U_n = U \sin \alpha$ and $U_t = U \cos \alpha$ and compute the force components $F_n = C_n \mu U_n b$ and $F_t = C_t \mu U_t b$. Then the total force on the body is the vector sum of F_n and F_t.

SOLUTIONS OF THE NEWTONIAN VISCOUS-FLOW EQUATIONS **173**

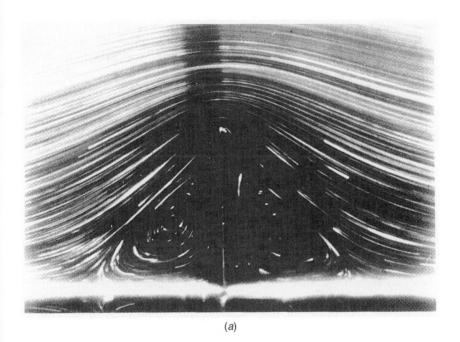

(a)

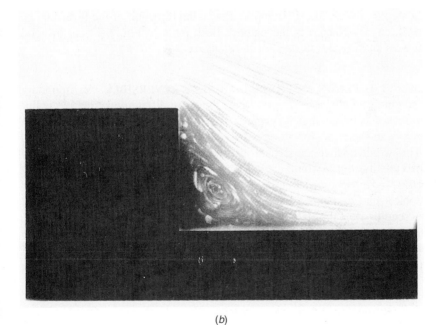

(b)

FIGURE 3-37
Separation occurs in creeping flow past sharp-cornered obstacles: (a) plane flow past a vertical fence at $Re = 0.014$; (b) plane flow past a step at $Re = 0.01$ (in either direction). [*Experimental visualization by aluminum particles in glycerin by Taneda (1979).*]

When falling unsymmetrically, such bodies behave in an interesting way. Consider the needle in Fig. 3-36b, oriented with its axis at an angle θ to the horizontal. It must fall such that its total fluid force F is vertical to balance the body weight. Since $C_n = 2C_t$, it falls at angle ϕ with respect to its axis such that $F_n/F_t = 2 \sin\phi/\cos\phi = \tan(90° - \theta)$, or

Falling rod or needle: $\qquad 2 \tan\phi = \cot\theta$

This is illustrated in Fig. 3-36b. For example, if $\theta = 20°$, then $\phi = 54°$, or the needle moves along a direction (20° + 54°) or 74° from the horizontal.

All of these forces are "in the ballpark" of the drag of a sphere of roughly the same size, i.e., shape effects do not vastly alter the drag. A remarkable theorem due to Hill and Power (1956) states that the Stokes drag of a body must be smaller than any circumscribed figure but larger than any inscribed figure. We can bound the drag of a particle of sand, for example, between inscribed and circumscribed spheres.

All of the immersed-body flows discussed have smooth streamlines near their surface with no separation. This is characteristic of rounded bodies in creeping flow. However, bodies with sharp corners or projecting appendages *do* show flow separation. Figure 3-37 illustrates two cases realized experimentally by Taneda (1979). In Fig. 3-37a, symmetric standing vortices form on either side of a plate or fence projecting into the flow. In Fig. 3-37b, a standing vortex forms in the corner region of a step in a wall; like all creeping flows, the direction of the flow may be *reversed* without any change in the pattern. Both parts of Fig. 3-37 are in good agreement with analytical solutions for the same cases.

3-9.3.1 CREEPING FLOWS ARE KINEMATICALLY REVERSIBLE. An important aspect of creeping flow is *reversibility*. Since the basic differential equation is linear, if a solution ψ is found, then its negative is also a solution. The streamlines can go either way. In Fig. 3-35 for Stokes flow, the arrows could be reversed with no penalty. This would change the signs of the viscosity-induced pressures and shear stresses, and the drag force would act to the left and still equal $6\pi\mu Ua$. If the body is symmetric, as in sphere flow, the fore-and-aft streamlines would be mirror images. Look ahead to another example in Fig. 3-50a for a perfectly symmetric, reversible creeping flow through an orifice.

Note that reversibility is also true of inviscid *potential flow*, which is itself governed by a linear relation, Laplace's Eq. (2-69), where ϕ denotes the velocity potential of the flow. Thus the arrows can also be reversed on the sphere potential flow in Fig. 3-35 and, in like manner, for the inviscid cylinder flow in Fig. 1-4.

3-9.4 Two-Dimensional Creeping Flow: Oseen's Improvement

As mentioned, the Stokes paradox is that a two-dimensional creeping flow cannot satisfy all boundary conditions without including inertia. Even three-dimensional flows are not rigorously valid in the far field. Oseen (1910) removed the paradox by adding ad hoc linearized convective acceleration to the momentum equation:

$$\rho U \frac{\partial \mathbf{V}}{\partial x} \approx -\nabla p + \mu \nabla^2 \mathbf{V} \tag{3-223}$$

where U is the stream velocity acting in the x direction. Equation (3-223) is linear and can be solved for a variety of flows, as discussed in the texts by Oseen (1927) and Lamb (1932). For an immersed body, the solutions are unsymmetrical and show a wake but no separation. For sphere flow, the Oseen approximation adds an additional term to Eq. (3-218):

$$C_{D_{\text{sphere}}} = \frac{24}{Re}\left(1 + \frac{3}{16} Re + \cdots\right) \tag{3-224}$$

Other workers have used asymptotic analyses to add to this expression. As reviewed by Proudman and Pearson (1957), the next term in parentheses should be $[9Re^2 (\ln Re)/160]$, but this diverges greatly for $Re > 3$. The idea of using creeping flow to expand into the higher Reynolds number region has not been successful.

The Stokes paradox is not really a fundamental barrier. It is a failure of the lowest order theory to match flow conditions. Oseen's ad hoc idea is worthy, but the "paradox" has now been truly resolved by the newer asymptotic methods that develop systematic analyses of the governing equations. Asymptotic methods are beyond the scope of this text and may be studied in specialized monographs such as Cebeci and Cousteix (1998) and Nayfeh (2000).

Figure 3-38b compares the Stokes Eq. (3-218) and Oseen Eq. (3-224) theories with experimental data for the drag of a sphere. For $Re > 1$, neither expression is accurate and the data seem to lie in between. Since the sphere geometry is important in engineering, we offer the following curve-fit formula for the (laminar-flow) data:

$$C_{D_{\text{sphere}}} \approx \frac{24}{Re} + \frac{6}{1 + \sqrt{Re}} + 0.4 \qquad 0 \le Re \le 2 \times 10^5 \tag{3-225}$$

which is also plotted in Fig. 3-38b. The accuracy is ± 10 percent up to the "drag crisis," $Re_D \approx 250{,}000$, where the boundary layer on the sphere becomes turbulent, markedly thinning the wake and reducing the drag. The drag crisis occurs slightly earlier with rough surfaces or a fluctuating freestream.

The Oseen Eq. (3-223) can also be solved for various two-dimensional bodies. Of special interest is the drag on a flat plate of length L placed parallel to the stream from Lewis and Carrier (1949):

Plate: $$C_D = \frac{2F'}{\rho U^2 L} = \frac{4\pi}{Re_L[1 - \Gamma + \ln(16/Re_L)]} \tag{3-226}$$

where F' is the drag per unit depth and $\Gamma = 0.577216\ldots$ is Euler's constant. Note that the drag depends upon density no matter how small the Reynolds number.

The Oseen solution for the drag of a cylinder in crossflow is given by Tomotika and Aoi (1951):

Cylinder: $$C_D = \frac{2F'}{\rho U^2 D} = \frac{8\pi}{Re_D[0.5 - \Gamma + \ln(8/Re_D)]} \tag{3-227}$$

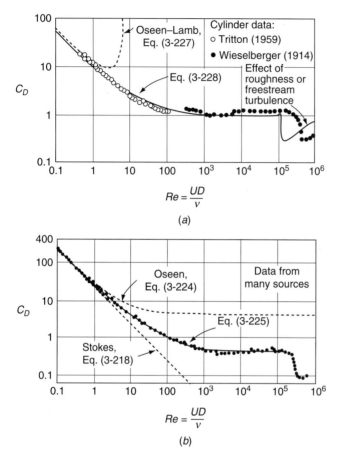

FIGURE 3-38
Comparison of experiment, theory, and empirical formulas for drag coefficients of a cylinder and a sphere (smooth walls): (a) cylinder; (b) sphere.

Figure 3-38a compares this expression with data for the drag of cylinders by Tritton (1959) and Wieselberger (1914). The formula fits up to about $Re \approx 1$ and then diverges. The author earlier offered the following simple curve-fit formula:

$$C_{D_{\text{cylinder}}} \approx 1 + \frac{10.0}{Re_D^{2/3}} \qquad (3\text{-}228)$$

which is in fair agreement up to the drag crisis, $Re \approx 250{,}000$. Sucker and Brauer (1975) offered a better curve-fit formula for the same data:

$$C_{D_{\text{cylinder}}} \approx 1.18 + \frac{6.8}{Re_D^{0.89}} + \frac{1.96}{Re_D^{1/2}} - \frac{0.0004 Re_D}{1 + 3.64\text{E}{-}7 Re_D^2} \qquad (3\text{-}229)$$

valid with good accuracy for $10^{-4} < Re_D < 2 \times 10^5$. Textbooks which cover particle drag and motion are by Clift et al. (1978) and Sirignano (1999).

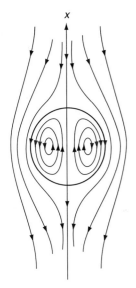

FIGURE 3-39
Streamlines for Stokes flow past a liquid droplet. See also Fig. 3-48.

3-9.5 Creeping Flow Past a Fluid Sphere

A large variety of creeping-motion solutions are given in the texts by Langlois (1964) and Happel and Brenner (1983). A particularly interesting solution is the flow past a spherical droplet of fluid. The outer stream has velocity U at infinity and viscosity μ_0, and the droplet has viscosity μ_i and a fixed interface. The boundary conditions at the droplet interface would be (1) zero radial velocities and (2) equality of surface shear and tangential velocity on either side of the interface. The solution was given by Rybczynski (1911) and independently by Hadamard (1911), and the drag force on a droplet is given by

$$F = 6\pi a \mu_0 U \frac{1 + 2\mu_0/3\mu_i}{1 + \mu_0/\mu_i} \quad (3\text{-}230)$$

For $\mu_i \gg \mu_0$, this simulates a solid sphere (Stokes solution), for which $F = 6\pi a \mu_0 U$. For $\mu_i \ll \mu_0$, this simulates a gas bubble in a liquid, for which $F = 4\pi a \mu_0 U$. A liquid droplet in another liquid would lie in between. Surface tension does not contribute to this drag force. Some streamlines of the flow are shown in Fig. 3-39. These patterns were verified in experiments by Spells (1952). For higher Reynolds numbers (>1.0), the flow changes character into a nearly irrotational outer flow about a droplet which distorts gradually into nonspherical, mushroom-like shapes.

3-9.6 Boundary-Element CFD Creeping-Flow Solutions

Naturally most of our analytical creeping-flow solutions are for simple body shapes and walls. For more complex geometries, the boundary-element method (BEM)—see

Beer (2001) or Wrobel (2002)—is ideal because the creeping-flow equations are linear. Computations are compact and economical because no internal nodes are needed. One sums elemental biharmonic or Stokesian solutions to make the element strengths match the boundary conditions (no slip or streaming flow or a porous wall, etc.). Here are some recent examples of BEM creeping flows.

Trogdon and Farmer (1991) compute unsteady creeping flow through an orifice. Keh and Chen (2001) consider creeping flow of a droplet between plane walls. Vainshtein et al. (2002) study creeping flow near a permeable spheroid. Richardson and Power (1996) report BEM results for flow past two porous bodies of arbitrary shape. Roumeliotis and Fulford (2000) compute interactions between droplets. Lin and Han (1991) report a variety of BEM simulations: a rectangular cavity, a square bank, flow over a fence, and flow past a cylindrical arc. All these results are in good agreement with known experiments and theories. We conclude that any sensible creeping flow can be computed and analyzed with reliability and accuracy.

3-9.7 Heat Transfer in Creeping Motion

Temperature and heat transfer in Stokes flow can be computed from the linearized energy equation, including an Oseen term:

$$\rho c_p U \frac{\partial T}{\partial x} \approx k \nabla^2 T \tag{3-231}$$

where U is the (constant) freestream velocity. This relation is entirely uncoupled from the companion Stokes–Oseen velocity distribution. Typical boundary conditions would be known temperatures at the wall, T_w, and in the stream, T_∞. Nondimensionalization of Eq. (3-231) would yield a single parameter, the Peclet number $Pe = RePr = \rho c_p UL/k$. The mean Nusselt number would be defined as

$$Nu_m = \frac{\overline{q}_w L}{k(T_w - T_\infty)}$$

and would vary only with the Peclet number and the geometry.

The solution for flow past a sphere ($L = 2a$) was given by Tomotika et al. (1953):

$$Nu_{m,\text{sphere}} = 2.0 + 0.5 PrRe + \mathcal{O}(Pr^2 Re^2) \ldots \tag{3-232}$$

The first term (2.0) is the Stokes theory, and the second term is the Oseen correction. When compared with sphere data in Fig. 3-40b, this formula is seen to be strictly qualitative. The curve-fit expression shown in the figure is the suggested formula

$$Nu_{m,\text{sphere}} \approx 2.0 + 0.3 Pr^{1/3} Re^{0.6} \tag{3-233}$$

which seems to agree with both liquid- and gas-flow data. If physical-property variations are important, it is suggested that μ, k, and c_p all be evaluated at the so-called film temperature $(T_w + T_\infty)/2$.

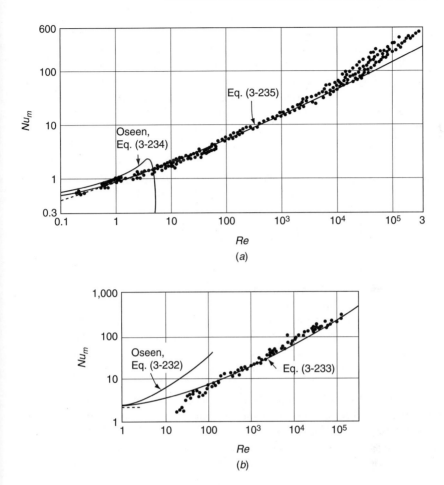

FIGURE 3-40
Comparison of theory and experiment for heat transfer from air to (*a*) cylinders and (*b*) spheres.

Tomotika et al. (1953) also obtained the Oseen solution for flow past a circular cylinder. The mean heat transfer ($L = 2a$) is given by

$$Nu_{m,\text{cylinder}} = B - \frac{Pr^2 Re^2}{12}(16 + B^2)$$

where
$$B = \frac{2}{\ln(8/PrRe) - \Gamma} \tag{3-234}$$

where $\Gamma = 0.577\ldots$, as before. This formula is compared with cylinder data for airflow by Hilpert (1933) and is seen to be of very limited utility. A better formula was suggested by Kramers (1946) as a curve fit

$$Nu_{m,\text{cylinder}} \approx 0.42 Pr^{0.2} + 0.57 Pr^{1/3} Re_D^{1/2} \tag{3-235}$$

valid for $0.1 < Re_D < 10^4$. This formula follows *King's law*, $q_w = a + b\sqrt{U}$, predicted in an earlier study by L. V. King in 1914. It may be used to design the hot-wire anemometer, a fundamental instrument for the study of turbulence [see, e.g., Goldstein (1996)]. A fine wire mounted between two needles is the "cylinder in crossflow." If a current I is passed through the wire and the wire placed normal to a moving stream U, then I will be a measure of U. If the wire is held at constant temperature (constant resistance), Eq. (3-235) predicts a relation of the form $I^2 = a + bU^{1/2}$, where a and b are constants. Equation (3-235) could be used to compute a and b, but it is better to actually calibrate the hot wire in a stream of known velocity. The good agreement of Kramer's formula suggests that for air, with a typical wire diameter of 0.001 in., a hot-wire anemometer should be accurate in the velocity range suggested by Hinze (1975):

$$4.0 \le U \le 4.0 \times 10^6 \text{ cm/s} \qquad (3\text{-}236)$$

which is a vast span, certainly sufficient for most experimental purposes.

Note that the cylinder data in Fig. 3-40a fall above Kramer's formula Eq. (3-235) for $Re > 10^4$. This is attributed to noisiness or "freestream" turbulence in the oncoming stream approaching the cylinder. It is thought that the stream fluctuations trigger concave streamwise "Görtler vortices" (see Fig. 5-23) near the front of the body, increasing the nose heat transfer. Stream turbulence also hastens the onset of the drag crisis (Fig. 3-38a), causing increased (turbulent) heat transfer at the rear of the cylinder. This subject is discussed in detail in the monograph by Zukauskas and Ziukzda (1985).

We close this section by noting that the basic idea of Stokes or creeping flow is valid for $Re \ll 1$ but has not been successfully extended or modified into an accurate simulation of immersed-body flows when $Re > 1$. We will give a few more Stokes flow examples as problem exercises.

3-9.8 Lubrication Theory

The lubrication or friction reduction of two bodies in near contact is generally accomplished by a viscous fluid moving through a narrow but variable gap between the two bodies. One or both bodies may be moving. The theory was developed by Reynolds (1886).

An idealization of this problem is the slipper-pad bearing shown in Fig. 3-41. The bottom wall moves at velocity U and creates a Couette flow in the gap. Let us assume for the present that the upper block is fixed. The gap is very narrow, $h(x) \ll L$, and it decreases from h_0 in the entrance to h_L at the exit. The problem is to find the pressure and velocity distribution.

Assume in Fig. 3-41 that the flow is two-dimensional, that is, $\partial/\partial z = 0$ into the paper. Further assume Stokes flow, i.e., negligible inertia. This requires that $\rho u\, \partial u/\partial x \ll \mu\, \partial^2 u/\partial y^2$, or, approximately,

$$\rho U \frac{U}{L} \ll \mu \left(\frac{U}{h}\right)^2$$

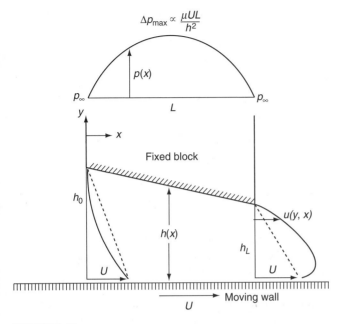

FIGURE 3-41
Low Reynolds number Couette flow in a varying gap: To maintain continuity, the gap pressure rises to a maximum and superimposes Poiseuille flow toward both ends of the gap.

or, finally,

$$\frac{\rho U L}{\mu} \frac{h^2}{L^2} \ll 1$$

Thus the Reynolds number Re_L can be large if the gap is very small. As a practical example, take $U = 10$ m/s, $L = 4$ cm, $h = 0.1$ mm, and SAE 50 lubricating oil, with $\nu \approx 7\text{E}{-}4$ m^2/s. Then $Re_L = 570$, but $Re_L h^2/L^2 = 0.004$. It is thus acceptable to neglect inertia in this case.

Now examine Fig. 3-41. The simple linear Couette flow profiles, shown as dashed lines at the entrance and exit, are impossible for this geometry because continuity is violated. The mass flow at the entrance would exceed the exit flow. To relieve this difficulty, high pressure develops in the gap and causes Poiseuille (parabolic) flow toward *both* ends of the gap, just sufficient to make the mass flow constant at every section x.

At any section in the gap, then, the local velocity profile is combined Couette–Poiseuille flow, from Eq. (3-42) modified for the new coordinates:

$$u = \frac{1}{2\mu} \frac{\partial p}{\partial x} y(y - h) + U\left(1 - \frac{y}{h}\right) \tag{3-237}$$

The correct distribution $p(x)$ is one which everywhere satisfies the continuity Eq. (3-1) for two-dimensional flow in the gap:

$$\int_0^h \frac{\partial u}{\partial x}\, dy = -\int_0^h \frac{\partial v}{\partial y}\, dy = -v(h) + v(0) \tag{3-238}$$

where in this particular case we are assuming that the vertical velocities $v(h)$ and $v(0)$ are zero at both walls.

Substituting for u from Eq. (3-237) and carrying out the integration in Eq. (3-238), we obtain a second-order differential equation for the pressure:

$$\frac{\partial}{\partial x}\left[h^3 \frac{\partial p}{\partial x}\right] = 6\mu U \frac{\partial h}{\partial x} \tag{3-239}$$

Here we assume U is constant and that the gap variation $h(x)$ is known. We then find $p(x)$ subject to the conditions $p(0) = p(L) = p_\infty$. Equation (3-42) is a simplified form of the Reynolds (1886) equation for lubrication. The development of Eq. (3-239) from Eqs. (3-237) and (3-238) is an excellent exercise, involving Leibnitz' rule, which we give as a problem assignment.

3-9.8.1 SOLUTION FOR A LINEARLY CONTRACTING GAP. Equation (3-239) may be integrated numerically for any gap variation $h(x)$. A closed-form solution is possible for a linear gap as in Fig. 3-41:

$$h = h_0 + (h_L - h_0)\frac{x}{L}$$

We may substitute in Eq. (3-239) and carry out the double integration. The algebra is quite laborious and we give only the final result:

$$\frac{p - p_\infty}{\mu U L / h_0^2} = \frac{6(x/L)(1 - x/L)(1 - h_L/h_0)}{(1 + h_L/h_0)[1 - (1 - h_L/h_0)x/L]^2} \tag{3-240}$$

This expression is plotted in Fig. 3-42 for various values of the gap contraction ratio h_L/h_0. When the contraction is only slight, the pressure distribution is nearly symmetric, with p_{max} at $x/L \approx 0.5$. As the degree of contraction increases, p_{max} increases and moves toward the exit plane. The maximum pressure rise is of the order of $(\mu U L / h_0^2)$ and can be amazingly high. For our previous example, SAE 50 oil with $U = 10$ m/s, $L = 4$ cm, and $h_0 = 0.1$ mm, we may compute that $(\mu U L / h_0^2) \approx 2.5\text{E}7$ Pa, or 250 atm. This provides a high force to the slipper block which can support a large load without the block touching the wall.

Recall that Stokes flows, being linear, are *reversible*. If we reverse the wall in Fig. 3-41 to move to the left, that is, $U < 0$, then the pressure change in Eq. (3-240) is *negative*. The fluid will not actually develop a large negative pressure but rather will cavitate and form a vapor void in the gap, as is well shown in the G. I. Taylor film "Low Reynolds Number Hydrodynamics" mentioned earlier on p. 167. Thus flow into an expanding narrow gap may not generally bear much load or provide good lubrication. This effect is unavoidable in a rotating journal bearing, where the gap contracts and then expands, and partial cavitation often occurs.

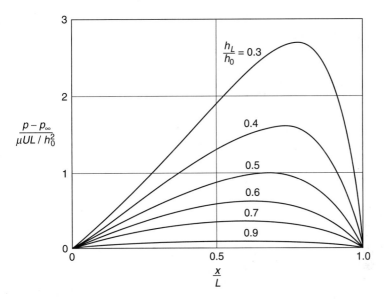

FIGURE 3-42
Pressure distribution in a two-dimensional linear-gap slipper-pad bearing, from Eq. (3-240).

3-9.8.2 THE GENERAL REYNOLDS EQUATION. In a general lubrication problem, both the upper and lower walls in Fig. 3-41 may be moving tangentially and normally, and the depth into the paper may be small, inducing a flow in the z direction. It is assumed there is no translation of the walls in the z direction. The complete derivation for this case is given, for example, in the text by Szeri (1998). The pressure now varies with both x and z and satisfies the following:

$$\frac{\partial}{\partial x}\left(h^3 \frac{\partial p}{\partial x}\right) + \frac{\partial}{\partial z}\left(h^3 \frac{\partial p}{\partial z}\right) = 6\mu \frac{\partial}{\partial x}[h\{U(0) + U(h)\}] + 12\mu[V(h) - V(0)]$$
(3-241)

where, in general, $h = h(x, z)$. This is the three-dimensional *Reynolds equation* for incompressible fluid lubrication. The pressure must be known on all four open sides of the gap.

3-10 COMPUTATIONAL FLUID DYNAMICS

This chapter has nearly exhausted the number of types of exact analytic solutions presently possible with viscous flows. The remainder of the text is primarily concerned with the boundary-layer (thin-shear) approximation. But there is still an important method of simulating viscous flows, often with nearly exact results: numerical analysis on a digital computer. One replaces the partial differential equations of continuity, momentum, and energy by algebraic approximations applied at a finite number of discrete mesh points. Subject to certain restrictions on the Reynolds number, mesh spacing, and time increment, the results can be remarkably accurate, detailed, and useful. Three-dimensional flows can be modeled, and there is no need to assume boundary-layer (parabolic) behavior.

Laminar flows—the subject of the present chapter—are well adapted to near-exact simulation, subject only to limitations of computer size and speed and possible irregularities in the shape of the flow region. Turbulent flows (Chap. 6) can be computed accurately by direct simulation at very low (transitional) Reynolds numbers, but may be computed only approximately at high Reynolds numbers, using a variety of turbulence-modeling assumptions. The accuracy of turbulence models is variable at present, being only fair in reverse-flow or "recirculating" situations, but modeling and database improvements are being published regularly.

The algebraic simulation of a partial differential equation has taken two paths: (1) finite differences and (2) finite elements. The finite-difference technique [Lapidus and Pinder (1999)] constructs algebraic mesh-point models of the partial derivatives, whereas finite-element methods [Huebner et al. (2001)] model the functions themselves over regions between mesh points. Both schemes have been liberally applied to viscous flows, but finite differences are somewhat more popular, perhaps due to a feeling that finite elements, though pleasingly adaptable to irregular geometries, are somewhat slower and require more computer storage. However, a convincing argument in favor of finite elements is given in the text by Pironneau (1989). A special case of finite elements, called the boundary-element method, is applicable to linear viscous flows [Beer (2001) and Wrobel (2002)] and requires no mesh points whatsoever in the interior of the flow.

There is a third CFD option, *spectral methods*, now becoming popular. These methods approximate the desired solution by sophisticated trial functions that cover a wide region, not just one element as with FDM and FEM methods. Typical approximating functions are Fourier, Chebyshev, or Legendre series. The exact solution is usually approached by minimizing the errors with the *method of weighted residuals* (MWR). Spectral methods have been the topic of at least four recent monographs, Guo and Kuo (1998), Karniadakis and Sherwin (1999), Trefethen (2001), and Peyret (2002). These ideas may be applied to the full Navier–Stokes equations.

There is, finally, a fourth CFD option that does not model the Navier–Stokes equations, but rather models the flow itself. These are *vortex methods*, wherein both inviscid and viscous flows are simulated by summing hundreds, perhaps thousands, of elementary vortices to construct a model of a flow. Details are given in the texts by Cottet and Koumoutsakos (1999) and by Kamemoto and Tsutahara (2000).

3-10.1 Overview of the Literature

It would be an understatement to say that the literature on computational fluid dynamics has exploded. The emergence of the digital computer has provided an irresistible tool for numerical investigations. Although Prandtl and others had developed pencil-and-paper marching methods for (parabolic) boundary layers prior to the computer era, very few workers had the patience, desire, or ability to attempt a hand computation of a fully elliptical viscous flow. One of these pioneers was Thom (1933), who calculated, by hand, a remarkably accurate solution of flow past a circular cylinder at $Re_D = 10$. The advent of the computer has created a new industry and a vast literature for numerical modeling.

The fluids engineering community is now blessed with a multitude of new monographs on computational fluid dynamics and/or heat transfer. Earlier classics, admired by this writer, were by Roache (1976), Patankar (1980), Peyret and Taylor (1983), Sod (1985), Jaluria and Torrance (1986), Minkowycz et al. (1988), Pironneau (1989), and Anderson (1995). Add to these a group of newer texts, in alphabetical order: Blazek (2001), Chung (2002), Ferziger and Peric (2001), Garg (1998), Hoffmann and Chiang (2000), Löhner (2001), Lomax et al. (2001), Tannehill et al. (1997), Versteeg and Malalasekera (1996), and Wesseling (2001). And there are more to come: Aref (2004), Caughey and Hafez (2004), Engquist and Rizzi (2004), and Morgan (2004). Some wider-ranging exact-solution studies are reported by Hamdan (1998), Profilo et al. (1998), and Bourchtein (2002). One should also understand the generation of numerical grids: Carey (1997), Thompson et al. (1998), and Miles and Farrashkhalvat (2003). The mathematics of numerical discretization is covered by Hirsch (1988). For readers new to numerical analysis, the general texts by Lucquin and Pironneau (1998) and Moin (2001) show how to handle basic numerical partial differential equations, including FORTRAN and BASIC codes and software. The present text cannot possibly compete with these monographs and will simply present a brief overview.

The traditional general fluid-flow and heat-transfer journals continue to publish numerical studies: *Journal of Fluid Mechanics, Journal of Fluids Engineering, Journal of Heat Transfer, Physics of Fluids, International Journal of Heat and Mass Transfer, Heat and Fluid Flow,* and the *AIAA Journal.* In addition, a number of numerically oriented, fluids-friendly journals have emerged, in alphabetical order: *Applied Mathematics and Computation, Applied Numerical Mathematics, Computational Mechanics, Computer Methods in Applied Mechanics and Engineering, Engineering Analysis with Boundary Elements, International Journal of Fluid Mechanics Research, International Journal for Numerical Methods in Engineering, International Journal for Numerical Methods in Fluids, Journal of Computational Physics, Numerical Heat Transfer (A) Applications* and *(B) Fundamentals, Numerical Methods in Heat Transfer, Numerical Methods for Partial Differential Equations,* and *SIAM Journal of Numerical Analysis.* One could use help in keeping up with such a burgeoning paper mass.

3-10.2 Finite-Difference Approximations

Finite differences simulate a partial derivative by an algebraic difference between discrete mesh points. Consider a two-dimensional flow property $Q(x, y)$, where Q could be velocity, pressure, temperature, etc. In numerical modeling, Q is computed only at a finite number of mesh points, usually equally spaced. A typical two-dimensional mesh is shown in Fig. 3-43. The spacings Δx and Δy need not be the

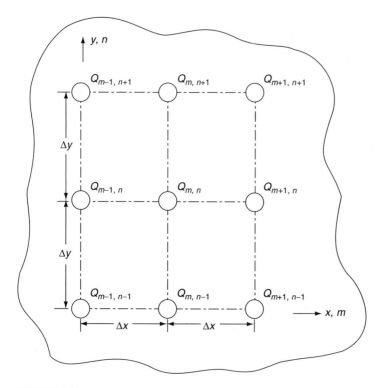

FIGURE 3-43
A typical rectangular finite-difference mesh, showing a discrete point property $Q_{m,n} = Q(x, y)$ surrounded by its eight neighbors.

same. To avoid a lot of parentheses, we adopt the subscript notation (m, n) shown for any particular location:

$$Q_{m,n} = Q(x, y) = Q(x_0 + m\Delta x, y_0 + n\Delta y)$$

The economy of description is immediately evident.

Partial derivatives are then approximated by using the mesh points in Fig. 3-43. For example, $(\partial Q/\partial x)$ at (x, y) may be simulated by a forward, backward, or central "difference":

$$\frac{\partial Q}{\partial x}(x, y) \approx \frac{Q_{m+1,n} - Q_{m,n}}{\Delta x} + \mathcal{O}(\Delta x) \qquad \text{forward}$$

$$\approx \frac{Q_{m,n} - Q_{m-1,n}}{\Delta x} + \mathcal{O}(\Delta x) \qquad \text{backward}$$

$$\approx \frac{Q_{m+1,n} - Q_{m-1,n}}{2\Delta x} + \mathcal{O}(\Delta x^2) \qquad \text{central} \qquad (3\text{-}242)$$

where $\mathcal{O}(\Delta x)$ means that the error in the approximation decreases linearly as the mesh spacing Δx decreases. Here the central difference is obviously more accurate, but it may lead to numerical problems in a full simulation. Similar approximations hold for $(\partial Q/\partial y)$ and $(\partial Q/\partial t)$.

By subtracting a backward from a forward difference in Eq. (3-242) and dividing by Δx, we simulate a second derivative:

$$\frac{\partial^2 Q}{\partial x^2}(x, y) \approx \frac{Q_{m+1,n} - 2Q_{m,n} + Q_{m-1,n}}{(\Delta x)^2} + \mathcal{O}(\Delta x^2) \qquad (3\text{-}243)$$

with an exactly analogous expression for $(\partial^2 Q/\partial y^2)$:

$$\frac{\partial^2 Q}{\partial y^2}(x, y) \approx \frac{Q_{m,n+1} - 2Q_{m,n} + Q_{m,n-1}}{(\Delta y)^2} + \mathcal{O}(\Delta y^2) \qquad (3\text{-}244)$$

Summing these two expressions gives a finite-difference approximation for the Laplacian of Q. For the common case of a square mesh, $\Delta x = \Delta y = h$,

$$\Delta^2 Q \approx \frac{Q_{m+1,n} + Q_{m-1,n} + Q_{m,n+1} + Q_{m,n-1} - 4Q_{m,n}}{h^2} + \mathcal{O}(h^2) \qquad (3\text{-}245)$$

The idea of this approximation is taken in at a glance by sketching what Milne (1953) calls a *stencil*, a term taken from the days when such computations were made by hand. Equation (3-245) is not unique. Here are three equivalent stencils for the Laplacian in a square mesh:

$$\nabla^2 \approx \frac{1}{h^2} \begin{bmatrix} & 1 & \\ 1 & -4 & 1 \\ & 1 & \end{bmatrix} \approx \frac{1}{2h^2} \begin{bmatrix} 1 & & 1 \\ & -4 & \\ 1 & & 1 \end{bmatrix} \approx \frac{1}{6h^2} \begin{bmatrix} 1 & 4 & 1 \\ 4 & -20 & 4 \\ 1 & 4 & 1 \end{bmatrix}$$

All have errors of $\mathcal{O}(h^2)$. The third stencil has a much smaller absolute error than the first two but is seldom used because of the extra computation needed and difficulties at the boundaries. The previous examples give an idea how finite-difference approximations are constructed. Texts on numerical methods give whole arrays of alternative difference models, e.g., Lucquin and Pironneau (1998), Moin (2001), or Lapidus and Pinder (1999).

3-10.3 Application to One-Dimensional Unsteady Laminar Flow

As an example of a finite-difference solution, consider one-dimensional unsteady flow with no pressure gradient, as discussed earlier in Sec. 3-5:

$$\frac{\partial u}{\partial t} = \nu \frac{\partial^2 u}{\partial y^2} \qquad (3\text{-}246)$$

subject to various boundary conditions such as the Stokes moving-wall problems of Fig. 3-18.

To include time in the notation, we use the superscript j as follows:
$$u^j_{m,n} = u(x, y, t) = u(x_0 + m\Delta x, y_0 + n\Delta y, t_0 + j\Delta t)$$
In the present case, Eq. (3-246), the m subscript is not needed. We adopt a forward difference for time and a central difference for the y derivative to model Eq. (3-246):
$$\frac{u^{j+1}_n - u^j_n}{\Delta t} \approx \nu \frac{u^j_{n+1} - 2u^j_n + u^j_{n-1}}{(\Delta y)^2}$$
The truncation error is $\mathcal{O}(\Delta t) + \mathcal{O}(\Delta y^2)$. Rearrange this to solve for the next nodal velocity at time $j + 1$:
$$u^{j+1}_n \approx \sigma(u^j_{n+1} + u^j_{n-1}) + (1 - 2\sigma)u^j_n \tag{3-247}$$
where $\sigma = \nu\Delta t/(\Delta y)^2$ is a dimensionless mesh-size parameter. Since the right-hand side contains known (previous) values, this method is called *explicit*, i.e., the new value of u^{j+1} is computed immediately. However, there is a penalty for this algebraic ease: The coefficients of u^j must all be positive for numerical stability. Thus we are restricted to
$$\sigma \leq \frac{1}{2}$$
or
$$\Delta t \leq \frac{\Delta y^2}{2\nu} \tag{3-248}$$

For a given mesh size Δy, then, the time step Δt is limited—sometimes severely—by Eq. (3-248). Given appropriate initial and boundary values for the mesh points, a complete numerical solution can be generated by repeated application of Eq. (3-247). For example, for Stokes' first problem, Fig. 3-18a, we would specify $u_{n>1} = 0$ at the initial time $j = 1$, plus $u_1 = U$ at the wall, and $u_{n_{\max}} = 0$ for all j. We would limit Δt by Eq. (3-248) and continually check $n_{\max}$ to see that the solution has not diffused that far.

To avoid the limitation on time step, we may change the model so that the second derivative is evaluated at the *new* time step:
$$\frac{u^{j+1}_n - u^j_n}{\Delta t} \approx \nu \frac{u^{j+1}_{n+1} - 2u^{j+1}_n + u^{j+1}_{n-1}}{(\Delta y)^2}$$
We may rearrange this as follows:
$$-\sigma u^{j+1}_{n+1} + (1 + 2\sigma)u^{j+1}_n - \sigma u^{j+1}_{n-1} \approx u^j_n \tag{3-249}$$

This an *implicit* method: When applied to each mesh point n, it results in a system of simultaneous algebraic equations. Since each equation contains only three unknowns centered around the diagonal of the coefficient matrix, the system may be solved rapidly and easily by the tridiagonal matrix algorithm (TDMA), which will be described later in Sec. 4-7.2.1. There is no limitation, in this implicit method, on the time step, which can typically be an order of magnitude higher than that required by Eq. (3-248) and still achieve the same numerical accuracy.

Slight complications arise in these one-dimensional models if the equations are non-linear, i.e., contain convective-acceleration terms. We will illustrate both explicit and implicit non-linear models in Sec. 4-7 by applying them to the laminar boundary-layer equations.

More substantial complications arise if there are two dimensions or more. For example, suppose we wish to model the two-dimensional linear diffusion or heat-conduction equation,

$$\frac{\partial u}{\partial t} = \nu \left(\frac{\partial^2 u}{\partial x^2} + \frac{\partial^2 u}{\partial y^2} \right) \tag{3-250}$$

The explicit algorithm, to be given as a problem exercise, has the stability limitation

$$\frac{\nu \Delta t}{\Delta x^2} + \frac{\nu \Delta t}{\Delta y^2} \leq \frac{1}{2} \tag{3-251}$$

This limitation is more restrictive than Eq. (3-248).

If we develop an implicit model for Eq. (3-250), there will be a system of equations each with five unknowns, corresponding to (m, n), $(m, n \pm 1)$, $(m \pm 1, n)$. Further, the matrix of coefficients is not pentadiagonal, i.e., there are nonzero terms far removed from the main diagonal. One must therefore contemplate lengthier schemes such as (1) iteration, (2) Gauss elimination, or (3) inversion of a sparse $N \times N$ matrix. In fact, what is often done is to split the procedure into two applications of the TDMA—one over a half time step in the x direction, then one over a second half time step in the y direction, each time using previous values for the additional two unknowns. This is the alternating-direction implicit (ADI) method developed by Peaceman and Rachford (1955). When applied to Eq. (3-250), the ADI model takes the form

Half-step 1: $\quad \dfrac{u_{m,n}^{j+1/2} - u_{m,n}^{j}}{\frac{1}{2}\Delta t} \approx \nu(\delta_x^2 u_{m,n}^{j+1/2} + \delta_y^2 u_{m,n}^{j}) \tag{3-252}$

Half-step 2: $\quad \dfrac{u_{m,n}^{j+1} - u_{m,n}^{j+1/2}}{\frac{1}{2}\Delta t} \approx \nu(\delta_x^2 u_{m,n}^{j+1/2} + \delta_y^2 u_{m,n}^{j+1})$

where δ^2 denotes the second partial derivative difference model from Eqs. (3-243) and (3-244). In half-step 1, the TDMA is used for the east–west neighbors ($m \pm 1$), while half-step 2 sweeps the north–south neighbors ($n \pm 1$). Further details of the ADI scheme are given by Ferziger (1998). There are also alternating-direction explicit (ADE) methods which half-step forward without using the TDMA [Tannehill et al. (1997)].

3-10.4 Modeling the Full Navier–Stokes Equations

The Navier–Stokes equations are non-linear, nonunique, and multidimensional, and solutions often contain very fine details of flow structure. To model them numerically requires considerably more sophistication than the one-dimensional explicit and implicit models discussed in Sec. 3-10.3. The last two decades have seen marked progress in creating "Navier–Stokes solvers," but three-dimensional complex-geometry flows remain challenging.

There is no single model in use. The handbook by Minkowycz et al. (1988) outlines, in detail, four different finite-difference models and two different finite-element models. Some of these—and related formulations—have been developed into commercial codes which a user can purchase, add boundary conditions, and get immediate results, especially for two-dimensional steady flows.

It is not the purpose—or hope—of this section to outline the latest advances at the frontiers of computational fluid dynamics. The reader is referred for this purpose to the review by Boris (1989), who discusses ideas beyond our scope: cellular automata, near-neighbor "atom" algorithms, adaptive gridding, spectral methods, and parallel processing for supercomputers. Here we choose only to give a brief overview of Navier–Stokes models.

3-10.4.1 ISSUES INVOLVED IN NAVIER–STOKES MODELING. There is considerable science, and not a little art, necessary to complete a successful Navier–Stokes computation for recirculating or "nonboundary-layer" flows, which have strong interactions between viscous and inviscid regions. Some of the issues involved are as follows:

1. *The coordinate-system viewpoint.* Most studies use the Eulerian or flow-field system. Unsteady flows, especially with material boundaries such as droplets or free surfaces, can benefit from a Lagrangian, or particle, formulation.
2. *The method of formulation.* The two main choices are finite differences (FDM) and finite elements (FEM)—usually developed and described independently. Workers rarely straddle both camps. The FDM formulation is localized by simple derivative approximations and leads to sparse matrices. The FEM method uses basis functions over a finite region and yields large, banded matrices. The FEM approach (not developed here) is very attractive for irregular geometries, but FDM methods are improving in this regard by the use of *boundary-fitted coordinates* [Thompson (1998)]. The two methods are compared by Pepper and Baker [Minkowycz et al. (1988, Chap. 13)].
3. *The method of discretization.* In FEM methods, the issue is the choice of basis function and the scheme for minimizing the integrated error. In FDM methods, the discretization may be accomplished either by (1) Taylor-series derivative truncations or by (2) control-volume (integral) techniques, the latter approach being a balance of physical terms popularized by Patankar (1980).

4. *The choice of dependent variables.* In two-dimensional flow, the stream function and vorticity can be convenient variables. In three dimensions, multiple stream functions are awkward, so the "primitive" variables of velocity and pressure are used.
5. *Grid selection.* The choice of grid is crucial to the performance of a numerical method. Plain, unadorned square meshes are rarely appropriate—grids should be finer in regions of high gradient. Poor gridding can result in instability or failure to converge. Grids can be generated in both two and three dimensions, using algebraic or differential transformations [Thompson et al. (1998)]. There are clustering techniques to pack grids into a high-gradient area. In unsteady flow, the gradients can move; hence the grids should be changeable or *adaptive*, using grid-speed algorithms to reduce the instantaneous numerical error [Tannehill et al. (1997, Sec. 10-7)].
6. *Simulation of boundary conditions.* Viscous flows depend strongly upon their boundary conditions, which should be modeled as accurately as possible. The effect of far-field boundary placement, such as "infinity," "upstream," or "downstream," must be carefully investigated. The mathematical requirements are discussed by Kreiss and Lorenz (1989).
7. *Solution techniques and numerical uncertainty.* Both explicit and implicit solutions are widely used in FDM methods. FEM methods, with large banded matrices coupled in space and time, are inherently implicit. Error estimates should be made for all numerical results. The availability of large and/or fast digital computers should eliminate untested single-fixed-grid computation. Since 1986, the *Journal of Fluids Engineering* will not accept any CFD articles unless their numerical accuracy is systematically tested by grid variation and far-field boundary placement.

3-10.4.2 THE STREAM FUNCTION-VORTICITY APPROACH. For many years, most CFD papers studied two-dimensional incompressible viscous flows, for which the stream function ψ exists and the vorticity ω has only a single component:

$$u = \frac{\partial \psi}{\partial y} \qquad v = -\frac{\partial \psi}{\partial x} \qquad \omega = \frac{\partial v}{\partial x} - \frac{\partial u}{\partial y} \qquad (3\text{-}253)$$

The vorticity transport Eq. (2-114) may be written as

$$\frac{\partial \omega}{\partial t} + \frac{\partial \psi}{\partial y}\frac{\partial \omega}{\partial x} - \frac{\partial \psi}{\partial x}\frac{\partial \omega}{\partial y} = \nu\left(\frac{\partial^2 \omega}{\partial x^2} + \frac{\partial^2 \omega}{\partial y^2}\right) \qquad (3\text{-}254)$$

This equation is parabolic in time and elliptical in space and may be modeled in FDM just as was done in Sec. 3-10.3 for Eq. (3-246). Meanwhile, the stream function may be computed by rewriting Eq. (3-253) in the form

$$\frac{\partial^2 \psi}{\partial x^2} + \frac{\partial^2 \psi}{\partial y^2} = -\omega \qquad (3\text{-}255)$$

This is an elliptical (Poisson) equation for ψ and may be modeled in FDM similar to Laplace's equation in Eq. (3-245). Computations for nodal values of ψ and ω may be either explicit or implicit. Note that pressure is not present, having been eliminated in Eq. (3-254) by taking the curl of the Navier–Stokes equation.

Once the stream function is known everywhere, the pressure is computed from the following Poisson equation:

$$\Delta^2 p = 2\rho \left(\frac{\partial u}{\partial x} \frac{\partial v}{\partial y} - \frac{\partial u}{\partial y} \frac{\partial v}{\partial x} \right) = 2\rho \left[\frac{\partial^2 \psi}{\partial x^2} \frac{\partial^2 \psi}{\partial y^2} - \left(\frac{\partial^2 \psi}{\partial x \partial y} \right)^2 \right] \quad (3\text{-}256)$$

This equation is derived by differentiation of the x- and y-momentum equations and is given as a problem exercise. There are many finite-difference models for the mixed partial derivative [Tannehill et al. (1997, Table 3-2)], of which a popular form is

$$\frac{\partial^2 \psi}{\partial x \partial y} \approx \frac{\psi_{m+1,n+1} - \psi_{m+1,n-1} - \psi_{m-1,n+1} + \psi_{m-1,n-1}}{4 \Delta x \Delta y}$$

The computations must satisfy appropriate boundary conditions. In the far-field "freestream," one usually knows the values of ψ, ω, and p. At a wall with no slip, $u = v = 0$, we would satisfy

$$\psi_w = \text{const} \qquad \omega_w \approx \frac{2(\psi_w - \psi_{w+1})}{\Delta n^2} \qquad \left. \frac{\partial p}{\partial s} \right|_w = -\mu \left. \frac{\partial \omega}{\partial n} \right|_w \quad (3\text{-}257)$$

where s and n denote coordinates parallel and normal to the wall, respectively. In this manner, beginning with Thom (1933), almost all of the classical two-dimensional solutions in the literature were computed. Plane-flow (ψ, ω) modeling is still popular today.

3-10.4.3 THE PRIMITIVE-VARIABLE APPROACH.
Presently, the most widely used CFD models of viscous flow use the direct or "primitive" variables (u, v, w, p, T). Consider, for simplicity, the two-dimensional, incompressible equations of motion with constant transport properties:

Continuity:
$$\frac{\partial u}{\partial x} + \frac{\partial v}{\partial y} = 0 \quad (3\text{-}258a)$$

x momentum:
$$\frac{\partial u}{\partial t} + u \frac{\partial u}{\partial x} + v \frac{\partial u}{\partial y} = -\frac{1}{\rho} \frac{\partial p}{\partial x} + \nu \Delta^2 u \quad (3\text{-}258b)$$

y momentum:
$$\frac{\partial v}{\partial t} + u \frac{\partial v}{\partial x} + v \frac{\partial v}{\partial y} = -\frac{1}{\rho} \frac{\partial p}{\partial y} + \nu \Delta^2 v \quad (3\text{-}258c)$$

Energy:
$$\rho c_p \left(\frac{\partial T}{\partial t} + u \frac{\partial T}{\partial x} + v \frac{\partial T}{\partial y} \right) = k \Delta^2 T + \Phi \quad (3\text{-}258d)$$

subject to suitable boundary conditions. The latter three have the same mathematical form—elliptical in space, parabolic in time, with non-linear convective terms—and may be modeled in a single generic manner [e.g., Patankar (1980)]. Thus the variables (u, v, T) are all computed in the same manner in a CFD model. The *odd* thing about the system Eq. (3-258) is that the fourth relation, continuity, does *not* define the fourth variable, pressure. We develop this oddity in the next section.

Each of Eqs. (3-258b) to (3-258d) consists of a time derivative, convective terms, a diffusion (Laplacian) term, and some remaining low-order terms which could be called "sources." For example, in Eq. (3-258b), the source term is the pressure gradient $(-\partial p/\partial x)/\rho$, and the equation could be modeled implicitly as

$$\frac{u_{m,n}^{j+1} - u_{m,n}^{j}}{\Delta t} + \text{convection}$$

$$\approx \text{source} + \nu \left(\frac{u_{m+1,n}^{j+1} + u_{m-1,n}^{j+1} + u_{m,n+1}^{j+1} + u_{m,n-1}^{j+1} - 4u_{m,n}^{j+1}}{\Delta x^2} \right)$$

where we have assumed a square mesh, $\Delta x = \Delta y$. For an explicit method, the superscripts on the right would be j, not $j + 1$.

At low Reynolds numbers, the convection terms could be modeled as central differences:

$$u\frac{\partial u}{\partial x} + v\frac{\partial u}{\partial y} \approx u_{m,n}^{j}\left(\frac{u_{m+1,n}^{j+1} - u_{m-1,n}^{j+1}}{2\Delta x} \right) + v_{m,n}^{j}\left(\frac{u_{m,n+1}^{j+1} - u_{m,n-1}^{j+1}}{2\Delta y} \right)$$

This expression causes increasing error as the Reynolds number increases. It is also physically wrong, implying that convection is equally dependent upon both upstream and downstream velocities. If we think of the "cell" surrounding point (m, n), its convection is *received* from upstream and *transmitted* to the next cell downstream.

The remedy to convection instability is simple and satisfying: Model convection only with the upstream contribution. For example,

$$u\frac{\partial u}{\partial x} = \frac{\partial}{\partial x}\left(\frac{u^2}{2}\right) \approx \frac{u_{m,n}^{j}(u_{m+1,n}^{j+1} + u_{m,n}^{j+1}) - u_{m-1,n}^{j}(u_{m,n}^{j+1} + u_{m-1,n}^{j+1})}{4\Delta x}$$

$$\text{if } u_{m,n}^{j} > 0$$

$$\approx \frac{u_{m+1,n}^{j}(u_{m+1,n}^{j+1} + u_{m,n}^{j+1}) - u_{m,n}^{j}(u_{m,n}^{j+1} + u_{m-1,n}^{j+1})}{4\Delta x}$$

$$\text{if } u_{m,n}^{j} < 0$$

This stabilizing procedure is called *upwind*, or donor-cell, differencing. Other upwind schemes are discussed by Tannehill et al. (1997). An interesting—and popular—alternative is given by Patankar (1980), who develops the equations of motion in control-volume (integral) form and then finds an exact exponential solution for the convective terms. Exponential convection, or its power-law approximation, avoids the need to check the signs of the velocities but requires more execution time.

The pressure-gradient source terms in the momentum equation would normally be written as central differences. For example,

$$\frac{\partial p}{\partial x} \approx \frac{p^j_{m+1,n} - p^j_{m-1,n}}{2\Delta x} \qquad \frac{\partial p}{\partial y} \approx \frac{p^j_{m,n+1} - p^j_{m,n-1}}{2\Delta y}$$

However, this formulation leads to a vexing artifact: no net pressure force if the nodal pressures simply alternate. For example, the "checkerboard" pattern in Fig. 3-44, consisting of four sets of different pressure values alternating vertically and horizontally, contributes no net force to the momentum model. This pattern is obviously unrealistic and should be avoided. A similar paradox occurs in the continuity equations, i.e., central-difference models for $(\partial u/\partial x)$ and $(\partial v/\partial y)$ yield zero net fluxes on a checkboard pattern. Therefore Patankar (1980), in particular, recommends a staggered grid, with (u, v, p) evaluated at three different places in a cell. Such a grid leads to differences between *adjacent* grid points and thus avoids the possibility of checkerboarding.

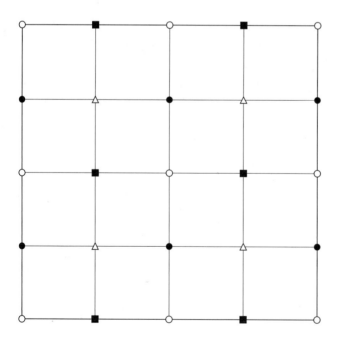

FIGURE 3-44
A checkerboard pressure pattern yields no net force yet is wildly unrealistic. Each symbol represents a different value of the pressure. Patankar (1980) recommends a staggered grid to eliminate this effect.

3-10.4.4 THE COMPUTATION OF FLUID PRESSURE.

Suppose that temperature is not a variable (isothermal flow). Then an FDM for Eqs. (3-258b) and (3-258c) yields the "next" values of the velocities $u_{m,n}$ and $v_{m,n}$. It remains to compute the "next" pressure, $p_{m,n}$. Somewhat to our surprise, the remaining relation, (3-258a), does not even *contain* the fluid pressure. It concerns only velocities; yet it is fundamental and must also be satisfied.

In the stream function-vorticity formulation, the pressure is broken out and satisfies its own Poisson relation, Eq. (3-256). With primitive variables, however, it is appropriate to *construct* a pressure equation, either artificially or by manipulation.

For plane flow, Roache (1976) recommends use of Eq. (3-256)—which uses continuity in its derivation—to compute the pressure. To avoid occasional instability or long convergence time, this relation has been modified to include an *artificial compressibility*:

$$\Delta^2 p = \mu \Delta^2 B - \rho \left[\frac{\partial B}{\partial t} + \frac{\partial^2(u^2)}{\partial x^2} + \frac{\partial^2(v^2)}{\partial y^2} + 2\frac{\partial^2(uv)}{\partial x \, \partial y} \right] \quad (3\text{-}259)$$

where $B = (\partial u/\partial x) + (\partial v/\partial y)$ is the *dilation* of the fluid and remains nonzero in the finite-difference approximation. Harlow and Welch (1965) demonstrate the stability of this type of computation. A similar relation is used for three-dimensional flow.

Other ideas for pressure computation are reviewed in the text by Tannehill et al. (1997). A popular method, called a "pressure-correction procedure," is given by Patankar (1980). The method presumes that initial guesses (u_0, v_0, p_0) are available, to be "corrected" by a convergent process to the actual (u, v, p):

$$u = u_0 + u' \qquad v = v_0 + v' \qquad p = p_0 + p'$$

After substitution in the momentum equations, the velocity corrections may be estimated by the relations

$$u' = -C \frac{\partial p'}{\partial x} \qquad v' = -C \frac{\partial p'}{\partial y} \quad (3\text{-}260)$$

where C is a coefficient related to the element area and the convective-acceleration model. Meanwhile, substitution in the continuity relation yields the following "pressure-correction" equation:

$$\Delta^2 p' = \frac{1}{C} \left(\frac{\partial u_0}{\partial x} + \frac{\partial v_0}{\partial y} \right) \quad (3\text{-}261)$$

As (u_0, v_0) approaches the correct velocity field (u, v), the right-hand side of Eq. (3-261) approaches zero; hence p' vanishes and p_0 is the desired pressure field.

Patankar (1980) calls this sequence the SIMPLE algorithm, meaning semi-implicit method for pressure-linked equations. The following steps constitute the algorithm:

1. Begin with the most recent pressure field p_0 as a guess.
2. Solve the discretized momentum equations for u_0 and v_0.
3. Solve for the corrections p' from a discretized Eq. (3-261).
4. Compute u' and v' from discretized Eqs. (3-260).
5. Correct the pressures and velocities everywhere.
6. Return to step 2 and repeat until converged.

In some cases, this procedure tends to overestimate the desired pressure correction, and convergence is slow. Patankar (1980) recommends an alternative SIMPLER algorithm, for SIMPLE-revised, which uses the pressure corrections only to change the velocity field. Pressures are then computed from a Poisson equation for total pressure. The SIMPLER method cuts computer time about one-third compared to SIMPLE.

There is no intrinsic difficulty in adding other variables (temperature, concentration, salinity) to a finite-difference model. These properties satisfy conservation laws of the same form as the Navier–Stokes equations and are modeled in the same manner.

3-10.4.5 SOME SUCCESSFUL COMPUTATIONS FROM THE LITERATURE. There have been thousands of publications in the literature of numerical analyses for various viscous flows. One of the earliest [Thom (1933)] hand-computed the still-popular problem of uniform crossflow past a circular cylinder. Figure 3-45 shows computer-generated streamlines and vorticity lines for $Re_D = 4$ [Keller and Takami (1966)] and for $Re_D = 40$ [Apelt (1961)].

The flow in Fig. 3-45a, for $Re_D = 4$, resembles an Oseen creeping-flow pattern, with slight asymmetry and streamlines slightly spread in the wake. At $Re_D = 40$, the asymmetry is striking, as vorticity lines are swept downstream by the strong convection. The Reynolds number can be thought of as the ratio of convection effects (ρU^2) to viscous effects $(\mu U/L)$. A standing eddy, whose size increases with Re, has formed behind the body. Even at $Re = 40$, the vorticity lines in front of the body have begun to concentrate in a narrow region, a portent of the "boundary layer" which becomes much thinner than the body at $Re = \mathcal{O}(1000)$. The calculations were performed by the stream function-vorticity $(\psi - \omega)$ formulation.

The standing eddy appears at about $Re = 7$ and grows linearly with Re to about $S/d = 2.5$ at $Re = 45$, as shown in Fig. 3-46 from the measurements of Taneda (1956). Above $Re = 60$, an unsteady Kármán vortex street forms, which can be suppressed up to $Re = 280$ by attaching a splitter plate aft of the cylinder [Grove et al. (1964)]. With a splitter plate, standing eddy growth continues to be linear, that is, $S/d \approx 0.065(Re - 7.0)$ up to $Re = 280$.

The computed wall shear stresses and surface pressures agree with experiments also, as does the total drag force. Figure 3-47 shows a compilation by Chang

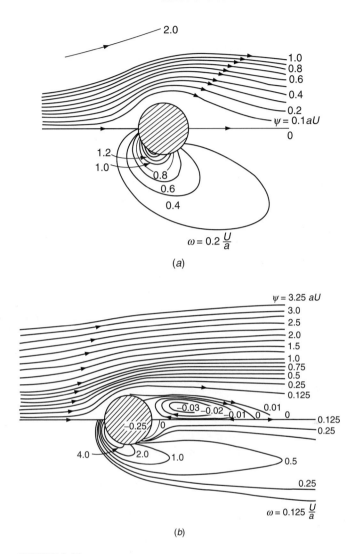

FIGURE 3-45
Numerical solutions for the flow past a circular cylinder: (a) streamlines (*upper half*) and lines of constant vorticity (*lower half*) in flow past a circular cylinder at $Re = 4$, calculated by Keller and Takami (1966); (b) streamlines (*upper half*) and lines of constant vorticity (*lower half*) in flow past a circular cylinder at $Re = 40$, calculated by Apelt (1961).

and Findlayson (1987) of numerical and experimental cylinder drag coefficients in the range $3 < Re < 100$. The results agree within experimental and numerical uncertainty. Our recommended curve fit, Eq. (3-229), is in fair agreement, being meant to fit well over a much wider range, $1E-4 < Re < 2E5$.

One can also compute the temperature field and heat transfer by adding a model of the energy equation to the CFD system. Chang and Findlayson (1987),

198 VISCOUS FLUID FLOW

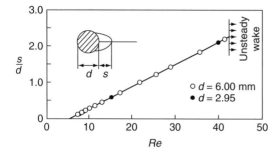

FIGURE 3-46
Observed lengths of the region of closed streamlines behind a circular cylinder. [*From Taneda (1956).*] Compare with Fig. 3-45.

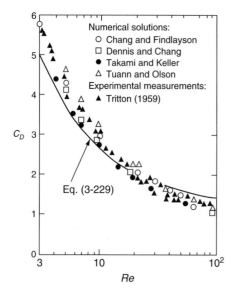

FIGURE 3-47
Comparison of experimental and numerical results for drag coefficient of a cylinder in crossflow. [*After Chang and Findlayson (1987).*]

for example, added heat transfer to their finite-element drag results in Fig. 3-47. Their numerical results, for $Re < 150$ and $0.01 \leq Pr \leq 1E4$, are in good agreement with our power-law correlation, Eq. (3-235).

3-10.4.6 FLOW PAST A LIQUID SPHERE. The related problem of flow past a solid sphere has been computed in many studies. The results are qualitatively similar to those of the cylinder in Fig. 3-46, with a standing eddy forming at $Re \approx 20$ and growing until $s/d \approx 1.1$ at $Re \approx 130$, above which the wake becomes unsteady.

A topic of current interest in multiphase flow of bubbles and droplets is flow past a *liquid* sphere of arbitrary viscosity ratio, $\lambda = \mu_{inner}/\mu_{outer}$. Analytically, only the creeping-flow solution of Rybczynski (1911) is known from Eq. (3-230). Numerical results, for $0.5 \leq Re \leq 1000$, were reported by Feng and Michaelides (2001) and summarized in the Freeman Scholar Lecture of Michaelides (2003). Their finite-difference method split the outer flow field into two matched layers and utilized 20,000 grid points. Figure 3-48 shows their computed streamlines and vorticity

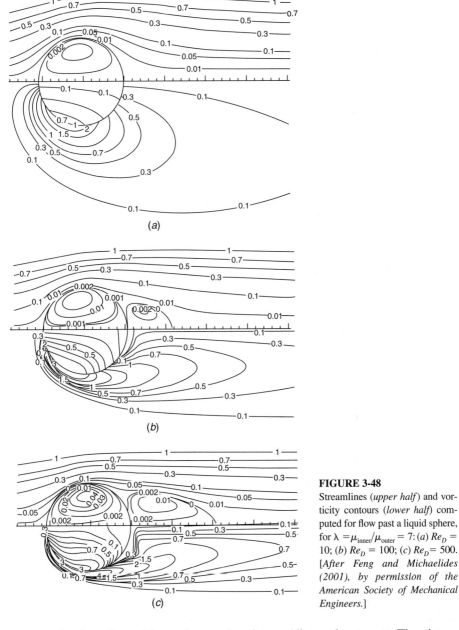

FIGURE 3-48
Streamlines (*upper half*) and vorticity contours (*lower half*) computed for flow past a liquid sphere, for $\lambda = \mu_{inner}/\mu_{outer} = 7$: (a) $Re_D = 10$; (b) $Re_D = 100$; (c) $Re_D = 500$. [*After Feng and Michaelides (2001), by permission of the American Society of Mechanical Engineers.*]

contours for three Reynolds numbers and an intermediate value $\lambda = 7$. The picture is similar to Fig. 3-45 for a cylinder. As *Re* increases, a standing eddy forms and lengthens, and the vorticity contours crowd together and are convected downstream.

Figure 3-49 shows the computed drag coefficients, based on outer fluid density, over the full viscosity range from $\lambda = 0$ (a bubble) to $\lambda = \infty$ (a solid sphere).

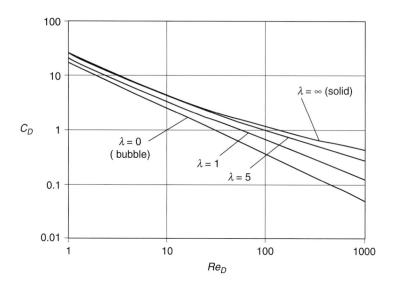

FIGURE 3-49
Drag coefficient of a liquid sphere for various values of the viscosity ratio $\lambda = \mu_{inner}/\mu_{outer}$. [*From calculations by Feng and Michaelides (2001), by permission of the American Society of Mechanical Engineers.*]

The bubble has a third less drag than a solid in the creeping-flow region, $C_D = 16/Re$, and at $Re = 1000$ has only one-ninth the drag of the solid. The solid curve in Fig. 3-49 has a simple and excellent curve fit

Solid sphere, $\lambda = \infty$, $Re \leq 1000$: $$C_D \approx \frac{24}{Re}\left(1 + \frac{Re^{2/3}}{6}\right) \qquad (3\text{-}262)$$

as recommended by Feng and Michaelides (2001) for $Re \leq 1000$. Note, however, that this formula is *not* accurate for $Re > 1000$.

Let us summarize some facts about these two examples, CFD for a cylinder and a sphere. They are steady *laminar* flows, with no unsteady shedding. They use reasonable grid sizes and could be extended to shedding situations with good results, if the unsteady downstream boundary conditions are properly handled. Laminar-flow modeling can thus serve as a "numerical laboratory" for a wide variety of geometries. The same accuracy and reliability cannot yet be assured for turbulence modeling at higher Reynolds numbers.

3-10.4.7 FLOW THROUGH AN ORIFICE. Internal flows, such as ducts and channels, have well-defined geometries of confined size which correspondingly limit the number of necessary grid points. Figure 3-50 shows the streamlines and vorticity contours computed by Mills (1968) for laminar flow through a circular orifice whose diameter ($D_0 = a$) is 50 percent of the pipe size $D = 2a$. The orifice Reynolds number is $Re_0 = 4aU_{av}/\nu$ where $U_{av} = Q/A$ is the average velocity in the pipe.

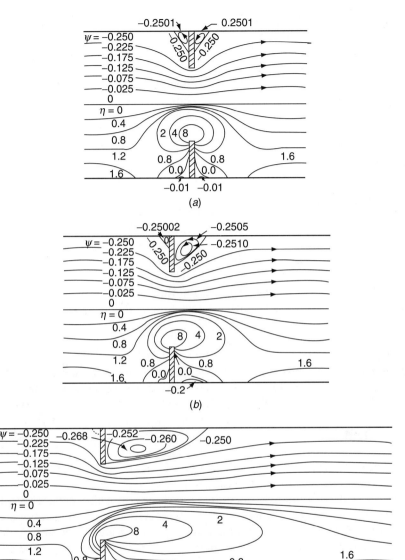

FIGURE 3-50
Numerical solutions for viscous flow through a circular orifice ($Re_0 = 4\,aU_{av}/\nu$). Streamlines and vorticity contours for (a) $Re_0 = 0$, $C_d = 0$; (b) $Re_0 = 10$, $C_d = 0.463$; (c) $Re_0 = 50$, $C_d = 0.690$.; [*From Mills (1968)*.]

The flow pattern for creeping motion, $Re_0 = 0$, in Fig. 3-50a is perfectly symmetrical, with small standing eddies in the corners. Detailed examination would reveal that such separated-flow corners possess an infinity of nested eddies, each successive eddy being an order of magnitude smaller. At $Re_0 = 10$, Fig. 3-50b, convection effects have caused a clear downstream asymmetry. In Fig. 3-50c, $Re_0 = 50$, the downstream eddy is elongated, and the flow through the orifice has the free-jet configuration characteristic of very high (laminar and turbulent) Reynolds numbers.

For the flows in Fig. 3-50, Mills (1968) computed the discharge coefficient, $C_d = Q/Q_{\text{ideal}}$, where Q_{ideal} is the flow rate computed for a one-dimensional inviscid approximation [White (2003, p. 416)]. For $Re_0 \leq 5$, $C_d \approx 0.15\sqrt{Re_0}$, and for $Re_0 > 5$, C_d rises slightly and levels off at the value of 0.7 expected for high Reynolds number thin-plate orifices. The computations are in excellent agreement with experiments.

As Re_0 increases beyond 50, orifice flow continues to resemble Fig. 3-50c, with discharge coefficients in the range of 0.6 to 0.7 expected for turbulent flow. One could continue the calculations with a turbulence model (Chap. 6). Since accurate discharge estimates are crucial for flowmeter applications, turbulent-flow correlations are provided, for example, by Scott et al. (1994) as a function of Re and $\beta = d_{\text{orifice}}/d_{\text{pipe}}$. Engineering formulas are provided in basic textbooks such as White (2003), Sec. 6.12. Recently a startling (to this writer) approach for orifice-data correlations was proposed by Morrison (2003), who replaced the Reynolds number with the *Euler* number, $Eu = 2\Delta p/(\rho U^2)$. This is quite a change from traditional practice.

3-10.4.8 CFD MODELING OF A MICROPUMP. There is now intense interest in microelectromechanical systems (MEMS), much of which involves flow at low Reynolds numbers in such systems. How can one pump a fluid through such tiny systems? Ordinary *rotodynamic* devices, such as compressors and centrifugal pumps, do not work at low Re. Positive-displacement pumps (PDPs) do work but involve complex fabrication and sealing problems. Sharatchandra et al. (1997) proposed, with testing and a CFD model, a simple rotating cylinder placed across a fluid between two plates. The cylinder picks up fluid due to the no-slip condition and drives it through the gap in the plates. Their CFD model was a finite-volume, two-dimensional computation using boundary-fitted coordinates and up to 8400 grid points.

Figure 3-51 shows the streamlines around their novel rotating-cylinder pump for $Re = 0.5$ as a function of the ratio s of plate spacing to cylinder diameter. The cylinder nearly touches the bottom plate. The quantity $\bar{u}$ is the dimensionless average velocity through the gap above the cylinder and is seen to be a maximum when $s \approx 1.5$. The gap is small, however, so the maximum volume flow occurs when $s \approx 2.0$. One can see in Fig. 3-51d that not much fluid is pumped when the gap is large. These CFD results are in good agreement with microflow experiments by the same group. For gas microflows, they investigated wall-slip effects and found no degradation in volume flow if the Knudsen number is less than 0.1. Finally, they investigated possible inertial effects (larger Reynolds number) and found no decrease in flow rate until $Re > 100$.

3-10.4.9 BOUNDARY-FITTING OF IRREGULAR GEOMETRIES. The CFD solutions illustrated in Figs. 3-45 to 3-51 have simple circular and/or rectangular geometries for which the equations of motion are well documented (App. B). If the

(a) $s = 1.25$ ($\bar{u} = 0.093$)

(b) $s = 1.5$ ($\bar{u} = 0.106$)

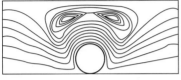

(c) $s = 2.5$ ($\bar{u} = 0.067$)

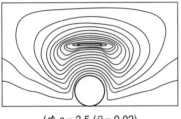

(d) $s = 3.5$ ($\bar{u} = 0.02$)

FIGURE 3-51
Streamlines computed for a rotating-cylinder micropump for $Re = 0.5$. Note the symmetry due to the low Reynolds number. The quantity s is the ratio of plate spacing to cylinder diameter, and $\bar{u}$ is the average velocity through the gap above the cylinder. [*After Sharatchandra et al. (1997), by permission of the American Society of Mechanical Engineers.*]

boundaries of the flow are irregular or asymmetrical, two approaches are suggested. Option 1 is to use a *finite-element* scheme, with triangular or quadrilateral elements that often fit irregular shapes very well. Details of the finite-element method are given in textbooks by, for example, Löhner (2001), Wrobel (2002), and Becker (2004), and commercial software is available.

Option 2 is to transform the coordinates and equations of motion to fit the irregular shapes. This is the method of *boundary-fitted coordinates*, popularized and now summarized by Thompson et al. (1998). First, one defines new (usually orthogonal) coordinates by solving the classical second-order partial differential equations that map the irregular region onto, say, a unit square. Numerical estimates of the coordinates' Jacobian are then used to transform the equations of motion so that conventional finite-difference schemes can be used on the unit square.

An example of a boundary-fitted solution is shown in Fig. 3-52, from Bramley and Sloan (1987). Laminar flow into a 45° half-angle Y branch is modeled by first defining orthogonal curvilinear coordinates for the region, using the method of Thompson et al. (1998). The resulting coordinates for the upper half of the flow

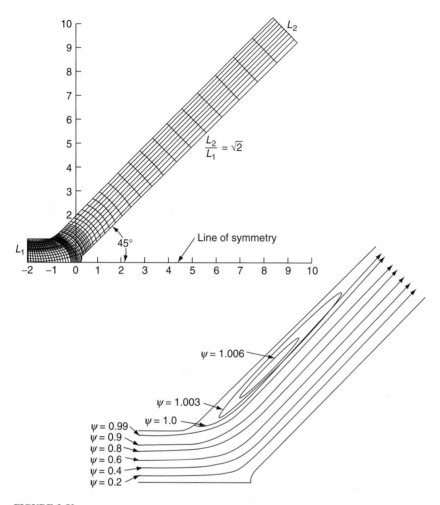

FIGURE 3-52
Computation of flow in a 45° Y branch: (a) boundary-fitted coordinates; (b) computed streamlines for $Re = 1000$. [*After Bramley and Sloan (1987)*.]

are shown in Fig. 3-52a. The transformed Navier–Stokes equations are then solved for the flow pattern, shown in Fig. 3-52b for $Re = 1000$. An elongated separation bubble, similar to those in Fig. 3-51, is formed on the outside (suction) side of the Y. The bubble size increases with Re, but not linearly: It is 6.5 inlet half-widths long at $Re = 1000$, rising to 7.8 at $Re = 2000$ and is nearly independent of the half-angle of the Y in the range of 30 to 60°. Though we are confident that such laminar-flow results are accurate, the same computation for turbulent flow would be uncertain because empirical turbulence models are very sensitive to geometry and flow conditions.

SUMMARY

It has been the intent of the chapter to give a fairly complete survey of the various types of exact or near-exact solutions presently known for incompressible Navier–Stokes equations. Where appropriate, solutions have included not only the velocity but also temperature and pressure distributions, and dimensionless variables have been emphasized. The various types of solutions discussed were

1. Couette flows with steadily moving surfaces
2. Poiseuille flow through ducts
3. Unsteady duct flows
4. Unsteady flows with moving boundaries
5. Asymptotic suction flows
6. Wind-driven Ekman flows
7. Similarity solutions: stagnation flow, rotating disk, wedge flow
8. Low Reynolds number (creeping) flows
9. Lubrication theory
10. Computational fluid dynamics (CFD)

This list seems rather substantial until we reflect that none of these types possess any degree of arbitrariness except the computer solutions, which themselves are limited by mesh size and Reynolds number considerations. As the Reynolds number increases arbitrarily, laminar flows become unstable, and no method exists for exact analysis of such problems. In many flows, whether laminar or turbulent, the condition of high Reynolds number makes certain terms in the Navier–Stokes equations become negligible. What remains is a far more exploitable field of analysis: boundary-layer theory for laminar (Chap. 4), transitional (Chap. 5), turbulent (Chap. 6), and compressible (Chap. 7) flows. For nonboundary-layer flows, the use of computational fluid dynamics (CFD) for both laminar and turbulent flow is an increasingly powerful tool.

PROBLEMS

3-1. Reconsider the problem of Couette flow between parallel plates, Fig. 3-1, for a power-law nonnewtonian fluid, $\tau_{xy} = K(du/dy)^n$, where $n \neq 1$. Assuming constant pressure and temperature, solve for the velocity distribution $u(y)$ between the plates of (a) $n < 1$ and (b) $n > 1$, and compare with the newtonian solution, Eq. (3-6). Comment on the results.

3-2. Consider the axial Couette flow of Fig. 3-3 with both cylinders moving. Find the velocity distribution $u(r)$ and plot it for (a) $U_1 = U_0$, (b) $U_1 = -U_0$, and (c) $U_1 = 2U_0$. Comment on the results.

3-3. Consider the axial Couette flow of Fig. 3-3 with the inner cylinder moving at speed U_0 and the outer cylinder fixed. Solve for the temperature distribution $T(r)$ in the fluid if the inner and outer cylinder walls are at temperatures T_0 and T_1, respectively.

3-4. A long thin rod of radius R is pulled axially at speed U through an infinite expanse of still fluid. Solve the Navier–Stokes equation for the velocity distribution $u(r)$ in the fluid and comment on a possible paradox.

3-5. A circular cylinder of radius R is rotating at steady angular rate ω in an infinite fluid of constant ρ and μ. Assuming purely circular streamlines, find the velocity and pressure distribution in the fluid and compare with the flow field of an inviscid "potential" vortex.

3-6. Assuming that the velocity distribution between rotating concentric cylinders is known from Eq. (3-22), find the pressure distribution $p(r)$ if the pressure is p_0 at the inner cylinder.

3-7. An open U tube of radius r_0 filled with a length L of a viscous fluid is displaced from rest and oscillates with amplitude $X(t)$, as shown in Fig. P3-7. Show with a one-dimensional integral analysis that the governing equation for $X(t)$ is

$$\frac{d^2X}{dt^2} + \frac{C_f}{r_0}\left|\frac{dX}{dt}\right|\frac{dX}{dt} + \frac{2g}{L}X = 0$$

where C_f is the wall friction coefficient. Find the natural frequency of oscillation and the time to damp to one-half amplitude for an assumed Poiseuille-type friction factor, $C_f = 16\mu/[\rho|dX/dt|d_0]$.

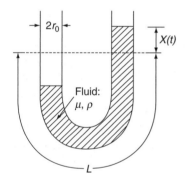

FIGURE P3-7

3-8. Air at 20°C and 1 atm is driven between two parallel plates 1 cm apart by an imposed pressure gradient (dp/dx) and by the upper plate moving at 20 cm/s. Find (a) the volume flow rate (in cm³/s per meter of width) if $dp/dx = -0.3$ Pa/m and (b) the value of (dp/dx) (in Pa/m) which causes the shear stress at the lower plate to be zero.

3-9. Derive the solution $u(y, z)$ for flow through an elliptical duct, Fig. 3-9, by solving Eq. (3-30). Begin with a guessed quadratic solution, $u = A + By^2 + Cz^2$, and work your way through to the exact solution.

3-10. Air at 20°C and approximately 1 atm flows at an average velocity of 1.7 m/s through a rectangular 1 × 4 cm duct. Estimate the pressure drop (in Pa/m) by (a) an exact calculation and (b) the hydraulic diameter approximation.

3-11. Consider a swirling motion superimposed upon a circular-duct flow by letting the velocity components take the form

$$v_r = 0 \qquad v_\theta = v_\theta(r, t) \qquad v_z = v_z(r)$$

Show that the axial flow v_z is unaffected by the swirl, so that an arbitrary v_θ can be added without changing the Poiseuille distribution.

3-12. It is desired to measure the viscosity of light lubricating oils ($\mu \approx 0.02$ to 0.1 Pa · s) by passing approximately 1 m³/h of fluid through an annulus of length 30 cm with inner and outer radii of 9 and 10 mm, respectively. Estimate the expected pressure drop through the device and an appropriate instrument for the pressure measurement.

3-13. Lubricating oil at 20°C [$\rho = 890$ kg/m³, $\mu = 0.8$ Pa · s, $k = 0.15$ W/(m · K), $c_p = 1800$ J/(kg · K)] is to be cooled by flowing at an average velocity of 2 m/s through a 3 cm diameter pipe whose walls are at 10°C. Estimate (a) the heat loss (in W/m²) at $x = 10$ cm and (b) the mean oil temperature at the pipe exit, $L = 2$ m. Comment on the results.

3-14. For plane polar coordinates with circular streamlines, show that the only nonzero vorticity component, $\omega = \omega_z(r)$, satisfies the equation

$$\frac{\partial \omega}{\partial t} = \nu \left(\frac{\partial^2 \omega}{\partial r^2} + \frac{1}{r} \frac{\partial \omega}{\partial r} \right)$$

Solve this equation for the decay of a line vortex initially concentrated at the origin with circulation Γ_0. Solve for $\omega(r, t)$ and show that

$$v_\theta = \frac{\Gamma_0}{2\pi r} \left[1 - \exp\left(-\frac{r^2}{4\nu t} \right) \right]$$

Sketch velocity profiles for a few representative times, including $t = 0$.

3-15. Consider a wide liquid film of constant thickness h flowing steadily due to gravity down an inclined plane at angle θ, as shown in Fig. P3-15. The atmosphere exerts constant pressure and negligible shear on the free surface. Show that the velocity distribution is given by

$$u = \frac{\rho g \sin \theta}{2\mu} y(2h - y)$$

and that the volume flow rate per unit width is $Q = \rho g h^3 \sin \theta / 3\mu$. Compare this result with flow between parallel plates, Eqs. (3-44) and (3-45).

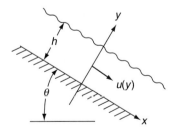

FIGURE P3-15

3-16. Consider a film of liquid draining at volume flow rate Q down the outside of a vertical rod of radius a, as shown in Fig. P3-16. Some distance down the rod, a fully developed region is reached where fluid shear balances gravity and the film thickness remains constant. Assuming incompressible laminar flow and negligible shear interaction with the atmosphere, find an expression for $v_z(r)$ and a relation between Q and film radius b.

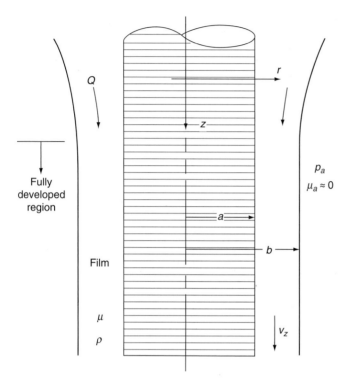

FIGURE P3-16

3-17. By extension of Prob. 3-15, consider a *double* layer of immiscible fluids 1 and 2, flowing steadily down an inclined plane, as in Fig. P3-17. The atmosphere exerts no shear stress on the surface and is at constant pressure. Find the laminar velocity distribution in the two layers.

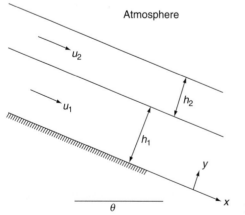

FIGURE P3-17

3-18. The surface of the draining liquid film in Prob. 3-15 will become wavy and unstable at a critical Reynolds number $Re^* = hu_{max}/\nu \approx \mathcal{O}(10)$ that depends upon slope θ, gravity g, density ρ, viscosity μ, and the surface tension $\mathcal{T}$. Rewrite this dependence in terms of two dimensionless parameters [Fulford (1964)]. For water at 20°C, what is the numerical value of the parameter which contains viscosity?

3-19. Derive Eq. (3-96) for start-up of flow in a circular pipe. Plot the instantaneous velocity profile for early times, $\nu t/r_0^2 = 0.005$ and 0.01.

3-20. Air at 20°C and 1 atm is at rest between two fixed parallel plates 2 cm apart. At time $t = 0$, the lower plate suddenly begins to move tangentially at 30 cm/s. Compute the air velocity in the center between plates after 2 s. When will the center velocity reach 14 cm/s?

3-21. A Couette pump consists of a rotating inner cylinder and a baffled entrance and exit, as shown in Fig. P3-21. Assuming zero circumferential pressure gradient and $(a - b) \ll a$, derive formulas for the volume flow and pumping power per unit depth. Illustrate for SAE 30 oil at 20°C, with $a = 10$ cm, $b = 9$ cm, and $\omega = 600$ rpm.

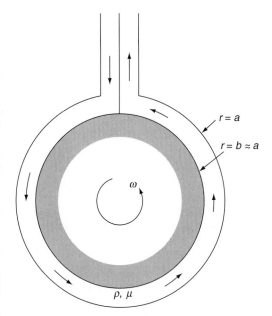

FIGURE P3-21

3-22. The Taylor vortex is defined by a purely circumferential flow

$$v_\theta = \text{const}\, \frac{r}{\nu t^2} \exp\left(-\frac{r^2}{\nu t}\right)$$

Determine whether this vortex is an exact solution of the incompressible Navier–Stokes equations with negligible gravity. Sketch a few instantaneous velocity profiles and compare to the Oseen vortex in Prob. 3-14.

3-23. Consider radial outflow between two parallel disks fed by symmetric entrance holes, as shown in Fig. P3-23. Assume that $v_z = v_\theta = 0$ and $v_r = f(r, z)$, with constant ρ and μ and $p = p(r)$ only. Neglect gravity and entrance effects at $r = 0$. Set up the appropriate differential equation and boundary conditions and solve as far as possible—numerical (e.g., Runge–Kutta) integration may be needed for a complete solution. Sketch the expected velocity profile shape.

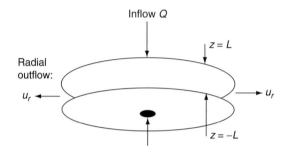

FIGURE P3-23

3-24. A long, uniformly porous cylinder of radius R exudes fluid at velocity U_0 into an unbounded fluid of constant ρ and μ. The pressure at the cylinder surface is p_0. Assuming purely radial outflow with negligible gravity, find the velocity and pressure distributions in the fluid.

3-25. Consider the problem of steady flow induced by a circular cylinder of radius r_0 rotating at surface vorticity ω_0 and having a wall-suction velocity $v_r(r = r_0) = -v_w = $ const. Set up the problem in polar coordinates assuming no circumferential variations $\partial/\partial\theta = 0$, and show that the vorticity in the fluid is given by

$$\omega = \frac{1}{r}\frac{\partial}{\partial r}(rv_\theta) = \omega_0 \left(\frac{r_0}{r}\right)^{Re}$$

where $Re = r_0 v_w/\nu$ is the wall-suction Reynolds number of the cylinder. Integrate this relation to find the velocity distribution $v_\theta(r)$ in the fluid and show that the character of the solution is quite different for the three cases of the wall Reynolds number Re less than, equal to, or greater than 2.0.

3-26. Consider laminar flow in a uniformly porous tube, similar to the porous-channel problem, Eqs. (3-130). Let the similarity variable be $\zeta = (r/r_0)^2$ and find a suitable expression for the axisymmetric stream function $\psi(x, \zeta)$, similar to Eq. (3-128) for the channel. Then show that the differential equation equivalent to (3-132) is

$$2\zeta g'''' + 4g''' + Re(g'g'' - gg''') = 0 \qquad \text{[Yuan and Finkelstein (1956)]}$$

where $Re = v_w r_0/\nu$ and $g = g(\zeta)$ only. This equation has no known solutions in the range $2.305 \leq Re \leq 9.105$ [White (1962a)]. Establish the proper boundary conditions on the function $g(\zeta)$.

3-27. The practical difficulty with the Ekman spiral solution, Eq. (3-144), is that it assumes laminar flow whereas the real ocean is turbulent. One approximate remedy is to replace kinematic viscosity ν everywhere by a (constant) turbulent or "eddy" viscosity correlated with wind shear and penetration depth using a suggestion by Clauser (1956):

$$\nu_{\text{turb}} \approx 0.04 D \left(\frac{\tau_0}{\rho}\right)^{1/2}$$

Repeat our text example, $V_{\text{wind}} = 6$ m/s over a 20°C air-water interface of 41°N latitude. Compute penetration depth D and surface velocity V_0.

3-28. Repeat the analysis of the Ekman flow, Sec. 3-7.2, for shallow water, that is, apply the bottom boundary condition. Eq. (3-143), at $z = -h$. Find the velocity components and show that the surface velocity is no longer at a 45° angle to the wind but rather satisfies the equation as follows for the surface angle θ:

$$\tan \theta = \frac{\sinh(2\pi h/D) - \sin(2\pi h/D)}{\sinh(2\pi h/D) + \sin(2\pi h/D)} \quad [\text{Ekman (1905)}]$$

Find the value of h/D for which $\theta = 20°$.

3-29. Air at 20°C and 1 atm flows at 1 m/s across a cylinder of diameter 2 cm and wall temperature 40°C. Using inviscid theory, Eq. (1-2), to establish the constant B, use the results of Sec. 3-8.1 to estimate (a) the shear layer thickness δ, (b) the wall shear stress, and (c) the heat-transfer rate q_w at the front stagnation point of the cylinder.

3-30. The opposite of von Kármán's rotating disk problem is Bödewadt's case [Rogers and Lance (1960)] of a fixed disk with a rotating outer stream, $v_\theta = r\omega$ as $z \to \infty$. Using the approach of Sec. 3-8.2, set up this problem and carry it out as far as possible, including numerical integration if a computer is available.

3-31. The rotating disk is sometimes called von Kármán's centrifugal pump, since it brings in fluid axially and throws it out radially. Consider one side of a 50 cm disk rotating at 1200 rpm in air at 20°C and 1 atm. Assuming laminar flow, compute (a) the flow rate, (b) the torque and power required, and (c) the maximum radial velocity at the disk edge.

3-32. Solve the Jeffery–Hamel wedge-flow relation, Eq. (3-195), for creeping flow, $Re = 0$ but $\alpha \neq 0$. Show that the proper solution is

$$f(\eta) = 1 + \frac{1}{2}\csc^2\alpha \left[\sin\left(\frac{\pi}{2} - 2\alpha\eta\right) - 1\right]$$

Show also that the constant $C = 4\alpha^2 \cot^2\alpha$ and sketch a few profiles. Show that backflow always occurs for $\alpha > 90°$.

3-33. In spherical polar coordinates, when the variations $\partial/\partial\phi$ vanish, an incompressible stream function $\psi(r, \theta)$ can be defined such that

$$u_r = \frac{\partial\psi/\partial\theta}{r^2 \sin\theta} \qquad u_\theta = -\frac{\partial\psi/\partial r}{r \sin\theta}$$

The particular stream function

$$\psi(r, \theta) = \frac{2\nu r \sin^2\theta}{1 + a - \cos\theta} \qquad a = \text{const}$$

is an exact solution of the Navier–Stokes equations and represents a round jet issuing from the origin. Sketch the streamlines in the upper two quadrants for a particular value of a between 0.001 and 0.1. (Various values could be distributed among a group.) Sketch the jet profile shape $u_r(1, \theta)$, and determine how the jet width δ (where $u_r = 0.01 u_{max}$) and jet mass flow vary with r.

3-34. A sphere of specific gravity 7.8 is dropped into oil of specific gravity 0.88 and viscosity $\mu = 0.15$ Pa·s. Estimate the terminal velocity of the sphere if its diameter is (a) 0.1 mm, (b) 1 mm, and (c) 10 mm. Which of these is a creeping motion?

3-35. Verify Eq. (3-230) for the drag of a liquid droplet by repeating the Stokes problem with boundary conditions of zero radial velocity, equal tangential velocity, and equal shear stress at the droplet surface (assumed to be a perfect sphere, $r = a$).

3-36. Repeat the analysis of radial outflow between parallel disks, as in Prob. 3-23, for creeping flow, i.e., negligible inertia. Find the velocity and pressure distributions for this case.

3-37. Analyze the problem of creeping flow between parallel disks of radius R and separation distance L. The lower disk ($z = 0$) is fixed and the upper disk ($z = L$) rotates at angular rate ω. Assuming that $v_\theta = rf(z)$, reduce the problem as far as possible and solve for the velocities.

3-38. Set up the method of *separation of variables* for finding $u(y, z)$ for a rectangular duct, Fig. 3-9 and Eq. (3-48), by analyzing Eq. (3-30). First note that the separation will not work until one defines a new variable $U = u - F$, where $\nabla^2 F = (1/\mu)(dp/dx)$, so that $\nabla^2 U = 0$. Then separate U into x and y parts and find the form of each part. Show how an infinite series would be required to satisfy the boundary conditions, but do not determine the series coefficients.

3-39. Repeat the analysis of Sec. 3-9.7.1 for a parabolically varying gap:

$$h = h_L + (h_0 - h_L)\left(1 - \frac{x}{L}\right)^2$$

Nondimensionalize and solve for the pressure distribution similar to Eq. (3-240). Numerical integration may be required. Plot the resulting pressures for various h_L/h_0 and compare with Fig. 3-42.

3-40. Set up Stokes' first problem of the impulsively started plane wall, Fig. 3-18a, for solution by the explicit numerical method of Eq. (3-247). Plot some typical velocity profiles and compare quantitatively with the exact solution, Eq. (3-107).

3-41. Repeat Prob. 3-40 using the implicit method of Eq. (3-249).

3-42. Set up Stokes' second problem of the (long-term) oscillating wall, Fig. 3-19a, for solution by the explicit numerical method of Eq. (3-247). Let the initial transient die out. Plot some typical velocity profiles and compare quantitatively with the exact solution, Eq. (3-111).

3-43. Repeat Prob. 3-42 using the implicit method of Eq. (3-249).

3-44. Develop an explicit numerical algorithm for the two-dimensional unsteady viscous diffusion relation, Eq. (3-250). Determine the appropriate stability limits on time step and mesh sizes.

3-45. Develop an implicit numerical algorithm for the two-dimensional unsteady viscous diffusion relation, Eq. (3-250). Comment on a possible solution procedure and possible instability.

3-46. Derive the two-dimensional Poisson relation for pressure, Eq. (3-256), assuming unsteady incompressible flow.

3-47. The solutions $f_n(r)$ for Eq. (3-76) are called *Graetz functions*, but they are not tabulated. Set up a numerical solution of Eq. (3-76), perhaps using the Runge–Kutta subroutine of App. C, and solve iteratively for the first three functions f_{1-2-3} and their eigenvalues λ_{1-2-3}. Compare with Table 3-1, but no fair using the table for your initial guesses.

3-48. Explain the physical significance of each of the boundary conditions Eq. (3-257) and show how they are derived.

3-49. Set up a finite-difference model for fully elliptical flow in the entrance between parallel plates a distance h apart. At the entrance ($x = 0$), assume uniform velocity U_0 and pressure p_0. Let the Reynolds number $U_0 h/\nu = 10$. Experiment with exit placement at ($x = 2h$), using a square mesh, $\Delta x = \Delta y = h/10$. Modify the mesh size and exit position if necessary for convergence or accuracy. Solve (iteratively) for the interior nodal velocities and pressures and compare the results with Fig. 4-33 and Table 4-7.

3-50. Extend Prob. 1-21, where we found only the Knudsen number, $Kn \approx 0.17$, by using all the data given there. (*a*) Find the Reynolds number and see if it is less than 2000 (laminar flow). (*b*) Estimate the required pressure gradient in Pa/m. (*c*) Estimate the flow rate in mm^3/s.

3-51. Starting from the axial momentum equation, derive Eqs. (3-40) for slip flow in tubes.

3-52. For the geometry of Fig. 3-1, assume a constant pressure gradient with both walls fixed. Solve continuity and x momentum for laminar *slip flow* between the plates. Find the velocity distribution and the volume flow rate per unit depth. Does the Knudsen number appear?

3-53. Hadjiconstantinou (2003) has updated a second-order slip theory by Cercignani (2000) to give new numerical coefficients for the wall-slip velocity, to be compared with Eq. (1-91):

$$u_w \approx 1.11\ell \left.\frac{\partial u}{\partial y}\right|_w - 0.61\ell^2 \left.\frac{\partial^2 u}{\partial y^2}\right|_w$$

Repeat Prob. 3-52 with this formulation to solve continuity and x momentum for laminar *slip flow* between the plates. Find the velocity distribution and the volume flow rate per unit depth. Does the Knudsen number appear?

3-54. Carry out the steps that lead, for Stokes sphere flow, from Eq. (3-211) to (3-213) by assuming a product solution $\psi(r, \theta) = f(r)g(\theta)$, separating the variables, and solving for f and g.

3-55. Hill and Power (1956) proved that the creeping-flow (Stokes) drag of a solid object is greater than the drag of any inscribed shape but less than the drag of any circumscribed shape. Verify this result for the spheroid of Fig. 3-36 by comparing it to inscribed and circumscribed spheres. Do the relative drag forces differ markedly or by only a few percent?

3-56. The elemental creeping-flow solution $\psi = r \ln r \sin \theta$ is called an *Oseenlet*. Is this a solution for plane flow, Eq. (3-206), or axisymmetric flow, Eq. (3-211), or both? How might Oseenlets be used in analysis of more complex creeping flows?

3-57. Using a numerical method such as the Runge–Kutta subroutine of App. C, solve the axisymmetric stagnation flow Eq. (3-165) for the function $F(\eta)$. Use an iterative scheme to determine the proper value of $F''(0)$.

3-58. The rotating disk of Fig. 3-28 acts rather like a centrifugal pump. Consider a disk 60 cm in diameter, rotating at 120 rev/min in air at 20°C and 1 atm. Is the flow at the disk tip laminar, transitional, or turbulent? Using Kármán's theory of Sec. 3-8.2, estimate the volume flow of air, in cm^3/s, pumped outward by one side of the disk.

CHAPTER
4

LAMINAR BOUNDARY LAYERS

4-1 INTRODUCTION

Several of the exact solutions considered in Chap. 3—notably moving-boundary flows, stagnation flow, the rotating disk, convergent-wedge flow, and the flat plate with asymptotic suction—have hinted strongly at boundary-layer behavior. That is, at large Reynolds numbers, the effects of viscosity become increasingly confined to narrow regions near solid walls. The digital-computer solutions of Sec. 3-10 also showed this tendency at large Re to sweep the vorticity downstream and leave the flow far from the walls essentially irrotational. Physically, this means that the rate of downstream convection is much larger than the rate of transverse viscous diffusion. Consider a flow at speed U past a thin body of length L. The time a fluid particle spends near the body is approximately L/U, while the time required for viscous effects to spread across the streamlines is of order $\sqrt{\nu L/U^3}$. Then the viscous region will be thin if the diffusion time is much shorter than the residence time:

$$\sqrt{\frac{\nu L}{U^3}} \ll \frac{L}{U}$$

or (4-1)

$$\sqrt{\frac{UL}{\nu}} = \sqrt{Re_L} \gg 1$$

Thus a *thin boundary layer* should exist if the flow Reynolds number Re_L is large. How large is difficult to specify and depends upon the geometry, the accuracy desired, and whether the quantity L is a streamwise or a transverse length scale. The boundary layer is likely to be laminar at first and then, as Re_L increases, undergoes transition to turbulence. However, a thin layer is not ensured on the rear or lee side of bluff bodies. As we saw for spheres and cylinders in Chap. 3, at moderate Re, a standing eddy forms behind the body. At higher Re, the eddies are shed and a broad wake develops. The thin-boundary-layer approximations to be discussed here do not apply in "separated" regions such as occur behind bluff bodies. Boundary-layer theory can, however, estimate where the point of separation occurs on the body. Analytical studies of separated flow are difficult and rare, but there is considerable experimentation and numerical modeling.

Even though standard boundary-layer analysis is not applicable to (1) low Reynolds numbers or (2) flow separation, it is a very important subject, especially for understanding viscous flows. The present chapter is a reasonable survey of the traditional approach. Low Reynolds numbers and separated flows can yield to analysis by more advanced boundary-layer techniques. For further study, there are monographs entirely devoted to boundary-layer theory: Cebeci and Cousteix (1998), Schlichting and Gersten (2000), Schetz (1992), Oleinik and Samokhin (1999), plus the classical monograph by Rosenhead (1963).

4-1.1 Flat-Plate Integral Analysis

Following an idea first put forth by Kármán (1921), we may gain much insight and a remarkable amount of quantitative information about boundary layers by making a broad-brush momentum analysis of the flow of a viscous fluid at high Re past a flat plate. The proposed flow is sketched in Fig. 4-1. The sharp edge of the plate is at $(x, y) = (0, 0)$. The fluid will shear against the plate due to the no-slip condition and cause a frictional drag force D, and the velocity distribution $u(y)$ at any particular downstream position x will show a smooth drop-off to zero at the wall, as sketched in the figure. To satisfy conservation of mass, the streamlines will be deflected away from the plate—but not too much, we hope, so that the fluid pressure remains

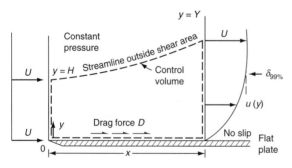

FIGURE 4-1
Definition of control volume for the analysis of flow past a flat plate.

approximately constant. We have labeled the shear-layer thickness as $\delta_{99\%}$; streamlines outside this shear layer will deflect an amount δ^* (the displacement thickness) which depends only upon x. Thus the streamline in the figure moves outward from $y = H$ at $x = 0$ at $y = Y = H + \delta^*$ at $x = x_1$. Finally, we assume that the velocity just upstream of the sharp edge is uniform and parallel, $u = U = \text{const}$; this would not be a good approximation at creeping-flow Reynolds numbers.

The control volume chosen for the analysis is enclosed by dashed lines in Fig. 4-1. This selection is not arbitrary but very clever. Since velocity distributions are known only at the inlet and exit, it is imperative that the other two sides on the control volume be streamlines, where no mass or momentum crosses. The lower side should be the wall itself; hence the drag force will be exposed. The upper side should be a streamline outside the shear layer, so that the viscous drag is zero along this line.

4-1.2 The Displacement Thickness

Conservation of mass for this control volume is obtained by applying Eq. (2-132) to this assumed steady flow:

$$\iint_{CS} \rho \mathbf{V} \cdot d\mathbf{A} = 0 = \int_0^Y \rho u \, dy - \int_0^H \rho U \, dy \qquad (4\text{-}2)$$

Assuming incompressible flow (constant density), this relation simplifies to

$$UH = \int_0^Y u \, dy = \int_0^Y (U + u - U) \, dy = UY + \int_0^Y (u - U) \, dy \qquad (4\text{-}3)$$

Rearranging this and noting that $Y = H + \delta^*$, we can express the mass-flow relation in the following simple manner:

$$U(Y - H) = U\delta^* = \int_0^Y (U - u) \, dy$$

or $\qquad\qquad\qquad\qquad\qquad\qquad\qquad\qquad\qquad\qquad\qquad\qquad\qquad\qquad$ (4-4)

$$\delta^* = \int_0^{Y \to \infty} \left(1 - \frac{u}{U}\right) dy$$

Equation (4-4) is the formal definition of the boundary-layer displacement thickness δ^* and holds true for any incompressible flow, whether laminar or turbulent, constant or variable pressure, constant or variable temperature. In other words, to define δ^* is simply to state conservation of mass in steady flow. Note that since y variations are integrated away, δ^* is a function only of x. Its exact value depends upon the distribution $u(y)$.

4-1.3 Momentum Thickness as Related to Flat-Plate Drag

Conservation of x momentum results by applying Eq. (2-133) to our control volume:

$$\sum F_x = -D = \iint_{CS} u(\rho \mathbf{V} \cdot d\mathbf{A}) = \int_0^Y u(\rho u \, dy) - \int_0^H U(\rho U \, dy)$$

or (4-5)

$$\text{Drag} = D = \rho U^2 H - \int_0^Y \rho u^2 \, dy$$

Again assuming constant ρ and introducing

$$H = \int_0^Y \frac{u}{U} \, dy$$

from Eq. (4-3), we obtain, per unit depth into the paper,

$$\text{Drag} = \rho \int_0^Y u(U - u) \, dy$$

or (4-6)

$$\frac{D}{\rho U^2} = \theta = \int_0^{Y \to \infty} \frac{u}{U}\left(1 - \frac{u}{U}\right) dy$$

Equation (4-6) is the defining relation for a second parameter, the momentum thickness θ, which, like δ^*, is clearly a function of x only. The definition holds true for any arbitrary incompressible boundary layer, but if we are not talking about flat-plate flow, θ is *not* equal to the drag divided by ρU^2. This means that momentum thickness, although interesting and useful in certain empirical correlations, is not as fundamental a quantity as δ^*.

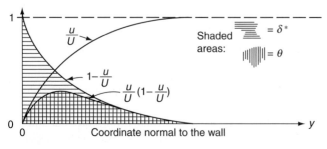

FIGURE 4-2
Momentum and displacement thicknesses.

The evaluation of δ^* and θ is illustrated graphically in Fig. 4-2. Clearly, δ^* is always the greater of the two. Their ratio, called the *shape factor*, is often used in boundary-layer analyses:

$$H = \frac{\delta^*}{\theta} > 1 \qquad (4\text{-}7)$$

If $\tau_w(x)$ is the wall shear stress on the plate, the total drag per unit width on one side of a plate of length L is the integral of the wall shear forces:

$$D = \int_0^L \tau_w(x)\, dx \qquad (4\text{-}8)$$

The wall shear and drag force may be nondimensionalized in the same manner as in Chap. 3:

Friction coefficient: $\quad C_f(x) = \dfrac{\tau_w(x)}{\frac{1}{2}\rho U^2}$

Drag coefficient: $\quad C_D = \dfrac{D}{\frac{1}{2}\rho U^2 L}$

$$(4\text{-}9)$$

These are general definitions. For a flat plate, by comparison with Eqs. (4-6) and (4-8), these take the specialized form

$$C_{f,\,\text{plate}} = \frac{d}{dx}[xC_D(x)] = 2\frac{d\theta}{dx}$$

$$C_{D,\,\text{plate}} = \frac{1}{L}\int_0^L C_f(x)\, dx = \frac{2\theta(L)}{L}$$

$$(4\text{-}10)$$

These formulas were derived by Kármán (1921) in his classic paper. They are valid for either laminar or turbulent flow. It is interesting that flat-plate friction and drag boil down entirely to the determination of the momentum thickness $\theta(x)$.

4-1.4 A Guessed Profile Yields Numerical Estimates

Assuming that the inlet profile is indeed a uniform velocity U, the previous relations are *exact* if one uses the correct profile $u(y)$. With the exact solution, though, we do not need integrals; we merely calculate the drag and wall shear and that is that. The beauty of Kármán's idea is that we can *guess* a reasonable form for $u(y)$ and get reasonable estimates because integration tends to wash out the positive and negative deviations of the assumed profile.

Let us therefore try a simple expression for $u(y)$ at the position x in Fig. 4-1. We will satisfy three physical conditions for the profile:

No slip at the wall: $\qquad u(0) = 0$

Smooth merging with the stream:

$$u(\delta) = U \qquad \left.\frac{\partial u}{\partial y}\right|_{y=\delta} = 0$$

Other constraints will be discussed later. We can satisfy these three conditions for laminar flow with a second-order polynomial:

$$u \approx U\left(\frac{2y}{\delta} - \frac{y^2}{\delta^2}\right) \tag{4-11}$$

Now insert this approximation into Eqs. (4-4) and (4-6) to estimate

$$\delta^* \approx \frac{\delta}{3} \qquad \theta \approx \frac{2\delta}{15} \qquad H \approx \frac{5}{2} \tag{4-12}$$

This is not enough. We must also relate wall shear to the assumed shape $u(y)$, which we do, following Kármán (1921), by differentiating the profile relation, Eq. (4-11):

$$\tau_w = \mu \left.\frac{\partial u}{\partial y}\right|_{y=0} \approx \frac{2\mu U}{\delta} \tag{4-13}$$

Substituting Eqs. (4-12) and (4-13) into the first of Eqs. (4-10), we obtain

$$C_f \approx \frac{2\mu U/\delta}{\frac{1}{2}\rho U^2} = 2\frac{d\theta}{dx} \approx 2\frac{d}{dx}\left(\frac{2\delta}{15}\right)$$

or

$$\delta\, d\delta \approx \frac{15\mu\, dx}{\rho U}$$

Integrate, assuming (for large Re) that the boundary layer begins at the leading edge, $\delta(0) = 0$. The result is

$$\delta^2 \approx \frac{30\mu x}{\rho U}$$

or

$$\frac{\delta}{x} \approx \frac{5.5}{\sqrt{Re_x}} \tag{4-14}$$

where $Re_x = \rho U x/\mu$. This is only 10 percent higher than the exact solution for laminar flat-plate flow given by Blasius (1908).

By substituting δ from Eq. (4-14) into Eqs. (4-12) and (4-13), we obtain the additional flat-plate estimates

$$\frac{\delta^*}{x} \approx \frac{1.83}{\sqrt{Re_x}}$$

$$\frac{\theta}{x} = C_f \approx \frac{0.73}{\sqrt{Re_x}} \tag{4-15}$$

which are also within 10 percent of the exact solutions. Finally, by integration of $C_f(x)$ from Eq. (4-10), we obtain the drag coefficient:

$$C_D \approx \frac{1.46}{\sqrt{Re_L}} = 2C_f(L) \qquad (4\text{-}16)$$

As Kármán (1921) pointed out, these results are quite easily obtained compared to a full-blown attack on the continuity and momentum partial differential equations.

All of the previous integral momentum estimates for flat-plate flow are collected in the following table and compared with the exact solution due to Blasius (1908). The error is 10 percent, which is typical of integral theories. The advantage is that, by integrating across the shear layer in the y direction, we are left only with an *ordinary* differential equation to solve in the x direction for parameters such as $\delta(x)$.

Parameter	u from Eq. (4-11)	Exact from Blasius (1908)	Error, %
$\frac{\theta}{x}\sqrt{Re_x}$	0.73	0.664	+10
$\frac{\delta^*}{x}\sqrt{Re_x}$	1.83	1.721	+6
$\frac{\delta_{99\%}}{x}\sqrt{Re_x}$	5.5	5.0	+10
$C_f\sqrt{Re_x}$	0.73	0.664	+10
$C_D\sqrt{Re_L}$	1.46	1.328	+10

4-1.5 Some Insight into Boundary-Layer Approximations

The essence of the "boundary-layer approximation" is that the shear layer is thin, $\delta \ll x$. From Eq. (4-14), we see that this is true if $Re_x \gg 1$. Other scale arguments also follow from the results. First, since the velocity ratio $v/u \leq d\delta^*/dx$, the outer streamline slope, it follows from Eq. (4-15) that $v \ll u$ if $Re_x \gg 1$. Second, by differentiating Eq. (4-11) we may determine that $\partial u/\partial x \ll \partial u/\partial y$ if, again, $Re_x \gg 1$, and a similar result holds for the derivatives of normal velocity v. In summary, a large Reynolds number creates the following strong inequalities:

$$Re_x \gg 1: \qquad \delta \ll x \qquad v \ll u \qquad \frac{\partial u}{\partial x} \ll \frac{\partial u}{\partial y} \qquad \frac{\partial v}{\partial x} \ll \frac{\partial v}{\partial y}$$

All of these approximations were used by Prandtl (1904) in deriving his celebrated boundary-layer equations.

4-1.6 Integral Analysis of the Energy Equation

In similar manner, the control-volume energy equation can be used to calculate approximate values for the mean and local heat transfer and the thickness of the

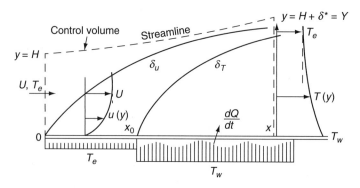

FIGURE 4-3
Sketch and control volume for the thermal boundary layer. Hot wall shown for convenience.

thermal boundary layer δ_T. The control volume chosen is shown in Fig. 4-3, where we propose that the rear portion of the plate, $x \geq x_0$, is heated (or cooled) to a temperature T_w different from the environment temperature T_e in the freestream (and on the first portion of the plate). Beginning at x_0, then, a thermal boundary layer will grow, assuming of course that the Reynolds number is large.

If we make sure that the upper streamline is outside the region of shear (the velocity boundary layer), there will be no shear work of significance at any boundary; therefore Eq. (2-139) will be valid and the shaft work will also be zero:

$$\frac{dQ}{dt} = \iint_{CS} \left(\frac{p}{\rho} + e + gy + \frac{1}{2}u^2 + \frac{1}{2}v^2\right) \rho \mathbf{V} \cdot d\mathbf{A} \qquad (4\text{-}17)$$

At large Re_x, $\delta^* \ll x$, so that $v \ll u$ and the change in potential energy will be very small. Hence we neglect the terms gy and $\frac{1}{2}v^2$. The quantity $p/\rho + e$ is the enthalpy h, and the equation reduces to integrals over the inlet and outlet of the stagnation enthalpy $h_0 = (h + \frac{1}{2}u^2)$:

$$\frac{dQ}{dt} \approx \int_0^Y \left(h + \frac{1}{2}u^2\right) \rho u \, dy - \int_0^H \left(h + \frac{1}{2}u^2\right) \rho u \, dy \qquad (4\text{-}18)$$

This expression applies even for compressible flow and will be studied again in Chap. 7. Here we assume constant density. Since the conditions at the inlet are uniform (h_e, T_e, U), we obtain

Inlet: $$\int_0^H \left(h + \frac{1}{2}u^2\right) \rho u \, dy = \left(h_e + \frac{1}{2}U^2\right) \rho U H$$

Now the mass-flow relation from Eq. (4-3) is still valid, so that we can eliminate H

from the energy relation, giving

$$\frac{dQ}{dt} = \rho \int_0^{Y \to \infty} \left[(h - h_e) - \frac{1}{2}(U^2 - u^2) \right] u\, dy \qquad (4\text{-}19)$$

This is the overall *integral-energy* equation. Since $dQ/dt = \int q_w\, dx$, it may be differentiated and written in *local* form:

$$q_w = \frac{d}{dx}\left[\int_0^\infty \rho u \left(h + \frac{1}{2} u^2 - h_e - \frac{1}{2} U^2 \right) dy \right] = -k \frac{\partial T}{\partial y}\bigg|_{y=0} \qquad (4\text{-}20)$$

For low-speed flow, it is customary to neglect kinetic energy $u^2/2$ (i.e., $Ec \ll 1$) and to use the approximation $h \approx c_p T$:

$$q_w \approx \frac{d}{dx}\left[\int_0^\infty \rho c_p u (T - T_e)\, dy \right] \qquad (4\text{-}21)$$

This may also be written in dimensionless form as a Stanton number, with a local length scale called the *enthalpy* thickness, δ_h:

$$C_h(x) = \frac{q_w}{\rho U c_p (T_w - T_e)} = \frac{d}{dx}(xC_H) = \frac{d\delta_h}{dx} \qquad (4\text{-}22)$$

where
$$\delta_h = \text{enthalpy thickness} \approx \int_0^\infty \frac{T - T_e}{T_w - T_e} \frac{u}{U}\, dy$$

Note the similarity to the local momentum integral relation, Eq. (4-11). Equation (4-22) is valid for either laminar or turbulent flow.

4-1.7 A Guessed Temperature Profile Yields the Heat-Transfer Rate

We can evaluate the heat transfer approximately for laminar flow by combining the velocity estimate Eq. (4-11) with a similar second-order polynomial temperature estimate:

$$T - T_e \approx (T_w - T_e)\left(1 - \frac{2y}{\delta_T} + \frac{y^2}{\delta_T^2} \right) \qquad (4\text{-}23)$$

Substitute these approximations, Eqs. (4-11) and (4-23), into the integral relation Eq. (4-20) to obtain

$$q_w \approx \frac{d}{dx}\left[\rho c_p U (T_w - T_e) \delta \left(\frac{\zeta^2}{6} - \frac{\zeta^3}{30} \right) \right] \approx \frac{2k(T_w - T_e)}{\zeta \delta} \qquad (4\text{-}24)$$

where $\zeta = \delta_T/\delta$ is the ratio of boundary-layer thicknesses. If we assume constant wall and stream temperatures, the term $(T_w - T_e)$ drops out. Also assume that ρ, μ, k, and c_p are constant.

Let us illustrate two solutions of Eq. (4-24). First, let the thermal boundary layer start from the plate leading edge, $\delta_T = 0$ at $x = 0$. Further assume, for convenience in the integration, that $\delta_T < \delta$. After introducing $\delta(x)$ from our momentum solution, Eq. (4-14), we obtain

$$\zeta^3 - \frac{\zeta^4}{5} \approx \frac{4\alpha}{5\nu} = \frac{0.8}{Pr} \tag{4-25}$$

An approximate solution to this polynomial, valid for Prandtl numbers not too far from unity, is

$$\zeta = \frac{\delta_T}{\delta} \approx Pr^{-1/3} \tag{4-26}$$

With ζ and δ known, the local Nusselt number may be obtained from Eq. (4-24):

$$Nu_x = \frac{q_w x}{k(T_w - T_e)} \approx 0.365 Re_x^{1/2} Pr^{1/3} \tag{4-27}$$

This is only 10 percent higher than the exact solution, for constant-T_w laminar flat-plate flow, given by Pohlhausen (1921)—the correct constant should be 0.332, not 0.365.

Second, consider an *unheated starting length*, where the wall temperature $T_w \neq T_e$ does not begin until $x = x_0$, as in Fig. 4-3. Here $\zeta(x)$ is not constant, and Eq. (4-24) must be solved:

$$\zeta\delta \frac{d}{dx}\left[\delta\left(\zeta^2 - \frac{\zeta^3}{5}\right)\right] \approx \frac{12\alpha}{U} \qquad \zeta = 0 \text{ at } x = x_0$$

This is readily solved numerically by, e.g., Subroutine RUNGE in App. C. To avoid this, a simple solution is possible if, noting that $\zeta < 1$ for $Pr > 1$, we neglect the term $(\zeta^3/5)$ and, concurrently, increase the constant 12 to 15. The differential equation then becomes

$$\zeta^3 + \frac{4}{3}x\frac{d}{dx}(\zeta^3) \approx \frac{1}{Pr}$$

or

$$\zeta^3 = \frac{1}{Pr} + \frac{C}{x^{3/4}}$$

To make $\zeta = 0$ at $x = x_0$, we require that the constant $C = -x_0^{3/4}/Pr$. The final solution may again be placed in the form of a Nusselt number for plate flow with an unheated starting length:

$$x > x_0: \qquad Nu_x \approx \frac{0.365 Re_x^{1/2} Pr^{1/3}}{\left[1 - (x_0/x)^{3/4}\right]^{1/3}} \tag{4-28}$$

Compare with Eq. (4-27) for $x_0 = 0$. If the constant 0.365 is reduced by 9 percent to 0.332, the result agrees with the formula recommended for flat-plate flow in most heat-transfer textbooks, e.g., Eq. (9-26) of Kays and Crawford (1993).

In summary, we see that simple integral techniques, using guessed velocity and temperature profiles, yield estimates for both mean and local friction and heat transfer with accuracies on the order of ± 10 percent. We shall return to these ideas in Sec. 4-6 for non-flat-plate flows.

4-2 LAMINAR-BOUNDARY-LAYER EQUATIONS

Integral analyses not only gave numbers but also yielded information about the sizes of various terms:

$$u, T, \text{ and } x = \mathcal{O}(\text{unity})$$
$$v \text{ and } y = \mathcal{O}(Re^{-1/2}) \tag{4-29}$$

These are solid estimates, not guesses, which can be used to derive the famous boundary-layer equations first propounded by Prandtl (1904). By redefining all variables in terms of these estimates, we can quickly spot which terms in the equations of motion are negligible if Re is large. Let us confine ourselves to two-dimensional incompressible flow, for which the relevant equations are

$$\frac{\partial u}{\partial x} + \frac{\partial v}{\partial y} = 0$$

$$\frac{\partial u}{\partial t} + u\frac{\partial u}{\partial x} + v\frac{\partial u}{\partial y} = -\frac{1}{\rho}\frac{\partial p}{\partial x} + g_x\beta(T - T_0) + \nu\left(\frac{\partial^2 u}{\partial x^2} + \frac{\partial^2 u}{\partial y^2}\right)$$

$$\frac{\partial v}{\partial t} + u\frac{\partial v}{\partial x} + v\frac{\partial v}{\partial y} = -\frac{1}{\rho}\frac{\partial p}{\partial y} + g_y\beta(T - T_0) + \nu\left(\frac{\partial^2 v}{\partial x^2} + \frac{\partial^2 v}{\partial y^2}\right) \tag{4-30}$$

$$\rho c_p\left(\frac{\partial T}{\partial t} + u\frac{\partial T}{\partial x} + v\frac{\partial T}{\partial y}\right) = k\left(\frac{\partial^2 T}{\partial x^2} + \frac{\partial^2 T}{\partial y^2}\right)$$
$$+ \mu\left[2\left(\frac{\partial u}{\partial x}\right)^2 + 2\left(\frac{\partial v}{\partial y}\right)^2 + \left(\frac{\partial u}{\partial y} + \frac{\partial v}{\partial x}\right)^2\right]$$

With our estimates from Eq. (4-29), we can now define dimensionless variables, all of which are sure to be of order unity if Re is large:

$$x^* = \frac{x}{L} \quad y^* = \frac{y}{L}\sqrt{Re} \quad t^* = \frac{tU}{L} \quad T^* = \frac{T - T_0}{T_w - T_0}$$
$$u^* = \frac{u}{U} \quad v^* = \frac{v}{U}\sqrt{Re} \quad p^* = \frac{p - p_0}{\rho U^2} \tag{4-31}$$

where U, L, p_0, and T_0 are reference values and $Re = UL/\nu$ is the Reynolds number characterizing the flow. Now substitute these variables into Eqs. (4-30) and take the limit of these equations as Re becomes very large. Many terms drop out, and we retain only

$$\frac{\partial u^*}{\partial x^*} + \frac{\partial v^*}{\partial y^*} = 0$$

$$\frac{\partial u^*}{\partial t^*} + u^*\frac{\partial u^*}{\partial x^*} + v^*\frac{\partial u^*}{\partial y^*} = -\frac{\partial p^*}{\partial x^*} + \frac{\beta(T_w - T_0)T^*}{Fr_x} + \frac{\partial^2 u^*}{\partial y^{*2}}$$

$$0 = \frac{\partial p^*}{\partial y^*} + \frac{\beta(T_w - T_0)}{Fr_y}T^* \qquad (4\text{-}32)$$

$$\frac{\partial T^*}{\partial t^*} + u^*\frac{\partial T^*}{\partial x^*} + v^*\frac{\partial T^*}{\partial y^*} = \frac{1}{Pr}\frac{\partial^2 T^*}{\partial y^{*2}} + Ec\left(\frac{\partial u^*}{\partial y^*}\right)^2$$

where $Ec = U^2/c_p(T_w - T_0)$ is the Eckert number and $Fr_i = U^2/Lg_i$ is the Froude number in each direction. All other terms had coefficients $1/Re$ or $(1/Re)^2$ and were dropped as asymptotically small. There are many things to notice about this simplified set of equations.

1. The continuity equation is unaffected by Reynolds number considerations.
2. The term $\partial^2 u/\partial x^2$ in the x-momentum equation has been neglected, and the buoyant term is also small if the Froude number is large (large U, small L), since $\beta(T_w - T_0)$ can only be of order unity at best (see Fig. 1-27). Therefore, except for small velocities and large sizes, free convection is negligible in the boundary-layer approximation.
3. The pressure gradient in the y direction is nearly zero, being affected only by a buoyant or stratification term that does not contribute to accelerations in the y direction. To all intents, then, in a boundary layer, the transverse pressure gradient in negligible.

$$\frac{\partial p}{\partial y} \approx 0 \qquad p = p(x) \text{ only} \qquad (4\text{-}33)$$

This splendid observation is due to Prandtl (1904), showing that pressure is a *known* variable in boundary-layer analysis, with $p(x)$ assumed to be *impressed* upon the boundary layer from without by an inviscid outer flow analysis. That is, the freestream outside the boundary layer, $U = U(x)$, where x is the coordinate parallel to the wall, is related to $p(x)$ by Bernoulli's theorem for incompressible flow. For steady flow, for example,

$$\frac{dp}{dx} = -\rho U\frac{dU}{dx} \qquad (4\text{-}34)$$

so that specifying $p(x)$ is equivalent to specifying $U(x)$ outside the boundary layer.

4. The energy equation shows that $\partial^2 T/\partial x^2$ is negligible and that only the $\partial u/\partial y$ portion of the dissipation is important. Further, it is clear that dissipation is entirely negligible if the Eckert number is small (small velocity and large temperature differences).
5. A very interesting observation is that the Reynolds number does not even appear in Eqs. (4-32). In this coordinate system, all thicknesses have been scaled to unit size, which is the only role of the Reynolds number in laminar flows.[†]
6. Perhaps most interesting, we note that all second derivatives with respect to x have been lost in the boundary-layer approximation. This has two consequences: (1) the equations are now *parabolic* instead of elliptic, so that x is now a marching variable and computer solutions are relatively easier than in Sec. 3-10; and (2) we have lost certain boundary conditions, notably those on v and x. The variable v has only one derivative left, $\partial v/\partial y$, with $\partial v/\partial x$ and the two second derivatives having been discarded. We now need only one condition on v at one y position. The obvious condition to retain is no slip: $v = 0$ at $y = 0$. We need not specify v at the outer edge of the layer. We have lost one condition each on u and T by discarding $\partial^2 u/\partial x^2$ and $\partial^2 T/\partial x^2$; therefore we disavow all knowledge of u and T at one x position, the best choice being the exit plane. The solutions will yield the correct values of u and T at the exit without our specifying them.

To sum up, the boundary-layer equations are far simpler than their parents, the Navier–Stokes equations. We can rewrite Eqs. (4-32) in the more common dimensional form for two-dimensional incompressible flow with constant properties

$$\frac{\partial u}{\partial x} + \frac{\partial v}{\partial y} = 0 \tag{4-35a}$$

$$\frac{\partial u}{\partial t} + u\frac{\partial u}{\partial x} + v\frac{\partial u}{\partial y} \approx \left(\frac{\partial U}{\partial t} + U\frac{\partial U}{\partial x}\right) + g_x \beta(T - T_0) + \nu\frac{\partial^2 u}{\partial y^2} \tag{4-35b}$$

$$\rho c_p \left(\frac{\partial T}{\partial t} + u\frac{\partial T}{\partial x} + v\frac{\partial T}{\partial y}\right) \approx k\frac{\partial^2 T}{\partial y^2} + \mu\left(\frac{\partial u}{\partial y}\right)^2 \tag{4-35c}$$

where $U = U(x, t)$, assumed known, denotes the freestream velocity just outside the boundary layer. The boundary conditions are

No slip: $u(x, 0, t) = v(x, 0, t) = 0 \quad T(x, 0, t) = T_w(x, t)$

Inlet condition: $u(x_0, y, t), v(x_0, y, t),$ and $T(x_0, y, t)$ known

Patching to the outer layer: $u(x, \infty, t) \rightarrow U(x, t)$ (4-36)

$T(x, \infty, t) \rightarrow T_e(x, t)$

Initial condition: $u(x, y, 0), v(x, y, 0),$ and $T(x, y, 0)$ known

---

[†]In turbulent flows (Sec. 6-4), the Reynolds number remains as a parameter because the turbulent inertia terms cannot be scaled by the square root of Re.

These equations, developed by Prandtl (1904), approximate the flow of a viscous fluid at high Reynolds numbers. They may be solved, at least numerically, for any practical distribution of stream velocity and temperature and of wall temperature. Some important solutions will be detailed here—for additional laminar boundary layers, see the text by Rosenhead (1963).

The limitations of the boundary-layer equations are

1. The Reynolds number must be large.
2. If the outer flow is decelerating ($dU/dx < 0, dp/dx > 0$), a point may be reached where wall shear approaches zero, the *separation point*. Beyond this point, the boundary-layer approximations are not accurate.
3. At some large $Re_x = \mathcal{O}(10^6)$, the laminar solutions become unstable and transition to turbulence occurs. The thin-layer approximations still hold then for the *turbulent* boundary layer.

The problem of separated flow is still frustrating because boundary-layer approximations fail. However, some newer techniques, to be discussed, have made headway on separated-flow analysis.

Here we present only two-dimensional incompressible-boundary-layer relations. They may be readily extended to three-dimensional and to compressible-flow conditions.

4-2.1 Orthogonal Curvilinear Coordinates

At first glance, Eqs. (4-35) seem to be valid only for a Cartesian system (x, y). However, they are also valid for flow along the curved wall shown in Fig. 4-4, subject only to the requirement that the boundary-layer thickness δ be much smaller than the radius of curvature $\mathcal{R}$ of the wall. The exact boundary-layer equations in such a curvilinear coordinate system were given by Tollmien (1931). After suitable order-of-magnitude assumptions, we find that the chief difference between the curvilinear and the Cartesian equations lies in the pressure gradient normal to the wall, which is no longer negligibly small for a curved body:

$$\frac{\partial p}{\partial y} \approx \frac{\rho u^2}{\mathcal{R}} \tag{4-37}$$

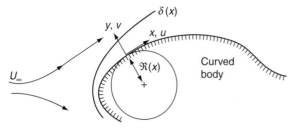

FIGURE 4-4
Boundary-layer flow over a curved body shape.

The gradient is of order unity. But if we integrate this from $y = 0$ to $y = \delta$, assuming, say, a linear distribution $u = Uy/\delta$, we obtain

$$p(\delta) - p(0) \approx \frac{\rho U^2 \delta}{3\mathcal{R}}$$

or (4-38)

$$\Delta p^* \approx \frac{\delta}{3\mathcal{R}}$$

which is negligibly small if $\delta \ll \mathcal{R}$. Therefore Eqs. (4-35) are valid for general curved-wall flows as long as the boundary-layer thickness is small compared to the wall radius of curvature. This would not be true at a sharp corner, but sharp corners invite immediate flow separation and are thus to be avoided.

4-2.2 General Remarks about Flow Separation

Before we attempt actual solutions, we can spot flow-separation effects from the boundary-layer equations themselves. If we apply the momentum equation at the wall, where $u = v = 0$, we find that

$$\left.\frac{\partial^2 u}{\partial y^2}\right|_{y=0} = \frac{1}{\mu}\frac{dp}{dx} \qquad (4\text{-}39)$$

Thus the wall curvature has the sign of the pressure gradient, whereas further out the profile must have *negative* curvature when it merges with the freestream. Profile curvature is an indicator of possible boundary-layer *separation*. Three examples are shown in Fig. 4-5a. For negative (favorable) pressure gradient, the curvature is negative throughout, and no flow separation can occur. For zero gradient, e.g., flat-plate flow, the curvature is zero at the wall and negative further out; there is no separation. For positive (adverse) gradient, the curvature changes sign and the profile is

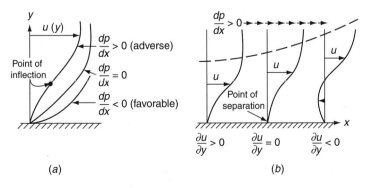

FIGURE 4-5
Geometric effects due to pressure gradient: (*a*) types of profile; (*b*) persistent adverse gradient.

S-shaped. The increasing downstream pressure slows down the wall flow and can make it go backward—flow separation.

Figure 4-5b illustrates the separation process. A persistent adverse gradient ($dp/dx > 0$) makes the profile more and more S-shaped, reducing the wall shear to zero (the separation point) and then causing backflow, while the boundary layer becomes much thicker. Laminar flows have poor resistance to adverse gradients and separate easily. Turbulent boundary layers can resist separation longer—at the expense of increased wall friction and heat transfer.

4-2.3 Shear Stress in a Boundary Layer

After the boundary-layer equations have been solved, a major interest is shear stress, which in two-dimensional flow would be given by

$$\tau = \mu\left(\frac{\partial u}{\partial y} + \frac{\partial v}{\partial x}\right)$$

But the order-of-magnitude analysis shows that the term $\partial v/\partial x$ is two orders smaller than $\partial u/\partial y$. Hence it is quite proper in boundary-layer theory to calculate the shear stress from the approximation

$$\tau \approx \mu\frac{\partial u}{\partial y} \qquad \tau_w \approx \mu\left(\frac{\partial u}{\partial y}\right)_{y=0} \tag{4-40}$$

The friction drag would of course be the integrated value of this local shear stress. The relation also shows that the approximate point of separation in Fig. 4-5b is the point of zero shear stress, i.e., the point where τ_w passes from positive to negative.

4-3 SIMILARITY SOLUTIONS FOR STEADY TWO-DIMENSIONAL FLOW

If we neglect buoyancy in Eqs. (4-35), the velocity is uncoupled from the temperature, and we can confine our attention temporarily to the momentum and continuity equations for steady flow:

$$\frac{\partial u}{\partial x} + \frac{\partial v}{\partial y} = 0 \tag{4-41a}$$

$$u\frac{\partial u}{\partial x} + v\frac{\partial u}{\partial y} = U\frac{dU}{dx} + \nu\frac{\partial^2 u}{\partial y^2} \tag{4-41b}$$

subject to $u(x, 0) = v(x, 0) = 0 \qquad u(x, \infty) = U(x)$

This is a system of partial differential equations of the parabolic type which can be solved with finite-difference techniques by marching downstream in the x direction. However, before trying that idea in Sec. 4-5, we study here the possibility of reducing these equations to *ordinary* differential equations by similarity transformation. Similarity solutions for the full Navier–Stokes equations were discussed in Sec. 3-8, and one of those successes, stagnation flow, is also a solution of the boundary-layer equations, as we shall see. Whole books are devoted to similarity techniques, e.g., Hansen (1964) and Sachdev (2000). We shall discuss here three similar boundary layers: the flat plate, the Falkner–Skan wedge flows, and the converging channel (sink flow). In each case the two-dimensional partial differential equations of continuity and momentum can be reduced to a single ordinary differential equation.

4-3.1 The Blasius Solution for Flat-Plate Flow

Ludwig Prandtl's first student, H. Blasius (1908), found a celebrated solution for laminar-boundary-layer flow past a flat plate, as in Fig. 4-1. If the displacement thickness is small ($Re \gg 1$), then $U = $ const and $dU/dx = 0$ in Eq. (4-41). There is a leading edge, $(x, y) = (0, 0)$, but no characteristic length "L." Therefore the local velocity profiles must all have the same dimensionless shape, $u/U = fcn(y/\delta)$. Since, from our integral analysis, Eq. (4-14), $\delta = \text{const}(\nu x/U)^{1/2}$, the appropriate dimensionless *similarity* variable should be

$$\eta = y\sqrt{\frac{U}{2\nu x}} \qquad (4\text{-}42)$$

The factor "2" is not necessary—Blasius himself did not use it—but it will avoid another factor "2" in the final differential equation.

The stream function of the flow, $\psi = \int u\, dy |_{x=\text{const}}$, should increase as δ, or $x^{1/2}$, and has the following nondimensional form:

$$\psi = \sqrt{2\nu U x}\, f(\eta) \qquad (4\text{-}43)$$

where f is a function to be determined. Note another factor "2" introduced for convenience. From the definition of stream function,

$$u = \frac{\partial \psi}{\partial y} = U f'(\eta)$$

$$v = -\frac{\partial \psi}{\partial x} = \sqrt{\frac{\nu U}{2x}} (\eta f' - f) \qquad (4\text{-}44)$$

where the prime denotes differentiation with respect to η. We see that, as expected, u is of order U while v is of smaller order, $U/\sqrt{Re_x}$.

Substitution of u and v from Eqs. (4-44) into the boundary-layer momentum relation (4-41) yields, after considerable manipulation useful as a student exercise, the following differential equation

$$f''' + f f'' = 0 \qquad (4\text{-}45)$$

Referring to Eqs. (4-44), the no-slip conditions $u(x, 0) = v(x, 0) = 0$, and the freestream-merge condition, $u(x, \infty) = U$, convert to

$$f'(0) = f(0) = 0 \qquad f'(\infty) = 1 \qquad (4\text{-}46)$$

Equation (4-45) is the celebrated non-linear *Blasius equation* for flat-plate flow. Note its similar appearance to the stagnation-flow relation, Eq. (3-151), whose extra terms are due to the (favorable) pressure gradient.

The Blasius equation has never yielded to exact analytic solution. Blasius himself (1908) gave matching inner and outer series solutions. Many other methods of attack are chronicled in the text by Rosenhead (1963). One method led to a general boundary-layer technique, now outdated, outlined in the text by Meksyn (1961). With the advent of the personal computer, it is now a simple matter to program Eqs. (4-45) and (4-46) using Subroutine RUNGE from App. C. Let $f'' = Y_1$, $f' = Y_2$, and $f = Y_3$. Then the proper FORTRAN relations are

$$F(1) = -Y(1) * Y(3)$$
$$F(2) = Y(1) \quad \quad (4\text{-}47)$$
$$F(3) = Y(2)$$

The problem is to find the correct value $f''(0) = Y_1\,(\eta = 0)$ which will make $f' = Y_2$ approach 1.0 as η approaches infinity. By a simple asymptotic analysis, we find that "infinity" is approximately $\eta = 10$, and we can run off a few solutions for various values of $f''(0)$, all of which behave beautifully. From these guesses, we can interpolate to find the value $f''(0)$ which makes $f'(\infty) \approx Y_2(10.0) \approx 1.0$. The accepted value, correct to six significant figures, is

$$f''(0) = 0.469600 \ldots \quad (4\text{-}48)$$

The complete numerical solution for the Blasius equation is given in Table 4-1. From this we can find all the flow parameters of interest to flat-plate flow.

Figure 4-6a shows a plot of the Blasius functions f, f', and f'', and Fig. 4-6b compares the profile $f' = u/U$ with the experiments of Liepmann (1943). The agreement is excellent, and we may regard this first test of the boundary-layer approximation as a success.

Other properties follow from the table. We note that $f' = 0.99$ at $\eta \approx 3.5$. Thus we have a value of the 99 percent boundary-layer thickness

$$\delta_{99\%} \approx 3.5\sqrt{\frac{2\nu x}{U}}$$

or $\quad\quad\quad\quad\quad\quad\quad\quad\quad\quad\quad\quad\quad\quad\quad\quad\quad\quad\quad\quad\quad\quad\quad\quad$ (4-49)

$$\frac{\delta_{99\%}}{x} \approx \frac{5.0}{\sqrt{Re_x}}$$

The momentum and displacement thickness are related to integrals of f' through their definitions from Eqs. (4-4) and (4-6):

$$\delta^* \sqrt{\frac{U}{2\nu x}} = \int_0^\infty (1 - f')\,d\eta = \lim_{\eta \to \infty}(\eta - f) = 1.21678$$

or $\quad\quad\quad\quad\quad\quad\quad\quad\quad\quad\quad\quad\quad\quad\quad\quad\quad\quad\quad\quad\quad\quad\quad\quad$ (4-50)

$$\frac{\delta^*}{x} = \frac{1.7208}{\sqrt{Re_x}}$$

TABLE 4-1
Numerical solution of the Blasius flat-plate relation, Eq. (4-45)

η	$f(\eta)$	$f'(\eta)$	$f''(\eta)$
0.0	0.0	0.0	0.46960
0.1	0.00235	0.04696	0.46956
0.2	0.00939	0.09391	0.46931
0.3	0.02113	0.14081	0.46861
0.4	0.03755	0.18761	0.46725
0.5	0.05864	0.23423	0.46503
0.6	0.08439	0.28058	0.46173
0.7	0.11474	0.32653	0.45718
0.8	0.14967	0.37196	0.45119
0.9	0.18911	0.41672	0.44363
1.0	0.23299	0.46063	0.43438
1.1	0.28121	0.50354	0.42337
1.2	0.33366	0.54525	0.41057
1.3	0.39021	0.58559	0.39598
1.4	0.45072	0.62439	0.37969
1.5	0.51503	0.66147	0.36180
1.6	0.58296	0.69670	0.34249
1.7	0.65430	0.72993	0.32195
1.8	0.72887	0.76106	0.30045
1.9	0.80644	0.79000	0.27825
2.0	0.88680	0.81669	0.25567
2.2	1.05495	0.86330	0.21058
2.4	1.23153	0.90107	0.16756
2.6	1.41482	0.93060	0.12861
2.8	1.60328	0.95288	0.09511
3.0	1.79557	0.96905	0.06771
3.2	1.99058	0.98037	0.04637
3.4	2.18747	0.98797	0.03054
3.6	2.38559	0.99289	0.01933
3.8	2.58450	0.99594	0.01176
4.0	2.78388	0.99777	0.00687
4.2	2.98355	0.99882	0.00386
4.4	3.18338	0.99940	0.00208
4.6	3.38329	0.99970	0.00108
4.8	3.58325	0.99986	0.00054
5.0	3.78323	0.99994	0.00026
5.2	3.98322	0.999971	0.000119
5.4	4.18322	0.999988	0.000052
5.6	4.38322	0.999995	0.000022
5.8	4.58322	0.999998	0.000009
6.0	4.78322	0.999999	0.000003

234 VISCOUS FLUID FLOW

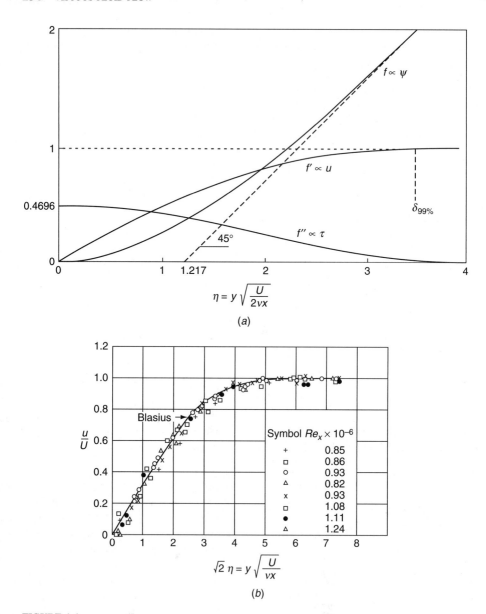

FIGURE 4-6
The Blasius solution for the flat-plate boundary layer: (a) numerical solution of Eq. (4-45); (b) comparison of $f' = u/U$ with experiments by Liepmann (1943).

The numerical value 1.21678 is shown in Fig. 4-6a to be the η intercept of a 45° line which is asymptotic to the curve $f(\eta)$ at large η. Since f is proportional to the stream function, the dashed 45° line represents an inviscid stream function displaced a dimensionless amount 1.21678 away from the plate. Similarly, we calculate

$$\theta\sqrt{\frac{U}{2\nu x}} = \int_0^\infty f'(1 - f')\,d\eta = f''(0) = 0.4696$$

or (4-51)

$$\frac{\theta}{x} = \frac{0.664}{\sqrt{Re_x}}$$

We can also calculate the wall shear stress

$$\tau_w = \mu \left.\frac{\partial u}{\partial y}\right|_w = \frac{\mu U f''(0)}{\sqrt{2\nu x/U}}$$

or (4-52)

$$C_f = \frac{2\tau_w}{\rho U^2} = \frac{0.664}{\sqrt{Re_x}} = \frac{\theta}{x}$$

Finally, the integrated drag coefficient on a plate of length L is

$$C_D(L) = \frac{1}{L}\int_0^L C_f\,dx = 2C_f(L) = \frac{1.328}{\sqrt{Re_L}} \quad (4\text{-}53)$$

This is the drag on one side of the plate. Experimental values of the drag coefficient of a plate in the range $1 \leq Re_L \leq 2000$ are shown in Fig. 4-7 and compared with

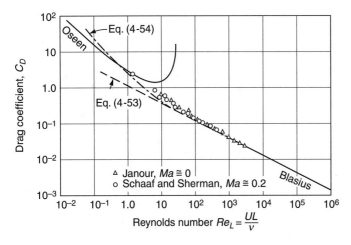

FIGURE 4-7
Theoretical and experimental drag of a flat plate.

Eq. (4-53) and also with the Oseen relation discussed in Chap. 3, Eq. (3-226). We see that the Blasius relation is accurate for $Re_L \geq 1000$, and the Oseen theory is valid for $Re_L \leq 1$.

The intermediate region $1 < Re_L < 1000$ has been the subject of many analytical and numerical studies. It may be fit reasonably well by the correction factor given in a perturbation theory by Imai (1957):

$$C_D \approx \frac{1.328}{\sqrt{Re_L}} + \frac{2.3}{Re_L} \qquad (4\text{-}54)$$

This relation is also plotted in Fig. 4-7 and can be seen to fall slightly lower than the data.

Numerical solutions of the full Navier–Stokes equations for $0.1 < Re_L < 1000$ by Dennis and Dunwoody (1966) and by Brauer and Sucker (1976) verify that the drag is higher than predicted by boundary-layer theory. Figure 4-8 shows the computations of Dennis and Dunwoody (1966) for (a) the local friction coefficient and (b) the local surface pressure. For lower Reynolds numbers, there are marked effects at both the leading and trailing edges, and the boundary-layer approximation is not realized until $Re_L \geq 1000$.

It is interesting that the normal velocity v is not zero at the edge of the boundary layer. From Eq. (4-44), we compute, as $\eta \to \infty$,

$$v(x, \infty) = \frac{0.8604 U}{\sqrt{Re_x}} \qquad (4\text{-}55)$$

There is a slight upwelling of the flow because of displacement of the outer stream. Panton (1996, Sec. 20.7) contains a good discussion of this effect. In favorable gradients, there can be downwelling toward the wall.

4-3.2 Flat-Plate Heat Transfer for Constant Wall Temperature

If T_w and T_e are constant, the temperature profiles also satisfy similarity relations. By analogy with the stagnation-flow relation, Eq. (3-167), we define a dimensionless temperature difference

$$\Theta(\eta) = \frac{T - T_e}{T_w - T_e} \qquad (4\text{-}56)$$

Assuming that u and v are known from the Blasius solution, we substitute Eq. (4-56) into the boundary-layer energy Eq. (4-35c), *neglecting dissipation*, that is, $Ec \ll 1$. The result is

$$\Theta'' + Pr f(\eta)\Theta' = 0 \qquad (4\text{-}57)$$

subject to $\qquad \Theta(0) = 1 \qquad \Theta(\infty) = 0$

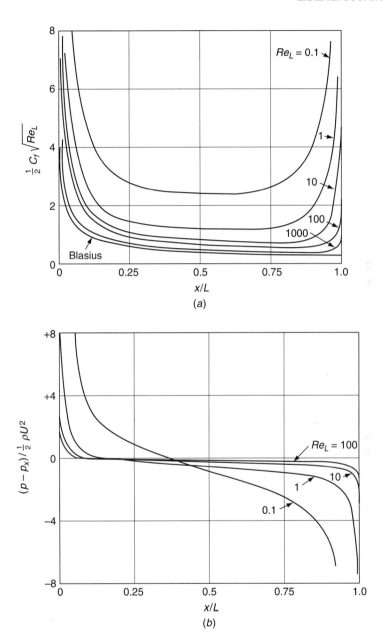

FIGURE 4-8
Numerical solution of the full Navier–Stokes equations for flat-plate flow at moderate Reynolds numbers: (*a*) local friction coefficient; (*b*) local surface pressure. [*After Dennis and Dunwoody* (*1966*).]

238 VISCOUS FLUID FLOW

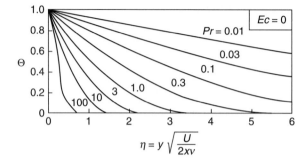

FIGURE 4-9
Flat-plate temperature profiles for zero dissipation.

This is identical to the stagnation-flow relation Eq. (3-168) with the boundary conditions reversed. The solution is, by analogy,

$$\Theta = \frac{\int_\eta^\infty d\eta \, \exp\left(-Pr \int_0^\eta f \, ds\right)}{\int_0^\infty d\eta \, \exp\left(-Pr \int_0^\eta f \, ds\right)} \tag{4-58}$$

This result was given by Pohlhausen (1921). The temperature profiles computed from Eq. (4-58) are shown in Fig. 4-9 for various Prandtl numbers. Since Pr is a ratio of viscous to conduction effects, the higher the Prandtl number, the thinner the thermal boundary layer. The thickness ratio may be approximated by

$$\frac{\delta_T}{\delta} \approx Pr^{-0.4} \tag{4-59}$$

The wall heat transfer may be computed from Fourier's law:

$$q_w = -k \left. \frac{\partial T}{\partial y} \right|_{y=0} = -k(T_w - T_e)\Theta'(0)\sqrt{\frac{U}{2\nu x}}$$

or, in dimensionless form,

$$Nu_x = \frac{-\Theta'(0)}{\sqrt{2}} Re_x^{1/2} \tag{4-60}$$

As seen in Fig. 4-9, the wall temperature slope $\Theta'(0)$ is a function of the Prandtl number and from Eq. (4-56) is given by

$$-\frac{1}{\Theta'(0)} = \int_0^\infty d\eta \, \exp\left(-Pr \int_0^\eta f \, ds\right) \tag{4-61}$$

By exact analogy with the stagnation-flow relation Eq. (3-172), this parameter may be evaluated by adding two FORTRAN statements to our Blasius analysis, Eq. (4-45):

$$F(4) = Y(3)$$

$$F(5) = EXP(-PR * Y(4))$$

The output $1/Y_5(\infty)$ is the desired wall slope $|\Theta'(0)|$. Some values of this slope may be tabulated as follows:

Pr	$-\Theta'(0)/\sqrt{2}$	Pr	$-\Theta'(0)/\sqrt{2}$
0.001	0.0173	10	0.7281
0.01	0.0516	100	1.572
0.1	0.1400	1000	3.387
1.0	0.3321	10,000	7.297

For Prandtl numbers, in the range $0.1 < Pr < 10{,}000$, a good curve fit to these tabulated values is $-\Theta'(0)/\sqrt{2} \approx 0.332 Pr^{1/3}$. Therefore Eq. (4-60) for low-speed laminar flat-plate heat transfer may be written as the familiar approximate relation

$$Nu_x \approx 0.332\, Re_x^{1/2} Pr^{1/3} \qquad (4\text{-}62)$$

This has the same form as our energy-integral result, Eq. (4-27). It is the accepted engineering approximation but predicts too high for liquid metals, $Pr < 0.1$.

The solution of the flat-plate energy equation when dissipation is *not* neglected is very interesting and leads to the concept of the recovery factor and the adiabatic-wall temperature. However, these concepts are realized in practice only by high-speed (compressible) boundary layers. Therefore discussion will be postponed until Chap. 7.

4-3.3 The Falkner–Skan Wedge Flows

The most famous family of boundary-layer similarity solutions was discovered by Falkner and Skan (1931) and later calculated numerically by Hartree (1937). Rather than merely "anticipating" the solution as we did with the Blasius problem, let us outline a search for similarity.

To begin with, we can eliminate v from the momentum equation by using the no-slip condition to solve continuity for v

$$v = -\frac{\partial}{\partial x}\int_0^y u\, dy \qquad (4\text{-}63)$$

the integration being carried out at constant x. The problem now reduces to a single integrodifferential equation for u alone:

$$u\frac{\partial u}{\partial x} - \frac{\partial u}{\partial y}\left(\frac{\partial}{\partial x}\int_0^y u\, dy\right) = U\frac{dU}{dx} + \nu\frac{\partial^2 u}{\partial y^2} \qquad (4\text{-}64)$$

We now inquire: Is there any possibility of combining x and y into a single variable $\eta(x, y)$ such that the above equation becomes an ordinary differential equation in a function of η only? If so, we have achieved similarity, but only for certain special cases of the freestream velocity distribution $U(x)$.

Let us generalize the Blasius solution to variable freestream velocity:

$$u(x, y) = U(x) f'(\eta) \qquad (4\text{-}65)$$

where $\eta = \eta(x, y)$ is dimensionless but is *not* the Blasius variable from Eq. (4-42). With careful chain-rule differentiation, we can substitute this in the momentum Eq. (4-64). For example, if $u = Uf'$, then

$$\frac{\partial u}{\partial x} = U'f' + Uf''\frac{\partial \eta}{\partial x}$$
$$\frac{\partial u}{\partial y} = Uf''\frac{\partial \eta}{\partial y}$$
(4-66)

and likewise for the second derivatives. An especially appealing form is linear in y, so that $(\partial^2 \eta/\partial y^2)$ vanishes:

$$\eta = yg(x) \tag{4-67}$$

Substitute Eq. (4-67) into the momentum relation, Eq. (4-64), evaluating the integral by Leibnitz' rule, using integration by parts. The result may be rearranged as follows:

$$f''' = ff''\left(\frac{Ug'}{\nu g^3}\right) + (f'^2 - ff'' - 1)\left(\frac{dU/dx}{\nu g^2}\right) \tag{4-68}$$

Similarity is achieved if each of the two coefficients in this relation is such that all x's disappear, leaving only constants. For Eq. (4-68), this implies that $\ln(U)$ is proportional to $\ln(g)$.

Falkner and Skan (1931) found that similarity was achieved by the variable $\eta = Cyx^a$, which is consistent with a power-law freestream velocity distribution:

$$U(x) = Kx^m \qquad m = 2a + 1 \tag{4-69}$$

The exponent m may be termed the Falkner–Skan *power-law parameter*. The constant C must make η dimensionless but is otherwise arbitrary. The best choice is $C^2 = K(1 + m)/2\nu$, which is consistent with its limiting case for $m = 0$, the Blasius variable from Eq. (4-42). Thus we choose

$$\eta = y\sqrt{\frac{m+1}{2}\frac{U(x)}{\nu x}} \tag{4-70}$$

Substituting this particular C into g_2 and g_3 in Eq. (4-67) gives the most common form of the Falkner–Skan equation for similar flows:

$$f''' + ff'' + \beta(1 - f'^2) = 0$$
(4-71)

where
$$\beta = \frac{2m}{1 + m}$$

The boundary conditions are exactly the same as for the flat plate:

$$f(0) = f'(0) = 0 \qquad f'(\infty) = 1 \tag{4-72}$$

The parameter β is a measure of the pressure gradient dp/dx. If β is positive, the pressure gradient is negative or favorable, and negative β denotes an unfavorable positive pressure gradient. Naturally, $\beta = 0$ denotes the flat plate. A priori, we expect the case $m = -1$ ($\beta = \pm\infty$) to be a trouble spot, but it is not: This case (U inversely proportional to x) is handled by a different choice of the constant C.

4-3.3.1 INVISCID FLOW PAST WEDGES AND CORNERS. The Falkner–Skan solution illustrates both favorable and adverse pressure gradients and is a realistic engineering flow pattern. The power-law freestream, $U = Kx^m$, is the exact solution to inviscid flow past a wedge or corner shape. In plane polar coordinates, Laplace's equation for the stream function $\psi(r, \theta)$ is

$$\frac{1}{r}\frac{\partial}{\partial r}\left(r\frac{\partial \psi}{\partial r}\right) + \frac{1}{r^2}\frac{\partial^2 \psi}{\partial \theta^2} = 0$$

As shown in texts covering potential flow, e.g., White (2003), p. 546, an exact solution is

$$\psi(r, \theta) = \text{const } r^{m+1} \sin[(m + 1)\theta] \tag{4-73}$$

This formula has certain radial streamlines that can be interpreted as the "walls" of a wedge or a corner, as in Fig. 4-10, depending upon the value of $\beta = 2m/(m + 1)$. The velocity along these walls has the form $U = Kx^m$ and represents the freestream driving the boundary layer on the wall, with $x = 0$ at the tip of the wedge. We may list the following cases:

$-2 \leq \beta \leq 0, -\frac{1}{2} \leq m \leq 0$: flow around an expansion corner of turning angle $\beta\pi/2$
$\beta = 0, m = 0$: the flat plate
$0 \leq \beta \leq +2, 0 \leq m \leq \infty$: flow against a wedge of half-angle $\beta\pi/2$
$\beta = 1, m = 1$: the plane stagnation point (180° wedge)
$\beta = +4, m = -2$: doublet flow near a plane wall
$\beta = +5, m = -\frac{5}{3}$: doublet flow near a 90° corner
$\beta = +\infty, m = -1$: flow toward a point sink [the boundary-layer version of the Jeffery–Hamel flow in a convergent wedge (Sec. 3-8)]

These are *similar* flows, i.e., for a given β the velocity profiles all look alike when scaled by $U(x)$ and $\delta(x)$. They may also be used, with modest success, to predict the behavior of nonsimilar flows.

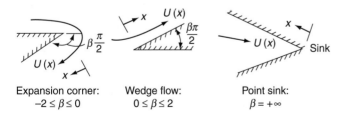

Expansion corner: $-2 \leq \beta \leq 0$
Wedge flow: $0 \leq \beta \leq 2$
Point sink: $\beta = +\infty$

FIGURE 4-10
Some examples of Falkner–Skan potential flows.

Equation (4-71) obviously yields to a digital-computer solution simply by adding the term $\beta(1 - f'^2)$ to the FORTRAN relations of Eq. (4-47) for the Blasius equation:

$$F(1) = -Y(1) * Y(3) - BETA * (1. - Y(2) * Y(2))$$
$$F(2) = Y(1) \qquad (4\text{-}74)$$
$$F(3) = Y(2)$$

As before, we select β and try to find the proper value of f_0'' that makes $f'(\infty)$ asymptotically approach unity.

Although we could easily attack Eq. (4-71) with our personal computers, in fact, the Falkner–Skan solutions have been well tabulated and charted. Extensive results are given in the text by Evans (1968). The most important results are plotted in Fig. 4-11, spanning the range from the stagnation point ($\beta = m = 1$) down through flat-plate flow ($\beta = m = 0$) to the separation point ($\beta = -0.19884$, $m = -0.09043$). Figure 4-11a shows the velocity profiles $u/U = f'$, which grow thicker as β decreases and, for $\beta < 0$, become S-shaped as in Fig. 4-5b and separate ($\tau_w = 0$) for $\beta = -0.19884$. Separation corresponds to an expansion angle in Fig. 4-10 of only 18°.

Figure 4-11b shows the shear-stress profiles $f''(\eta)$. Note that shear, in accelerated (favorable) flows, falls away from the wall value but instead rises from the wall in decelerated (adverse) flows. This is a consequence of the momentum-equation condition $\partial \tau / \partial y |_{\text{wall}} \, dp/dx$ and is true also in turbulent boundary layers. Note in passing that the solution for $\beta = \frac{1}{2}, m = \frac{1}{3}$ corresponds to the axisymmetric stagnation flow from Sec. 3-8.1.3.

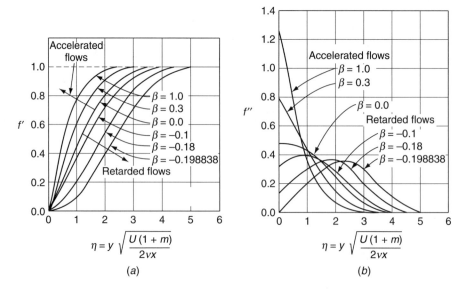

FIGURE 4-11
(a) Velocity profiles and (b) shear-stress profiles for the Falkner–Skan equation.

TABLE 4-2
Numerical values of the streamwise velocity $f'(\eta)$ for Falkner–Skan similarity flows

η	β f_0'' η^* θ^*	−0.19884 0.0 2.35885 0.58544	−0.18 0.12864 1.87157 0.56771	0.0 0.46960 1.21678 0.46960	0.3 0.77476 0.91099 0.38574	1.0 1.23259 0.64790 0.29235	2.0 1.68722 0.49743 0.23079	10.0 3.67523 0.24077 0.11523
0.0		0.0	0.0	0.0	0.0	0.0	0.0	0.0
0.1		0.00099	0.01376	0.04696	0.07597	0.11826	0.15876	0.31843
0.2		0.00398	0.02933	0.09391	0.14894	0.22661	0.29794	0.54730
0.3		0.00895	0.04668	0.14081	0.21886	0.32524	0.41854	0.70496
0.4		0.01591	0.06582	0.18761	0.28569	0.41446	0.52190	0.81043
0.5		0.02485	0.08673	0.23423	0.34938	0.49465	0.60964	0.87954
0.6		0.03578	0.10937	0.28058	0.40988	0.56628	0.68343	0.92414
0.7		0.04868	0.13373	0.32653	0.46713	0.62986	0.74496	0.95259
0.8		0.06355	0.15975	0.37196	0.52107	0.68594	0.79587	0.97057
0.9		0.08038	0.18737	0.41672	0.57167	0.73508	0.83767	0.98185
1.0		0.09913	0.21651	0.46063	0.61890	0.77787	0.87172	0.98888
1.2		0.14232	0.27899	0.54525	0.70322	0.84667	0.92142	0.99591
1.4		0.19274	0.34622	0.62439	0.77425	0.89681	0.95308	0.99856
1.6		0.24982	0.41691	0.69670	0.83254	0.93235	0.97269	0.99957
1.8		0.31271	0.48946	0.76106	0.87906	0.95683	0.98452	0.99998
2.0		0.38026	0.56205	0.81669	0.91509	0.97322	0.99146	0.99999
2.2		0.45097	0.63269	0.86330	0.94211	0.98385	0.99542	
2.4		0.52308	0.69942	0.90107	0.96173	0.99055	0.99761	
2.6		0.59460	0.76048	0.93060	0.97548	0.99463	0.99879	
2.8		0.66348	0.81449	0.95288	0.98480	0.99705	0.99940	
3.0		0.72776	0.86061	0.96905	0.99088	0.99842	0.99972	
3.2		0.78578	0.89853	0.98037	0.99471	0.99919	0.99987	
3.4		0.83635	0.92854	0.98797	0.99704	0.99959	0.99995	
3.6		0.87882	0.95138	0.99289	0.99840	0.99980	0.99998	
3.8		0.91315	0.96805	0.99594	0.99916	0.99991	0.99999	
4.0		0.93982	0.97975	0.99777	0.99958	0.99996		
4.5		0.97940	0.99449	0.99957	0.99994	0.99999		
5.0		0.99439	0.99997	0.99994	0.99999			

Table 4-2 lists the values of $u/U = f'(\eta)$ for a variety of solutions. Also tabulated are the proper initial conditions $f''(0)$, plus the dimensionless displacement and momentum thicknesses:

$$\eta^* = \int_0^\infty (1 - f')\, d\eta = (\eta - f)|_{\eta \to \infty}$$

$$\theta^* = \int_0^\infty f'(1 - f')\, d\eta = \frac{f_0'' - \beta \eta^*}{1 + \beta} \tag{4-75}$$

The ratio of these two, the shape factor H, will be especially useful in some approximate theories to be discussed.

In keeping with our constant reminders that the Navier–Stokes equations are nonunique, the boundary-layer equations also show multiple solutions. Stewartson (1954) pointed out the following pathology of the Falkner–Skan equation for negative β:

1. For $-0.19884 \leq \beta \leq 0$, there are (at least) two solutions of Eq. (4-71) for any given β, one of which is of the type shown in Fig. 4-11a and the second of which always shows a backflow at the wall. The two types of solution are identical at $\beta = -0.19884$ but are entirely different at $\beta = 0$.
2. For $\beta < -0.19884$, a multiple (probably an infinity) of solutions to Eq. (4-71) exists for any given value of the wall gradient f_0''. For example, Fig. 4-12b shows a family of separating profiles ($f_0'' = 0$) calculated by Libby and Liu (1967);

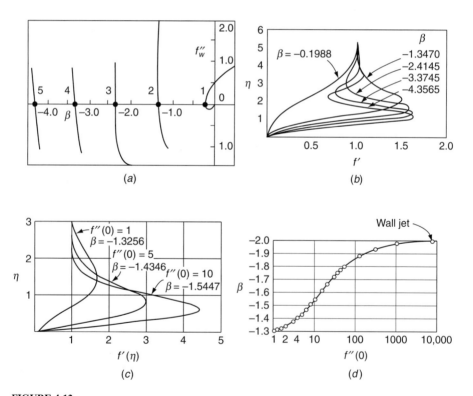

FIGURE 4-12
The multiplicity of Falkner–Skan solutions for negative β: (a) five branches of solutions for negative β, as found by Libby and Liu (1967); (b) five separating profiles corresponding to the five heavily marked intercepts in (a); (c) three overshoot profiles without backflow, calculated moving up along branch 2 [*After Steinheuer (1968)*]; (d) solution sets along upper branch 2 as calculated by Steinheuer (1968), showing asymptotic approach to the wall jet solution found by Glauert (1956).

presumably there are even more of this same family; for $\beta \approx -5.3, -6.3$, etc., Stewartson (1954) has proved that all solutions in this range of β must show velocity overshoot, that is, f' greater than 1.0, at some point in the boundary layer.

The gist of these remarks is shown in Fig. 4-12a, which gives a polar plot of β vs. f_0'' for the first five branches of solutions for negative β found by Libby and Liu (1967). Figure 4-12c shows some profiles for positive f_0'' along branch 2 as computed by Steinheuer (1968); these, like the separating profiles, show overshoot but no backflow. Libby and Liu (1967) liken these profiles to streamwise blowing into a moving stream—a wall jet. Indeed, as $f''(0)$ becomes very large along branch 2 of Fig. 4-12a, β approaches -2.0 and, as pointed out by Steinheuer (1968), the velocity profile becomes identical with the pure laminar-wall-jet solution found by Glauert (1956). This effect is shown in Figs. 4-12c and d.

4-3.4 Heat Transfer for the Falkner–Skan Flows

If we neglect dissipation and assume constant wall and stream temperature, the flat-plate analysis of Sec. 4-3.2 holds in exactly the same form:

$$\Theta'' + Prf(\eta, \beta)\Theta' = 0 \qquad (4\text{-}76)$$

where $\Theta = (T - T_e)/(T_w - T_e)$ as before. Here $f(\eta, \beta)$ is the Falkner–Skan stream function—the output Y_3 from Eq. (4-74)—defined as

$$\psi(x, y) = f(\eta)\sqrt{\frac{2}{m+1}\nu x U(x)} \qquad (4\text{-}77)$$

Compare with Eq. (4-43). The solution is given by Eq. (4-58), and the local Nusselt number may be written in the form

$$Nu_x = \sqrt{\frac{m+1}{2}} G(Pr, \beta) Re_x^{1/2} \qquad Re_x = \frac{xU(x)}{\nu} \qquad (4\text{-}78)$$

where

$$\frac{1}{G(Pr, \beta)} = \int_0^\infty \exp\left(-Pr \int_0^\eta f\, ds\right) d\eta$$

A plot of $G(Pr, \beta)$ for various β is given in Fig. 4-13, and tabulated values are given in Table 4-3. For a given β, the variation with Pr is nearly a power law, as in Eq. (4-62). Three examples would be

$$\frac{Nu_x}{Re_x^{1/2}} \approx \begin{cases} 0.22\, Pr^{0.27} & (\text{separation}, \beta = -0.1988) \\ 0.332\, Pr^{1/3} & (\text{flat plate}, \beta = 0) \\ 0.57\, Pr^{0.4} & (\text{stagnation}, \beta = 1) \end{cases} \qquad (4\text{-}79)$$

Note that separation flow, where skin friction is *zero*, supports considerable heat transfer. In fact, heat transfer in general (nonsimilar) separated-flow regions is substantial for both laminar and turbulent flows, e.g., the rear of bluff bodies.

TABLE 4-3
Numerical values of the heat-transfer parameter $G(Pr, \beta)$ from Eq. (4-78)

	β	−0.19884	−0.18	0.0	0.3	1.0	2.0	10.0
	f_0''	0.0	0.12864	0.46960	0.77476	1.23259	1.68722	3.67523
Pr	η^*	2.35885	1.87157	1.21678	0.91099	0.64790	0.49743	0.24077
	0.001	0.02383	0.02410	0.02449	0.02467	0.02483	0.02492	0.02508
	0.003	0.03967	0.04047	0.04154	0.04206	0.04252	0.04278	0.04325
	0.006	0.05409	0.05555	0.05759	0.05859	0.05947	0.05999	0.06091
	0.01	0.06745	0.06972	0.07296	0.07455	0.07597	0.07681	0.07831
	0.03	0.10547	0.11109	0.11935	0.12353	0.12734	0.12972	0.13385
	0.06	0.13666	0.14619	0.16050	0.16791	0.17480	0.17903	0.18693
	0.1	0.16339	0.17709	0.19803	0.20908	0.21950	0.22600	0.23843
	0.3	0.23180	0.25971	0.30371	0.32783	0.35147	0.36681	0.39801
	0.6	0.28318	0.32498	0.39168	0.42892	0.46633	0.49130	0.54459
Air	0.72	0.29777	0.34400	0.41786	0.45929	0.50113	0.52928	0.59054
	1.0	0.32581	0.38112	0.46960	0.51952	0.57047	0.60520	0.68219
	2.0	0.39145	0.47090	0.59723	0.66905	0.74372	0.79599	0.91815
	3.0	0.43478	0.53224	0.68596	0.77344	0.86522	0.93036	1.0872
	6.0	0.51896	0.65591	0.86728	0.98727	1.1147	1.2069	1.4396
	10.0	0.59054	0.76545	1.02974	1.1791	1.3388	1.4557	1.7597
	30.0	0.77839	1.0703	1.4873	1.7198	1.9706	2.1577	2.6682
	60.0	0.92602	1.3260	1.8746	2.1776	2.5054	2.7520	3.4395
	100.0	1.0523	1.5550	2.2229	2.5892	2.9863	3.2863	4.1332
	400.0	1.4885	2.4098	3.5292	4.1331	4.7894	5.2890	6.7332
	1000.0	1.8717	3.2319	4.7901	5.6230	6.5291	7.2212	9.2401
	4000.0	2.6471	5.0631	7.6039	8.9481	10.4112	11.5320	14.8312
	10,000.0	3.3285	6.8289	10.3201	12.1577	14.1583	15.6928	20.2262

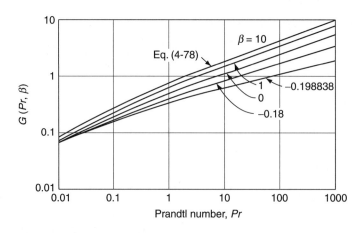

FIGURE 4-13
Heat-transfer parameter $G(Pr, \beta)$ for Falkner–Skan similarity flows with constant wall temperature and negligible dissipation.

4-3.5 The Reynolds Analogy as a Function of Pressure Gradient

We saw in Sec. 3-8.1.5 that there is a proportionality between friction and heat transfer in stagnation flow. A comparable Reynolds analogy holds in flat-plate flow. If we divide Eq. (4-52) by Eq. (4-62), we obtain

$$\frac{C_f}{C_h} \text{ (flat plate)} = 2Pr^{2/3} \quad (4\text{-}80)$$

This is a pure proportionality, valid for laminar or turbulent flow, independent of the Reynolds number. It is often used as an approximation in other (nonsimilar, non-flat-plate) situations, such as duct flow. But Eq. (4-79) alerts us that there is a pressure gradient effect.

Let us see if the Reynolds analogy holds for the Falkner–Skan solutions. The skin friction for these flows is given by

$$C_f(x) = \frac{2\mu(\partial u/\partial y)_w}{\rho U^2(x)} = f_0'' \sqrt{\frac{2(1+m)\nu}{Ux}} \quad (4\text{-}81)$$

and the Stanton number C_h is given by Eq. (4-78). If we adopt the usual power-law approximation near Prandtl number unity

$$G(Pr, \beta) \approx G(1, \beta) Pr^{1/3} \quad (4\text{-}82)$$

we find that the ratio of skin friction to heat transfer for the Falkner–Skan flows is given by

Gases:
$$\frac{C_f(x)}{C_h(x)} \approx 2Pr^{2/3} \frac{f_0''}{G(1, \beta)} \quad (4\text{-}83)$$

and the simple Reynolds analogy will be valid only if the factor $f_0''/G(1, \beta)$ is unity, which indeed it is for $\beta = 0$. However, this "analogy factor" varies strongly with β, as shown in Fig. 4-14. It is zero at the separation point, where C_f is zero but C_h

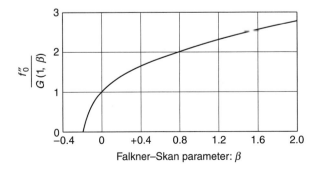

FIGURE 4-14
Variation of the Reynolds analogy from Eq. (4-83) with pressure gradient β for the Falkner–Skan solutions.

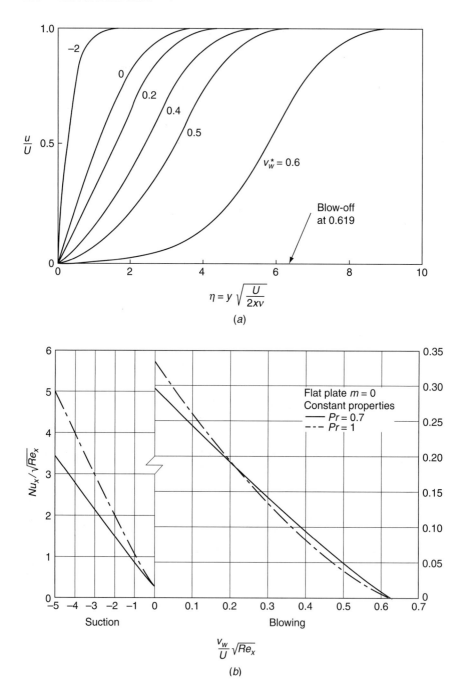

FIGURE 4-15
Flat-plate flow with suction or blowing: (a) velocity profiles; (b) local heat-transfer rates [*After Hartnett and Eckert (1957).*]

is finite, and increases without bound as β becomes large. The analogy is thus reliable only for modest, near-zero pressure gradients. It breaks down completely for nonsimilar flows or if the wall temperature varies.

4-3.6 The Flat Plate with Wall Suction or Blowing

The Blasius solution can be extended to nonzero wall velocity, $v_w \ll U$, either positive (blowing) or negative (suction). The streamwise wall velocity, u_w, is still zero from the no-slip condition. This has practical application to many problems: mass transfer, drying, ablation, transpiration cooling, and boundary-layer control. For similarity, only a certain variation $v_w(x)$ is allowed. From Eq. (4-44), at $\eta = 0$, the wall velocity $v_w = -f(0)\sqrt{\nu U/2x}$. Therefore suction and blowing can be simulated by a nonzero value of the Blasius stream function, $f(0)$, and v_w must vary as $x^{-1/2}$. We solve the Blasius Eq. (4-45) with

$$f'(0) = 0 \quad f'(\infty) = 1 \quad f(0) \neq 0 \qquad (4\text{-}84)$$

The results will vary with the *suction-blowing parameter*, v_w^*.

$$v_w^* = \frac{v_w}{U}\sqrt{Re_x} = \frac{-f(0)}{\sqrt{2}} \qquad (4\text{-}85)$$

The momentum problem was studied by Schlichting and Bussmann (1943), with heat-transfer results added by Hartnett and Eckert (1957).

Figure 4-15 shows the basic results. The velocity profiles in Fig. 4-15a are strongly affected by v_w^*. Suction thins the boundary layer and greatly increases the wall slope (friction, heat transfer). The suction profiles have strong negative curvature, like a favorable gradient, and are very stable and delay transition to turbulence (see Fig. 5-12). Blowing thickens the boundary layer and makes the profile S-shaped, less stable, and prone to transition to turbulence (look again at Fig. 5-12). At a finite value $v_w^* = +0.619$, the solution yields $\partial u/\partial y = 0$ at $y = 0$, with $u = 0$ for all finite y. The boundary layer is said to be "blown off" by the wall effect, and both the heat transfer and friction are zero. (The boundary-layer approximations fail, of course, for this extreme case.) Figure 4-15b shows the heat transfer $Nu_x/\sqrt{Re_x}$ versus the suction-blowing parameter. The effect of the Prandtl number is seen to be slight, at least for gases. The dash-dot lines in Fig. 4-15b also represent the skin friction $C_f/2$, since the Reynolds analogy is exactly valid if $Pr = 1$ in flat-plate flow.

4-3.7 Flow toward a Point Sink

Figure 4-10 indicated that the limiting case $\beta = +\infty, m = -1$ corresponds to flow toward a point sink. However, the Falkner–Skan approach is inappropriate,

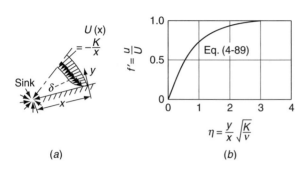

FIGURE 4-16
Boundary-layer similar solution for a point sink: (*a*) geometry; (*b*) solution.

since the similarity variable in Eq. (4-70) is squeezed to nothing when $(1 + m) = 0$. The remedy is to redefine the similarity variable for this case. For the point-sink coordinates in Fig. 4-16*a*,

$$U(x) = -\frac{K}{x} \quad (4\text{-}86)$$

Comparison with Eq. (4-68) reveals that similarity is achieved if η is proportional to y/x. For the nicest constants, we choose

$$\eta_{\text{point sink}} = \frac{y}{x}\sqrt{\frac{K}{\nu}} \quad (4\text{-}87)$$

Substitution in Eq. (4-67) gives the following differential equation for flow into a convergent channel:

$$f''' - f'^2 + 1 = 0 \quad (4\text{-}88)$$

subject to $f(0) = f'(0) = 0$ and $f'(\infty) = 1$. An exact solution may be found if one resists the temptation to retreat into Subroutine RUNGE:

$$f' = \frac{u}{U} = 3\tanh^2\left(\frac{\eta}{\sqrt{2}} + \tanh^{-1}\sqrt{\tfrac{2}{3}}\right) - 2 \quad (4\text{-}89)$$

where the constant $\tanh^{-1}\sqrt{\tfrac{2}{3}} = 1.146$. A plot of this relation is given in Fig. 4-16*b*. We have $f' = 0.99$ at $\eta \approx 3.4$, so that once again we have the customary inverse square-root relationship

Point sink: $\quad\dfrac{\delta_{99\%}}{x} \approx \dfrac{3.4}{\sqrt{-Ux/\nu}} \quad (4\text{-}90)$

the minus sign merely denoting that U is opposite to x, as shown in Fig. 4-16*a*. It is interesting that the sink-flow profile in Fig. 4-16*b* is the same as the Jeffery–Hamel wedge-flow profiles of Sec. 3-8.3.1 for large negative $Re\alpha$. See, e.g., the profile for $Re\alpha = -100$ in Fig. 3-34. Note, however, that η is defined quite differently for Jeffery–Hamel flow.

4-4 FREE-SHEAR FLOWS

Free-shear layers are unaffected by walls and develop and spread in an open ambient fluid. They possess velocity gradients, created by some upstream mechanism, that they try to smooth out by viscous diffusion in the presence of convective deceleration. Three examples are (1) the free-shear layer between parallel moving streams, (2) a jet, and (3) the wake behind a body immersed in a stream.

Let the dominant free-shear velocity be u in the x direction. Then, if the Reynolds number is large, the boundary-layer approximations will hold: $v \ll u, \partial u/\partial x \ll \partial u/\partial y, \partial p/\partial y \approx 0$. Further, if there are no confining walls, the pressure gradient dp/dx will be essentially zero. Then plane free-shear flows satisfy the flat-plate equations

$$\frac{\partial u}{\partial x} + \frac{\partial v}{\partial y} = 0$$

$$u\frac{\partial u}{\partial x} + v\frac{\partial u}{\partial y} \approx \nu \frac{\partial^2 u}{\partial y^2}$$

(4-91)

except of course there are no walls to enforce a no-slip condition.

Just downstream of the disturbance that caused the velocity gradients (the meeting point of the two parallel streams, the jet exit, the rear of the immersed body), the flow will be *developing* and nonsimilar. Further downstream, the flow will be *similar* and the velocity profiles will all look alike when suitably scaled. Here we discuss only the similar solutions for the shear layer, the jet, and the wake.

4-4.1 The Free-Shear Layer between Two Different Streams

Figure 4-17a shows two parallel uniform streams, U_1 (upper) and U_2 (lower), meeting at $x = 0$. As we move downstream, the discontinuity between U_1 and U_2 is smoothed out by viscosity into an S-shaped *free-shear layer* between the two. The simplest application would be for $U_2 = 0$, such as a plane airflow emerging from a slot into ambient air at rest. Lock (1951) generalized this into two different fluids with physical properties (ρ_1, μ_1) and (ρ_2, μ_2), respectively—also shown in Fig. 4-17a. He defined a Blasius-type similarity variable for each stream:

$$\eta_j = y\sqrt{\frac{U_1}{2x\nu_j}}$$

$$f'_j = \frac{u_j}{U_1} \qquad j = 1, 2$$

(4-92)

Note that U_1, not U_j, is specified in both variables. Substitution in Eqs. (4-92) yields a Blasius-type equation for each layer:

$$f'''_j + f_j f''_j = 0 \qquad j = 1, 2$$

(4-93)

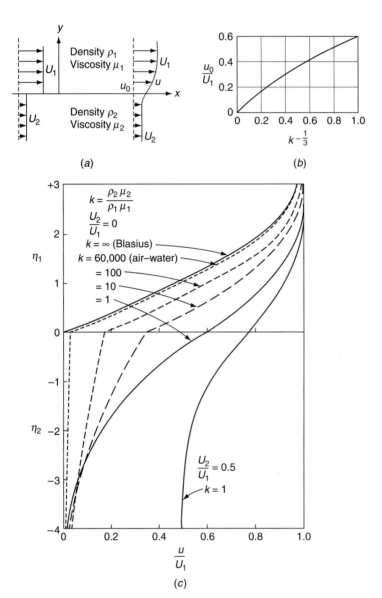

FIGURE 4-17
Velocity distribution between two parallel streams of different properties: (*a*) geometry; (*b*) velocities at the interface ($U_2 = 0$) (*c*) representative velocity profiles. [*After Lock (1951)*.] (*By permission of The Clarendon Press, Oxford.*)

The boundary conditions are of three types. First, an asymptotic approach to the two stream velocities:

$$f'_1(+\infty) = 1 \qquad f'_2(-\infty) = \frac{U_2}{U_1} \qquad (4\text{-}94)$$

Second, there should be kinematic equality, $u_1 = u_2$ and $v_1 = v_2$, at the interface, taken to be at $\eta_j = 0$:

$$f_1(0) = f_2(0) = 0$$
$$f'_1(0) = f'_2(0) \neq 0 \qquad (4\text{-}95)$$

Third, there should be equality of shear stress at the interface:

$$\mu_1 \frac{\partial u_1}{\partial y}(0) = \mu_2 \frac{\partial u_2}{\partial y}(0)$$

or (4-96)

$$f''_1(0) = k^{1/2} f''_2(0)$$

where $k = (\rho_2 \mu_2 / \rho_1 \mu_1)$. The most practical cases, since we are neglecting mass transfer between the two fluids, are $k = 1$ (identical fluids) or $k \gg 1$ (a gas flowing over a liquid). For the air–water interface, $k \approx 60{,}000$ or $k^{1/2} \approx 245$.

Some solutions computed by Lock (1951) for various k are shown in Fig. 4-17c. As k increases, the lower layer moves more slowly. The air–water case, $k = 60{,}000$, gives a good physical picture of slow "wind-driven" flow in the surface layer of a lake or ocean—although *that* large-scale flow would likely be turbulent, not laminar. The interface velocity when $U_2 = 0$ is shown in Fig. 4-17b as a function of k.

The classic case $k = 1$, $U_2 = 0$ in Fig. 4-17c has two interesting facets. First, it is not antisymmetric. The interface velocity is greater than $0.5\,U_1$ because the two layers have different convective deceleration. Second, the asymptotic value $f_2(-\infty)/\sqrt{2} = -0.619$, which represents a flat plate at $(-\infty)$ with its boundary layer "blown off" as in Fig. 4-15a.

4-4.2 The Plane Laminar Jet

Consider a plane jet emerging into a still ambient (identical) fluid from a (two-dimensional) slot at $x = 0$, as shown in Fig. 4-18. Since the jet spreads at constant pressure and there are no bounding walls, it satisfies Eqs. (4-91) and must also have constant momentum flux across any ($x = $ const) cross section:

$$J = \rho \int_{-\infty}^{+\infty} u^2 \, dy = \text{const} \qquad (4\text{-}97)$$

which is the zero-drag, zero-freestream version of Eq. (4-5) for a constant-pressure control volume. Schlichting (1933a) showed that if boundary-layer approximations

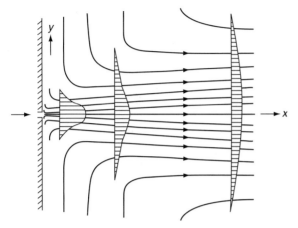

FIGURE 4-18
Definition sketch for a two-dimensional laminar free jet. [*After Schlichting (1933a)*.]

are valid, the jet entrainment spreads as the cube root of x, and the proper stream function is

$$\psi = \nu^{1/2} x^{1/3} f(\eta) \tag{4-98}$$

where
$$\eta = \frac{y}{3\nu^{1/2} x^{2/3}}$$

The corresponding velocity components are

$$u = \frac{f'(\eta)}{3x^{1/3}}$$
$$v = \frac{-\nu^{1/2}}{3x^{2/3}}(f - 2f'\eta) \tag{4-99}$$

Substitution in Eq. (4-91) gives the following relation to be solved:

$$f''' + ff'' + f'^2 = 0 \tag{4-100}$$

The boundary conditions are symmetry about the x axis ($v = 0$ and $\partial u/\partial y = 0$ at $y = 0$) and a quiescent ambient fluid ($u = 0$ at $y = \infty$). This translates in the similarity variables to

$$f(0) = f''(0) = 0$$
$$f'(\infty) = 0 \tag{4-101}$$

With all zero boundary conditions, it looks as if we have no driving force for the equation. Upon reflection, we see that the driving potential is the momentum flux J, rather analogous to the Jeffery–Hamel wedge flows of Chap. 3.

Equation (4-100) contains the same two non-linearities as the Falkner–Skan equation (4-69), thus issuing a seductive call for Subroutine RUNGE. In fact, however, Schlichting (1933a) deduced the exact analytic solution, which is strikingly simple:

$$f(\eta) = 2a \tanh(a\eta)$$

or (4-102)

$$f'(\eta) = 2a^2 \operatorname{sech}^2(a\eta)$$

The jet velocity profile thus has the symmetrical $\operatorname{sech}^2 y$ shape, reminiscent of a Gaussian probability distribution. The constant a is determined by evaluating the momentum flux J from Eq. (4-97):

$$J = \rho \int_{-\infty}^{+\infty} \left(\frac{2a^2}{3x^{1/3}} \operatorname{sech}^2 a\eta\right)^2 3\nu^{1/2} x^{2/3} d\eta = \tfrac{16}{9}\rho\nu^{1/2}a^3$$

or (4-103)

$$a = \left(\frac{9J}{16\sqrt{\rho\mu}}\right)^{1/3} \approx 0.8255 \frac{J^{1/3}}{(\rho\mu)^{1/6}}$$

Since $\operatorname{sech} 0 = 1$, the maximum or centerline velocity is seen to be

$$u_{\max} = \frac{2a^2}{3x^{1/3}} = \frac{2}{3}\left(\frac{9}{16}\right)^{2/3} \frac{J^{2/3}}{(\rho\mu x)^{1/3}} \approx 0.4543 \left(\frac{J^2}{\rho\mu x}\right)^{1/3} \quad (4\text{-}104)$$

Thus the jet spreads so that the centerline velocity drops off as $x^{-1/3}$. The velocity distribution is

$$u(x, y) = u_{\max} \operatorname{sech}^2 a\eta = u_{\max} \operatorname{sech}^2 \left[0.2752 \left(\frac{J\rho}{\mu^2 x^2}\right)^{1/3} y\right] \quad (4\text{-}105)$$

We may define the width of the jet as twice the distance y where $u = 0.01 u_{\max}$. Noting that $\operatorname{sech}^2 3 \approx 0.01$, we have

$$\text{Width} = 2y|_{1\%} = b \approx 21.8 \left(\frac{x^2 \mu^2}{J\rho}\right)^{1/3} \quad (4\text{-}106)$$

and thus the jet spreads as $x^{2/3}$. The mass rate of flow across any vertical plane is given by

$$\dot{m} = \rho \int_{-\infty}^{+\infty} u \, dy = (36 J\rho\mu x)^{1/3} \approx 3.302(J\rho\mu x)^{1/3} \quad (4\text{-}107)$$

which is seen to increase with $x^{1/3}$ as the jet entrains ambient fluid by dragging it along. This result is correct at large x but implies falsely that $\dot{m} = 0$ at $x = 0$, which is the slot where the jet issues. The reason is that the boundary-layer approximations fail if the Reynolds number is small, and the appropriate Reynolds number here is $\dot{m}/\mu \sim (J\rho x/\mu^2)^{1/3}$. Thus the solution is invalid for small values of $J\rho x/\mu^2$, meaning that we cannot ascertain any details of the flow near the jet outlet with a boundary-layer theory.

Since jet velocity profiles are S-shaped (i.e., have a point of inflection), they are unstable and undergo transition to turbulence early—at a Reynolds number of about 30, based on exit slot width and mean slot velocity. Thus although there is further analysis of laminar jets in the literature [Pai (1954)], jets are more likely to be turbulent. Textbooks treating turbulent-jet analysis and experiments are by Abramovich (1963), Schetz (1980), and Morris et al. (2002).

The analysis of axisymmetric (round) jets and wakes will be discussed in Sec. 4-10.

4-4.3 The Plane Laminar Wake: Far-Field Approximation

A wake is the *defect* in stream velocity behind an immersed body in a flow, as sketched in Fig. 4-19. A slender plane body with zero lift, such as the airfoil parallel to the stream in Fig. 4-19, usually produces a smooth wake whose velocity defect u_1 decays monotonically downstream. A blunt body, such as a cylinder, has a wake distorted by an alternating shed vortex structure—the picture in Fig. 4-19 would be a time-average wake.

Immediately downstream of the body in Fig. 4-19, the wake is developing and nonsimilar. About three body-lengths downstream, the wake becomes developed, with similar profiles. Here we make only a simple far-field approximation for a developed wake by assuming that the velocity defect is very small:

$$u_1(x, y) = U_0 - u(x, y) \ll U_0 \qquad (4\text{-}108)$$

in which case the convective acceleration can be linearized in the manner of the Oseen approximation, Sec. 3-9.4. The momentum Eq. (4-91) simplifies to

$$U_0 \frac{\partial u_1}{\partial x} \approx \nu \frac{\partial^2 u_1}{\partial y^2} \qquad (4\text{-}109)$$

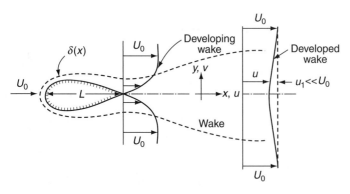

FIGURE 4-19
Flow in the wake of a body immersed in a stream.

subject to $u_1(x, \pm\infty) = 0$ and $\partial u_1/\partial y = 0$ at $y = 0$. This is the linear heat-conduction equation, and the solution is

$$u_1 = BU_0 x^{-1/2} \exp\left(-\frac{U_0 y^2}{4x\nu}\right) \qquad (4\text{-}110)$$

where B is a constant. In the far field, then, the wake has a Gaussian velocity distribution whose centerline value drops off as $x^{-1/2}$. The constant B is evaluated from the condition that the body drag force F equals the momentum flux defect in the wake:

$$F = \int_{-\infty}^{\infty} \rho u u_1 \, dy \approx \rho U_0 \int_{-\infty}^{\infty} u_1 \, dy = 2\rho U_0^2 B \sqrt{\frac{\pi \nu}{U_0}} \qquad (4\text{-}111)$$

Meanwhile, this force is correlated by the body drag coefficient:

$$F = C_D \tfrac{1}{2} \rho U_0^2 L$$

per unit depth into the paper. Equate this to Eq. (4-111) and solve for B, after which the wake velocity may be written in the form

$$\frac{u_1}{U_0} = C_D \left(\frac{Re_L}{16\pi}\right)^{1/2} \left(\frac{L}{x}\right)^{1/2} \exp\left(-\frac{U_0 y^2}{4x\nu}\right) \qquad (4\text{-}112)$$

where $Re_L = U_0 L/\nu$ is the body Reynolds number. The wake defect is thus proportional to the body drag coefficient. For a flat plate wetted on both sides, $C_D = 2.656/\sqrt{Re_L}$ from Eq. (4-53), the centerline velocity defect becomes

$$\left.\frac{u_1(x, 0)}{U_0}\right|_{\text{flat plate}} = \frac{0.664}{\sqrt{\pi}} \left(\frac{L}{x}\right)^{1/2} \qquad (4\text{-}113)$$

a result given by Tollmein (1931) and valid for $x > 3L$.

A complete review of laminar wakes, including near-field and three-dimensional geometries and compressible flows, is given in the monograph by Berger (1971). Like jets, wakes are unstable and are more likely in practice to be turbulent than laminar.

4-5 OTHER ANALYTIC TWO-DIMENSIONAL SOLUTIONS

It is clear that the similarity solutions of Secs. 4-3 and 4-4 are very special, in that their profile shapes remain the same (in the dimensionless sense) as we move downstream. It is far more likely that typical flow in practice is nonsimilar and changes from adverse to favorable gradients and perhaps back again. For example, the freestream flow near a cylinder has a favorable gradient near the nose which changes continuously to an adverse gradient (with separation) as we move toward the rear. No similarity technique can attack such a flow.

In the first 50 years after Prandtl's development of boundary-layer theory, nonsimilar flows were vexing and had to be computed laboriously by hand using analytic methods. Computed results, such as the boundary layer over a cylinder, were cherished and used as examples over and over. Now, however, in this personal-computer era, any practical laminar freestream distribution $U(x)$ can be computed swiftly and accurately by numerical methods which we illustrate in Sec. 4-7. (The same computational power is also available for turbulent boundary layers, but in that case the equations themselves are less reliable.)

Three basic approaches have been used to analyze nonsimilar boundary layers:

1. Analytic continuation by series expansion, e.g., Howarth (1938), Görtler (1957), and Meksyn (1961). These methods are now obsolete.
2. Approximate integral methods—an extension of Sec. 4-1.1 An amazing number of these techniques are discussed by Rosenhead (1963). We will discuss here the simple and accurate laminar-flow correlation of Thwaites (1949), which seems to serve well for any $U(x)$.
3. Numerical modeling on a digital computer. Scores of these numerical techniques, both finite-difference and finite-element, are reported in recent literature, e.g., Schetz (1992) and Cebeci and Cousteix (1998). We will outline a simple finite-difference method easily implemented on a personal computer.

Although approaches 2 and 3 above can handle any new problem with proper care, two classic results are worth a brief description here.

4-5.1 The Linearly Retarded Flow of Howarth

A simple decelerating nonsimilar freestream distribution,

$$U(x) = U_0\left(1 - \frac{x}{L}\right) \qquad (4\text{-}114)$$

was studied by Howarth (1938) to illustrate adverse gradients and laminar-boundary-layer separation. Howarth expanded the stream function into a power series in x/L whose coefficients were Blasius-type functions of $[y\sqrt{U_0/\nu x}]$. He kept seven terms—each a complicated function—and plotted the velocity profiles shown in Fig. 4-20. The curve for $x/L = 0$ is the Blasius profile, downstream of which the profiles become increasingly S-shaped and finally separation occurs at $x/L \approx 0.125$, according to Howarth's estimate.

The series-expansion approach is definitely obsolete now, but even the most accurate numerical solution of boundary-layer equations is invalid and inappropriate at separation and beyond. No matter how the external potential flow $U(x)$ varies, the wall shear stress at the separation point comes in vertically to zero, that is, $\partial \tau_w / \partial n = 0$ at separation.[†] This is known as the *Goldstein singularity*, after Goldstein (1948), who showed that, in boundary-layer theory, the wall shear stress

---

[†]Look ahead to see similar singular wall shear stress behavior on a cylinder, Fig. 4-24b.

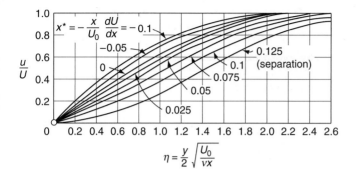

FIGURE 4-20
Velocity profiles from a series expansion of linearly retarded flow $U(x) = U_0(1 - x/L)$. [*After Howarth (1938)*.]

has square-root behavior near separation:

$$\mu \frac{\partial u}{\partial y}\bigg|_{y=0} = \text{const}(x_{\text{sep}} - x)^{1/2} \qquad (4\text{-}116)$$

Though interesting mathematically, this is unrealistic physically and is a fundamental limitation of boundary-layer theory. The dilemma was resolved by Sychev (1972), who showed that a *free-streamline potential theory*, discussed, e.g., in Milne–Thomson's text (1968), plus a rescaled boundary-layer approximation in the immediate vicinity of separation, removes the singularity. The wall shear stress then varies smoothly through zero. Another example of this rescaling technique is given later in Sec. 4-11.5. In the separated region, higher order theory and, indeed, the full Navier–Stokes equations should replace boundary-layer approximations. Additional detailed calculations of the separation region structure are given by Smith (1977). The Sychev–Smith theory is most appropriate for bluff-body flows, where the free streamline produces an adverse pressure gradient just sufficient to cause flow separation. For further discussion, see Chap. 14 of the monograph by Schlichting and Gersten (2000).

4-5.2 The Flat Plate with Uniform Wall Suction

In another series solution of an important case, Iglisch (1944) studied boundary-layer flow past a flat plate with uniform suction, as in Fig. 4-21. Wall suction is a very practical means of delaying boundary-layer transition to turbulence. Iglisch solved the boundary-layer Eqs. (4-91) for constant pressure, $dU/dx = 0$, and a wall-suction boundary condition:

$$u(x, 0) = 0 \qquad u(x, \infty) = U_\infty \qquad v(x, 0) = v_0 = \text{const} < 0 \qquad (4\text{-}117)$$

The flow is nonsimilar, and the velocity profiles in Fig. 4-21 gradually change shape from near-Blasius flow at the leading edge to the asymptotic exponential

260 VISCOUS FLUID FLOW

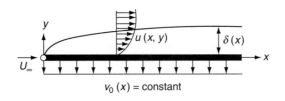

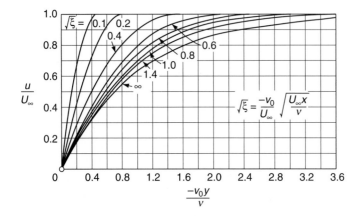

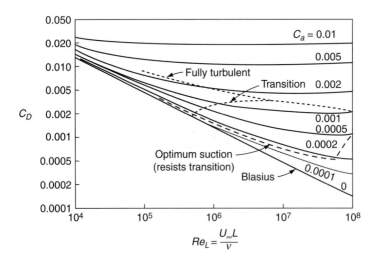

FIGURE 4-21
Velocity profile and drag coefficient of a flat plate with uniform suction. [*After Iglisch (1994)*.]

profile of Eq. (3-121) far downstream. The asymptotic condition is reached at about $(-v_0/U_\infty)\sqrt{Re_x} \approx 2.0$. If the suction rate is a typical value, $(-v_0/U_\infty) = 0.001$, in air at 22°C with $U_\infty = 10$ m/s, this corresponds to a distance $x = 6$ m. But the third chart in Fig. 4-21 shows that increasing suction raises the plate drag, possibly even above the turbulent-flow drag that we are trying to avoid with suction. The reader may show as an exercise (from a control volume similar to Fig. 4-1) that the drag coefficient on the plate is increased by suction, due to swallowing tangential momentum through the wall:

$$C_D = \frac{1}{L}\int_0^L C_f(x)\,dx + 2C_Q$$

$$C_Q = -\frac{v_0}{U_0}$$

(4-118)

Here C_f is the local viscous shear, $2\tau_w/\rho U_0^2$. Iglisch's drag computations are shown in Fig. 4-21. The dashed line shows the optimum suction, the locus of points of minimum suction required to maintain laminar flow, according to the transition theory of Chap. 5. For further study, the reader is referred to the excellent review by Wuest in Lachmann (1961).

4-5.3 Other Classical Analytic Solutions

There are quite a few other published classical solutions of boundary-layer equations for special $U(x)$. Some of these are theoretical:

1. Tani (1949): $U = U_0(1 - x^{*n})$, $n = 2, 4, 8,$ and $x^* = x/L$
2. Görtler (1957): $U/U_0 = (1 - x^*)^n$, $n = \frac{1}{2}, 2$; $(1 + x^*)^m$, $m = -1, -2, \cos(x^*)$
3. Curle (1958): $U = U_0(x^* - x^{*3} + ax^{*5})$, $a = 0, +0.07885, -0.12156$
4. Terrill (1960): $U = U_0 \sin(x^*)$

Two of them are experiments used for comparison with theory:

5. Hiemenz (1911): Flow past a circular cylinder
6. Schubauer (1935): Flow past a slender elliptical cylinder

All of these laminar flows are driven to separation and make good test cases for an alternate method. The reader will appreciate that they also make good problem exercises for integral or finite-difference methods.

4-6 APPROXIMATE INTEGRAL METHODS

The control-volume approach of Sec. 4-1 is exact in principle—if exact velocity and temperature distributions are used—but is approximate in practice, using guessed profiles and other correlations. A wide variety of these "integral methods" are discussed in the text edited by Rosenhead (1963). Here we present only this writer's favorite method, due to Thwaites (1949).

4-6.1 The Boundary-Layer Integral Equations

All integral methods use partially integrated forms of the continuity, momentum, and energy equations, which for two-dimensional incompressible flow are Eqs. (4-35). Continuity is eliminated during the derivation, so that the two basic results are (1) the *momentum* integral relation [Kármán (1921)] and (2) the *thermal-energy* integral relation [Frankl (1934)]. Some workers also use a third, called the *mechanical-energy* integral relation [Leibenson (1935)]. The mechanical-energy equation is the x-momentum equation multiplied by u, that is, it changes forces into the rate of work done by those forces. It is not a fundamental relation but can be useful as a correlation tool, especially for turbulent boundary layers.

We may derive the integral relations by direct integration, in the y direction, of the boundary-layer equations, repeated here for convenience:

Continuity: $$\frac{\partial u}{\partial x} + \frac{\partial v}{\partial y} = 0 \qquad (4\text{-}35a)$$

Momentum: $$\frac{\partial u}{\partial t} + u\frac{\partial u}{\partial x} + v\frac{\partial u}{\partial y} \approx \left(\frac{\partial U}{\partial t} + U\frac{\partial U}{\partial x}\right) + \frac{1}{\rho}\frac{\partial \tau}{\partial y} \qquad (4\text{-}35b)$$

Thermal energy: $$\rho c_v\left(\frac{\partial T}{\partial t} + u\frac{\partial T}{\partial x} + v\frac{\partial T}{\partial y}\right) \approx -\frac{\partial q}{\partial y} + \tau\frac{\partial u}{\partial y} \qquad (4\text{-}35c)$$

We have neglected buoyant forces and have used τ and q to represent shear stress and heat transfer, because then the equations are valid, as we shall see, for turbulent boundary layers also. In the laminar case, $\tau = \mu\, \partial u/\partial y$ and $q = -k\, \partial T/\partial y$.

4-6.2 The Momentum-Integral Relation

To obtain the momentum-integral relation, we first multiply continuity by $u - U$ and subtract from momentum, with the result

$$-\frac{1}{\rho}\frac{\partial \tau}{\partial y} = \frac{\partial}{\partial t}(U - u) + \frac{\partial}{\partial x}(uU - u^2)$$

$$+ (U - u)\frac{\partial U}{\partial x} + \frac{\partial}{\partial y}(vU - vu) \qquad (4\text{-}119)$$

We allow unsteady flow and the possibility of a porous wall with normal velocity $v_w(x)$ which is positive for injection. We then integrate from the wall to infinity, noting that

τ vanishes at infinity in the boundary-layer approximation. The result is

$$\frac{\tau_w}{\rho} = \frac{\partial}{\partial t}\int_0^\infty (U - u)\,dy + \frac{\partial}{\partial x}\int_0^\infty u(U - u)\,dy$$

$$+ \frac{\partial U}{\partial x}\int_0^\infty (U - u)\,dy - Uv_w \qquad (4\text{-}120)$$

This a fairly general form of the momentum-integral relation, often called the Kármán integral relation after T. von Kármán, who first suggested this approach in boundary-layer analysis (1921). As might be expected, the integrals of $U - u$ and $u(U - u)$ are equivalent to the displacement and momentum thicknesses, Eqs. (4-4) and (4-6). Hence we can rewrite the momentum relation in the more compact form

$$\frac{\tau_w}{\rho U^2} = \frac{C_f}{2} = \frac{1}{U^2}\frac{\partial}{\partial t}(U\delta^*) + \frac{\partial \theta}{\partial x} + (2\theta + \delta^*)\frac{1}{U}\frac{\partial U}{\partial x} - \frac{v_w}{U} \qquad (4\text{-}121)$$

For steady flow with an impermeable wall, this reduces to

$$\frac{C_f}{2} = \frac{d\theta}{dx} + (2 + H)\frac{\theta}{U}\frac{dU}{dx} \qquad H = \frac{\delta^*}{\theta} \qquad (4\text{-}122)$$

which is the most heavily analyzed and most commonly seen form of the Kármán integral relation. The so-called shape factor H is always greater than unity, as seen from the geometry of Fig. 4-2, and varies for laminar flow from about 2.0 at the stagnation point to about 3.5 at the separation point. In turbulent flow, the variation is even less (about 1.3 to 2.5), but, paradoxically, determining the exact value of H in turbulent flow is critical to an accurate analysis of this type.

4-6.3 The Thermal-Energy Integral Relation

The thermal-energy integral is most easily derived by first multiplying the momentum Eq. (4-35b) by u and adding this to the thermal-energy Eq. (4-35c). The result is

$$\rho\frac{\partial h_0}{\partial t} + \rho\left(u\frac{\partial h_0}{\partial x} + v\frac{\partial h_0}{\partial y}\right) = \frac{\partial}{\partial y}(-q + u\tau) \qquad (4\text{-}123)$$

where $h_0 = c_p T + u^2/2$ is the total, or stagnation, enthalpy of the flow—neglecting $v^2/2$, naturally. This equation will have important consequences in laminar compressible flow (Chap. 7) of gases. Presently, we merely wish to integrate the equation from zero to infinity, including porous walls and unsteady flow. The result is

$$q_w = \frac{\partial}{\partial t}\int_0^\infty \rho c_p T\,dy + \frac{\partial}{\partial x}\int_0^\infty \rho u(h_0 - h_{0e})\,dy - \rho c_p v_w(T_w - T_e) \qquad (4\text{-}124)$$

This is the general form of the thermal-energy integral relation, derived by Frankl (1934). For steady flow with impermeable walls and negligible dissipation, we have the simpler form

$$q_w = \frac{d}{dx}\int_0^\infty \rho c_p u(T - T_e)\, dy \qquad (4\text{-}125)$$

which was used earlier in our flat-plate analysis, Eq. (4-21), and in fact is valid for any low-speed laminar or turbulent boundary layer.

4-6.4 The Mechanical-Energy Integral Relation

To derive the mechanical-energy integral relation, we multiply continuity by $u^2 - U^2$ and momentum by $2u$, subtract, and integrate as before from the wall to infinity. The result is

$$\frac{2}{\rho}\int_0^\infty \tau \frac{\partial u}{\partial y}\, dy = \frac{\partial}{\partial t}\int_0^\infty u(U - u)\, dy + U^2 \frac{\partial}{\partial t}\int_0^\infty \left(1 - \frac{u}{U}\right) dy$$

$$+ \frac{\partial}{\partial x}\int_0^\infty u(U^2 - u^2)\, dy - U^2 v_w \qquad (4\text{-}126)$$

The integral on the left-hand side is often called the *dissipation integral*:

$$\mathcal{D} = \int_0^\infty \tau \frac{\partial u}{\partial y}\, dy \qquad (4\text{-}127)$$

On the right-hand side, we recognize the momentum thickness, the displacement thickness, and a third integral related to the so-called kinetic-energy thickness (sometimes called the dissipation thickness):

$$\delta_3 = \int_0^\infty \frac{u}{U}\left(1 - \frac{u^2}{U^2}\right) dy \qquad (4\text{-}128)$$

With this notation, Eq. (4-126) can be rewritten in the compact form

$$C_\mathcal{D} = \frac{2\mathcal{D}}{\rho U^3} = \frac{1}{U}\frac{\partial}{\partial t}(\theta + \delta^*) + \frac{2\theta}{U^2}\frac{\partial U}{\partial t} + \frac{1}{U^3}\frac{\partial}{\partial x}(U^3 \delta_3) - \frac{v_w}{U} \qquad (4\text{-}129)$$

This is the mechanical-energy integral relation, first derived by Leibenson (1935) and valid for laminar or turbulent flow. The relation has been used in laminar flow but is more appropriate for turbulent flow, where an extra correlation can help make up for uncertain modeling laws.

4-6.5 One-Parameter Integral Methods

The three integral relations derived before are exact but are used in *approximate* methods, and many have been published. There are two basic categories: (1) guessed velocity and temperature profiles and (2) empirical correlations among the integral parameters.

For momentum analysis, Eq. (4-122), one could approximate u by a one-parameter family of velocity profiles:

$$u(x, y) \approx U(x)f[\eta, P(x)] \quad (4\text{-}130)$$

where $\eta = y/\delta$ and P is a suitable dimensionless parameter. From such a profile we could then compute θ, δ^*, H, and τ_w and substitute in Eq. (4-122) to obtain a first-order differential equation for, say, $\delta(x)$.

The guessed-profile idea was dominated for decades by the first such method, developed by Pohlhausen (1921). Guided by the concurrent paper of Kármán (1921), Pohlhausen proposed a fourth-order polynomial:

$$\frac{u}{U} \approx 2\eta - 2\eta^3 + \eta^4 + \frac{\Lambda}{6}[\eta(1-\eta)^3] \quad (4\text{-}131)$$

where $\Lambda = \delta^2(dU/dx)/\nu$ varies with the local pressure gradient and is now called the *Pohlhausen parameter*. Equation (4-131) looks good when plotted, fits five realistic profile boundary conditions, and, when substituted into the momentum-integral relation (4-122), produces a differential equation that is easy to solve on a personal computer. Unfortunately, it is *not* very accurate and thus should be discarded.

There have been many other methods using profile assumptions, some involving multiple parameters [Rosenhead (1963)]. But this writer believes that the correlation idea of Thwaites (1949) is excellent and sufficient for laminar flow and need not be supplemented by any other approximation.

4-6.6 The Correlation Method of Thwaites

Thwaites (1949) modified and improved an idea of Holstein and Bohlen (1940), who cleverly rewrote the momentum-integral Eq. (4-122) in terms of a better parameter λ, defined as

$$\lambda = \frac{\theta^2 U'}{\nu} = \left(\frac{\theta}{\delta}\right)^2 \Lambda \quad (4\text{-}132)$$

Their simple but inspired idea was to multiply the momentum-integral relation, Eq. (4-122), by $U\theta/\nu$, with the result

$$\frac{\tau_w \theta}{\mu U} = \frac{U\theta}{\nu}\frac{d\theta}{dx} + \frac{\theta^2 U'}{\nu}(2 + H) \quad (4\text{-}133)$$

Now H and the left-hand side of this equation are dimensionless boundary-layer functions and thus, by assumption, are correlated reasonably by a single parameter (λ in this case). Thus we assume, after Holstein and Bohlen (1940), that

$$\frac{\tau_w \theta}{\mu U} \approx S(\lambda) \quad \text{shear correlation}$$

$$H = \frac{\delta^*}{\theta} \approx H(\lambda) \quad \text{shape-factor correlation} \quad (4\text{-}134)$$

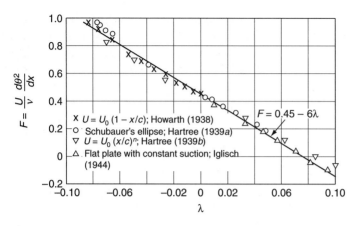

FIGURE 4-22
Empirical correlation of the boundary-layer function $F(\lambda)$ in Eq. (4-135). [*After Thwaites (1949)*.]

and further note that $\theta \, d\theta = d(\theta^2/2)$. Equation (4-133) may thus be rewritten in the very neat form

$$U \frac{d}{dx}\left(\frac{\lambda}{U'}\right) \approx 2[S(\lambda) - \lambda(2 + H)] = F(\lambda) \qquad (4\text{-}135)$$

Whereas earlier workers would have proposed a family of profiles to evaluate the parametric functions in Eq. (4-135), Thwaites (1949) abandoned the favorite-family idea and looked at the entire collection of known analytic and experimental results to see if they could be fit by a set of average one-parameter functions. As shown in Fig. 4-22, he found excellent correlation for the function F and proposed a simple linear fit

$$F(\lambda) \approx 0.45 - 6.0\lambda \quad \text{[Thwaites (1949)]} \qquad (4\text{-}136)$$

If $F = a - b\lambda$, Eq. (4-135) has a closed-form solution which the reader may verify as an exercise:

$$\frac{\theta^2}{\nu} = aU^{-b}\left(\int_{x_0}^{x} U^{b-1} dx + C\right) \qquad (4\text{-}137)$$

If x_0 is a stagnation point, the constant C must be zero to avoid an infinite momentum thickness where $U = 0$. Thus Thwaites has shown that $\theta(x)$ is predicted very accurately (± 3 percent), for all types of laminar boundary layers, by the simple quadrature

$$\theta^2 \approx \frac{0.45\nu}{U^6} \int_0^x U^5 \, dx \qquad (4\text{-}138)$$

TABLE 4-4
Shear and shape functions correlated by Thwaites (1949)

λ	H(λ)	S(λ)	λ	H(λ)	S(λ)
+0.25	2.00	0.500	−0.056	2.94	0.122
0.20	2.07	0.463	−0.060	2.99	0.113
0.14	2.18	0.404	−0.064	3.04	0.104
0.12	2.23	0.382	−0.068	3.09	0.095
0.10	2.28	0.359	−0.072	3.15	0.085
+0.080	2.34	0.333	−0.076	3.22	0.072
0.064	2.39	0.313	−0.080	3.30	0.056
0.048	2.44	0.291	−0.084	3.39	0.038
0.032	2.49	0.268	−0.086	3.44	0.027
0.016	2.55	0.244	−0.088	3.49	0.015
0.0	2.61	0.220	−0.090	3.55 (Separation)	0.000
−0.016	2.67	0.195			
−0.032	2.75	0.168			
−0.040	2.81	0.153			
−0.048	2.87	0.138			
−0.052	2.90	0.130			

Having found θ from this relation, one calculates $\lambda = \theta^2 U'/\nu$ and calculates the skin friction and displacement thickness from the assumed one-parameter correlations

$$\tau_w = \frac{\mu U}{\theta} S(\lambda)$$
$$\delta^* = \theta H(\lambda)$$
(4-139)

Thwaites' suggested correlations for $S(\lambda)$ and $H(\lambda)$ are listed in Table 4-4 and plotted in Fig. 4-23. A simple and accurate curve fit to the shear function, shown in Fig. 4-23b, is

$$S(\lambda) \approx (\lambda + 0.09)^{0.62} \qquad (4\text{-}140)$$

Being unable to divine a comparably simple formula for the shape-factor function $H(\lambda)$, this writer offers the following desperation polynomial:

$$H(\lambda) \approx 2.0 + 4.14z - 83.5z^2 + 854z^3 - 3337z^4 + 4576z^5 \qquad (4\text{-}141)$$

where $z = (0.25 - \lambda)$. The fit in Fig. 4-23a is good, though laborious.

The accuracy of the Thwaites method is about ±5 percent for favorable or mild adverse gradients but may be as much as ±15 percent near the separation point. Nevertheless, since the method is an average of many exact solutions, it can be regarded as a best available one-parameter method. If more accuracy is desired, the finite-difference computer method of Sec. 4-7 is recommended.

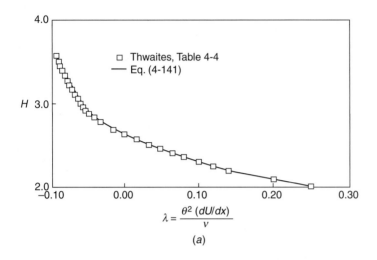

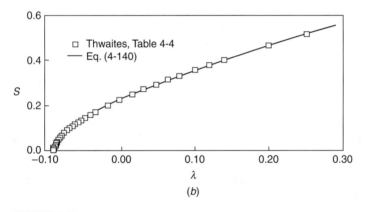

FIGURE 4-23
The laminar-boundary-layer correlation functions suggested by Thwaites (1949): (*a*) shape factor; (*b*) shear stress, with curve fits.

4-6.7 Application to the Howarth Decelerating Flow

To illustrate the simplicity and accuracy of Thwaites' method, we apply it to the Howarth linearly decelerating flow of Eq. (4-114) and Fig. 4-20, with $dU/dx = -U_0/L = $ const. The momentum thickness is computed, approximately, from Eq. (4-138):

$$\theta^2 = \frac{0.45\nu}{U_0^6(1 - x/L)^6} \int_0^x U_0^5 \left(1 - \frac{x}{L}\right)^5 dx = 0.075 \frac{\nu L}{U_0}\left[\left(1 - \frac{x}{L}\right)^{-6} - 1\right]$$

from which, by definition,

$$\lambda = \frac{\theta^2}{\nu}\frac{dU}{dx} = -0.075\left[\left(1 - \frac{x}{L}\right)^{-6} - 1\right] \quad (4\text{-}142)$$

With $\lambda(x)$ given by this (approximate) expression, we could then compute the wall shear $\tau_w(x)$ from the function $S(\lambda)$ in Table 4-4 or Fig. 4-23b. Let us save that until Sec. 4-7 as a comparison.

Given $\lambda(x)$, the separation point is predicted by

$$\lambda_{\text{sep}} \approx -0.09$$

or

$$\frac{x_{\text{sep}}}{L} = 1 - (2.2)^{-1/6} = 0.123$$

This is within 3 percent of the exact finite-difference result $x_{\text{sep}} = 0.120\,L$, achieved with very modest computational effort. This is better accuracy than the ± 10 percent that one might expect in any given case. Further, if $U(x)$ is complicated algebraically, it may be necessary to carry out the integral of U^5 in Eq. (4-138) numerically—which is still much simpler than a full-blown finite-difference computer attack.

4-6.8 Application to Laminar Flow Past a Circular Cylinder

Both the accuracy and the dilemma of a bluff-body boundary-layer calculation are illustrated by the circular cylinder. In terms of the dimensionless arc length $x^* = x/a$, where a is the cylinder radius, the potential-flow velocity distribution is

$$\frac{U}{U_\infty} = 2\sin x^* = 2.0 x^* - 0.333 x^{*3} + 0.0167 x^{*5} \ldots \quad (4\text{-}143)$$

from which we can easily generate boundary-layer solutions of any type (integral, series, or digital computer). Calculations of this type have been made. Separation is predicted at an angle $x^* = \phi = 104.5°$ in numerical results by Terrill (1960), which we might (mistakenly) think would be reproduced in an experimental laminar cylinder flow.

Unfortunately, as discussed earlier in Sec. 1-2, the broad wake caused by bluff-body separation is a *first-order effect*, i.e., it is so different from potential flow (see Figs. 1-5 and 1-6) that it alters $U(x)$ greatly everywhere, even at the stagnation point. For example, the experiment of Hiemenz (1911) for a cylinder at a Reynolds number $Re_a = U_\infty a/\nu = 9500$ fit the polynomial

$$\frac{U}{U_\infty} \approx 1.814 x^* - 0.271 x^{*3} - 0.0471 x^{*5} \quad (4\text{-}144)$$

which is quite different from potential flow. Even the stagnation velocity gradient

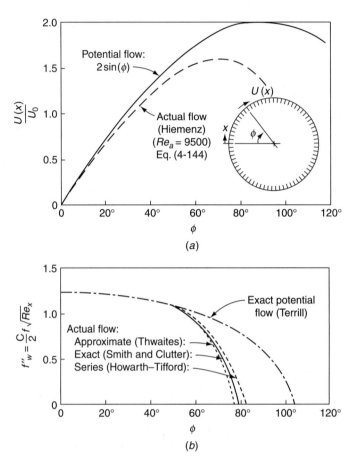

FIGURE 4-24
Comparison of potential-flow and actual-flow computations for a circular cylinder: (a) potential and actual velocity distributions; (b) potential-flow and actual-flow skin friction.

(1.814) is 9.3 percent less than the potential-flow value (2.0), and the maximum velocity ($1.595U_\infty$) occurs at $\phi = 71.2°$, instead of $2.0U_\infty$ at $\phi = 90°$. The two distributions are illustrated in Fig. 4-24a. It is clear that the potential flow is not a suitable input for the boundary-layer calculation. Once the actual $U(x)$ is known, the various theories are accurate, as seen in Fig. 4-24b. The finite-difference method of Smith and Clutter (1963) places ϕ_{sep} at $80°$, whereas Thwaites' integral method, Eq. (4-138), predicts $78.5°$ and a Howarth–Tifford series (including 19 terms) predicts $83°$. All three are in reasonable agreement with the Hiemenz (1911) experimental observation $\phi_{sep} \approx 80.5°$.

All laminar-boundary-layer computations hinge upon knowing the correct $U(x)$, however. It is presently a very active area of research to develop coupled methods in which a separating boundary layer interacts with and strongly modifies the external inviscid flow.

TABLE 4-5
Laminar-separation-point prediction by Thwaites' method

		Thwaites	
$U(x)$	x_{sep} (exact)	x_{sep}	Error, %
Howarth (1938)			
$1 - x$	0.120	0.123	+2.5
Tani (1949)			
$1 - x^2$	0.271	0.268	−1.1
$1 - x^4$	0.462	0.449	−2.8
$1 - x^8$	0.640	0.621	−3.0
Terrill (1960)			
$\sin(x)$	1.823	1.800	−1.3
Curle (1958)			
$x - x^3$	0.655	0.648	−1.1
Görtler (1957)			
$\cos(x)$	0.389	0.384	−1.3
$(1 - x)^{1/2}$	0.218	0.221	+1.3
$(1 - x)^2$	0.0637	0.0652	+2.4
$(1 + x)^{-1}$	0.151	0.158	+4.6
$(1 + x)^{-2}$	0.0713	0.0739	+3.6

4-6.9 Laminar-Separation-Point Prediction

One test of the general effectiveness of Thwaites' theory Eq. (4-138) is to see how well it predicts laminar separation for a variety of $U(x)$ freestream velocity distributions. Apply Eq. (4-138) to the distribution, compute $\theta(x)$, and note the position where $\lambda_{sep} = \theta^2 (dU\,dx)/\nu \approx -0.09$. Some results are shown in Table 4-5 for eleven different flows with adverse pressure gradients. The separation position error is less than 4 percent, but this success is tempered by our previous observation that large-scale separation greatly changes the external velocity and pressure distribution. The boundary-layer approximation is inadequate in the vicinity of separated flow.

4-7 DIGITAL-COMPUTER SOLUTIONS

The integral methods discussed in Sec. 4-6 are easy to use but limited in accuracy. The similarity solutions of Sec. 4-3 are accurate but limited in applicability. If one is faced with a heavy-duty study of nonsimilar boundary layers and needs great accuracy, then a digital-computer numerical model is required.

For convenience, let us repeat the two-dimensional, steady, laminar, boundary-layer equations that we wish to solve:

Continuity: $\quad \dfrac{\partial u}{\partial x} + \dfrac{\partial v}{\partial y} = 0 \quad$ (4-41a)

Momentum: $\quad u\dfrac{\partial u}{\partial x} + v\dfrac{\partial u}{\partial y} = U\dfrac{dU}{dx} + \nu\dfrac{\partial^2 u}{\partial y^2} \quad$ (4-41b)

These equations are *parabolic* in x, so the numerical models are all of the downstream-marching type. The usual inputs for any method are (1) known upstream profiles $u(0, y)$ and $v(0, y)$, (2) known freestream velocity $U(x)$, and (3) known wall conditions $u(x, 0) = 0$ and $v(x, 0) = v_w(x)$.

The laminar-boundary-layer equations are *parabolic* or "marching" in character and thus relatively easy to model with CFD. The literature contains scores of numerical models, many of which are summarized in three boundary-layer texts: Schetz (1992), Cebeci and Cousteix (1998), and Schlichting and Gersten (2000). The first two of these texts include FORTRAN codes for various boundary-layer models. Moreover, Schetz' work has been extended, as sponsored by the National Science Foundation, in the form of Internet Boundary Layer Applets,[†] Devenport and Schetz (2002), which anyone can access and run on their own PC. These excellent Applets are JAVA codes that include laminar and turbulent flows, both incompressible and compressible. With such resources available, the present text will not attempt to become a CFD monograph. We will simply present a simple, direct boundary-layer model that is effective and useful. The reader may wish to learn from this simple method and then graduate to more advanced codes.

There are two types of marching schemes: *explicit* and *implicit*. In the explicit methods, the downstream profiles $u(x + \Delta x, y)$ and $v(x + \Delta x, y)$ are calculated immediately from the known upstream profiles $u(x, y)$ and $v(x, y)$ by direct application of an algebraic model of Eqs. (4-41). Explicit models are simple but become unstable numerically unless small step sizes Δx are used, which may mean excessive computer time.

An implicit method is also an algebraic model of Eqs. (4-41), but points on the downstream profile $u(x + \Delta x)$ must be *solved simultaneously* by either iteration or matrix inversion. Computation time per step Δx is larger than for explicit schemes, but there is no numerical instability. The step size may be as large as one likes, subject only to the normal truncation errors that do not diverge or oscillate. We will illustrate both an explicit and an implicit model.

The methods we illustrate are *finite-difference* schemes, modeling derivatives by differences between nodal points on a rectangular (x, y) mesh. There are also numerous schemes using the *finite-element* method Beer (2001) or Löhner (2001) that is quite different in principle—simulating whole fields or "elements" of flow bounded by mesh points. We shall not discuss finite-element models here.

4-7.1 An Explicit Finite-Difference Model for Plane Flow

For both of our methods, we use the rectangular finite-difference mesh shown in Fig. 4-25. Subscripts are convenient: $u_{m,n}$ means the velocity u at mesh point

---

[†]An Applet is an Internet software program that a browser can download and use interactively. Not all Applets are excellent. Poorly constructed Applets can slow or crash PCs and are known as "craplets."

LAMINAR BOUNDARY LAYERS   **273**

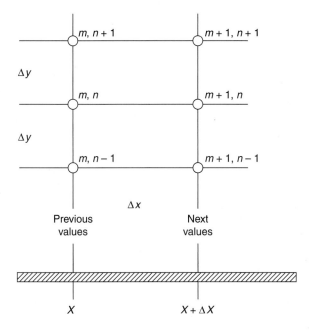

FIGURE 4-25
Finite-difference mesh for a two-dimensional boundary layer.

(m, n), located at $(x, y) = [(m - 1) \Delta x, (n - 1) \Delta y]$. Thus $(m, n) = (1, 1)$ locates the origin, $(x, y) = (0, 0)$. The step sizes Δx and Δy need not be equal. We also locate vertical velocity $v_{m,n}$ and (later) temperature $T_{m,n}$ at the same mesh points. All values at station m are assumed known or "previous," and we use them to march downstream and predict the "next" station values at $(m + 1)$. Let the total mesh be $M \times N$ large, where $m = M$ is the exit, $n = N$ is the freestream, $m = 1$ is the inlet, and $n = 1$ is the wall.

Let us begin by writing a finite-difference model of the momentum Eq. (4-41b) representing level n of the mesh:

$$u_{m,n} \frac{u_{m+1,n} - u_{m,n}}{\Delta x} + v_{m,n} \frac{u_{m,n+1} - u_{m,n-1}}{2 \Delta y}$$

$$\approx \frac{U_{m+1}^2 - U_m^2}{2 \Delta x} + \nu \frac{u_{m,n+1} - 2u_{m,n} + u_{m,n-1}}{\Delta y^2} \qquad (4\text{-}145)$$

Now, moving from left to right, let us remark about each term. The first term uses a *forward* difference to model $\partial u/\partial x$. The second term uses a *central* difference to represent $\partial u/\partial y$, thus keeping the model at level "n." The third term is the pressure gradient, using the form $U(dU/dx) = d(U^2/2)/dx$ and writing this as a forward difference—note that U_{m+1} is *known* because $U(x)$ is the freestream velocity. Finally, the fourth term is a central-difference model of $\partial^2 u/\partial y^2$. All terms center about level "n," which is good practice for numerical accuracy. The only unknown velocity in the model is $u_{m+1,n}$, which we may solve for and write as follows:

$$u_{m+1,n} \approx (\alpha - \beta)u_{m,n+1} + (1 - 2\alpha)u_{m,n} + (\alpha + \beta)u_{m,n-1} + \frac{U_{m+1}^2 - U_m^2}{2u_{m,n}}$$
(4-146)

where $\quad \alpha = \dfrac{\nu \Delta x}{u_{m,n} \Delta y^2} \quad$ and $\quad \beta = \dfrac{v_{m,n} \Delta x}{2 u_{m,n} \Delta y}$

The right-hand terms in Eq. (4-146) are all known from the "previous" station. Therefore we may solve for $u_{m+1,n}$ immediately—the model is *explicit*. However, algebraic recurrence relations such as Eq. (4-146) require, for numerical stability, that all the coefficients of the previous values u_m be positive. In this case, $\alpha < \frac{1}{2}$ and $\beta < \alpha$. The restraints limit both step sizes in the mesh:

$$\Delta x < \frac{u_{\min} \Delta y^2}{2\nu} \qquad \Delta y < \frac{2\nu}{|v_{\max}|} \qquad (4\text{-}147)$$

We need the absolute value on $v_{\max}$ because v is positive in adverse gradients but may be negative in favorable gradients. Usually $u_{\min}$ occurs near the wall and $v_{\max}$ near the freestream. During this explicit computation, we have to monitor $\alpha(y)$ and $\beta(y)$ to make sure that Eqs. (4-147) are satisfied.

With $u_{m+1,n}$ known from Eq. (4-146), we find $v_{m+1,n}$ by modeling the continuity Eq. (4-41a). The simplest way would be a forward difference for both terms:

$$\frac{u_{m+1,n} - u_{m,n}}{\Delta x} + \frac{v_{m+1,n} - v_{m+1,n-1}}{\Delta y} \approx 0$$

We start with $n = 2$ to compute $v_{m+1,2}$ because $v_{m+1,1} = v_w$ is known. However, let's not bother: The first term is at level n, and the second is at level $n - \frac{1}{2}$; numerical accuracy will be poor. Moving $\partial v/\partial y$ up to a central difference $[v_{m+1,n+1} - v_{m+1,n-1}]/2\Delta y$ at level n is self-defeating because we skip the value we wanted, $v_{m+1,n}$. Wu (1961) suggested that we move $\partial u/\partial x$ *down* to level $n - \frac{1}{2}$ by using an average value:

$$\left. \frac{\partial u}{\partial x} \right|_{\text{avg}} \approx \frac{1}{2} \left[\frac{u_{m+1,n} - u_{m,n}}{\Delta x} + \frac{u_{m+1,n-1} - u_{m,n-1}}{\Delta x} \right]$$

We use this expression instead in the continuity model and solve explicitly for the next vertical velocity:

$$v_{m+1,n} \approx v_{m+1,n-1} - \frac{\Delta y}{2\Delta x}[u_{m+1,n} - u_{m,n} + u_{m+1,n-1} - u_{m,n-1}] \qquad (4\text{-}148)$$

There is no instability in this relation other than that already required by Eqs. (4-147). We begin at $n = 2$ and move upward.

Equations (4-146) and (4-148) constitute a satisfactory, explicit, laminar, boundary-layer model which may easily be programmed on a personal computer.

Since we are "marching," there is really no need to use the "m" subscript, e.g., to dimension $u(1000, 50)$ and $v(1000, 50)$ for 1000 downstream steps. One can merely denote station m as $up(50)$ and $up(50)$ for "previous" and station $m + 1$ as $un(50)$ and $vn(50)$ for "next," keep the results long enough to print them, reinitialize $up(n) = un(n)$ and $vp(n) = vn(n)$, and move on. During the computation, one should constantly monitor the stability conditions Eqs. (4-147) and also check to see that the outer values $u_{m, N-1}$ and $u_{m, N}$, etc., merge smoothly with the freestream velocity U_m, since the boundary-layer thickness is *a priori* unknown.

4-7.1.1 APPLICATION TO HOWARTH LINEARLY DECELERATING FLOW. For illustration, Eqs. (4-146) and (4-148) were programmed to solve the Howarth problem $U = U_0(1 - x/L)$ from $x = 0$ until separation. Arbitrary values can be taken for U_0, L, and ν, since the results will be nondimensionalized. Figure 4-26 shows the computed values of normalized skin friction $C_f/C_f(0)$ compared with the Howarth series solution, and Thwaites' method using $S(\lambda)$ when λ is given by

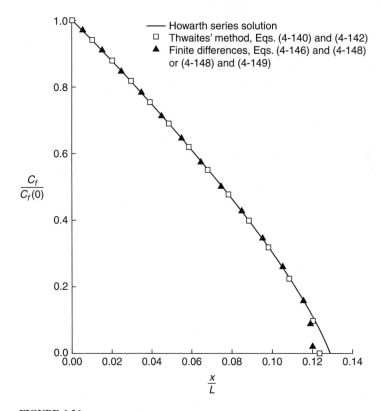

FIGURE 4-26

Comparison of finite-difference, integral, and series solutions for wall friction in the Howarth linearly decelerating flow of Eq. (4-114).

Eq. (4-142). We see that the computer solution equals the Howarth series for $x/L \leq 0.1$ and then is more accurate near separation, $x_{sep}/L \approx 0.120$. Thwaites' method is quite good (± 5 percent) over the entire range. The explicit computer method required 3000 steps, with Δx being extremely restricted by Eqs. (4-147) near separation.

4-7.2 An Implicit Finite-Difference Model

In an implicit model, more of the "next" terms are used to approximate the derivatives, resulting in simultaneous algebraic equations. In the present application, we model the viscous or second-derivative term in the momentum equation at the next station, $m + 1$:

$$\frac{\partial^2 u}{\partial y^2} \approx \frac{u_{m+1, n+1} - 2u_{m+1, n} + u_{m+1, n-1}}{\Delta y^2}$$

When this replaces the "previous" second derivative used in the explicit model, the result is that each nth-level equation has three unknowns:

$$-\alpha u_{m+1, n+1} + (1 + 2\alpha) u_{m+1, n} - \alpha u_{m+1, n-1}$$
$$\approx u_{m, n} - \beta(u_{m, n+1} - u_{m, n-1}) + \frac{U_{m+1}^2 - U_m^2}{2 u_{m, n}} \quad (4\text{-}149)$$

where α and β have the same meaning as before, Eq. (4-146). Now we must solve simultaneously for the $u_{m+1, n}$—the method is *implicit*. The benefit is that the results are unconditionally stable, regardless of the step sizes Δx and Δy—no need to use conditions Eqs. (4-147). We should select Δy small enough that, say, 20 or more points are within the boundary layer, and Δx should be small enough that changes in $u_{m, n}$ from station to station are small, say, less than 5 percent.

One way to solve Eq. (4-149) is by Gauss–Seidel iteration: Solve for the dominant term $u_{m+1, n}$, moving $u_{m+1, n \pm 1}$ over, and sweep the right-hand side repeatedly until the values of u do not change. Convergence is not guaranteed unless Δx and Δy are relatively small. A second method is direct solution by Gauss elimination or matrix inversion.

4-7.2.1 INVERSION OF A TRIDIAGONAL MATRIX.
Assuming that $n = 1$ is the wall and $n = N$ is the freestream, Eqs. (4-149) represent $(N - 2)$ equations, each with three unknowns: The matrix is thus *tridiagonal*. Such matrices are easy to invert by a procedure, called the tridiagonal matrix algorithm (TDMA), very nicely described in Sec. 4.2-7 of the text by Patankar (1980). It works because there are only two unknowns at the bottom, $n = 2$, where $u_{n=1} = 0$ (no slip), and only two at the top, $n = N$, where $u_N = U(x)$. Thus we can begin at the bottom and eliminate one variable at a time until we reach the top, where u_{N-1} is immediately found.

We then work our way back to the bottom ("back substitution"), picking up u_n in terms of u_{n+1} until we secure the final value, u_2.

The TDMA is outlined as follows. Dropping the "m" subscripts as superfluous, Eq. (4-149) may be written, at any n, as

$$(1 + 2\alpha_n)u_n = \alpha_n u_{n+1} + \alpha_n u_{n-1} + C_n$$

where C_n denotes the entire right-hand side of Eq. (4-149). The desired back-substitution recurrence relation has the form

$$u_n = P_n u_{n+1} + Q_n \qquad (4\text{-}150)$$

First calculate P_n and Q_n. Begin at the bottom, $u_1 = 0$, by computing

$$P_2 = \frac{\alpha_2}{1 + 2\alpha_2}$$

$$Q_2 = \frac{C_2}{1 + 2\alpha_2} \qquad (4\text{-}151)$$

Then calculate the remaining P's and Q's by recurrence relations:

$$P_n = \frac{\alpha_n}{1 + 2\alpha_n - \alpha_n P_{n-1}}$$

$$Q_n = \frac{C_n + \alpha_n Q_{n-1}}{1 + 2\alpha_n - \alpha_n P_{n-1}} \qquad (4\text{-}152)$$

At the top, since $u_N = U$, the value of velocity follows immediately:

$$u_{N-1} = \frac{\alpha_{N-1} U + \alpha_{N-1} Q_{N-2} + C_{N-1}}{1 + 2\alpha_{N-1} - \alpha_{N-1} P_{N-2}} \qquad (4\text{-}153)$$

where U is the freestream velocity at that station. With u_{N-1} known, plus all the P's and Q's, we work our way downward using Eq. (4-150) until we reach the final unknown, u_2. The algorithm works extremely well for a well-behaved matrix such as Eq. (4-149). Remember that the coefficient α itself varies with n because it contains $u_{m,n}$.

The implicit method was also applied to the Howarth distribution and results of accuracy equal to the explicit method, plotted in Fig. 4-26, were obtained with $\Delta x/L = 0.001$. Run time on a personal computer was less than the explicit method. There is no instability and no special need for decreased Δx near separation.

Finally, to illustrate the yin and yang of nonsimilar boundary layers, the method was run for a circular cylinder with the Hiemenz experimental velocity distribution, Eq. (4-144). Boundary-layer profiles are plotted in Fig. 4-27 for every $\Delta\theta = 10°$, measured from the stagnation point. Separation was predicted at $\theta = 81°$ (see Fig. 4-24b). The nominal Reynolds number was taken as $U_0 R/\nu = 10^5$ to show the actual boundary-layer thickness, which is about 1 percent of the cylinder radius. In the favorable gradient for θ up to $60°$, the profiles are strongly curved and have increasing wall shear stress. From $60°$ on, as the freestream levels off and begins to decelerate, the profiles become thicker and then S-shaped, with an early separation.

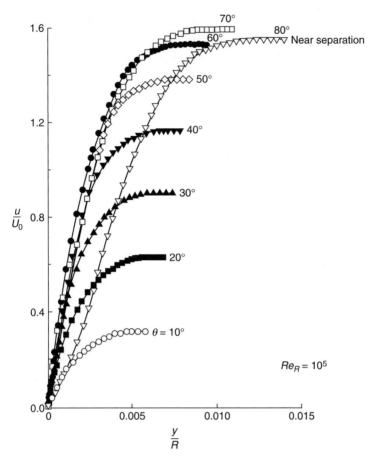

FIGURE 4-27
Laminar boundary layers on a cylinder computed by the implicit numerical method of Eqs. (4-148) and (4-149). Freestream velocity given by Eq. (4-144).

4-8 THERMAL-BOUNDARY-LAYER CALCULATIONS

The previous three sections were devoted to velocity calculations, because the assumption of incompressible flow with constant viscosity uncouples the velocity and enables it to be calculated independent of temperature. Afterward, the temperature can be calculated from the thermal-energy equation (4-35c) and the known velocity solution. We now review some of the methods of making temperature calculations under fairly arbitrary conditions.

Almost all analyses of incompressible flow assume that dissipation is negligible and work with the simplified thermal-energy equation

Differential form: $\quad u\dfrac{\partial T}{\partial x} + v\dfrac{\partial T}{\partial y} \approx \dfrac{k}{\rho c_p}\dfrac{\partial^2 T}{\partial y^2}$ \hfill (4-154a)

Integral form: $\quad q_w = \dfrac{d}{dx}\left[\displaystyle\int_0^\infty \rho u c_p (T - T_e)\, dy\right]$ \hfill (4-154b)

It turns out that temperature is somewhat of a poor relation of velocity, in that not nearly so much work has been done on Eqs. (4-154). This is partly because the thermal-energy equation is easier, i.e., it is linear. If, for a given $U(x)$, two solutions $T_1(x, y)$ and $T_2(x, y)$ are found, then their sum $T_1 + T_2$ is also a solution; this of course would not be true for velocity solutions u_1 and u_2.

As with momentum analyses, there are three basic approaches:

1. Series expansion, e.g., Meksyn (1961). These are now obsolete.
2. Integral-energy (approximate) methods—13 of these were discussed in an interesting review by Spalding and Pun (1962). We will select one of these which is simple yet gives good accuracy.
3. Numerical methods—many have been proposed, including well-developed user-friendly computer codes Tannehill et al. (1997). We will apply the same primitive-variable technique of Sec. 4-7.1 to the present heat-transfer problem. Finite differences are indicated when there are arbitrary variations of $U(x)$, $T_e(x)$, and $T_w(x)$.

4-8.1 One-Parameter Integral Method: Variable $U(x)$, Constant $(T_w - T_e)$

One approach to an integral analysis of heat transfer is to guess both velocity and temperature profiles and substitute them in Eq. (4-154b). However, good accuracy is obtained by a simple laminar-correlation scheme in the same spirit as Thwaites' momentum method, Sec. 4-6.6.

Consider any thickness measure of a thermal boundary layer, Δ, such as the enthalpy thickness from Eq. (4-22). Eckert (1942) theorized that the rate of growth of such a thickness should depend only on local stream parameters and fluid properties:

$$\frac{d\Delta}{dx} \approx fcn\left(\Delta, U, \frac{dU}{dx}, \nu, \frac{k}{\rho c_p}\right)$$

If temperature difference $(T_w - T_e)$ is constant, it would have no effect, since every term in the linear Eq. (4-154a) is proportional to T. By dimensional analysis we can rewrite the above expression as

$$\frac{U}{\nu}\frac{d\Delta^2}{dx} = fcn\left(\frac{\Delta^2}{\nu}\frac{dU}{dx}, Pr\right) \qquad (4\text{-}155)$$

Note the close resemblance of this relation to the Holstein–Bohlen formulation, Eq. (4-135). If one chooses Δ as the enthalpy thickness δ_h, a lot of integration is involved and the method bogs down into excessive charts and formulas.

An elegant simplification of this procedure was proposed by Smith and Spalding (1958). With an eye toward direct computation of the heat transfer, they suggested using instead the so-called conduction thickness

$$\delta_c = \frac{k(T_w - T_e)}{q_w} \qquad (4\text{-}156)$$

Thus the correlation now desired is of the form

$$\frac{U}{\nu}\frac{d}{dx}(\delta_c^2) = f\left(\frac{\delta_c^2 U'}{\nu}, Pr\right) \qquad (4\text{-}157)$$

and numerical values are selected from the Falkner–Skan solutions for $U(x) = Cx^m$. In terms of m and the heat-transfer parameter $G(Pr, \beta)$ from Eq. (4-78), we can determine that our desired new parameters are

$$\frac{U}{\nu}\frac{d}{dx}(\delta_c^2) = \frac{2(1-m)}{(1+m)G^2} \qquad \frac{\delta_c^2 U'}{\nu} = \lambda_c = \frac{2m}{(1+m)G^2} \qquad (4\text{-}158)$$

Figure 4-28 shows a typical plot of Eq. (4-157), using these relations for $Pr = 1$. We can share Smith and Spalding's delight that the curve is nearly linear over the wide range ($-0.18 \leq \beta \leq +2.0$). This is true for all Prandtl numbers, and thus Smith and Spalding propose a general approximation

$$\frac{U}{\nu}\frac{d}{dx}(\delta_c^2) \approx a(Pr) - b(Pr)\lambda_c \qquad (4\text{-}159)$$

where a and b are tabulated in Table 4-6. Equation (4-159) is the exact analog of Thwaites' linear fit, Eq. (4-136), and thus the exact analytic solution is

$$\frac{\delta_c^2}{\nu} = aU^{-b}\left(\int_{x_0}^{x} U^{b-1}dx + C\right) \qquad (4\text{-}160)$$

where, again, $C = 0$ if x_0 is the stagnation point. Since δ_c is directly related to q_w, we may readily convert Eq. (4-160) to a Nusselt number based on the local heat

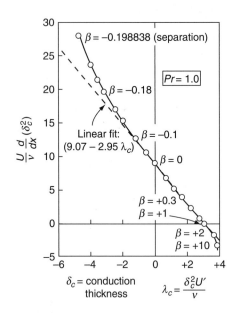

FIGURE 4-28
The correlation of Eq. (4-157) from the Falkner–Skan solutions (Table 4-3) compared with the linear approximation proposed by Smith and Spalding (1958).

TABLE 4-6
Constants $a(Pr)$ and $b(Pr)$ for use in Eq. (4-161)

	Pr	$a^{-1/2}$	b
	0.0	0.0	2.0
	0.001	0.0173	2.06
	0.01	0.0760	2.17
	0.1	0.140	2.46
	0.3	0.215	2.68
Air	0.72	0.296	2.88
	1.0	0.332	2.95
	2.0	0.422	3.10
	3.0	0.485	3.18
	10.0	0.728	3.38
	100.0	1.572	3.61
	1000.0	3.387	3.72
	10,000.0	7.297	3.76
	∞	∞	3.81

conductance $h = q_w/(T_w - T_e)$, a reference length L, and a velocity U_0:

$$\frac{Nu_L}{\sqrt{Re_L}} = \frac{hL/k}{\sqrt{U_0 L/\nu}} = \frac{a^{-1/2}}{\left[\left(\dfrac{U}{U_0}\right)^{-b} \displaystyle\int_0^x \left(\dfrac{U}{U_0}\right)^{b-1} \dfrac{dx}{L}\right]^{1/2}} \quad (4\text{-}161)$$

This is an accurate and remarkably simple expression, involving only a single quadrature of the velocity distribution. From Table 4-6, the terms $a(Pr)$ and $b(Pr)$ can be approximated in the range $0.1 \leq Pr \leq 10.0$ by the power laws

$$a^{-1/2} \approx 0.332 \, Pr^{0.35}$$
$$b \approx 2.95 \, Pr^{0.07} \quad (4\text{-}162)$$

For air, the most common case, we have $a^{-1/2} = 0.296$ and $b = 2.88$ ($Pr = 0.72$).

4-8.2 A Finite-Difference Method: Explicit or Implicit

The finite-difference momentum method outlined in Sec. 4-7.1 may readily be extended to temperature calculations. By defining temperature at each mesh point, $T_{m,n}$, we may write the energy Eq. (4-154a) in the difference form

$$u_{m,n} \frac{T_{m+1,n} - T_{m,n}}{\Delta x} + v_{m,n} \frac{T_{m,n+1} - T_{m,n-1}}{2\Delta y} \approx \frac{k}{\rho c_p} \frac{T_{m,n+1} - 2T_{m,n} + T_{m,n-1}}{\Delta y^2}$$

where $u_{m,n}$ and $v_{m,n}$ are presumed known already from the momentum model of Sec. 4-7.1. All differences are centered about level "n." We can rearrange this into an explicit model for downstream temperature:

$$T_{m+1,n} \approx (\gamma - \beta)T_{m,n+1} + (1 - 2\gamma)T_{m,n} + (\gamma + \beta)T_{m,n-1} \quad (4\text{-}163)$$

where $\quad \gamma = \dfrac{k\Delta x}{\rho c_p u_{m,n} \Delta y^2} \quad$ and, as before, $\quad \beta = \dfrac{v_{m,n}\Delta x}{2u_{m,n}\Delta y}$

The stability of this explicit model is similar to the momentum formulation, that is, $\gamma < \tfrac{1}{2}$ and $\beta < \gamma$, or step size is further limited by

$$\Delta x < \dfrac{\rho c_p u_{\min}\Delta y^2}{2k} \qquad \Delta y < \dfrac{2\nu}{|v_{\max}|} \quad (4\text{-}164)$$

If $Pr < 1$, this condition on Δx is more restrictive than Eq. (4-147).

Equation (4-163) is to be *combined with* Eqs. (4-146) and (4-148) to form a complete explicit numerical model for laminar-boundary-layer velocity and temperature.

An implicit model can be constructed, as before, in Eq. (4-149), by using a downstream central-difference model for the conduction term:

$$\dfrac{\partial^2 T}{\partial y^2} \approx \dfrac{T_{m+1,n+1} - 2T_{m+1,n} + T_{m+1,n-1}}{\Delta y^2}$$

The result is a set of simultaneous equations for the new temperatures:

$$-\gamma T_{m+1,n+1} + (1 + 2\gamma)T_{m+1,n} - \gamma T_{m+1,n-1} \approx T_{m,n} - \beta(T_{m,n+1} - T_{m,n-1})$$
$$(4\text{-}165)$$

Note the strong resemblance to the implicit momentum model, Eq. (4-149). Since γ and β involve $u_{m,n}$ and $v_{m,n}$, Eq. (4-165) must be combined with Eqs. (4-148) and (4-149) for the complete model.

Equation (4-165) can also be solved by the TDMA algorithm explained in Sec. 4-7.2.1, Eqs. (4-150) to (4-153), with several modifications. Since the computation is coupled with the momentum TDMA, let us just rewrite all the recurrence relations with new notation, and these terms will be added to the momentum analysis:

$$D_n = \text{right-hand side of Eq. (4-165)} \quad (4\text{-}166a)$$

$$T_n = R_n T_{n+1} + S_n \quad (4\text{-}166b)$$

$$R_2 = \dfrac{\gamma_2}{1 + 2\gamma_2} \qquad S_2 = \dfrac{\gamma_2 T_w + D_2}{1 + 2\gamma_2} \quad (4\text{-}166c)$$

$$R_n = \dfrac{\gamma_n}{1 + 2\gamma_n - \gamma_n R_{n-1}} \qquad S_n = \dfrac{D_n + \gamma_n S_{n-1}}{1 + 2\gamma_n - \gamma_n R_{n-1}} \quad (4\text{-}166d)$$

$$T_{N-1} = \dfrac{\gamma_{N-1} T_e + \gamma_{N-1} S_{N-1} + D_{N-1}}{1 + 2\gamma_{N-1} - \gamma_{N-1} R_{N-1}} \quad (4\text{-}166e)$$

where T_e is the freestream temperature. Calculate the R's and S's from the bottom up, then back-calculate the temperature with Eq. (4-166b) from the top, T_{N-1} down to T_2. As with the momentum analysis, the algorithm works extremely well, and results are accurate if changes in temperature between steps Δx and Δy are no more than a few percent.

4-8.3 Experimental Thermal Boundary Layer: The Circular Cylinder

Heat transfer from a circular cylinder has been the subject of many experiments because of its importance in heat-exchanger design. Here we concentrate on the *local* Nusselt number in the laminar boundary layer on the front of a cylinder, as measured by Schmidt and Wenner (1941) and by Giedt (1949).

The freestream velocity near the nose is approximated by

$$\frac{U}{U_0} \approx 1.82 \frac{x}{R} - 0.4 \frac{x^3}{R^3} \qquad (4\text{-}167)$$

where R is the cylinder radius and U_0 the velocity approaching the cylinder. The maximum velocity is $U/U_0 = 1.49$ at $x/R = 1.23$ ($\phi = 71°$). The temperature difference ($T_w - T_e$) was constant, so that the integral method of Smith and Spalding, Eq. (4-161), applies.

Figure 4-29 compares the Smith–Spalding formula and our implicit finite-difference technique, Eq. (4-165), with the local heat-transfer data. Both theories are in good agreement with Schmidt and Wenner's low Reynolds number data, with boundary-layer separation occurring at $\phi \approx 81°$. However, Giedt's data at $Re_D = 140{,}000$ are 10 percent higher. In fact, there is a systematic increase of heat transfer with the Reynolds number, which defies our contention in Sec. 4-2 that laminar boundary layers should scale directly with $\sqrt{Re}$. The reason was traced by Kestin et al. (1961) to turbulence or noisiness in the stream approaching the cylinder. This turbulence increased with the Reynolds number and caused streamwise vortices in the boundary layer that increased the heat transfer. A review of such freestream turbulence effects is given in the monograph by Zukauskas and Ziukzda (1985). Barrett and Hollingsworth (2003) review freestream turbulence effects in flat-plate flows.

4-8.4 Variable Wall Temperature: Superposition of Indicial Solutions

Our previous analyses have been for constant driving-temperature difference ($T_w - T_e$), a common design assumption. For variable ΔT, it is seriously inadequate to use our previous formulas so that Nu is related to the *local* temperature difference. For example, if T_w drops as we move downstream, q_w will change sign even before ($T_w - T_e$) changes sign, because the fluid in the local boundary layer has been warmed by the hotter upstream wall. We will show this effect with an example.

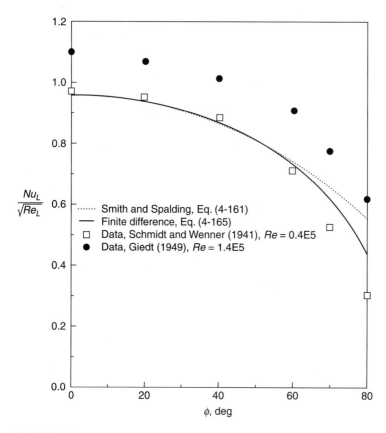

FIGURE 4-29
Comparison of theory and experiment for local heat transfer in airflow ($Pr = 0.72$) past a cylinder, constant $(T_w - T_e)$.

The thermal-energy equation (4-154a) is linear in temperature. Therefore one solution for variable $(T_w - T_e)$ is to use superposition of "indicial" or step-wall-change formulas for that particular $U(x)$. Let us illustrate with the most common case, a flat plate with variable $\Delta T(x)$. In Eq. (4-28), we solved the problem stated by Fig. 4-3: heat transfer with a discontinuity in wall temperature at $x = x_0$. The heat transfer downstream of the discontinuity was found to be

$$q_w(x_0, x) = (T_w - T_e)h(x_0, x)$$

where
$$h(x_0, x) = \frac{0.332k}{x} Pr^{1/3} Re_x^{1/2} \left[1 - \left(\frac{x_0}{x}\right)^{3/4} \right]^{-1/3} \qquad (4\text{-}168)$$

is the local heat-transfer coefficient.

Imagine now that $T_w - T_e$ changes continuously with x, so that we have a continuous series of starting problems at changing x_0. If $d(T_w - T_e)$ is the change which occurs between x_0 and $x_0 + dx_0$, the wall temperature at any x is given by using x_0 as a dummy variable and superimposing all these infinitesimal changes:

$$q_w(x) = \int_0^x h(x_0, x) \frac{d(T_w - T_e)}{dx_0} dx_0 \qquad (4\text{-}169)$$

the integral being evaluated with x held constant. This is a straightforward application of the superposition theorem; the justification is, of course, the fact that the thermal-energy equation is linear.

Suppose further, for generality, that $T_w - T_e$ changes discontinuously at a certain number of points $x_0(i)$ by the amount ΔT_i. Then we must add on to our previous solution a summation of these discrete starting solutions. The complete solution for arbitrarily variable wall temperature is, then,

$$q_w(x) = \int_0^x h(x_0, x) \frac{d}{dx_0}(T_w - T_e) dx_0 + \sum_{i=1}^N h[x_0(i), x] \Delta T_i \qquad (4\text{-}170)$$

We must remember that the indicial solution $h(x_0, x)$ must be appropriate for the particular velocity distribution $U(x)$ of the problem.

Now let us apply this theorem to the flat plate with a polynomial distribution of $T_w - T_e$. Suppose that

$$T_w - T_e = \sum_{j=0}^N a_j x^j = a_0 + a_1 x + a_2 x^2 + a_3 x^3 + a_4 x^4 + \cdots \qquad (4\text{-}171)$$

There are no discontinuities, hence Eq. (4-169) applies:

$$q_w(x) = \frac{0.332k}{x} Pr^{1/3} Re_x^{1/2} \int_0^x \left[1 - \left(\frac{x_0}{x}\right)^{3/4}\right]^{-1/3} \left(a_0 + \sum_{j=1}^N j a_j x_0^{j-1}\right) dx_0$$

The integrals are readily evaluated in terms of gamma functions, and the result is

$$q_w(x) = \frac{0.332k}{x} Pr^{1/3} Re_x^{1/2} \left[a_0 + \sum_{j=1}^N \frac{4}{3} j a_j x^j \frac{\Gamma(4j/3)\Gamma(2/3)}{\Gamma(4j/3 + 2/3)}\right] \qquad (4\text{-}172)$$

All the gamma functions can be evaluated numerically from the three known values $\Gamma(\frac{1}{3}) = 2.6789385$, $\Gamma(\frac{2}{3}) = 1.3541179$, and $\Gamma(1.0) = 1.0$, plus the recurrence relation $\Gamma(1 + k) = k\Gamma(k)$. The first five terms in the square brackets from Eq. (4-172) are

$$a_0 + 1.6123 a_1 x + 1.9556 a_2 x^2 + 2.2091 a_3 x^3 + 2.4151 a_4 x^4 + \cdots \qquad (4\text{-}173)$$

This is a far cry from q_w proportional to local $\Delta T = a_0 + a_1 x + a_2 x^2 + a_3 x^3 + a_4 x^4 + \cdots$; hence the warning. To illustrate how drastically things are changed from what our intuition might tell us, consider the linear distribution

$$T_w - T_e = \Delta T_0 \left(1 \pm \frac{x}{L}\right) \qquad (4\text{-}174)$$

286 VISCOUS FLUID FLOW

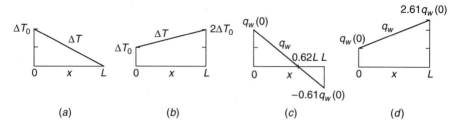

FIGURE 4-30
Local heat transfer with a linearly varying wall temperature on a flat plate, Eq. (4-175): (*a*, *c*) falling temperature difference; (*b*, *d*) rising temperature difference.

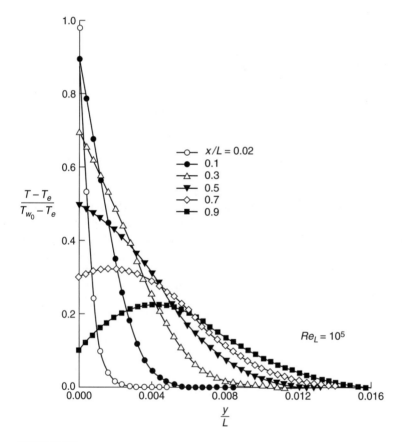

FIGURE 4-31
Finite-difference solution (Sec. 4-8.2) for temperature profiles in flat-plate flow with a linearly decreasing wall temperature. The wall heat transfer changes sign at $x/L = 0.62$.

From Eq. (4-173), the Nusselt number based upon ΔT_0 is

$$Nu_x = \frac{q_w}{k \Delta T_0} = 0.332 Pr^{1/3} Re_x^{1/2} \left(1 \pm \frac{1.6123x}{L}\right) \quad (4\text{-}175)$$

showing strong effects due to the linear contribution. For the rising temperature difference, $1 + x/L$, the heat transfer at the trailing edge is 31 percent greater than what one might expect from the local difference $2\Delta T_0$. For the falling case, $1 - x/L$, the heat transfer actually changes *sign* at the trailing edge, and the effect is not negligible, being 61 percent of the leading-edge heat transfer. These effects are sketched in Fig. 4-30.

The finite-difference method of Sec. 4-8.2 is valid for any $U(x)$ and/or $\Delta T(x)$ and was applied to this particular problem of falling $\Delta T_0(1 - x/L)$. The resulting temperature profiles are shown in Fig. 4-31 and illustrate warming of the local fluid by the hot upstream wall.

4-9 FLOW IN THE INLET OF DUCTS

The developing laminar flow in the entrance of a duct has captured the imaginations of scores of workers. Extensive reviews are given by Schmidt and Zeldin (1969) and by Shah and London (1978).

The entrance-flow geometry for a pipe of radius a is shown in Fig. 4-32. In a well-rounded entrance, the velocity profile at $x = 0$ is nearly uniform, $u = \bar{u} = $ const, and this assumption is common in theoretical studies. A boundary layer begins at the entrance whose initial behavior, when $\delta \ll a$, is like the Blasius flat-plate solution. However, due to continuity requirements, a retardation near the wall must cause the "inviscid-core" center flow to speed up (favorable gradient), thinning the boundary layer more than the Blasius estimate Eq. (4-49). Some distance downstream, the shear layers meet, and the duct is filled with boundary layer. Shortly thereafter, at $x = x_L$, the flow is essentially *developed* into the Poiseuille paraboloid. We term the region $0 \leq x \leq x_L$ the *entrance region*. Of particular engineering interest are (1) the excess pressure drop compared to Poiseuille flow and (2) the entrance length x_L/D. Similar considerations occur in turbulent entrance flow, where the initial boundary layers are likely to be laminar and then undergo transition to turbulence before they meet downstream.

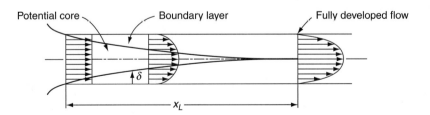

FIGURE 4-32
Laminar flow in the entrance of a duct.

The excess pressure drop is due to both increased shear in the entrance boundary layers and the acceleration of the core flow. Let the inlet pressure at $x = 0$ be p_0, and let $\tau_P = 4\mu u_p/a$ be the shear stress in the developed Poiseuille flow region, where $u_p = 2\bar{u}(1 - r^2/a^2)$. A control-volume momentum analysis of the region between $x = 0$ and any $x \geq x_L$ gives the following balance:

$$(p_0 - p_x)\pi a^2 = \tau_P 2\pi a x + \int_0^a \rho(u_p^2 - \bar{u}^2) 2\pi r\, dr + \int_0^x (\tau - \tau_P) 2\pi a\, dx$$

This may be written in dimensionless form as an "apparent" friction:

$$\frac{p_0 - p_x}{\tfrac{1}{2}\rho\bar{u}^2} = C_{f,\text{app}} \frac{4x}{D} = C_{fP} \frac{4x}{D} + K \qquad (4\text{-}176)$$

where

$$K = \frac{2}{3} + \int_0^{x/a} \frac{4(\tau - \tau_P)}{\rho\bar{u}^2} \frac{dx}{a}$$

Here $C_{fP} = 16/Re_D$ is the Poiseuille friction factor from Eq. (3-40), with $D = 2a$ the pipe diameter. The term K is the *excess* pressure drop due to the entrance region, whose component $2/3$ represents the pressure drop necessary to accelerate a uniform flow into a Poiseuille paraboloid. The integral term in K is the excess-shear contribution, approximately 0.6. Note that, for a long pipe, $x \gg D$, the term $C_{fP}(4x/D)$ dominates the total friction loss.

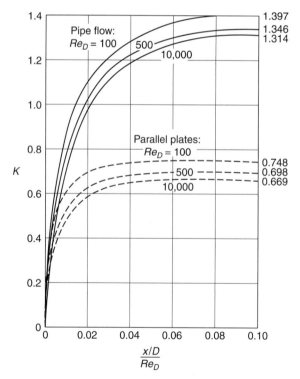

FIGURE 4-33
Pressure-drop parameter K [Eq. (4-176)] in the entrance of a duct for laminar flow. From finite-difference solutions of the full Navier–Stokes equations by Schmidt and Zeldin (1969).

If we extend Eq. (4-176) into the entrance region, $K = K(x)$, rising from zero at $x = 0$ to an asymptotic constant value K_∞ in the developed region. Figure 4-33 shows the finite-difference computations of Schmidt and Zeldin (1969) for $K(x)$ in pipes and between parallel plates. Note that there is only a slight Reynolds number effect, verifying the boundary-layer character of the entrance flow. Although $K_\infty = 1.31$ to 1.40 for pipe-flow computations in Fig. 4-33, the experiments reviewed by Shah (1978) indicate $K_\infty \approx 1.25$ is more realistic.

The overall friction for any section of the pipe correlates with a Graetz-type variable $\zeta = (x/D)/Re_D$. Shah (1978) suggests an interpolation formula that is valid within ± 2 percent for many duct shapes:

$$C_{f,\text{app}} Re \approx \frac{3.44}{\sqrt{\zeta}} + \frac{C_{fP}Re + K_\infty/4\zeta - 3.44/\sqrt{\zeta}}{1 + c/\zeta^2} \quad (4\text{-}177)$$

where C_{fP}, K_∞, and c are appropriate to the geometry listed in Table 4-7. The noncircular sections use hydraulic diameter D_h and Re_{D_h}.

The "entrance length" x_L can be defined in various ways. Figure 4-33 shows that development could be considered to occur when K approaches its final value K_∞ at $(x/D)/Re_D \approx 0.08$—an upper bound. A lower bound could be when the shear layers "meet" in a boundary-layer entrance theory. If the Blasius theory from Eq. (4-49) were accurate, we could set this meeting at $\delta = a = 5.0\sqrt{\nu x/u}$, or $(x/D)/Re_D \approx 0.01$. This is far too low because the accelerating core thins the shear layer. [There is a problem assignment to analyze the entrance as boundary layers patched to a flat inviscid core $u_c(x)$ which satisfies Bernoulli's equation.]

Shah and London (1978) define x_L as the point where the developing centerline velocity equals 99 percent of the Poiseuille value u_max and recommend the

TABLE 4-7
Constants to be used in Eq. (4-177)

b/a	$C_{fP}Re$	K_∞	c
	Pipe or Concentric Annulus		
0.0	16.00	1.25	0.000212
0.05	21.57	0.830	0.000050
0.10	22.34	0.784	0.000043
0.50	23.81	0.688	0.000032
0.75	23.97	0.678	0.000030
1.00	24.00	0.674	0.000029
	Rectangular Duct		
1.00	14.23	1.43	0.00029
0.50	15.55	1.28	0.00021
0.20	19.07	0.931	0.000076
0.00	24.00	0.674	0.000029
	Equilateral Triangle		
—	13.33	1.69	0.00053

following correlation for entrance length:

$$\frac{x_L}{D} \approx \frac{0.6}{1 + 0.035Re_D} + 0.056Re_D \qquad (4\text{-}178)$$

The entrance length does not vanish as Re approaches zero. It still takes about 0.6 diameters for noninertial creeping flow to change from a uniform into a parabolic profile.

4-10 ROTATIONALLY SYMMETRICAL BOUNDARY LAYERS

So far, all our boundary-layer solutions have been for plane flow. Let us now look at the rotationally symmetric case, as a preliminary to more general three-dimensional boundary layers. By rotationally symmetric, we mean flow in cylindrical polar coordinates (r, θ, z) where none of the variables $(v_r, v_\theta, v_z, p, T)$ depends upon θ. The circumferential velocity $v_\theta(r, z)$ is called the *swirl*. If $v_\theta = 0$ everywhere, the flow reduces to the special case of *axisymmetric flow*, where the streamlines remain in meridian planes.

We have already considered several rotationally symmetric flows in Chap. 3: pipe flow (including uniform swirl), Couette flow between rotating cylinders, axisymmetric stagnation flow, and flow near an infinite rotating disk. The boundary-layer equations under rotational symmetry were first given by Mangler (1945) in the form used today. Mangler chose the curvilinear coordinates (x, y, θ), with corresponding velocity components (u, v, w), shown in Fig. 4-34. The boundary-layer equations then become

Continuity: $\qquad \dfrac{\partial}{\partial x}(r_0 u) + r_0 \dfrac{\partial v}{\partial y} = 0 \qquad (4\text{-}179a)$

x momentum: $\quad \dfrac{\partial u}{\partial t} + u\dfrac{\partial u}{\partial x} + v\dfrac{\partial u}{\partial y} - \dfrac{w^2}{r_0}\dfrac{dr_0}{dx} = -\dfrac{1}{\rho}\dfrac{\partial p}{\partial x} + \nu\dfrac{\partial^2 u}{\partial y^2} \qquad (4\text{-}179b)$

y momentum: $\qquad \dfrac{\partial p}{\partial y} = 0 \qquad (4\text{-}179c)$

θ momentum: $\quad \dfrac{\partial w}{\partial t} + u\dfrac{\partial w}{\partial x} + v\dfrac{\partial w}{\partial y} + \dfrac{uw}{r_0}\dfrac{dr_0}{dx} = \nu\dfrac{\partial^2 w}{\partial y^2} \qquad (4\text{-}179d)$

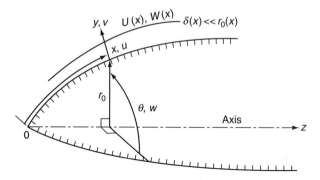

FIGURE 4-34
Coordinate system for rotationally symmetric flow, Eqs. (4-179).

Here $r_0(x)$ is the local surface radius, measured from the axis (*not* the radius of curvature of the surface). The derivation further assumes that r_0 is much larger than the boundary-layer thickness δ, so that variations in r_0 through the boundary layer can be neglected. At the edge of the boundary layer, the freestream velocity is $U(x)$, and the swirl is $W(x)$, which must be related to the pressure by the potential-flow relations

$$\frac{\partial U}{\partial t} + U\frac{\partial U}{\partial x} - \frac{W^2}{r_0}\frac{dr_0}{dx} = -\frac{1}{\rho}\frac{\partial p}{\partial x} \qquad (4\text{-}180)$$

$$\frac{\partial W}{\partial t} + U\frac{\partial W}{\partial x} + \frac{UW}{r_0}\frac{dr_0}{dx} = 0$$

The boundary conditions could include, for generality, a porous surface rotating about its axis at angular speed $\Omega(t)$:

At $y = 0$: $\quad u = 0 \quad v = v_w(x, t) \quad w = r_0\Omega(t)$

At $y \to \infty$: $\quad u = U(x, t) \quad w = W(x, t)$

At $x = 0$: $\quad$ known $u(0, y, t)$, $v(0, y, t)$, $w(0, y, t)$

At $t = 0$: $\quad$ known $u(x, y, 0)$, $v(x, y, 0)$, $w(x, y, 0)$

$\hfill (4\text{-}181)$

As with two-dimensional boundary-layer theory, we have lost the condition on v as $y \to \infty$, through neglect of the normal pressure gradient.

4-10.1 Axisymmetric Boundary Layers: Similarity Solutions

A much simpler special case is that of steady axisymmetric flow ($w = 0$), for which a great deal of work has been reported. Equations (4-179) reduce to

$$\frac{\partial}{\partial x}(r_0 u) + r_0 \frac{\partial}{\partial y}(v) = 0$$

$$u\frac{\partial u}{\partial x} + v\frac{\partial u}{\partial y} = U\frac{dU}{dx} + \nu\frac{\partial^2 u}{\partial y^2} \qquad (4\text{-}182)$$

with $\quad u(x, 0) = 0 \quad v(x, 0) = v_w(x) \quad u(x, \infty) \to U(x)$

It is assumed that the body shape $r_0(x)$ is known.

A family of similarity solutions to Eqs. (4-182) may be found if the freestream velocity is of the power-law form

$$U(x) = Cx^n \qquad (4\text{-}183)$$

These correspond to potential flow at zero angle of attack past a cone of half-angle ϕ. An extensive discussion is given by Evans (1968), and the half-angles which relate to various n are given in Table 4-8. It is not a simple relationship such as the Falkner–Skan plane wedge flows, where $\phi_{\text{wedge}} = m\pi/(1 + m)$.

TABLE 4-8
Cone half-angles ϕ vs. velocity parameter n

n	ϕ, deg	n	ϕ, deg
0.0	0.0	1.2	97.01
0.05	19.10	1.4	102.99
0.1	27.73	1.6	108.12
0.15	34.52	1.8	112.61
0.2	40.33	2.0	116.58
0.3	50.11	2.5	124.60
0.4	58.22	3.0	130.89
0.5	65.20	4.0	139.90
0.6	71.31	5.0	146.12
0.7	76.84	6.0	150.71
0.8	81.60	7.0	154.12
0.9	86.00	8.0	156.86
1.0	90.00	9.0	159.70

The proper similarity variables for these cone flows are

$$u(x, y) = U(x)f'(\eta)$$

where
$$\eta = y \left[\frac{(3 + n)U(x)}{2x\nu} \right]^{1/2} \qquad (4\text{-}184)$$

Substitution in Eqs. (4-182) gives the ordinary differential equation

$$f''' + ff'' + \frac{2n}{3 + n}(1 - f'^2) = 0 \qquad (4\text{-}185)$$

with
$$f(0) = f'(0) = 0 \quad \text{and} \quad f'(\infty) = 1$$

But this is identical to the Falkner–Skan relation, Eq. (4-71), except that the equivalent value of β is different. If we set the two equal, we find that any given cone flow is equivalent mathematically to a certain wedge flow:

$$\beta_{\text{cone}} = \frac{2n}{3 + n} = \beta_{\text{wedge}} = \frac{2m}{1 + m}$$

or
$$m_{\text{wedge}} = \frac{1}{3} n_{\text{cone}} \qquad (4\text{-}186)$$

Thus the cone flow $U = Cx^n$ has properties identical to the wedge flow $U = C' x'^{n/3}$, and the Falkner–Skan solution determines both. Note that C and x for the cone are not the same as C' and x' for the wedge, i.e., the pressure gradients are quite different for the two flows. The equivalence is only through the similarity solutions, e.g., the solution for axisymmetric stagnation flow ($n = 1$) can be taken directly from the known similar solution for a 90° wedge ($m = 1/3$).

The cone-wedge equivalence is no accident. Mangler (1948) deduced an extraordinary transformation,

$$x' = \frac{1}{L^2}\int_0^x r_0^2\,dx \qquad y' = \frac{r_0 y}{L}$$

$$u' = u \quad U'(x') = U(x) \qquad v' = \frac{L}{r_0}\left(v + \frac{yu}{r_0}\frac{dr_0}{dx}\right) \qquad (4\text{-}187)$$

where L is a reference length, which converts Eqs. (4-182) directly into an equivalent plane-boundary-layer flow, Eqs. (4-41). The advantage is that plane-flow methods may be used for (x', y', u', v'), but in turn the transformed pressure gradient $U'(x')$ may be a complicated function. Sometimes the Mangler transformation is welcome. Consider the cone of Eq. (4-184), where $r_0 = x\sin\phi$. The transformation Eqs. (4-187) gives $x' = (\text{const})x^3$. Thus $U = Cx^n$ for the cone is equivalent to plane flow $U' = C'x'^{n/3}$, as shown in Eq. (4-186). Since the Falkner–Skan results are already known, no new calculations are needed for cone flow.

4-10.2 A Thwaites-Type Method for Axisymmetric Flow

We may eschew the Mangler transformation and convert Eqs. (4-182) directly into integral form. The derivation is similar to Sec. 4-6.2 for plane flow, and the result is

$$\frac{C_f}{2} = \frac{1}{U^2}\frac{\partial}{\partial t}(U\delta^*) + \frac{2\theta + \delta^*}{U}\frac{\partial u}{\partial x} + \frac{\partial \theta}{\partial x} + \frac{\theta}{r_0}\frac{dr_0}{dx} - \frac{v_w}{U} \qquad (4\text{-}188)$$

The only difference from plane flow is the term involving dr_0/dx. The derivation assumes that $\delta \ll r_0$. For steady axisymmetric flow with an impermeable wall, this reduces to

$$\frac{\tau_w}{\rho} = \frac{1}{r_0}\frac{d}{dx}(r_0 U^2\theta) + U\frac{dU}{dx}\delta^* \qquad (4\text{-}189)$$

By exact analogy with Eq. (4-135) for plane flow, we regroup this relation in terms of the Holstein–Bohlen parameter, $\lambda = \theta^2(dU/dx)/\nu$. The result is

$$\frac{U}{r_0^2}\frac{d}{dx}\left(\frac{r_0^2\lambda}{dU/dx}\right) = 2S - 2\lambda(2 + H) = F(\lambda) \approx 0.45 - 6\lambda \qquad (4\text{-}190)$$

where $S(\lambda)$ and $H(\lambda)$ have the same meanings and correlations as in plane flow, Sec. 4-6.6. Again, by exact analogy with plane flow, Eq. (4-138), we can integrate Eq. (4-190) immediately:

$$\theta^2 \approx \frac{0.45\nu}{r_0^2 U^6}\int_0^x r_0^2 U^5\,dx \qquad (4\text{-}191)$$

with separation again assumed to occur at $\lambda = -0.09$. This expression was developed by Rott and Crabtree (1952). If r_0 is constant or much larger than x, Eq. (4-191) clearly reduces to the Thwaites' plane-flow method. The expected accuracy is ±5 percent.

4-10.3 Axisymmetric Finite-Difference Methods

Since the only difference between plane and axisymmetric boundary layers is the presence of $r_0(x)$ in the continuity Eq. (4-182), very little modification is needed to our numerical techniques of Sec. 4-7.

For axisymmetric flow, both the explicit x-momentum model, Eq. (4-146), and its implicit sibling, Eq. (4-149), are still valid. The stability constraints for the explicit model, Eqs. (4-147), are still valid. Only the continuity relation Eq. (4-148) must be changed by introducing $r_0(x)$, which we denote as r_m. The result of this modification is

$$v_{m+1,n} \approx v_{m+1,n-1} - \frac{\Delta y}{2\, \Delta x\, r_{m+1}}$$
$$\times [r_{m+1}(u_{m+1,n} + u_{m+1,n-1}) - r_m(u_{m,n} + u_{m,n-1})] \quad (4\text{-}192)$$

This relation is used to compute the next column of v's once the new u's are known from either the explicit or implicit models.

4-10.4 Application to Flow Past a Sphere

Consider a sphere of radius a immersed in a stream of velocity U_0. If ϕ is the angle measured from the nose, the potential-flow surface-velocity distribution is [White (2003), Sec. 8.8]

$$U = \tfrac{3}{2} U_0 \sin \phi = \tfrac{3}{2} U_0 \sin\left(\frac{x}{a}\right) \quad (4\text{-}193)$$

We may test this theoretical freestream distribution against our two methods: (1) the Rott–Crabtree modification of Thwaites' method, Eq. (4-191); and (2) a finite-difference method, Eqs. (4-149) and (4-192). The results for predicted skin friction are shown in Fig. 4-35. The Rott–Crabtree method predicts separation at $\phi = 103.6°$, and the most exact finite-difference computation gives $\phi_{\text{sep}} = 105.5°$.

Like the earlier cylinder example, the potential flow is not a good approximation because of first-order flow-separation effects. The actual velocity distribution measured by Fage (1936) at $Re_a \approx 200{,}000$ fits the curve

$$\frac{U}{U_0} \approx 1.5\frac{x}{a} - 0.4371\left(\frac{x}{a}\right)^3 + 0.1481\left(\frac{x}{a}\right)^5 - 0.0423\left(\frac{x}{a}\right)^7$$
$$\text{for } 0 \leq \frac{x}{a} \leq 1.48 \quad (4\text{-}194)$$

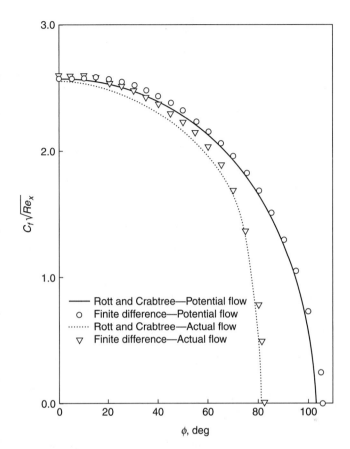

FIGURE 4-35
Comparison of the Rott–Crabtree (1952) integral method and a finite-difference computation for skin friction in sphere flow.

This drops off much faster than the potential flow, reaching a maximum of $U/U_0 = 1.274$ at $x/a = 1.291$ ($\phi = 74°$), whereas the potential flow has a maximum of $U/U_0 = 1.5$ at $\phi = 90°$.

Figure 4-35 also compares the two methods for this experimental $U(x)$ distribution. Again the two are acceptably close—the separation angles are 81.1° for the Rott–Crabtree method and 82.4° for the (nearly exact) finite-difference computations. The numerical method used was implicit and used a step size $\Delta\theta = 0.5°$.

4-10.5 Thick Axisymmetric Boundary Layers

This section has so far considered only "thin" boundary layers, where $\delta \ll r_0$. If a cylindrical body is long enough, however, there must be some point downstream where this assumption fails, and the boundary-layer thickness becomes of the order of the body diameter or greater. A long towed cable is an important practical example. The main effect of a thick axisymmetric boundary layer is that the cylindrical geometry must be taken into account; hence the term *transverse-curvature effect* is often used to describe these flows.

We now consider the problem of the axisymmetric laminar boundary layer along an extremely long cylinder of constant radius a. The pressure gradient is negligible ($U = U_0 \approx$ const), and the boundary-layer approximations apply if the square root of the Reynolds number Re_x is large. Using the "wall" coordinate $y = r - a$, $dy = dr$, the axisymmetric boundary-layer equations become

$$\frac{\partial u}{\partial x} + \frac{1}{a+y}\frac{\partial}{\partial y}[(a+y)v] = 0 \qquad (4\text{-}195a)$$

$$u\frac{\partial u}{\partial x} + v\frac{\partial u}{\partial y} = \frac{\nu}{a+y}\frac{\partial}{\partial y}\left[(a+y)\frac{\partial u}{\partial y}\right] \qquad (4\text{-}195b)$$

Note that if $y \ll a$, these equations reduce to the flat-plate equations, so that the boundary layer on a short cylinder is the Blasius solution.

There have been many studies of this problem, notably Glauert and Lighthill (1955), who gave an approximate solution valid over the whole range from short to very long cylinders. They pointed out that, for $\delta \gg a$, the convective acceleration is negligible and the momentum Eq. (4-195b) may be solved for

$$u(x, y) \approx \frac{a}{\mu}\tau_w(x)\ln\left(1 + \frac{y}{a}\right) \qquad \delta \gg a \qquad (4\text{-}196)$$

Interestingly, this formula should be valid for either laminar or turbulent flow along an extremely long cylinder. Glauert and Lighthill then extended this formula with a parameter $\alpha(x) = \mu U_0/(a\tau_w)$ to define the approximate profiles

$$u = \begin{cases} \dfrac{U_0}{\alpha}\ln\left(1 + \dfrac{y}{a}\right) & \text{for } y \leq \delta = a(e^\alpha - 1) \\ U_0 & \text{for } y \geq \delta \end{cases} \qquad (4\text{-}197)$$

The steady-flow cylindrical momentum integral relation is

$$\tau_w = \frac{d}{dx}\left[\int_0^\delta \rho u(U_0 - u)\left(1 + \frac{y}{a}\right)dy\right] \qquad (4\text{-}198)$$

Substitution of u from Eq. (4-197) in this relation gives an algebraic expression for $\alpha(x)$:

$$\frac{4\nu x}{U_0 a^2} = e^{2\alpha} + 3 - \frac{2}{\alpha}(e^{2\alpha} - 1) + \int_0^{2\alpha}\frac{e^z - 1}{z}dz \qquad (4\text{-}199)$$

The exponential integral is tabulated—or easily evaluated numerically—so Eq. (4-199) relates local shear $\alpha(x)$ to the *long-cylinder* parameter:

$$\xi = \sqrt{\frac{\nu x}{U_0 a^2}} = \frac{x/a}{\sqrt{Re_x}} \qquad (4\text{-}200)$$

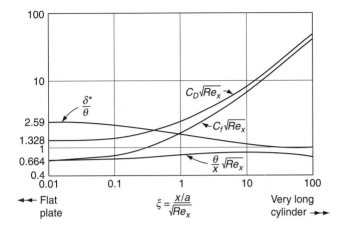

FIGURE 4-36
Boundary-layer parameters for steady laminar flow past the outside of an extremely long cylinder. Computed from the theory of Glauert and Lighthill (1955).

With $\xi(\alpha)$ known from Eq. (4-199), Glauert and Lighthill were able to tabulate shear and drag and thickness parameters, which are plotted in Fig. 4-36. For $\xi > 100$, a "very long" cylinder, the skin friction and drag approach the asymptotic result $C_f \sqrt{Re_x} \approx 2\xi/\ln(2\xi)$. For $\xi < 0.01$, the Blasius solution obtains.

Figure 4-36 shows approximate results, but they are accurate when compared with finite-difference computations. Equations (4-195) are parabolic and easily amenable to the numerical techniques of Sec. 4-7. In fact, the finite-difference programming of these equations is suggested—yes—as a problem assignment for this chapter.

4-10.6 The Narrow Axisymmetric Jet

If a round jet emerges from a circular hole with sufficient momentum, it remains narrow and grows slowly, the radial changes $\partial/\partial r$ being much larger than axial changes $\partial/\partial x$. An exact solution to the Navier–Stokes equations for a "point source of momentum," found by Squire (1951), was given as a problem assignment in Chap. 3. Schlichting (1933a) found a solution for a narrow jet, which satisfies the axisymmetric boundary-layer equations,

Continuity: $$\frac{\partial u}{\partial x} + \frac{1}{r}\frac{\partial}{\partial r}(rv) = 0 \qquad (4\text{-}201a)$$

x momentum: $$u\frac{\partial u}{\partial x} + v\frac{\partial u}{\partial y} = \frac{\nu}{r}\frac{\partial}{\partial r}\left(r\frac{\partial u}{\partial r}\right) \qquad (4\text{-}201b)$$

Schlichting reasoned that the jet thickness grew linearly, so that the proper similarity variable was r/x. He defined a stream function

$$\psi(r, x) = \nu x F(\eta) \qquad \eta = \frac{r}{x} \qquad (4\text{-}202)$$

from which the axisymmetrical a velocity components are

$$u = \frac{1}{r}\frac{\partial \psi}{\partial r} = \frac{\nu F'}{r}$$
$$v = -\frac{1}{r}\frac{\partial \psi}{\partial x} = \frac{\nu}{r}(\eta F' - F)$$
(4-203)

Substitution in the x-momentum equation (4-201b) gives the following third-order non-linear differential equation

$$\frac{d}{d\eta}\left(F'' - \frac{F'}{\eta}\right) = \frac{1}{\eta^2}(FF' - \eta F'^2 - \eta FF'')$$
(4-204)

This is devoid of x and r, therefore similarity has been achieved. The boundary conditions are $F(0) = F'(0) = F'(\infty) = 0$. Again we resist the temptation to retreat into Subroutine RUNGE in App. C, for Schlichting (1933a) found the exact solution:

$$F = \frac{(C\eta)^2}{1 + (C\eta/2)^2}$$
(4-205)

where the constant C is determined from the momentum of the jet, as was done for the plane jet in Sec. 4-4.2:

$$J = \rho \int_0^\infty u^2 2\pi r\, dr = \frac{16\pi}{3}\rho C^2 \nu^2$$

or
(4-206)

$$C = \left(\frac{3J}{16\pi\rho\nu^2}\right)^{1/2}$$

The axial jet velocity, then, from Eq. (4-203), is

$$u = \frac{3J}{8\pi\mu x}\left(1 + \frac{C^2\eta^2}{4}\right)^{-2}$$
(4-207)

The term in parentheses is the shape of the jet profile, which differs somewhat from the (sech^2) shape of the plane jet, Eq. (4-102). The jet centerline velocity drops off as x^{-1}. The mass rate of flow across any axial section of the jet is

$$\dot{m} = \rho \int_0^\infty u 2\pi r\, dr = 8\pi\mu x$$
(4-208)

The mass (or volume) flow increases with x and is independent of the jet momentum J, which affects only the width of the jet through C.

Figure 4-37 shows the jet streamlines from Eq. (4-205) for $C = 1$ and $C = 4$. The former case is not "narrow" and the boundary-layer approximations are marginal. The case $C \geq 4$ is clearly a narrow ($\partial/\partial r \gg \partial/\partial x$) jet, corresponding from Eq. (4-206) to a Reynolds number $J/\rho\nu^2 > 250$.

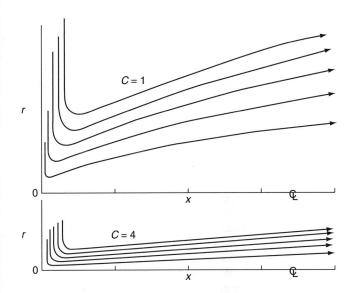

FIGURE 4-37
Streamlines of a round jet for two cases, from the boundary-layer theory of Eq. (4-205).

4-10.7 Axisymmetric Wakes

The plane wake was discussed in Sec. 4-4.3, using an Oseen-type linearization for the convective acceleration. Refer to Fig. 4-19 for the wake geometry, where the wake velocity defect $u_1 = (U_0 - u)$ is assumed small. For axisymmetric flow, the momentum Eq. (4-201b) is linearized to

$$U_0 \frac{\partial u_1}{\partial x} \approx \frac{\nu}{r} \frac{\partial}{\partial r}\left(r \frac{\partial u_1}{\partial r}\right) \qquad (4\text{-}209)$$

subject to $(\partial u_1/\partial r)(0, x) = 0$, $u_1(\pm\infty, x) = 0$, and $u_1(r, \infty) = 0$.

The solution is a Gaussian wake-velocity distribution:

$$u_1 = \frac{C}{x} U_0 \exp\left(-\frac{U_0 r^2}{4x\nu}\right) \qquad (4\text{-}210)$$

where C is a constant related to the drag force on the body

$$F \approx \rho U_0 \int_0^\infty u_1 2\pi r \, dr = \rho U_0^2 C \frac{4\pi\nu}{U_0} = C_D \tfrac{1}{2} \rho U_0^2 L^2$$

from which $C = C_D U_0 L^2/8\pi\nu$, where L is a body reference length. The wake defect velocity thus is

$$\frac{u_1}{U_0} = C_D \frac{U_0 L}{8\pi\nu} \frac{L}{x} \exp\left(-\frac{U_0 r^2}{4x\nu}\right) \qquad (4\text{-}211)$$

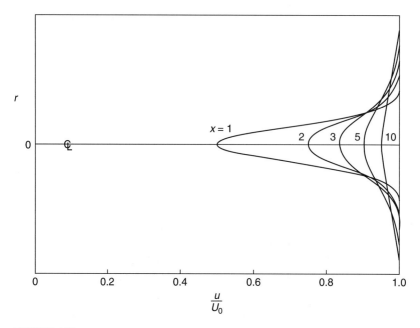

FIGURE 4-38
Velocity profiles illustrating the decay of a round wake from Eq. (4-211). Units on x and r are arbitrary.

These similar profiles occur in the far-field wake, at least three body-lengths downstream of the trailing edge.

Figure 4-38 shows some typical wake-velocity profiles plotted from Eq. (4-211), illustrating the decay of the centerline velocity defect and the spreading of the wake. For further details on laminar wakes, consult the monograph by Berger (1971).

4-11 ASYMPTOTIC EXPANSIONS AND TRIPLE-DECK THEORY

Boundary-layer theory is a marvelous and useful simplification of the full Navier–Stokes equations. It may be regarded as a first approximation to analysis of near-wall flows with no slip. However, the theory has several vexing limitations, even for incompressible flows:

1. High Reynolds numbers $= \mathcal{O}(1000)$
2. Not valid in separated-flow regions
3. Not valid too near the leading edge (where $Re_x \ll Re_L$)
4. Not valid at the trailing edge, where no slip gives way to a wake
5. Not able to predict the effect of viscous streamline displacement on the outer stream

6. Not able to predict the effect of wall curvature, whether transverse or longitudinal
7. Not able to predict the effect of finite shear (vorticity) in the outer stream

For compressible flows, additional vexations would be the effects of entropy and stagnation temperature gradients in the outer stream.

It is not surprising, then, that many workers have attempted to extend boundary-layer theory to higher approximations. Prandtl himself studied the modification of the freestream by correcting the potential-flow result for boundary-layer displacement thickness $\delta^*(x)$, iterating this correction to a (not always successful) conclusion. The general study of higher order boundary-layer approximations is reviewed by van Dyke (1969), and the use of these and other perturbation methods in fluid mechanics is treated in his text [van Dyke (1964)] and in the text by Nayfeh (2000). Probably the most successful method is triple-deck theory, initiated by Stewartson in 1969 and generalized in his review article [Stewartson (1974)]. The "upper deck" is controlled by the freestream interaction and the "lower deck" reacts to wall conditions. The "main deck" serves as a merging layer for the upper and lower decks. The present section is a brief discussion of such enhanced boundary-layer approximations.

4-11.1 Friedrichs' Model of the Boundary Layer

One trouble with the boundary-layer momentum equation is that it is a *singular perturbation problem:* The highest order derivative, $\partial^2 u/\partial y^2$, is multiplied by a very small coefficient, μ. In the limit as μ approaches zero (inviscid flow), we drop back to first-order derivatives and the no-slip condition is lost. Thus high Reynolds number flow is not merely a simple "correction" to inviscid flow; it gives a singular change in the near-wall flow pattern. Attempts to expand a near-wall solution in, say, powers of μ (or $1/Re$), have not been entirely successful.

Friedrichs (1942) gave a model ordinary differential equation which illustrates these difficulties and the asymptotic methods which resolve them. One form of Friedrichs' model is

$$\epsilon \frac{d^2 u}{dy^2} + \frac{du}{dy} = a \qquad \epsilon \ll 1 \qquad (4\text{-}212)$$

with
$$u(0) = 0 \qquad u(1) = 1$$

There is a rough resemblance to the boundary-layer momentum Eq. (4-41), with $\epsilon = 1/Re$ and $a < 1$ crudely modeling the x-derivative terms.

Although a (non-linear) boundary-layer problem would not have a closed-form solution, we can easily determine the exact solution of Eq. (4-212):

$$u = ay + (1-a)\frac{1 - \exp(-y/\epsilon)}{1 - \exp(-1/\epsilon)} \qquad (4\text{-}213)$$

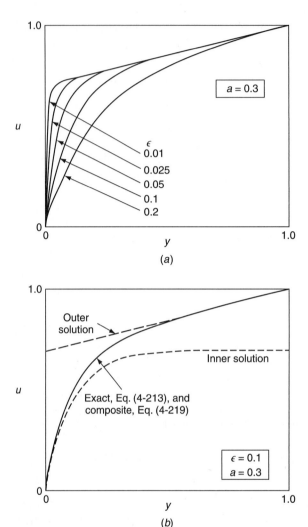

FIGURE 4-39
Friedrichs' model of the boundary layer: (*a*) exact solutions for various ϵ; (*b*) the inner and outer solutions from Eqs. (4-218).

Some exact profiles are plotted in Fig. 4-39*a* for $a = 0.3$ and various (small) values of ϵ. There is a "boundary layer," whose thickness is proportional to ϵ, rising from no slip at the wall to merge with a linear "outer stream." Similar profile shapes would occur for other values of the constant a.

In the limit as $\epsilon \to 0$, we lose the highest order derivative in Eq. (4-212), and the remaining "outer layer" equation, $u' = a$, has the solution $u = ay + C$. We can satisfy only one boundary condition, appropriately at the outer edge, $u(1) = 1$. Thus the "outer" solution of Friedrichs' model is

$$u_{\text{outer}} = ay + 1 - a \qquad (4\text{-}214)$$

This does not satisfy the no-slip condition at the wall.

The inner region is thin, of $\mathcal{O}(\epsilon)$, where $\epsilon \ll 1$. To enlarge it to unit size, we define a magnified variable and rename the function:

$$Y = \frac{y}{\epsilon} \qquad F(Y) = f(y)$$

Making this change in Friedrichs' model Eq. (4-212) gives

$$\frac{d^2F}{dY^2} + \frac{dF}{dY} = a\epsilon \approx 0 \tag{4-215}$$

The solution of this "inner" equation which satisfies no slip, $F(0) = 0$, is

$$F_{\text{inner}} = C(1 - e^{-Y}) \tag{4-216}$$

where C is arbitrary because the formula is not valid in the outer region.

4-11.2 The Matching Principle

The inner and outer solutions are valid only in their respective regions. How can we match the two to approximate the desired solution? One idea would be set the two expressions equal at some intermediate point, say, $y = \epsilon$. Accuracy will generally be poor for this oversimplified scheme.

Van Dyke (1964) proposes the following heuristic matching philosophy. Write the two solutions in terms of the same variable—either y or Y in the present case—and compare terms containing like powers of the perturbation parameter—ϵ in the present case. This comparison gives the best-match values of any undetermined constants.

Let us leave Eq. (4-214) alone and rewrite Eq. (4-216) in terms of y:

$$F = C\left(1 - e^{-y/\epsilon}\right) \approx C \tag{4-217}$$

since, as $\epsilon \to 0$, $\exp(-y/\epsilon)$ is vanishingly small. Comparing Eqs. (4-217) and (4-214), we see that the best-match value of C is $(1 - a)$. By writing the inner solution F in terms of the outer variable y, we find its strongest effect on the outer region.

This first-order analysis of Friedrichs' model problem then gives

$$\begin{aligned} F_{\text{inner}} &\approx (1 - a)(1 - e^{-Y}) \\ f_{\text{outer}} &\approx (1 - a) + ay \end{aligned} \tag{4-218}$$

These are plotted together in Fig. 4-39b for $\epsilon = 0.1$. Each is seen to be an excellent approximation in its respective region, although clearly they do not merge to form a complete approximation to the problem.

We may use the inner and outer solutions to form a composite function that should be a good approximation over the entire region. A pragmatic approach is suggested in Sec. 5.10 of van Dyke (1964): Sum the two solutions and subtract the terms common to both, which would otherwise be counted twice. Thus $u \approx [F(Y) + f(y) - (\text{common terms})]$. In the simple first-order solutions of

Eqs. (4-218), we see that the only common term is $(1 - a)$, which we subtract from the sum and thus obtain the approximation

$$u_{\text{composite}} \approx (1 - a)(1 - e^{-y/\epsilon}) + ay \tag{4-219}$$

This composite estimate is within ± 1 percent of the exact solution, Eq. (4-213), for all $\epsilon \leq 0.2$. It is plotted in Fig. 4-39b for $\epsilon = 0.1$ and is nearly coincident with the exact solution.

4-11.3 Higher Order Terms: Matched Asymptotic Expansions

Van Dyke generalizes the previous simple first-order procedure into a method of matching inner and outer expansions of any order. The two solutions are expanded in terms of a small parameter (ϵ in Friedrichs' model) and their respective variables. Then each expansion is rewritten in terms of its opposite variable, and terms of similar order are compared so that arbitrary constants can be determined. This is an *asymptotic matching principle* [van Dyke (1964)]:

The m-term inner expansion of (the n-term outer expansion)
= the n-term outer expansion of (the m-term inner expansion)

In general, m and n may be different integers, but it is convenient to let them be equal. Although exceptions to this principle may exist, the matching procedure works well for boundary-layer problems.

We illustrate this procedure by applying it to Friedrichs' model Eq. (4-212). The outer solution is expanded in powers of ϵ:

$$f_{\text{outer}} = f0(y) + \epsilon f1(y) + \epsilon^2 f2(y) + \cdots \tag{4-220}$$

subject to the single outer boundary condition $f(1) = 1$, with the first-order expansion taking the whole burden: $f0(1) = 1$, $f1(1) = 0$, $f2(1) = 0$.

Now substitute Eq. (4-220) in the basic differential equation (4-212) and collect like powers of ϵ. The resulting equations and solutions are

$$\begin{aligned} f0' &= a & \text{or} & & f0 &= (1 - a) + ay \\ f0'' + f1' &= 0 & \text{or} & & f1 &= C_1 y \quad C_1 = 0 \\ f1'' + f2' &= 0 & \text{or} & & f2 &= C_2 y \quad C_2 = 0 \end{aligned} \tag{4-221}$$

For this particular simple model, then, the outer solution to third order is exactly what we found in a preliminary analysis, Eq. (4-214).

Now expand the inner solution in a similar manner in terms of $Y = y/\epsilon$:

$$F_{\text{inner}} = F0(Y) + \epsilon F1(Y) + \epsilon^2 F2(Y) + \cdots \tag{4-222}$$

subject to the single inner condition $F(0) = 0 = F0(0) = F1(0) = F2(0)$. Substitute this expansion also in Eq. (4-212) and collect like powers of ϵ. The resulting outer equations and solutions are

$$F0'' + F0' = 0 \quad \text{or} \quad F0 = C_0(1 - e^{-Y})$$
$$F1'' + F1' = a \quad \text{or} \quad F1 = aY + C_1(1 - e^{-Y}) \quad (4\text{-}223)$$
$$F2'' + F2' = 0 \quad \text{or} \quad F2 = C_2(1 - e^{-Y})$$

This is quite different from the simple inner solution found earlier in Eq. (4-216).

To apply the matching principle, rewrite these expansions in terms of their opposite-layer variables and compare. Since the outer solution is so simple, we merely convert the inner solution from Y to y/ϵ and compare:

$$u_{\text{outer}} = (1 - a) + ay + \epsilon(0) + \epsilon^2(0)$$
$$u_{\text{inner}} = C_0 + ay + \epsilon C_1 + \epsilon^2 C_2$$

The comparison yields $C_0 = (1 - a)$ and $C_1 = C_2 = 0$. Thus, for this particular simple model, expansion to third order does not add any more significant terms, and the proper inner–outer solution is Eq. (4-218). This is one reason why the first-order composite function Eq. (4-219) was such a smashing success. In boundary-layer theory, increasing the order of the asymptotic expansions does indeed yield new terms and improved results.

4-11.4 Application to a Flat-Plate Boundary Layer

A number of matched asymptotic expansion solutions to boundary-layer problems are given in the texts by van Dyke (1964) and Nayfeh (2000). Here we briefly discuss results for incompressible flow past a flat plate that improve upon the "first-order" Blasius solution. Matched asymptotic expansions were used to correct for (1) the leading-edge (low Reynolds number) effect and (2) the displacement of the outer streamlines. A third effect, at the trailing edge, which changes suddenly from no slip to a growing wake, is resolved by the "triple-deck" theory of the next section. All three of these effects can be visualized in the numerical calculations of Dennis and Dunwoody (1966) in Fig. 4-8.

Second-order boundary-layer theory for the flat plate, as reviewed by van Dyke (1964, 1969) does not include the trailing edge. The appropriate small parameter for expansion is

$$\epsilon = Re^{-1/2}$$

where
$$Re = \frac{UL}{\nu} \quad (4\text{-}224)$$

The stream function is expanded up to second order in terms of variables nondimensionalized by U and L:

$$\psi(x, y, \epsilon) = \psi_1(x, y) + \epsilon \psi_2(x, y) + \cdots$$

Substitution in the continuity and Navier–Stokes equations reveals that the inner expansion has $\psi_2 = 0$ and ψ_1 equals the Blasius solution from Sec. 4-3.1. After application of the asymptotic matching principle from Sec. 4-11.3, a two-term outer expansion results for small y:

$$\psi = y - \epsilon(1.7208)x^{1/2} \qquad (4\text{-}225)$$

Thus the second-order correction of an undisturbed freestream, $\psi = y$, is the (parabolic-shaped) effect of the displacement thickness δ^* from Eq. (4-50).

The local skin friction is computed to second order by differentiating the inner solution at the wall:

$$C_f = \frac{0.664}{\sqrt{Re_x}} + \frac{0}{Re_x} \qquad (4\text{-}226)$$

There is no second-order correction of local friction. However, as shown by Imai (1957), the integrated friction drag does indeed show an effect:

$$C_D = \frac{1.328}{\sqrt{Re_L}} + \frac{2.326}{Re_L} + \cdots \qquad (4\text{-}227)$$

The correction term was stated without proof in Eq. (4-54) and compared to low Reynolds number data in Fig. 4-7. Interestingly, Imai (1957) extends the whole analysis to *third* order and finds that the results involve new integration constants which cannot be resolved without supplementary information such as a finite-difference solution. For further details, consult van Dyke (1964, 1969).

4-11.5 Trailing-Edge Flow: The Triple-Deck Structure

Near the trailing edge of a body immersed in a stream, either there is boundary-layer separation (blunt-body flows) or an unseparated wake (flat-plate flow). In either case, the two-layer inner–outer expansions are insufficient to resolve the flow field accurately. It was discovered by Stewartson (1969) and independently by Messiter (1970) that the flow near the trailing edge of a finite-length flat plate has a *triple-deck* structure, illustrated in Fig. 4-40. The triple-deck region is of size $Re^{-3/8}$, centered about the trailing edge, merging upstream with the two-layer Blasius flow of Sec. 4-11.4 and downstream with the two-layer wake of Goldstein (1930). Near the wall, dominated by viscous effects, is the *lower* deck, of size $Re^{-5/8}$, which enforces the no-slip (or wake-slip) condition at the inner boundary. This is patched by velocity and pressure to the *main* deck, of size $Re^{-1/2}$, in which viscous effects are secondary. Finally, there is an *upper* deck, of purely potential flow, which is controlled by changes in the main deck. Centered in a radius of size $Re^{-3/4}$ about the trailing edge is a small region where the full equations of motion are needed, as shown in Fig. 4-40.

The friction drag from triple-deck theory predicts a new contribution from the trailing edge:

$$C_D = \frac{1.328}{Re_L^{1/2}} + \frac{2.661}{Re_L^{7/8}} + \mathcal{O}(Re_L^{-1}) \qquad (4\text{-}228)$$

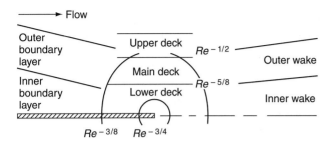

FIGURE 4-40
Sketch of the triple-deck region at the trailing edge of a flat plate, merging into two-layer upstream and downstream regions. [*After Stewartson (1969) and Messiter (1970).*]

This formula is both more rigorous and more accurate than Eq. (4-227). It is in excellent agreement with both the data of Fig. 4-6 and the flat-plate finite-difference calculations in Table 12.1 of the text by Cebeci and Cousteix (1998).

The Navier–Stokes equations reduce to a different form in each part of a triple deck. The complete equations and matching principles are given by Stewartson (1974) and are beyond the scope of this text. Triple-deck analysis applies not only to low-speed flat-plate flow but also to various body shapes and compressible flows. In fact, the example used in Stewartson's review (1974) is the interaction between a supersonic boundary layer and an impinging oblique shock wave.

In more complicated geometries, triple-deck theory requires numerical evaluation by, say, finite-difference methods. For example, Ragab and Nayfeh (1982) analyze low-speed boundary-layer flow near a small hump in the wall and compare the triple-deck results with an interacting boundary-layer theory. Theories are satisfactory, but the complexity is such that the effort required is not too far removed from full Navier–Stokes modeling. For further discussion, see Sec. 14.4 of the monograph by Schlichting and Gersten (2000).

4-12 THREE-DIMENSIONAL LAMINAR BOUNDARY LAYERS

This text is primarily concerned with two-dimensional boundary layers—including axisymmetric flows, which can readily be interpreted using two-dimensional concepts. However, many real-world flows are three-dimensional: swept finite wings, corners, turbine blades, slender bodies at an angle of attack, spinning projectiles, curved ducts, and wing–body junctions. The theory of laminar three-dimensional boundary layers is well developed but limited, of course, to unseparated flows. Flow separation in three dimensions, when it occurs, is a complex topological problem, as reviewed by Tobak and Peake (1982) and by Wang (1997). Progress in analysis of turbulent three-dimensional flows has been slower because of the lack of detailed data. For example, the 1980–81 Stanford Conference on Complex

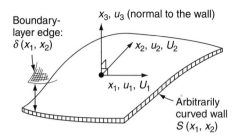

FIGURE 4-41
Orthogonal coordinate system for three-dimensional boundary layer.

Turbulent Flows [Kline et al. (1982)] had very few test cases that exhibited three-dimensionality.

There have been several recent reviews of three-dimensional boundary-layer physics and analysis, primarily for turbulent flows: Eichelbrenner (1973), Cebeci (1984), Cousteix (1986), and Bradshaw (1987). Experimental work is broadly represented in the conference edited by Fernholz and Krause (1982). Meanwhile, Dwyer (1981) specifically reviews the equations, boundary conditions, and numerical analysis of laminar three-dimensional flows.

In general, the boundary layer flows over an arbitrarily curved surface S, as in Fig. 4-41. We may choose orthogonal curvilinear coordinates (x_1, x_2, x_3) with x_3 everywhere normal to S. The boundary-layer velocities are (u_1, u_2, u_3); there are *two* freestream velocity components, U_1 and U_2; and the boundary layer is presumed thin, so that $u_3 \ll u_1, u_2$. In orthogonal coordinates, the element of arc length ds is related to (x_1, x_2, x_3) through scale factors h_1, h_2, and h_3, just as discussed in Sec. 2-7:

$$(ds)^2 = (h_1 dx_1)^2 + (h_2 dx_2)^2 + (h_3 dx_3)^2 \qquad (4\text{-}229)$$

Here, however, with x_3 everywhere normal to the wall, we have $h_3 \equiv 1$ everywhere. If the scale factors are not known, they can be found by relating (x_1, x_2, x_3) to any convenient Cartesian system (x, y, z), from which it follows that

$$h_1^2 = \left(\frac{\partial x}{\partial x_1}\right)^2 + \left(\frac{\partial y}{\partial x_1}\right)^2 + \left(\frac{\partial z}{\partial x_1}\right)^2 = \left[\left(\frac{\partial x_1}{\partial x}\right)^2 + \left(\frac{\partial x_1}{\partial y}\right)^2 + \left(\frac{\partial x_1}{\partial z}\right)^2\right]^{-1}$$

$$(4\text{-}230)$$

with exactly similar expressions for h_2 and h_3. Again, we are choosing the particular case $h_3 = 1$.

In these coordinates, the incompressible continuity equation becomes

$$\frac{1}{h_1 h_2}\left[\frac{\partial}{\partial x_1}(h_2 u_1) + \frac{\partial}{\partial x_2}(h_1 u_2) + \frac{\partial}{\partial x_3}(h_1 h_2 u_3)\right] = 0 \qquad (4\text{-}231)$$

which is exact. By the boundary-layer approximations, the third term is approximately equal to simply $\partial u_3/\partial x_3$.

As hoped for, the thin boundary layer leads to the result that the normal pressure gradient $\partial p/\partial x_3$ can be neglected. The two remaining boundary-layer momentum equations then are

$$\frac{Du_1}{Dt} + \frac{u_1 u_2}{h_1 h_2}\frac{\partial h_1}{\partial x_2} - \frac{u_2^2}{h_1 h_2}\frac{\partial h_2}{\partial x_1} \approx -\frac{1}{\rho h_1}\frac{\partial p}{\partial x_1} + \nu\frac{\partial^2 u_1}{\partial x_3^2} \quad (4\text{-}232)$$

$$\frac{Du_2}{Dt} - \frac{u_1^2}{h_1 h_2}\frac{\partial h_1}{\partial x_2} + \frac{u_1 u_2}{h_1 h_2}\frac{\partial h_2}{\partial x_1} \approx -\frac{1}{\rho h_2}\frac{\partial p}{\partial x_2} + \nu\frac{\partial^2 u_2}{\partial x_3^2}$$

where

$$\frac{D}{Dt} = \frac{\partial}{\partial t} + \frac{u_1}{h_1}\frac{\partial}{\partial x_1} + \frac{u_2}{h_2}\frac{\partial}{\partial x_2} + u_3\frac{\partial}{\partial x_3}$$

The boundary-layer thermal-energy equation can be written in terms of the total enthalpy $H \approx c_p T + \tfrac{1}{2}(u_1^2 + u_2^2)$:

$$\frac{DH}{Dt} \approx \frac{1}{\rho}\frac{\partial p}{\partial t} + \frac{\alpha}{Pr}\frac{\partial^2}{\partial x_3^2}\left[H + \frac{(Pr-1)}{2}(u_1^2 + u_2^2)\right] \quad (4\text{-}233)$$

where Pr is the Prandtl number and $\alpha = k/\rho c_p$. From this equation, we notice that in steady flow with Prandtl number unity, the total enthalpy H is constant everywhere in the boundary layer if the wall is insulated, i.e., Crocco's relation is not affected by three-dimensionality.

The pressure gradients in Eqs. (4-232) must be matched to the freestream velocity, which has two components $U_1(x_1, x_2, t)$ and $U_2(x_1, x_2, t)$. The proper matching is given by Euler's relations for inviscid flow

$$-\frac{1}{\rho h_1}\frac{\partial p}{\partial x_1} = \frac{\partial U_1}{\partial t} + \frac{U_1}{h_1}\frac{\partial U_1}{\partial x_1} + \frac{U_2}{h_2}\frac{\partial U_1}{\partial x_2} + \frac{U_1 U_2}{h_1 h_2}\frac{\partial h_1}{\partial x_2} - \frac{U_2^2}{h_1 h_2}\frac{\partial h_2}{\partial x_1}$$

$$-\frac{1}{\rho h_2}\frac{\partial p}{\partial x_2} = \frac{\partial U_2}{\partial t} + \frac{U_1}{h_1}\frac{\partial U_2}{\partial x_1} + \frac{U_2}{h_2}\frac{\partial U_2}{\partial x_2} + \frac{U_1 U_2}{h_1 h_2}\frac{\partial h_2}{\partial x_1} - \frac{U_1^2}{h_1 h_2}\frac{\partial h_1}{\partial x_2} \quad (4\text{-}234)$$

Finally, the boundary conditions, including a porous wall, are

At $x_3 = 0$: $\quad u_1 = u_2 = 0 \quad u_3 = v_w(x_1, x_2, t) \quad T = T_w$

At $x_3 \to \infty$: $\quad u_1 \to U_1 \quad\quad u_2 \to U_2 \quad\quad\quad\quad T = T_e$ $\quad (4\text{-}235)$

Another item of interest is that there are two components of shear stress in a three-dimensional boundary layer:

$$\tau_1 = \mu\frac{\partial u_1}{\partial x_3}$$

$$\tau_2 = \mu\frac{\partial u_2}{\partial x_3} \quad (4\text{-}236)$$

This holds at the wall also, and the resultant wall shear direction may not be in the same direction as the resultant freestream velocity and probably is not.

4-12.1 Cartesian Coordinates: Secondary Flow and Skewing

In Cartesian coordinates, $h_1 = h_2 = 1$ and we may choose y as normal to the wall. The boundary-layer continuity and momentum equations, for steady flow with constant density and viscosity, become

$$\frac{\partial u}{\partial x} + \frac{\partial v}{\partial y} + \frac{\partial w}{\partial z} = 0$$

$$u\frac{\partial u}{\partial x} + v\frac{\partial u}{\partial y} + w\frac{\partial u}{\partial z} = U\frac{\partial U}{\partial x} + W\frac{\partial U}{\partial z} + \nu\frac{\partial^2 u}{\partial y^2} \quad (4\text{-}237)$$

$$u\frac{\partial w}{\partial x} + v\frac{\partial w}{\partial y} + w\frac{\partial w}{\partial z} = U\frac{\partial W}{\partial x} + W\frac{\partial W}{\partial z} + \nu\frac{\partial^2 w}{\partial z^2}$$

where $U(x, z)$ and $W(x, z)$ are the freestream velocity components. These equations, like their curvilinear counterparts Eqs. (4-232), are parabolic in x and z and can be marched downstream from known initial values, if U and W are known. The initial conditions are not limited to a single "entrance," $x = 0$, but extend along the *sides* of the computational field. Such side conditions, for example, $(x, y, z) = (x, y, 0)$ and (x, y, L), are often not well known or well modeled. Dwyer (1981) takes care to point out that the "region of influence" of a point in a three-dimensional boundary layer is affected by both convection and diffusion and is especially strong when there is large crossflow convection, as in a spinning projectile. Thus numerical and other analytical methods are greatly influenced by the details of the region of influence.

Suppose that U is locally aligned with the mainstream direction. Then $w(x, y, z)$ is called the crossflow or *secondary flow* and is strongly dependent on the curvature of the main streamlines and the associated crossflow pressure gradients. Two different types of flow are shown in Fig. 4-42. In Fig. 4-42a, the crossflow is unidirectional, implying a uniformly positive crossflow pressure gradient, similar to the rotating disk flow $v_r(z)$ in Fig. 3-7. An interesting way to visualize this three-dimensional motion is a hodograph plot of w vs. u, as in Fig. 4-42c. Johnston (1960) suggests that a triangular hodograph shape is suitable for many unidirectional crossflows. Thus, if we model the mainflow U by, say, a two-dimensional velocity profile and correlate W with U through the hodograph, we may approximate such problems with integral methods.

A second type of crossflow is bidirectional, as shown in Fig. 4-42b. This could be caused, for example, by a reversal in outer streamline curvature. The associated hodograph, Fig. 4-42d, is S-shaped, and the triangular approximation fails. Since such a hodograph shape is not well correlated with the outer stream conditions, integral analysis is less successful.

Note that, in Figs. 4-42c and d, the surface streamline is at an angle β with respect to the main freestream and is related to the surface shear stresses by

$$\tan \beta = \frac{\tau_{zy,\text{wall}}}{\tau_{xy,\text{wall}}} \quad (4\text{-}238)$$

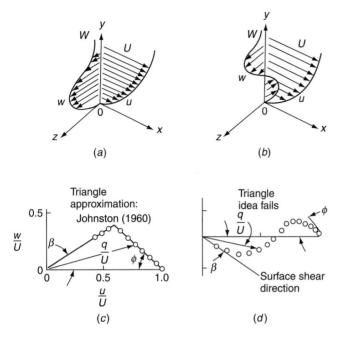

FIGURE 4-42
Unidirectional and bidirectional skewing in three-dimensional pressure-driven boundary layers: (*a*) unidirectional skewing; (*b*) bidirectional skewing; (*c*) unidirectional hodograph; (*d*) bidirectional hodograph.

In our numerical solution for the rotating disk, Table 3-5, the surface streamline lay at an angle $\beta = 39.6°$ compared to purely circumferential "mainstream" motion.

4-12.2 Flat-Plate Flow with a Parabolic Freestream

A simple and beautiful example of secondary flow was given by Loos (1955), who postulated a sharp flat plate immersed in a freestream that moves in parabolic curves across the plate, as shown in Fig. 4-43. The velocity components are[†]

$$U = \text{const} \qquad W = U(a - bx)$$

Here a is the initial slope, and b is the curvature of the outer streamlines, as we can demonstrate from the streamline geometric relationship

$$\left.\frac{dz}{dx}\right|_{\text{streamline}} = \frac{W}{U} = a - bx \quad \text{or} \quad z = z_0 + ax - \frac{1}{2}bx^2 \quad (4\text{-}239)$$

Hence, z_0 is where the streamline intercepts the plate leading edge ($x = 0$), a is the leading-edge streamline slope, $a = \tan \phi_0$, and $-b = z''_{\text{streamline}}$ is the curvature of the parabolic streamlines. All these are shown in Fig. 4-43. Since the plate has such

---

[†]Note that this freestream is a rotational flow, that is, $\partial U/\partial z - \partial W/\partial x \neq 0$. This can be ignored; the illustration of secondary flow is what matters.

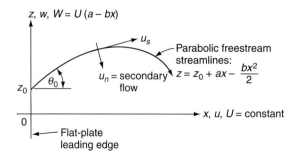

FIGURE 4-43
Definition sketch for parabolic flow over a flat plate.

a simple geometry, the best choice of coordinates is the Cartesian system (x, y, z) with velocities u, v, w. For this parabolic freestream, Eqs. (4-237) reduce to

$$\frac{\partial u}{\partial x} + \frac{\partial v}{\partial y} = 0$$

$$u\frac{\partial w}{\partial x} + v\frac{\partial w}{\partial y} = -bU^2 + \nu\frac{\partial^2 w}{\partial y^2} \qquad (4\text{-}240)$$

$$u\frac{\partial u}{\partial x} + v\frac{\partial u}{\partial y} = \nu\frac{\partial^2 u}{\partial y^2}$$

since the flow is everywhere independent of z. The boundary conditions are

At $y = 0$: $\quad u = v = w = 0$
At $y \to \infty$: $\quad u \to U \quad w \to W$ $\qquad (4\text{-}241)$

The flow reduces to the Blasius solution if $b = 0$, since the term $-bU^2$ is the curvature effect that induces the secondary motion. Even if $a \neq 0$, the solution is still a Blasius flow along the slanted straight lines $dz/dx = a = \tan\phi_0$. In other words, the flow along a flat plate is not affected by the direction with which the freestream crosses the leading edge.

Loos (1955) obtained the solution using a Blasius similarity variable,

$$\eta = \frac{y}{2}\left(\frac{U}{\nu x}\right)^{1/2}$$

from which the velocities are related to three functions f, g, and h:

$$u = \frac{1}{2}Uf'(\eta)$$

$$v = \frac{1}{2}(\eta f' - f)\sqrt{\frac{U\nu}{x}} \qquad (4\text{-}242)$$

$$w = U[ag(\eta) - cxh(\eta)]$$

Substitution in Eqs. (4-240) gives three ordinary differential equations

$$f''' + ff'' = 0 \qquad (4\text{-}243a)$$

$$g'' + fg' = 0 \qquad (4\text{-}243b)$$

$$h'' + fh' - 2f'h = 0 \qquad (4\text{-}243c)$$

with the associated boundary conditions

$$f(0) = f'(0) = g(0) = h(0) = 0$$
$$f'(\infty) = 2 \tag{4-244}$$
$$g(\infty) = h(\infty) = 1$$

Equation (4-243a) is the Blasius equation for two-dimensional flat-plate flow. Equation (4-243b) can be integrated to give

$$g = \tfrac{1}{2}f' \tag{4-245}$$

This is again the Blasius solution for that portion of the spanwise flow w not associated with curvature—another verification of the independence principle. Finally, with $f(\eta)$ known, Eq. (4-243c) is linear but with no known closed-form solution. Sowerby (1954) computed a numerical solution to complete the problem.

To illustrate the secondary flow, we can resolve the flow into velocity components parallel and normal to the outer streamlines (see Fig. 4-43):

$$\frac{u_s}{U_0} = \tfrac{1}{2}f' + (a - bx)\left(\tfrac{1}{2}af' - cxh\right)$$
$$\frac{u_n}{U_0} = cx\left(h - \tfrac{1}{2}f'\right) \tag{4-246}$$

where
$$U_0 = \frac{U}{[1 + (a - bx)^2]^{1/2}}$$

Here u_n is positive if directed toward the center of curvature of the streamlines. Since $h - f'/2$ is always positive, it follows that the secondary flow is in fact always toward the center of curvature, like the tea leaves in a cup.

Figure 4-44 shows a numerical example for $a = +1$, $bx = +2$, where the resultant freestream velocity is $U_\infty = U\sqrt{2}$. It is seen that the secondary flow is substantial, with a maximum of 64 percent of U_∞. The streamwise component u_s has an overshoot, another aspect of curving boundary layers. The angle of deviation of the resultant velocity from the outer streamline is also substantial, being 38° at the wall. The equivalent wall angle for a rotating disk (Chap. 3) is 39.6°.

4-12.3 Boundary Layer on a Yawed Infinite Cylinder

The practical problem of designing swept-back wings for aircraft has led to the need for analysis of flow over a body whose leading edge is not normal to the oncoming stream. It is known that fluid in the boundary layer near the trailing edge moves outward along the wing axis toward the rearward part of the wing. This is another example of secondary flow.

For finite-span swept (possibly tapered) wings, no similarity relations obtain, analysis is possible only through fully three-dimensional finite-difference modeling of the Navier–Stokes equations on a supercomputer. However, some

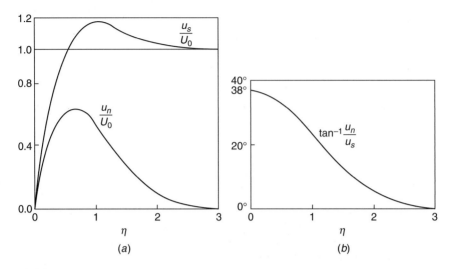

FIGURE 4-44
Laminar-boundary-layer solution for parabolic flow over a flat plate for $a = +1$, $bx = +2$: (a) velocity components; (b) deviation from external streamlines. [*After Sowerby (1954)*.]

illumination is possible here if we assume an infinite-span wing of constant cross section, as sketched in Fig. 4-45. This flow is three-dimensional, but its velocities are independent of the spanwise coordinate z. It is convenient to split the freestream into components $U(x)$ and $W(x)$ normal and parallel to the leading edge, respectively. Since the velocities do not vary with z, the boundary-layer Eqs. (4-231) and (4-232) reduce to

$$\frac{\partial u}{\partial x} + \frac{\partial v}{\partial y} = 0 \qquad (4\text{-}247a)$$

$$u\frac{\partial u}{\partial x} + v\frac{\partial u}{\partial y} = U\frac{dU}{dx} + \nu\frac{\partial^2 u}{\partial y^2} \qquad (4\text{-}247b)$$

$$u\frac{\partial w}{\partial x} + v\frac{\partial w}{\partial y} = \nu\frac{\partial^2 w}{\partial y^2} \qquad (4\text{-}247c)$$

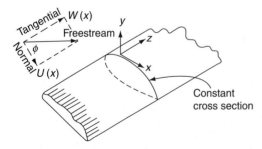

FIGURE 4-45
Coordinate system for a yawed infinite cylinder.

The boundary conditions are no slip at the wall and velocity matching in the freestream:

$$u(x, 0) = v(x, 0) = w(x, 0) = 0$$
$$u(x, \infty) = U(x) \tag{4-248}$$
$$w(x, \infty) = W(x)$$

These equations were first deduced by Prandtl (1945b) in a pioneering attempt to consider three-dimensional boundary layers.

Equations (4-247a) and (4-247b) are independent of w and can be solved for u and v in a two-dimensional-type analysis. The spanwise flow $w(x, y)$ is then analogous to convection and diffusion of temperature when dissipation is negligible and the Prandtl number is unity. This splitting of u and v from w is the *independence principle,* first noted by Prandtl (1945b). If separation occurs, it is due to adverse gradients in the normal component $U(x)$ through a two-dimensional analysis. The separation "line" is one of the cylinder generators, as illustrated in the two particular examples computed in Fig. 4-46. Using approximate two-dimensional methods, Wild (1949) computed the flow over a 45° yawed 6:1 elliptical cylinder at $\alpha = 7°$, and Sears (1948) computed the case $U = C(x - x^3)$. In both cases, the surface streamline turns asymptotically toward the separation line. Such spanwise surface flow occurs for finite wings also, as shown in the photographs of Fig. 1-4. It is customary to provide chordwise boundary-layer "fences" on swept wings to avoid loss of aileron effectiveness from this crossflow.

Another interesting example of three-dimensional boundary layers is the corner flow between intersecting bodies. Even the limiting case of intersecting sharp flat plates fails to reduce below three independent velocity components. Gersten (1959) has studied flat-plate corners and reports their friction drag to be less than

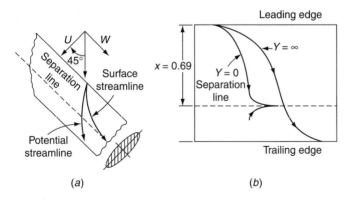

FIGURE 4-46
Two examples of outer potential flow and limiting surface streamlines for separating flow over an infinite yawed cylinder: (*a*) elliptical cylinder (1:6), $\phi = 45°$ [*after* Wild (1949)]; (*b*) $U = C(x - x^3)$, $W = W_0$ [*after* Sears (1948)].

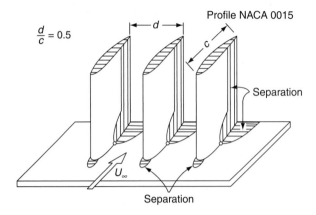

FIGURE 4-47
Separation regions in corner flow between airfoils. [*After Gersten (1959).*]

two independent plates of the same area:

$$C_{D,\text{corner}} \approx \frac{1.328}{Re_L^{1/2}} - \frac{C(\alpha)}{Re_L} \tag{4-249}$$

where α is the corner angle. Gersten reports $C(90°) \approx 5.76$. Experimental measurements on flat-plate corners are typically under turbulent-flow conditions, e.g., Nakayama and Rahai (1984).

Gersten also studied the flow along the roots of an airfoil cascade, as illustrated in Fig. 4.47. The double retardation of the walls causes the low-momentum corner flow to separate almost immediately in adverse gradients, as shown. Also shown is a separation zone which forms in the front of such round-nosed junctions and, as nose bluntness increases, grows into a horseshoe-shaped "junction vortex."

4-12.4 Separation Geometry in Three-Dimensional Flow

Two-dimensional separation has a simple geometry, as in Fig. 4-5*b*. The entire boundary-layer flow breaks away at the point of zero wall shear stress and, having no way to diverge left or right, has to go up and over the resulting separation bubble or wake. The ability to separate in three dimensions allows many more options. For example, the surface streamlines in Fig. 4-46 move tangentially toward the separation line, and the outer flow moves up and over the bubble.

There is now active research in the mathematical topology of three-dimensional separation zones, as reviewed by Tobak and Peake (1982) and by Wang (1997). They distinguish four different special points in separation:

1. *A nodal point*, where an infinite number of surface streamlines ("skin-friction lines") merge tangentially to the separation line

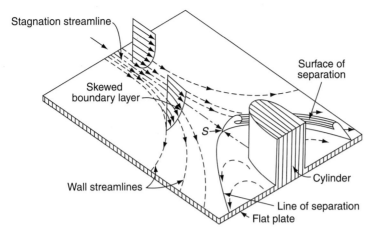

FIGURE 4-48
Three-dimensional separation in flat-plate flow against a cylindrical obstacle. [*After Johnston (1960)*.]

2. A *saddle point*, where only two surface streamlines intersect and all others divert to either side
3. A *focus*, or spiral node, which forms near a saddle point and around which an infinite number of surface streamlines swirl
4. A *three-dimensional singular point*, not on the wall, where the velocity is zero, generally serving as the center for a horseshoe vortex

A classic example is the boundary-layer flow on the wall near a round-nosed obstacle in Fig. 4-48, as studied by Johnston (1960). Because the obstacle creates an adverse gradient on the wall in front of it, the flow must separate, and the resulting bubble wraps itself around the obstacle in a U-shaped junction or horseshoe vortex. In the Tobak–Peake notation, the point S in Fig. 4-48 is a nodal point of separation. Inside the bubble, on the center plane between S and the obstacle, is a three-dimensional singular point where the vortex center forms. In the rear of such obstacles (not shown), a nodal attachment point and twin foci may also form. Baker (1979) reports experiments on the laminar horseshoe vortex. Since most practical applications (turbomachinery blade roots, wing–body junctions, control surfaces on ships) are at high Reynolds numbers, research on turbulent junction vortices continues, e.g., Menna and Pierce (1988).

A second example of laminar three-dimensional separation is the flow past a round-nosed body at an angle of attack, sketched in Fig. 4-49 and first described by Legendre (1965). In the Tobak–Peake (1981) notation, point A is a nodal attachment point, point S is a saddle point, and point F (always near S) is a focus of separation. In the sketch, Fig. 4-49, instead of forming a bubble or streamwise vortex, the separated flow has broken away from the surface in what

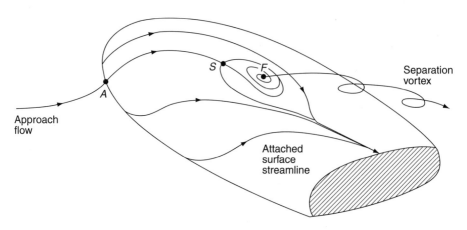

FIGURE 4-49
Three-dimensional separation on a round-nosed body at an angle of attack, first described by Legendre (1965). Point A is a nodal attachment point, point S is a saddle point, and point F is a focus of separation.

Wang (1997) calls a *tornado-like vortex*. Contrast this behavior with the "open" or nonbubble separation on the prolate spheroid discussed earlier in Fig. 1-13*b*, where flow from upstream may enter the separated region. Though experimentation has been the key to such complex flow descriptions, mega-mesh Navier–Stokes CFD models, such as Fig. 1-13*a*, can now simulate three-dimensional flow separation.

4-13 UNSTEADY BOUNDARY LAYERS: SEPARATION ANXIETY

The majority of boundary-layer applications are steady, but unsteady flows also occur, primarily due to (1) start-up of a flow or (2) periodic flow. Chapter 3 presented some exact unsteady Navier–Stokes solutions: in Sec. 3-4, pipe flow with a pressure gradient suddenly applied (start-up flow) or oscillating (periodic flow). Section 3-5 illustrated Stokes' solutions for a suddenly accelerated plate (start-up flow) and an oscillating plate (periodic flow). None of these examples encountered flow separation, but many unsteady boundary layers do separate.

4-13.1 Start-Up Flows

Consider the start-up flow that occurs when a cylinder at rest suddenly begins to move at speed U. An interesting sequence ensues, as in Fig. 4-50. Initially, Fig. 4-50*a*, the flow is inviscid and matches the classical potential-flow pattern of Fig. 1-4.

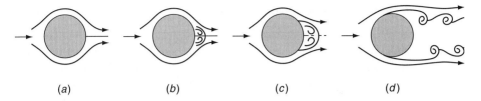

FIGURE 4-50
Start-up of viscous flow past a cylinder accelerated from rest: *(a)* Initial flow is nearly inviscid and resembles Fig. 1-4; *(b)* later, separation begins on the rear surface; *(c)* still later, separation extends up the rear surface in a symmetric vortex pattern; *(d)* finally, the double vortex becomes unstable and alternate shedding begins, with the separation point on the front of the body.

At $Ut/R \approx 0.35$, the shear stress vanishes at the rear stagnation point and separation begins there, Fig. 4-50*b*. However, boundary-layer theory predicts no Goldstein-type singularity at this time. Later, at about $Ut/R \approx 1.5$, a large standing double-vortex layer has formed. Much later, the alternating Kármán vortex street has formed, Fig. 4-50*d*, as the final flow pattern, with separation occurring on the front of the cylinder at about 81° from the front stagnation point.

The profound flow-pattern changes in Fig. 4-50 prompt the questions (1) how is separation defined in unsteady flow, and (2) can boundary-layer theory still be useful? In the 1950s, Moore, Rott, and Sears [in Moore (1958)] independently proposed that unsteady separation is signaled by a singular point $x_s(t)$, $y_s(t)$ in the flow where the velocity relative to the moving point *s*, the shear stress, and the vorticity all vanish:

$$u_s = \frac{dx_s}{dt} = 0 \qquad \tau_s = \omega_s = \left.\frac{\partial u}{\partial y}\right|_s = 0$$

This is now known as the MRS criterion and is the unsteady version of the Goldstein steady-flow singularity, Eq. (4-116). The point *s* need not be at the wall, and there need not be any reverse flow. Later, in the 1980s, the exact behavior of the boundary layer on, for example, the impulsively started cylinder of Fig. 4-50, became controversial, as discussed in the monographs by Cebeci (1982) and by Telionis (1981). Van Dommelen and Shen (1981, 1982) give a good explanation in a careful numerical study. Using Lagrangian coordinates and a fine mesh, they calculated a non-Goldstein singularity, at about 111° from the front, in the form of a nearly inviscid "peel-off" of a vorticity layer from the wall. The wall shear stress was not necessarily zero at this point. In Van Dommelen's words, "the boundary layer is ejected away from the wall after a finite time" of order $Ut/R = 1.5$. Thus boundary-layer theory is also fundamentally limited in unsteady separated flow. Using multivortex methods, Koumoutsakos and Leonard (1995) report detailed CFD calculations of start-up flow around a circular cylinder.

4-13.2 Periodic Flows: Acoustic Streaming

The second important type of unsteady viscous flow is an oscillatory boundary layer. An illustrative case occurs when a cylindrical body of length L oscillates in a fluid at rest. Further assume a small amplitude, $A \ll L$. If, in steady freestream flow, the body has the potential flow distribution $U_o(x)$, the oscillating body will create a potential flow given by

$$U(x, t) = U_o(x)\cos(\omega t)$$

Thus the boundary layer is driven by an oscillating freestream, much like the flat wall in Stokes' second problem, Sec. 3-5.1 and Fig. 3-19b. For a circular cylinder oscillating at high frequency, Schlichting (1932) gave a solution in series form. The first-order solution is identical to Eq. (3-112) for Stokes' problem:

$$u_o(x, t) = U_o(x)[\cos(\omega t) - e^{-\eta}\cos(\omega t - \eta)]$$

where

$$\eta = y\sqrt{\frac{\omega}{2\nu}}$$

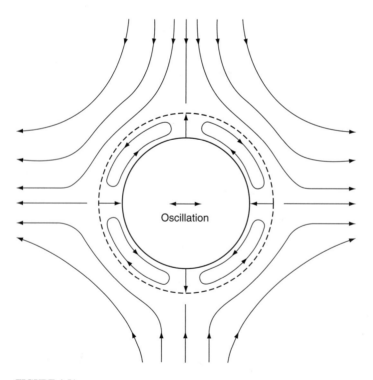

FIGURE 4-51
Streamlines for acoustic streaming in the outer flow about an oscillating circular cylinder. [*After Schlichting (1932).*]

The second-order solution $u_1(x, t)$ turns out to have a nonzero value in the freestream itself:

$$u_1(x, \infty) = -\frac{3}{4}\frac{U_o}{\omega}\frac{dU_o}{dx}$$

This term occurs because the convective acceleration terms, such as $u_o(\partial u_o/\partial x)$, have a nonzero mean value when they oscillate. The term $u_1(x, \infty)$ is thus called *acoustic streaming*, a flow toward decreasing freestream velocity $U_o(x)$, created by the body oscillation. Schlichting's solution explains the formation of dust patterns in a Kundt tube when standing waves occur, as first studied by Lord Rayleigh in 1883. Figure 4-51 shows Schlichting's sketches of acoustic streaming caused by oscillation of a cylinder. This concept of periodic motions causing nonzero mean convective terms will be very important in flow stability (Chap. 5) and turbulent flow (Chap. 6). Further studies of unsteady viscous flows are given in the monograph by Telionis (1981).

4-14 FREE-CONVECTION BOUNDARY LAYERS

Although we included body forces in the derivation of the basic equations, we have not yet considered any solutions of flow with buoyancy effects. The purpose of this section is to discuss some examples of natural-convection boundary layers. There are many practical cases of buoyant flow in both engineering (heat exchangers, computer chips, fires, and plumes) and geophysics (the ocean and atmosphere). For further reading, see the monographs by Jaluria (1980) and by Kakaç et al. (1985).

4-14.1 Velocity Scales: The Grashof Number

We consider purely buoyant flows, i.e., with no superimposed freestream or "forced" motion. With no stream velocity U_∞ as a reference, we must scale the flow velocities with the body force. We further restrict ourselves to single-phase flows, neglecting the important cases of condensation and melting.

If a fluid contains density differences $\Delta\rho$, arising from temperature or concentration differences, a gravitational body force $g\,\Delta\rho$ will drive the motion. For example, in flow next to a hot vertical wall, the average buoyant force in the near-wall layers would be $\frac{1}{2}g(\rho_\infty - \rho_w)$. If the lighter fluid rises, from rest, a distance L, it will (approximately) lose potential energy and gain kinetic energy or motion. Thus we have a crude energy balance

$$\frac{1}{2}\Delta\rho\, gL \approx \frac{1}{2}\rho V^2$$

or (4-250)

$$V \approx \left(gL\frac{\Delta\rho}{\rho}\right)^{1/2}$$

The quantity V is the appropriate velocity scale for free convection. Thus we may form an equivalent Reynolds number squared for buoyant flows:

$$Re^2_{\text{effective}} = \left(\frac{VL}{\nu}\right)^2 = \frac{gL^3}{\nu^2}\frac{\Delta\rho}{\rho} = Gr_L \qquad (4\text{-}251)$$

This new parameter Gr, the *Grashof number*, takes the place of the Reynolds number (squared) in free convection. In the common case of buoyancy caused by temperature differences, we relate $\Delta\rho$ to ΔT through the coefficient of thermal expansion β and a linear approximation:

$$\Delta\rho \approx \rho\beta\,\Delta T$$

or $\qquad\qquad\qquad\qquad\qquad\qquad\qquad\qquad\qquad\qquad\qquad\qquad\qquad$ (4-252)

$$Gr_L = \frac{g\beta\,\Delta T\,L^3}{\nu^2}$$

By comparing the relative sizes of Grashof and Reynolds numbers, we may determine when free-convection effects are important:

$Gr \ll Re^2$: $\qquad$ forced convection dominates

$Gr \gg Re^2$: $\qquad$ free convection dominates

$Gr = \mathcal{O}(Re^2)$: $\qquad$ "mixed" free and forced convection

For a given size and temperature difference the fluid buoyancy parameter $(g\beta/\nu^2)$ determines the strength of the Grashof number. Some values are listed in Table 4-9. We see that viscous oils have poor buoyancy, gases are moderate to good, and light liquids and liquid metals have high buoyancy potential.

TABLE 4-9
Buoyancy parameter of various fluids at 20°C and 1 atm in earth gravity

Fluid	$g\beta/\nu^2$, $K^{-1}m^{-3}$
Glycerin	3.2E3
Engine oil	7.9E3
Helium	2.4E6
Hydrogen	3.1E6
Air	1.5E8
Carbon dioxide	5.3E8
Water	2.0E9
Ethyl alcohol	4.5E9
Mercury	1.4E11

4-14.2 Two-Dimensional Steady Free Convection

Let us now reconsider the boundary-layer equations from Eqs. (4-35) for steady flow with constant properties and nonnegligible buoyancy:

$$\frac{\partial u}{\partial x} + \frac{\partial v}{\partial y} = 0$$

$$u\frac{\partial u}{\partial x} + v\frac{\partial u}{\partial y} = g_x\beta(T - T_0) + \nu\frac{\partial^2 u}{\partial y^2} \quad (4\text{-}253)$$

$$u\frac{\partial T}{\partial x} + v\frac{\partial T}{\partial y} = \alpha\frac{\partial^2 T}{\partial y^2}$$

where $\alpha = k/\rho c_p$ is the thermal diffusivity, assumed constant. Note that we have neglected dissipation. We have also assumed no forced motion and have dropped the external pressure gradient $U\,dU/dx$. Note that buoyancy is proportional to g_x, the component of gravity parallel to the solid boundary. If the surface is curved or inclined, we must account for geometric changes in g_x along the surface.

The usual boundary conditions are no slip or temperature jumps at the wall and a fluid at rest under ambient temperature at infinity:

$$u(x, 0) = v(x, 0) = u(x, \infty) = 0$$
$$T(x, 0) = T_w(x) \quad T(x, \infty) = T_\infty = \text{const} \quad (4\text{-}254)$$

There are many analyses of these equations in the literature for various geometries and applications. An excellent treatment, both broad and deep, is given in the monograph by Gebhart et al. (1988). These writers also discuss combined buoyancy and species diffusion, mixed convection, flow instability, turbulent flows, porous media, nonnewtonian fluids, and variable fluid properties (such as water near freezing).

4-14.3 Free Convection along a Vertical Isothermal Plate

For a vertical plate, $g_x = g$, and we assume constant T_w. If we ignore the anomalous case of water near freezing, where β may be negative, all common fluids have low density at a hot wall. The buoyancy and free convection are thus upward along a hot vertical plate, and the leading edge ($x = 0$) is at the bottom. For a cold plate, the buoyancy is downward and ($x = 0$) is at the top. The coordinate y is normal to the plate.

Since the local Grashof number Gr_x plays the role of Re_x^2 in buoyant flow, we could guess (correctly) that the boundary-layer thickness (δ/x) would be proportional to $Gr_x^{-1/4}$. Indeed, the "scale analysis" recommended in the text by Bejan (1994) predicts that δ is proportional to $x^{1/4}$. For vertical plate flow, velocity and thickness scales were deduced by Schmidt et al. (1930), who defined the following similarity variables:

$$\eta = \left(\frac{Gr_x}{4}\right)^{1/4} \frac{y}{x} \qquad Gr_x = \frac{\beta g(T_w - T_\infty)x^3}{\nu^2}$$

$$u = 2\sqrt{x\beta g(T_w - T_\infty)}\, f'$$

$$v = \left[\frac{\beta g(T_w - T_\infty)\nu^2}{4x}\right]^{1/4} (\eta f' - 3f) \qquad (4\text{-}255)$$

$$\Theta = \frac{T - T_\infty}{T_w - T_\infty}$$

The reader should verify that u and v satisfy continuity exactly and that these variables will reduce momentum and energy to the two coupled non-linear ordinary differential equations

$$f''' + 3ff'' - 2f'^2 + \Theta = 0$$
$$\Theta'' + 3Prf\Theta' = 0 \qquad (4\text{-}256)$$

subject to the boundary conditions

$$f(0) = f'(0) = f'(\infty) = 0$$
$$\Theta(0) = 1 \quad \Theta(\infty) = 0 \qquad (4\text{-}257)$$

Note that Eqs. (4-256) are coupled and must be solved simultaneously, which is always the case in free-convection problems. No analytic solution is known, so numerical integration is necessary. There are two unknown initial values at the wall. One must find the proper values of $f''(0)$ and $\Theta'(0)$ which cause the velocity and temperature to vanish for large η. The Prandtl number is a parameter.

The original approximate solutions given by Pohlhausen in 1930 were improved with digital-computer solutions provided by Ostrach (1953) and others. Some accurate initial values are listed in Table 4-10. Velocity and temperature profiles are shown in Fig. 4-52 for various Prandtl numbers. Except for $Pr \ll 1$, the velocity layers are thicker than the temperature layers. These theoretical profiles are in good agreement with experimental laminar free-convection data on vertical surfaces.

The most important results of the analysis are the dimensionless heat-transfer rates, or Nusselt numbers:

Local: $$Nu_x = \frac{xq_w}{k(T_w - T_\infty)}$$

Mean: $$Nu_m = \frac{1}{L}\int_0^L Nu_x\, dx \qquad (4\text{-}258)$$

From the definition of Θ, we have $Nu_x(Gr_x/4)^{-1/4} = -\Theta'(0) = fcn(Pr)$, where Gr_x is the local Grashof number defined in Eq. (4-255). Since q_w varies as $x^{-1/4}$, it follows that the two Nusselt numbers are related simply by $Nu_x(L) = \frac{3}{4}Nu_m$. Some numerical values of $Nu_x/Gr_x^{1/4}$ are listed in Table 4-10. For gases, with Pr near unity, a

TABLE 4-10
Computed parameters from Eqs. (4-256) for free convection on a vertical isothermal plate

Pr	$f''(0)$	$-\Theta'(0)$	$Nu_x/Gr_x^{1/4}$
0.01	0.9873	0.0807	0.0571
0.1	0.8591	0.2301	0.1627
0.72	0.6760	0.5046	0.3568
1.0	0.6422	0.5671	0.4010
2.0	0.5713	0.7165	0.5066
3.0	0.5309	0.8155	0.5767
6.0	0.4649	1.0075	0.7124
10.0	0.4192	1.1693	0.8268
30.0	0.3312	1.5891	1.1237
100.0	0.2517	2.1913	1.5495
1000.0	0.1449	3.9650	2.8037

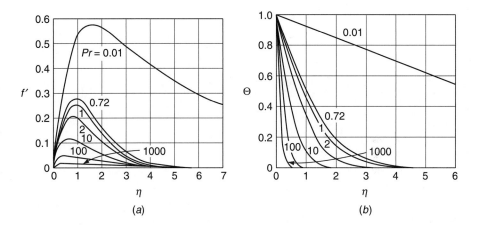

FIGURE 4-52
(a) Velocity and (b) temperature profiles for free convection on a vertical plate. Computed by Ostrach (1953) from the differential Eqs. (4-256) of Pohlhausen.

simple power law will suffice:

$$Nu_x \approx 0.4\, Ra_x^{1/4} \qquad Ra_x = Gr_x Pr \qquad (4\text{-}259)$$

where Ra_x is called the *Rayleigh number* of the flow. Meanwhile, Churchill and Usagi (1972) proposed the following curve-fit formula which is valid with ± 0.5 percent accuracy over the entire range of Prandtl numbers:

$$Nu_x \approx \frac{0.503\, Ra_x^{1/4}}{\left[1 + (0.492/Pr)^{9/16}\right]^{4/9}} \qquad (4\text{-}260)$$

This formula is limited to isothermal walls and the laminar free-convection flow regime, $10^5 < Ra_x < 10^9$. It predicts that Nu_x varies as $(Gr_x Pr^2)^{1/4}$ for very small Pr and as $(Gr_x Pr)^{1/4}$ for very large Pr, which can be proved by an asymptotic analysis.

The stability and transition of free-convection flows are reviewed by Gebhart and Majahan (1982) and also in Chap. 11 of the monograph by Gebhart et al. (1988). Boundary-layer instability begins for Grashof numbers as low as 400, and fully turbulent flow generally develops by $Gr_x \approx 10^9$. The local Nusselt numbers in turbulent flow then vary approximately as $Gr_x^{1/3}$. Although local heat flux may change sharply during transition, the overall heat transfer increases gradually and is well represented by the curve-fit formula of Churchill and Chu (1975):

$$Nu_L^{1/2} \approx 0.825 + \frac{0.387 \, Ra_L^{1/6}}{\left[1 + (0.492/Pr)^{9/16}\right]^{8/27}} \qquad (4\text{-}261)$$

valid for isothermal walls and any Prandtl number and for $Ra_L \leq 10^{12}$.

The related problem of laminar free convection on a vertical surface with uniform heat flux was solved by Sparrow and Gregg (1956). Their digital-computer results for local heat transfer were fit by Churchill and Ozoe (1973) into the formula

$$Nu_x \approx \left(\frac{Gr_x^* Pr^2}{4 + 9Pr^{1/2} + 10Pr}\right)^{1/5} \qquad (4\text{-}262)$$

valid for any Prandtl number and for $10^5 < Gr_x^* < 10^{12}$. The quantity $Gr_x^* = Gr_x Nu_x = g\beta q_w x^4/k\nu^2$ is called the modified or "heat-flux" Grashof number. For overall heat transfer in both laminar and turbulent flow, Churchill and Chu (1975) proposed the curve-fit correlation

$$Nu_L^{1/2} \approx 0.825 + \frac{0.387(Gr_L Pr)^{1/6}}{\left[1 + (0.437/Pr)^{9/16}\right]^{8/27}} \qquad (4\text{-}263)$$

valid for any Prandtl number and for $1 < Gr_L Pr < 10^{11}$. Curiously, this formula uses the traditional rather than the modified Grashof number and therefore must be iterated with the relation $Gr_L = Gr_L^*/Nu_L$ since the temperature difference varies over the surface.

4-14.4 Other Geometries

If a plane surface in free convection is inclined from the vertical by an angle ϕ, the first-order theoretical effect is to decrease effective gravity from g to $g \cos \phi$. The second-order effect depends upon whether the lateral buoyancy component is toward or away from the plate surface. In the latter case, the flow tends to separate away from the plate. The overall conclusion of many experiments [Gebhart et al. (1988, Sec. 5.2)] is that, if $|\phi| \leq 60°$, the heat transfer may be computed by evaluating the vertical-surface formulas at the effective Grashof number $Gr_x \cos \phi$.

FIGURE 4-53
Comparison of isotherms in free convection near a heated horizontal cylinder at $Ra_D = 10^5$: (*right*) experiment; (*left*) fully elliptic Navier–Stokes computations. [*From Kuehn and Goldstein (1980), courtesy of the authors.*]

The laminar-flow solution for a vertical cylinder was given by Sparrow and Gregg (1956). The heat transfer is increased if the boundary layer becomes thick compared to the cylinder radius. The appropriate parameter is $\zeta = (L/D)Ra_L^{-1/4}$. An approximate correction relates the overall heat transfer to a vertical plate of the same length:

$$Nu_{L,\text{cylinder}} \approx Nu_{L,\text{plate}}(1 + 1.3\zeta^{0.9}) \qquad (4\text{-}264)$$

where the plate Nusselt number is taken from Eq. (4-261).

Perhaps the most practical shape is the horizontal cylinder, which occurs so often in heat-exchanger design. The first theory was by Hermann (1936), using the boundary-layer Eqs. (4-253) with gravity varying as $g_x = g\sin(x/R)$. More recently, Kuehn and Goldstein (1980) reported fully elliptic Navier–Stokes computations for this case. Their computed and experimental isotherms for $Ra_D = 10^5$ are shown in Fig. 4-53. Note that, unlike forced motion past a cylinder, there is no separation at the rear of the body. Rather, the flow at the rear gradually spreads and turns to form a plume above the cylinder. At lower Rayleigh numbers, $Ra_D < 10^4$ [Kuehn and Goldstein (1980)], the entire flow resembles a plume near a hot "source" and is much larger in extent than the body radius.

The free-convection heat transfer on a horizontal cylinder could be approximated by a vertical plate of length $(\pi D/2)$, the "length of travel" of a particle in the boundary layer. Churchill and Chu (1975) recommend the following general curve-fit correlation formula for cylinders:

$$Nu_D^{1/2} \approx 0.60 + \frac{0.387 Ra_D^{1/6}}{\left[1 + (0.559/Pr)^{9/16}\right]^{8/27}} \qquad (4\text{-}265)$$

valid for any Prandtl number and $10^{-5} < Ra_D < 10^{12}$. Note the similarity to Eq. (4-261) for a vertical plate.

For further study of buoyant free- or mixed-convection flows, the reader is referred to the monograph by Gebhart et al. (1988).

SUMMARY

This chapter reviews in some detail the many different types of laminar-boundary-layer flows, with emphasis upon both velocity and temperature distributions. The simple integral momentum and energy analyses of Sec. 4-1 lead into the full boundary-layer equations of Sec. 4-2. A variety of similarity solutions are discussed in Sec. 4-3, followed by free-shear flows (mixing layers, jets, and wakes) in Sec. 4-4 and a few nonsimilar flows in Sec. 4-5. Methods of analysis of general boundary-layer flows are introduced in Sec. 4-6 (integral techniques) and Sec. 4-7 (finite-difference computer methods). Thermal boundary layers are analyzed in Sec. 4-8 by both integral and finite-difference methods. Section 4-9 is a brief treatment of flow in the inlets of ducts, and Sec. 4-10 presents a variety of axisymmetric boundary-layer flows, including round jets and wakes. Section 4-11 gives a brief introduction to higher order boundary-layer theories by matched asymptotic expansions. The chapter ends with discussions of three-dimensional boundary layers (Sec. 4-12), unsteady boundary layers (Sec. 4-13), and free-convection flows (Sec. 4-14).

We should always emphasize that laminar flows inevitably break down at some finite critical Reynolds number and the ensuing flow will be turbulent (Chap. 6). It is the purpose of the next chapter to examine the stability and transition of laminar flows. Perhaps we should not dwell too long on laminar flows, when turbulence is so often the actual state of affairs. However, laminar theory is far more advanced than turbulent-boundary-layer theory, and the many phenomena we have just exhibited—similarity, separation, computer techniques, favorable gradients, free convection, three-dimensional flow—are representative of the general behavior of boundary layers. Thus almost everything we learn in this chapter will serve us well in the study of turbulent flow. Often the only differences will be in the exact numerical values.

PROBLEMS

4-1. Repeat the integral momentum analysis of the flat plate in Sec. 4-1 for the assumed velocity profile

$$\frac{u}{U} = \frac{3}{2}\left(\frac{y}{\delta_u}\right) - \frac{1}{2}\left(\frac{y}{\delta_u}\right)^3$$

where δ_u is the velocity boundary-layer thickness. Is this profile any more (or less) realistic than the approximation of Eq. (4-11)? For the above profile, compute (a) $(\theta/x)\sqrt{Re_x}$; (b) $(\delta^*/x)\sqrt{Re_x}$; (c) $(\delta/x)\sqrt{Re_x}$; (d) $C_f\sqrt{Re_x}$; (e) $C_D\sqrt{Re_x}$.

Answers: (a) 0.646, (b) 1.740, (c) 4.64, (d) 0.646, (e) 1.293.

4-2. Repeat the integral heat-transfer analysis of Sec. 4-1.7 by replacing Eq. (4-23) by the quartic temperature profile approximation

$$T - T_e \approx (T_w - T_e)(1 - 2\zeta + 2\zeta^3 - \zeta^4)$$

What boundary conditions does this profile satisfy? Use the velocity profile from Eq. (4-11). Compare your results with Eqs. (4-25) and (4-26).

4-3. Schlichting (1979, p. 206) points out that the simple flat-plate velocity profile approximation

$$u \approx U \sin\left(\frac{\pi y}{2\delta}\right)$$

gives much better accuracy for c_f, θ, and $\delta^*(\pm 2$ percent) than the parabolic profile of Eq. (4.11). Verify this by computing c_f. Does this sine-wave shape satisfy any additional boundary conditions compared to Eq. (4-11)?

4-4. Air at 20°C and 1 atm flows past a smooth flat plate as in Fig. P4-4. A pitot stagnation tube, placed 2 mm from the wall, develops a water manometer head $h = 21$ mm. Use this information with the Blasius solution, Table 4-1, to estimate the position x of the pitot tube. Check to see if the flow is laminar.

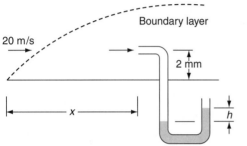

FIGURE P4-4

4-5. In the unheated-starting-length analysis that led to Eq. (4-28), a term ($\zeta^3/5$) was neglected in the integral-energy equation. Without neglecting this term, solve the differential equation numerically (e.g., with Runge–Kutta) for $1 < x < 5$ for the special case $Pr = 1$ and $x_0 = 1$. Compare your numerical results with Eq. (4-28).

4-6. Develop a numerical solution, perhaps with Subroutine RUNGE and Eqs. (4-47), that will iterate the Blasius equation from an initial guess $f''(0) = 0.3$ and converge to the exact value $f''(0) = 0.4696$.

4-7. Consider a long flat plate emerging from a wall at velocity U, as in Fig. P4-7. There is no freestream. Show that the Blasius Eq. (4-45) holds for this case, with $f(0) = 0$,

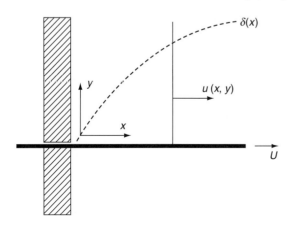

FIGURE P4-7

$f'(0) = 1$, and $f'(\infty) = 0$. Solve the equation numerically and show that $C_f \approx 0.444/Re_x^{1/2}$. Also evaluate $v(\infty)$ and discuss. [Hint: Note that $f''(0)$ is negative.]

4-8. Develop a numerical solution, perhaps with Subroutine RUNGE and Eqs. (4-74), which will iterate the Falkner–Skan Eq. (4-71), for any value of $\beta \neq 0$, from an initial guess for $f''(0)$ which converges to the exact value of $f''(0)$, for that β. Compare your results with Fig. 4-11.

4-9. The Blasius Eq. (4-45) must be iterated to find the value of $f''(0)$ which causes $f'(\infty)$ to equal 1.0. Goldstein (1938)—see also Panton (1996)—suggests the following to avoid iteration: Define

$$f(\eta) = \alpha F(\alpha \eta), \quad \text{where } \alpha \text{ is a constant}$$

(*a*) Show that the function F also satisfies the Blasius equation. (*b*) If we arbitrarily set an initial condition $F''(0) = 1.0$, explain how α can immediately be found without iteration. (*c*) Even with α found, explain why the solution for F is still awkward for filling out Table 4-1. (*d*) If an integration (not required of you) then yields $F'(\infty) = 1.6552$, what is the proper value of α? (*e*) Show that the value of α from part (*d*) leads to the result $f''(0) = 0.4696$.

4-10. The quantity $(\delta^*/\tau_w)(dp/dx)$ is called Clauser's parameter. It compares an external pressure gradient to wall friction and is very useful for turbulent boundary layers (Chap. 6). Show that this parameter is a constant for a given laminar Falkner–Skan wedge-flow boundary layer. What value does this parameter have at the separation condition?

4-11. If, instead of Eq. (4-70), we choose the Falkner–Skan similarity variable $\eta = y(|U|/\nu x)^{1/2}$, the Falkner–Skan equation becomes

$$f''' + \tfrac{1}{2}(m + 1)ff'' + m(f^2 - 1) = 0$$

subject to the same boundary conditions Eq. (4-72). Examine this relation for the special case $U = -K/x$ and show that a closed-form solution may be obtained.

4-12. A thin equilateral triangle plate is immersed parallel to a 12 m/s stream of air at 20°C and 1 atm, as in Fig. P4-12. Assuming laminar flow, estimate the drag of this plate (in N).

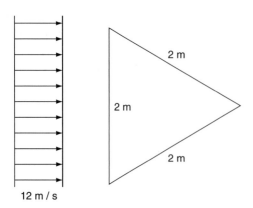

FIGURE P4-12

4-13. Flow straighteners consist of arrays of narrow ducts placed in a flow to remove swirl and other transverse (secondary) velocities. One element can be idealized as a square box with thin sides as in Fig. P4-13. Using laminar flat-plate theory, derive a formula for the pressure drop Δp across an $N \times N$ bundle of such boxes.

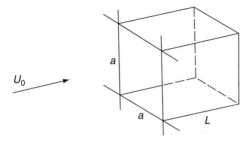

FIGURE P4-13

4-14. Develop a numerical solution for the Blasius equation with wall suction or blowing, for a particular value $v_w^* \neq 0$, and compare your results with Fig. 4-15.

4-15. Derive a relation for skin-friction coefficient C_f as a function of local Reynolds number Re_x for boundary-layer flow toward a point sink, Eq. (4-89). Compare your result with the Falkner–Skan relations.

4-16. Develop a numerical solution for a laminar mixing layer between parallel streams, for a particular value of k, and compare with Fig. 4-17.

4-17. Air at 20°C and 1 atm issues from a narrow slot and forms a two-dimensional laminar jet. At 50 cm downstream of the slot the maximum velocity is 20 cm/s. Estimate, at this position, (a) the jet width, (b) the jet mass flow per unit depth, and (c) an appropriate Reynolds number for the jet.

4-18. Air at 20°C and 1 atm flows at 1 m/s past a slender two-dimensional body, of length $L = 30$ cm, whose drag coefficient is 0.05 based on "plan" area (bL). Assuming laminar flow at a point 3 m downstream of the trailing edge, estimate (a) the maximum wake velocity defect (in cm/s), (b) the "one percent" wake thickness (in cm), and (c) the wake-thickness Reynolds number.

4-19. Derive the steady-flow version of the integral momentum relation, Eq. (4-121) or equivalent, including wall suction or blowing, by making a mass and force balance on the differential boundary-layer control volume shown in Fig. P4-19.

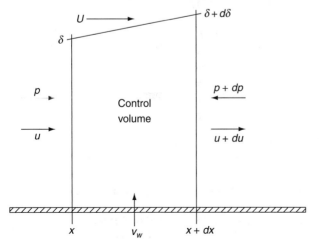

FIGURE P4-19

4-20. Solve the integral relation Eq. (4-121), with $v_w \neq 0$ and a velocity profile given by Eq. (4-11), by assuming that both δ and $1/v_w$ are proportional to $x^{1/2}$. Compare the computed wall friction with Fig. 4-15b for $Pr = 1$, assuming that $Nu_x/Re_x = c_f/2$.

4-21. Improve Prob. 4-20 by developing a parametric polynomial velocity profile which accounts for blowing and suction in the manner of Fig. 4-15a. Match your profile at the wall to the boundary-layer equations.

4-22. Modify Prob. 4-20 by using the same profile Eq. (4-11) but letting the suction or blowing v_w be constant. Solve for $c_f(x)$ and $\delta(x)$ and compare the asymptotic results with Fig. 4-21 and Eq. (2-123).

4-23. Apply the method of Thwaites, Sec. 4-6.6, to boundary-layer flow on a cylinder, using either the inviscid Eq. (4-143) or measured Eq. (4-144) freestream velocity distributions. Compare the computed local wall friction with Fig. 4-24b.

4-24. Apply the Thwaites' integral method to one of the laminar-flow test cases in Table 4-5 (for best results have each member of the class take a different case). Compute and plot the local friction distribution $c_f\sqrt{Re_x}$ and compare the predicted separation point with Table 4-5.

4-25. Consider a two-dimensional flat-walled diffuser, as in Fig. P4-25. Assume incompressible flow with a one-dimensional freestream velocity $U(x)$ and entrance velocity $U_o(x)$. The entrance height is W and the constant depth into the paper is b. Using Thwaites' method, find an expression for the angle θ at which separation will occur at $x = L$. What is the value of θ if $L = 1.5\ W$?

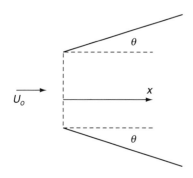

FIGURE P4-25

4-26. Apply the explicit finite-difference model, Sec. 4-7.1, to boundary-layer flow on a cylinder, using either the inviscid, Eq. (4-143), or measured, Eq. (4-144), freestream velocity distributions. Compare the computed local wall friction with Fig. 4-24b.

4-27. Modify Prob. 4-26 by instead using the implicit finite-difference model of Sec. 4-7.2.

4-28. Investigate the use of the Crank–Nicolson (1947) method for computer analysis of a laminar boundary layer. What are its numerical advantages and disadvantages?

4-29. Apply the explicit finite-difference method of Sec. 4-7.1 to one of the laminar-flow test cases in Table 4-5 (for best results have each member of the class take a different case). Compute and plot the local friction distribution $c_f\sqrt{Re_x}$ and compare the predicted separation point with Table 4-5.

4-30. Sherman (1990) gives a CFD solution for laminar flow due to a freestream U_o approaching a parabolic cylinder, as in Fig. P4-30. The cylinder surface is defined by $y/R = (2x/R)^{1/2}$, where R is the cylinder nose radius. The arc length s along the surface

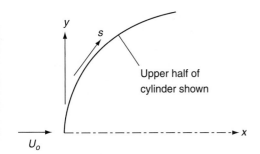

FIGURE P4-30

is defined by $ds = R(1 + \zeta^2)^{1/2}d\zeta$, where $\zeta = y/R$. From potential theory, the surface velocity is $U = U_o\zeta/(1 + \zeta^2)^{1/2}$. (a) Show that the surface velocity approaches the stream velocity U_o as one moves up the surface. (b) Using Thwaites' method, estimate the distance s/R along the surface where $\tau\theta/(\mu U)$ is within 10 percent of the flat-plate value of 0.22.

4-31. Apply the Smith and Spalding thermal-integral method of Eq. (4-161) for $Pr = 1$ to the Howarth velocity distribution, Eq. (4-114), with constant wall temperature, computing the local heat-transfer rate up to the point of separation. Also compute the Reynolds analogy factor, $C_f(x)/2C_h(x)$, and relate this to Fig. 4-14.

4-32. Modify Prob. 4-31 by using instead the explicit or implicit finite-difference method of Sec. 4-8.2.

4-33. Apply the Smith and Spalding thermal-integral method of Eq.(4-161) for $Pr = 0.72$ to flow past an isothermal circular cylinder, Eq. (4-167), and compare your results with Fig. 4-29.

4-34. In Eq. (4-144) for $Re_a = 9500$, $U/U_\infty \approx 1.814 \, x/a$ near the front of the cylinder. For air at 20°C and 1 atm, this corresponds approximately to $a = 5$ cm and $U_\infty = 2.85$ m/s. Using the Falkner–Skan theory, Tables 4-2 and 4-3, and a temperature difference $(T_w - T_\infty) = 12$°C, estimate (a) the momentum thickness in mm and (b) the heat-transfer rate in W/m² at the front of this cylinder.

4-35. For a flat plate, $U = U_0$, and a wall temperature distribution $T_w - T_e = \Delta T_0[1 - (x/L)^3]$, use the superposition method of Sec. 4-8.4 to compute the value of x at which the local heat transfer q_w changes sign.

4-36. Modify Prob. 4-35 by using instead the explicit or implicit finite-difference method of Sec. 4-8.2.

4-37. Make an integral analysis of laminar flow in the entrance between impermeable parallel plates a distance $(2H)$ apart, analogous to Fig. 4-32 [Sparrow (1955)]. At any x, let the velocity profile consist of (a) a potential core $U(x)$ that satisfies Bernoulli's equation and (b) a parabolic boundary layer profile, Eq. (4-11), extending out to distance $\delta \leq H$. The flow enters with a flat profile $U = U_0$. Apply the steady-flow integral momentum relation Eq. (4-120), plus mass conservation across the entire channel, to compute $\delta(x)$ and $U(x)$. Find the entrance length x_L.

4-38. For potential freestream flow past a sphere, $U = 1.5U_0 \sin(x/a)$, use the Rott–Crabtree integral method, Eq. (4-191), to compute the point of laminar-boundary-layer separation. Compare with Fig. 4-35.

4-39. In the spirit of Eq. (4-146) for two-dimensional flow, develop an explicit finite-difference model for the thick axisymmetric flow momentum relation, Eq. (4-195b). Use the same mesh as shown in Fig. 4-25, with y_n as the radial coordinate, but do not analyze the axisymmetric continuity Eq. (4-195a). Do the parameters α and β still appear? Note that there is no pressure gradient, $U_{m+1} = U_m$.

4-40. Air at 20°C and 1 atm issues from a circular hole and forms a round laminar jet. At 20 cm downstream of the hole, the maximum velocity is 35 cm/s. Estimate, at this position, (a) the "one percent" jet thickness, (b) the jet mass flow, and (c) an appropriate Reynolds number for the jet.

4-41. Air at 20°C and 1 atm flows at 1 m/s past a slender body of revolution, of length $L = 15$ cm, whose drag coefficient is 0.008 based on area (L^2). Assuming laminar flow at a point 3 m downstream of the trailing edge, estimate (a) the maximum wake velocity defect (in cm/s), (b) the "one percent" wake thickness (in cm), and (c) the wake-thickness Reynolds number.

4-42. In a lecture in 1932, Prandtl [Schlichting (1979, p. 80)] gave the following model equation analogous to Friedrichs' problem, Eq. (4-212):

$$\epsilon \frac{d^2 u}{dy^2} + \frac{du}{dy} + y = 0 \qquad \epsilon \ll 1$$

with $\qquad u(0) = 0 \qquad u(\infty) = 0$

This problem exhibits "boundary-layer" behavior. Find an exact solution of this problem and compare with a matched inner and outer solution.

4-43. Panton (1995) suggests the following modification of Friedrichs' model Eq. (4-212):

$$\epsilon \frac{d^2 u}{dy^2} + \frac{du}{dy} = -\frac{3}{2}(1 - 3\epsilon)e^{-3y} \qquad \epsilon \ll 1$$

with $\qquad u(0) = 0 \qquad u(\infty) = 1$

Find the exact solution of this problem and compare with a matched inner and outer solution.

4-44. The rotating-disk laminar boundary layer of Sec. 3-8.2 is fully three-dimensional. By changing the frame of reference of v_θ so that it is zero at the wall (i.e., subtracting v_θ from $r\omega$), plot the results of Table 3-5 in the form of a hodograph, with v_θ as the "streamwise" flow. Find the angle β and compare the results with Fig. 4-42c.

4-45. Verify that the similarity variables of Eq. (4-255) do indeed lead to the coupled ordinary differential Eqs. (4-256).

4-46. Develop a numerical solution, e.g., with Subroutine RUNGE in App. C, for the vertical free-convection Eqs. (4-256) and (4-257). Take any Prandtl number not listed in Table 4-10 (each class member could take a different case). The iteration is quite challenging because there are two unknown conditions, $f''(0)$ and $\Theta'(0)$. Compute the Nusselt number and compare with Eq. (4-260).

4-47. A vertical isothermal plate 40 cm high and 30 cm wide is immersed in air at 20°C and 1 atm. Each side of the plate is to dissipate 100 W of heat to the air. What is the wall temperature (in °C)?

4-48. A horizontal pipe of outer diameter 5 cm is immersed in air at 20°C and 1 atm. If the cylinder surface is at 300°C, how much heat (in W) is lost to the air per meter of pipe length?

4-49. A model two-dimensional airfoil has the following theoretical potential-flow surface velocities on its upper surface at a small angle of attack:

x/C	0.0	0.025	0.05	0.1	0.2	0.3	0.4	0.6	0.8	1.0
V/U_∞	0.0	0.97	1.23	1.28	1.29	1.29	1.24	1.14	0.99	0.82

The stream velocity U_∞ is variable. The chord length C is 30 cm. The fluid is air at 20°C and 1 atm. Assume that x is a good approximation of the arc length along the upper surface. Using any laminar-boundary-layer method of your choice, find the predicted separation point, if any, for (a) $Re_C = 1E6$; and (b) $Re_C = 4E5$.

4-50. Air at about 1 atm and 20°C flows through a 12 cm square duct at 0.4 m³/s, as in Fig. P4-50. Two hundred thin flat plates of 1 cm chord length are stretched across the duct at random positions. They do not interfere with each other. How much additional pressure drop do these plates contribute to the duct flow loss?

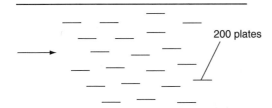

FIGURE P4-50

4-51. Nondimensionalize Thwaites' method, using U_o and L as reference values, for laminar-boundary-layer flow with an external potential-flow distribution $U(x)$. Show that the predicted separation point is independent of the Reynolds number $U_o L/\nu$. Apply this conclusion to discuss the expected results for Prob. 4-49.

4-52. A conical diffuser of initial radius R expands at a uniform angle θ, as in Fig. P4-52. The flow enters at uniform velocity U_o. Assuming a one-dimensional freestream, use any laminar-boundary-layer method of your choosing to estimate the angle θ for which flow separation occurs at $x = 2R$.

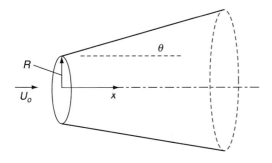

FIGURE P4-52

4-53. Show that the point-sink boundary-layer solution of Eqs. (4-86) and (4-89) may be interpreted through Thwaites' method as a constant value of Thwaites' parameter, $\lambda = 9/80$. How does this value of λ compare to two-dimensional stagnation flow?

4-54. Show that the exact Falkner–Skan solutions for two-dimensional stagnation flow are equivalent to Thwaites' method parameters of $\lambda = 0.0855$, $S = 0.360$, and $H = 2.216$. How do these values compare with values from the use of Thwaites' method for the freestream $U = Bx$?

4-55. When a jet is formed parallel to a wall, it is called a *wall jet*, as shown in Fig. P4-55. The jet origin can be thought of as a source of momentum, and further down the wall the velocity profiles are similar. Glauert (1956) showed that both plane and axisymmetrical wall jets satisfy the same ordinary differential equation for laminar flow:

$$\frac{d^3f}{d\eta^3} + f\frac{d^2f}{d\eta^2} + 2\left(\frac{df}{d\eta}\right)^2 = 0$$

where $\eta \propto y$, $f \propto \psi$, $f' \propto u$, and $f'' \propto \tau$. The boundary-layer thickness δ is proportional to $x^{3/4}$. (a) Verify Glauert's choice of boundary conditions: $f(0) = f'(0) = f'(\infty) = 0$. (b) To enforce the normalized condition $f(\infty) = 1$, Glauert showed that $f''(0) = 2/9$. For this value, numerically integrate the similarity equation out to $\eta = 8$ and plot both $f(\eta)$ and $f'(\eta)$.

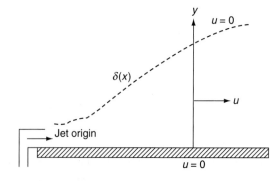

FIGURE P4-55

CHAPTER
5
THE STABILITY OF LAMINAR FLOWS

5-1 INTRODUCTION: THE CONCEPT OF SMALL-DISTURBANCE STABILITY

Chapters 3 and 4 have attacked a wide variety of problems in laminar viscous flow. Chapter 3 considered the full Navier–Stokes equations and a host of simple flow geometries. Chapter 4 introduced the Prandtl boundary-layer approximation that allowed us to look at a wider class of more realistic flows, subject only to accurate knowledge of the inviscid outer flow distribution. One way or another, it seemed that we could resolve, to reasonable accuracy, almost any problem in laminar flow.

But this text is not over, because, as the reader probably knows, laminar flows have a fatal weakness: poor resistance to high Reynolds numbers. For any given laminar flow, there is a finite value of its Reynolds number which threatens its very existence. Since this *critical Reynolds number*, as we shall call it, has only a modest value, being of the order of 1000 when referred to a transverse thickness, it follows that laminar flows are the exception rather than the rule in most engineering situations. At higher Reynolds numbers, the flow is always turbulent, i.e., disorderly, randomly unsteady, usually impossible to analyze exactly, but fortunately amenable to study of its average values (Chap. 6).

Thus laminar flow is found to be unstable, and its critical Reynolds numbers are of such a magnitude that flows of low-viscosity fluids (water, mercury, ammonia, gases, . . .) are normally turbulent, not laminar. Coffee stirred in a cup mixes turbulently. Smoke rises turbulently from a chimney. Water in a bathroom shower pipe

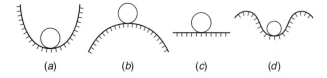

FIGURE 5-1
Relative stability of a ball at rest: (*a*) stable; (*b*) unstable; (*c*) neutral stability; (*d*) stable for small disturbances but unstable for large ones.

flows turbulently.* The boundary layer on a commercial jet airplane wing is turbulent. Any river worthy of its name flows turbulently. Meanwhile, laminar flows should not be disregarded, because many practical situations arise that are indeed laminar, such as low-speed flows, small-scale bodies, very viscous fluids, or leading-edge problems.

The two key words in this chapter are *stability* and *transition*. The general concept of stability has been discussed many times, probably never more eloquently than by Cunningham (1963). The discussion always boils down to one question: Can a given physical state withstand a disturbance and still return to its original state? If so, it is *stable*. If not, that particular state is unstable. It is the job of the stability analyst to test the effect of a particular disturbance.

A simple example is shown in Fig. 5-1, where a ball lies at rest under various conditions. In Fig. 5-1*a*, its position is unconditionally stable, because it would return to its initial position even if disturbed by a large displacement. Conversely, Fig. 5-1*b* shows an unstable state, since any slight disturbance would topple the ball, never to return. A flat tabletop, as in Fig. 5-1*c*, is an example of neutral stability, since the ball will rest anywhere it is displaced. Finally, Fig. 5-1*d* illustrates a more complicated case, where the ball is stable for small displacements but will diverge if disturbed far enough to drop over the edge. This is often the case in boundary-layer flow, where a large trip wire can cause an otherwise stable laminar flow to become turbulent. Note that stability requires simply a yes or no answer. One can prove that a physical state is unstable without being able to determine to what true stable state the disturbance will lead. In viscous flow, we can prove that laminar flow is unstable above certain Reynolds numbers, but that is all. The analysis does *not* predict turbulence. Turbulence is an experimentally observed fact. It has never been proved mathematically that turbulent flow is the proper stable state at high Reynolds numbers. Thus we can discuss only qualitatively our second key word, *transition*, which is defined as the change, over space and time and a certain Reynolds number range, of a laminar flow into a turbulent flow.† Although small-disturbance stability theory has been widely accepted for 40 years and many transition data have been accumulated, there is still no theory of transition‡ although

---

* The shower spray is turbulent also but not continuous. When viewed under stroboscopic light, it shows a striking pattern of irregular droplets.
† The opposite effect of *reverse transition* or *relaminarization* is also possible under certain conditions, as we shall see in Chap. 6.
‡ It can be stated, however, that transition never occurs for $Re < Re_{\text{crit}}$.

there is, as we shall see, a modestly successful empirical prediction of transition based on the spatial amplification rates of the linearized stability theory. It is also now possible, at least for moderate Reynolds numbers, to use computational fluid dynamics (CFD) to predict flow instability and transition. This new field is called *direct numerical simulation* (DNS) and uses the size and speed of the newest computers to simulate all important scales of a flow, including turbulent fluctuations. Thus DNS has become a numerical laboratory for investigating stability, transition, and low Reynolds number turbulence, as reviewed by Moin and Mahesh (1998).

The present chapter is a reasonably comprehensive review of the subject, but there are several complete monographs for further reading. Two classic analytical texts are by Lin (1955) and by Chandrasekhar (1961). Betchov and Criminale (1967) cover the linear stability of parallel flows, and Joseph (1976) emphasizes non-linear theories. Drazin and Reid (1981) cover the subject broadly, including numerical methods. The most recent texts are by Godreche and Manneville (1998), Riahi (2000), and Schmid and Henningson (2001). For pure enjoyment of this very interesting subject, the writer recommends the short text by Philip Drazin (2002), regrettably the last of his many excellent textbooks.

5-1.1 Outline of a Typical Stability Analysis

All small-disturbance stability analyses follow the same general line of attack, which may be listed in seven steps.

1. We seek to examine the stability of a basic solution to the physical problem, Q_0, which may be a scalar or vector function.
2. Add a disturbance variable Q' and substitute $(Q_0 + Q')$ into the basic equations which govern the problem.
3. From the equation(s) resulting from step 2, subtract the basic terms that Q_0 satisfies identically. What remains is the *disturbance equation*.
4. Linearize the disturbance equation by assuming *small* disturbances, that is, $Q' \ll Q_0$, and neglect terms such as Q'^2 and Q'^3, etc.
5. If the linearized disturbance equation is complicated and multidimensional, it can be simplified by assuming a form for the disturbances, such as a traveling wave or a perturbation in only one direction.
6. The linearized disturbance equation should be homogeneous and have homogeneous boundary conditions. It can thus be solved only for certain specific values of the equation's parameters. In other words, it is an *eigenvalue* problem.
7. The eigenvalues found in step 6 are examined to determine when they grow (are unstable), decay (are stable), or remain constant (neutrally stable). Typically the analysis ends with a chart showing regions of stability separated from unstable regions by *neutral curves*.

Let us illustrate this process with an example more than a century old.

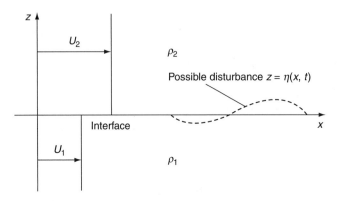

FIGURE 5-2
Sketch and nomenclature for the Kelvin–Helmholtz interfacial instability.

5-1.2 Wind-Generated Waves: The Kelvin–Helmholtz Instability

Our first example is very important yet not too difficult mathematically, with no digital computer required. The reader knows from experience that a wind of sufficient velocity will form waves on a calm water surface. The air–water interface is *unstable* under certain conditions. Hermann von Helmholtz posed the problem physically in 1868, and his friend, Lord Kelvin, set it up mathematically and gave the solution in 1871.

The mathematical model is shown in Fig. 5-2. A horizontal interface divides two uniform flows of differing velocity and density. Kelvin assumed that both the lower (region 1) and upper (region 2) flows were incompressible, irrotational, and inviscid.† Thus both *basic flows* (Step 1 of Sec. 5-1.1) possess a velocity potential and a hydrostatic pressure distribution:

$$z < 0: \quad \phi_1 = U_1 x \quad p_1 = p_o - \rho_1 g z$$
$$z > 0: \quad \phi_2 = U_2 x \quad p_2 = p_o - \rho_2 g z \tag{5-1}$$

We allow tangential slip at the interface, which is a *vortex sheet* with a discontinuity in velocity. Step 2 of Sec. 5-1.1 then asks us to apply a disturbance to both variables, denoted by a "hat":

$$\phi_1 = U_1 x + \hat{\phi}_1(x, z, t)$$
$$\phi_2 = U_2 x + \hat{\phi}_2(x, z, t) \tag{5-2}$$

The interface is disturbed also, to $z = \eta(x, t)$, as shown in Fig. 5-2. The pressure becomes unsteady when the flow is disturbed and satisfies the unsteady Bernoulli

---

†More recent work has added viscous effects to the analysis. See the text by Phillips (1978).

relation, Eq. (2-95):

$$p_i = C_i - \rho_i \frac{\partial \phi_i}{\partial t} - \frac{\rho_i}{2}|\nabla \phi_i|^2 - \rho_i g z \quad \text{for regions } i = 1, 2 \quad (2\text{-}95)$$

For this to be valid at $z = 0$ when there is no disturbance, it must be that

$$C_1 - \frac{\rho_1}{2} U_1^2 = C_2 - \frac{\rho_2}{2} U_2^2$$

One boundary condition is that the pressure must be continuous across the interface:

At $z = \eta$:
$$C_1 - \rho_1 \frac{\partial \phi_1}{\partial t} - \frac{\rho_1}{2}|\nabla \phi_1|^2 - \rho_1 g \eta$$
$$= C_2 - \rho_2 \frac{\partial \phi_2}{\partial t} - \frac{\rho_2}{2}|\nabla \phi_2|^2 - \rho_2 g \eta$$

A second condition is that the vertical velocities must each match the interfacial motion:

At $z = \eta$: $\quad w_i = \dfrac{\partial \phi_i}{\partial z} = \dfrac{d\eta}{dt} = \dfrac{\partial \eta}{\partial t} + \dfrac{\partial \phi_i}{\partial x}\dfrac{\partial \eta}{\partial x} \quad$ for $i = 1, 2$

The other two conditions are that the disturbance dies out far from the interface:

$$\nabla \hat{\phi}_2 \to 0 \quad \text{as} \quad z \to +\infty$$

and (5-3)

$$\nabla \hat{\phi}_1 \to 0 \quad \text{as} \quad z \to -\infty$$

Step 3 of Sec. 5-1.1 is to subtract the basic-flow equation to leave the *disturbance equation*. This is easy for our inviscid flow because the disturbance terms also satisfy Laplace's equation:

$$\nabla^2 \hat{\phi}_1 = 0$$

and (5-4)

$$\nabla^2 \hat{\phi}_2 = 0$$

Step 4 is to linearize by assuming that the disturbances are much weaker than the basic flow. First, the interface is assumed to have small displacements and small slopes:

$$g\eta \ll U_i^2 \quad \frac{\partial \eta}{\partial x} \ll 1$$

This means that the interfacial conditions can be approximated at $z = \eta \approx 0$. Second, the disturbance velocities are assumed small:

$$|\nabla \hat{\phi}_i| \ll U_i$$

This enables us to linearize the interfacial conditions:

Pressure: $\rho_1 \left(U_1 \dfrac{\partial \hat{\phi}_1}{\partial x} + \dfrac{\partial \hat{\phi}_1}{\partial t} + g\eta \right) \approx \rho_2 \left(U_2 \dfrac{\partial \hat{\phi}_2}{\partial x} + \dfrac{\partial \hat{\phi}_2}{\partial t} + g\eta \right)$

at $z \approx 0$

Kinematic: $\dfrac{\partial \hat{\phi}_1}{\partial x} \approx \dfrac{\partial \eta}{\partial t} + U_1 \dfrac{\partial \eta}{\partial x}$

and (5-5)

$$\dfrac{\partial \hat{\phi}_2}{\partial x} \approx \dfrac{\partial \eta}{\partial t} + U_2 \dfrac{\partial \eta}{\partial x} \quad \text{at} \quad z \approx 0$$

Although the equations and boundary conditions are now linear, they are still too difficult to handle for an arbitrary disturbance $\eta(x, t)$. Thus step 5 of Sec. 5-1.1 suggests that we assume a simple form for the disturbances, namely, two-dimensional traveling waves or *normal modes*:

$$\eta = \eta_o e^{i(\alpha x - \sigma t)}$$
$$\hat{\phi}_j = \phi_j'(z) e^{i(\alpha x - \sigma t)} \quad \text{for} \quad j = 1, 2$$

where $i = \sqrt{(-1)}$. The wave number α is real, but the frequency σ is a complex number, and the disturbance will be *unstable* (grow without bound) if its imaginary part σ_i is positive.

Step 6 of Sec. 5-1.1: Solve the Laplacian disturbance Eqs. (5-4) to yield exponential solutions $e^{\pm kz}$, and the far-field boundary conditions Eqs. (5-3) reduce them to

$$\phi_1' = A_1 e^{kz}$$

and (5-6)

$$\phi_2' = A_2 e^{-kz}$$

Substitution back in the interfacial conditions Eqs. (5-5) gives three linear homogeneous equations for the three constants η_0, A_1, and A_2. The results are

$$A_1 = i\eta_o \left(U_1 - \dfrac{\sigma}{\alpha} \right)$$

$$A_2 = -i\eta_o \left(U_2 - \dfrac{\sigma}{\alpha} \right)$$

This specifies the coefficients in the disturbance *eigenfunctions* of Eqs. (5-6). They are proportional to η_o, which is arbitrary but small. The *eigenvalues* σ are found from the pressure condition, Eq. (5-5), which is a quadratic equation with the following solution:

$$\sigma = \alpha \frac{\rho_1 U_1 + \rho_2 U_2}{\rho_1 + \rho_2} \pm \left[\frac{\alpha^2 \rho_1 \rho_2 (U_1 - U_2)^2}{(\rho_1 + \rho_2)^2} - \frac{\alpha g(\rho_1 - \rho_2)}{(\rho_1 + \rho_2)}\right]^{1/2}$$

The key to stability is the square bracket []. The disturbance is stable/neutral/unstable as the bracketed term is negative/zero/positive. The unstable condition is thus

$$\alpha \rho_1 \rho_2 (U_1 - U_2)^2 > g(\rho_1^2 - \rho_2^2) \qquad \alpha = \frac{2\pi}{\lambda} \qquad (5\text{-}7)$$

where λ is the wavelength of the disturbance. This condition is always true for large enough α, or small enough λ, so, at this level of approximation, vortex sheets are always unstable.

A more realistic result is obtained by adding linearized surface (interfacial) tension to the analysis, using Eq. (1-107). The final result is

$$\sigma = \alpha \frac{\rho_1 U_1 + \rho_2 U_2}{\rho_1 + \rho_2} \pm \left[\frac{\alpha^2 \rho_1 \rho_2 (U_1 - U_2)^2}{(\rho_1 + \rho_2)^2} - \frac{\alpha[g(\rho_1 - \rho_2) + \alpha^2 \mathcal{T}]}{(\rho_1 + \rho_2)}\right]^{1/2} \qquad (5\text{-}8)$$

where $\mathcal{T}$ is the surface-tension coefficient. The bracketed term [] is positive (unstable) when

$$(U_1 - U_2)^2 > \frac{[g(\rho_1 - \rho_2) + \alpha^2 \mathcal{T}](\rho_1 + \rho_2)}{\alpha \rho_1 \rho_2} \qquad (5\text{-}9)$$

Unlike Eq. (5-8), which neglects surface tension, Eq. (5-9) requires a finite velocity difference to cause waves to form. The reader may show as an exercise that the right-hand side of Eq. (5-9) is a minimum when $\alpha = [g(\rho_1 - \rho_2)/\mathcal{T}]^{1/2}$. As an example, take air at 20°C and 1 atm blowing over a fresh water surface. From Tables A-1 and A-2 in App. A, $\rho_{\text{water}} = 998$ kg/m^3, $\rho_{\text{air}} = 1.20$ kg/m^3, and $\mathcal{T} = 0.0727$ N/m. The minimum value of $(U_1 - U_2)^2$ occurs at $\alpha \approx 367$ m^{-1}. Then Eq. (5-9) predicts that waves will begin to form at a critical value $|U_1 - U_2| \approx 6.67$ m/s ≈ 15 mph, for which the critical wavelength is 0.0171 m. This calculation is merely an estimate, ±50 percent. The actual critical velocity difference will depend upon viscosity, wind nonuniformity, and turbulence.

The final step 7 of Sec. 5-1.1 suggests drawing a chart showing regions of stability and instability for the flow. This is appropriate for the complex computer solutions of the next section, but here we derived a formula, Eq. (5-9), which makes a chart unnecessary.

Although modified by viscosity, Kelvin–Helmholtz instability is common in the atmosphere and the oceans. Wind-generated surface waves are visible. Deep in the ocean, there are internal waves at density interfaces, as explained by Phillips (1978). In the atmosphere, wind shear, the bane of pilots, generates waves that are generally invisible. But sometimes cloud and light conditions reveal the waves.

FIGURE 5-3
Kelvin–Helmholtz breaking waves outlined by a billow-cloud formation near Laramie, Wyoming. [*Courtesy of Brooks Martner, NOAA Environmental Technology Laboratory.*]

Figure 5-3 shows beautiful rolled-up waves photographed by Brooks Martner for the NOAA Environmental Technology Laboratory.

5-2 LINEARIZED STABILITY OF PARALLEL VISCOUS FLOWS

We now use the same method of Sec. 5-1 to study viscous-flow instability. Here, both the basic solution and its assumed disturbances will in general be three-dimensional unsteady functions. The same seven steps apply:

1. Select a basic solution flow.
2. Add a disturbance.
3. Find the disturbance equations.
4. Linearize.
5. Simplify (in this case to a traveling wave).
6. Solve for the eigenvalues.
7. Interpret the stability conditions and draw a chart showing the neutral curves and the growth and decay rates.

5-2.1 Derivation of the Orr–Sommerfeld Equation

Let us now carry out the seven steps for incompressible laminar flow with constant (ρ, μ, k, c_p) and no buoyancy effects. Thus we analyze the stability of the continuity and Navier–Stokes relations for the two variables $\mathbf{V}$ and p:

$$\nabla \cdot \mathbf{V} = 0$$
$$\frac{D\mathbf{V}}{Dt} = -\frac{1}{\rho}\nabla p + \nu \nabla^2 \mathbf{V} \qquad (5\text{-}10)$$

Let us assume that we have found, by the methods of Chap. 3 or 4, a laminar-flow solution to these equations: $\mathbf{V}_0 = (U, V, W) = \mathbf{V}_0(\mathbf{x}, t)$ and $p_0(\mathbf{x}, t)$. To determine whether these are stable solutions, we add on small-disturbance variables $\mathbf{v}(\mathbf{x}, t) = (\hat{u}, \hat{v}, \hat{w})$ and $\hat{p}(\mathbf{x}, t)$. Substitute the superimposed variables $\mathbf{V}_0 + \mathbf{v}$ and $p_0 + \hat{p}$ in Eqs. (5-10), subtract the original $\mathbf{V}_0$ and p_0 equalities, and neglect higher powers and products of $\mathbf{v}$, which occur only in one place, the (non-linear) convective acceleration. The reader may verify the following linearized disturbance equations, written out in full:

$$\frac{\partial \hat{u}}{\partial x} + \frac{\partial \hat{v}}{\partial y} + \frac{\partial \hat{w}}{\partial z} = 0$$

$$\frac{\partial \hat{u}}{\partial t} + U\frac{\partial \hat{u}}{\partial x} + \hat{u}\frac{\partial U}{\partial x} + V\frac{\partial \hat{u}}{\partial y} + \hat{v}\frac{\partial U}{\partial y} + W\frac{\partial \hat{u}}{\partial z} + \hat{w}\frac{\partial U}{\partial z}$$

$$\approx -\frac{1}{\rho}\frac{\partial \hat{p}}{\partial x} + \nu \nabla^2 \hat{u}$$

$$\frac{\partial \hat{v}}{\partial t} + U\frac{\partial \hat{v}}{\partial x} + \hat{u}\frac{\partial V}{\partial x} + V\frac{\partial \hat{v}}{\partial y} + \hat{v}\frac{\partial V}{\partial y} + W\frac{\partial \hat{v}}{\partial z} + \hat{w}\frac{\partial V}{\partial z} \quad (5\text{-}11)$$

$$\approx -\frac{1}{\rho}\frac{\partial \hat{p}}{\partial y} + \nu \nabla^2 \hat{v}$$

$$\frac{\partial \hat{w}}{\partial t} + U\frac{\partial \hat{w}}{\partial x} + \hat{u}\frac{\partial W}{\partial x} + V\frac{\partial \hat{w}}{\partial y} + \hat{v}\frac{\partial W}{\partial y} + W\frac{\partial \hat{w}}{\partial z} + \hat{w}\frac{\partial W}{\partial z}$$

$$\approx -\frac{1}{\rho}\frac{\partial \hat{p}}{\partial z} + \nu \nabla^2 \hat{w}$$

These are formidable but *linear* partial differential equations for $\hat{u}, \hat{v}, \hat{w}, \hat{p}$, since U, V, and W are known functions and thus no more than variable coefficients.

Equations (5-11) can be systematically reduced to a single ordinary differential equation by assuming a locally parallel basic flow. If y is the coordinate normal to the wall or across the shear layer, we assume that the component V across the layer is negligibly small, as in duct flow, and further assume that $U \approx U(y)$ and $W \approx W(y)$. This will eliminate 10 convective terms on the left-hand sides. The disturbances are also assumed to be parallel flows. Then the most general form of three-dimensional disturbance $(\hat{u}, \hat{v}, \hat{w})$ is that of a traveling wave whose amplitude varies with y and which moves along the wall at an angle ϕ with respect to the x axis. Taking advantage of the complex notation, we specify the disturbances

$$(\hat{u}, \hat{v}, \hat{w}, \hat{p}) = [u(y), v(y), w(y), p(y)]\exp[i\alpha(x\cos\phi + z\sin\phi - ct)] \quad (5\text{-}12)$$

where $i = \sqrt{-1}$. All disturbances have wave number α, propagation speed c, and frequency $\omega = \alpha c$. They are historically referred to as Tollmien–Schlichting waves, which are the first (infinitesimal) indications of laminar-flow instability. If we substitute Eqs. (5-12) in Eqs. (5-11), we obtain the following linear ordinary differential equations with complex coefficients:

$$i\alpha u \cos\phi + v' + i\alpha w \sin\phi = 0 \tag{5-13a}$$

$$i\alpha u(U\cos\phi + W\sin\phi - c) + U'v = -\frac{i}{\rho}\alpha p \cos\phi + \nu(u'' - \alpha^2 u) \tag{5-13b}$$

$$i\alpha v(U\cos\phi + W\sin\phi - c) = -\frac{1}{\rho}p' + \nu(v'' - \alpha^2 v) \tag{5-13c}$$

$$i\alpha w(U\cos\phi + W\sin\phi - c) + W'v = -\frac{i}{\rho}\alpha p \sin\phi + \nu(w'' - \alpha^2 w) \tag{5-13d}$$

where the primes denote differentiation with respect to y. As expected, they are second order in u, v, w and first order in p. We may assume that the disturbances grow either spatially (α complex and αc real) or else temporally (α real and c complex).

A remarkable property lurks within Eqs. (5-13). To see this, we multiply (5-13b) by cos ϕ and (5-13d) by sin ϕ and add them together. Further, introduce the compact notation

$$\begin{aligned} u_0 &= u\cos\phi + w\sin\phi \\ U_0 &= U\cos\phi + W\sin\phi \end{aligned} \tag{5-14}$$

We then find ourselves left with only three equations in three variables, u_0, v, and p:

$$i\alpha u_0 + v' = 0 \tag{5-15a}$$

$$i\alpha u_0(U_0 - c) + U_0'v = -\frac{i}{\rho}\alpha p + \nu(u_0'' - \alpha^2 u_0) \tag{5-15b}$$

$$i\alpha v(U_0 - c) = -\frac{1}{\rho}p' + \nu(v'' - \alpha^2 v) \tag{5-15c}$$

These are two-dimensional equations whose components u_0 and U_0 are parallel to the direction of propagation ϕ of the traveling waves, as seen from Eqs. (5-14). Thus the stability of any parallel flow in any direction ϕ can be found from a two-dimensional analysis for the effective basic flow $U_0(y)$ in that direction. This powerful result was first deduced by Dunn and Lin (1955). It has been used by Gregory et al. (1955) to derive general results about the stability of three-dimensional flows, particularly the rotating disk flow of Sec. 3-8.

Now suppose further that W is identically zero, i.e., the basic flow is purely two-dimensional. Then the shape of the profile $U_0(y)$ is independent of ϕ, which means that the stability computation is also independent of ϕ, except for the scale factor cos ϕ. Therefore, for an oblique disturbance, the basic flow $U\cos\phi$ is relatively slower and thus more stable than when the disturbance propagates parallel to U(y). This result was deduced by Squire (1933) in his now famous theorem.

Squire's theorem. For a two-dimensional parallel flow U(y), the minimum critical unstable Reynolds number occurs for a two-dimensional disturbance propagating in the same direction ($\phi = 0$).

Thus here we consider only two-dimensional disturbances. Squire's theorem is concerned with the minimum Reynolds number, which occurs when $\phi = 0$. However, in his review, Stuart (1987) points out that maximum wave *amplification* occurs for $\phi \neq 0$, so that three-dimensional disturbances do indeed affect the growth and shape of an unstable disturbance.

Let us now drop the subscript 0 in Eqs. (5-15). It is possible to eliminate any two of the three variables u, v, p. If we eliminate u, we obtain a simple equation for the pressure fluctuation.

$$p'' - \alpha^2 p = -2i\alpha\rho U' v \qquad (5\text{-}16)$$

Thus the pressure satisfies a Poisson equation whose pressure-source term is proportional to the product

$$\frac{dU}{dy}\frac{\partial \hat{v}}{\partial x}$$

which is to say the product of the mean shear and the longitudinal variation of the normal velocity fluctuation. In the outer freestream, where $U' \approx 0$, the solution must be of the form

$$p = ae^{-\alpha y} + be^{+\alpha y} \qquad (5\text{-}17)$$

Since the disturbance is assumed to vanish at infinity, it follows that $b = 0$ and the pressure fluctuation damps exponentially.

It is relatively difficult to eliminate v and p from these equations; the reader is invited to try it. The result will be the same mathematically if we eliminate u and p, which results in the celebrated Orr–Sommerfeld relation:

$$(U - c)(v'' - \alpha^2 v) - U''v + \frac{i\nu}{\alpha}(v'''' - 2\alpha^2 v'' + \alpha^4 v) = 0 \qquad (5\text{-}18)$$

The secrets of infinitesimal laminar-flow instability lie within this fourth-order linear homogeneous equation, first derived independently by Orr (1907) and Sommerfeld (1908). The boundary conditions are that the disturbances u and v must vanish at infinity and at any walls (no slip). The continuity relation Eq. (5-15a) shows that $u = 0$ implies $v' = 0$. Hence the proper conditions on the Orr–Sommerfeld equation are of the following types:

Duct flows: $\quad v(\pm h) = v'(\pm h) = 0$

Boundary layers: $\quad v(0) = v'(0) = 0 \quad v(\infty) = v'(\infty) = 0 \qquad (5\text{-}19)$

Free-shear layers: $\quad v(\pm\infty) = v'(\pm\infty) = 0$

Since the equation and its boundary conditions are both homogeneous, it follows that what we are talking about is an eigenvalue problem. The Orr–Sommerfeld equation has three parameters: α, c, and ν (or, better, a Reynolds number $Re = U\delta/\nu$). For a given profile $U(y)$ and $U''(y)$, only a certain continuous but limited sequence of these parameters (the eigenvalues) will satisfy the relations

Eqs. (5-18) and (5-19). The mathematical problem is to find this sequence, which has a different functional form for the spatial versus temporal growth of disturbances:

Temporal growth: $\quad f(Re, \alpha, c_r, c_i) = 0 \quad$ (5-20a)

Spatial growth: $\quad g(Re, \alpha_r, \alpha_i, \omega) = 0 \quad$ (5-20b)

where the subscripts r and i mean real and imaginary, respectively. Of particular interest is the case of neutral stability: $c_i = 0$ for the temporal case and $\alpha_i = 0$ for spatial neutral growth. The locus of these neutral points forms the boundary between stability and instability. The solutions $v(y)$ that generate the eigenvalues in Eqs. (5-20) are called *eigenfunctions*.

5-2.2 Inviscid-Stability Theory

A limiting but very instructive case is that of an infinite Reynolds number or negligible viscosity, for which we can drop the term involving ν in the Orr-Sommerfeld relation, Eq. (5-18). This gives an inviscid-disturbance relation

$$v'' - \left(\frac{U''}{U-c} + \alpha^2\right)v = 0 \quad (5\text{-}21)$$

that is named in honor of Lord Rayleigh, who first studied this equation in a series of papers between 1878 and 1915, all of which are now collected in six volumes of his "Scientific Papers." The Rayleigh equation is second order, and thus we can no longer maintain conditions on the tangential velocity u (or v'). We retain two conditions on the normal velocity v, for example,

Boundary layer: $\quad v(0) = v(\infty) = 0 \quad$ (5-22)

There are many inviscid analyses reported in the literature, and an excellent discussion is given in the text by Drazin and Reid (1981). The theory gives good insight into profile shape effects on stability, and some workers have used the two "inviscid" solutions to help generate two additional "viscous" solutions of the full Orr–Sommerfeld equation.

The Rayleigh Eq. (5-21) can be readily solved either analytically or numerically. There is a singularity on the real axis, $c = c_r$, at the point where the velocity $U = c_r$. The analytical behavior of v near this singularity was established in pioneering work by Tollmien (1929).

Five important theorems on inviscid stability are as follows:

Theorem 1 [Rayleigh (1880)]. It is necessary for instability that the velocity profile have a point of inflection.

Theorem 2 [Fjørtoft (1950)]. It is further necessary for instability that the numerical value $|U'|$ of the vorticity be a maximum at the point of inflection.

Theorem 3 [Fjørtoft (1950)]. If a point of inflection exists, it is further necessary that $U''(U - U_{PI}) < 0$ somewhere on the profile, where U_{PI} is the velocity at the point of inflection.

Theorem 4 [Lin (1955)]. If $U(y)$ possesses an inflection point at $y = y_c$, a neutral disturbance ($c_i = 0$) may exist whose phase velocity is $c_r = U(y_c)$.

Theorem 5 [Rayleigh (1880)]. The phase velocity c_r of an amplified disturbance must always lie between the minimum and maximum values of $U(y)$.

Rayleigh's result, Theorem 1, led workers for many years to believe that real (viscous) profiles without a point of inflection, such as channel flows and boundary layers with favorable pressure gradients, would be stable. It remained for Prandtl (1921) to show that viscosity can indeed be destabilizing for certain wave numbers at finite Reynolds numbers.

Figure 5-4 shows four different velocity profiles to be evaluated for inviscid instability. The first three are unconditionally stable; only the fourth is possibly unstable by virtue of Fjørtoft's Theorem 3. Meanwhile, all four should possess viscous instability at a finite Reynolds number.

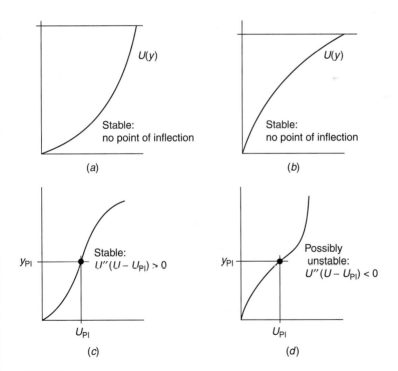

FIGURE 5-4
Four candidate inviscid velocity profiles evaluated for stability by Theorems 1 and 3 of Sec. 5-2.2.

Some actual profiles which resemble Fig. 5-4d—and thus are unstable even in the inviscid limit—are boundary layers with adverse pressure gradients, the shear layer between parallel streams, jets, and wakes.

5-2.3 Viscous Stability: Solution of the Orr–Sommerfeld Equation

The Rayleigh equation reveals the point of inflection to be a source of possible instability as Re approaches infinity. But it is clear from experiments that all types of laminar profiles become unstable at finite Reynolds numbers, as predicted physically by Prandtl (1921) and first shown mathematically by Tollmien (1929). These instabilities are manifested in the eigenvalues of the Orr–Sommerfeld equation, written here in dimensionless form:

$$(U^* - c^*)(\phi'' - \alpha_\delta^2 \phi) - U^{*\prime\prime}\phi + \frac{i}{\alpha_\delta Re_\delta}(\phi'''' - 2\alpha_\delta^2 \phi'' + \alpha_\delta^4 \phi) = 0 \quad (5\text{-}23)$$

where

$$\phi = \frac{v}{U_e} \quad U^* = \frac{U}{U_e} \quad c^* = \frac{c}{U_e} \quad \alpha_\delta = \alpha\delta \quad Re_\delta = \frac{U_e \delta}{\nu}$$

The independent variable is $\eta = y/\delta$. The boundary conditions have the same homogeneous form as in Eq. (5-19).

For a given velocity profile $U(y)$, Eq. (5-23) can be solved for its eigenvalues α_δ and c^*, which vary with Re_δ. The problem can be solved for either temporal stability (real α_δ) or spatial stability (real $\omega^* = \alpha_\delta c^*$). The solution will be unstable if we find eigenvalues such that

Temporal instability: $\qquad c_i^* > 0$ \hfill (5-24a)

Spatial instability: $\qquad \alpha_{\delta i} < 0$ \hfill (5-24b)

To satisfy the freestream boundary conditions $\phi(\infty) = \phi'(\infty) = 0$, we set $U^{*\prime\prime} = 0$ and $U' = U_e^*$, so that Eq. (5-23) can be solved for damped exponential behavior:

Freestream, $\eta \gg 1$: $\qquad \phi = (\text{const})e^{-\alpha_\delta \eta} + (\text{const})e^{-\zeta \eta}$ \hfill (5-25)

where $\qquad \zeta^2 = \alpha_\delta^2 + i\alpha_\delta Re_\delta (U_e^* - c^*)$

The "constants" are not important since the differential equation and boundary conditions are homogeneous. This asymptotic behavior reveals a formidable numerical difficulty in the solution of the Orr–Sommerfeld equation. Since α_δ and c^* are of order unity and $Re_\delta \gg 1$, the two rates of change in Eq. (5-25) are quite different, that is, generally speaking, regardless of the velocity profile shape,

$$|\zeta| \gg |\alpha_\delta| \quad (5\text{-}26)$$

Thus, of the four independent solutions of Eq. (5-23), two of them (associated with $e^{\alpha_\delta \eta}$) grow slowly, whereas two of them (associated with $e^{\zeta \eta}$) grow rapidly. The fast-growing solutions tend to contaminate and smear the slower pair, so that accurate eigenvalues cannot easily be found.

5-2.3.1 NUMERICAL METHODS. The Rayleigh inviscid-stability Eq. (5-21) is only moderately difficult to model numerically and thus is suitable for problem assignments. However, in the writer's opinion, the viscous Orr–Sommerfeld Eq. (5-23), with competing small and large solutions, is too difficult for homework. Orr–Sommerfeld programs are lengthy and sophisticated. The pioneers were Tollmien (1929) and Schlichting (1933b), who developed analytical (noncomputer) methods. By the 1950s, digital-computer methods were tried but proved inaccurate. Kaplan (1964) made a breakthrough with a "purification" scheme to keep slow- and fast-moving solutions separate, as described in the text by Betchov and Criminale (1967). Wazzan et al. (1968a) used Gram–Schmidt orthonormalization to ensure accuracy up to $Re_{\delta*} \approx 10^5$. Modern methods use either spectral techniques or finite differences, as described in the text by Drazin and Reid (1981). Sherman (1990) and Cebeci and Cousteix (1998) list FORTRAN programs. The recent text by Schmid and Henningson (2001) includes MATLAB programs for stability calculations. There is also commercial software available. The writer will abstain from Orr–Sommerfeld assignments. The transition-prediction methods of Sec. 5-4 will also avoid Orr–Sommerfeld computation.

The viscous theories predict a finite region at low Reynolds numbers (of the order of $Re_\delta \approx 1000$) where infinitesimal disturbances are amplified (c_i greater than zero). The boundary of this region is the so-called *neutral curve*, which is the locus ($c_i = 0$). From their shapes, they are called *thumb curves*, and two examples are sketched in Fig. 5-5. Outside the thumb, all disturbances are damped. If the profile $U(y)$ has no point of inflection, the thumb disappears at a large Reynolds number, in accordance with the Rayleigh criterion. In contrast, if $U(y)$ has a point of inflection, the thumb curve remains open at infinity. The lowest Reynolds number to which the

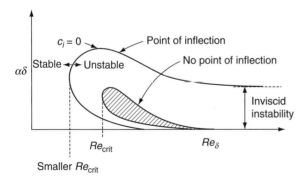

FIGURE 5-5
Neutral curves of the Orr–Sommerfeld equation.

thumb protrudes is called Re_{crit}, and typically the inflection profile has a smaller Re_{crit}, and a bigger thumb all around. The point Re_{crit} denotes nothing more than the smallest Reynolds number at which disturbances can be amplified. It *should not* be confused with the point of transition to turbulence, which occurs some 10 to 20 times the distance x_{crit} further downstream in the boundary layer. More important

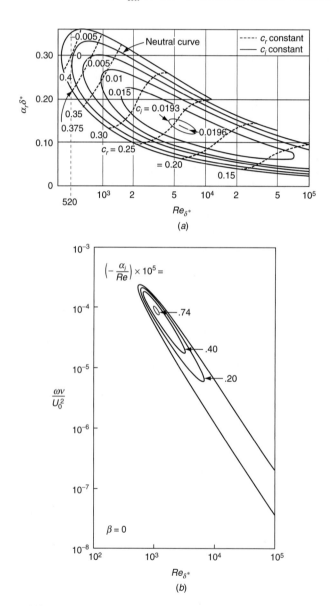

FIGURE 5-6
Amplification curves for the Blasius flat-plate boundary layer, from the point of view of (*a*) temporal stability; (*b*) spatial stability. [*After Wazzan et al. (1968a).*]

than Re_{crit} are the values of c_i and $\alpha\delta$ within the thumb because these determine the spatial and temporal growth rates which lead eventually to turbulent flow.

5-2.4 Stability of the Blasius and Falkner–Skan Profiles

Most of the classical velocity profiles have now been studied and their eigenvalues tabulated and charted. Early work concentrated on *temporal* stability, Eq. (5-20a). In the 1960s, emphasis shifted to *spatial stability*, Eq. (5-20b), which is a more realistic model of disturbances in a boundary layer growing and propagating downstream. Review articles such as Wazzan (1975) and Arnal (1984) discuss a wide variety of profiles such as the Falkner–Skan flows, the plane jet, the plane wake, the free-shear layer, Poiseuille pipe and channel flow, and many nonsimilar flows.

Accurate computations for the Blasius profile, assuming parallel flow, are shown in Figs. 5-6a and b, from Wazzan et al. (1968a). Note the difference in viewpoint between *temporal* instability (Fig. 5-6a), where unstable contours of $c_i > 0$ are plotted, and *spatial* instability (Fig. 5-6b), where values of $\alpha_1 < 0$ are contoured. The neutral curves are the same for both although shown in different coordinates. The following details are listed:

1. The minimum or *critical* Reynolds number for initial instability is $Re_{\delta^*,\,crit} = 520$, or $Re_{x,\,crit} \approx 91{,}000$. (Nonparallel-flow effects reduce these further to about 400 and 54,000, respectively.)
2. At Re_{crit}, the wave parameters, as calculated by Jordinson (1970), are $\alpha\delta^* = 0.3012$, $c_r/U_0 = 0.3961$, and $\omega\nu/U_0^2 = 2.29\text{E}{-}4$.
3. The maximum wave number for instability is $\alpha\delta^* \approx 0.35$, hence the smallest unstable wavelength is $\lambda_{min} = 2\pi\delta^*/0.35 \approx 18\delta^* \approx 6\delta$. Thus unstable Tollmien–Schlichting waves are long compared to boundary-layer thickness.
4. The maximum temporal growth rate is $c_i/U_0 \approx 0.0196$.
5. The maximum spatial growth rate is $\alpha_i\nu/U_0 \approx -0.74\text{E}{-}5$.
6. The maximum phase velocity of unstable waves is $c_r/U_0 \approx 0.4$; hence Tollmien–Schlichting waves travel rather slowly and tend to arise *near the wall*.
7. Compared to $Re_{x,\,crit} \approx 91{,}000$, the point of final transition to turbulence is at about $Re_{x,\,tr} \approx 3 \times 10^6$, or about 30 times further downstream.

Figure 5-7 shows neutral curves for Falkner–Skan wedge-flow profiles as computed by Wazzan (1968b). The parameter β has the same meaning as in Eq. (4-71). We see a very strong effect of pressure gradient, with, e.g., stagnation flow ($\beta = 1.0$) hundreds of times more stable than separating flow ($\beta = -0.1988$). This is true not only of Re_{crit} but also of the maximum spatial amplification rates, as shown in Table 5-1. Note how the curves for $\beta < 0$ (adverse gradients) remain open at large Reynolds numbers, since they have a point of inflection and thus have inviscid instability.

TABLE 5-1
Spatial stability parameters for Falkner–Skan profiles

β	$Re_{\delta^*, \text{crit}}$	$Re_{\theta, \text{crit}}$	$c_{i, \text{max}}$	$\left(-\dfrac{\alpha\delta^*}{Re_{\delta^*}}\right)_{\text{max}} \times 10^7$
+1.0	12,490	5,636	0.0065	1.14
0.8	10,920	4,874	0.0070	1.35
0.6	8,890	3,909	0.0075	1.67
0.5	7,680	3,344	0.0080	1.92
0.4	6,230	2,679	0.0085	2.42
0.3	4,550	1,927	0.0095	3.45
0.2	2,830	1,174	0.0104	6.0
0.1	1,380	556	0.0129	15.7
0.05	865	342	0.0154	32
0.0	520	201	0.0196	74
−0.05	318	119	0.0275	186
−0.1	199	71	0.0388	450
−0.14	138	47	0.0525	963
−0.1988	67	17	0.12	5,600

Source: Computations by Wazzan et al. (1968b).

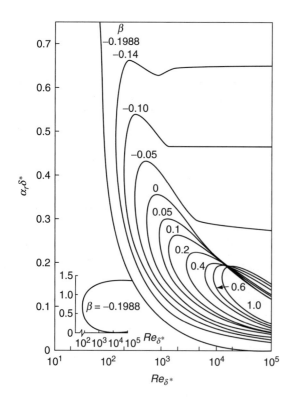

FIGURE 5-7
Neutral stability curves for Falkner–Skan boundary-layer profiles. [*After Wazzan et al. (1968b).*]

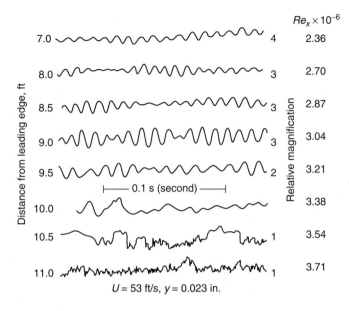

FIGURE 5-8
Hot-wire oscillograms showing natural transition from laminar to turbulent flow on a flat plate. [*After Schubauer and Skramstad (1947)*.]

5-2.5 Comparison of Stability Theory with Experiment

Although the concepts of instability and transition were well established from the work of Reynolds (1883), stability theory did not receive wide acceptance at first. Early wind tunnel transition experiments, such as Burgers (1924), reported $Re_{x,\text{tr}} \approx 350{,}000$ in flat-plate flow and did not detect any Tollmien–Schlichting waves. Even after the hot-wire anemometer was invented, careful measurements by Dryden (1934) confirmed the onset of turbulent fluctuations but did not detect Tollmien–Schlichting waves. These experiments were contaminated by the very noisy turbulence and the acoustics in the wind tunnels of the day.

Then, in 1940, the U.S. National Bureau of Standards, under the direction of H. L. Dryden, constructed a new wind tunnel with the extremely small freestream turbulence level of 0.02 percent[†] [compared to about 1.2 percent for Burgers' (1924) tunnel]. In this tunnel, in 1940–1941 (but not reported until after World War II), the classic experiment of Schubauer and Skramstad (1947) was performed. First, they did hot-wire studies to determine flat-plate transition in the presence of natural random fluctuations in the freestream. These results are shown in Fig. 5-8. At x less

---

[†]The turbulence level is defined as

$$T = \frac{1}{U_0}\left[\frac{1}{3}(\overline{u'^2} + \overline{v'^2} + \overline{w'^2})\right]^{1/2}$$

where u', v', and w' are the velocity fluctuations.

than 7 ft from the leading edge for this run, the velocity oscillations are negligibly small. At 8 ft, selective amplification clearly occurs. At 9 ft, the amplitudes of certain sinusoidal components have become quite large. At 10 ft, non-linear processes have begun to dominate, and the fluctuations are becoming more random. At 11 ft, the flow is fully turbulent and the scale of motion has changed drastically. Relevant frequencies are now in thousands instead of tens of hertz. This information would be lost if the freestream were noisy. The large "natural" fluctuations would cause almost immediate transition on the final (non-linear) scale of the process. This of course was the case in the early experiments, for which the local Reynolds number was much lower.

Figure 5-8 shows the effect of natural disturbances. To determine the effect of specific sinusoidal components, Schubauer and Skramstad placed a metal ribbon 0.1 in. wide and 0.002 in. thick across the flat plate at a distance of 0.005 in. from the wall and oscillated it electromagnetically. In this way they could introduce disturbances of any frequency into the boundary layer. For example, at a given Reynolds number, one can start from zero and slowly increase the oscillation frequency of the ribbon. At first, the downstream hot wire shows only insignificant natural oscillations, until a frequency is reached where definite sinusoidal components appear. This is the lower boundary of the neutral curve. As the frequency increases, the sinusoidal oscillations grow to a maximum amplitude and then decrease, finally disappearing at a higher finite frequency, the upper boundary of the thumb curve. In this way, one can establish the entire neutral curve by varying the Reynolds number of the flow.

The neutral curve measured in this classic controlled experiment is shown in Fig. 5-9 and compared to stability theories assuming (1) parallel flow [Wazzan et al. (1968a)] and (2) nonparallel flow [Saric and Nayfeh (1975)]. The effect of

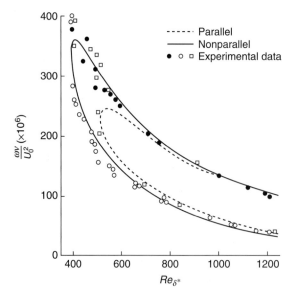

FIGURE 5-9

Comparison of theory and experiment for the neutral curves of the Blasius flat-plate boundary layer: □ data of Schubauer and Skramstad (1947); ● upper branch and ○ lower branch data by Ross and Barnes (1970). [*After Saric and Nayfeh (1975)*.]

nonparallel flow—boundary-layer growth $U(x, y)$ and $V(x, y)$—is seen to be significant at the lower Reynolds numbers and is in good agreement with the experiment. Saric and Nayfeh (1975) also used their "method of multiple scales" to investigate nonparallel flow with pressure gradients. As expected, favorable gradients ($\beta > 0$) grow slowly and show small effects on the neutral curve. Rapidly growing adverse gradients ($\beta < 0$) have large effects and are difficult to assess accurately.

The experimental stability measurements are also in good agreement with theory for the phase velocities and eigenfunctions—see, e.g., Kaplan (1964). Thus the basic correctness of linear instability theory is now well established.

5-3 PARAMETRIC EFFECTS IN THE LINEAR STABILITY THEORY

The success of the stability theory means that we can analytically predict the effect of various parameters such as Mach number, wall temperature, and wall suction on laminar-flow stability. A great many such analyses have now appeared, mostly using the newer digital-computer techniques of handling the Orr–Sommerfeld equation. In Fig. 5-7, we saw the computed effect of pressure gradient on Falkner–Skan profiles, showing that favorable (adverse) gradients increase (decrease) stability. This effect was qualitatively verified in the same experiment of Schubauer and Skramstad (1947), as shown in Fig. 5-10. Moving down from the top, we see that an established unstable flow is completely damped by a favorable gradient (downstream pressure drop) and then the instabilities appear even stronger when the pressure rises (adverse gradient). A pressure gradient influences not only the initial linear instability but also the final breakdown into turbulence, as reviewed in detail by Tani (1969). In a near-separating flow, this final process is abrupt.

5-3.1 Classical Laminar-Flow Profiles

The classical Poiseuille and Couette flow profiles from Chap. 3 have been analyzed for linear stability in a number of papers.

The parabolic Poiseuille channel flow due to a pressure gradient (Fig. 3-8) has the form $U = U_0(1 - y^2/L^2)$ and was studied as one of the last of the asymptotic analyses by Shen (1954). His computed amplification curves, shown in Fig. 5-11, are accurate to about 5 percent. A subsequent machine calculation by Nachtsheim (1964) gave $Re_{\text{crit}} = 5767$ and $\alpha L = 1.02$, which are labeled in the figure. Apparently linear instability does *not* control this flow because experiments in channel flow typically show transition to turbulence at about $Re_L \approx 1000$, far less than the critical value in Fig. 5-11. The most probable explanation is that *large* disturbances cause the transition, but it may also be possible that the developing-flow profiles in the channel entrance become unstable.

Even more paradoxical is axisymmetric Poiseuille pipe flow, with the paraboloidal profile $U(r) = U_0(1 - r^2/r_0^2)$. All papers known to this writer report the same result: no apparent linear instability at any Reynolds number. Yet we know

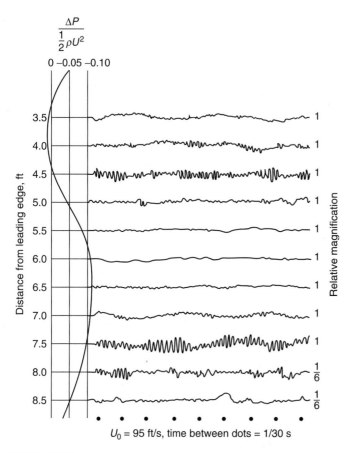

FIGURE 5-10
Effect of pressure gradient on laminar-boundary-layer oscillations. [*After Schubauer and Skramstad (1947)*.]

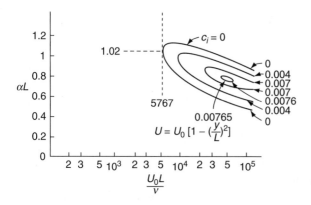

FIGURE 5-11
Stability diagram for plane Poiseuille flow. [*After Shen (1954)*.]

since the work of Reynolds (1883) that transition to turbulence typically occurs at about $Re_D \approx 2000$. Here again it is thought that the instability is due to large disturbances in the entrance, since with proper care one can maintain laminar flow in a smooth pipe up to $Re_D \approx 10^5$. Tatsumi (1952) shows that the developing pipe entrance profiles become unstable at about $Re_D \approx 10^4$. Alternately, Meseguer (2003) analyzes the *streak breakdown* process that, in conjunction with finite disturbances, is a possible path toward pipe-flow transition. For further details, see Drazin and Reid (1981) or Sec. 9.3 of Schmid and Henningson (2001).

The case of the linear Couette profile $U = U_0 y/L$ generated between a fixed and a moving plate (Fig. 3-1) is also apparently completely stable for all infinitesimal disturbances. Yet we know from experiment that turbulent flow ensues if $Re = U_o L/\nu > 1500$ [see Fig. 19.3 of Schlichting (1979)]. Once again it is thought that Couette flow becomes unstable due either to (1) large disturbances or (2) Tollmien–Schlichting waves in the developing transient profiles.

5-3.1.1 PSEUDORESONANCE. As mentioned earlier, no unstable normal-mode traveling waves have been found for either plane Couette flow or Poiseuille pipe flow. The linear theory is stable, and large disturbances are necessary to trigger turbulence. Meanwhile, Trefethen et al. (1993, 1999) reported an entirely different linear stability approach that predicts large-amplitude growth, if not actual instability. The theory employs spectral methods to generate a continuous set of linear three-dimensional perturbations. Part of this *pseudospectrum* is highly amplified, especially at higher Reynolds numbers, and certain small disturbances can grow by three or four orders of magnitude. Presumably some of these amplified disturbances spur non-linear effects that lead to instability.

5-3.2 Instability of Suction Profiles: A Shape Factor Correlation

The asymptotic (exponential) suction profile of Fig. 4-21 has the greatest stability of any profile known to this writer. Hughes and Reid (1965) report for this condition that

$$U(y) = U_0(1 - e^{-y}) \qquad Re_{\text{crit}} = \frac{U_0 \delta^*}{\nu} = 46{,}000 \qquad (5\text{-}27)$$

This is 90 times greater than the Blasius value. The profile also has the smallest laminar shape factor, $H = \delta^*/\theta = 2.0$, known to this writer. Suction profiles, with no inflection point and U'' positive everywhere, are very stable and are used to delay transition in aerodynamic and hydrodynamic applications. The dashed line labeled "optimum suction" on the drag curve in Fig. 4-21 was computed from the suction stability theory. Obviously, there is a large decrease from the turbulent-flow drag curve.

If the values of $Re_{\delta^*,\text{crit}}$ from Table 5-1 were plotted versus the boundary-layer shape factor $H = \delta^*/\theta$, a smooth curve would result, as shown in Obremski et al. (1969). Even more interesting, stability results for nonsimilar profiles fall close to this same curve. This is shown in Fig. 5-12 from Wazzan et al. (1979). The points on the curve include Falkner–Skan wedge flows from Table 5-1, flat-plate suction

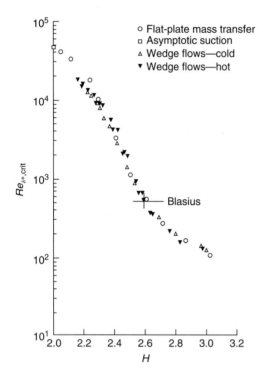

FIGURE 5-12
A correlation of critical Reynolds number versus boundary-layer shape factor which may be used as an engineering estimate of initial instability. [*After Wazzan et al. (1979)*.]

and blowing stability computations from Tsou and Sparrow (1970), and wedge flows with hot walls and variable viscosity from Wazzan et al. (1979). Wazzan thus proposes that Fig. 5-12 is a universal correlation, so that the shape factor H is all that is needed for an engineering estimate of boundary-layer instability. One can compute $H(x)$ from Thwaites' method, Table 4-4.

5-3.3 Wakes, Jets, and Shear Layers

When no wall or pressure gradient is present, the so-called free-shear layers always possess strong curvature and one or more points of inflection, e.g., the shear layer (Fig. 4-17 with $k = 1$), the laminar jet (Fig. 4-18), and the developed wake (Fig. 4-41). Thus they are relatively unstable.

The two-dimensional jet has the profile shape $U = U_0 \text{sech}^2 y$ and has been analyzed by many workers. Figure 5-13 shows the jet eigenvalues computed by Kaplan (1964). Instability occurs very early:

Jet: $\qquad Re_{\text{crit}} \approx 4 \quad \text{at} \quad \alpha \approx 0.2 \qquad (5\text{-}28)$

Similarly, the two-dimensional wake has a reverse-jet appearance and has been analyzed by, among others, Mattingly and Criminale (1972). The velocity profile has the approximate form given by $U = 1 + (U_0 - 1)\text{sech}^2 y$ and, like the jet, the critical Reynolds number is approximately 4.

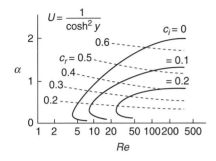

FIGURE 5-13
Eigenvalues for a two-dimensional jet. [*After Kaplan (1964)*.]

The two-dimensional shear layer has the form $U = U_0 \tanh(y/L)$ and was analyzed by Betchov and Szewczyk (1963):

$$Re_{\text{crit}} = 0$$

and (5-29)

$$C_{i,\text{max}} = 0.19 \quad \text{at} \quad \alpha L = 0.5$$

The shear layer between streams of different velocity is thus unconditionally unstable for all Reynolds numbers. However, these jet, wake, and shear-layer calculations assume parallel flow and should be corrected for nonparallel effects that clearly occur at such low Reynolds numbers. These flows are discussed in detail by Betchov and Criminale (1967) and by Drazin and Reid (1981).

5-3.4 Effect of Wall Temperature

Real fluids have viscosities that vary with temperature. Therefore the viscosity near the wall—and thereby the near-wall profile shape—can be influenced by heat transfer. Generally speaking, boundary-layer stability is enhanced by increasing the near-wall negative curvature $U''(0)$. As pointed out by Reshotko (1987), if $\mu = \mu(T)$, the boundary-layer equation at the wall, with no slip, yields the curvature expression

$$\frac{\partial^2 u}{\partial y^2}(0) = \frac{1}{\mu_w}\left[\frac{dp}{dx} - \frac{d\mu}{dT}\frac{\partial T}{\partial y}(0)\frac{\partial u}{\partial y}(0)\right] \tag{5-30}$$

As already seen, we can make $U''(0)$ more negative by imposing a negative (favorable) pressure gradient $dp/dx < 0$. Equation (5-30) shows an additional effect due to heat transfer. Since $\partial u/\partial y > 0$, the second term on the right will be negative under two different conditions:

Gases: $\quad \dfrac{d\mu}{dT} > 0 \quad \dfrac{dT}{dy} > 0:$ Cold wall (5-31a)

Liquids: $\quad \dfrac{d\mu}{dT} < 0 \quad \dfrac{dT}{dy} < 0:$ Hot wall (5-31b)

Thus gases are stabilized by cooling and liquids by wall heating.

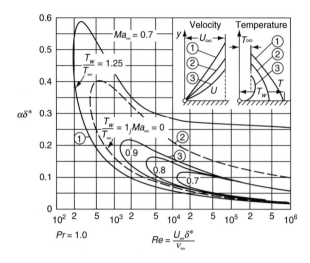

FIGURE 5-14
Effect of wall temperature on stability of a subsonic gas boundary layer. [*After Lees and Lin (1946)*.]

If we continue to assume parallel, incompressible flow but allow variable viscosity $\mu(y)$ due to temperature variations, the Orr–Sommerfeld Eq. (5-23) is modified to

$$(U - c)\left(\phi'' - \alpha_\delta^2 \phi\right) - U''\phi = \frac{-i}{\alpha_\delta Re_\delta}\left[\mu\left(\phi'''' - 2\alpha_\delta^2 \phi'' + \alpha_\delta^4 \phi\right) \right.$$
$$\left. + 2\mu'\left(\phi''' - \alpha_\delta^2 \phi'\right) + \mu''\left(\phi'' + \alpha_\delta^2 \phi\right)\right] \quad (5\text{-}32)$$

where the two added terms involve μ' and μ''. Parallel flow is still assumed. When applied to a gas in a subsonic flow, Eq. (5-32) reveals a very strong effect on both the neutral curves and the amplification factors. Figure 5-14 shows the computations of Lees and Lin (1946) at a freestream Mach number of 0.7. Reduction in wall temperature to $T_w = 0.7T_e$ gives a 30-fold increase in the critical Reynolds number and a similar decrease in the growth rate c_i.

Exactly the opposite effect is reported by Wazzan et al. (1968a) in studies of flat-plate flow with water, whose viscosity drops sharply with temperature. Their computations, assuming $T_e = 60°F$ and varying T_w, are given in Table 5-2. The most stable condition is for a hot wall with $T_w = 135°F$, which has Re_{crit} 22 times larger and a spatial amplification factor 63 times smaller than the Blasius (isothermal) boundary layer.

Further stability computations involving liquids are discussed in the review article by Wazzan et al. (1979).

5-3.5 Stability of Compressible Flows

The addition of gas compressibility to a stability problem greatly increases its complexity. Since pressure is now coupled to temperature and density through the

TABLE 5-2
Stability of flat-plate flow with water at $T_e = 60°F$

T_w, °F	$Re_{\delta,\text{crit}}$	$\left(\dfrac{\alpha_i \delta^* c_i}{U_e}\right)_{\max}$	$\left(-\dfrac{\alpha_r \delta^*}{Re_{\delta^*}}\right)_{\max} \times 10^5$
32	240	0.0131	4.80
45	280	0.00780	2.50
60	520	0.00355	0.740[†]
75	1680	0.00135	0.105
90	5200	0.00080	0.028
105	8925	0.00067	0.016
120	10,750	0.00060	0.0125
135	11,400	0.00057	0.0117[‡]
140	11,440	0.00057	0.0120
150	11,300	0.00058	0.0130
175	9600		0.0166
200	8200		0.0200
250	5500		0.0295
300	3700		0.0405

Source: Computations by Wazzan et al. (1968a).
[†] Blasius.
[‡] Most stable.

equation of state, one must add fluctuations $\rho(y)$ and $T(y)$ to the quartet (u, v, w, p) considered previously in Eq. (5-12). It is also necessary to consider variations of the fluid transport properties (μ, k, c_p). For parallel flow, the disturbances are again taken as traveling waves or "normal modes" as in Eq. (5-12), resulting in a sixth-order set of differential equations. The complete analysis is beyond our scope here, but a good discussion is given in the text by Betchov and Criminale (1967, Chap. 8) and in the extensive notes by Mack (1969). Mack (1984, 1987) has also reviewed the subject.

Lees and Lin (1946) developed the inviscid compressible stability theory and Brown (1962) began the first direct numerical solutions of the viscous theory. Numerical studies since that time have revealed a variety of interesting phenomena, such as unstable supersonic waves. The field has progressed to the point that laminar-flow control can now be estimated in compressible-flow design studies using computer codes.

Let us briefly discuss some significant results. Figure 5-15 shows experimental data compared with theories for the neutral curve on an adiabatic flat plate in air at $Ma = 2.2$. We see that the approximate analysis of Dunn and Lin (1955), either in its asymptotic or numerically integrated form, is not in good agreement with the data. The full multivariable stability equations are needed for accuracy. Some further calculations of Mack (1969) for an adiabatic plate are shown in Fig. 5-16. At first glance, it appears that compressibility is stabilizing, since $Re_{x,\text{crit}}$ is increasing with the Mach number. Actually, great changes are taking place in the character of the instabilities. The amplification rates and the general shape of the curves are changing greatly with both the Mach number and the Reynolds number. Note that the curve for $Ma_e = 1.6$ is of the incompressible thumb-curve type. At $Ma_e = 2.2$, a local

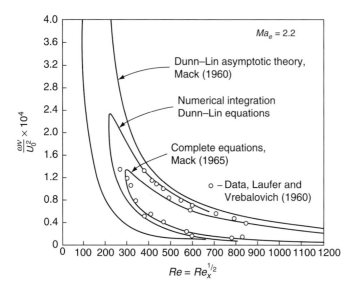

FIGURE 5-15
Comparison of three stability theories for the neutral curve on an adiabatic flat plate at $Ma_e = 2.2$ [*After Mack (1969)*.]

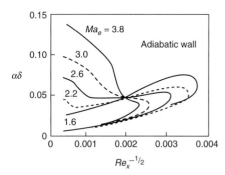

FIGURE 5-16
Computer solutions by Mack (1969) for the neutral curves in supersonic adiabatic flow over a flat plate (perfect gas). (*Courtesy of the Jet Propulsion Laboratory, California Institute of Technology.*)

minimum appears in the upper branch of the curve, indicating that viscosity has become stabilizing in the higher range of Reynolds numbers. This minimum moves further to the right as Ma_e increases and disappears at about $Ma_e = 3.0$, above which viscosity is stabilizing for all Reynolds numbers. At $Ma_e = 3.8$, the neutral curve has the general shape of the separation profile in Fig. 5-7, with the maximum instability in the inviscid limit. The temporal amplification rates for $Ma_e = 3.8$ are given in Fig. 5-17, showing that inviscid rates are the highest.

For Ma_e greater than about 3.0, a new phenomenon appears, higher modes. These modes occur when the disturbance phase velocity is supersonic relative to the wall $c_r > a_w$. The second mode appears at very high frequency and is insignificantly weak until about $Ma_e = 3.7$. It becomes stronger and moves downward

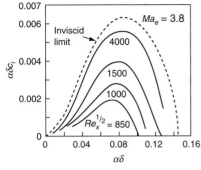

FIGURE 5-17
Effect of Reynolds number on first-mode amplification rate at $Ma_e = 3.8$ on an adiabatic flat plate. [*Courtesy of the Jet Propulsion Laboratory, California Institute of Technology, after Mack (1969).*]

with increasing Mach number, merging with the first mode at about $Ma_e = 4.8$. This effect is illustrated by the neutral curves for $Ma_e = 4.5$ and 4.8 in Fig. 5-18. The merging of modes is even more complete at higher Mach numbers, and the two modes are indistinguishable at $Ma_e = 7$. Also, vexingly, the higher modes are not damped by wall cooling, unlike the first-mode effect shown in Fig. 5-14. Calculations by Mack (1969) show that amplification rates of the higher modes are slightly increased by wall cooling.

Still another interesting facet of compressibility is that Squire's theorem no longer holds: The most unstable waves need not be parallel to the freestream. In fact, as dissipation becomes important at high Mach numbers, it is not quite correct to treat a three-dimensional disturbance analysis by an equivalent two-dimensional method, as was done in Eq. (5-15) for the incompressible case. This was first shown by Dunn and Lin (1955), who demonstrated that viscous and conduction terms transform properly to an equivalent two-dimensional system but that dissipation terms do not. Some results of Mack (1969) for amplification rates and the most

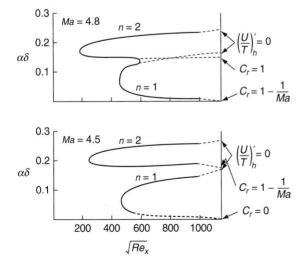

FIGURE 5-18
Neutral curves for supersonic, adiabatic flat-plate flow illustrating the appearance of higher modes. [*After Mack (1969).*]

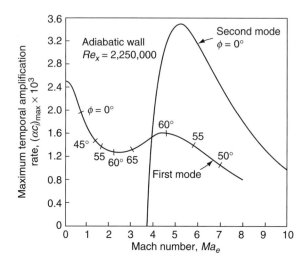

FIGURE 5-19
Temporal amplification rates and the most unstable wave direction ϕ for adiabatic flow past a flat plate. [*Courtesy of the Jet Propulsion Laboratory, California Institute of Technology, after Mack (1969).*]

unstable wave angle ϕ [from Eq. (5-12)] on a flat plate are shown in Fig. 5-19. Squire's theorem ($\phi = 0$) holds for the first mode up to about $Ma_e = 0.7$, after which the critical angle jumps to $\phi = 45$ to $65°$. The second mode satisfies Squire's theorem throughout. Note that the second mode quickly assumes prominence at $Ma_e = 4$ and becomes the more unstable mode.

The stability of high Mach number (hypersonic) flows, both theoretical and experimental, is the subject of many studies, notably Stetson (1988), Lachowicz et al. (1996), and Cassel et al. (1996). The dominant instability is the second mode, whose growth rates compared reasonably well with linear theory. High-frequency disturbances were detected in the experiments and imply strong non-linear effects in hypersonic stability.

5-3.6 Stability with Compliant Boundaries

All results discussed so far are for rigid walls, where a simple no-slip condition holds. A number of important examples occur when the wall is flexible: (1) panel flutter, (2) motion of liquid films, (3) growth of ocean waves, (4) rubber-coated instruments such as sonar domes, (5) cardiovascular flow, and (6) the skin of a swimming fish. The thesis of Kaplan (1964), which led to the purification breakthrough in numerical analysis, was in fact a study of compliant boundaries. His work was triggered by some intriguing experiments by Kramer (1957), which indicated a reduction in drag on bodies towed in water if their surfaces were coated with a compliant material such as natural rubber. Kramer speculated that the drag reduction was due to a delay in transition to turbulence caused by compliant boundaries. Subsequently, Benjamin (1960) showed analytically that compliant walls were potentially stabilizing in some cases and that there were three types of unstable waves: class A (Tollmien–Schlichting waves); class B (aeroelastic boundary flutter), and class C (Kelvin–Helmholtz instability). Class A waves are typically damped and stabilized by the compliant wall, but class B waves are often

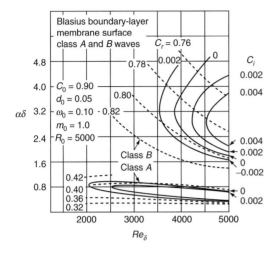

FIGURE 5-20
Amplification rates and phase speeds for flow over a compliant membrane surface. [*After Kaplan (1964).*]

destabilized by compliance and become the critical case. Class C waves are analogous to inviscid instability and can cause problems in the fabrication and maintenance of the coating.

Kaplan (1964) reported numerical stability solutions for some specific compliant walls. His results for a membrane surface with a Blasius boundary layer are shown in Fig. 5-20. The constants $(C_0, d_0, \omega_0, m_0, R_0)$ are characteristics of the particular membrane. Both class A and B waves appear, and the critical Reynolds number for class A waves is 2150, about 40 percent higher than for the rigid wall case (Fig. 5-6). Class B waves become unstable at about $Re_{\text{crit}} \approx 3800$ and have higher growth rates and phase speeds than class A waves.

Although Kramer's experiments were provocative and were supported qualitatively by theory, the promise of compliant-surface transition delay and drag reduction has not been too fruitful, and no reliable engineering designs have been reported.

5-3.7 Stability of Free-Convection Flows

Free or buoyant convection is an excellent medium for studying laminar instability. The varying temperature field allows for flow visualization through the interferometric techniques pioneered by Eckert and Soehngen [see the text by Eckert and Drake (1972)]. An excellent review of the subject is given by Gebhart (1969) and in Chap. 11 of the text by Gebhart et al. (1988).

The free-convection equivalent to the Orr–Sommerfeld relation was derived by Plapp (1957). Like the boundary-layer relations themselves [Eqs. (4-253)], these equations are coupled between velocity and temperature—but because of the small-disturbance assumption, they are linear. Computer solutions of Plapp's full coupled equations were given by Nachtsheim (1963), and the results shown in Fig. 5-21 agree well with experiment. Instability occurs very early because the velocity

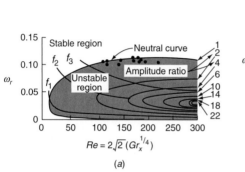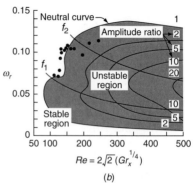

FIGURE 5-21
Amplification curves for free-convection flow on a vertical plate: (a) water: $Pr = 6.7$; ● = data from Knowles and Gebhart (1968); (b) gases: $Pr = 0.7$; ● = data from Polymeropoulos and Gebhart (1967).

profiles have a point of inflection (see Fig. 4-50a). In Fig. 5-21, we see that the critical Grashof numbers are about 2500 for water and 50,000 for gases, which correspond to equivalent values of $Re_{crit} \approx 50$ and 700, respectively. Growth rates are moderate, so that final transition to turbulence does not occur until $Gr_x \approx 10^9$, or about 30 to 70 times further downstream. Note from Fig. 5-21b that the neutral curve for gases has two modes, the upper and lower humps corresponding to velocity and temperature fluctuations, respectively.

5-3.8 Centrifugal Instability

The previous stability examples have all been parallel viscous flows. A somewhat different but related case is that of centrifugal instability of rotating flows. Thus the classic Couette flow between rotating cylinders, Eq. (3-22), breaks down for a certain finite configuration of rotational speeds and cylinder radii. The flow that then ensues is *not* turbulent but rather is a stable laminar flow of rows of circumferential toroidal vortices, shown in Fig. 5-22a, called *Taylor vortices*, after G. I. Taylor (1923). Turbulence does not appear until much higher rotational rates.

A criterion for inviscid rotational instability was deduced by Rayleigh (1916): "An inviscid rotating flow is unstable if the square of its circulation decreases outward." In other words, stability is ensured if

$$\frac{d}{dr}(rv_\theta)^2 > 0$$

For Couette flow between cylinders, Eq. (3-22), this reduces to

Stability: $$\Omega_o r_o^2 > \Omega_i r_i^2 \qquad (5\text{-}33)$$

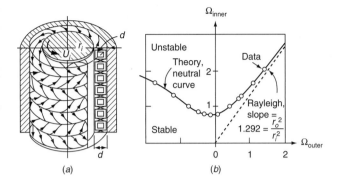

FIGURE 5-22
Theory and experiment by Taylor (1923) for the instability of Couette flow between rotating cylinders: (a) Taylor vortices; (b) theory and experiment for $r_o = 4.035$ cm, $r_i = 3.55$ cm.

where the subscripts i and o denote the inner and outer surfaces, respectively. From this we deduce the following intriguing inviscid predictions:

1. Inner cylinder rotating, outer cylinder at rest: *unstable*
2. Inner cylinder at rest, outer cylinder rotating: *stable*
3. Cylinders rotating in opposite directions: *unstable*

Actually, viscosity tends to stabilize and smudge these boundaries somewhat. Viscous instability is studied by allowing the basic flow, Eq. (3-22), to be perturbed by small disturbances in v_r, v_θ, v_z, p. This analysis was performed by Taylor (1923) in a classic experimental and analytical study. By assuming that the clearance, $C = (r_o - r_i)$, is small compared to r_i, Taylor simplified the problem so that stability is dependent only on Ω_o/Ω_i and a single parameter, now called the *Taylor number*:

$$Ta = \frac{r_i C^3 \left(\Omega_i^2 - \Omega_o^2\right)}{\nu^2} \qquad (5\text{-}34)$$

Then small-gap stability theory, for $0 \leq \Omega_o/\Omega_i \leq 1$, predicts that instability occurs at

$$Ta_{\text{crit}} \approx 1708$$

and (5-35)

$$\alpha C \approx 3.12$$

The critical wave number of 3.12 is nearly π, that is, the toroidal vortices that form in Fig. 5-22a are nearly square. These results are in good agreement with the careful experiments of Coles (1965). As Ta increases, the laminar vortices begin to develop circumferential waves but remain laminar. At $Ta \approx 160{,}000$, turbulence

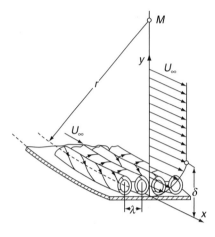

FIGURE 5-23
Longitudinal vortices in concave-wall flow. [*After Görtler (1955).*]

ensues. Photographs of these toroidal vortices are given in the photo album of van Dyke (1982).

Experimental results for turbulent Couette flow at very high Taylor numbers, $8 \times 10^7 < Ta < 4 \times 10^9$, are reported by Townsend (1984).

In a related example, Görtler (1955) pointed out the existence of longitudinal vortex instability in boundary-layer flow along a concave wall, as illustrated in Fig. 5-23. For small $\delta/r \approx 0.02$–0.1, these paired streamwise vortices begin to form at about $U_\infty \delta/\nu \approx 5$. They are thought to have a strong influence on early transition to turbulence on concave walls.

Still another example of centrifugal instability is the flow over a rotating disk, whose laminar solution was discussed in Sec. 3-8.2. The experiments of Gregory et al. (1955) showed unstable logarithmic-spiral vortices in a finite-thickness ring on the disk, as in Fig. 3-31. The three-dimensional disturbance equations were found by Malik (1986) to be unstable at a Reynolds number $Re = r(\omega/\nu)^{1/2} \approx 285$. Lingwood (1995) then found a point of *absolute* (multidirectional, nonconvective) instability at $Re \approx 510$. Her experiments, Lingwood (1996), then showed that absolute instability is immediately followed by transition to turbulence.

Excellent general reviews of the subject of centrifugal instability are given by Drazin and Reid (1981) and by Koschmieder (1993).

5-4 TRANSITION TO TURBULENCE

The linearized stability theory of the previous two sections predicts the demise of laminar flow at some finite Reynolds number. It *does not* predict the onset of turbulence. Following the initial breakdown of laminar flow through amplification of infinitesimal disturbances, the flow passes through a complicated sequence of spatial changes; the end result is that unsteady and disorderly but strangely rational and marvelously stable phenomenon known as turbulence. The whole process of change from laminar to turbulent flow is termed transition.

After a century of research on the transition process, there has been significant progress, especially in visualization of unstable waves, but the mechanisms are still not completely understood. Historically, opinions about transition have shifted from one concept to another. The Rayleigh inviscid theories of 1880 obscured the real problem in boundary layers: viscous instability of infinitesimal disturbances. Then, when the viscous Tollmien–Schlichting waves were predicted in 1929, researchers doubted them because the (noisy) experiments of the day pointed clearly to transition as a three-dimensional "explosion" into turbulence. Twenty years later, when Schubauer and Skramstad and also Liepmann documented the Tollmien–Schlichting waves, the rush of opinion to embrace the two-dimensional cause was so great that Liepmann's clear indication of spanwise fluctuations was entirely ignored. A decade later, Emmons (1951) accidentally noticed sporadic turbulent spots on shallow running water; two-dimensionality was abandoned, and suddenly, as Morkovin (1969) wryly notes, "everyone was seeing spots." It was another decade before Klebanoff et al. (1962) clarified the essential intermediate process involving longitudinal vorticity and spanwise energy exchange. Added to this is the annoying unit Reynolds number effect, by which transition processes differ among the various experimental facilities, indicating a strong role for the initial disturbance spectrum (freestream turbulence, radiated noise, surface roughness) on the actual paths towards turbulence. The complexities of these parameters and their interactions continue to discourage our hopes for a definitive picture of the transition process.

A dramatic example of our limited knowledge is that fact that the original transition experiment (pipe flow) is still not well understood. Reynolds (1883) conducted a classic experiment of introducing dye into the entrance of a circular pipe, as in Fig. 5-24. By varying the flow speed at constant diameter and viscosity, Reynolds was in effect varying the dimensionless parameter $Re = \bar{u}d/\nu$, named for him. At low speed $Re < 2000$, the flow remained laminar, and the dye filament

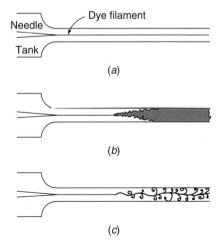

FIGURE 5-24
The classic pipe-flow dye experiment of Reynolds (1883): (*a*) low speed: laminar flow; (*b*) high speed: turbulent flow; (*c*) spark photograph of condition (*b*).

stayed along a nearly straight and distinct streamline, as in Fig. 5-24a. As speed was increased to a value of Re between 2000 and 13,000, depending upon the smoothness of entrance conditions, the filament broke up somewhere downstream and mixed rapidly with surrounding water to color the tube completely, as in Fig. 5-24b. The flow in the well-mixed region is turbulent, but a spark photograph (Fig. 5-24c) reveals that the instantaneous filament is actually still confined to a relatively distinct set of curls and eddies. These eddies and associated fluid parcels are in rapid three-dimensional motion at frequencies up to thousands of hertz, thus producing a color-blurred effect to the naked eye (Fig. 5-24b). It was also found that no transition occurs for Re less than 2000, no matter how rough and noisy the entrance conditions are made.

Some other facts regarding Fig. 5-24 are worth noting: (1) the transition occurs downstream, not in the pipe entrance, the actual point tending to move toward the entrance with increasing Reynolds number; (2) there is a clear region of amplification from laminar flow into turbulence, and far downstream the turbulence reaches a fully developed equilibrium state of balance between production and decay which is independent of the initial disturbance; (3) in the beginning turbulent zone, the flow is only *intermittently* turbulent, interspersed with laminar-flow regions. For example, in the measurements of Rotta (1956), at $Re = 2500$, the flow is about 50 percent turbulent (*intermittency factor* $\gamma = 0.5$) at $x/d = 200$ and still only about 85 percent turbulent ($\gamma = 0.85$) at $x/d = 500$. The same phenomenon of intermittency obtains in boundary-layer flow, where it delineates a rather sharp interface between turbulent-boundary-layer flow and nearly nonturbulent outer flow (Fig. 6-5).

5-4.1 Development of Spanwise Vorticity

If the initial disturbance spectrum is nearly infinitesimal and random (with no discrete frequency peaks), the theory predicts and experiment verifies that the initial instability will occur as two-dimensional Tollmien–Schlichting waves, traveling in the mean flow direction if compressibility is not important. However, three-dimensionality soon appears as the Tollmien–Schlichting waves rather quickly begin to show spanwise variations. This tendency toward rapid achievement of three-dimensionality was not predicted or expected until the experiments of the 1960s. As shown by Klebanoff et al. (1962), a shear layer in the unstable region has a strong ability to amplify any slight three-dimensionality which surely must be present in any natural-disturbance spectrum. Figure 5-25 shows the rapid development of spanwise peaks and valleys in the streamwise velocity fluctuations downstream of a vibrating ribbon containing spacers. The spanwise variation is slight 7 cm downstream, but 11 cm later, the peaks are 4 times as strong as the valleys. Similar variations (not shown) have developed in the lateral and vertical fluctuations, and the flow is approaching a condition of thorough three-dimensionality so characteristic of fully turbulent flow. The flow in Fig. 5-25, though, is still laminar.

The 1980s have produced some excellent visualizations of the growth process of Tollmien–Schlichting waves. There are many excellent discussions by various

THE STABILITY OF LAMINAR FLOWS   373

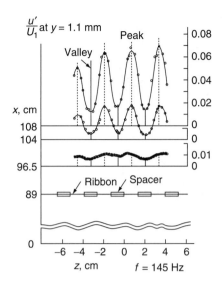

FIGURE 5-25
Development of spanwise variations in the streamwise velocity fluctuation downstream of a vibrating ribbon with spacers. [*After Klebanoff et al.* (*1962*).]

authors in the monograph edited by Tatsumi (1984). In particular, Saric and Thomas (1984) have used smoke-wire visualization of flat-plate flow to demonstrate three different types of Tollmien–Schlichting wave breakdown. Whereas Klebanoff et al. (1962) used a ribbon with spacers, Saric and Thomas (1984) used an unadorned ribbon and varied the rms amplitude and frequency of its vibrations. The results are shown in Fig. 5-26 and are called Λ vortices.

Figure 5-26a shows in-phase Λ vortices—i.e., peak-following-peak—which have the same wavelength as a Tollmien–Schlichting wave and arise at fluctuation amplitudes of approximately 1 percent U_∞. These waves are the same as those shown in Fig. 5-25 and are thus called K-type vortices, after Klebanoff. The spanwise wavelength λ_z is about one-half of the streamwise value, λ_x.

At lower amplitudes (≈ 0.3 percent U_∞), a structure of staggered waves, with peak-following-valley, appears as in Fig. 5-26b. The streamwise wave number and frequency are approximately one-half of a Tollmien–Schlichting wave, and $\lambda_z \approx 1.5\lambda_x$. This is a *subharmonic* unstable wave, now called a C-type vortex after Craik (1971), who first explained them as a superposition of a wave triad.

Finally, Fig. 5-26c shows a second type of staggered (subharmonic) vortex pattern with $\lambda_x \approx 2\lambda_{TS}$ but with short span, $\lambda_z \approx 0.7\lambda_x$. These waves occur at an intermediate fluctuation amplitude of about 0.6 percent U_∞ and are called H waves, after Herbert (1983), who showed that they arise from secondary instability of Tollmien–Schlichting waves to three-dimensional disturbances.

Thus there are now known to be many paths from Tollmien–Schlichting waves to turbulence, and many of them are explained by the non-linear theories of flow instability, as reviewed by Stuart (1987) and in Chap. 5 of the text by Schmid and Henningson (2001). The non-linear theories are beyond the scope of the present text.

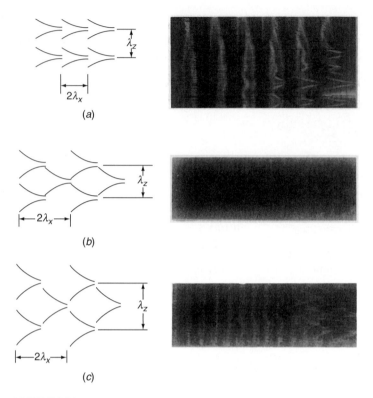

FIGURE 5-26
Patterns of Λ-type unstable vortex breakdown in a boundary layer: (a) K-type ($u'/U_\infty \approx 1$ percent) aligned in-phase and similar to a Tollmien–Schlichting wave; (b) C-type (0.3 percent) staggered subharmonic with $\lambda_z \approx 1.5\lambda_x$; (c) H-type (0.6 percent) staggered subharmonic with $\lambda_z \approx 0.7\lambda_x$. [*After Saric and Thomas (1984)*.]

5-4.2 Turbulent Spots: The Final Transition Process

We have seen that a shear layer develops viscous instability and forms Tollmien–Schlichting waves that grow, while still laminar, into finite amplitude (1 to 2 percent U_∞) three-dimensional fluctuations in velocity and pressure. From then on, the process is more a breakdown than a growth. The longitudinally stretched vortices begin a cascading breakdown into smaller units, until the relevant frequencies and wave numbers are approaching randomness. Then, in this diffusely fluctuating state, intense local changes occur at random times and locations in the shear layer near the wall. Turbulence bursts forth in the form of growing and spreading *spots*, first noticed by Emmons (1951) on the surface of a shallow-water channel. Since then, many workers have been able to generate artificial spots with, e.g., an electric spark.

Figure 5-27 shows a compilation of various observations of the growth and shape of turbulent spots. Viewed from above—Figs. 5-27a and c—a spot looks

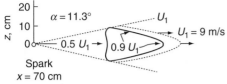
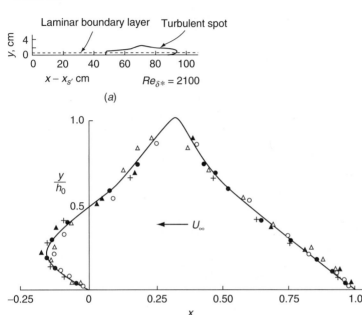
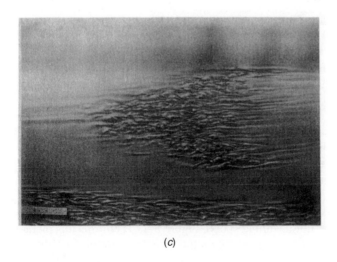

FIGURE 5-27
Observations of turbulent spot growth: (a) plan and elevation sketches [*after Schubauer and Klebanoff* (*1955*)]; (b) normalized measured shape [*after Wygnanski et al.* (*1976*)]; (c) visualization by aluminum flakes in a water channel [*after Cantwell et al.* (*1978*)].

like an arrowhead moving downstream and spreading at a half-angle variously reported from 8 to 12°. The leading edge travels at about $0.9U_\infty$ whereas the trailing edge is much slower, about $0.5U_\infty$. Thus the spot grows in size but its normalized shape remains the same, as shown in Fig. 5-27b for five different x positions. The disorderly details in Fig. 5-27c show that the spot is fully turbulent. Its detailed structure becomes finer as the Reynolds number increases—see, e.g., photos 111 in van Dyke (1982). For Fig. 5-27c, $Re_x \approx 400{,}000$ at the center of the spot.

Carlson et al. (1982) used a visualization method that showed the traveling spot itself, unlike smoke or dye releases that remain fixed to the same particles. Their photos show strong oblique waves at both the front and the rear "wings" of the arrowhead. This indicates that unstable wave breakdown continues to play a role even after the spot is formed.

Since the spots grow and remain intensely turbulent, they must entrain the surrounding laminar fluid to maintain themselves. Wygnanski et al. (1976) report that the strongest entrainment is at the leading edge and on the upper surface of the trailing edge.

In boundary-layer flow, spots form randomly and naturally in what Dhawan and Narasimha (1958) report is a narrow transverse band Δx which is small compared to the total length of the transition zone. Since they spread downstream, they inevitably coalesce into a region where spots continually exist and prosper. The flow downstream is then said to be fully turbulent and, as will be seen in Chap. 6, is scaled by modeling laws entirely different from laminar flow.

Our overall picture of the transition process in quiet boundary-layer flow past a smooth surface thus consists of the following processes as one moves downstream:

1. Stable laminar flow near the leading edge
2. Unstable two-dimensional Tollmien–Schlichting waves
3. Development of three-dimensional unstable waves and hairpin eddies
4. Vortex breakdown at regions of high localized shear
5. Cascading vortex breakdown into fully three-dimensional fluctuations
6. Formation of turbulent spots at locally intense fluctuations
7. Coalescence of spots into fully turbulent flow

These phenomena are sketched as an idealized flat-plate flow in Fig. 5-28a, which also illustrates the contamination effect of the lateral edges of the plate. The edges simulate a finite disturbance which, if severe enough, can cause immediate transition to turbulent flow (without spots). Figure 5-28b confirms this picture with a smoke-flow visualization of transition on the straight cylindrical portion of a body of revolution.

Qualitative similar effects occur when there are mild pressure gradients. A strong adverse gradient, being both unstable and also subject to flow separation,

THE STABILITY OF LAMINAR FLOWS **377**

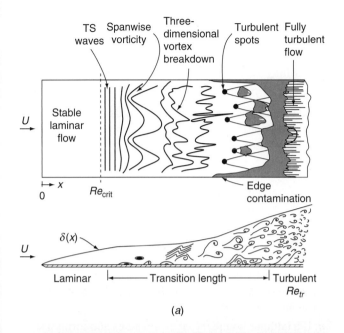

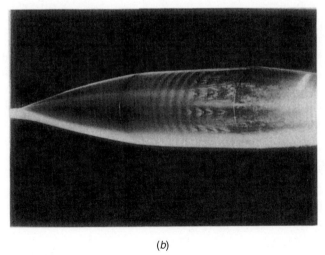

FIGURE 5-28
Description of the boundary-layer transition process: (*a*) idealized sketch of flat-plate flow; (*b*) smoke-flow visualization of flow with transition induced early by acoustic input at $Re_L = 814{,}000$ and 500 Hz. (*Courtesy of J.T. Kegelman and T. J. Mueller, University of Notre Dame.*)

may short-circuit some steps and instead form a separation bubble, with the reattached flow downstream being fully turbulent.

5-4.3 Classification of Boundary Layer Transition Processes

There are three types of wall-bounded transition processes in fluids engineering. An interesting chart was given by Morkovin (1969b), and the subject was reviewed by Mayle (1991) in his IGTI Scholar Lecture related to gas turbine engines. The three scenarios are

1. *Natural transition.* This is the gradual process of Fig. 5-28, caused by infinitesimal disturbances. The flow changes from TS waves to three-dimensional waves to vortex breakdown to turbulent spots to fully turbulent flow. The stream should be quiet and the walls smooth.
2. *Bypass transition.* If the walls are rough, the freestream noisy, the surface vibrating, or the stream subjected to acoustic waves, the flow may skip the early stages of natural instability and go directly to vortex breakdown or turbulent spot production. Large disturbances cause the bypass.
3. *Separated-flow transition.* If a laminar boundary layer separates and forms a separation bubble, it will likely reattach as a turbulent flow. Some or all of the natural-transition processes of Fig. 5-28 may occur in the region of the bubble.

5-4.3.1 BOUNDARY-LAYER RECEPTIVITY. Bypass transition is not just a sudden change. It may take a number of different paths, usually called *transient growth*, on the way to transition. Finding these paths by theory, CFD simulations, or experiment is a field of research called *boundary-layer receptivity*, reviewed by Saric et al. (2002). Results are incomplete and beyond the scope of this text.

5-5 ENGINEERING PREDICTION OF TRANSITION

There is no fundamental *theory* of transition, but there are experiments and correlations which try to predict the final onset of fully turbulent flow, such as $Re_{x,\,tr}$ or $Re_{\theta,\,tr}$, as a function of the following parameters:

1. Pressure gradient
2. Freestream turbulence
3. Wall roughness
4. Mach number
5. Wall suction or blowing
6. Wall heating or cooling

Understandably, most of the approximations deal with only one or two of these parameters.

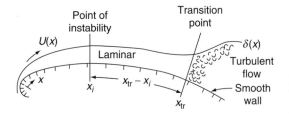

FIGURE 5-29
The two-step calculation of the transition point.

5-5.1 Effect of Pressure Gradient

Consider the problem sketched in Fig. 5-29. We assume that $U(x)$ is known, with zero freestream turbulence, and that the walls are smooth, impermeable, and unheated. The boundary layer will be initially laminar and will become unstable at point x_i, where undamped Tollmien–Schlichting waves first appear. These waves will grow and distort, in the manner of Fig. 5-28, until the point of *transition* or fully turbulent flow, x_{tr}, is reached. We wish to predict x_{tr}, using x_i as an input if necessary. We briefly look at four methods.

5-5.1.1 THE TWO-STEP METHOD OF GRANVILLE. An early and still popular method of Granville (1953) requires the computation of x_i first, perhaps by following $H(x)$ from Thwaites' method until it hits the Re_{crit} correlation of Fig. 5-12. Then, while continuing to monitor $Re_\theta(x)$, Granville suggests computing a mean Thwaites' parameter,

$$\lambda_m = \frac{1}{(x_{tr} - x_i)} \int_{x_i}^{x_{tr}} \lambda(x)\, dx \qquad (5\text{-}36)$$

where $\lambda = \theta^2 (dU/dx)/\nu$. We then assume that x_{tr} occurs when Re_θ strikes Granville's transition data, which can be fit to the following formula:

$$Re_\theta(x_{tr}) \approx Re_\theta(x_i) + 450 + 400 e^{60\lambda_m} \qquad (5\text{-}37)$$

for $\lambda < 0.04$. For an adverse gradient ($\lambda \approx -0.1$), the last term is negligible and transition is very near to x_i. For a favorable gradient, the last term is very large and transition moves far downstream.

5-5.1.2 THE ONE-STEP METHOD OF MICHEL. An extremely simple but effective method was proposed by Michel (1952) in the form of a correlation based on local values of momentum thickness and position. One ignores x_i and simply computes $\theta(x)$ from, say, Thwaites' method, Eq. (4-138), until we strike Michel's "transition line"

$$Re_{\theta,\,tr} \approx \frac{U(x)\,\theta(x)}{\nu} \approx 2.9 Re_{x,\,tr}^{0.4} \qquad (5\text{-}38)$$

The computed $Re_\theta(x)$ should approach this curve unambiguously from below.

5-5.1.3 THE ORR–SOMMERFELD METHOD OF JAFFE ET AL.

In the same paper in which Eq. (5-38) was proposed, Michel (1952) noted that the transition points in his data compilation seemed in all cases to correspond to total amplification of Tollmien–Schlichting waves equal to about $A/A_0 \approx 10^4$. This fact inspired workers in computational stability to evaluate the eigenvalues of various boundary-layer profiles and then to compute the total growth of waves of a given frequency. In this way Smith and Gamberoni (1956) and, independently, van Ingen (1956) verified that temporal stability theory applied to these experiments would give a total growth

$$\frac{A}{A_0} = \exp\left[\int_{x_i}^{x_{tr}} \alpha c_i \, dt\right] \approx e^9 \quad (5\text{-}39)$$

Thereafter this method became known as the "e^9" method. Later, Jaffe et al. (1970) showed that a more realistic procedure would be to use *spatial* stability theory to evaluate the overall growth. Their computations with the exact velocity profiles, rather than with local-similarity approximations, gave good agreement with transition measurements when

$$\frac{A}{A_0} = \exp\left[\int_{x_i}^{x_{tr}} (-\alpha_i) \, dx\right] \approx e^{10} \quad (5\text{-}40)$$

Thus one also hears of the "e^{10}" method. An example of their calculations of constant-frequency growth curves for the Blasius profile is shown in Fig. 5-30. Although Gaster (1962) proposed a transformation to relate temporal and spatial growth rates and propagation speeds, the accuracy is only fair and thus Eq. (5-40) is preferred.

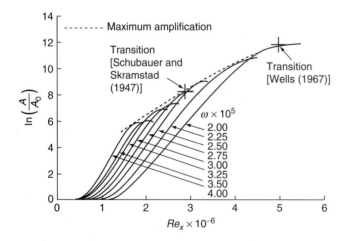

FIGURE 5-30
Constant-frequency amplification curves for flat-plate flow from linear stability theory, Eq. (5-40). Each curve represents a slice through the Blasius thumb curves in Fig. 5-6b. [*After Jaffe et al. (1970).*]

Also, in the spirit of Michel's formula Eq. (5-38), Cebeci and Smith (1974) suggested that the e^9 computations could be correlated by the formula

$$Re_{\theta,\,tr} \approx 1.174\left[1 + \left(\frac{22{,}400}{Re_{x,\,tr}}\right)\right]Re_{x,\,tr}^{0.46} \qquad (5\text{-}41)$$

which they claim is somewhat more accurate than Michel's relation.

5-5.1.4 THE ONE-STEP METHOD OF WAZZAN ET AL.
The success of Fig. 5-12 in correlating Re_{crit} vs. H for widely different cases led Wazzan et al. (1979, 1981) to propose a similar correlation for the transition Reynolds number. They found that the e^9 method gave the same result for parameters such as pressure gradient, suction and blowing, or heating and cooling, if plotted in terms of the transition shape factor. Their correlation is shown as the upper curve in Fig. 5-31. The lower curve gives Re_{crit} computed from Fig. 5-12 by converting Re_{δ^*} to Re_x. To use Fig. 5-31, one computes $H(x)$ by any laminar-boundary-layer method, e.g., Thwaites. Transition is predicted when this locus $H(Re_x)$, rising from below, hits the upper curve. Wazzan et al. (1981) suggest the following curve fit for the upper curve:

$$\log_{10}(Re_{x,\,tr}) \approx -40.4557 + 64.8066H - 26.7538H^2 + 3.3819H^3 \qquad (5\text{-}42)$$

for $2.1 < H < 2.8$. This method is not well verified versus experiment, but it is attractive nonetheless because of the basic acceptance of the e^9 method.

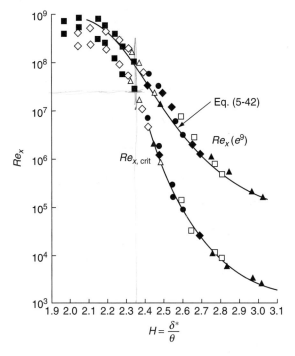

FIGURE 5-31
Correlation of critical and transition Reynolds numbers versus boundary-layer shape factor. [*After Wazzan et al.* (*1981*).]

5-5.1.5 COMPARISON FOR FLAT-PLATE FLOW. As a preliminary comparison, we can compute $Re_{x,\,tr}$ for flat-plate flow by each of the four previous methods and tabulate them as follows:

Method	Equation used	$Re_{x,\,tr}$ (predicted)
Granville (1953)	(5-37)	2,505,000
Michel (1952)	(5-38)	2,525,000
Smith and Gamberoni (1956)	(5-41)	2,027,000
Wazzan et al. (1981)	(5-42)	4,753,000

Obviously there is a sharp difference between the older methods and Wazzan et al. (1981). One reason might be seen in the two flat-plate data points placed quietly but provocatively in Fig. 5-30. The older methods were turned toward the experiment of Schubauer and Skramstad (1947), which reported $Re_{x,\,tr} \approx 2.8 \times 10^6$ at 0.02 percent turbulence. Later, Wells (1967) extended the transition point to $Re_{x,\,tr} \approx 4.9 \times 10^6$ by eliminating acoustic disturbances present in the Schubauer–Skramstad data. The Wazzan estimate reflects this quieter experiment. Because transition is so sensitive to various kinds of disturbances, it is doubtful if a simple theory can resolve these effects.

5-5.1.6 COMPARISON FOR THE FALKNER–SKAN FLOWS. As a further comparison of pressure gradient effects, in the absence of freestream turbulence or wall roughness and heat transfer, the four theories above, plus two others to be discussed, were applied to the Falkner–Skan wedge flows. Where necessary, $Re_\theta(x)$ was computed by Thwaites' method, Eq. (4-138). The results are plotted in Fig. 5-32. Mindful of the log scale, we see that all six methods are about the same for $\beta = 0$ and fairly consistent for adverse gradients ($\beta < 0$). However, they differ by five orders of magnitude for strong favorable gradients ($\beta > 0.5$). One surmises that these methods were not well calibrated for favorable gradients, yet many such cases occur in fluids engineering. It is recommended that Wazzan's relation Eq. (5-42) be used in such cases.

5-5.1.7 DIRECT NUMERICAL SIMULATION OF TRANSITION. The formulas recommended in this section are correlations, not theories. There is no pure theory of transition, but numerical methods are making inroads. Direct numerical simulation (DNS), reviewed by Moin and Mahesh (1998), solves the full time-dependent Navier–Stokes equations, without "closure" models, accounting for all important scales of the flow. As the Reynolds number increases, the number of CFD mesh points must increase rapidly. To avoid a googol of nodes, the foreseeable limit is a low-turbulent Reynolds number. But this is sufficient for a DNS study of transition, simulating disturbances such as vibrating ribbons, freestream turbulence, and wall roughness. Kleiser and Zang (1991) review results through 1990. Rist and Fasel (1995) use an improved spatial (as opposed to temporal) computation and replicate transition experiments. Bake et al. (2002) model a vibrating-ribbon disturbance, and

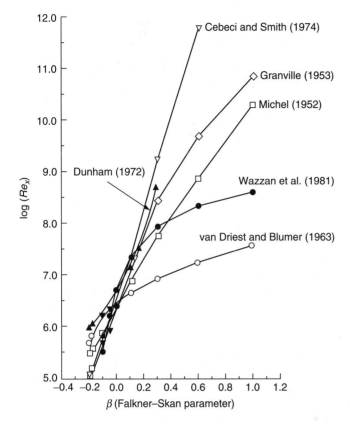

FIGURE 5-32
Transition Reynolds numbers for the Falkner–Skan wedge flows as predicted by six different methods.

their results agree well with Klebanoff (K-type) transition experiments (Fig. 5-26a). In the near future, DNS computations could predict transition for a variety of geometries and disturbances.

5-5.2 Effect of Freestream Turbulence

We have already discussed the fact that a noisy freestream hastens the transition process and, in the early days, masked the initial role of Tollmien–Schlichting waves. The parameter characterizing freestream turbulence level is defined as

$$T = \frac{q}{U}$$

where
$$q = \left[\frac{1}{3}(\overline{u'^2} + \overline{v'^2} + \overline{w'^2})\right]^{1/2} \tag{5-43}$$

in which U is the mean freestream velocity and u', v', w' are the fluctuating velocities in the freestream. The notation $\overline{u'^2}$ is the mean square fluctuation, averaged over a long period of time (see Sec. 6-1 for the rules of time-averaging). The effect of T on transition is very strong: At $T = 0.35$ percent, Re_{tr} has dropped 50 percent from its "quiet" value in the experiment of Schubauer and Skramstad (1947).

The term "freestream disturbances" is ambiguous. Different effects occur when the disturbances are due to (1) grid-generated turbulence, (2) acoustic noise, (3) excited standing waves, and (4) excited traveling waves. This is illustrated in Fig. 5-33 for flat-plate flow experiments. The experiment of Wells (1967), which was free of acoustic noise but contained standing waves, gave a high "quiet" value $Re_{tr} \approx 4.9 \times 10^6$. When grids were added to produce freestream turbulence, Wells' data dropped down to meet those of Schubauer and Skramstad (1947), whose data are thought to contain acoustic noise. The latter workers report a quiet Re_{tr} of only 2.8×10^6. Spangler and Wells (1968) then used Wells' facility to stimulate the boundary layer with excited acoustic waves of specific frequencies known to be unstable from the Orr–Sommerfeld equation, Fig. 5-9. Standing waves were suppressed. Traveling waves at 82 Hz (above the thumb curve) caused no effect, but waves at 27, 43, and 76 Hz caused the decreases shown in Fig. 5-33.

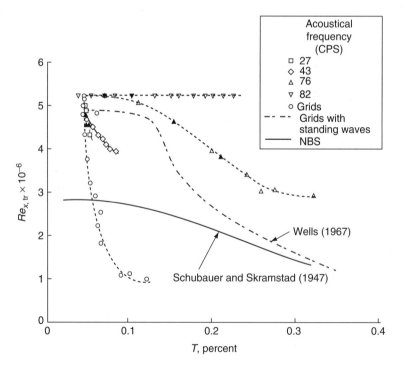

FIGURE 5-33
Measured effect of various types of freestream disturbances on the flat-plate transition point. [*After Spangler and Wells (1968).*]

Grids (open circles) were far more effective than in Wells' (1967) experiment. Spangler and Wells also report that the spectral density distribution of the disturbances was significant. Clearly, as Arnal (1984) remarks, "There is no universal $Re_{x,\,tr}(T)$ curve."

5-5.2.1 THE CORRELATION OF VAN DRIEST AND BLUMER.

Following a suggestion given by Liepmann in 1943, van Driest and Blumer (1963) theorized that transition occurs when the Reynolds number associated with maximum vorticity in the boundary layer,

$$Re_\omega = \left(\frac{\omega y^2}{\nu}\right)_{max}$$

reaches a critical value to be correlated with freestream turbulence. By relating Re_ω to the shape of the profile and fitting its critical value to the data of Schubauer and Skramstad (1947), van Driest and Blumer derived the following formula for flat-plate transition:

$$Re_{x,\,tr}^{1/2} = \frac{-1 + \sqrt{1 + 132{,}500T^2}}{39.2T^2} \quad (5\text{-}44)$$

where T is to be taken as a fraction, not a percentage. Equation (5-44) is in excellent agreement with the solid line of Fig. 5-33 and in fact was "tuned" to the Schubauer and Skramstad data.

To illustrate the combined effect of freestream turbulence and pressure gradient, van Driest and Blumer evaluated Re_ω for the Falkner–Skan wedge flows, $U = Cx^m$, and arrived at the following formula:

$$\frac{1690}{Re_{x,\,tr}^{1/2}} \approx 0.312(m + 0.11)^{-0.528} + 1.6\eta_\delta^2 Re_{x,\,tr}^{1/2} T^2 \quad (5\text{-}45)$$

where η_δ is the 99 percent boundary-layer thickness in terms of the similarity variable from Eq. (4-70). Equation (5-45) includes the flat-plate correlation, Eq. (5-44), as the special case $m = 0$ and $\eta_\delta = 3.5$.

Figure 5-34a is a plot of Eq. (5-45), showing that both freestream turbulence and pressure gradient have strong effects on transition. This formula, with $T = 0$, is also displayed for comparison in Fig. 5-32 and is seen to predict early transition in strong favorable gradients.

5-5.2.2 THE CORRELATION OF DUNHAM.

In studying boundary-layer transition on turbomachinery blades, Dunham (1972) collected data on combined freestream turbulence and pressure gradient effects in the following correlation:

$$Re_{\theta,\,tr} \approx (0.27 + 0.73 e^{-80T})\left[550 + \frac{680}{(1 + 100T - 21\lambda_{tr})}\right] \quad (5\text{-}46)$$

for $\lambda < 0.04$, where $\lambda = \theta^2(dU/dx)/\nu$ is the Thwaites parameter. This formula is plotted in Fig. 5-34b and is seen to predict a strong stabilizing effect of favorable

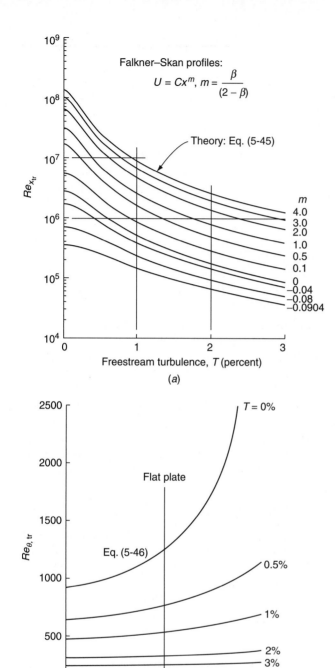

FIGURE 5-34
Combined effects of freestream turbulence and pressure gradient on transition: (*a*) the correlation of van Driest and Blumer (1963) for wedge flows; (*b*) the correlation of Dunham (1972).

gradients if T is small. Note also that Dunham's correlation predicts very little effect of pressure gradient at high turbulence levels. Equation (5-46), for $T = 0$, is converted to Re_x from the Falkner–Skan relations and also displayed in Fig. 5-32 as a comparison. If plotted in Fig. 5-33 for $\lambda = 0$, Eq. (5-46) would lie about 10 percent higher than the curve labeled Schubauer and Skramstad (1947).

5-5.2.3 MACK'S MODIFICATION OF THE e^9 METHOD. One difficulty with the correlations Eqs. (5-45) and (5-46) is that one gets no feeling for the effect of freestream turbulence on the spectrum of disturbances or their subsequent amplification. Mack (1977) proposed that turbulence is related to the initial level of disturbances, which would then modify the e^9 method into an "e^N" prediction, where $N = N(T) \leq 9$. He suggested the following correlation:

$$N \approx -8.43 - 2.4 \ln(T)$$

valid for $0.0007 \leq T \leq 0.0298$. For example, for $T = 0.0015$, $N = 7.2$. For $T = 0.0298$, $N \approx 0$, indicating that transition occurs right at Re_{crit}. However, to use this idea, one has to integrate the amplitude ratios from the Orr–Sommerfeld results using Eq. (5-39) or (5-40). Alternately, one could modify Wazzan's method in Fig. 5-31 to include lower levels of e^N—see, e.g., Fig. 25 of Wazzan et al. (1979).

5-5.2.4 HIGH FREESTREAM TURBULENCE: A BYPASS FORMULA. A glance at Fig. 5-34b shows that, for freestream turbulence > 1 percent, there is very little effect of pressure gradient. The disturbance is large, and the flow tends to bypass the natural transition. Using an optimal-growth criterion, proportional to the Reynolds number, Schmid and Henningson (2001) derive a simple but elegant formula for bypass transition:

$$T(Re_{x,\text{tr}})^{1/2} \approx \text{dimensionless constant} \approx 1200 \pm 200 \qquad (5\text{-}47)$$

where the freestream turbulence T is expressed in percent. Schmid and Henningson point out that this concept is analogous to the method of van Driest and Blumer, Eq. (5-45). However, Eq. (5-45) shows a substantial pressure gradient effect, even at T as high as 3 percent, Fig. 5-34a.

5-5.3 Effect of Surface Roughness on Transition

The previous results are for smooth walls. The introduction of a roughness or projection at the wall will generally cause earlier transition because of the additional disturbances it feeds into the boundary layer. We should distinguish between two different geometries: (1) two-dimensional roughness, a wire or cylinder stretched across the flow; and (2) three-dimensional roughness, a sphere or spike or a single grain of sand. These are *single* roughnesses. There is also the possibility of *distributed* roughness such as sandpaper or rows of cylinders or multiple rivets.

Two- and three-dimensional roughnesses have quite different effects. Figure 5-35a shows an idealized two-dimensional element—a transverse wire on a

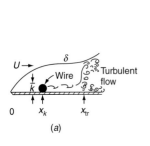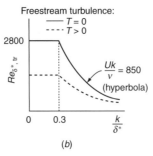

FIGURE 5-35
Idealized effect of two-dimensional roughness on transition: (a) flat plate with trip wire; (b) transition data.

flat plate. As discussed by Klebanoff and Tidstrom (1972), the wire wake introduces disturbances to raise the level of the Tollmien–Schlichting waves growing downstream. If the wire height k is much smaller than the local displacement thickness $\delta^*(x_k)$, there is little effect and transition occurs at the smooth-wall location x_{tr} from, e.g., Fig. 5-33. As U or k increases until k/δ^* exceeds 0.3, the point x_{tr} moves forward toward the roughness element. The value of $Re_{\delta^*,\,tr}$ drops along a hyperbolic curve, as shown in Fig. 5-35b. Thus a criterion for a wire to "trip" a flow into turbulence is

$$\frac{Uk}{\nu} \approx 850 \qquad [\text{Gibbings (1959)}] \qquad (5\text{-}48)$$

When there is freestream turbulence, both the smooth-wall and roughness-affected transition values are lower, as shown in the dashed line of Fig. 5-35b. The value $Re_{\delta^*,\,tr}(k=0)$ can be estimated from Fig. 5-33.

As Dryden (1953) pointed out, the effect of freestream turbulence in Fig. 5-35b can be removed by normalizing the abscissa with its "quiet" value. The result, as correlated by Dryden (1953), is shown in Fig. 5-36a to be a nearly universal curve.

At higher Mach numbers, a two-dimensional trip is much less effective. Figure 5-36b shows normalized transition data for $Ma = 0$ (from Fig. 5-36a), $Ma = 3.1$ [Brinich (1954)], and $Ma = 5.8$ [Korkegi (1956)]. Much larger roughness heights are required for tripping, listed as follows by Gibbings (1959):

Flat-plate Mach number	$\frac{Uk}{\nu}$
0.0	850
2.0	2000
4.0	10,000
6.0	∞(?)

A single three-dimensional roughness element causes a zone of spreading disturbances in its wake. There is little effect on the transition point until $k/\delta^* \approx 0.6$, after which there is a sharp drop in $Re_{\delta^*,\,tr}$. At a local roughness Reynolds number $u(k)k/\nu \approx 600$, a low-speed boundary layer is considered "tripped," and the flow downstream of the element is a wedge of continuous turbulence. As the Mach number increases, the element remains an effective trip until $Ma > 4$, after which the

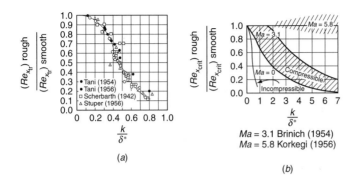

FIGURE 5-36
Flat-plate, two-dimensional roughness transition data normalized to eliminate freestream turbulence effects: (a) incompressible flow [after Dryden (1953)]; (b) compressible flow.

necessary roughness height increases markedly. Whitfield and Iannuzzi (1969) conducted experiments at hypersonic Mach numbers. They review the literature and give a general transition correlation for spherical elements from $Ma = 0$ to $Ma = 16$.

Distributed roughness has been less studied. Results for sand-grain roughness reported by Feindt in 1957 [see Fig. 17.45 of Schlichting (1979)] show no effect on transition until $Uk/\nu \approx 120$, in both favorable and adverse pressure gradients. For $Re_k > 120$, the transition point x_{tr} decreases markedly with Re_k.

5-5.4 Transition of Unsteady Flows

All the discussions in this section have described transition in steady shear flows. Additional parameters appear in unsteady flows, to which the symposium edited by Dwoyer and Hussaini (1987) was devoted. Usually a dimensionless time or frequency is added to the correlation.

5-5.4.1 OSCILLATING BOUNDARY LAYERS. Obremski and Fejer (1967) experimented with flat-plate airflow whose stream velocity varied approximately sinusoidally about a mean:

$$U = U_0(1 + N_A \sin \omega t) \tag{5-49}$$

The amplitude ratio N_A varied from 0.014 to 0.27, and the frequency ω (4.5 to 62 Hz) was kept below the range of expected unstable Tollmien–Schlichting waves. Initial instability began in the low-velocity trough of the cycle ($\omega t \approx 3\pi/2$). In addition to the usual $Re_{x,tr}$, Obremski and Fejer defined a "nonsteady Reynolds number," $Re_{ns} = N_A U_0^2/\omega\nu$, and found that $Re_{ns} \approx 26{,}000$ was a critical point. As shown in Fig. 5-37a, for $Re_{ns} > 25{,}000$, there was no effect of the oscillations on the transition point $Re_{x,tr} \approx 1.6 \times 10^6$. However, for $Re_{ns} > 27{,}000$, the transition point moves rapidly forward as N_A increases.

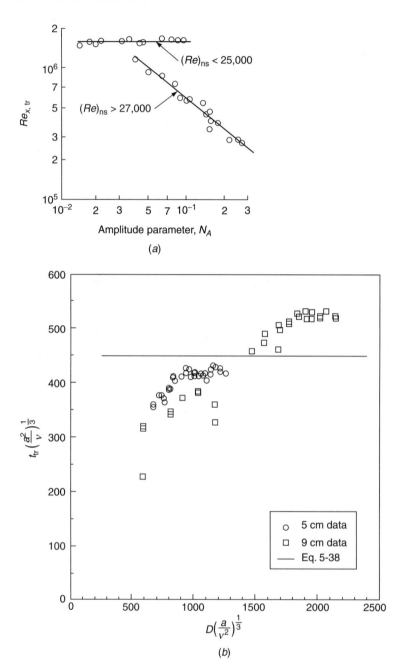

FIGURE 5-37
Transition in unsteady shear flows: (a) on a flat plate with oscillating freestream [*after Obremski and Fejer (1967)*]; (b) in pipe flow accelerated linearly from rest [*after Lefebvre and White (1991)*].

Two additional results, characteristic of unsteady instability, were reported by Obremski and Fejer: (1) unstable wave packets were formed at the velocity trough for all values of Re_{ns}, but only for $Re_{ns} > 27{,}000$ did these packets burst into turbulence; and (2) at certain times during transition, the wave packet interface appeared to move upstream—something that is not observed in steady-flow instability. By evaluating the Orr–Sommerfeld eigenvalues of the instantaneous velocity profiles and following their paths, Obremski and Morkovin (1969) were able to construct a satisfactory quasi-steady theory which explained these results. The subject of time-periodic flow stability is reviewed by Davis (1976).

5-5.4.2 ACCELERATING FLOW IN A PIPE. Lefebvre and White (1989, 1991) experimented with water flow accelerated linearly from rest in a 30 m long pipe with $D = 5$ cm and (later) 9 cm. The flow began as laminar, underwent transition abruptly and almost globally (within 30 ms), then became fully turbulent. The laminar velocity profiles were in agreement with the classic solution of Szymanski (1932) from Sec. 3-4 (Fig. 3-16). Compared to steady pipe flow, transition occurred at very high Reynolds numbers, $2 \times 10^5 \le Re_{D,\,tr} \le 1.1 \times 10^6$, for accelerations from 0.2 to 11.8 m/s^2. The comparison is misleading because the velocity profiles are not parabolas but rather resemble the earliest profile in Fig. 3-16, with thin boundary layers and a flat core. The laminar profiles burst globally into turbulence at a transition time t_{tr} which is a function of pipe diameter D, acceleration a, and kinematic viscosity ν. By dimensional analysis,

$$t_{tr}\left(\frac{a^2}{\nu}\right)^{1/3} = fcn\left[D\left(\frac{a}{\nu^2}\right)^{1/3}\right] \qquad (5\text{-}50)$$

The transition-time data for two different pipe sizes are shown in Fig. 5-37b. As an approximate comparison, Lefebvre and White (1991) noted that Michel's steady-flow criterion, Eq. (5-38), gave the simple result $t_{tr}(a^2/\nu)^{1/3} \approx 450$. This is also shown in Fig. 5-37b. The data are more closely fit by a square-root relation, $t_{tr} \propto D^{1/2}$.

5-5.5 Control of Boundary-Layer Transition

Although existing theories are certainly not definitive, we know enough about transition to make some serious attempts at designing systems that *control* transition. This subject is reviewed by Reshotko (1987), who also reviews the concept of *receptivity*, the means by which freestream disturbances enter the boundary layer and excite its "normal modes" or Tollmien–Schlichting waves. For example, with reference to Fig. 5-33, why is the boundary layer so *receptive* to certain types of grid turbulence?

Basically, boundary layers are stabilized and transition delayed by increasing the negative curvature of the profile. Reshotko (1987) extends Eq. (5-30) to include a nonzero wall velocity:

$$\mu_w \frac{\partial^2 u}{\partial y^2}(0) = \frac{dp}{dx} + \left[\rho v_w - \frac{d\mu}{dT}\frac{\partial T}{\partial y}(0)\right]\frac{\partial u}{\partial y}(0) \qquad (5\text{-}51)$$

From this we see that the effects that make the right-hand side more negative (stable) are

1. Favorable pressure gradient: $dp/dx < 0$
2. Wall suction: $v_w < 0$
3. Cooling of gases: $d\mu/dT > 0, \partial T/\partial y > 0$
4. Heating of liquids: $d\mu/dT < 0, \partial T/\partial y < 0$

Delaying transition (maintaining laminar flow) will lower the drag of vehicles—about 60 percent of the drag of aircraft and 90 percent of the drag of underwater bodies are due to wall friction. Reshotko (1987) reviews work to date on the above effects, which we summarize here.

A favorable pressure gradient (effect 1) is achieved by "shaping" the body so that its point of maximum velocity moves far aft. One is then faced with designing the (short) tail of the body to avoid extensive flow separation.

Wall suction (effect 2) has long been known to keep the boundary layer laminar, reducing drag, and delaying separation (thus achieving higher lift). Pfenninger and Reed (1966) summarize research on airfoil suction. Suction is often less effective for swept wings, where turbulence from the fuselage tends to spread out along the leading edge. Wagner and Fischer (1984) summarize efforts to keep leading edges free from such contamination.

Cooling of gases (effect 3) is very effective (Fig. 5-14) but expensive, requiring cryogenic equipment for application to aircraft. Nevertheless, Reshotko (1979) shows that drag reductions of 20 to 25 percent might be a practical and achievable goal.

Surface heating of water flows (effect 4) definitely stabilizes the boundary layer. By overheating a body of revolution in a water flow to 25°C above ambient, Lauchle and Gurney (1984) showed that $Re_{x,\text{tr}}$ increased from 4.5×10^6 to 3.64×10^7. Heating power requirements are large (approximately 100 kW/m^2) because of the high heat capacity and thermal conductivity of water.

There is much research now on *active control* of boundary layers to delay transition or otherwise improve a flow. A general review is given by Gad-el-Hak (1996). One method is *wave cancellation,* wherein growing disturbances are detected and the control system generates waves 180° out of phase to cancel them. This idea is surveyed by Joslin et al. (1996).

A second active-control proposal, useful in both transitional and turbulent boundary layers, is to cover the surface with an array of *piezoelectric actuators* that generate counterrotating motions to cancel the streamwise vortices that feed turbulence. This idea is reviewed by Jacobson and Reynolds (1998).

All of the above six laminar-flow-control techniques suffer from environmental factors—such as ice crystals, rain, dirt, and sediment or small organisms in water—which may bypass the control and cause disturbances sufficient to trigger turbulence.

5-5.6 Chaos and Turbulence

In 1963, E. N. Lorentz, while analyzing a system of first-order differential equations simulating atmospheric behavior, discovered that these apparently deterministic equations led to disorderly solutions, now termed *chaos*. These findings have triggered an explosion of chaos research by physicists, mathematicians, and engineers. Many physical systems are found to be chaotic.

For example, consider the *logistic equation*, a simple algebraic formula for predicting the growth of a population x of a biological species:

$$x_{\text{new}} = rx_{\text{old}}(1 - x_{\text{old}}) \quad (5\text{-}52)$$

where $1 < r < 4$ is a parameter. Begin with any initial guess $0 < x < 1$ and iterate this equation for a given value of r. For $1 < r < 3$, the iterates converge to a single stable curve. At $r = 3$, the iterates bifurcate into two curves. Two more bifurcations occur at $r = 3.5$ and 3.6. Then, for $r > 3.6$, the iterates spread chaotically throughout the region between the highest and lowest bifurcation. All this behavior is shown graphically in Fig. 5-38 and is characteristic of many seemingly deterministic dynamic systems.

The chief key to chaos in a dynamic system is its *extreme sensitivity to initial conditions*. Slight changes in initial data will create a trajectory which eventually diverges completely from neighbors with nearly the same starting conditions. Often the trajectories will circle near a point, not periodically (i.e., a limit cycle) but rather wandering through changing layers that, in time, pass near every part of the region. These points are called *strange attractors* and have a "fractal" dimension greater than the perceived geometric dimensions of the region.

Chaos theory is now a mature science and the subject of dozens of books. There is a strong similarity to transition, as one can imagine by viewing Figs. 5-28a and 5-38. Both processes involve multiple bifurcations and lead toward random content, but, as discussed in Chap. 9 of Drazin (2002), it is an oversimplification to

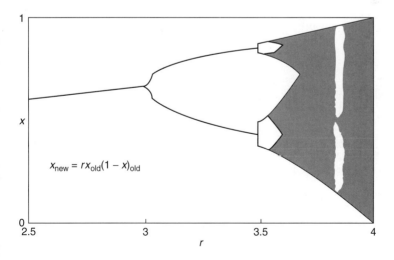

FIGURE 5-38
Iterates of the logistic formula, Eq. (5-52). Bifurcations occur at $r = 3.0$, 3.5, and 3.6. For $r > 3.6$, solutions are chaotic.

match the two. Transition is too diverse, with too many parameters and too many paths toward turbulence. Chaos and transition were first compared in books by Tatsumi (1984), Swinney and Gollub (1985), and Chevray (1989). It is known that certain laminar flows—Couette flow, Bénard convection, and fluid mixing—exhibit chaotic behavior. Certain turbulent structures—the edges of jets and shear layers—have a fractal (noninteger) dimension. Chaos and transition are not directly analogous, but the parallel remains intriguing and is the subject of recent books by Favre (1995), Branover et al. (1999), and Debnath and Riahi (2000).

SUMMARY

As far as we know, all laminar flows eventually become unstable and develop or "transition" into turbulence at a sufficiently high velocity parameter appropriate to the flow: the Reynolds, Grashof, or Taylor numbers.

The stability of any flow can be studied by superimposing small disturbances on a basic laminar-flow condition. For boundary-layer flows, the linearized disturbance relation is the Orr–Sommerfeld equation that has been solved for a variety of flows. Many parametric studies are given here, and parameters such as favorable pressure gradient, wall suction, wall compliance, heating of liquids, and cooling of gases are found to be stabilizing.

Experiments show that the unstable, plane traveling Tollmien–Schlichting waves, predicted by the Orr–Sommerfeld analysis, grow downstream and become three-dimensional, causing vortex breakdown and formation of turbulent spots that coalesce into fully turbulent flow. There are several empirical correlations for the prediction of this final *transition point*, and an attempt is made here to describe and compare them. Transition is strongly affected by freestream turbulence, acoustic excitations, and wall roughness. After a brief discussion of transition in unsteady shear flows, the chapter ends with a description of some possible design criteria for controlling the onset of transition.

PROBLEMS

5-1. While holding $(g, \rho_1, \rho_2, \mathcal{T})$ constant, show that the right-hand side of Eq. (5-9) has a minimum at the wave number $\alpha = [g(\rho_1 - \rho_2)/\mathcal{T}]^{1/2}$. Find experimental data somewhere and estimate this "critical" wavelength and velocity difference for air blowing over gasoline. For further discussion, see Drazin (2002), p. 57.

5-2. Show that, if the upper and lower velocities in Fig. 5-2 are negligible and if surface tension is neglected, a disturbance of the interface will propagate at the phase speed

$$c = \sqrt{\frac{g\lambda(\rho_1 - \rho_2)}{2\pi(\rho_1 + \rho_2)}}$$

where λ is the wavelength of the disturbance. Discuss what might happen if $\rho_1 < \rho_2$. Estimate this propagation speed for an air–water interface when the wavelength is 3 m.

5-3. Derive a linearized disturbance equation for the celebrated (non-linear) van der Pol equation,

$$\frac{d^2X}{dt^2} + C(X^2 - 1)\frac{dX}{dt} + X = 0$$

where C is a constant. Assume that $X_0(t)$ is a known exact solution. Comment upon the disturbance equation but do not solve.

5-4. Verify the Orr–Sommerfeld Eq. (5-18) by eliminating p and u from Eqs. (5-15).

5-5. Consider Rayleigh's inviscid stability Eq. (5-21). The perturbation amplitude $v(y)$ is complex, and, for temporal stability analysis, α is real and c is complex. Show that Rayleigh's equation may be split into real and imaginary parts, as follows:

$$v_r'' - \left[\alpha^2 + \frac{U''(U - c_r)}{(U - c_r)^2 + c_i^2}\right]v_r + \frac{U''c_i}{(U - c_r)^2 + c_i^2} = 0$$

$$v_i'' - \left[\alpha^2 + \frac{U''(U - c_r)}{(U - c_r)^2 + c_i^2}\right]v_i - \frac{U''c_r}{(U - c_r)^2 + c_i^2} = 0$$

Explain, in words, a possible method for solving this system numerically for a given $U(y)$.

5-6. For an approximate quartic-polynomial flat-plate velocity distribution.

$$\frac{u}{U} \approx 2\eta - 2\eta^3 + \eta^4 \qquad \eta = \frac{y}{\delta}$$

solve the Orr–Sommerfeld Eq. (5-18) numerically for the inviscid case, $\nu = 0$. Begin at $y = 2\delta$ by assuming that $v \approx e^{-\alpha y}$ and integrate inward to satisfy the wall ($y = 0$) conditions $v = v' = 0$. Assuming temporal amplification, find some (damped) eigenvalues and plot c_i vs. α in dimensionless form.

5-7. For stagnation boundary-layer flow, $U = Kx$, estimate the position Re_x where instability first occurs.

5-8. For the separating Falkner–Skan wedge-flow boundary layer, $\beta = -0.19884$, estimate the position Re_x where instability first occurs.

5-9. For the Howarth freestream velocity $U = U_0(1 - x/L)$, if $U_0 L/\nu = 10^6$, estimate the point (x/L) where boundary-layer instability first occurs. Assume a low subsonic Mach number.

5-10. Generalize Prob. 5-9 to calculate and plot the instability point $(x/L)_{crit}$ as a function of $U_0 L/\nu$.

5-11. For potential freestream flow across a cylinder, $U = 2U_0 \sin(x/a)$, if $Re_D = 10^6$, estimate the position $(x/a)_{crit}$ where boundary-layer instability first occurs.

5-12. Using the guidance of Probs. 5-5 and 5-6 and any numerical method of your choosing, solve the Rayleigh Eq. (5-21) for simplified boundary-layer flow, $U = \tanh(y)$, $0 \le y \le \infty$, with a typical value of $\alpha \approx 0.2$ to 0.3. Do you expect any inviscid instability? If time permits, plot some computed values of c_r versus α. [Hint: Begin at large $y \ge 4$ with the exponential approximation of Eq. (5-6) and integrate backward to the wall.]

5-13. Using the guidance of Probs. 5-5 and 5-6 and any numerical method of your choosing, solve the Rayleigh Eq. (5-21) for the Blasius boundary-layer flow, which you should generate from Eqs. (4-45) and (4-46). Do you expect any inviscid instability? Select a value of α in the range 0.2 to 0.3. [Hint: Begin at large $\eta \geq 5$ with the exponential approximation of Eq. (5-6) and integrate backward to the wall.]

5-14. For stagnation boundary-layer flow, $U = Kx$, estimate the position Re_x where transition first occurs, using the method of Michel, Eq. (5-38). What makes the correlation of Granville (Sec. 5-5.1.1) inappropriate? Assume negligible freestream turbulence.

5-15. For the separating Falkner–Skan wedge-flow boundary layer, $\beta = -0.19884$, use any appropriate correlation to estimate the position Re_x where transition first occurs. Neglect freestream turbulence. Compare your result with Fig. 5-32.

5-16. For the Howarth freestream velocity $U = U_0(1 - x/L)$, if $U_0 L/\nu = 4 \times 10^6$, use the correlation of Michel, Eq. (5-38), to estimate the point (x/L) where boundary-layer transition occurs. Neglect freestream turbulence. Compare your result with Fig. 5-31.

5-17. Generalize Prob. 5-16 into a parametric computer study to compute and plot $(x/L)_{tr}$ vs. $U_0 L/\nu$.

5-18. For potential freestream flow across a cylinder, $U = 2U_0 \sin(x/a)$, if $Re_D = 2 \times 10^6$, use the correlation of Michel, Eq. (5-38), to estimate the position $(x/a)_{tr}$ where boundary-layer transition first occurs. Neglect freestream turbulence. Compare your result with Fig. 5-31.

5-19. Air at 20°C and 1 atm flows quietly toward a wedge of half-angle 36°, resulting in a power-law freestream and a laminar boundary layer along the surface. Use Wazzan's method Eq. (5-42) to estimate the transition Reynolds number $Re_{x,\,tr}$.

5-20. Modify Prob. 5-14 for a freestream turbulence level of 1 percent.

5-21. Modify Prob. 5-15 for a freestream turbulence level of 1 percent.

5-22. Modify Prob. 5-16 for a freestream turbulence level of 1 percent.

5-23. Modify Prob. 5-18 for a freestream turbulence level of 1 percent.

5-24. For a pipe flow started from rest with acceleration a, as in Fig. 5-37b, the momentum thickness initially grows according to the formula $\theta \approx 0.35(\nu t)^{1/2}$. Apply this relation to Michel's steady-flow transition correlation, Eq. (5-38), assuming that "V" and "x" are given by constant-acceleration formulas. Show that the result is a constant value of the dimensionless transition time $t_{tr}(a^2/\nu)^{1/3}$. Why does the diameter D not appear?

5-25. Air at 20°C and 1 atm flows at $U = 12$ m/s past a smooth flat plate. It is desired to trip the boundary layer to turbulence by stretching a 1 mm diameter wire across the plate at the wall. Where will transition occur if the wire is placed at $x = 1$ m? What wire location x will cause the earliest transition?

5-26. Repeat Prob. 5-25 if the freestream turbulence level is 1 percent.

5-27. The narrow vertical white band in the chaotic area of the logistic map in Fig. 5-38 lies in the region $3.825 < r < 3.865$. Beginning at $r = 3.825$ with an initial guess $x = 0.5$, make repeated computer iterations, for small increment $\Delta r \leq 0.0005$, of the logistic relation Eq. (5-52) and plot the results on an expanded abscissa for this region. Comment on the remarkable pattern you find.

5-28. Repeat Prob. 5-19 if the freestream has a turbulence level of 4 percent. Find the estimated transition Reynolds number $Re_{x,\,tr}$ by two different methods and compare.

5-29. The famous neutral curve of Taylor (1923), for Couette flow between rotating cylinders, is in Fig. 5-22b. The region above the curve is simply labeled *unstable*. Some amazingly diverse flow regimes lie in this region, as shown in a wonderful chart by Andereck et al. (1986). Report to the class on this chart and its many unstable flow patterns.

5-30. For two-dimensional inviscid flow, the vorticity Eq. (2-116) may be written as

$$\frac{\partial \omega}{\partial t} + u \frac{\partial \omega}{\partial x} + v \frac{\partial \omega}{\partial y} = 0$$

where

$$\omega = \frac{\partial v}{\partial x} - \frac{\partial u}{\partial y} = -\nabla^2 \psi$$

Defining a basic flow (Ψ, U, V, Ω) and disturbances (ψ', u', v', ω'), derive a linearized disturbance equation for this flow. Then assume normal modes as traveling waves:

$$(\psi', u', v', \omega') = [\hat{\psi}'(y), \hat{u}(y), \hat{v}(y), \hat{\omega}(y)] \exp[i\alpha(x - ct)]$$

Derive the disturbance equations and, if possible, combine them to obtain a single differential equation for a single disturbance amplitude.

5-31. Use the airfoil surface-velocity data of Prob. 4.49 and Michel's method, Eq. (5-38), to estimate the position of transition to turbulence if $Re_c = 2 \times 10^6$. Assume air at 20°C and 1 atm. If you note an ambiguity in the results, please criticize them.

CHAPTER 6

INCOMPRESSIBLE TURBULENT MEAN FLOW

6-1 PHYSICAL AND MATHEMATICAL DESCRIPTION OF TURBULENCE

Chapter 5 showed that smooth, orderly laminar flow is strictly limited to finite values of a critical parameter—Reynolds number, Grashof number, Taylor number, Richardson number. Beyond that, laminar flow is unstable and will evolve to a new flow regime if the critical parameter is high enough. That new regime, not predicted by stability theory but nevertheless inevitable, is a fluctuating, disorderly motion called *turbulence*. Because turbulence is so complex, its complete analysis and quantification will probably never be achieved. Turbulent flow will be the subject of research in the foreseeable future, and hundreds of papers and articles are being published every year.

Much is known now about the structure of turbulence, due to excellent experimental techniques. First, there is advanced flow visualization, as described in the monographs by van Dyke (1982), Merzkirch (1987), Nakayama (1988), Yang (1989, 1994), Nakayama and Tanida (1996), and Smits and Lim (2000). In addition to still images, there is an outstanding list of fluid-flow videotapes and movies given by Carr and Young (1996). The flow-visualization community meets regularly and just had its 11th conference, Mueller (2004). Second, there is superb modern miniature instrumentation: hot-wires, Bruun (1995); laser-Doppler systems, Durst (1992), particle image velocimetry, Raffel et al. (1998), and other measurement techniques, Goldstein (1996).

The present chapter is a modest survey of the field of incompressible turbulent flow. There are many advanced monographs on the subject, in chronological order: Hinze (1975), Libby (1996), Schmitt (1997), Wilcox (1998), Mathieu and Scott (2000), Pope (2000), Launder and Sandham (2001), Bernard and Wallace (2002), Cebeci (2003), and Tzabiras (2003).

6-1.1 Physical Description

A provocative visualization of transitional and turbulent flat-plate flow may be seen in Fig. 6-1. Air ($\nu \approx 1.5\text{E}-5 \text{ m}^2/\text{s}$) flows at 3.3 m/s past a plate 2.4 m long and 1.2 m wide. The shear-layer motion is visualized by smoke introduced at the leading edge. Spots form and merge, and transition to turbulence occurs at about $x = 90$ cm,

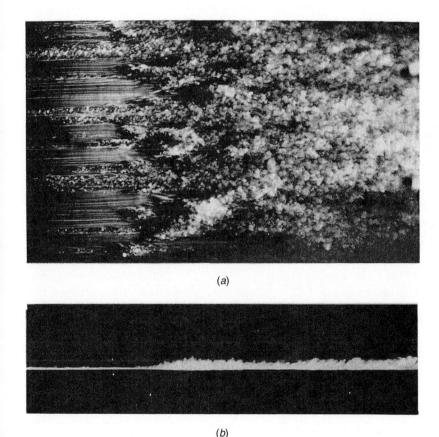

FIGURE 6-1
Smoke visualization of airflow at 3.3 m/s past a flat plate: (*a*) top view; (*b*) side view. Transition is at $x \approx$ 90 cm, or $Re_x \approx 2\text{E}5$. [*From Nakayama (1988), courtesy of S. Taneda.*]

or $Re_x \approx$ 2E5. After this point, the boundary layer is much thicker and quite disorderly. The disorder is not merely white noise. It clearly has spatial structure and may be described by the following characteristics:

1. *Fluctuations* in pressure and velocity (and also temperature when there is heat transfer). Velocity fluctuates in all three directions. Fluctuations are superimposed upon a mean value of each property.
2. *Eddies* or fluid packets of many sizes that intermingle and fill the shear layer. Eddy size varies continuously from a shear-layer thickness δ (about 40 mm in this case) down to the so-called Kolmogorov length scale, $L = (\nu^3 \delta / U^3)^{1/4}$, or about 0.05 mm in this case.
3. *Random* variations in fluid properties which have a particular form (not white noise). Each property has a specific continuous energy spectrum that drops off to zero at high wave numbers (small eddy size).
4. *Self-sustaining* motion. Once triggered, turbulent flow can maintain itself by producing new eddies to replace those lost by viscous dissipation. This is especially so in wall-bounded flows. Turbulence production is not generally related to the original instability mechanisms such as Tollmien–Schlichting waves.
5. *Mixing* that is much stronger than that due to laminar (molecular) action. Turbulent eddies actively move about in three dimensions and cause rapid diffusion of mass, momentum, and energy. Ambient fluid from nonturbulent zones will be strongly entrained in a turbulent flow. Heat transfer and friction are greatly enhanced compared to laminar flow. Turbulent mixing is associated with a gradient in the time-mean flow.

These are typical descriptors for turbulence: a spatially varying mean flow with superimposed three-dimensional random fluctuations that are self-sustaining and enhance mixing, diffusion, entrainment, and dissipation.

It is not necessary to have a nearby wall to generate turbulence. Boundary-free motions—jets, wakes, mixing layers—may also be turbulent and have all the characteristics mentioned above. In addition, turbulent flows that interface with nonturbulent fluid may contain clearly visible patterns called *coherent structures*. An example is shown in Fig. 6-2, turbulent flow in the mixing layer between two parallel streams of unequal velocity. Embedded within the disorderly small-eddy motions is a coherent pattern of paired vortices growing linearly downstream, associated in this case with the Kelvin–Helmholtz instability of an inviscid mixing layer [see Fig. 5-3]. Related coherent structures may also be found at the edges of turbulent jets, wakes, and boundary layers [Hussain (1986), Holmes et al. (1998)].

Since turbulence involves a wide spectrum of properties with random length and timescales, describing it is difficult. The present chapter will concentrate on engineering estimates of time-mean velocity, pressure, temperature, shear stress, and heat transfer. There are many other worthy measures of turbulent flow. Chapter 2 of Bernard and Wallace (2002), for example, details the properties of integral microscales, Reynolds stress, turbulent kinetic energy, turbulent dissipation, turbulent transport, pressure–strain correlations, vorticity, and enstrophy.

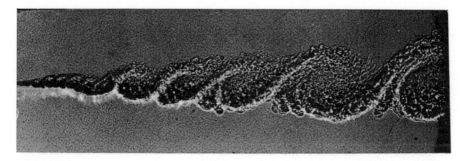

FIGURE 6-2
Spark shadowgraph of turbulent flow in the mixing layer between unequal parallel streams. A large coherent vortical structure is embedded within the random turbulent eddies. [*After Brown and Roshko (1974)*.]

6-1.2 Mathematical Description

Visualizations such as Fig. 6-1 are illuminating but fail to give much quantitative information. At a minimum, one wishes in a turbulent-flow analysis to be able to predict (1) velocity and/or temperature profiles and (2) wall friction and/or heat transfer. Additional information on the statistical properties of the turbulence is desirable, but it is essentially impossible to *predict* Fig. 6-1 with a theory.

As far as we know, the Navier–Stokes equations from Chap. 3 do apply to turbulent flow, and these equations can be modeled on a digital computer by finite differences or finite elements. Such frontal attacks on turbulence form the new research field of *direct numerical simulation* [Moin and Mahesh (1998)]. Because of the wide range of flow scales involved, the solutions require supercomputers and even then are limited to very low Reynolds numbers. A flow such as Fig. 6-1 is out of the question—the smallest eddy size is about 0.04 mm, hence simulation of a shear layer 10 cm high on a 1.2 m × 2.4 m plate would require 5 trillion mesh points. Such a mesh is far beyond the capacity of present computers. The largest meshes achieved thus far contain about 100 million nodes. This is sufficient to model transitional and low-turbulent flows ($Re \leq 10^4$) and yields much useful information for modeling higher Reynolds numbers.

At one time, DNS turbulence computation was thought to be impossible. Emmons (1970) estimated that prediction of fine details of turbulent pipe flow would require 10^{22} numerical operations, far beyond the capability of any single supercomputer. Since then, *parallel* computation, combining the efforts of a multitude of single computers, has enabled researchers to exceed Emmons' estimate by several orders of magnitude.

6-1.3 Fluctuations and Time Averaging

Since actual computation of a raw velocity component $u(x, y, z, t)$ is not possible in high Reynolds number flow, the standard analysis of turbulence separates the fluctuating property from its time-mean value.

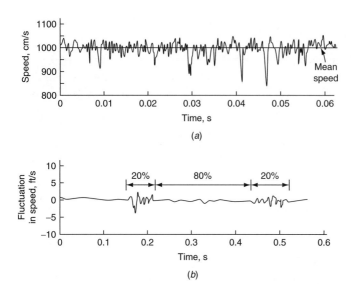

FIGURE 6-3
Hot-wire measurements showing turbulent velocity fluctuations: (*a*) typical trace of a single velocity component in a turbulent flow; (*b*) trace showing intermittent turbulence at the edge of a jet.

Figure 6-3 shows time traces, taken by a hot-wire, of a single velocity component in a turbulent flow. Trace *a* was taken deep within a shear layer and shows continuous random fluctuations of 5 to 10 percent about a mean value. Trace *b* was taken near the edge of a jet flow and shows intermittent regimes of turbulent and nonturbulent flow. The flow is seen to be turbulent about 20 percent of the time—hence we define the *intermittency factor,* γ = (turbulent time)/(total time) = 0.2. Note that the mean flow has been subtracted for convenience in trace *b*. The intermittency occurs because the edge of a turbulent jet is sharp and ragged and regions of turbulent and nonturbulent flow alternately pass over the probe.

When analyzed for energy content, the frequency spectrum of trace *a* is found to be continuous (random), with significant energy in the range of 1 to 10^5 Hz. The same is true of the turbulent portions of trace *b*. Other properties—pressure, temperature, and the remaining two velocity components—will show more or less the same behavior as in Fig. 6-3.

Suppose that trace *a* in Fig. 6-3 represents $u(t)$ at a particular spot (x, y, z). Then the time average of u is defined as

$$\bar{u} = \frac{1}{T} \int_{t_0}^{t_0+T} u \, dt \qquad (6\text{-}1)$$

where the integration interval T is chosen to be larger than any significant period of the fluctuations in u. We may then define the fluctuation, u', as what remains when the mean flow is subtracted.

$$u' = u - \bar{u} \tag{6-2}$$

Only the fluctuation is shown in trace b of Fig. 6-3. By definition, the mean fluctuation $\bar{u}' = 0$. Therefore, to characterize the magnitude of the fluctuation, we work with its mean-square value:

$$\overline{u'^2} = \frac{1}{T}\int_{t_0}^{t_0+T} u'^2 \, dt \tag{6-3}$$

The root-mean-square value of u' is defined as $u'_{\text{rms}} = (\overline{u'^2})^{1/2}$. If the integrals in Eqs. (6-1) and (6-3) are independent of starting time t_0, the fluctuations are said to be statistically *stationary*.

There are two paths to turbulent-flow analysis: (1) a statistical theory of turbulent correlation functions and (2) a semiempirical modeling of turbulent mean quantities. The former studies the statistical properties of the fluctuations: their frequency correlations, space–time correlations, and interactions with each other. The latter approach emphasizes the turbulent properties of most engineering significance: mean velocity and temperature profiles, wall friction and heat transfer, shear-layer thickness parameters, and rms fluctuation profiles. With the increase in sophistication in recent turbulent-modeling research, the two paths are beginning to overlap. But this text will primarily emphasize path 2, semiempirical modeling of engineering problems. The statistical theory is treated in several excellent texts: Monin and Yaglom (1972), Tennekes and Lumley (1972), Hinze (1975), and Heinz (2003).

6-1.4 Illustration: Turbulence Measurements in Flat-Plate Flow

Throughout boundary-layer theory we shall adopt coordinates (u, v, w) such that u is parallel to the freestream, v is normal to the wall, and w is lateral to the freestream. Figure 6-4 shows measured values of $(\sqrt{\overline{u'^2}}, \sqrt{\overline{v'^2}}, \sqrt{\overline{w'^2}})$ and the covariance $\overline{u'v'}$, which is called the *turbulent shear stress* for reasons we shall soon see, in a turbulent boundary layer on a flat plate at a Reynolds number $Re_x \approx 10^7$, after Klebanoff (1955). Note that the fluctuations are quite large, up to 11 percent of the freestream speed. The presence of the wall makes the fluctuations different in magnitude (anisotropy, typical of all shear flows) for clear geometric and physical reasons. The longitudinal fluctuation u' is largest, being unimpeded by the wall and slightly reinforced by the freestream. The term v' is smallest, being directly impeded by the presence of the wall, and reaches its maximum much farther out than u' or w'. The lateral component w' is intermediate and quite large. The reader may perhaps be struck by this evidence that even the most "two-dimensional" of turbulent boundary layers has a thoroughly three-dimensional set of velocity fluctuations. However, w' does not directly influence the mean flow in this case. Finally,

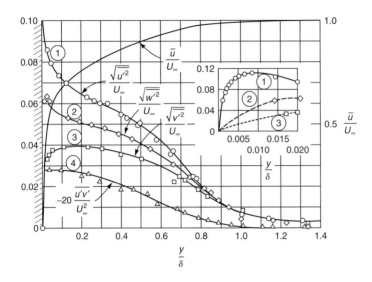

FIGURE 6-4
Flat-plate measurements of the fluctuating velocities u' (streamwise), v' (normal), and w' (lateral) and the turbulent shear $\overline{u'v'}$. [*After Klebanoff* (1955).]

the turbulent shear $\overline{u'v'}$ is much smaller but of fundamental importance to later analysis. Determining its characteristics is the central problem in our analysis of turbulent shear flows.

Figure 6-4 has other interesting features. First, near the wall, all fluctuations (u', v', w') should vanish due to the no-slip condition. Yet the large-scale graph ($0 < y/\delta < 1$) shows the mean flow $\bar{u}$ dropping to zero but not the fluctuations, which require an expanded inset scale to demonstrate no slip. Turbulence is quite resistant to wall damping, and measurements in thick boundary layers show significant fluctuations, even at $y/\delta = 0.0001$.

Second, at large y, we see that the fluctuations extend *outside* the point normally designated at the boundary-layer "edge" $y = \delta$ where $\bar{u} \approx U_\infty$. This is associated with the intermittency of the outer layer. Klebanoff (1955) demonstrated that a sharp interface or *superlayer* exists between turbulent and nonturbulent regions of the flow. This interface has a ragged shape and undulates while traveling downstream, as sketched in Fig. 6-5b. The superlayer varies between about 0.4δ and 1.2δ and propagates downstream at about $0.94U_\infty$. The interface exhibits *fractal* behavior, i.e., its length and surface area rise at a faster rate than its basic size would predict [see, e.g., Chevray (1989)].

Klebanoff measured the intermittency (percentage of time the flow is turbulent) in the boundary layer, with results shown in Fig. 6-5a. The oddity of data points where $\gamma > 1$ (near the wall) is explained by Klebanoff's use of a statistical formula, rather than the more recent technique of conditioned sampling with an instrument that discriminates between laminar and turbulent flow

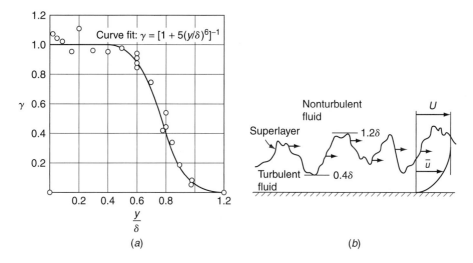

FIGURE 6-5
The phenomenon of intermittency in a turbulent boundary layer: (*a*) measured intermittency factors [*after Klebanoff* (*1955*)]; (*b*) the superlayer interface between turbulent and nonturbulent fluid.

[Kovaznay et al. (1970)]. The curve-fit formula in Fig. 6-5*a* is often used in empirical theories—actually, Klebanoff showed that the intermittency distribution is almost exactly Gaussian in the range between $y/\delta = 0.4$ and 1.2. Similar measurements by Corrsin and Kistler (1955) show the same Gaussian intermittency for turbulent jets and wakes. Klebanoff estimated the dominant wavelength of the superlayer to be about 2δ. The mean position of this interface (where $\gamma \approx 0.5$) is at $y = 0.78\delta$.

Note from Fig. 6-4 that the rms fluctuations (u', v', w') become approximately equal for $y/\delta \geq 0.8$, where the turbulence becomes *isotropic* [see, e.g., Bernard and Wallace (2002)]. Near the wall, however, the flow is strongly anisotropic, and this is the region where most of the production and dissipation of turbulence energy takes place.

Finally, Klebanoff (1955) also measured the energy content or *spectrum* of the turbulence. His results for the wave-number spectrum of the streamwise fluctuation u' are shown in Fig. 6-6. The area under these curves is a measure of the total mean-square fluctuation. Thus

$$\overline{u'^2} = \int_0^\infty F_u(k_u)\, dk_u \qquad (6\text{-}4)$$

We see that, for positions near the wall, there is more energy at a high wave number (small eddies), whereas away from the wall, the large eddies dominate the spectrum.

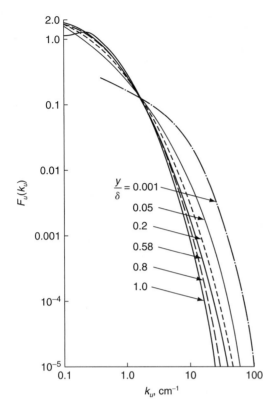

FIGURE 6-6
The wave-number spectrum of the streamwise turbulent velocity fluctuation in flat-plate flow. [*Adapted from Klebanoff* (1955).]

6-2 THE REYNOLDS EQUATIONS OF TURBULENT MOTION

Although statistical theory and numerical simulation are viable options, most of the research on turbulent-flow analysis in the past century has used the concept of time averaging. Applying time averaging to the basic equations of motion yields the Reynolds equations, which involve both mean and fluctuating quantities. One then attempts to model the fluctuation terms by relating them to mean properties or their gradients. This approach may now be yielding diminishing returns. Lumley (1989) gives a stimulating discussion of how time averaging might outlive its usefulness. The Reynolds equations are far from obsolete, however, and form the basis of most engineering analyses of turbulent flow.

Thus, following the original idea of Reynolds (1895), we assume that the fluid is in a randomly unsteady turbulent state and work with the time-averaged or mean equations of motion. Any variable Q is resolved into a mean value $\overline{Q}$ plus a fluctuating value Q', where, by definition,

$$\overline{Q} = \frac{1}{T}\int_{t_0}^{t_0+T} Q\, dt \qquad (6\text{-}5)$$

where T is large compared to the relevant period of the fluctuations. The mean value $\overline{Q}$ itself may vary slowly with time, as sketched in Fig. 6-7, and we speak of this

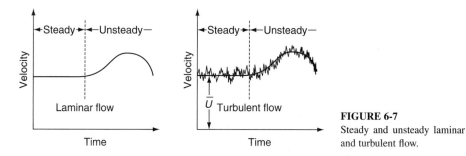

FIGURE 6-7
Steady and unsteady laminar and turbulent flow.

case as an *unsteady* turbulent flow. For example, in a tidal estuary, the velocities would have turbulent components in the 1 Hz range superimposed on a tidal variation with a 12 or 24 h period. After time averaging in the sense of Eq. (6-5), the mean flow would still oscillate with tidal periods.

Let us consider only incompressible turbulent flow with constant transport properties but with possible significant fluctuations in velocity, pressure, and temperature:

$$u = \bar{u} + u' \qquad p = \bar{p} + p'$$
$$v = \bar{v} + v' \qquad T = \bar{T} + T' \qquad (6\text{-}6)$$
$$w = \bar{w} + w'$$

Before substituting these functions in the basic equations, we can verify from the basic integral relation, Eq. (6-5), that the following rules of averaging apply for any two turbulent quantities f and g:

$$\overline{f'} = 0 \qquad \overline{\bar{f}} = \bar{f} \qquad \overline{\bar{f}g} = \bar{f}\bar{g}$$
$$\overline{f'\bar{g}} = 0 \qquad \overline{f+g} = \bar{f} + \bar{g} \qquad \overline{\frac{\partial f}{\partial s}} = \frac{\partial \bar{f}}{\partial s} \qquad (6\text{-}7)$$
$$\overline{fg} = \bar{f}\bar{g} + \overline{f'g'} \qquad \overline{\int f\,ds} = \int \bar{f}\,ds$$

Now consider the incompressible continuity equation

$$\nabla \cdot \mathbf{V} = 0 \qquad (6\text{-}8)$$

substitute u, v, and w from Eqs. (6-6), and take the time average of the entire equation. The result is

$$\frac{\partial \bar{u}}{\partial x} + \frac{\partial \bar{v}}{\partial y} + \frac{\partial \bar{w}}{\partial z} = 0 \qquad (6\text{-}9)$$

Now subtract Eq. (6-9) from Eq. (6-8) but do not take the time average. This gives

$$\frac{\partial u'}{\partial x} + \frac{\partial v'}{\partial y} + \frac{\partial w'}{\partial z} = 0 \qquad (6\text{-}10)$$

Thus the mean and fluctuating velocity components each separately satisfy an equation of continuity. This would not be true if we accounted for density fluctuations, because terms involving $\overline{\rho'u_i'}$ would couple the two relations. Our interest will center on the mean Eq. (6-9), not the fluctuations Eq. (6-10).

Now attempt the same procedure with the non-linear Navier–Stokes equations

$$\rho \frac{D\mathbf{V}}{Dt} = \rho\mathbf{g} - \nabla p + \mu\nabla^2\mathbf{V} \tag{6-11}$$

Before substituting for u, v, w, p, we can avoid unnecessary churning by noting the following rather clever rearrangement of the convective-acceleration term:

$$\mathbf{V}\cdot\nabla Q = u\frac{\partial Q}{\partial x} + v\frac{\partial Q}{\partial y} + w\frac{\partial Q}{\partial z} \equiv \frac{\partial}{\partial x}(uQ) + \frac{\partial}{\partial y}(vQ) + \frac{\partial}{\partial z}(wQ)$$

an identity which follows from the continuity relation, Eq. (6-8). If we now use this idea to substitute in Eq. (6-11) and take the time average, we obtain

$$\rho\frac{D\overline{\mathbf{V}}}{Dt} + \rho\frac{\partial}{\partial x_j}(\overline{u_i'u_j'}) = \rho\mathbf{g} - \nabla\overline{p} + \mu\nabla^2\overline{\mathbf{V}} \tag{6-12}$$

Thus the mean momentum equation is complicated by a new term involving the turbulent inertia tensor $\overline{u_i'u_j'}$. This new term is never negligible in any turbulent flow and is the source of our analytic difficulties because its analytic form is not known a priori. In essence, the time-averaging procedure has introduced nine new variables (the tensor components) that can be defined only through (unavailable) knowledge of the detailed turbulent structure. The components of $\overline{u_i'u_j'}$ are related not only to fluid physical properties but also to local flow conditions (velocity, geometry, surface roughness, and upstream history), and no further physical laws are available to resolve this dilemma. In a two-dimensional turbulent boundary layer ($\overline{w} = 0, \partial/\partial z = 0$), the only significant term reduces to $\overline{u'v'}$, but even this single term requires extensive scratching about to achieve an analytic correlation which is semiempirical at best. Some of the empirical approaches have been quite successful, though rather thinly formulated from nonrigorous postulates.

A slight amount of illumination is thrown upon Eq. (6-12) if it is rearranged to display the turbulent inertia terms as if they were stresses, which of course they are not. Thus we write

$$\rho\frac{D\overline{\mathbf{V}}}{Dt} = \rho\mathbf{g} - \nabla\overline{p} + \nabla\cdot\tau_{ij} \tag{6-13}$$

where
$$\tau_{ij} = \underbrace{\mu\left(\frac{\partial u_i}{\partial x_j} + \frac{\partial u_j}{\partial x_i}\right)}_{\text{Laminar}} - \underbrace{\rho\overline{u_i'u_j'}}_{\text{Turbulent}}$$

Mathematically, then, the turbulent inertia terms behave as if the total stress on the system were composed of newtonian viscous stresses plus an additional or apparent turbulent-stress tensor $-\rho\overline{u_i'u_j'}$. The turbulent stresses are still unknown, so that little has been gained but conceptual value. In a boundary layer, the dominant term $-\rho\overline{u'v'}$ is called *turbulent shear*.

Now consider the energy equation (first law of thermodynamics) for incompressible flow with constant properties

$$\rho c_p \frac{DT}{Dt} = k\nabla^2 T + \Phi \qquad (6\text{-}14)$$

where Φ is the dissipation function given by Eq. (2-46). Taking the time average, we obtain the mean-energy equation

$$\rho c_p \frac{D\overline{T}}{Dt} = -\frac{\partial}{\partial x_i}(q_i) + \overline{\Phi} \qquad (6\text{-}15)$$

where

$$\overline{\Phi} = \frac{\mu}{2}\overline{\left(\frac{\partial \overline{u}_i}{\partial x_j} + \frac{\partial u'_i}{\partial x_j} + \frac{\partial \overline{u}_j}{\partial x_i} + \frac{\partial u'_j}{\partial x_i}\right)^2}$$

and

$$q_i = \underbrace{-k\frac{\partial \overline{T}}{\partial x_i}}_{\text{Laminar}} + \underbrace{\rho c_p \overline{u'_i T'}}_{\text{Turbulent}}$$

By analogy with our rearrangement of the momentum equation, we have collected conduction and turbulent convection terms into a sort of total-heat-flux vector q_i which includes molecular flux plus the *turbulent flux* $\rho c_v \overline{u'_i T'}$. The total-dissipation term $\overline{\Phi}$ is obviously complex in the general case. In two-dimensional turbulent-boundary-layer flow (the most common situation), the dissipation reduces approximately to

$$\overline{\Phi} \approx \frac{\partial \overline{u}}{\partial y}\left(\mu \frac{\partial \overline{u}}{\partial y} - \rho \overline{u'v'}\right)$$

Equations (6-9), (6-13), and (6-15) are the Reynolds-averaged basic differential equations for turbulent mean continuity, mean momentum, and mean thermal energy. They contain many new unknowns involving time-mean correlations of fluctuating velocities and temperature. Therefore solutions cannot be attempted without additional relations or empirical modeling ideas.

6-2.1 The Turbulence Kinetic-Energy Equation

Many attempts have been made to add "turbulence conservation" relations to the time averaged continuity, momentum, and energy equations above. The most obvious single addition would be a relation for the *turbulence kinetic energy* K of the fluctuations, defined by

$$K = \tfrac{1}{2}(\overline{u'u'} + \overline{v'v'} + \overline{w'w'}) = \tfrac{1}{2}\overline{u'_i u'_i} \qquad (6\text{-}16)$$

Here we have introduced for convenience the Einstein summation notation, where $u_i = (u_1, u_2, u_3) = (u, v, w)$ and a repeated subscript implies summation. For example, $u_i u_i = u_1^2 + u_2^2 + u_3^2$. If turbulence is to be described by only one velocity scale, it should be $K^{1/2}$.

A conservation relation for K can be derived by forming the mechanical-energy equation, i.e., the dot product of u_i and the ith momentum equation. Then, following Prandtl (1945a), we subtract the instantaneous mechanical-energy equation from its time-averaged value. The result is the *turbulence kinetic-energy* relation for an incompressible fluid:

$$\underset{\text{I}}{\frac{DK}{Dt}} = \underset{\text{II}}{-\frac{\partial}{\partial x_i}\left[\overline{u'_i\left(\tfrac{1}{2}u'_j u'_j + \frac{p'}{\rho}\right)}\right]} \underset{\text{III}}{- \overline{u'_i u'_j}\frac{\partial \bar{u}_j}{\partial x_i}}$$

$$+ \underset{\text{IV}}{\frac{\partial}{\partial x_i}\left[\overline{\nu u'_j\left(\frac{\partial u'_i}{\partial x_j} + \frac{\partial u'_j}{\partial x_i}\right)}\right]} \underset{\text{V}}{- \nu\overline{\frac{\partial u'_j}{\partial x_i}\left(\frac{\partial u'_i}{\partial x_j} + \frac{\partial u'_j}{\partial x_i}\right)}} \quad (6\text{-}17)$$

We have labeled this relation with Roman numerals to state the relation in words. The rate of change (I) of turbulent energy is equal to (II) its convective diffusion, plus (III) its production, plus (IV) the work done by turbulent viscous stresses, plus (V) turbulent viscous dissipation. The terms in this relation are so complex that they cannot be computed from first principles. Therefore modeling ideas are needed, as will be discussed.

6-2.2 The Reynolds Stress Equation

From Eq. (6-13), the turbulent or "Reynolds" stresses have the form $S_{ij} = (-\rho\overline{u'_i v'_j})$. From this point of view, turbulence kinetic energy is actually proportional to the sum of the three turbulent normal stresses, $K = -S_{ii}/2\rho$. Of more importance to the engineer are the turbulent shear stresses, where $i \neq j$. It is possible to develop a conservation equation for a single Reynolds stress. The derivation involves subtracting the time-averaged momentum Eq. (6-12) from its instantaneous value, for both the i and j directions. The ith result is multiplied by u_j and added to the jth result multiplied by u_i. This relation is then time averaged to yield the *Reynolds stress equation*:

$$\underset{\text{I}}{\frac{D\overline{u'_i u'_j}}{Dt}} = \underset{\text{II}}{-\left[\overline{u'_j u'_k}\frac{\partial \bar{u}_i}{\partial x_k} + \overline{u'_i u'_k}\frac{\partial \bar{u}_j}{\partial x_k}\right]} \underset{\text{III}}{- \left[\&2\nu\overline{\frac{\partial u'_i}{\partial x_k}\frac{\partial u'_j}{\partial x_k}}\right.}$$

$$\underset{\text{IV}}{+ \overline{\frac{p'}{\rho}\left(\frac{\partial u'_i}{\partial x_j} + \frac{\partial u'_j}{\partial x_i}\right)}} \underset{\text{V}}{- \frac{\partial}{\partial x_k}\left[\overline{u'_i u'_j u'_k} - \nu\frac{\partial \overline{u'_i u'_j}}{\partial x_k} + \overline{\frac{p'}{\rho}(\delta_{jk}u'_i + \delta_{ik}u'_j)}\right]}$$

$$(6\text{-}18)$$

Here the Roman numerals denote (I) rate of change of Reynolds stress, (II) generation of stress, (III) dissipation, (IV) pressure–strain effects, and (V) diffusion of Reynolds stress.

In their full three-dimensional form, Eqs. (6-17) and (6-18) are extremely complex, with many unknown correlations to model. We will confine our attention in subsequent sections to the much simpler two-dimensional boundary-layer forms of these relations.

If buoyancy and temperature dependence of fluid properties are neglected, we may uncouple the mean-energy Eq. (6-15) and solve for mean temperature later. The mean velocity and pressure $(\bar{u}, \bar{v}, \bar{w}, \bar{p})$ may be solved from the continuity relation Eq. (6-9) and momentum Eq. (6-13) plus whatever turbulence modeling assumptions are made. Recall that this temperature uncoupling was also the case in laminar viscous flow.

6-3 THE TWO-DIMENSIONAL TURBULENT-BOUNDARY-LAYER EQUATIONS

We must face the fact that our chief success in turbulent-flow analysis lies with two-dimensional boundary layers. If we define a boundary layer as one in which there are large lateral changes and slow longitudinal changes in flow properties, this would include not only wall flows but also pipe flow, channel flow, wakes, and jets.

Let x be parallel to the freestream and y be normal to the wall, as usual. We assume the boundary-layer thickness $\delta(x) \ll x$, from which follow the same approximations as in laminar-boundary-layer analysis:

$$\bar{v} \ll \bar{u} \qquad \frac{\partial}{\partial x} \ll \frac{\partial}{\partial y} \tag{6-19}$$

In addition, we assume that the mean flow structure is two-dimensional

$$\bar{w} = 0 \qquad \frac{\partial}{\partial z} = 0 \tag{6-20}$$

The mean lateral turbulence is actually not zero, $\overline{w'^2} \neq 0$, but its z derivative is assumed to vanish. Then we can easily verify that the basic turbulent Eqs. (6-9), (6-13), and (6-15) reduce to the following boundary-layer approximations for incompressible flow:

Continuity:
$$\frac{\partial \bar{u}}{\partial x} + \frac{\partial \bar{v}}{\partial y} = 0 \tag{6-21a}$$

x momentum:
$$\bar{u}\frac{\partial \bar{u}}{\partial x} + \bar{v}\frac{\partial \bar{u}}{\partial y} \approx U_e \frac{dU_e}{dx} + \frac{1}{\rho}\frac{\partial \tau}{\partial y} \tag{6-21b}$$

Thermal energy:
$$\rho c_p \left(\bar{u}\frac{\partial \bar{T}}{\partial x} + \bar{v}\frac{\partial \bar{T}}{\partial y} \right) \approx \frac{\partial q}{\partial y} + \tau \frac{\partial \bar{u}}{\partial y} \tag{6-21c}$$

where $U_e(x)$ is the freestream velocity and where we have adopted the short notation

$$\begin{aligned} \tau &= \mu \frac{\partial \bar{u}}{\partial y} - \rho \overline{u'v'} \\ q &= k \frac{\partial \bar{T}}{\partial y} - \rho c_p \overline{v'T'} \end{aligned} \tag{6-22}$$

We see that Eqs. (6-21a) to (6-21c) closely resemble the laminar-flow equations from Chap. 4, except that q and τ contain a turbulent heat flux and turbulent shear stress, respectively, which must be modeled.

The y-momentum equation reduces to

$$\frac{\partial p}{\partial y} \approx -\rho \frac{\partial \overline{v'^2}}{\partial y}$$

which may be integrated across the boundary layer to yield

$$p \approx p_e(x) - \rho \overline{v'^2} \tag{6-23}$$

Thus, unlike laminar flow, there is a slight variation of pressure across the boundary layer due to velocity fluctuations normal to the wall. From Fig. 6-4, the rms v' fluctuations are no more than 4 percent of the stream velocity. Therefore the pressure differs from stream pressure by no more than about 0.4 percent of the freestream dynamic pressure, a negligible variation. Note that, due to the no-slip condition, the wall pressure equals the stream pressure.

The Bernoulli relation is assumed to hold in the (inviscid) freestream:

$$dp_e \approx -\rho U_e \, dU_e \tag{6-24}$$

This relation has already been used in Eq. (6-21b). It is assumed that the freestream conditions $U_e(x)$ and $T_e(x)$ are known, and the boundary conditions then become

No slip, no jump: $\quad \bar{u}(x, 0) = \bar{v}(x, 0) = 0 \quad \bar{T}(x, 0) = T_w(x)$

Freestream matching: $\quad \bar{u}(x, \delta) = U_e(x) \quad \bar{T}(x, \delta_T) = T_e(x)$ (6-25)

The velocity and thermal boundary-layer thicknesses (δ, δ_T) are not necessarily equal but depend upon the Prandtl number, as in laminar analyses. Equations (6-21a) and (6-21b) can be solved for $\bar{u}$ and $\bar{v}$ if a suitable correlation for the total shear τ is known; subsequently, the temperature $\bar{T}$ can be found from Eq. (6-21c) if the turbulent heat flux q can be correlated.

6-3.1 Turbulent Energy and Reynolds Stress

Since Eqs. (6-21) contain the two new unknowns, q and τ, they need supplementing for mathematical closure. One avenue is through the boundary-layer forms of the turbulence equations.

The turbulence kinetic-energy Eq. (6-17) has the two-dimensional boundary-layer form

$$\bar{u}\frac{\partial K}{\partial x} + \bar{v}\frac{\partial K}{\partial y} \approx -\frac{\partial}{\partial y}\left[\overline{v'\left(\tfrac{1}{2}u_i'u_i' + \frac{p'}{\rho}\right)}\right] + \frac{\tau}{\rho}\frac{\partial \bar{u}}{\partial y} - \epsilon \tag{6-26}$$

where $\quad \epsilon = -\nu \overline{\dfrac{\partial u_i'}{\partial x_j}\dfrac{\partial u_j'}{\partial x_i}} =$ turbulent dissipation

Unfortunately, this introduces two additional turbulence parameters, the pressure–strain term and the dissipation. It can be successfully modeled, however, as we will see later in this chapter.

The two-dimensional boundary-layer form of the Reynolds stress Eq. (6-18) is

$$\bar{u}\frac{\partial \overline{u'v'}}{\partial x} + \bar{v}\frac{\partial \overline{u'v'}}{\partial y} \approx 2\overline{u'v'}\frac{\partial \bar{u}}{\partial y} - \frac{\partial}{\partial y}\overline{\left(u'v'^2 + \frac{p'u'}{\rho}\right)} - 2\nu\,\overline{\frac{\partial u'}{\partial y}\frac{\partial v'}{\partial x}}$$

$$+ \overline{\frac{p'}{\rho}\left(\frac{\partial u'}{\partial y} + \frac{\partial v'}{\partial x}\right)} + \nu\frac{\partial^2 \overline{u'v'}}{\partial y^2} \qquad (6\text{-}27)$$

Again several new turbulence correlations are introduced that will need to be—and have been—successfully modeled.

6-3.2 Turbulent-Boundary-Layer Integral Relations

The integral-momentum, energy, etc., relations for turbulent flow may be derived by using continuity Eq. (6-21a) to eliminate $\bar{v}$ in favor of $\bar{u}$ and then integrating the result across the entire boundary layer. Alternately, one may use a control volume of width dx and height δ.

The integral-momentum equation was first derived by Kármán (1921) and has a form identical to the laminar-flow relation Eq. (4-122):

$$\frac{d\theta}{dx} + (2+H)\frac{\theta}{U_e}\frac{dU_e}{dx} = \frac{\tau_w}{\rho U_e^2} = \frac{C_f}{2} \qquad (6\text{-}28)$$

where
$$\theta = \text{momentum thickness} = \int_0^\infty \frac{\bar{u}}{U_e}\left(1 - \frac{\bar{u}}{U_e}\right)dy$$

$$H = \text{momentum shape factor} = \frac{\delta^*}{\theta}$$

$$\delta^* = \text{displacement thickness} = \int_0^\infty \left(1 - \frac{\bar{u}}{U_e}\right)dy$$

This equation contains three variables, θ, H, and C_f. In laminar flow, we can relate the three reasonably well with one-parameter velocity profile approximations such as Eq. (4-131). The turbulent-flow profile is, however, more complicated in shape, and many different correlations or additional relations have been proposed to effect closure of Eq. (6-28). There are over 50 different ideas in the literature.

In like manner, the two-dimensional turbulent integral-energy equation is identical to the laminar form, Eq. (4-124). For steady flow with impermeable walls, it may be written as

$$q_w = \frac{d}{dx}\left[\int_0^\infty \rho\bar{u}(\bar{h}_0 - \bar{h}_{0e})\,dy\right] \qquad (6\text{-}29)$$

where
$$\bar{h}_0 = \text{mean stagnation enthalpy} \approx c_p\bar{T} + \tfrac{1}{2}\bar{u}^2$$

Equation (6-29) can be used to develop approximate theories for turbulent heat convection by making suitable assumptions about the form of the (turbulent) velocity and temperature profiles.

Since the momentum-integral relation Eq. (6-28) contains unknown functions, it has often been supplemented by addition integral relations. One of these uses "mechanical energy" and is formed by multiplying Eq. (6-21b) by $\bar{u}$ and integrating across the layer. The result is the two-dimensional *mechanical-energy integral relation*, which takes the same form for either laminar or turbulent flow:

$$\frac{1}{2}\rho \frac{d}{dx}\left[U_e \int_0^\infty \bar{u}(U_e^2 - \bar{u}^2)\,dy\right] = \int_0^\infty \tau \frac{\partial \bar{u}}{\partial y}\,dy \qquad (6\text{-}30)$$

The right-hand side is called the *dissipation integral* and, for turbulent flow, contains new information that can be used for closure of a theory. Kline et al. (1968) discuss seven integral-boundary-layer methods that use this relation.

Finally, some workers use the integrated from of the continuity relation Eq. (6-21a), which becomes, for an impermeable wall,

$$v_e = -\int_0^\infty \frac{\partial \bar{u}}{\partial x}\,dy = -\frac{d}{dx}\left(\int_0^\infty \bar{u}\,dy\right) + U_e \frac{d\delta}{dx}$$

where Leibnitz' rule is used to evaluate the integral. By introducing δ^* from the definition of displacement thickness in Eq. (6-28), we may rewrite this as

$$\frac{d\delta}{dx} - \frac{v_e}{U_e} = \frac{1}{U_e}\frac{d}{dx}[U_e(\delta - \delta^*)] \qquad (6\text{-}31)$$

This expression represents the rate at which outer fluid is brought into the boundary layer. It is called the *entrainment relation*, valid for either laminar or turbulent flow, and was first derived by Head (1958). Kline et al. (1968) outline four turbulent-flow methods that use this relation.

Turbulent-boundary-layer integral methods were once the mainstay of design calculations, but now they have mostly been replaced by commercial CFD turbulence-modeling codes, assisted by large eddy simulation (LES) and direct numerical simulation (DNS). Integral methods are reviewed by Cousteix in Kline et al. (1982, vol. 2, pp. 650–671). Some excellent integral methods are also included in the Internet Applets of Devenport and Schetz (2002). The advantage of integral methods is easy implementation on a simple spreadsheet.

6-4 VELOCITY PROFILES: THE INNER, OUTER, AND OVERLAP LAYERS

Now let us consider velocity profiles $\bar{u}(x, y)$ for turbulent flow. First, what do they look like? Figure 6-8 shows some experimental profiles of $\bar{u}/U_e$ vs. y/δ for various pressure gradients, in turbulent flow of course. The profiles have a distinctive look about them that seems to spell analytic trouble. They are about as unlaminar as they can possibly be. They seem to smash up against the wall as if there were velocity

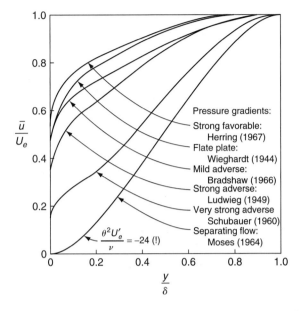

FIGURE 6-8
Experimental turbulent-boundary-layer velocity profiles for various pressure gradients. [*Data from Coles and Hirst (1968)*.]

slip at the wall; actually, they drop linearly to zero within a thickness too small to be seen. For the uppermost profile, for example, the linear (viscous) drop-off is in the region $0 \leq y/\delta < 0.002$. The profiles also seem to have a characteristic concavity near $y/\delta \approx 0.2$, as if a gremlin were sitting upon them. This is the beginning of the wakelike behavior of the outer (fully turbulent) layer.

For the key to profile shape, we are indebted to the physical insight of Ludwig Prandtl and Theodore von Kármán. They deduced that the profile consists of an inner and an outer layer, plus an intermediate overlap between the two:

Inner layer: viscous (molecular) shear dominates
Outer layer: turbulent (eddy) shear dominates
Overlap layer: both types of shear important; profile smoothly connects inner and outer regions

For the inner law, Prandtl reasoned in 1933 that the profile would depend upon wall shear stress, fluid properties, and distance y from the wall, but not upon freestream parameters:

Inner law: $\quad \bar{u} = fcn\,(\tau_w, \rho, \mu, y)$ (6-32)

Conversely, for the outer layer, Kármán deduced in 1930 that the wall acts merely as a source of retardation, reducing local velocity $\bar{u}(y)$ below the stream velocity U_e in a manner independent of viscosity μ but dependent upon wall shear stress, layer thickness, and freestream pressure gradient:

Outer law: $\quad U_e - \bar{u} = f\left(\tau_w, \rho, y, \delta, \dfrac{dp_e}{dx}\right)$ (6-33)

Finally, for the overlap layer, we simply specify that the inner and outer functions merge together smoothly over some finite intermediate region:

Overlap law: $\quad\quad\quad\quad\quad\quad \bar{u}_{\text{inner}} = \bar{u}_{\text{outer}} \quad\quad\quad\quad\quad\quad$ (6-34)

Interestingly, the physics of the problem determines the mathematical form of this overlap profile.

6-4.1 Dimensionless Profiles

The functional forms in Eqs. (6-32) to (6-34) are determined from experiment *after* use of dimensional analysis. These relations contain three primary dimensions (mass, length, and time). Therefore Eq. (6-32), with five variables, reduces to $5 - 3 = 2$ dimensionless parameters. The reader may show as an exercise that the proper dimensionless inner law is

Inner law: $\quad\quad \dfrac{\bar{u}}{v^*} = f\left(\dfrac{yv^*}{\nu}\right) \quad\quad v^* = \left(\dfrac{\tau_w}{\rho}\right)^{1/2} \quad\quad$ (6-35)

The variable v^* has units of velocity and is called the *wall-friction velocity*. We shall use v^* over and over again in our turbulent-flow analyses.

In like manner, Eq. (6-33) may be nondimensionalized as follows:

Outer law: $\quad\quad \dfrac{U_e - \bar{u}}{v^*} = g\left(\dfrac{y}{\delta}, \xi\right) \quad\quad \xi = \dfrac{\delta}{\tau_w}\dfrac{dp_e}{dx} \quad\quad$ (6-36)

This is often called the *velocity-defect law*, with $(U_e - \bar{u})$ being the "defect" or retardation of the flow due to wall effects. At any given position x, the defect shape $g(y/\delta)$ will depend upon the local pressure gradient ξ.

Let ξ have some particular value. Then the overlap function follows immediately by setting the inner and outer profiles equal in this intermediate region:

Overlap law: $\quad\quad \dfrac{\bar{u}}{v^*} = f\left(\dfrac{\delta v^*}{\nu}\dfrac{y}{\delta}\right) = \dfrac{U_e}{v^*} - g\left(\dfrac{y}{\delta}\right) \quad\quad$ (6-37)

This is supposedly an *identity*, yet the function f contains a multiplicative constant and the function g an additive constant. This is an elementary exercise in functional analysis. It can be true only if both f and g are *logarithmic* functions. Thus we conclude that in the overlap layer

Inner variables: $\quad\quad \dfrac{\bar{u}}{v^*} = \dfrac{1}{\kappa}\ln\dfrac{yv^*}{\nu} + B \quad\quad$ (6-38a)

Outer variables: $\quad\quad \dfrac{U_e - \bar{u}}{v^*} = -\dfrac{1}{\kappa}\ln\dfrac{y}{\delta} + A \quad\quad$ (6-38b)

where κ and B are near-universal constants for turbulent flow past smooth, impermeable walls and A varies with the pressure gradient ξ and perhaps with other parameters also. The original pipe-flow measurements in 1930 by Prandtl's student J. Nikuradse suggested that $\kappa \approx 0.40$ and $B \approx 5.5$, but later data correlations, e.g.,

Coles and Hirst (1968), use the values

$$\kappa \approx 0.41 \qquad B \approx 5.0 \qquad (6\text{-}39)$$

which we will adopt for use in this text.

The validity of the inner law, including the logarithmic overlap, is quickly established by replotting the velocity profiles from Fig. 6-8 in terms of the inner variables $u^+ = \bar{u}/v^*$ and $y^+ = yv^*/\nu$. This is done in Fig. 6-9, with smashing success. With the exception of the separating flow, all the curves, which looked so different in Fig. 6-8, now collapse into a single logarithmic relation in the overlap region $35 \leq y^+ \leq 350$, corresponding roughly to the range 2 percent $\leq y/\delta \leq 20$ percent, after which the curves either turn upward in the outer (wakelike) layer or turn down in the inner (viscous) layer.

The near-separating flow profile in Fig. 6-9 zooms to the left and up off the chart because v^* approaches zero and hence u^+ becomes very large and y^+ very small. Near separation, then, scaling based on wall shear stress is inappropriate. The outer law holds, but the inner and overlap layers become vanishingly small.

Note also from Fig. 6-9 that there is little to choose between (0.40, 5.5) or (0.41, 5.0) for the logarithmic-law constants κ and B. The latter values will be used here when numerical results are needed.

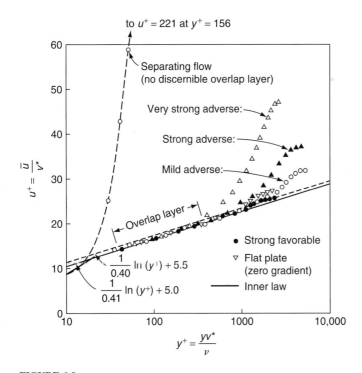

FIGURE 6-9
Replot of the velocity profiles of Fig. 6-8 using inner law variables y^+ and u^+.

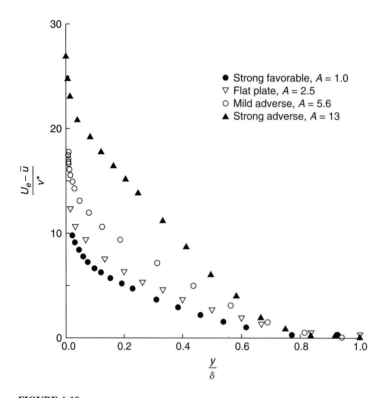

FIGURE 6-10
Replot of the velocity profiles of Fig. 6-8 using outer law variables from Eq. (6-36). Success is not evident because each profile has a different value of the parameter ξ.

The validity of the outer law, Eq. (6-36), is less quickly established by replotting the profiles from Fig. 6-8 using the outer variables of velocity defect vs. y/δ. The result is shown as Fig. 6-10. We are momentarily distressed by the fact that the profiles do not collapse into a single "universal" curve. The reason is that each profile has a different value of the pressure gradient parameter, ξ:

Strong favorable:	-4.8		Mild adverse:	$+6.3$
Flat plate:	0.0		Strong adverse:	$+29$

Therefore each shape is different and, as shown in Fig. 6-10, each has a different value of the constant A, so the overlap laws will not collapse together either. However, for a given ξ, a nearly unique profile will occur. For example, all flat-plate data would fall near the open-triangle points in Fig. 6-10.

6-4.2 Inner Layer Details: The Law of the Wall

From Fig. 6-9, we see that the inner law, Eq. (6-35), rises from no slip at the wall to merge smoothly, at about $y^+ \approx 30$, with the overlap log-law, Eq. (6-38a). Not shown in the figure is what happens very near the wall, where turbulence is damped out and the boundary layer is dominated by viscous shear. At very small y, the velocity profile is linear:

$y^+ \leq 5$: $\quad\quad\quad\quad\quad \tau_w = \dfrac{\mu \bar{u}}{y}$

or $\quad\quad\quad\quad\quad\quad\quad\quad\quad\quad\quad\quad\quad\quad\quad\quad\quad\quad\quad\quad$ (6-40)

$$u^+ = y^+$$

This (very thin) region near the wall is called the *viscous sublayer*. Its thickness is, by general agreement, $\delta_{sub} = 5\nu/v^*$, and the quantity (ν/v^*) is often called the *viscous length scale* of a turbulent boundary layer. For Wieghardt's flat-plate airflow data in Fig. 6-8, for example, with $v^* = 1.24$ m/s and $\nu_{air} \approx 1.51\text{E}{-}5$ m^2/s, $\delta_{sub} = 5(1.51\text{E}{-}5)/(1.24) \approx 0.06$ mm, or 500 times less then the 3 cm boundary-layer thickness.

Between $5 \leq y^+ \leq 30$, sometimes called the *buffer layer*, the velocity profile is neither linear nor logarithmic but is instead a smooth merge between the two. For decades, separate formulas were given for the sublayer, the buffer layer, and the log layer, until Spalding (1961) deduced a single composite formula that covered the entire wall-related region:

$$y^+ = u^+ + e^{-\kappa B}\left[e^{\kappa u^+} - 1 - \kappa u^+ - \dfrac{(\kappa u^+)^2}{2} - \dfrac{(\kappa u^+)^3}{6} \right] \quad (6\text{-}41)$$

This expression is an excellent fit to inner law data all the way from the wall to the point (usually for $y^+ > 100$) where the outer layer begins to rise above the logarithmic curve.

For decades, also, there were no mean-velocity data close enough to the wall to test the inner law, until Lindgren (1965), testing a smooth pipe with distilled water, provided the data shown in Fig. 6-11. The agreement with Spalding's formula Eq. (6-41) is excellent even after, for $y^+ > 300$, the outer law commences. This outer "wake" is very slight because fully developed pipe flow is a favorable gradient with $\xi \approx -2$, or about halfway between the "favorable" and "flat-plate" data in Fig. 6-9. Note the (acceptable) use in Fig. 6-11 of the older log constants $(\kappa, B) = (0.40, 5.5)$.

6-4.3 Outer Layer Details: Equilibrium Turbulent Flows

This writer regards the inner law as a complete success for smooth-wall turbulent flow, namely, Eq. (6-41) or its equivalent. The outer law, though, is sensitive to the pressure gradient parameter $\xi = (\delta/\tau_w)(dp_e/dx)$ and its variation with x, as shown by Eq. (6-36) or Fig. 6-10.

In two classic papers, Clauser (1954, 1956) developed the idea of specific cases, for which ξ is constant, which he termed *equilibrium turbulent flows*. Clauser replaced the fuzzily defined thickness δ by the rigorously defined displacement

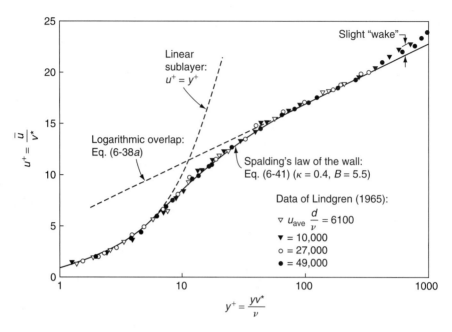

FIGURE 6-11
Comparison of Spalding's inner law expression with the pipe-flow data of Lindgren (1965).

thickness, so that the equilibrium pressure gradient parameter is now

$$\beta = \frac{\delta^*}{\tau_w}\frac{dp_e}{dx} \qquad \text{Clauser's equilibrium parameter} \qquad (6\text{-}42)$$

It turns out that constant-β flows correspond to a power-law freestream distribution $U_e = (\text{const})x^m$, exactly analogous to the laminar Falkner–Skan similarity flows of Sec. 4-3.4.

With considerable experimental effort, Clauser (1954) showed clearly that a boundary layer with variable $p_e(x)$ but constant β is in turbulent equilibrium in the sense that all the gross properties of that boundary layer can be scaled with a single parameter. The most relevant thickness parameter for equilibrium flow was determined by Clauser to be the *defect thickness* Δ

$$\Delta = \int_0^\infty \frac{U_e - \bar{u}}{v^*}\,dy = \delta^* \lambda \qquad (6\text{-}43)$$

where $\lambda = \sqrt{2/C_f}$ is a measure of the local skin friction (λ will be very useful in some approximate analyses which follow). Velocity profiles could be scaled with y/Δ, and a shape factor G that would remain constant in an equilibrium boundary layer was also defined by Clauser

$$G = \frac{1}{\Delta}\int_0^\infty \left(\frac{U_e - \bar{u}}{v^*}\right)^2 dy \qquad (6\text{-}44)$$

The ordinary Kármán-type shape factor, $H = \delta^*/\theta$, can be related to G as follows:

$$H = \left(1 - \frac{G}{\lambda}\right)^{-1} \qquad (6\text{-}45)$$

Since the skin friction varies with x, it follows that H is *not* constant in an equilibrium boundary layer.

There is no longer any doubt about the validity of Clauser's outer layer structural assumptions. Figure 6-12 shows the outer layer defect profile for a flat plate ($\beta = 0$), including data for rough walls. The data collapse beautifully about the eddy viscosity calculations of Mellor and Gibson. Also shown are the two equilibrium adverse gradients generated in the original experiments by Clauser (1954), again with excellent agreement. The two distributions are now commonly called Clauser **I** ($\beta \approx 1.8$, $G \approx 10.1$) and Clauser **II** ($\beta \approx 8.0$, $G \approx 19.3$). The equilibrium concept is even valid at the ultimate limit ($\beta \to \infty$), as measured, for example, by Stratford (1959). Also, the case of negative β (favorable gradient) is in equally good agreement with experiment, as shown by Herring and Norbury (1967), for example.

6-4.4 An Alternate View: Coles' Law of the Wake

From an application point of view, there are two difficulties with the outer layer approach of Clauser: (1) nonequilibrium flows deviate in shape from the "similarity" profiles of Fig. 6-12, and (2) even the equilibrium shapes have no simple analytical form to use in an engineering theory. These points were resolved by Coles (1956), who noted that the *deviations* or excess velocity of the outer layer above the log layer (see Fig. 6-9) have a wakelike shape when viewed from the freestream. If normalized by the maximum deviation at $y = \delta$, the data nearly collapse into a unique function of y/δ. In other words, Coles proposed that

$$\frac{u^+ - u^+_{\text{log-law}}}{U_e^+ - u^+_{\text{log-law}}(y = \delta)} \approx f\left(\frac{y}{\delta}\right)$$

where the *wake function* f is normalized to be zero at the wall and unity at $y = \delta$. Two popular curve fits are used for the S-shape we expect for the wake function:

$$f(\eta) \approx \sin^2\left(\frac{\pi}{2}\eta\right) \approx 3\eta^2 - 2\eta^3 \qquad (6\text{-}46)$$

The polynomial fit is easier to use in integral theories.

By adding the wake to the log-law, we have an accurate approximation of both the overlap and outer layers:

$$u^+ \approx \frac{1}{\kappa}\ln(y^+) + B + \frac{2\Pi}{\kappa}f\left(\frac{y}{\delta}\right) \qquad (6\text{-}47)$$

The quantity Π, called *Coles' wake parameter*, is directly related to the previously defined outer variable constant A, $\Pi = \kappa A/2$. It is also approximately related to

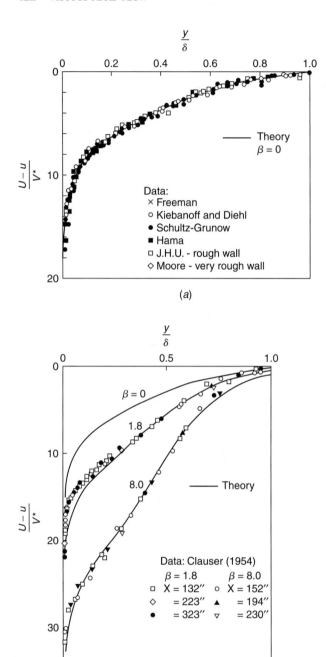

FIGURE 6-12
Equilibrium-defect profiles, as correlated by the Clauser parameter β and the theory of Mellor and Gibson (1966): (*a*) flat-plate data; (*b*) equilibrium adverse gradients.

Clauser's equilibrium parameter β (see Fig. 6-27). For equilibrium flows, Π should vary only with β.

The beauty of Eq. (6-47) is that it is a complete and reasonably accurate expression for any two-dimensional turbulent-boundary-layer profile, whether in equilibrium or not. If $y^+ < 30$, one should omit the wake and compute u^+ from Eq. (6-41). The efficacy of Eq. (6-47) is seen from the representative profiles plotted in Fig. 6-13, demonstrating the idea of adding an S-shaped wake function to the pure law of the wall ($\Pi = 0$). These are very realistic shapes and easy to compute. If normalized to u/U_e and replotted, they closely resemble the profiles shown in Fig. 6-8.

The wall–wake composite profile can be used in several types of turbulent shear-flow theories. Its admirable simplicity, for example, results in concise formulas for integral parameters. By integrating Eq. (6-47) across the boundary layer, we obtain

$$\frac{\delta^*}{\delta} \approx \frac{1+\Pi}{\kappa\lambda} \qquad \lambda = \left(\frac{2}{C_f}\right)^{1/2}$$

$$\frac{\theta}{\delta} \approx \frac{\delta^*}{\delta} - \frac{2 + 3.2\Pi + 1.5\Pi^2}{\kappa^2\lambda^2} \tag{6-48}$$

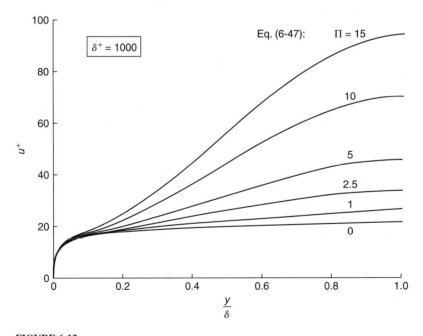

FIGURE 6-13
Turbulent velocity profiles computed from the Coles' wall–wake formula, Eq. (6-47), assuming $\delta^+ = 1000$. The curve for $\Pi = 0$ is the pure law of the wall from Eq. (6-41).

Similarly, the local skin-friction coefficient, $C_f = 2\tau_w/\rho U_e^2$, may be related to Π and local Reynolds number, $Re_\delta = U_e\delta/\nu$, by evaluating the wall–wake law Eq. (6-47) at the edge of the boundary layer:

$$\frac{U_e}{v^*} = \lambda = \left(\frac{2}{C_f}\right)^{1/2} = \frac{1}{\kappa}\ln\left(\frac{Re_\delta}{\lambda}\right) + B + \frac{2\Pi}{\kappa} \qquad (6\text{-}49)$$

We will use this simple algebraic approach for some practical turbulent shear-flow problems in the next sections. Huang and Bradshaw (1995) discuss the effect of pressure gradient.

6-4.5 The Alternative Power-Law Overlap: A Controversy

Fluids engineering is a straightforward field of study. Researchers usually make incremental gains and seldom argue, except about calculation methods. Thus, when Baranblatt et al. (1997) proposed a new nonlogarithmic formula for the overlap law, it became surprisingly controversial. Newspapers carried the headline "Law of the Wall Toppled." By this they meant the logarithmic-law of the wall, Eq. (6-38a), which had been accepted for 70 years. The reporter, Kathleen Stein, quoted the authors that "Engineering and aerodynamics textbooks, and design software will have to be rewritten." George and Castillo (1997) and others, notably Zagarola and Smits (1998), seconded the demise of the log-law. The alternative overlap formula, proposed for both pipe flow and flat-plate flow, is a *power law:*

$$u^+ \approx C(y^+)^\alpha$$

where $\qquad C \approx 3 + 0.62\ln(Re_\theta)$ and $\alpha \approx \dfrac{1.24}{\ln(Re_\theta)} \qquad (6\text{-}50)$

Unlike the classical log-law, with $\kappa \approx 0.41$ and $B \approx 5.0$, the constants C and α were acknowledged to vary with Reynolds number. The idea recalls the original observation in 1921 by Prandtl (1961, vol. II, pp. 620–626) that pipe-flow velocities approximated a one-seventh power law.

One might assume that the fluids community would now accept two viable overlap laws? Instead, the controversy escalated into fierce arguments at several international conferences. A good effort to resolve the debate was finally offered by Buschmann and Gad-el-Hak (2003), who analyzed 109 different zero-pressure-gradient data sets by both log and power formulas. They included newer data by Österlund (1999) which extended to $Re_\theta = 27{,}000$, much higher than data for which the log-law had been matched. They concluded that the two laws gave comparable agreement and neither was statistically superior. The power law is poorer near the wall, and the log-law is poorer near the wake region. Figure 6-14 shows the best fit values of log-law and power-law "constants" as a function of Reynolds number. We see that all coefficients vary slightly with Re_θ. It is generally agreed that the log-law should be valid as $Re_\theta \to \infty$, where Buschmann and Gad-el-Hak (2003) suggest the limiting values $\kappa \approx 0.38$ and $B \approx 4.08$. It is not known yet whether this neutral-observer

INCOMPRESSIBLE TURBULENT MEAN FLOW **425**

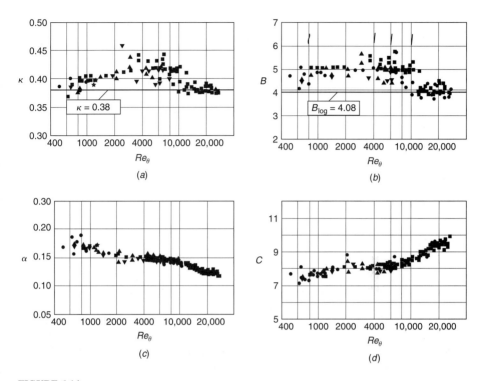

FIGURE 6-14
Overlap law constants as correlated by Buschmann and Gad-el-Hak (2003): (a, b) log-law, Eq. (6-38a); (c, d) power law, Eq. (6-50). [*Reprinted by permission of the American Institute of Aeronautics and Astronautics.*]

arbitration will satisfy both log-lovers and power-law lovers. The writer has enjoyed being a log-lover and remains so, encouraged by the wide ranging logarithm-friendly boundary-layer data compiled by Fernholz and Finley (1996).

6-5 TURBULENT FLOW IN PIPES AND CHANNELS

It is interesting that, for all the discussions and graphs given in Sec. 6-4, there was actually no theory. No equations were solved. All of the velocity profile formulas were correlations, i.e., the results of inspired and physically meaningful dimensional analysis. We can now proceed from these algebraic expressions to solve simple problems, without any extensive use of "theory." Here we discuss developed flow in ducts.

6-5.1 The Circular Pipe

Consider pipe flow first, assumed far downstream of the entrance, where velocity does not depend upon axial distance x. The laminar-flow analysis, Sec. 3-3.1, gave a parabolic velocity distribution, Eq. (3-34), plus a friction factor $C_f = 16/Re_D$, which was illustrated in Fig. 3-7. Laminar flow becomes unstable at about $Re_D = 2000$ and fully turbulent flow ensues at $Re_D \approx 4000$. At higher Reynolds numbers, the data in Fig. 3-7 follow a curve labeled "Blasius," which is not a theory but a curve fit to smooth-wall data collected in 1913 by Prandtl's student H. Blasius:

$$C_f(\text{pipe}) = \frac{2\tau_w}{\rho u_{av}^2} \approx 0.0791 \, Re_D^{-1/4} \qquad 4000 < Re_D < 10^5 \qquad (6\text{-}51)$$

The Blasius formula was one of the first modern applications of the technique of dimensional analysis. The formula is seen to have a limited range of applicability.

A better result follows from the wall law, Eq. (6-41) or (6-38a). Pipe flow, being a favorable gradient, should resemble the upper velocity profile in Fig. 6-8. Let the pipe radius be a, and let the wall coordinate be

$$y = a - r \qquad dy = -dr$$

With the profile known, we may evaluate the average pipe velocity:

$$u_{av} = \frac{Q}{A} = \frac{1}{\pi a^2} \int_0^a \bar{u} \, 2\pi \, r \, dr = \frac{1}{a^2} \int_0^a \bar{u} \, 2(a - y) \, dy \qquad (6\text{-}52)$$

Figure 6-11 shows clearly that turbulent pipe flow has very little wake, that is, $\Pi \approx 0$. Therefore, the law of the wall is accurate all the way across the pipe. Further, neglect the (very thin) viscous sublayer and substitute the simple log-law, Eq. (6-38a), with the result

$$u_{av} = v^* \left(\frac{1}{\kappa} \ln \frac{av^*}{\nu} + B - \frac{3}{2\kappa} \right) \qquad (6\text{-}53)$$

Now, from the definition of pipe-friction factor, $C_f = 2\tau_w/\rho u_{av}^2$, the following identities hold:

$$\frac{u_{av}}{v^*} = \left(\frac{2}{C_f} \right)^{1/2} \qquad \frac{av^*}{\nu} = Re_D \left(\frac{C_f}{8} \right)^{1/2} \qquad Re_D = \frac{2au_{av}}{\nu}$$

This piques our interest, revealing that Eq. (6-53) is actually a friction factor relation. Introducing base-10 logarithms (a tradition) plus $\kappa = 0.41$ and $B = 5.0$, we may clean up Eq. (6-53) as follows:

$$\frac{1}{\Lambda^{1/2}} = 1.99 \log_{10}(Re_D \Lambda^{1/2}) - 1.02$$

where $\Lambda = 4C_f$ is the Darcy friction factor. This formula was derived by Prandtl in 1935. Since he neglected the sublayer and the wake, Prandtl slightly adjusted the constants to fit pipe-friction data better, especially at lower Reynolds numbers. The final formula is

$$\frac{1}{\Lambda^{1/2}} = 2.0 \log_{10}(Re_D \Lambda^{1/2}) - 0.8 \qquad (6\text{-}54)$$

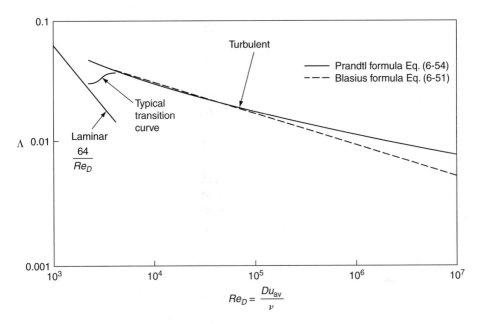

FIGURE 6-15
Analytical friction factor formulas for fully developed pipe flow. Compare with the data in Fig. 3-7.

valid for smooth-wall turbulent pipe flow for any Reynolds number greater than 4000. It supplants the Blasius correlation Eq. (6-51).

Figure 6-15 shows the laminar and turbulent friction factor formulas plotted versus Reynolds number. Note that the Blasius formula Eq. (6-51) falls too low for $Re_D > 10^5$. However, we do learn something about trends from the Blasius formula by rewriting it in terms of wall shear stress:

$$\tau_w \approx 0.0396 \rho^{3/4} u_{av}^{7/4} \mu^{1/4} D^{-1/4} \qquad (6\text{-}55)$$

Thus, in turbulent pipe flow, wall shear rises nearly linearly with density, nearly quadratically with velocity, very weakly with viscosity, and drops weakly with pipe size.

The transition curve in Fig. 6-15 is typical of the change from laminar to turbulent flow between $Re_D = 2000$ and 4000. This region is rather uncertain, and the flow pulsates as turbulent "slugs" propagate through the pipe. It should be avoided in the design of a pipe-flow system.

6-5.2 Channel Flow between Parallel Plates

Developed laminar flow between parallel plates was solved as Eq. (3-44), resulting in a parabolic velocity distribution $u(y)$ and a friction factor $C_f = 6/Re_h$, where h is the half-width between plates. The geometry is shown in Fig. 6-16 and illustrates the contrast between laminar and turbulent velocity profiles. Also shown is the total

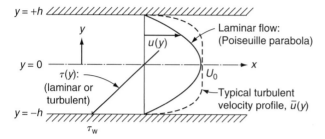

FIGURE 6-16
Fully developed laminar and turbulent flow in a channel.

shear-stress distribution $\tau(y)$, which, like pipe flow, drops linearly from the wall value to zero at the centerline.

Again the wake is small and the sublayer negligible, so it is satisfactory to use the log-law Eq. (6-38a) to represent the velocity profile across the entire channel. The wall variable is $Y = (h - y)$, and the average velocity may be computed approximately from the log-law:

$$u_{av} = \frac{1}{h}\int_0^h \bar{u}\, dY \approx v^*\left(\frac{1}{\kappa}\ln\frac{hv^*}{\nu} + B - \frac{1}{\kappa}\right)$$

Since, by definition, $u_{av}/v^* = (8/\Lambda)^{1/2}$, this is a friction factor relation. But a channel is not "round," so what is the proper Reynolds number? One obvious answer is to use the hydraulic diameter concept from Eq. (3-55):

$$Re_{D_h} = \frac{u_{av}D_h}{\nu} \qquad D_h = \frac{4A}{P} = 4h \qquad (6\text{-}56)$$

Introducing this into the expression for u_{av}, using base-10 logarithms, and cleaning it all up, we obtain, for smooth-wall turbulent channel flow,

$$\frac{1}{\Lambda^{1/2}} = 2.0\log_{10}(Re_{D_h}\Lambda^{1/2}) - 1.19 \qquad (6\text{-}57)$$

This is quite close to the pipe relation, Eq. (6-54), but predicts Λ slightly higher, +7 percent at $Re = 10^5$ decreasing to +4 percent at $Re = 10^8$.

Note that, in using the law of the wall to analyze pipe and channel flow, evaluation of u_{av} immediately yields the final result. No differential equations are solved, and no real "theory" is involved.

6-5.3 The Effective Diameter for Turbulent Noncircular Duct Flow

The channel-flow relation Eq. (6-57) is close enough to the pipe-flow result that it substantiates the common engineering practice [e.g., White (2003, p. 376)] of computing noncircular duct friction by using the hydraulic diameter D_h and the pipe-friction relation Eq. (6-54). Indeed, experiments with turbulent flow through triangular, square, rectangular, and annular ducts [see Schlichting (1979, Fig. 20.12)] show only a few percent error when this scheme is adopted. Thus the hydraulic diameter approximation is much better for turbulent than for laminar flow (recall Fig. 3-13 for comparison).

A further improvement is obtained by modifying D_h using laminar duct theory for the same cross section. For example, Eq. (6-57) for the channel will resemble the circular pipe law Eq. (6-54) if rewritten as

$$\frac{1}{\Lambda^{1/2}} \text{(channel)} = 2.0 \log_{10}(0.64 Re_{D_h} \Lambda^{1/2}) - 0.8 \qquad (6\text{-}58)$$

Best agreement between channel and pipe is predicted when one uses $(0.64 D_h)$ as the *effective diameter* of the channel. Now, in laminar flow, C_f (pipe) $= 16/Re_D$, and C_f(channel) $= 24/Re_{D_h}$, and the ratio of these two is $16/24 = 0.667$, quite near the prediction of 0.64. This is no accident. A general rule for estimating turbulent friction in noncircular ducts is to use the pipe-friction law Eq. (6-54) based on an effective Reynolds number

$$Re_{D_{\text{eff}}} = \frac{u_{\text{av}} D_{\text{eff}}}{\nu}$$

where $$D_{\text{eff}} = D_h \frac{16}{(C_f Re_{D_h})_{\text{laminar}}}$$ (6-59)

This idea was proposed and proven experimentally by O. C. Jones in tests on rectangular [Jones (1976)] and concentric annular ducts [Jones and Leung (1981)]. One could take $(C_f Re)_{\text{laminar}}$ for the given section from the results in Sec. 3-3.3. The method should work well for any squatty or blocky cross section, i.e., with no unusually thin regions. An experiment by Eckert and Irvine (1957) on a 12° isoceles triangle gave poorer results, because the flow in the 12° corner remained laminar up to surprisingly high Reynolds numbers. Obot (1988) reviews the entire subject of noncircular ducts, including small apex angles and rough walls. While agreeing with the effective diameter scheme, Obot goes further and proposes a "critical friction" or transition-oriented method for correlating the data in any duct.

6-5.4 Turbulent Flow in Rough Pipes

The previous formulas are valid only for smooth-wall turbulent duct flow. Wall roughness has little influence upon laminar flow. In turbulent flow, however, even a small roughness will break up the thin viscous sublayer and greatly increase the wall friction. Denote the average roughness height by k. Add this to our list of parameters and the wall law and friction law become, respectively,

$$u^+ = fcn(y^+, k^+)$$

where $$k^+ = \frac{kv^*}{\nu}$$

$$\Lambda = fcn\left(Re_D, \frac{k}{D}\right)$$

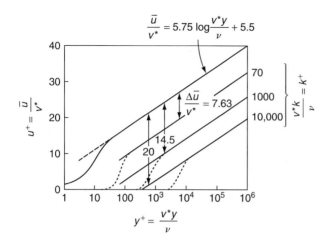

FIGURE 6-17
Experimental rough-pipe velocity profiles, showing the downward shift ΔB of the logarithmic overlap layer.

The logarithmic overlap layer still exists but, as k^+ increases, the intercept B begins to move downward monotonically and the effective "wall" position begins to move outward:

$$u^+_{\text{overlap}} = \frac{1}{\kappa} \ln \frac{yv^*}{\nu} + B - \Delta B(k^+) \qquad (6\text{-}60)$$

This is illustrated in Fig. 6-17; the downward shift is quite large, with a correspondingly large friction increase. Kármán's constant $\kappa \approx 0.41$ does not change with roughness. Unfortunately, ΔB is not a unique function of k^+ but varies somewhat with the *type* of roughness (uniform sand, sand mixtures, rivets, threads, spheres, etc.). This is seen in Fig. 6-18, taken from a compilation by Clauser (1956). There is quite a variation at small k^+, but in all cases the asymptotic variation at large k^+ is logarithmic with slope $1/\kappa$. The *outer* or defect layer is not affected by wall roughness (the data in Fig. 6-12a include rough and very rough surfaces!).

The dashed line in Fig. 6-18 is the classic Prandtl–Schlichting sand-grain roughness curve. For $k^+ < 4$, there is no roughness effect, and for $k^+ > 60$, ΔB is logarithmic. This defines the three roughness regimes:

$$\begin{aligned} k^+ < 4: &\quad \text{hydraulically smooth wall} \\ 4 < k^+ < 60: &\quad \text{transitional-roughness regime} \\ k^+ > 60: &\quad \text{fully rough flow (no } \mu \text{ effect)} \end{aligned} \qquad (6\text{-}61)$$

To derive a pipe-friction formula with sand-grain roughness, we may curve fit the dashed curve in Fig. 6-18 with the following expression:

$$\Delta B_{\text{sand grains}} \approx \frac{1}{\kappa} \ln(1 + 0.3k^+) \qquad (6\text{-}62)$$

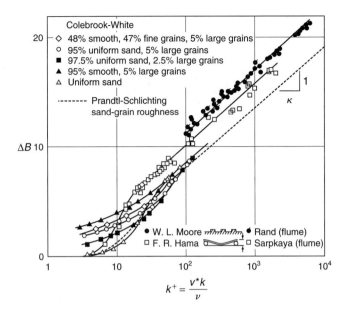

FIGURE 6-18
Composite plot of the profile-shift parameter $\Delta B(k^+)$ for various roughness geometries, compiled by Clauser (1956).

When this is substituted in Eq. (6-60), it reduces for large $k^+ > 60$ to

Fully rough flow: $$u^+ = \frac{1}{\kappa} \ln\left(\frac{y}{k}\right) + 8.5 \quad (6\text{-}63)$$

In this condition, there is no effect of viscosity on the velocity profile.

Similarly, the introduction of $(B - \Delta B)$ into the pipe-friction formula Eq. (6-53) yields, after cleaning up and introducing base-10 logarithms,

$$\frac{1}{\Lambda^{1/2}} \approx 2.0 \log\left[\frac{Re_D\sqrt{\Lambda}}{1 + 0.1(k/D)Re_D\sqrt{\Lambda}}\right] - 0.8 \quad (6\text{-}64)$$

This formula should be a good representation of sand-grain pipe friction over the entire turbulent-flow regime. The denominator term shows that if

$$\left(\frac{k}{D}\right)Re_D < 10: \quad \text{roughness unimportant}$$

$$\left(\frac{k}{D}\right)Re_D > 1000: \quad \text{fully rough (independent of } Re_D)$$

Commercial pipes have roughness somewhat different from sand-grain behavior, so C. F. Colebrook in 1939 devised a formula, different from Eq. (6-64), which

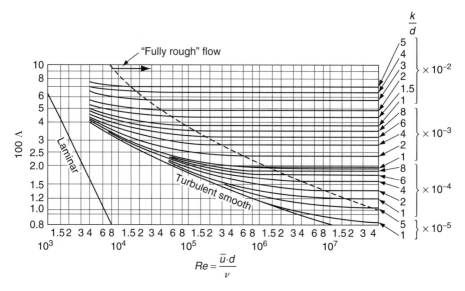

FIGURE 6-19
Friction factors for turbulent flow in rough pipes. [*After Moody (1944)*.]

was plotted by Moody (1944) and shown in Fig. 6-19. The Colebrook interpolation formula is

$$\frac{1}{\Lambda^{1/2}} \approx -2.0 \log_{10}\left(\frac{k/D}{3.7} + \frac{2.51}{Re_D \sqrt{\Lambda}}\right) \quad (6\text{-}65)$$

The plot of this expression, now celebrated as the *Moody chart* for commercial pipe friction, is shown in Fig. 6-19. The pipe roughness height may be estimated from the list given in Table 6-1.

TABLE 6-1
Roughness of commercial pipes

Type of pipe	Equivalent roughness k_s, ft
Glass	0.000001
Drawn tubing	0.000005
Steel, wrought iron	0.00015
Asphalted cast iron	0.0004
Galvanized iron, new	0.0005
3 years old	0.0009
Cast iron	0.00085
Wood stave	0.0006–0.003
Concrete	0.001–0.01
Riveted steel	0.003–0.03

The Colebrook–Moody formula is implicit, i.e., if Re_D and k/D are known, one has to iterate to compute Λ. This annoyance is avoided by instead using a clever explicit formula proposed by Haaland (1983):

$$\frac{1}{\Lambda^{1/2}} \approx -1.8 \log_{10}\left[\frac{6.9}{Re_D} + \left(\frac{k/D}{3.7}\right)^{1.11}\right] \qquad (6\text{-}66)$$

This function may be solved immediately for the friction factor and varies by less than ±2 percent from Eq. (6-65).

6-6 THE TURBULENT BOUNDARY LAYER ON A FLAT PLATE

The problem of flow past a sharp flat plate at high Reynolds numbers has been extensively studied, and numerous formulas have been proposed for the friction factor. The studies vary in sophistication from curve fits of data, use of the Kármán integral relation and/or the law of the wall, up to finite-difference computer studies using models of turbulent shear (Sec. 6-7).

6-6.1 Momentum-Integral Analysis

An integral approach is entirely adequate for this problem. The pressure gradient is zero, and the momentum-integral relation Eq. (6-28) reduces to

$$C_f = 2\frac{d\theta}{dx} \qquad (6\text{-}67)$$

Since this is an equilibrium flow ($\beta = 0$), the velocity profile is well approximated by the wall–wake law Eq. (6-47) with $\Pi \approx 0.45$. Evaluation of the profile function Eq. (6-47) at $y = \delta$ then gives

$$\frac{U_e}{v^*} = \left(\frac{2}{C_f}\right)^{1/2} = \frac{1}{\kappa}\ln\frac{\delta v^*}{\nu} + B + \frac{2\Pi}{\kappa} \approx 2.44 \ln\left[Re_\delta\left(\frac{C_f}{2}\right)^{1/2}\right] + 7.2$$

By analogy with Eq. (6-53), this is a friction relation between C_f and Re_δ. It is messy algebraically, so we plug in a few values for C_f from 0.001 to 0.005 and come up with an excellent power-law curve-fit approximation:

$$C_f \approx 0.020 Re_\delta^{-1/6} \qquad (6\text{-}68)$$

This takes care of the left-hand side of Eq. (6-67).

To evaluate momentum thickness without undue algebra, Prandtl suggested in 1921 [Prandtl (1961, vol. II, pp. 620–626)] that a simple one-seventh power-law profile, taken from pipe data, would suffice:

$$\frac{\bar{u}}{U_e} \approx \left(\frac{y}{\delta}\right)^{1/7}$$

hence
$$\frac{\theta}{\delta} \approx \frac{7}{72} \qquad (6\text{-}69)$$

Substitution of Eqs. (6-68) and (6-69) in Eq. (6-67) leads to the simple first-order ordinary differential equation

$$0.020 Re_\delta^{-1/6} \approx 2\frac{d}{dx}\left(\frac{7\delta}{72}\right) = \frac{7}{36}\frac{dRe_\delta}{dRe_x}$$

and, after integration assuming $\delta = 0$ at $x = 0$, the solution is

$$Re_\delta \approx 0.16 Re_x^{6/7} \qquad \frac{\delta}{x} \approx \frac{0.16}{Re_x^{1/7}} \qquad C_f \approx \frac{0.027}{Re_x^{1/7}} \qquad (6\text{-}70)$$

These simple power-law expressions are in good agreement with turbulent flat-plate data and are recommended for general use.

Slightly more detailed would be the direct use of the wall–wake momentum thickness result, Eq. (6-48), with $\kappa = 0.41$ and $\Pi \approx 0.45$:

$$\frac{\theta}{\delta} \approx \frac{3.54}{\lambda} - \frac{22.21}{\lambda^2}$$

where

$$\lambda = \left(\frac{2}{C_f}\right)^{1/2} \qquad (6\text{-}71)$$

Use of this expression in Eq. (6-67) leads to results that are not any more accurate than Eq. (6-70). We give this analysis as a problem exercise.

Meanwhile, Prandtl in 1927 gave the power-law result

$$\frac{\delta}{x} \approx \frac{0.37}{Re_x^{1/5}} \qquad C_f \approx \frac{0.058}{Re_x^{1/5}} \qquad (6\text{-}72)$$

These formulas though often quoted in the literature, were developed from very limited low Reynolds number data and are not very accurate.

6-6.2 Flat-Plate Computation Using Inner Variables

A very interesting alternate approach was suggested by Kestin and Persen (1962). The idea is to assume that the streamwise velocity $\bar{u}$ in the boundary layer is correlated by inner variables:

$$\frac{\bar{u}(x,y)}{v^*(x)} = u^+ \approx f(y^+) \qquad y^+ = \frac{yv^*(x)}{\nu} \qquad (6\text{-}73)$$

The wall function $f(y^+)$ could be Spalding's formula Eq. (6-41)—it is not necessary to specify at this point. The wake component $(\Pi W/\kappa)$ is neglected, but this is not strictly necessary. Then the normal velocity follows by straightforward integration of the continuity equation:

$$\bar{v} = -\int_0^y \frac{\partial \bar{u}}{\partial x} dy = -\frac{\nu}{v^*}\int_0^{y^+} \frac{\partial}{\partial x}(v^* u^+)\, dy^+ = -\frac{\nu}{v^*}\frac{dv^*}{dx} u^+ y^+ \qquad (6\text{-}74)$$

Kestin and Persen (1962) substituted these velocity assumptions directly in the streamwise boundary-layer momentum equation, with $dU_e/dx = 0$:

$$\overline{u}\frac{\partial \overline{u}}{\partial x} + \overline{v}\frac{\partial \overline{u}}{\partial y} = \frac{1}{\rho}\frac{\partial \tau}{\partial y}$$

The result can be expressed entirely in terms of inner variables:

$$v^* \frac{dv^*}{dx} u^{+2} \approx \frac{v^*}{\mu} \frac{\partial \tau}{\partial y^+}$$

Now integrate this expression across the entire boundary layer, from $y^+ = 0$ to $y^+ = \delta^+$, noting that $\tau = 0$ at $y = \delta$. The result is

$$\tau_w = \rho v^{*2} = -\mu \frac{dv^*}{dx} G(\lambda)$$

(6-75)

where
$$G = \int_0^{\delta^+} u^{+2} dy^+$$

where, as in Eq. (6-71), $\lambda = (2/C_f)^{1/2}$. This is a first-order differential equation for the distribution of wall-friction velocity $v^*(x)$, or, equivalently, $\tau_w(x)$. Before integrating, replace v^* by the dimensionless variable $\lambda = U_e/v^*$. The result can be integrated once again:

$$\frac{U_e}{\nu} = G(\lambda) \frac{d\lambda}{dx}$$

or
(6-76)

$$Re_x = \frac{U_e x}{\nu} = \int_0^\lambda G(\zeta) d\zeta$$

where we have assumed that the turbulent boundary layer begins at $x = 0$. Kestin and Persen evaluated the integral from Spalding's wall formula Eq. (6-41), with a somewhat algebraically involved but very accurate result:

$$G(\lambda) = \int_0^\lambda u^{+2} dy^+ = \tfrac{1}{3}\lambda^3 + \frac{e^{-\kappa B}}{\kappa^2}\left[e^Z(Z^2 - 2Z + 2) - 2 - \frac{Z^3}{3} - \frac{Z^4}{4}\right]$$

$$Re_x = \int_0^\lambda G\, d\lambda$$

(6-77)

$$= \frac{1}{12}\lambda^4 + \frac{e^{-\kappa B}}{\kappa^3}\left[e^Z(Z^2 - 4Z + 6) - 6 - 2Z - \frac{Z^4}{12} - \frac{Z^5}{20}\right]$$

where $Z = \kappa\lambda$. Since $\lambda = (2/C_f)^{1/2}$, Eq. (6-77) is the desired flat-plate wall-friction law for turbulent flow. However, it is implicit in C_f and therefore awkward to evaluate. White (1969) noted that the function G, in its practical range $20 < \lambda < 40$, is well approximated by an exponential:

$$G(\lambda) \approx 8.0 e^{0.48\lambda}$$

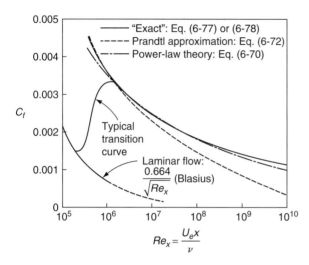

FIGURE 6-20
Local skin friction on a smooth flat plate for turbulent flow, showing several theories.

Introducing this approximation into Eq. (6-76), integrating, and rearranging, we obtain the desired explicit formula:

Flat plate:
$$C_f \approx \frac{0.455}{\ln^2(0.06 Re_x)} \qquad (6\text{-}78)$$

This simpler expression is within ±1 percent of Eq. (6-77) and is recommended as a more or less "exact" relation for flat-plate turbulent skin friction.

Figure 6-20 shows laminar and turbulent theories for $C_f(Re_x)$, with transition occurring around $Re_x \approx 5\text{E}5$. We see that Prandtl's traditional power-law expression Eq. (6-72) is not accurate. The newer power-law theory Eq. (6-70) is quite accurate—so why bother with the more complicated inner variable theory? The answer is that this latter theory can be readily extended to variable pressure gradients, as we shall see.

Other formulas for flat-plate skin friction, of comparable accuracy, abound in the literature. See, for example, Chap. XXI, part a, of Schlichting (1979). Note that the formulas above assume turbulent flow beginning at $x = 0$. If there is laminar flow and then transition, one should interpolate to find the "effective" turbulent point of origin.

6-6.3 Total Friction Drag of a Flat Plate

Let F be the total drag force per unit width on one side of a plate of length L and width b. As in laminar flow, F may be calculated by integration of the local wall shear stress:

$$F = \int_0^L \tau_w b\, dx$$

or
$$C_D = \frac{F}{\frac{1}{2}\rho U_e^2 L b} = \int_0^1 C_f\, d\!\left(\frac{x}{L}\right) \qquad (6\text{-}79)$$

where C_D is called the *drag coefficient*. In this particular case of a thin plate at zero incidence, there is no *form drag* because all pressure forces are normal to the freestream.

Introduction of our power law, Eq. (6-70), into Eq. (6-79) and subsequent integration reveals that C_D is one-sixth larger than the trailing-edge value of C_f:

$$C_{D,\text{plate}} \approx 0.031 Re_L^{-1/7} = \tfrac{7}{6} C_f(L) \tag{6-80}$$

This is a good approximation over the whole Reynolds number range in turbulent smooth-wall flow.

Integration of the log-squared law Eq. (6-78) is analytically vexing but numerically satisfying. The result can be approximated as

$$C_{D,\text{plate}} \approx 1.15 C_f(L) \approx \frac{0.523}{\ln^2(0.06 Re_L)} \tag{6-81}$$

This relation is quite accurate and approximately equivalent to Eq. (6-80). Again we can find a plethora of alternate formulas in the literature, but we will take Eq. (6-81) as our standard.

6-6.4 Turbulent Flow Past a Rough Plate

Suppose now that the plate has a uniform average roughness height k, and we wish to know the wall-friction variation $C_f(Re_x, k/x)$. The inner variable approach works for this case also, if we assume that, say, the sand-grain correlation Eq. (6-62) holds locally in the boundary layer:

$$u^+ = \frac{\bar{u}(x,y)}{v^*(x)} \approx \frac{1}{\kappa}\ln(y^+) + B - \frac{1}{\kappa}\ln(1 + 0.3k^+)$$

Substitution in the streamwise boundary-layer momentum equation yields

$$v^*\frac{dv^*}{dx}\left[u^{+2} - \frac{0.3k^+}{\kappa(1+0.3k^+)}\left(u^+ - \frac{1}{\kappa}\right)\right] = \frac{v^*}{\mu}\frac{\partial \tau}{\partial y^+}$$

Following the same procedure as Sec. 6-6.3, we integrate this relation twice, nondimensionalize, and rearrange as the final result:

$$Re_x \approx 1.73(1 + 0.3k^+)e^Z\left[Z^2 - 4Z + 6 - \frac{0.3k^+}{1+0.3k^+}(Z-1)\right] \tag{6-82}$$

where $\quad Z = \kappa\lambda \quad \lambda = \left(\frac{2}{C_f}\right)^{1/2} \quad$ and $\quad k^+ = \frac{kv^*}{\nu} = \frac{Re_x(k/x)}{\lambda}$

This result, although implicit in all three variables (Re_x, C_f, k/x), is valid for uniform sand-grain roughness over the complete range of hydraulically smooth, transitional, and fully rough walls in turbulent flat-plate flow. Some numerical values of C_f are plotted in Fig. 6-21 for various (x/k) and Re_x. These values may not be accurate for other types of roughness elements such as spheres or rivets.

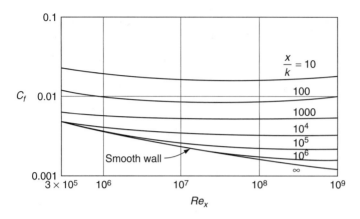

FIGURE 6-21
Local skin friction on a uniformly rough plate, from Eq. (6-82).

As k^+ becomes large, the effect of the Reynolds number vanishes and the flow becomes fully rough. An approximation of Eq. (6-82) for this case is

Fully rough: $\quad C_f \approx \left[1.4 + 3.7\log_{10}\left(\dfrac{x}{k}\right)\right]^{-2} \quad$ for $\dfrac{x}{k} > \dfrac{Re_x}{1000}$ \hfill (6-83)

This is in good agreement with experiments on sand-roughened plates by Nikuradse (1933), for which Schlichting (1979, p. 654) recommends the following curve-fit relations:

Fully rough: $\quad C_f \approx \left[2.87 + 1.58\log_{10}\left(\dfrac{x}{k}\right)\right]^{-2.5}$ \hfill (6-84a)

$$C_D \approx \left[1.89 + 1.62\log_{10}\left(\dfrac{L}{k}\right)\right]^{-2.5} \hfill (6\text{-}84b)$$

These formulas would not be too accurate for other than sand-grain types of roughness.

A general review of rough-wall effects is given by Raupach et al. (1991). The roughness log-law shift ΔB has been employed in turbulence modeling but is correctly criticized by Patel (1998), who points out "the uncertainty in the dependence of ΔB on the size and type of roughness and the effective location of the fictitious wall." Schlichting and Gersten (2000, Chap. 17), classify a number of roughness element shapes. For uniform periodic roughness, such as ribs and grooves, CFD results by Grégoire et al. (2003) show that the fictitious wall ($y = 0$) is close to the crests of the elements. An alternate approach called the *discrete-element model*, Taylor et al. (1985, 1988), makes direct calculations of the form drag and blockage of individual elements. Finally, although the computational effort is large and the Reynolds numbers low, it is now possible to simulate rough-wall flows by large-eddy simulation (LES), Lee (2002), and by direct numerical simulation, Miyake et al. (2001).

6-6.5 Turbulent Flow with Continuous Wall Suction or Blowing

An effect as strong as roughness is that of nonzero normal velocity at the wall: $v_w < 0$ (suction) or $v_w > 0$ (blowing). This adds a strong streamwise convective acceleration $v_w(\partial \overline{u}/\partial y)$ to the near-wall boundary layer. If we assume zero pressure gradient (flat-plate) flow, the momentum equation very near the wall becomes

$$\rho v_w \frac{\partial \overline{u}}{\partial y} \approx \frac{\partial \tau}{\partial y}$$

or (6-85)

$$\tau \approx \tau_w + \rho v_w \overline{u}$$

We see that wall transpiration changes the shear distribution significantly. By matching Eq. (6-85) to an "eddy-viscosity" model of turbulent shear (see Sec. 6-7), Stevenson (1963) derived the following modification of the logarithmic-law of the wall with suction or blowing:

$$\frac{2}{v_w^+}\left[(1 + v_w^+ u^+)^{1/2} - 1\right] \approx \frac{1}{\kappa}\ln(y^+) + B \quad (6\text{-}86)$$

where $v_w^+ = v_w/v^*$. We assign this derivation as a problem exercise. When $v_w = 0$ this expression reduces to the impermeable wall law Eq. (6-38a). The typical range of v_w^+ is ±0.06.

Figure 6-22 shows some law of the wall profiles plotted from Eq. (6-86)—the sublayer is not shown. Recalling that $C_f = 2v^{*2}/U_e^2$, we see that a small amount of wall transpiration causes a large friction change. For example, if $\delta^+ = 1000$, the impermeable wall friction is $C_{f0} \approx 0.0042$. For moderate blowing, $v_w^+ = +0.02$, $C_f \approx 0.0034$, or 19 percent less; for equivalent suction, $v_w^+ \approx -0.02$, $C_f \approx 0.0053$, or 26 percent more.

Figure 6-22, though satisfying physically, is not an exact correlation. Schetz (1984, pp. 151–155) criticizes both Stevenson's law Eq. (6-86) and a comparable law proposed by Simpson (1968) on the grounds that they do not collapse all available porous flat-plate data. One problem is that real porous walls are rough, at least slightly, so multiple effects occur. However, Schetz' own data suggest that even a very smooth porous wall will *increase* skin friction over the impermeable case. Thus, if one is trying to reduce friction (or heat transfer) by blowing, a certain amount of blowing is needed just to bring one back to the nonporous condition.

Suction and blowing have little effect on the outer defect law, so one can add, say, the Coles' wake function Eq. (6-47) to Eq. (6-86) to simulate the full velocity profile of a transpired boundary layer.

Stevenson's algebraic formula Eq. (6-86) is quite adequate for engineering estimates of turbulent wall suction and blowing, and also of course for homework assignments. Further details of suction/blowing effects can be found in CFD studies. Turbulence modeling yields good results, as reported, for example, by Sofialidis and Primos (1996, 1997). Direct numerical simulation (DNS), limited to

440 VISCOUS FLUID FLOW

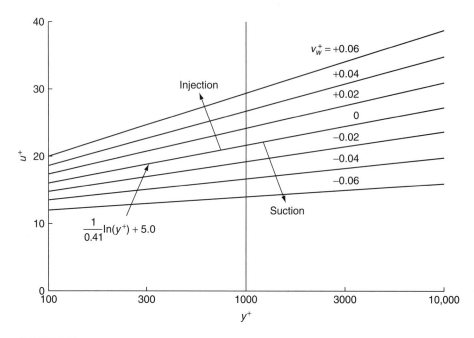

FIGURE 6-22
Illustration of the effect of suction and blowing on the law of the wall, using the correlation of Stevenson (1963), Eq. (6-86).

low-turbulence Reynolds numbers, can assist in model improvement and flow visualization, Sumitani and Kasagi (1995) and Kim et al. (2002).

6-7 TURBULENCE MODELING

Very quickly we run out of simple turbulent shear-flow cases for which we can use the wall–wake law Eq. (6-47), plus a bit of algebra, to wrap up the whole problem. What are we to do for arbitrarily variable (1) pressure gradient, (2) wall roughness, or (3) blowing and suction? What can we do if asked for more details than just the wall friction, for example: (1) the complete velocity profile, (2) the turbulent shear stress, or (3) the rms turbulent fluctuations? The answers lie in turbulence modeling.

Turbulence modeling has become a mature field and has spawned many recent monographs, notably Wilcox (1998), Chen and Jaw (1997), Durbin and Pettersson (2001), Piquet et al. (2001), Launder and Sandham (2001), Gross and Burchard (2002), and Cebeci (2003). The Sixth International Symposium on Engineering Turbulence Modeling and Measurements will be held in May 2005. Many commercial CFD turbulence codes are now available and are being (slowly) merged with computer-aided design (CAD) systems, Thilmany (2003). The present chapter merely highlights modeling concepts and leaves CFD programs for advanced reading.

There is a hierarchy of turbulence models, ranging from the simplest algebraic correlations to full-blown unsteady Navier–Stokes treatment. Wilcox (1998) gives an excellent discussion. Here the term "*n-Equation*" model means that, in addition to the time-mean continuity and momentum relations, *n* time-mean partial differential equations have been added. The six different classifications, in order of increasing complexity, are as follows:

1. *Zero-equation* model: simply adds algebraic eddy-viscosity formulas to the system.
2. *One-equation* model: adds the time-mean equation of either (1) turbulence kinetic energy or (2) eddy viscosity to the system, plus some algebraic formulas to model its various terms. This idea has not been too successful but is still used today.
3. *Two-equation* model: adds turbulent kinetic energy and a second partial differential equation, usually involving time-mean turbulence dissipation, to the system, plus more algebraic modeling formulas.
4. *Second-moment closure* model: This is the most complex time-mean turbulence model. It skips the turbulent kinetic-energy equation, keeps the dissipation equation, and models the full Reynolds-stress relation, Eq. (6-18). Proper modeling here is still a research topic.
5. *Large-eddy simulation* (LES): Carries out full unsteady Navier–Stokes calculations for motion scales of the order of the grid size or larger, then uses a turbulence model for the smaller sub-grid-scale motions. The system is computationally intensive and limited to moderate Reynolds numbers. See, for example, the review by Lesieur and Métais (1996) or the monographs by Sagaut and Germano (2002), Geurts (2003), or Volker (2003).
6. *Direct numerical simulation* (DNS): Attacks the full unsteady Navier–Stokes equations for all turbulence scales and thus uses no model at all. It is limited to low-turbulent Reynolds numbers. A very important use of DNS is to provide details for improving turbulence models. See the review by Moin and Mahesh (1998) or the monographs by Baritaud (1996) or Geurts (2003).

6-7.1 Zero-Equation Models: The Eddy Viscosity

In two-dimensional turbulent-boundary-layer flow, the only additional unknown in the momentum equation is turbulent shear $(-\rho \overline{u'v'})$. The traditional modeling assumption, following J. Boussinesq in 1877, is to make this a gradient diffusion term, analogous to molecular shear:

$$\tau_t = -\rho \overline{u'v'} = \mu_t \frac{\partial \overline{u}}{\partial y} \qquad (6\text{-}87)$$

where μ_t = eddy viscosity

The eddy viscosity μ_t has the same dimensions as μ, but it is *not* a fluid property, varying instead with flow conditions and the geometry. (It depends upon turbulent eddies, to put it colloquially.) In like manner, we will define an eddy conductivity later for turbulent heat transfer.

The first thing to notice is that μ_t is positive, i.e., the shear correlation $(\overline{u'v'})$ is negative. To explain this, consider a position y in shear flow where $(\partial \overline{u}/\partial y)$ is positive. An eddy coming down ($v' < 0$) to position y will generally bring with it a higher streamwise velocity ($u' > 0$). Similarly, a slower eddy moving up from below ($v' > 0$) will bring a decrement in velocity ($u' < 0$). Thus the majority of eddy motions are associated with a negative correlation $(\overline{u'v'})$.

The second point is that gradient diffusion of eddies is only an approximation, but it is a good one. In a typical boundary layer, $\overline{u}(y)$ rises monotonically to U_e, hence $(\partial \overline{u}/\partial y)$ is positive throughout and approaches zero in the freestream, where τ_t approaches zero also. The whole profile supports a plausible (positive) correlation between τ_t and $(\partial \overline{u}/\partial y)$. There are some profiles, however, that have a local maximum (or minimum), e.g., a wall jet (or an asymmetrical wake). The turbulent shear changes sign, and the measured values of τ_t are not necessarily exactly zero where $(\partial u/\partial y)$ is zero. The trend is right, though, and the use of the eddy-viscosity concept should not cause any serious problem in such cases.

6-7.1.1 MIXING-LENGTH THEORY. A very popular approach is the *mixing-length* concept of Prandtl (1925), who, by analogy with kinetic theory, proposed that each turbulent fluctuation could be related to a length scale and a velocity gradient:

$$-\overline{u'v'} \approx (\text{const})u'_{\text{rms}}v'_{\text{rms}} \approx (\text{const})\left(l_1 \frac{\partial \overline{u}}{\partial y}\right)\left(l_2 \frac{\partial \overline{u}}{\partial y}\right)$$

where the scales l_1 and l_2 are called *mixing lengths*, which represent some mean eddy size much larger than the fluid's mean-free path. Now, for convenience, replace $(\text{const})(l_1 l_2)$ by a single scale ℓ^2, and compare Prandtl's idea with Eq. (6-87). The result is

Mixing-length model: $\quad \mu_t \approx \rho \ell^2 \left|\dfrac{\partial \overline{u}}{\partial y}\right| \quad$ (6-88)

The model is complete if we can relate the mixing length ℓ to the flow conditions. The primary effect is the distance y from the wall. Prandtl and Kármán took turns with these estimates and arrived at the following:

In the sublayer: $\quad \ell \approx y^2 \quad$ (6-89a)

In the overlap layer: $\quad \ell \approx \kappa y \quad$ (6-89b)

In the outer layer: $\quad \ell \approx \text{constant} \quad$ (6-89c)

It is possible to merge all three of these conditions into one single magnificent composite function. However, most workers merge only (*a*) and (*b*) and then transfer to (*c*) at some matching condition. The most popular composite for (*a*) and (*b*) is due to van Driest (1956a), who added a "damping factor" to condition (b) to take care of (a):

$$\ell_{a,b} \approx \kappa y \left[1 - \exp\left(-\frac{y^+}{A}\right) \right] \qquad A \approx 26 \text{ for flat-plate flow} \qquad (6\text{-}90)$$

where, as usual, $y^+ = yv^*/\nu$. The van Driest *damping factor* in brackets [ ] is derived from oscillating laminar flow near a wall, Eq. (3-111). The dimensionless damping constant A varies with flow conditions such as pressure gradient, wall roughness, and blowing or suction. For a smooth, impermeable wall with zero pressure gradient, $A \approx 26$.

In the outer layer, condition Eq. (6-89c) is satisfied by relating mixing length to boundary-layer thickness. The most popular model is

$$\ell_{\text{outer layer}} \approx 0.09\delta \qquad (6\text{-}91)$$

This same idea of mixing length proportional to layer thickness is used in the analysis of jets, wakes, and free-shear layers. The crossover point between Eqs. (6-90) and (6-91) occurs where $\kappa y \approx 0.09\delta$, or $y_{\text{match}} \approx 0.22\delta$.

These models are very effective, that is, introducing

$$\tau = (\mu + \mu_t)\frac{\partial \overline{u}}{\partial y}$$

into the momentum Eq. (6-21b), with μ_t calculated from Eqs. (6-90) and (6-91), will give an excellent prediction for the complete velocity profile: linear sublayer, buffer layer, log layer, and wavelike outer layer.

6-7.1.2 MODELING WITHOUT THE MIXING LENGTH. It is possible to attack eddy viscosity directly without the mixing-length concept, in both inner and outer layers. For example, the inner layer model which Spalding (1961) used to derive Eq. (6-41)—see Fig. 6-11—was

$$\mu_t \approx \mu \kappa e^{-\kappa B}\left[e^Z - 1 - Z - \frac{Z^2}{2} \right] \qquad Z = \kappa u^+ = \frac{\kappa \overline{u}}{v^*} \qquad (6\text{-}92)$$

There is no mixing length or velocity gradient, yet the agreement with the data in Fig. 6-11 is excellent.

Similarly, a popular outer layer formulation by Clauser (1956) is

$$\mu_{t,\text{outer}} \approx C\rho U_e \delta^* \qquad C \approx 0.016 \qquad (6\text{-}93)$$

or, equivalently, $\mu_t/\mu \approx 0.016 Re_{\delta^*}$. This type of model is also used in computation of jets, wakes, and free-shear layers.

One can mix and match also: Eqs. (6-90) and (6-93) are the basis of the widely used Cebeci–Smith model outlined in their text [Cebeci and Smith (1974)] and at the Stanford 1968 Conference [Kline et al. (1968, pp. 346–355)]. They modified the outer law to account for intermittency near the edge of the boundary layer (see Fig. 6-5a):

$$\mu_{t,\text{outer}} \approx \frac{0.016\rho U_e \delta^*}{[1 + 5.5(y/\delta)^6]} \qquad (6\text{-}94)$$

Figure 6-23 illustrates these composite models, Eqs. (6-92) plus (6-93) or Eq. (6-94), for three different local Reynolds numbers. We see that μ_t is linear in the inner region—except for a nearly invisible region near the origin where damping reduces μ_t to a cubic function of y. Although Eqs. (6-88) and (6-92) have completely different formulations and parameters, both reduce in the inner layer to the linear relation

$$\mu_t \approx \kappa \rho v^* y$$

or $\qquad (6\text{-}95)$

$$\frac{\mu_t}{\mu} \approx \kappa y^+$$

except, as noted, in the sublayer region.

In the outer layer, Fig. 6-23 shows that, depending upon the local Reynolds number, μ_t is tens or even hundreds of times greater than the molecular viscosity.

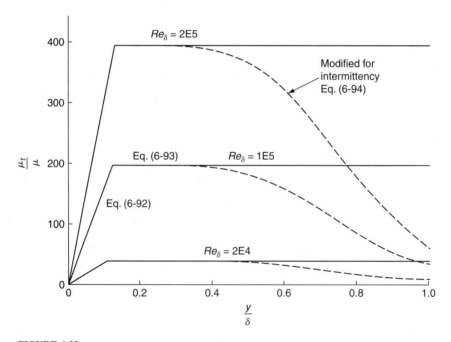

FIGURE 6-23
Eddy-viscosity distribution in a turbulent boundary layer computed from the inner law Eq. (6-92) and outer law Eq. (6-93). An expanded view near the origin ($y/\delta < 0.01$) would be needed to show the cubic damping effect.

A turbulent shear layer basically has a high outer and low inner effective viscosity, which is why turbulent velocity profiles are so steep at the wall and so flat further out.

A display of μ_t by the mixing-length formulas Eqs. (6-90) and (6-91) would be similar, but not identical, to Fig. 6-23. Either approach—or a mixture of the two—gives a good prediction of a turbulent-boundary-layer velocity profile.

6-7.1.3 APPLICATION TO FLAT-PLATE FLOW.
When the pressure gradient is zero, the total shear stress is constant near the wall, and one can immediately compute the overlap layer from the definition of eddy viscosity and, say, the mixing-length model Eq. (6-90):

$$\tau \approx \tau_w = (\mu + \mu_t)\frac{\partial \overline{u}}{\partial y} \approx \rho \kappa^2 y^2 \left|\frac{\partial \overline{u}}{\partial y}\right|\frac{\partial \overline{u}}{\partial y} = \rho v^{*2}$$

where we have neglected $\mu \ll \mu_t$. Taking the square root, separating and integrating this relation, we obtain

$$\frac{\overline{u}}{v^*} = \frac{1}{\kappa}\ln\left(\frac{yv^*}{\nu}\right) + B$$

The overlap law Eq. (6-38a) is thus reproduced faithfully by the mixing-length theory, which was probably Prandtl's goal in proposing the model. We have no "boundary condition" for the constant $B \approx 5.0$, so an experiment would be needed. If we add in the molecular viscosity term and the van Driest damping term from Eq. (6-90), then we may integrate all the way from the wall, using the no-slip condition. The result, after considerable algebra, is

$$u^+ = \int_0^{y^+} \frac{2\,dy^+}{1 + \{1 + 4\kappa^2 y^{+2}[1 - \exp(-y^+/A)]^2\}^{1/2}} \quad (6\text{-}96)$$

If we choose the particular value $A = 26$, the entire inner law—sublayer, buffer layer, and overlap—is reproduced accurately with the correct log-law constant $B \approx 5.0$. If we choose the wrong A, we get the wrong B. If we add other effects, e.g., pressure gradient, to the analysis, we have to modify A as in the next section.

In like manner, if we choose, instead, Spalding's expression Eq. (6-92) for μ_t and keep the molecular viscosity term, integration will yield Spalding's law of the wall Eq. (6-41), which is also a complete expression for the inner law. Both approaches are very successful for impermeable flat-plate flow.

6-7.1.4 MODIFICATION FOR PARAMETRIC EFFECTS.
The eddy-viscosity model has to be modified for pressure gradients or blowing or suction. Early efforts centered about changing only the damping constant A in Eq. (6-90). Later it was realized that the linear variation (κy) itself needed changing because the total shear

stress was no longer constant near the wall. With a pressure gradient or nonzero wall velocity, the variation of shear stress near the wall is, to first order,

$$\tau \approx \tau_w + \frac{dp_e}{dx} y + \rho v_w \bar{u} + \cdots \quad (6\text{-}97)$$

The changes in slope near the wall are substantial. Thomas and Hasani (1989) extended Eq. (6-97) into an interpolation formula for $\tau(y)$ which is reasonably accurate across the entire boundary layer:

$$\frac{\tau}{\tau_w} \approx 1 + \xi\eta + \varphi\frac{\bar{u}}{U_e} - (3 + 2\xi + 3\varphi)\eta^2 + (2 + \xi + 2\varphi)\eta^3 \quad (6\text{-}98)$$

where $\quad \eta = \dfrac{y}{\delta} \quad \xi = \dfrac{\delta}{\tau_w}\dfrac{dp_e}{dx} \quad$ and $\quad \varphi = v_w^+ U_e^+$

This relation is plotted in Fig. 6-24 for an impermeable wall ($\varphi = 0$) and three simple cases: (1) flat plate, $\xi = 0$; (2) a moderate adverse pressure gradient, $\xi = +4$; and (3) a strong favorable gradient, $\xi = -3$. The point of the figure is to illustrate the strong effect of pressure gradient on shear stress. To account for these effects, Galbraith et al. (1977) show convincingly that Eq. (6-90) should be modified by the square root of the shear:

$$\ell_{inner} \approx \kappa y \left(\frac{\tau}{\tau_w}\right)^{1/2}\left[1 - \exp\left(-\frac{y^+}{A}\right)\right] \quad (6\text{-}99)$$

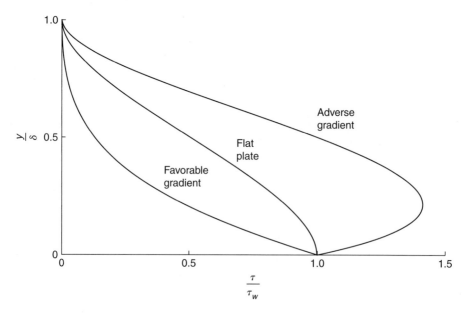

FIGURE 6-24
Illustration of the effect of a pressure gradient on the shear-stress distribution in a turbulent boundary layer. Plotted from Eq. (6-98) for an impermeable wall ($\varphi = 0$).

This gets the profile started well for any pressure gradient. To merge into the log layer smoothly, one needs to modify A. Granville (1989) suggested that the smoothest merge (with the log-law intercept *and* its slope) was accomplished by the following damping-constant variation:

$$A \approx \frac{26}{(1 + bp^+)^{1/2}} \qquad p^+ = \frac{\nu}{\rho v^{*3}} \frac{dp_e}{dx} \qquad (6\text{-}100)$$

where $b = 12.6$ if $p^+ > 0$ and $b = 14.76$ if $p^+ < 0$. Equations (6-99) and (6-100) are recommended for general use in turbulent boundary layers with impermeable walls.

6-7.1.5 THE BALDWIN–LOMAX MODEL. When using the previous eddy-viscosity models, iteration is always required, even when marching downstream with, say, an explicit finite-difference procedure. The chief reason is that the inner law is correlated with wall shear stress τ_w or its equivalent, v^*, which is unknown. A second reason is that the outer law formulas also contain unknowns: δ for the mixing-length model Eq. (6-91) and δ^* for the Clauser model Eq. (6-93). To avoid this second iteration, Baldwin and Lomax (1978) proposed an entirely different model for the outer law, eliminating δ or δ^* in favor of a certain maximum function occurring in the boundary layer. In their notation,

$$\mu_{t,\text{outer}} \approx \frac{0.016 C_{cp} \rho y_{\max} F_{\max}}{1 + 5.5(C_{\text{kleb}} y/y_{\max})^6} \qquad (6\text{-}101)$$

where

$$F_{\max} = \max\left[y \left| \frac{\partial \bar{u}}{\partial y} \right| (1 - e^{-y^+/A}) \right]$$

and where $y_{\max}$ is the value of y corresponding to $F_{\max}$. The parameters C_{cp} and C_{kleb} were suggested to be constants of the order of unity. This model has become especially popular in the aerospace literature. However, it results in discrepancies with other generally accepted outer layer relations, such as Coles' law of the wake.

Granville (1987) suggested that the "constants" should really be variable and fit them to known properties of Coles' wake law and equilibrium pressure gradients. His suggested correlations are as follows:

$$C_{\text{kleb}} \approx \frac{2}{3} - \frac{0.01312}{0.1724 + \beta^*} \qquad \beta^* = \frac{y_{\max}}{v^*} \frac{dU_e}{dx}$$

$$C_{cp} \approx \frac{3 - 4 C_{\text{kleb}}}{2 C_{\text{kleb}}\left(2 - 3 C_{\text{kleb}} + C_{\text{kleb}}^3\right)} \qquad (6\text{-}102)$$

It is thought that this modification will give the Baldwin–Lomax model accuracy comparable to the mixing-length and Clauser iterative models.

6-7.2 Higher Order Modeling

The eddy-viscosity approach adds *zero* equations and is a direct algebraic correlation for Reynolds stresses in the momentum equation. It can be pruned and tuned for various types of flows, but it has only one type of output: the mean velocities $\bar{u}$ and $\bar{v}$ and the turbulent shear, $(\overline{-u'v'})$. It cannot compute the turbulent energy K or any of its fluctuating components $(u', v', w')_{rms}$. Additional partial differential equations are needed. Modeling of the fluctuations began with Prandtl in 1945 and has taken many paths: kinetic energy, turbulent dissipation, turbulence length scale, vorticity fluctuations, dissipation timescale, pressure–strain correlations, and various stress relations. There are excellent monographs on turbulence modeling, such as Wilcox (1998), Piquet et al. (2001), and Cebeci (2003).

Turbulence modeling with CFD was in its infancy at the seminal 1968 Stanford Conference on Turbulent Boundary Layers, Kline et al. (1968). Now, nearly 40 years later, hundreds of modeling papers and their applications are published each year. The most popular two-equation models are available in commercial CFD codes and are widely used. One can contrast the many CFD vendors by perusing any issue of *Mechanical Engineering Magazine*. Eight different commercial codes are compared in an interesting paper by Freitas (1995).

6-7.2.1 ONE-EQUATION MODEL: TURBULENT KINETIC ENERGY. For convenience let us repeat here the two-dimensional boundary-layer form of the turbulent kinetic-energy equation:

$$\bar{u}\frac{\partial K}{\partial x} + \bar{v}\frac{\partial K}{\partial y} \approx -\frac{\partial}{\partial y}\left[\overline{v'\left(\frac{1}{2}u_i'u_i' + \frac{p'}{\rho}\right)}\right] + \frac{\tau}{\rho}\frac{\partial \bar{u}}{\partial y} - \epsilon \quad (6\text{-}26)$$

where $K = \overline{(u_i' u_i')}/2$ is the kinetic energy and ϵ is the turbulent dissipation with units of power per unit mass or (velocity)^3/(length).

Let L be a turbulence length scale or effective eddy size. Such an eddy would have a velocity scale $K^{1/2}$. Its dissipation then should scale, by dimensional reasoning, as

$$\epsilon \approx (\text{const})\frac{K^{3/2}}{L} \quad (6\text{-}103)$$

This makes sense physically. If an eddy of size L is moving with speed $K^{1/2}$, its power dissipated per unit mass should be proportional to

$$\frac{\text{drag} \times \text{velocity}}{\text{mass}} \approx \frac{[(\text{const})\rho K L^2]K^{1/2}}{(\text{const})\rho L^3} \approx (\text{const})\frac{K^{3/2}}{L}$$

The second or "production" term on the right-hand side of Eq. (6-26) has already been modeled by eddy viscosity, $\tau = (\mu + \mu_t)(\partial \bar{u}/\partial y)$. The first or "convective diffusion" term on the right-hand side is vexing because it cannot be measured in a boundary layer. The reason is that, although small flush-mounted sensors do measure *wall* pressure fluctuations, there is presently no instrument which measures fine-scale fluctuations p' in the fluid field itself. By analogy with turbulent shear stress, it is reasonable to assume that this term is akin to gradient diffusion, i.e.,

$$\overline{-v'\left(\frac{1}{2}u_i'u_i' + \frac{p'}{\rho}\right)} \approx (\text{const})\frac{\partial K}{\partial y}$$

Then our generic model of the turbulent kinetic-energy equation is

$$\bar{u}\frac{\partial K}{\partial x} + \bar{v}\frac{\partial K}{\partial y} \approx \frac{\partial}{\partial y}\left[(\text{const})\frac{\partial K}{\partial y}\right] + \nu_t\left(\frac{\partial \bar{u}}{\partial y}\right)^2 - (\text{const})\frac{K^{3/2}}{L} \quad (6\text{-}104)$$

This relation is to be coupled with the continuity and momentum Eqs. (6-21a) and (6-21b). Closure is not obtained until we model the length scale L. Three of the differential methods at the 1968 Stanford Conference [Kline et al. (1968)] used Eq. (6-104) with algebraic correlations for L. The results were satisfactory but, apparently, no better than the best zero-equation methods, which merely used a model for eddy viscosity. This fact, plus the extreme difficulty of extending a length-scale correlation to complex flows, has meant that a one-equation model is no longer popular.

6-7.2.2 TWO EQUATIONS: THE K–ϵ MODEL. The turbulent kinetic-energy Eq. (6-104) performs better if coupled with a second equation modeling the rate of change of either (1) dissipation ϵ or (2) turbulent length scale L. Dissipation is very popular, perhaps because of a seminal paper by Jones and Launder (1972), which—to this writer, at least—opened up the field of turbulence modeling to users as opposed to specialists. They modeled the complete dissipation equation [found in Tennekes and Lumley (1972)] in a manner similar to Eq. (6-104). The resulting two models are given for fully elliptical (nonboundary-layer) high Reynolds number flow as

Energy: $\quad \dfrac{DK}{Dt} \approx \dfrac{\partial}{\partial x_j}\left(\dfrac{\nu_t}{\sigma_K}\dfrac{\partial K}{\partial x_j}\right) + \nu_t \dfrac{\partial \bar{u}_i}{\partial x_j}\left(\dfrac{\partial \bar{u}_i}{\partial x_j} + \dfrac{\partial \bar{u}_j}{\partial x_i}\right) - \epsilon \quad (6\text{-}105a)$

Dissipation: $\quad \dfrac{D\epsilon}{Dt} \approx \dfrac{\partial}{\partial x_j}\left(\dfrac{\nu_t}{\sigma_\epsilon}\dfrac{\partial \epsilon}{\partial x_j}\right) + C_1\nu_t \dfrac{\epsilon}{K}\dfrac{\partial \bar{u}_i}{\partial x_j}\left(\dfrac{\partial \bar{u}_i}{\partial x_j} + \dfrac{\partial \bar{u}_j}{\partial x_i}\right) - C_2\dfrac{\epsilon^2}{K} \quad (6\text{-}105b)$

where σ_K and σ_ϵ are effective "Prandtl numbers" which relate eddy diffusion of K and ϵ to the momentum eddy viscosity: $\sigma_K = \nu_t/\nu_K$ and $\sigma_\epsilon = \nu_t/\nu_\epsilon$. The eddy viscosity itself is modeled as follows:

$$\nu_t \approx \frac{C_\mu K^2}{\epsilon} \quad (6\text{-}106)$$

The five empirical constants in these model relations have the following recommended values for attached boundary-layer calculations:

$$C_\mu = 0.09 \quad C_1 = 1.44 \quad C_2 = 1.92 \quad \sigma_K = 1.0 \quad \sigma_\epsilon = 1.3 \quad (6\text{-}107)$$

These values are, unfortunately, *not universal* but have to be modified for other problems such as jets and wakes and recirculating flows.

Equations (6-105) to (6-107), combined with continuity and momentum Eqs. (6-21a, b), form the complete K–ϵ model for analysis of two-dimensional turbulent shear flow. Except for flat-plate flow, to which the constants in Eq. (6-107) were tuned, accuracy is often only fair. Thus other researchers have proposed different, but related, two-equation models:

(1) The $K-\epsilon$ model using Renormalization Group methods, Yakhot and Smith (1992).
(2) The $K-\omega$ model, Prandtl (1945a), Wilcox (1998).
(3) The $K-L$ model, Rotta (1986).
(4) The $K-\omega^2$ model, Wilcox and Rubesin (1980).

In the same spirit is a $K-\epsilon$ model with two additional equations, i.e., a *four-equation* model:

(5) The $K-\epsilon-v^2-f$ model, sometimes simply called the v'^2-f model, Durbin (1995).

In this fifth model, v^2 is the fluctuating normal velocity, and f is its rate of production. Recall that L is the turbulence length scale, and ω is defined as the dissipation per unit turbulent kinetic energy, as modeled by the relation $\omega = \epsilon/(KC_\mu)$. Both L and ω also have their own partial differential equations to join the equation for K. Details are given in the text by Wilcox (1998). The five models above are generally more accurate than the standard $K-\epsilon$ model for adverse pressure gradients, low Reynolds numbers, and separated flows, as shown by Wilcox (1998). But the standard model is still widely used and is qualitatively satisfactory.

A stringent case for testing these models was reported by Iaccarino (2001) for flow through an asymmetrical two-dimensional diffuser with a flat upper wall and a lower wall expanding at a 10° angle. The Reynolds number, based on inlet channel height, was 20,000, for which experimental data are available from Obi et al. (1993). Iaccarino used a grid of 124 streamwise and 65 transverse nodes, packed especially closely near the walls. Figure 6-25 shows the computed isovelocity contours for (1) the v'^2-f model, and (2) the $K-\epsilon$ model with low Reynolds number corrections. The v'^2-f model shows a long thin separation bubble in the corner (the dashed lines)

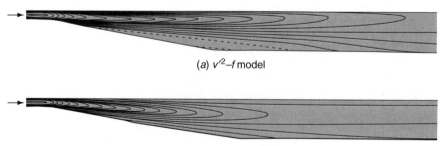

FIGURE 6-25
Turbulence modeling isovelocity contours of an asymmetrical diffuser with a 10° lower wall expansion at $Re \approx 20{,}000$: (*a*) the v'^2-f model correctly predicts the experimental separation bubble (dashed lines); (*b*) the $K-\epsilon$ model fails to predict separation but is qualitatively reasonable. [*From Iaccarino (2001), with permission of the American Society of Mechanical Engineers.*]

in excellent agreement with the experiment. The K–ϵ model shows no separation whatever, a recurrent defect, but is otherwise qualitatively correct.

6-7.2.3 WALL FUNCTIONS AND LOW REYNOLDS NUMBER MODIFICATIONS.
Most standard two-equation models are for "high Reynolds numbers." Note that the K–ϵ model Eqs. (6-105 to 6-106) do not contain the molecular viscosity μ. The model is meant to be applied in the fully turbulent region away from any viscosity-dominated near-walls. Two options are possible for the near-wall regions: (1) use *wall-function* inner law formulas that patch into the first computational node in the overlap region, or (2) add viscosity and damping terms to the model itself and use a fine mesh right up to the wall.

The wall-function formulas are essentially log-law relations tailored to the K–ϵ–ω parameters. For example, if y_P denotes the grid point nearest the wall, we could specify

$$\frac{\bar{u}_P}{v^*} = \frac{1}{\kappa} \ln\left(\frac{v^* y_P}{\nu}\right) + B$$

$$K_P = \frac{v^{*2}}{C_\mu^{1/2}}$$

$$\epsilon_P = \frac{v^{*3}}{\kappa y_P} \quad (6\text{-}108)$$

$$\omega_P = \frac{C_\mu^{3/4} K^{3/2}}{\kappa y_P}$$

This is generally satisfactory, certainly for qualitative calculations. Unfortunately, the results are numerically sensitive to the position of the match point y_P and are often inaccurate in separated-flow and corner-flow regions. The advantage, of course, is the gratifying increase in grid size. A substantial improvement was recently reported by Craft et al. (2002), who derive new analytical wall-function formulas for velocity, temperature, and dissipation rate. Their approach accounts for buoyancy, pressure gradient, and molecular transport property variations.

The second near-wall option is to add molecular viscosity and damping terms and use a fine mesh that accounts for local dynamics. Jones and Launder (1973) added these terms directly to both the K and ϵ differential equations. The fine grid should reach into the viscous sublayer with points as close as $y^+ < 1$. Jones and Launder suggest the following modifications to Eqs. (6-105):

$$\bar{u}\frac{\partial K}{\partial x} + \bar{v}\frac{\partial K}{\partial y} \approx \frac{\partial}{\partial y}\left[\left(\nu + \frac{\nu_t}{\sigma_K}\right)\frac{\partial K}{\partial y}\right] + \nu_t\left(\frac{\partial \bar{u}}{\partial y}\right)^2 - \epsilon \quad (6\text{-}109a)$$

$$\bar{u}\frac{\partial \epsilon}{\partial x} + \bar{v}\frac{\partial \epsilon}{\partial y} \approx \frac{\partial}{\partial y}\left[\left(\nu + \frac{\nu_t}{\sigma_\epsilon}\right)\frac{\partial \epsilon}{\partial y}\right] + C_1\frac{\epsilon \nu_t}{K}\left(\frac{\partial \bar{u}}{\partial y}\right)^2 - C_2\frac{\epsilon^2}{K} \quad (6\text{-}109b)$$

We have added viscosity μ and now the "constants" C_1, C_2, and C_μ are variable near the wall. In addition to the Jones and Launder (1973) formulation, Wilcox

(1998) discusses three other low Reynolds number models and compares them to data for 12 different boundary-layer experiments from Coles and Hirst (1968). All four models have defects, whereas the K–ω model performed quite well for all 12 flows. The review by Patel et al. (1985) tests seven low Reynolds number models, plus the K–ω^2 model of Wilcox and Rubesin (1980), and finds that all have inaccuracies, especially for favorable pressure gradients. Wilcox (1998) speculates that the inaccuracy occurs because the ϵ relation Eq. (6-105b) is improperly formulated.

A variation on low Reynolds number modifications is the v'^2–f model of Durbin (1995) that uses no damping or wall functions. Rather, it uses the fact that the normal velocity v'^2 is a natural damper, as one approaches the wall. Modeled correctly, it reproduces the sublayer and inner layer quite well. The equation modeled by Durbin is as follows:

$$\bar{u}\frac{\partial v'^2}{\partial x} + \bar{v}\frac{\partial v'^2}{\partial y} \approx K f_{22} - v'^2 \frac{\epsilon}{K} + \frac{\partial}{\partial y}\left[\left(\nu + \frac{\nu_t}{\sigma_K}\right)\frac{\partial v'^2}{\partial y}\right] \quad (6\text{-}110)$$

where the term f_{22} represents the *source*, or rate of production, of v'^2 and is itself modeled by a relaxation-type differential equation. In addition to (C_1, C_2, C_μ, σ_K, σ_ϵ) from the K–ϵ equations, Durbin must introduce three more tuned constants to complete the model. But Durbin's examples and those of Iaccarino (2001) show accuracy and great promise, for separation bubbles, backstep flows, and even vortex shedding from bluff bodies.

6-7.2.4 REYNOLDS STRESS MODELS. The modeling of Reynolds stresses ($-\rho\overline{u'v'}$) is a level higher than the previous schemes and is usually called a *second-order closure*. The eddy viscosity and velocity gradient approach are discarded and the stresses computed directly by either (1) an algebraic stress model or (2) a differential equation for the rate of change of stress. The modeling itself is at a higher level—rather complex and computationally intensive.

The subject is reviewed by Rodi (1984), Launder et al. (1975), and Wilcox (1998). Following the discussion by Rodi [Kline et al. (1982, pp. 681–690)], break the Reynolds stress equation into separate terms:

$$\frac{D}{Dt}(\overline{u'v'}) = D_{ij} + P_{ij} + \pi_{ij} - \epsilon_{ij} + \nu\nabla^2(\overline{u'v'}) \quad (6\text{-}111)$$

where D_{ij} = turbulent transport = $-\dfrac{\partial}{\partial x_k}\left(\overline{u'_i u'_j u'_k} + \dfrac{\overline{p'u'_i}}{\rho}\delta_{jk} + \dfrac{\overline{p'u'_j}}{\rho}\delta_{ki}\right)$

P_{ij} = production = $-\overline{u'_i u'_k}\dfrac{\partial \bar{u}_j}{\partial x_k} - \overline{u'_j u'_k}\dfrac{\partial \bar{u}_i}{\partial x_k}$

π_{ij} = pressure–strain = $\dfrac{\overline{p'}}{\rho}\left(\dfrac{\partial u'_i}{\partial x_j} + \dfrac{\partial u'_j}{\partial x_i}\right)$

ϵ_{ij} = dissipation = $2\nu\dfrac{\overline{\partial u'_i}\,\partial u'_j}{\partial x_k\,\partial x_k}$

where summation notation is assumed. This equation is to be used in concert with relations for K and ϵ and does not replace them

The production term P_{ij} is already in Reynolds stress form, but one must model the other three terms: dissipation, turbulent transport, and pressure–strain. We observe that all three are complex statistical correlations, difficult or even impossible (pressure–strain) to measure. What is needed is inspired modeling, backed up now and in the future by direct numerical simulation. Wilcox (1998) discusses these subtleties nicely, and the future is bright for second-order closure.

Dissipation is isotropic away from the wall and anisotropic near the wall. Hanjalic and Launder (1976) suggest the following model:

$$\epsilon_{ij} = \frac{2\epsilon}{3}\delta_{ij} + f_s\epsilon\left(\frac{-\overline{u'v'}}{2K} + \frac{1}{3}\delta_{ij}\right) \tag{6-112}$$

where the overall dissipation and turbulence intensity, ϵ and K, continue to be calculated from their own differential relations, such as Eqs. (6-105). The coefficient f_s is a damping function.

The turbulent transport term $D_{ij} = -\partial/\partial x_k(\beta_{ijk})$ has been studied by DNS calculations, e.g., Moin and Mahesh (1998), which support a model by Hanjalic and Launder (1976). For convenience, denote $(-\overline{u_i'v_j'}) = S_{ij}$. Then their model for the three-term quantity β_{ijk} is

$$\beta_{ijk} = C_s\frac{K}{\epsilon}\left(S_{im}\frac{\partial S_{jk}}{\partial x_m} + S_{jm}\frac{\partial S_{ik}}{\partial x_m} + S_{km}\frac{\partial S_{ij}}{\partial x_m}\right) \tag{6-113}$$

They suggest that the model constant $C_s \approx 0.11$. This model is used in the very popular Launder–Reece–Rodi (LRR) (1975) and Speziale–Sarkar–Gatski (SSG) (1991a) approaches.

Finally, the pressure–strain term π_{ij} is still under intensive evolution because it is (1) unmeasurable and (2) extremely difficult to model. The SSG version, for example, contains 11 terms and seven empirical constants. The Wilcox and Rubesin (1980) model encompasses eight terms and five constants. A simpler model is due to Rodi (1984):

$$\pi_{ij} \approx -C_1\frac{\epsilon}{K}\left(\overline{u_i'u_j'} - \frac{2}{3}K\delta_{ij}\right) - C_\gamma\left(P_{ij} - \frac{1}{3}P_{ii}\delta_{ij}\right) \tag{6-114}$$

These terms are collected into the Reynolds stress relation Eq. (6-111) and solved simultaneously with continuity, momentum, and rate equations for dissipation ϵ. The eddy-viscosity model Eq. (6-106) is also discarded, since turbulent shear is computed directly. Modifications are necessary to use the model very near the wall.

The constants suggested for use in this model are as follows:

$$C_s \approx 0.25 \quad C_1 \approx 1.5 \quad C_\gamma \approx 0.6 \tag{6-115}$$

Finally, we note that Reynolds stress modelers often use K–ϵ models different from those presented here in Sec. 6-7.2.2—see Rodi (1984) for details.

6-7.2.5 ALGEBRAIC STRESS MODELS. Reynolds stress closure models are an improvement, in principle, over one- and two-equation models, but they involve three to six additional partial differential equations and are thus computationally intensive. Therefore, beginning with Hanjalic and Launder (1972), many workers have proposed replacing the *differential* Eq. (6-111) with an algebraic stress model (ASM) that mimics the dissipation, turbulent transport, and pressure–strain effects of Eq. (6-111). For example, Rodi (1976) proposes the following non-linear relation:

$$\frac{\tau_{ij}}{\rho K}\left(\tau_{mn}\frac{\partial \bar{u}_m}{\partial x_n} - \rho \epsilon\right) \approx -\tau_{ik}\frac{\partial \bar{u}_j}{\partial x_k} - \tau_{jk}\frac{\partial \bar{u}_i}{\partial x_k} + \epsilon_{ij} - \pi_{ij} \quad (6\text{-}116)$$

to be supplemented by extremely complex three-dimensional correlations for dissipation ϵ_{ij} and pressure–strain π_{ij} terms. This idea is similar to a non-linear eddy-viscosity model. Presently, ASMs are not popular, especially for separating and reattaching flows.

It should be clear that the present Sec. 6-7 is only a brief overview of the very broad research area of turbulence modeling. For a detailed discussion, see Wilcox (1998), Bernard and Wallace (2002), or Cebeci (2003).

6-8 ANALYSIS OF TURBULENT BOUNDARY LAYERS WITH A PRESSURE GRADIENT

Using the tools outlined earlier in this chapter, a variety of methods have been developed to analyze turbulent boundary layers under fairly arbitrary conditions. Twenty-nine methods were studied at the seminal 1968 Stanford Conference [Kline et al. (1968)], and newer and more complex methods were added at the second Stanford Conference [Kline et al. (1982)]. We select a few of these for discussion here.

There are two basic groups of methods with subcategories:

1. *Integral* methods, using (1) the Kármán integral relation or (2) inner variables.
2. *Finite-difference* (differential) methods, using (1) eddy viscosity, (2) turbulence kinetic energy, (3) two equations, e.g., energy (K) and dissipation (ϵ), or (4) Reynolds stress equations.

The integral methods, being averaged across the boundary layer, are ordinary differential equations and yield gross parameters such as skin friction and momentum thickness. The differential methods yield those same parameters and also the complete profiles of velocity, turbulent shear, and whatever else is modeled. Integral methods are easy to program by the average reader of this text. Differential methods usually require a complex (commercially available) computer code. Both types of methods can give good accuracy for a variety of problems.

Let us discuss these methods for variable pressure gradients but limit ourselves to two-dimensional steady turbulent flow with smooth impermeable walls.

6-8.1 Methods Using the Kármán Integral Relation

Most integral methods in the literature use the famous momentum relation of Kármán (1921), which we repeat here for convenience:

$$\frac{d\theta}{dx} + (2 + H)\frac{\theta}{U_e}\frac{dU_e}{dx} = \frac{C_f}{2} \tag{6-28}$$

With $U_e(x)$ assumed known, there are three unknowns—θ, H, and C_f—hence at least two additional relations are needed. One of these is a purely algebraic relation that arises because the wall–wake law Eq. (6-47) fits the profile data so well. Ludwieg and Tillmann (1949) found the following classic correlation simply by correlating measured values of (θ, H, C_f) in a variety of flows:

Ludwieg–Tillmann relation:

$$C_f \approx 0.246\left(\frac{U_e\theta}{\nu}\right)^{-0.268} 10^{-0.678H} \tag{6-117}$$

When compared to data, the accuracy is ± 10 percent. Separation is not predicted, that is, C_f does not tend toward zero for finite H. Many similar correlations have since been proposed, as reviewed by Cousteix [Kline et al. (1982, p. 657)]. For example, Felsch et al. [Kline et al. (1968, p. 170)] propose

$$C_f \approx 0.058 Re_\theta^{-0.268}(0.93 - 1.95 \log_{10} H)^{1.705} \tag{6-118}$$

Separation ($C_f = 0$) is predicted at $H = 3.0$, which is quite reasonable.

The validity of all such correlations rests upon the excellence of the Coles wall–wake law, whose integral parameters were given as Eqs. (6-48). Let us rearrange those formulas as follows:

$$\lambda = a(\Pi)\frac{H}{H-1}$$

where

$$a(\Pi) = \frac{2 + 3.179\Pi + 1.5\Pi^2}{\kappa(1+\Pi)} \tag{6-119a}$$

$$Re_\theta = \frac{1+\Pi}{\kappa H}\exp(\kappa\lambda - \kappa B - 2\Pi) \tag{6-119b}$$

If we eliminate Π between these two formulas (in our hearts, if not actually by algebraic manipulation), we obtain a unique relation among $C_f = 2/\lambda^2$, H, and Re_θ. The results are plotted in Fig. 6-26 and are curve fit quite accurately (± 3 percent) by the following formula:

$$C_f \approx \frac{0.3 e^{-1.33H}}{(\log_{10} Re_\theta)^{1.74+0.31H}} \tag{6-120}$$

This writer, admittedly biased, proposed this as the most accurate correlation to use in Kármán-based methods.

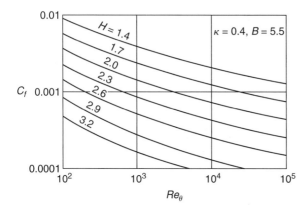

FIGURE 6-26
Computations from the law of the wake, Eq. (6-119), for skin friction as a function of momentum thickness and shape factor.

6-8.1.1 SIMPLE KÁRMÁN-TYPE CLOSURE WITH A $\Pi - \beta$ CORRELATION. A simple way to close the Kármán-type integral method is an approximate correlation between Coles' wake parameter Π and the Clauser parameter β, as discussed earlier in Sec. 6-4.4. The first edition of this text proposed the simple formula $\Pi \approx 0.8(\beta + 0.5)^{3/4}$, which fit a limited number of experimental equilibrium flows. Since then, Das (1987) has correlated hundreds of data points from the 1968 Stanford Conference [Coles and Hirst (1968)] into the following polynomial correlation:

$$\beta \approx -0.4 + 0.76\Pi + 0.42\Pi^2 \qquad (6\text{-}121)$$

This formula is compared with equilibrium and nonequilibrium data in Fig. 6.27. The considerable scatter of data does not invalidate the formula's quite serviceable

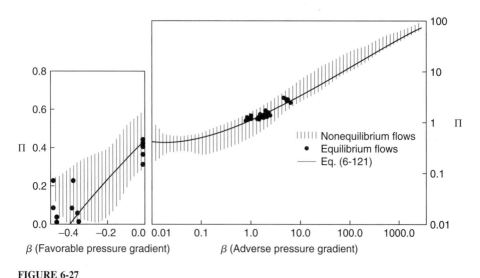

FIGURE 6-27
Correlation of the Coles' wake parameter with the Clauser parameter for equilibrium and nonequilibrium boundary layers. [*After Das (1987)*.]

use in an integral method. The large scatter for negative β (favorable gradients) is not experimental error but rather results from subtracting numbers that are nearly equal when Π is very small.

Equation (6-121) is the desired "third relation" but, as often happens, it introduces new parameters, Π and β, to be related to (C_f, H, Re_θ). The resolution is straightforward: Π is related to C_f and H by Eq. (6-119a), and β yields easily by virtue of its definition:

$$\beta = \frac{\delta^*}{\tau_w}\frac{dp_e}{dx} = -\lambda^2 H \frac{\theta}{U_e}\frac{dU_e}{dx} \qquad (6\text{-}122)$$

This closes the method. One solves Eqs. (6-28), (6-119a), (6-120), (6-121), and (6-122) simultaneously using, say, a Runge–Kutta method to integrate $d\theta/dx$. The initial conditions are known $C_f(0)$ and $\theta(0)$ at, say, $x = 0$. At each step $(x + \Delta x)$, iteration is required to compute H, C_f, etc., among the four algebraic relations, after which one can step forward to the next position $(x + 2\Delta x)$. Though this method has no pedigree, it makes quite adequate predictions for both favorable and adverse gradients.

6-8.1.2 ENTRAINMENT INTEGRAL METHODS.
Many workers believe that the "third relation" should not be an algebraic correlation but rather a second differential equation, namely, the entrainment relation discussed earlier:

$$E = \frac{1}{U_e}\frac{d}{dx}[U_e(\delta - \delta^*)] \qquad (6\text{-}31)$$

To be useful, E needs to be correlated with other integral parameters. The original method of Head (1958) defines a new shape factor

$$H_1 = (\delta - \delta^*)/\theta$$

and writes Eq. (6-31) as a curve-fit correlation in terms of this factor:

$$\frac{d}{dx}(U_e\theta H_1) = U_e F(H_1) \qquad F(H_1) \approx 0.0306(H_1 - 3.0)^{-0.6169} \qquad (6\text{-}123)$$

Head then showed the H_1 is related to the standard shape factor H, and his correlation was later curve-fit to the following formulas:

$$H_1 \approx \begin{cases} 3.3 + 0.8234(H - 1.1)^{-1.287} & \text{for } H \leq 1.6 \\ 3.3 + 1.5501(H - 0.6778)^{-3.064} & \text{for } H \geq 1.6 \end{cases} \qquad (6\text{-}124)$$

Equations (6-28), (6-120), (6-123), and (6-124) are a closed system to be solved for (θ, C_f, H, H_1) with known freestream velocity $U_e(x)$. The agreement is good except possibly for separating flows.

For strong adverse gradients and separating flows, Ferziger et al. (1982) suggest reformulating the entrainment in terms of a shape factor

$$\zeta = \frac{\delta^*}{\delta}$$

Their entrainment correlation takes the form

$$\frac{d}{dx}\left[U_e H\theta\left(\frac{1}{\zeta} - 1\right)\right] = U_e f(\zeta) \quad f(\zeta) \approx 0.0083(1 - \zeta)^{-2.5} \quad (6\text{-}125)$$

In addition, Ferziger et al. (1982) eschew the Ludwieg–Tillmann relation in favor of a skin-friction correlation directly related to ζ:

$$C_f \approx 0.1017|1 - 2\zeta|^{1.732}\left(\frac{\zeta}{HRe_\theta}\right)^{0.268} \text{sgn}(1 - 2\zeta) \quad (6\text{-}126)$$

Separation ($C_f = 0$) occurs at $\zeta = 1/2$. The formula can be used even *past* the separation point, where $C_f < 0$ and the wall flow is reversed. Finally, they relate ζ to C_f and H by manipulating the Coles' wall–wake law Eq. (6-49) into the following correlation:

$$\frac{H-1}{H} \approx 1.5\zeta + 0.309 C_f^{1/2} + 0.955\frac{C_f}{\zeta} \quad (6\text{-}127)$$

Equations (6-28), (6-125), (6-126), and (6-127) may be solved simultaneously for (θ, C_f, H, ζ). The agreement is good for adverse gradients but, unfortunately, not so good for favorable gradients.

6-8.2 An Inner Variable Integral Method

The inner variable method that led to Eq. (6-76) for a flat plate can be extended to pressure gradient conditions by adding a wake component. The result is a first-order ordinary differential equation, with known algebraic coefficients, for the skin friction $C_f(x)$.

An early attempt by White (1969), using a wake varying linearly with y, was reasonably effective—even for hand calculation—and is described on pp. 522–530 of the first edition of this text. It has been supplanted by a more accurate model due to Das (1988), who used the Coles' wall–wake velocity profile in the following form:

$$\frac{\overline{u}(x, y)}{v^*(x)} \approx \frac{1}{\kappa}\ln y^+ + B + \frac{2\Pi}{\kappa}(3\eta^2 - 2\eta^3) \quad (6\text{-}128)$$

where $\quad v^* = \left(\dfrac{\tau_w}{\rho}\right)^{1/2} \quad \eta = \dfrac{y^+}{\delta^+} \quad$ and $\quad y^+ = \dfrac{yv^*}{\nu}$

The polynomial wake, suggested by Moses [Kline et al. (1968, p. 78)], replaces the sine-squared law of Eq. (6-46) to simplify the integration.

With $\overline{u}$ known from Eq. (6-128), the normal velocity component follows from the equation of continuity, Eq. (6-21a):

$$\overline{v} \approx -\frac{\nu}{v^*}\int_0^{y^+}\frac{\partial \overline{u}}{\partial x}\,dy^+ \quad (6\text{-}129)$$

In similar—but algebraically more complicated—fashion to Sec. 6-6.2, $\overline{u}$ and $\overline{v}$ are substituted in the momentum Eq. (6-21b) and integrated across the boundary layer from the wall ($y^+ = 0, \tau = \tau_w$) to the edge of the layer ($y^+ = \delta^+, \tau = 0$). The result of this intermediate step is

$$A_1 \frac{dv^*}{dx} + A_2 \frac{d\Pi}{dx} + A_3 \frac{d\delta^+}{dx} \approx \delta^+ U_e \frac{dU_e}{dx} - \frac{v^* \tau_w}{\mu} \quad (6\text{-}130)$$

where coefficients ($A_{1,2,3}$) are given by Das (1988) but are not strictly needed in the final method.

To eliminate ($d\delta^+/dx$) from Eq. (6-130), we use the differential form of the wall–wake friction law Eq. (6-49) and rearrange:

$$\frac{d\delta^+}{dx} \approx \frac{\kappa \delta^+}{v^*} \frac{dU_e}{dx} - \frac{\delta^+}{v^*}(\ln \delta^+ + \kappa B + 2\Pi)\frac{dv^*}{dx} - 2\delta^+ \frac{d\Pi}{dx} \quad (6\text{-}131)$$

To eliminate ($d\Pi/dx$) from Eq. (6-130), we first introduce the definition of the Clauser parameter:

$$\beta = \frac{\delta^* }{\tau_w} \frac{dp_e}{dx} = -\frac{\delta U_e}{v^{*2}} \frac{dU_e}{dx} \frac{\delta^*}{\delta} \quad \frac{\delta^*}{\delta} \approx \frac{(1+\Pi)v^*}{\kappa U_e} \quad (6\text{-}132)$$

where the latter relation follows from Eq. (6-48). Finally, we related β to Π, approximately, by the polynomial curve fit of Eq. (6-121), as shown in Fig. 6-27. In this way, ($d\Pi/dx$) is also eliminated from Eq. (6-130). The result is a (dimensional) differential equation containing only (dv^*/dx). Finally, Das (1988) nondimensionalizes this relation by defining

$$x^* = \frac{x}{L} \quad V = \frac{U_e}{U_0} \quad \zeta = \frac{v^*(x)}{U_e(x)} \quad (6\text{-}133)$$

where L and U_0 are any convenient reference length and velocity, respectively.

The final differential equation has the form

$$\frac{d\zeta}{dx^*} = -\frac{\zeta}{V}\frac{dV}{dx^*} + \frac{1}{V}\frac{T_1 - T_2 + R_L T_3(T_4 - R_L T_5)}{R_L T_3(T_6 + T_7) + T_8 + T_9} \quad (6\text{-}134)$$

where $R_L = U_0 L/\nu$ and where T_{1-9} are dimensionless coefficients given in Table 6-2. A known initial value $\zeta(0)$ is necessary. Once $\zeta(x^*)$ is computed, the skin friction follows from the identity $C_f = 2\zeta^2$. If desired, one can calculate estimates of $\delta^*(x)$, $H(x)$, and $\theta(x)$ from the wall–wake law:

$$\frac{1}{\kappa}\ln \delta^+ + B + \frac{2\Pi}{\kappa} = \frac{1}{\zeta}$$

$$\beta = -\frac{\delta^+(1+\Pi)}{\kappa \zeta^2 R_L} V \frac{dV}{dx^*} \quad (6\text{-}135)$$

The method requires a reasonably smooth function for the freestream velocity $V(x^*)$ and its first two derivatives but otherwise is free from pathology.

TABLE 6-2
The coefficient functions T_i for use in Eq. (6-134)
[After Das (1988).]

$$T_1 = \delta^+[F_1]\left[\frac{1}{V}\left(\frac{dV}{dx^*}\right)^2\right]$$

where $F_1 = \zeta(6.25 \ln \delta^+ + 8.54 + 6.42\Pi^2) + \zeta\Pi(12.5 \ln \delta^+ + 18.97)$

$$T_2 = 2.5\delta^+\zeta[F_1]\left(-\frac{d^2V}{dx^{*2}}\right)$$

$$T_3 = \frac{[F_2]\zeta^2}{(1+\Pi)^2} - \frac{5\delta^+}{R_L V^2}\left(\frac{dV}{dx^*}\right)$$

where $F_2 = 1.16 + 0.84\Pi + 0.42\Pi^2$

$$T_4 = [\zeta(1.28\Pi^2) + \zeta\Pi(2.5 \ln \delta^+ + 2.58) + 1]\left(V\frac{dV}{dx^*}\right)$$

$$T_5 = \frac{\zeta^3 V^3}{\delta^+}$$

$$T_6 = [\zeta(6.25 \ln^2 \delta^+ + 8.75 \ln \delta^+ + 14.0)$$
$$+ \zeta\Pi^2(3.21 \ln \delta^+ + 7.06)$$
$$+ \zeta\Pi(6.25 \ln^2 \delta^+ + 20.2 \ln \delta^+ + 21.48)]V$$

$$T_7 = [\zeta\Pi(12.5 \ln \delta^+ + 15.0 + 6.42\Pi^2)$$
$$+ \zeta(6.25 \ln \delta^+ + 1.25) + \zeta\Pi^2(12.5 \ln \delta^+ + 22.18)]V$$

$$T_8 = \delta^+(2.5 \ln \delta^+ + 5.5)[F_1]\left(\frac{1}{V}\frac{dV}{dx^*}\right)$$

$$T_9 = 5\delta^+(\Pi + 1)[F_1]\left(\frac{1}{V}\frac{dV}{dx^*}\right)$$

A related, and slightly simpler, polynomial-shear integral method was proposed by Thomas and Kadry (1990). Also, the method of Das, Eq. (6-134), was extended to wall transpiration by Oljaca and Sucec (1997). Their results, straightforward but even more lengthy than Table 6-2, are in excellent agreement with blowing/suction experiments.

6-8.2.1 RELAXATION EFFECTS: LAG EQUATIONS. Most integral methods use correlations developed from equilibrium turbulent flows such as the flat plate. They should work well in nonequilibrium flows if the changes are gradual. In rapid changes, the actual integral parameters will lag behind the equilibrium prediction. For example, Fig. 6-28 shows a *relaxing* boundary layer, where a strong adverse pressure gradient is suddenly removed. The wall region reacts quickly, but the outer wake slowly relaxes toward equilibrium over a distance of five to 10 boundary-layer thicknesses. A theory tied directly to the sudden change from large positive β to, say, $\beta = 0$ will give premature predictions that do not actually occur until further downstream. To combat premature changes, some workers use an exponential-type "lag equation":

$$\frac{dQ}{dx} = \frac{C}{\delta}(Q_{eq} - Q) \qquad (6\text{-}136)$$

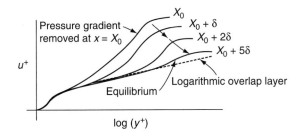

FIGURE 6-28
Schematic of the relaxation zone following a sudden removal of a pressure gradient.

where Q is any integral property (H, Π, etc.), C is a dimensionless constant, and Q_{eq} is the downstream asymptote. For example, in their entrainment method from Sec. 6-8.1.2, Ferziger et al. (1982) suggest that Eq. (6-136) be used in certain rapidly changing pressure gradients, where Q is taken to be E and $C \approx 0.025$.

6-8.3 Finite-Difference Analysis of Boundary Layers

The integral methods discussed above are ordinary differential equations solved in the streamwise direction for parameters such as $C_f(x)$ and $\theta(x)$. Details of the flow field can be found only from algebraic approximations. In contrast, *differential* methods attack the full partial differential equations of motion and compute flow-field details: $\bar{u}(x, y)$, $\bar{v}(x, y)$, $K(x, y)$, $\epsilon(x, y)$, etc. In principle, they are more accurate than integral methods. In practice, being based on modeling assumptions rather than pure physics, they often show deviations from experimental data.

All of these methods require a digital computer and a two-dimensional (or three-dimensional where necessary) nodal mesh. Such a mesh must be finer near the wall, where changes in turbulent flow are more rapid. Although the finite-element method [Löhner (2001)] is applicable to this problem, such as the method of Hytopoulos et al. (1993), in fact most schemes presently use either finite differences or the finite-volume method popularized by Patankar (1980). A variety of boundary-layer computation schemes are discussed in the texts by Schetz (1992), Cebeci and Cousteix (1998), and Schlichting and Gersten (2000). Also, the boundary-layer Applet codes of Devenport and Schetz (2002) are available for direct use over the Internet.

6-8.3.1 ZERO-EQUATION MODELS. These methods use the eddy-viscosity concept to solve the continuity and momentum Eqs. (6-21a) and (6-21b) without additional differential equations. In a method discussed in detail in their text, Cebeci and Smith (1974) use the van Driest inner layer model Eq. (6-90) and the Clauser outer layer model Eq. (6-93) for eddy viscosity. The damping constant A in Eq. (6-90) is adjusted for pressure gradient in a manner similar to Eq. (6-100). Cebeci and Smith use a Blasius-type coordinate across the boundary layer,

$$d\eta = \left(\frac{U_e}{\nu x}\right)^{1/2} dy$$

which makes the same method very convenient for laminar boundary layers. For turbulent flow, they use a mesh size $\Delta \eta$ which is very fine near the wall, becoming coarser in the outer layer. Their computer code has been tested for a variety of parametric conditions—rough walls, suction and blowing, low Reynolds numbers, free convection—and has been popular since its inception.

6-8.3.2 TWO-EQUATION MODELS. The $K-\epsilon$ method of Sec. 6-7.2.2 is the best known two-equation model, but a dozen other formulations were used at the 1981 Stanford Conference [Kline et al. (1982)]. Finite differences are necessary. Jones and Launder (1972) used the Patankar–Spalding stream function variable with 100 cross-stream steps $\Delta \omega$ and a streamwise step $\Delta x \approx 0.3\delta$, with good results for boundary layers. It is less accurate for recirculating flows, i.e., flow with a separated (backflow) region.

Wilcox (1998) has shown, by comparison to numerous data sets, that the $K-\omega$ model is superior to the $K-\epsilon$ model for boundary layers, especially with adverse gradients. Two-equation models can be extended to separated flow, wall blowing/suction, and compressibility.

6-8.3.3 REYNOLDS STRESS MODELS. Even in 1981 at the Stanford Conference [Kline et al. (1982)], six of the participants used Reynolds stress models (RSM). Half of these also used additional differential equations for K, ϵ, or L, and half did not. Clearly the best of these RSM models are superior for complex flows such as separation and reattachment, and even for attached boundary layers, they surpass the $K-\epsilon$ model. The RSM can easily account for gravity and wall curvature and apply readily to three-dimensional flows, albeit with much computational effort. Since large-eddy simulation (LES) and direct numerical simulation (DNS) are limited to low Reynolds numbers, RSM will probably become the dominant future CFD approach, as reviewed by Speziale (1991*b*), Hanjalic (1994), and Wilcox (1998).

6-8.4 Turbulent Flow Past a Flat Plate

A simple but important test of any model, used at both Stanford conferences, is the flat-plate flow measured by Wieghardt and Tillmann (1951). Air with $\nu \approx 1.51\text{E}-5 \text{ m}^2/\text{s}$ flowed at $U_e \approx 33$ m/s past a 5 m long waxed plywood plate. The flow was tripped at the leading edge so that the first measurement station, $x = 0.087$ m, was barely turbulent ($Re_x \approx 1.9\text{E}5$, $Re_\theta \approx 460$) and very thin ($\delta \approx 4$ mm); so only five velocity profile data points were taken. In fact, for all stations with $x < 0.8$ m, there is considerable uncertainty in both the velocity profiles and the skin-friction coefficient. In spite of this defect, the experiment is an excellent preliminary test for any prediction method.

Figure 6-29 is a comparison of data for skin friction and momentum thickness with four methods described earlier: Head (1958) and Ferziger et al. (1982) from Sec. 6-8.1.2, Das (1988) from Sec. 6-8.2, and Cebeci and Smith (1974) from Sec. 6-8.3.1. To avoid some of the data uncertainty, the prediction was begun at the fourth station, $x = 0.387$ m. All four methods show good accuracy except that

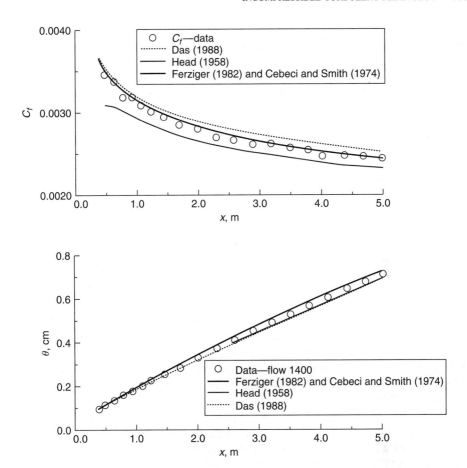

FIGURE 6-29
Comparison of four boundary-layer prediction methods with the flat-plate data of Wieghardt [flow 1400 of Coles and Hirst (1968)].

Head (1958) predicts C_f too low, the problem being that the given (uncertain) initial values of C_f, H, and θ are incompatible with the algebraic correlations of Eq. (6-117) or (6-120). When compared with Eq. (6-78), the experimental value $C_f(0) = 0.00364$ is compatible with an effective origin $x(0) = -0.15$ m. When compared with Eq. (6-70), the experimental value $\theta(0) = 0.092$ cm is compatible with a different origin, $x(0) = -0.09$ m, another indicator of data uncertainty.

Not shown in Fig. 6-29, to avoid clutter, are the Kármán-based method of Sec. 6-8.1.1 [results similar to Das (1988)] or the K–ϵ and Reynolds stress models [see Kline et al. (1982, pp. 1249–1250)], all of which show good agreement. The differential models, being field methods, can also predict the velocity profiles $\bar{u}(x, y)$, with good results. This test shows that all the previously mentioned integral and differential methods are reasonably accurate and well behaved for flat-plate flow.

464 VISCOUS FLUID FLOW

6-8.4.1 VISUALIZATION AND SIMULATION OF FLAT-PLATE FLOW. The results shown in Fig. 6-29 are typical of "boundary-layer analyses" that predict time-mean properties such as $C_f(x)$, $\theta(x)$ plus, in the case of the differential methods, $\bar{u}(x, y)$ or $K(x, y)$ or $\overline{u'v'}(x, y)$. The time-varying fluctuations themselves—or the unsteady flow field—are *not* computed. But the instantaneous turbulent flow field can be photographed and now, with the advent of supercomputers, it can be simulated numerically.

Figure 6-30*a–e* shows hydrogen bubble visualizations of water flow past a flat plate, taken by Hirata and colleagues at the University of Tokyo [see also Nakayama (1988, p. 27)]. Figure 6-30*f* shows the mean velocity profile and the five locations *a–e*. Figure 6-30*a* and *b* is in the sublayer and shows a characteristic spanwise variation of fast and slow "streaks" of streamwise vorticity. In the buffer layer, Fig. 6-30*c*, the flow mixes better and the streaks become less pronounced. In the log layer, Fig. 6-30*d*, mixing is thorough and the streaks disappear. In the outer (wake) layer, Fig. 6-30*e*, the scale of the eddies becomes much larger

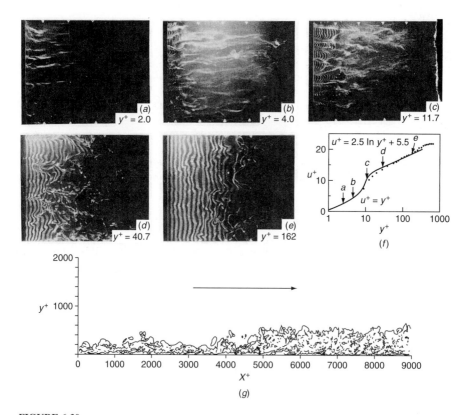

FIGURE 6-30
Visualization of flat-plate flow: (*a–e*) hydrogen bubbles released at five different locations in the boundary layer (*courtesy of K. Hirata, University of Tokyo*); (*f*) velocity profile showing locations (*a–e*); (*g*) supercomputer simulation of vorticity contours at $Re_\theta = 1410$. [*From Spalart (1988)*.]

and the flow is slightly intermittent. All this structure is lost when we take the time-mean as in Fig. 6-30f.

In a DNS using the NASA–Ames Laboratory supercomputer with *10 million* mesh points, Spalart (1988) solved the three-dimensional time-dependent Navier–Stokes equations with a spectral method for flat-plate flow at Re_θ = 225, 300, 670, and 1410. Figure 6-30g shows the computed instantaneous vorticity contours at the highest Reynolds number. When these 10 million point values were correlated over time and space, Spalart was able to verify the following results:

1. Friction coefficients and shape factors accurate to ±5 percent;
2. Velocity profiles that verify the inner law, the log-law with $\kappa \approx 0.41$ and $B \approx 5.0$, and an outer wake with Π slightly less than Coles (1956);
3. Accurate estimates of Reynolds stress and rms fluctuations of both velocity and pressure.
4. Reasonably accurate frequency spectra of both velocity and pressure whose trends agree with the statistical theories [Hinze (1975)].

Although the computed Reynolds numbers are small, of low turbulence—and may remain so unless mega-supercomputers are developed—the value of such simulations in providing a wealth of space–time information about turbulent flows is striking and is bringing engineering analysis of turbulent flow to a new level.

6-8.5 Equilibrium Flow with an Adverse Pressure Gradient

A more difficult case than flat-plate flow is flow 2200 of Coles and Hirst (1968), from the experiment of Clauser (1954) which first demonstrated the properties of equilibrium turbulent boundary layers. The flow was meant to have constant $\beta \approx 1.8$ [see Eq. (6-42)], but in fact the measured β increased from about 1.4 to 2.3 over a 25 ft distance. The flow was a thick boundary layer in a narrow wind tunnel and almost certainly contains three-dimensional effects. It was rated "most difficult" at the 1968 Stanford Conference, and none of the 28 methods presented there gave accurate predictions.

Figure 6-31a shows the measured freestream velocity, which fits a power-law distribution, expected for equilibrium flows, quite well. Figure 6-31b compares the measured skin-friction coefficients with the same four theories used in Fig. 6-29 plus the Kármán-based method of Sec. 6-8.1.1. The Kármán-based and Das predictions give the best agreement, as might be expected since both use an equilibrium correlation, Eq. (6-121). The other methods consistently predict friction far too low, for reasons unknown to this writer.

Figure 6-31c compares the computed and measured momentum thickness variation. All five methods give essentially the same (too-high) predictions—perhaps another indication of unknown effects in the data. However, an excellent prediction of $\theta(x)$ for flow 2200 at the 1968 Stanford Conference resulted from a one-equation

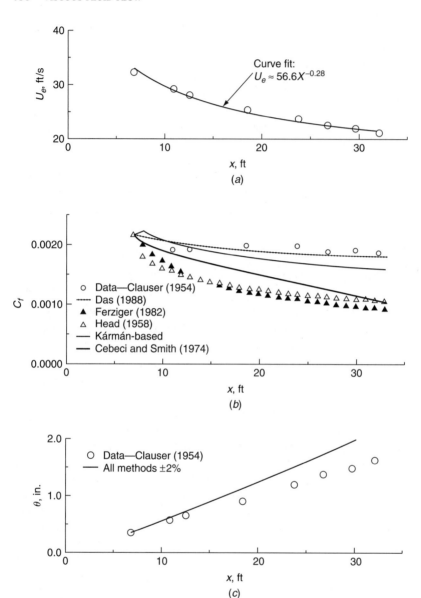

FIGURE 6-31
Comparison of theory and experiment for the equilibrium flow of Clauser (1954), flow 2200 of Coles and Hirst (1968).

turbulent kinetic-energy method by Bradshaw and Ferriss [Kline et al. (1968, pp. 264–274)].

6-8.6 Nonequilibrium Separating Flow: The Newman Airfoil

Newman (1951) made a wind tunnel study of flow at 120 ft/s past an airfoil section. The measurements on the upper (suction) surface proceed along an adverse pressure gradient toward separation at $x \approx 5$ ft. Figure 6-32a shows the measured freestream velocity, which fits a least-squares cubic polynomial quite well. Approximately one-third of the methods used in Kline et al. (1968) gave good predictions for this nonequilibrium flow.

Figure 6-32b shows the skin-friction data compared to our five methods. The equilibrium-oriented methods, Das (1988) and Kármán-based, show (unrealistic) relaxation of C_f before separation. Ferziger et al. (1982) show separation too early. Head (1958) is reasonable but too low. Only the Cebeci and Smith (1974) predictions are in very good agreement.

Figure 6-32c shows the prediction and data for momentum thickness. Das (1988) is too high, but the other four methods give reasonable agreement. This trend of methods being better able to predict $\theta(x)$, than $C_f(x)$ was a general characteristic of the 1968 Stanford Conference. We conclude that, although the prediction techniques presented here are "adequate" for engineering use in boundary-layer prediction, no single method is in excellent agreement with all available data.

6-8.6.1 SEPARATING FLOWS: INVERSE TECHNIQUES. Near separation, C_f approaches zero, β and Π approach infinity, and most boundary-layer methods become inaccurate—primarily because of sensitivity to the freestream velocity $U_e(x)$. After separation, τ_w and v^* are negative, displacement thickness becomes very large, and the boundary-layer approximations become inaccurate; normal velocity and streamwise gradients become much larger. Under these conditions, the most fundamental approach is to solve, with finite differences, the full equations of motion in the whole flow field—freestream and wall regions—a computer-intensive procedure. Yet boundary-layer theory can still be used effectively in an "inverse" mode.

The appropriate interaction near separation is to assume that the displacement thickness distribution $\delta^*(x)$ is known and the freestream velocity $U_e(x)$ is unknown—hence the term *inverse* method. With this modification, satisfactory agreement with measured separating flows can be obtained by both finite-difference methods, Pletcher (1978), and integral methods, Moses et al. (1979) and Strawn and Kline (1983).

Das (1987) modifies the "direct" inner variable method of Sec. 6-8.2 into an inverse technique as follows. First the wall–wake law Eq. (6-128) is written in terms of a wake velocity $v_\beta = \Pi v^*$:

$$\bar{u} = \frac{v^*}{\kappa} \ln\left(\frac{yv^*}{\nu}\right) + Bv^* + \frac{v_\beta}{\kappa}(6\eta^2 - 4\eta^3) \qquad \eta = \frac{y}{\delta} \qquad (6\text{-}137)$$

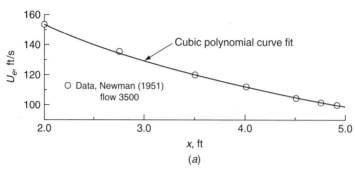
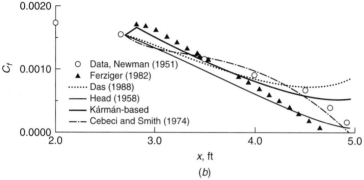
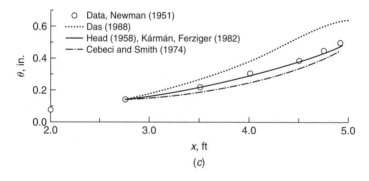

FIGURE 6-32
Comparison of theory and experiment for the separating airfoil flow of Newman (1951), flow 3500 of Coles and Hirst (1968).

The advantage is that v_β remains finite and positive through separation, whereas v^* changes sign and Π, no longer used, becomes singular. The new version of the momentum-integral relation Eq. (6-130) becomes

$$A_{11}\frac{dv^*}{dx} + A_{12}\frac{dv_\beta}{dx} + A_{13}\frac{d\delta^+}{dx} + A_{14}\frac{dU_e}{dx} = A_{15} \qquad (6\text{-}138)$$

where A_{ij} are given by Das (1987). The derivative form of the wall–wake law Eq. (6-131) also takes the new form of Eq. (6-138):

$$\left(\frac{1}{\kappa}\ln\delta^+ + B\right)\frac{dv^*}{dx} + \frac{2}{\kappa}\frac{dv_\beta}{dx} + \frac{v^*}{\kappa\delta^+}\frac{d\delta^+}{dx} + \frac{dU_e}{dx} = 0 \qquad (6\text{-}139)$$

Equation (6-132) is combined with the Π–β correlation Eq. (6-121) and retained as a third relation for dU_e/dx, except that $(1 + \Pi)v^*$ is replaced by $(v^* + v_\beta)$, in the following rearrangement:

$$\frac{dU_e}{dx} = -\frac{\kappa|v^*|\left[-0.4v^{*2} + 0.76v_\beta|v^*| + 0.42v_\beta^2\right]}{\nu\delta^+(v_\beta + v^*)} \qquad (6\text{-}140)$$

Finally, the derivative form of Eq. (6-132) yields a fourth differential equation:

$$\frac{dv_\beta}{dx} = B_1\frac{dv^*}{dx} + B_2\frac{d\delta^+}{dx} + B_3\frac{dU_e}{dx} + B_4 \qquad (6\text{-}141)$$

where B_i are given by Das (1987).

Equations (6-138) to (6-141) are four first-order ordinary differential equations to be solved for U_e, v^*, v_β, and δ^+, assuming a known input distribution $\delta^*(x)$. For example, when this inverse scheme is applied to the Newman airfoil flow of Fig. 6-32 (for which the direct inner variable method is not very accurate), the results agree well with the measurements for U_e, C_f, and $\theta(x)$—even in the separated-flow region. Other satisfactory examples are given by Das (1987). In principle, such inverse methods can be used to design "tailored" flows to, say, avoid separation or reduce body drag.

6-8.7 Complex Turbulent Flows

Although the emphasis in this section has been upon developing relatively simple turbulent-boundary-layer methods for engineering use, there are complex flows for which such methods fail. In the late 1970s, many computer-oriented methods were developed to analyze such complicated flows. The 1981 Stanford Conference on Complex Turbulent Flows [Kline et al. (1982)] was intended to explore this second generation of prediction techniques by providing a trustworthy set of complex-flow experiments against which computer methods could be tested in an open and adjudicated forum.

An overview of the 1981 Stanford Conference is given on pp. 1–22 of Kline et al. (1982) that we summarize here. In one sense, a "complex" turbulent flow is simply one that is more complicated than the steady, incompressible, two-dimensional attached thin shear layers featured in the 1968 conference [Kline et al. (1968)]. Three-dimensionality is the most obvious complexity. Other parameters contributing to complex flows are

1. Fluid effects: polymer additives, chemical reactions, multiple phases, compressibility, surface tension, and nonnewtonian shear
2. Fluctuations or turbulence or periodicity in the freestream

3. Wall effects: irregular geometry, suction and blowing, roughness, compliance, moving walls
4. Body forces, especially due to streamline curvature
5. Strain interactions, especially in three dimensions

One straightforward example is an *irregular geometry*. As illustrated in Fig. 6-33, right-angle changes in the wall or body shape will immediately cause the flow to separate and, if wall length permits, to reattach further downstream. The separation "bubble" is not of a boundary-layer character and it interacts strongly with the freestream. A full flow-field computation is required to predict each of the seven examples in Fig. 6-33.

The question of whether enough trustworthy data sets could be assembled for the 1981 Stanford Conference turned out to be moot. A total of 26 incompressible- and 18 compressible-flow experiments were included in vol. II of Kline et al. (1982), many of which treated multiple cases. There were 66 test cases in all. Among the incompressible-flow experiments were the following:

1. Corner flows
2. Diffuser flows
3. Freestream turbulence
4. Curved streamlines
5. Suction or blowing
6. Separated flows
7. Wall jets
8. Relaminarizing flows
9. Wakes—near and far
10. Decay of grid turbulence
11. Backward-facing step (Fig. 6-33a)
12. Secondary flows

Needless to say, no single method could successfully attack the complete spectrum of this rich variety of cases.

A total of 35 computer groups participated, using 67 methods in all, including (1) integral methods, (2) eddy viscosity, (3) two-equation models, (4) algebraic stress models, and (5) differential Reynolds stress models. To quote from the Evaluation Committee report [Kline et al. (1982, pp. 979–986)]: "No method had any significant universality. Likewise no method proved to be universally bad." Because different methods worked in different cases, they recommended that all methods receive further study and refinement. Best progress toward predictive accuracy occurred in separated flows, shock-wave–boundary-layer interactions, calculation of fluctuating velocities (u', v', w'), decay of turbulence toward laminar flow, and transonic flow.

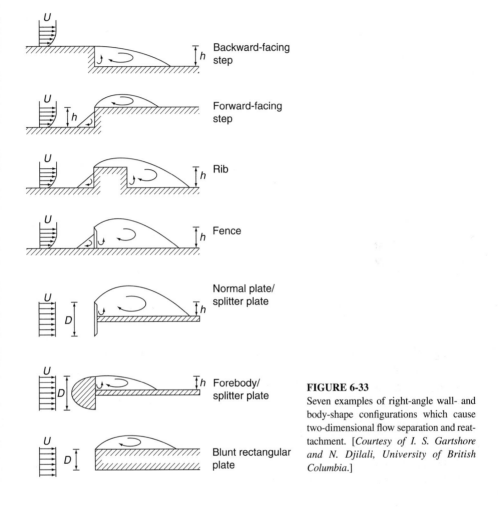

FIGURE 6-33
Seven examples of right-angle wall- and body-shape configurations which cause two-dimensional flow separation and reattachment. [*Courtesy of I. S. Gartshore and N. Djilali, University of British Columbia.*]

Some further points made by the Evaluation Committee are

1. The weakest part of two-equation models is the ϵ relation. Agreement in particular cases could be found by "tweaking its constants," but in general a better dissipation model is needed.
2. Algebraic stress models, though expected to be better than eddy-viscosity closures, did not give significant improvement on average.
3. Differential methods that integrate right down to the wall give better results than "wall functions."
4. Grid spacing and numerical uncertainty were often inadequate.
5. In some difficult cases—transonic airfoils, curved walls, and diffusers—the simplest (integral) methods were, vexingly, the most accurate. In general, there was

no correlation between the complexity of the models used and their actual predictive capability.
6. Prediction accuracy for separated flows, e.g., Fig. 6-33, was significantly worse than for the corresponding attached flows.

The 1968 and 1981 Stanford conferences were outstanding mergers between cooperation and controversy, and each has reached a new plateau in turbulence prediction. No doubt, when it occurs, the third such conference will be even better. The present status is one of uncertainty as to whether we will ever achieve "universal" models that handle all (or most) turbulent flows, or whether instead we should assemble groups of *zonal* models that are each effective for certain regions or classes or types of flow.

6-8.7.1 RELAMINARIZATION. An intriguing test case in the 1981 Stanford Conference—not computed by many participants—is flow that undergoes reverse transition or *relaminarization*. About 30 experiments in the literature, reviewed by Narasimha and Sreenivasan (1979), Sreenivasan (1982), and Warnack (1998), have studied this interesting phenomenon.

Disregarding special effects due to buoyancy and electromagnetic forces, there are two fluid-flow mechanisms causing reversion of turbulent to laminar flow. These are shown in Fig. 6-34a, strong convective acceleration of a boundary layer, and Fig. 6-34b, gradual expansion of fully developed duct flow.

In Fig. 6-34a, "weak" laminarization occurs, in the sense that the turbulence does not disappear, but mean parameters such as the velocity profile and skin friction approach laminar values. Turbulence becomes smaller only in an absolute sense: $(-\overline{u'v'})$ remains almost constant during the laminarization process but is a sharply decreasing fraction of the stream energy U_e^2. Although it is thought that turbulence cannot exist for Reynolds numbers $Re_\theta < 300$, this critical value does *not* correlate laminarization in Fig. 6-34a. The proposed critical parameter is a dimensionless acceleration:

$$K_{\text{crit}} = \frac{\nu}{U_e^2} \frac{dU_e}{dx} \approx 3\text{E}{-}6 \qquad (6\text{-}142)$$

Typically, this value is reached about 20 boundary-layer thicknesses after acceleration begins. However, Sreenivasan (1982) points out that K cannot truly be a fundamental parameter since it contains no boundary-layer information.

The case of Fig. 6-34b may be called "strong" laminarization since, sufficiently far downstream, the turbulence disappears entirely. For a small (1–2°) expansion angle, the duct Reynolds number drops gradually until turbulent flow can no longer be sustained. Turbulent shear decreases faster than the rms fluctuations, but both decay slowly over tens or even hundreds of diameters. Dissipation exceeds turbulent production during this expansion, so Fig. 6-34b is a "viscous" laminarization, whereas in Fig. 6-34a, where dissipation is small, the reversion is thought to be due to the stabilizing effect of acceleration on the flow field. At the time of the second Stanford Conference [Kline et al. (1982, p. 1165)], few methods could predict

INCOMPRESSIBLE TURBULENT MEAN FLOW   473

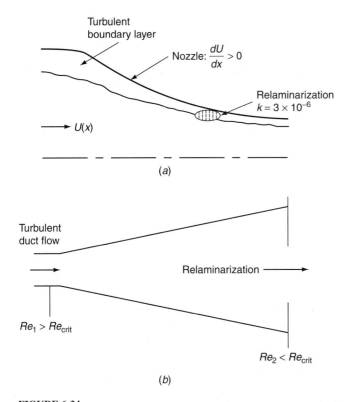

FIGURE 6-34
Two examples of reverse transition experiments: (*a*) convective acceleration of an initially turbulent boundary layer; (*b*) reduction of duct Reynolds number due to gradual (unstalled) expansion.

relaminarization, although the original K–ϵ model of Jones and Launder (1972) was developed for this purpose. More recently, however, it has been possible to predict this effect, both with a second-moment closure, Shima (1993), and also with direct numerical simulation, Iida and Nagano (1998).

6-9 FREE TURBULENCE: JETS, WAKES, AND MIXING LAYERS

The previous sections have been concerned with wall-bounded flows, where the interaction between an inner and outer layer is crucial to rational analysis. Now let us consider *free turbulence*: high Reynolds number shear flow in an open ambient fluid, unconfined or uninfluenced by walls. There are several monographs and reviews about free turbulent flows, notably Pai (1954), Abramovich (1963), List (1982), Ho and Huerre (1984), Middleman (1995), Morris et al. (2002), and Lee and Chu (2003). There is also a very nice discussion in Sec. 14.7 of the text by Sherman (1990).

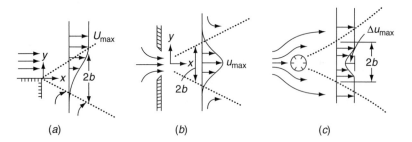

FIGURE 6-35
Three types of free turbulent flow: (*a*) mixing layer; (*b*) free jet; (*c*) wake of a body.

Figure 6-35 shows the three most common types of free turbulence: (1) a mixing layer between two streams of different velocity, (2) a jet issuing into a still (or moving) stream, and (3) a wake behind a body. In all three cases, there is a characteristic velocity scale, $U_{max}(x)$ or $\Delta u_{max}(x)$, and a characteristic shear-layer width, $b(x)$. Since these flows are "free," or unconfined, the pressure is approximately constant throughout the flow, except for (small) turbulent fluctuations within the layer. (A confined shear layer not discussed here, such as from a jet pump, might have an impressed pressure gradient.)

In Fig. 6-35, we are looking at the asymptotic downstream behavior of the free turbulence, traditionally assumed to be independent of the exact type of source which created the flow. The source, thus ignored, is also assumed to be symmetrical, so that the shear layer is not *skewed* in shape. One then analyzes the asymptotic behavior of width and velocity scales and the velocity profile, $\bar{u}(x, y)$ for plane flow or $\bar{u}(x, r)$ for axisymmetric flow. Actually, there are certain effects, to be discussed, of the exact form of the jet source or the body creating the wake.

Figure 6-36 shows the details of the initial formation of a jet, assuming a still ambient fluid. The figure is valid only for similar jet and ambient fluids, e.g., air-into-air, not water-into-air. Velocity profiles are shown as thick dark lines across the flow. Typically the jet issues at a nearly flat, fully developed, turbulent velocity U_{exit}. Mixing layers form at the lip of the exit, as in Fig. 6-35*a*, growing between the still ambient and the nearly inviscid *potential core* flowing at velocity U_{exit}. The potential core vanishes quickly at a distance of about one diameter from the exit, where the velocity profile loses its mixing-layer–flat-core shape. Downstream of the core, the flow begins to develop into the distinctive Gaussian-type shape we think of as a "jet"—recall Fig. 4-18 for a plane laminar jet. Finally, at about 20 diameters downstream of the exit, the velocity profile reaches and maintains a *self-preserving* shape,

$$\frac{\bar{u}}{U_{max}} \approx f\left(\frac{y}{b}\right) \quad \text{or} \quad f\left(\frac{r}{b}\right) \qquad (6\text{-}143)$$

depending upon whether the jet is plane or axisymmetric. It is this asymptotic self-similar form of free turbulent flows that we wish to study here. Equation (6-143) is

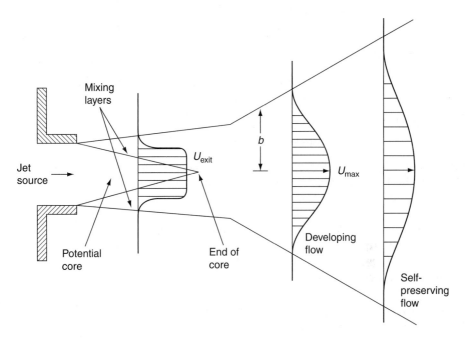

FIGURE 6-36
Details of the early development of a real jet.

satisfying physically but hardly surprising. Most curves of this shape—falling from a maximum through a point of inflection to approach zero—will collapse together reasonably well when normalized with their maximum values and their characteristic widths. The developed or self-similar region seems to grow from an "apparent origin" in front of the exit, as sketched in Fig. 6-36.

Note that the velocity profiles in Fig. 6-36 have the same momentum but *not* the same mass flow. Fluid is entrained in the jet from the ambient region and the jet mass flow increases downstream. One should also emphasize that Fig. 6-36 is a schematic of the time-mean flow field—the instantaneous flow might resemble the structure in Fig. 6-2.

6-9.1 Self-Similar Analysis of Turbulent Jets

Assume that we are sufficiently far downstream for the jet velocity profiles to become self-similar, as in Eq. (6-143). Since there is no pressure gradient, the jet momentum J must remain constant at each cross section [recall Eq. (4-97) for laminar jets]:

$$J = \int_{-\infty}^{+\infty} \rho \bar{u}^2 \, dA = \text{const} = \text{const } \rho b u_{\max}^2 \quad \text{(plane jet)}$$

$$= \text{const } \rho b^2 u_{\max}^2 \quad \text{(axisymmetric)} \quad (6\text{-}144)$$

In the self-similar region, the centerline velocity and jet width should depend only upon jet momentum, density, and distance, but *not* upon molecular viscosity since there are no walls:

$$U_{max} = fcn(x, J, \rho)$$
$$b = fcn(x, J, \rho)$$

By dimensional analysis, the width can be only a linear growth:

Turbulent jet: $b = (\text{const}) \, x$ (plane or axisymmetric)

Recall from Chap. 4 that this was also true for a laminar circular jet, Fig. 4-37, but there the constant depended upon jet momentum and molecular viscosity. Here the constant is unique: a single growth rate for all self-similar turbulent jets, regardless of the Reynolds number. Recall also that the growth of a plane laminar jet, Eq. (4-106), is $b = Cx^{2/3}$.

Dimensional analysis of U_{max} leads to the following two relations:

Plane flow: $\quad U_{max} = (\text{const}) \left(\dfrac{J}{\rho}\right)^{1/2} x^{-1/2}$

Axisymmetric flow: $\quad U_{max} = (\text{const}) \left(\dfrac{J}{\rho}\right)^{1/2} x^{-1}$

Again, unlike laminar flow, the constants are unique, independent of the Reynolds number. They are determined from the experimental data. An eloquent discussion of the restrictions on these self-preserving assumptions in given by George (1989).

6-9.1.1 THEORETICAL VELOCITY PROFILE FOR A PLANE JET.
With the forms of $b(x)$ and $U_{max}(x)$ revealed by dimensional analysis, the jet profile shape is found by solving the turbulent-boundary-layer continuity and momentum relations for zero pressure gradient:

$$\frac{\partial \overline{u}}{\partial x} + \frac{\partial \overline{v}}{\partial y} = 0$$
$$\overline{u}\frac{\partial \overline{u}}{\partial x} + \overline{v}\frac{\partial \overline{u}}{\partial y} = \frac{1}{\rho}\frac{\partial \tau}{\partial y} \tag{6-145}$$

As suggested by Prandtl in 1926, shear stress can be modeled by an eddy viscosity which depends only upon x and has the form of Clauser's outer wake model for a boundary layer, Eq. (6-93):

$$\mu_t(\text{jet}) \approx K\rho U_{max} b = (\text{const})\, x^{1/2} \qquad K \approx 0.016 \tag{6-146}$$

where the variation $x^{1/2}$ follows from our dimensional analysis. Görtler (1942) made the same assumption for plane-jet flow and defined the following similarity variables:

$$\overline{u} = U_0 \left(\frac{x_0}{x}\right)^{1/2} F'(\eta)$$

$$\nu_t = KU_0 b_0 \left(\frac{x}{x_0}\right)^{1/2} \qquad \eta = \frac{\sigma y}{x} \tag{6-147}$$

where σ is a free constant and (U_0, b_0) are the values of $(U_{\max}, b)$ at an initial reference point x_0. Substitution of Eqs. (6-146) and (6-147) in Eqs. (6-145) yields a similarity differential equation for the function F:

$$\tfrac{1}{2} F''' + FF'' + F'^2 = 0 \tag{6-148}$$

where the coefficient $\tfrac{1}{2}$ has been chosen for convenience by specifying that

$$\frac{4Kb_0\sigma^2}{x_0} = 1 \tag{6-149}$$

Equation (6-148) is subject to symmetry at the centerline, $\bar{v} = \partial \bar{u}/\partial y = 0$ at $y = 0$, and a vanishing velocity $\bar{u}$ as y becomes very large:

$$F(0) = F''(0) = F'(\infty) = 0 \tag{6-150}$$

plus the condition $F'(0) = 1$ from the definition of velocity in Eq. (6-147). Comparing these with their laminar-jet equivalents, Eqs. (4-100) and (4-101), we see that the solution for a turbulent plane jet is

$$F = \tanh(\eta)$$

$$\frac{\bar{u}}{U_{\max}} = \mathrm{sech}^2(\eta) \tag{6-151}$$

This distribution is in very good agreement with experiments for plane jets. Görtler (1942) matched the data at the *half-velocity point*, $y_{1/2}$, where $\bar{u} = U_{\max}/2$ and $\eta_{1/2} = \mathrm{sech}^{-2}(0.5) = 0.88$, thus obtaining the estimate

$$\sigma_{\text{plane jet}} \approx 7.67$$

The "width" of the jet is ill-defined, since the velocity drops asymptotically to zero at large y. If we defined $b = 2y_{1/2}$, then the jet grows as

$$\frac{b}{x} = \frac{1.76}{7.67} = \tan 13°$$

with

$$K = \frac{x}{4b\sigma^2} = 0.018$$

Thus a turbulent plane jet grows at a half-angle of 13°, independent of the Reynolds number, and the eddy viscosity is in reasonable agreement with Clauser's formula Eq. (6-93).

6-9.1.2 SOLUTION FOR A CIRCULAR JET.
The analysis for a round jet is quite similar, except that one takes $U_{\max}$ proportional to x^{-1} and uses the axisymmetric boundary-layer relations, Eqs. (4-201). The Görtler (1942) theory for this case gives the profile

$$\frac{\bar{u}}{U_{\max}} \approx \left(1 + \frac{\eta^2}{4}\right)^{-2} \qquad \eta \approx 15.2 \frac{y}{x} \qquad U_{\max} \approx 7.4 \frac{(J/\rho)^{1/2}}{x} \tag{6-152}$$

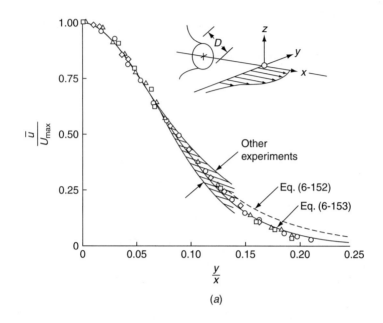

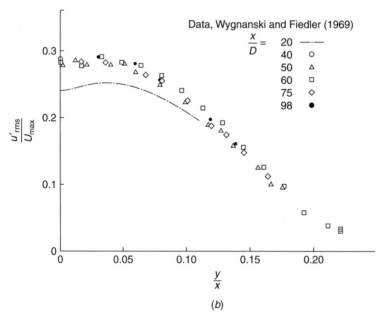

FIGURE 6-37
Experimental data for a circular jet, verifying the approach to self-similarity: (*a*) mean velocity; (*b*) streamwise fluctuation. [*After Wygnanski and Fiedler (1969).*]

This has the same form as the laminar round jet, Eq. (4-207). This solution is compared with experiment in Fig. 6-37a. It agrees reasonably well with the data but is too high in the outer regions of the jet, a discrepancy possibly due to intermittency of turbulence near the jet edge [Corrsin and Kistler (1955)].

A better formula for the round jet is found simply by carrying over the plane-jet solution with a different value of σ:

Round jet: $$\frac{\bar{u}}{U_{max}} \approx \text{sech}^2\left(10.4\frac{y}{x}\right) \quad (6\text{-}153)$$

This expression is seen in Fig. 6-37a to be in excellent agreement with the data of Wygnanski and Fiedler (1969). Self-similarity in the velocity profile occurs for $x/D > 20$, where D is the source diameter. At the centerline, U_{max} decreased as x^{-1} as if the flow began from a virtual origin approximately seven diameters in front of the actual source.

The turbulence components take longer to develop than the mean velocity. Figure 6-37b shows the measured streamwise velocity fluctuation, which is nonsimilar at $x/D = 20$ and does not become self-similar until $x/D = 50$. Note the very high levels of turbulence (30 percent) compared to boundary layers (Fig. 6-4). The transverse components, v'_{rms} and w'_{rms}, do not become similar until $x/D = 70$, at which point the round jet is truly *self-preserving*. Even at $x/D = 100$, the components v' and w' are smaller then u', so that the turbulence has not become isotropic.

Turbulence in free-shear flows can be predicted by higher order theories, such as the K–ϵ model of Sec. 6-7.2.2, but the "constants" in Eqs. (6-107) have to be modified to improve the agreement, as reviewed by Rodi (1984) and Taulbee (1989). The models predict mean velocity very well, growth rate and shear stress moderately well, and are only fair to poor for the turbulence compoments. See Wilcox (1998) for further discussion.

If the jet issues at velocity U_j into a *co-flowing* stream of velocity U_e, the development of self-similarity is strongly affected, as shown by the data of Antonia and Bilger (1973). The mean-velocity difference scales well with $y/y_{1/2}$, similar to Fig. 6-37a, and is well fit by

$$\frac{\bar{u} - U_e}{U_j - U_e} \approx \cos^2\left(\frac{\pi}{4}\frac{y}{y_{1/2}}\right) \quad 0 < y < 2y_{1/2}$$

but the growth rate $y_{1/2}(x)$ is smaller and decidedly non-linear, as U_e/U_j increases. For example, from Fig. 6-37a for $U_e = 0$, $y_{1/2}/D \approx 25$ at $x/D = 300$. However, from the measurements of Antonia and Bilger (1973) for $U_e/U_j = 1/3$ and at $x/D = 300$, $y_{1/2}D \approx 3.2$, or eight times smaller. The mean velocity becomes self-similar for $x/D > 40$, but the streamwise turbulence u'_{rms}/U_{max} is not similar even at $x/D = 266$.

6-9.2 The Plane Mixing Layer

Using an eddy-viscosity analysis, Görtler (1942) also solved for the velocity profile in a mixing layer between parallel streams, including the case where the upper

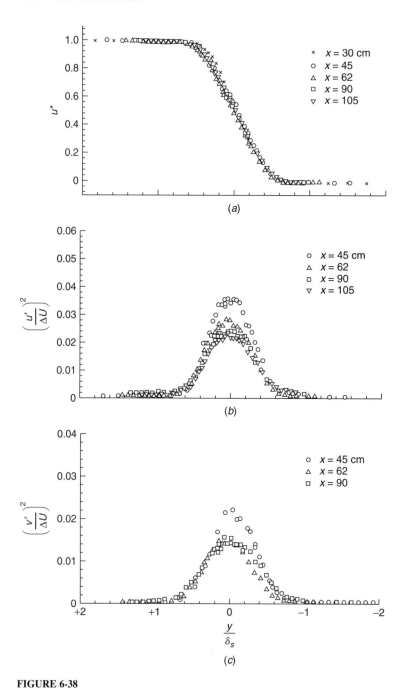

FIGURE 6-38
Measurements in a plane mixing layer: (*a*) mean velocity; (*b*) longitudinal fluctuation; (*c*) transverse fluctuation. [*After Plesniak and Johnston* (*1988*).]

stream moves at U_2 and the lower stream at U_1. The flow is thus approximately antisymmetrical about the midpoint, with boundary conditions

$$\bar{u}(-\infty) = U_1 \quad \bar{u}(+\infty) = U_2 \quad \bar{u}(0) = \tfrac{1}{2}(U_1 + U_2)$$

The solution given by Görtler (1942) is

$$u^* = \frac{\bar{u} - U_1}{U_2 - U_1} = \frac{1}{2}\left[1 + \mathrm{erf}\left(\frac{\sigma y}{x}\right)\right] \quad \sigma \approx 13.5 \quad (6\text{-}154)$$

If we define the layer half-thickness as the point where $u^* = 0.99$, then $\sigma b/x \approx 1.64$, or $b/x \approx 0.121 = \tan(7°)$. Equation (6-154) is in good agreement with experiments such as Plesniak and Johnston (1988), whose data shown in Fig. 6-38a were taken in a water channel with $U_1/U_2 = 0.5$. The abscissa in Fig. 6-38 is not y/x but rather y nondimensionalized by a "shear thickness" δ_s equal to $(U_2 - U_1)$ divided by the slope $(\partial \bar{u}/\partial y)_{\max}$ at the center point ($y = 0$).

Note in Fig. 6-38, as with jet flow, the staged development of self-similarity. The mean velocity in Fig. 6-38a develops at $x = 30$ cm, whereas u'_{rms} in Fig. 6-38b is not self-similar until $x \approx 70$ cm. Finally, v'_{rms} in Fig. 6-38c is not similar within the distance where the data were taken and is not as large as the streamwise fluctuation. Plesniak and Johnston (1988) further show that the mixing layer is strongly affected by streamline curvature, being stabilized or destabilized depending upon the angular momentum ratio of the upper and lower streams.

The plane mixing layer at low Reynolds members is an excellent test case for direct simulation of turbulence by supercomputers. Sandham and Reynolds (1987) used a spectral method, with 1024 longitudinal and 256 transverse mesh points, to compute a turbulent mixing layer with $U_1/U_2 = 0.5$ and an inlet profile with $Re_{\theta 1} = 100$. Vortex pairing and turbulence were triggered by forcing the inlet profile with three of its unstable frequencies. Strictly periodic forcing gave erratic, anomalous results. However, by "jittering" the inlet forcing, i.e., randomly varying the phases of the exciting modes, very realistic results were obtained, as shown in Fig. 6-39. Note how these "numerical snapshots" resemble the visualization of a real mixing layer in Fig. 6-2.

When time-averaged, the simulations in Fig. 6-39 yield a mean-velocity profile almost identical to Fig. 6-38a or Eq. (6-154). However, the agreement is not as good for turbulence components, and the mean scalar fluxes seem to have the wrong sign near the layer edges. It is clear that direct numerical simulation of turbulence is a fruitful research area.

6-9.3 Turbulent Wakes

The wake in Fig. 6-35c looks like a jet carved out of a uniform stream. Since a wake is a "defect" in a moving stream, it has a much stronger effect due to convective acceleration than a jet, and the resulting formulas are different. Far downstream, we assume self-similarity:

$$\frac{\Delta u}{\Delta u_{\max}(x)} = fcn\left[\frac{y}{b(x)}\right] \quad (6\text{-}155)$$

FIGURE 6-39
Three instantaneous snapshots of passive-scalar contours in a supercomputer simulation of a turbulent mixing layer. Compare with Fig. 6-2. [*After Sandham and Reynolds (1987)*.]

As with jet flow, the pressure in the wake is nearly constant—except for turbulent-fluctuation effects—because of the open environment. This time, the momentum theorem states that the drag force F associated with the wake profile is independent of x:

$$F = \int_{-\infty}^{+\infty} \rho \bar{u}\, \Delta u\, dA = \text{const} \approx (\text{const})\, \rho U \Delta u_{\max} b \quad \text{(plane wake)}$$

$$\approx (\text{const})\, \rho U \Delta u_{\max} b^2 \quad \text{(round wake)} \quad (6\text{-}156)$$

where U is the stream velocity outside the wake, assumed uniform. The last two results in Eq. (6-156) follow from the small-defect assumption, $\Delta u \ll U$. Thus, unlike the jet, $\Delta u_{\max}$ is proportional to b^{-1} (plane) and to b^{-2} (circular wake). When these facts are substituted in the boundary-layer equations with the small-defect assumption $u(\partial u/\partial x) \approx U(\partial u/\partial x)$, we find that similarity cannot be achieved unless

Plane wake: $b = \text{const } x^{1/2}$ $\Delta u_{\max} = \text{const } x^{-1/2}$

Circular wake: $b = \text{const } x^{1/3}$ $\Delta u_{\max} = \text{const } x^{-2/3}$ (6-157)

This information enables us to solve for the velocity defect similarity profiles. If we use the Clauser-type eddy-viscosity distribution, Eq. (6-146), and take "b" as the half-velocity point, the solution is a Gaussian distribution [recall Eq. (4-211)]:

$$\frac{\Delta u}{\Delta u_{\max}} \approx \exp\left(\frac{-0.693 y^2}{y_{1/2}^2}\right) \qquad (6\text{-}158)$$

The constants in Eqs. (6-157) that determine the variations of $\Delta u_{\max}$ and $y_{1/2}$ must be established by experiment. This is difficult because of the large-scale structures, e.g., Kármán vortex streets, in typical wakes—similarity may not develop until up to 1000 diameters downstream. Using a dual-plate "small-disturbance" wake generator, Sreenivasan and Narasimha (1982) proposed the following growth-rate expressions for a plane wake:

$$y_{1/2} \approx 0.30 (x\theta)^{1/2}$$

$$\Delta u_{\max} \approx 1.63 U \left(\frac{\theta}{x}\right)^{1/2} \qquad (6\text{-}159)$$

where θ is the momentum thickness of the wake

$$\theta = \int_{-\infty}^{+\infty} \frac{\Delta u}{U}\left(1 - \frac{\Delta u}{U}\right) dy = \text{const}$$

that is independent of x. This writer knows of no comparable correlations for a circular or a three-dimensional wake.

Wygnanski et al. (1986) measured the plane wakes behind various types of bodies—cylinders, high-solidity screens, strips, and airfoils—with the results shown in Fig. 6-40. The flows did not become self-similar until hundreds of momentum thicknesses downstream of the bodies.

The mean-velocity defect data in Fig. 6-40a are in good agreement with the Gaussian formula, Eq. (6-158), (not shown) except near the edge of the wake. However, the growth rates for cylinder wakes fit constants different from Eq. (6-159), that is, (0.275, 1.75) rather than (0.30, 1.63), an 8 percent deviation. Moreover, Fig. 6-40b shows that the streamwise fluctuation behind each reaches self-similarity, but the distributions do not agree with each other. Even the normalized turbulent shear, $(-\overline{u'v'})/\Delta u_{\max}^2$, was not the same behind the various body shapes. Wygnanski et al. (1986) conclude that there is no "universal" state of similarity for two-dimensional wakes except for the mean-velocity defect.

The various cases of free turbulent flows just discussed have their growth rates and velocity decay rates summarized in Table 6-3.

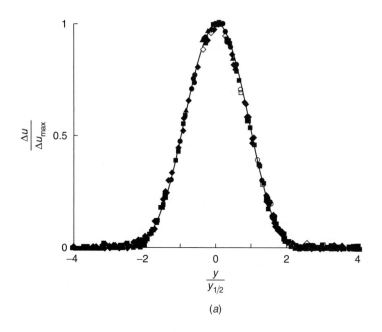

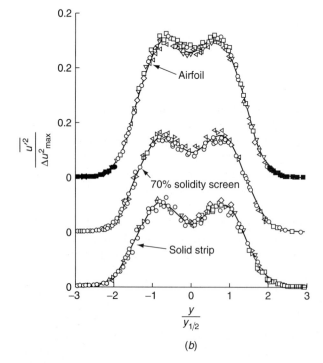

FIGURE 6-40
Measurements of plane wakes behind various-shaped bodies: (a) mean-velocity defect; (b) streamwise turbulent fluctuation. The data are taken for $x/\theta = 200$ to 700. [*After Wygnanski et al.* (*1986*).]

TABLE 6-3
Power laws for jets and wakes

	Laminar flow		Turbulent flow	
	Growth of width b	Velocity decay of $\bar{u}$ or Δu	Growth of width b	Velocity decay of $\bar{u}$ or Δu
Mixing zone	$x^{1/2}$	1.0	x	1.0
Plane jet	$x^{2/3}$	$x^{-1/3}$	x	$x^{-1/2}$
Circular jet	x	x^{-1}	x	x^{-1}
Plane wake	$x^{1/2}$	$x^{-1/2}$	$x^{1/2}$	$x^{-1/2}$
Circular wake	$x^{1/2}$	x^{-1}	$x^{1/3}$	$x^{-2/3}$

6-10 TURBULENT CONVECTIVE HEAT TRANSFER

As in laminar incompressible flow, the heat transfer seems to arise as an afterthought, because the velocity profiles, skin friction, etc., can be calculated essentially independent of temperature. Then, with velocities known, the temperature can, in principle, be calculated from the thermal-energy equation for turbulent-boundary-layer flow

$$\rho c_p \left(\bar{u} \frac{\partial \bar{T}}{\partial x} + \bar{v} \frac{\partial \bar{T}}{\partial y} \right) = \frac{\partial \bar{q}}{\partial y} + \tau \frac{\partial \bar{u}}{\partial y} \qquad (6\text{-}160)$$

where $\quad \tau = \mu \dfrac{\partial \bar{u}}{\partial y} - \rho \overline{u'v'} \quad$ and $\quad \bar{q} = k \dfrac{\partial \bar{T}}{\partial y} - \rho c_p \overline{v'T'}$

The difficulty, as usual, is the correlation of the turbulent-inertia terms.

This section is a brief overview of turbulent convective heat transfer, a subject popular with many textbook and monograph authors. In chronological order, they are Arpaci (1984), Kays and Crawford (1993), Burmeister (1993), Bejan (1994), Kakac and Yener (1994), Oosthuizen (1999), and Kaviany (2001). Some authors specialize in the computational or CHT aspects of the subject: Jaluria and Torrance (1986), Nakayama (1995), Tannehill et al. (1997), and Comini and Sunden (2000). All types of turbulence models, from zero- and one- and two-equation models to second-moment closures, large-eddy simulation (LES), and direct numerical simulation (DNS), have been applied to turbulent heat transfer.

6-10.1 Turbulent Eddy Conductivity

We formally write the Boussinesq analogy for eddy viscosity and eddy conductivity:

$$\tau_{\text{total}} = (\mu + \mu_t) \frac{\partial \bar{u}}{\partial y}$$

$$q_{\text{total}} = -(k + k_t) \frac{\partial \bar{T}}{\partial y} \qquad (6\text{-}22)$$

Now neither μ_t nor k_t is a property of a fluid, but they can be formed into a dimensionless ratio called the *turbulent Prandtl number*:

$$Pr_t = \frac{c_p \mu_t}{k_t} \quad (6\text{-}161)$$

Since the turbulent-flux terms $\overline{v'u'}$ and $\overline{v'T'}$ are due to the same mechanism of time-averaged convection, it follows that their ratio Pr_t ought to be of order unity. This idea is essentially that of Reynolds (1874), who postulated that turbulent momentum flux and heat flux should be equivalent phenomena. Hence

$$Pr_t = f(Pr) = \mathbb{O}(1)$$

This is one form of the celebrated Reynolds analogy for turbulent flow. Even today, over 100 years later, the turbulent Prandtl number, which surely must vary somewhat with local conditions, is taken to be a constant or at most a function of the molecular Prandtl number Pr. Even the latest digital-computer programs for computing turbulent heat transfer take advantage of this assumption.

Much of our experimental database for turbulent forced convection has come from the Stanford University group, reviewed by Moffat and Kays (1984). Figure 6-41 shows the data of Blackwell (1973) for the turbulent Prandtl number

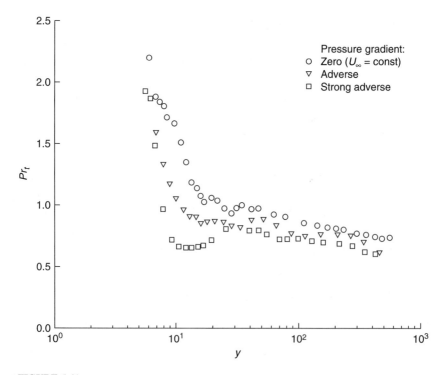

FIGURE 6-41
Experimental measurements of the turbulent Prandtl number in a boundary layer with pressure gradients. [*After Blackwell (1973).*]

of air in positive, zero, and negative pressure gradients. In all cases, Pr_t decreases from about 1.5 in the sublayer to 0.7 at the outer edge of the boundary layer. The increase near the wall is not too important, since both μ_t and k_t are small there; so it is not necessary to curve-fit the data. Instead, it is common to assume a constant value:

$$Pr > 0.7: \qquad Pr_t \approx 0.9 \text{ or } 1.0 \qquad (6\text{-}162)$$

Additional experiments with wall suction and blowing [Simpson et al. (1970)] indicate no effect, so that Eq. (6-162) still holds. Presumably the data uncertainty in Fig. 6-41 is large, since numerical differentiation of both velocity and temperature profiles is involved. Use of a constant Pr_t has been very successful in theoretical models. A lower limit on the molecular Prandtl number is indicated because Pr_t seems to be higher, approximately 2 to 4, for liquid metals, $Pr < 0.1$. Also, in free turbulence (jets, wakes, mixing layers), one should take $Pr_t \approx 0.7$, as Fig. 6-41 indicates.

With Pr_t known, $k_t = c_p \mu_t / Pr_t$ and μ_t can be correlated by the usual eddy viscosity models, e.g., Eqs. (6-92) and (6-93). Therefore the mean temperature profiles and wall heat transfer can be computed without any further assumptions. All turbulence models can be readily extended to thermal calculations.

6-10.2 The Temperature Law of the Wall

Although a pressure gradient affects the near-wall shear stress $\tau(y)$—see Eq. (6-98) and Fig. 6-24—it does not affect, to first order, the near-wall heat flux. To see this, consider the energy Eq. (6-21c) for small y, with $\bar{u} \approx 0$ but with suction or blowing, $\bar{v} \approx v_w \neq 0$. We may integrate this "Couette flow" approximation as follows:

$$\frac{\partial q}{\partial y} \approx \rho c_p v_w \frac{\partial \overline{T}}{\partial y}$$

or
$$q \approx q_w + \rho c_p v_w (\overline{T} - T_w) + \cdots \qquad (6\text{-}163)$$

Near an impermeable wall, then, $q \approx q_w \approx \rho c_p (\nu / Pr + \nu_t / Pr_t)(\partial \overline{T} / \partial y)$, and we may separate and integrate using the law-of-the-wall variable y^+:

$$T^+ = \frac{T_w - \overline{T}}{T^*} = \int_0^{y^+} \frac{dy^1}{1/Pr + \mu_t / \mu Pr_t} \qquad (6\text{-}164)$$

where $T^* = q_w / (\rho c_p v^*)$ is a wall-conduction temperature exactly analogous to the wall-friction velocity v^*. This is the *temperature law of the wall*, first proposed by Kármán (1939).

Assuming constant Pr and Pr_t, Eq. (6-164) may be integrated numerically using, say, Eq. (6-92) for the eddy viscosity μ_t / μ. The results are shown in Fig. 6-42 for $Pr_t = 1.0$ and various molecular Prandtl numbers. Very near the wall,

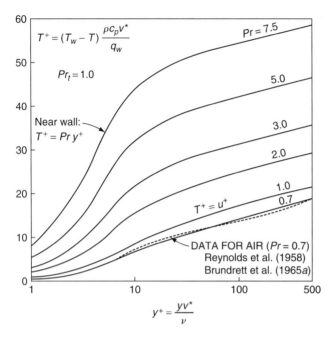

FIGURE 6-42
The temperature law of the wall for fluids with a Prandtl number greater than 0.7, from Eqs. (6-92) and (6-164).

the thermal sublayer has the form $T^+ = Pr\, y^+$. Further away from the wall, a logarithmic layer is obtained:

$$T^+ \approx \frac{Pr_t}{\kappa} \ln y^+ + A(Pr) \qquad (6\text{-}165)$$

the intercept $A(Pr)$ varies strongly with Pr as shown in Fig. 6-42, and an easy-to-remember curve-fit expression is

$$A(Pr) \approx 13 Pr^{2/3} - 7 \qquad (6\text{-}166)$$

valid for $Pr \geq 0.7$. Kader (1981) gives a more complicated curve fit that is valid even for liquid metals ($Pr < 0.03$):

$$A(Pr) \approx (3.85 Pr^{1/3} - 1.3)^2 + 2.12 \ln(Pr) \qquad (6\text{-}167)$$

Kader found this fit to be in good agreement with his survey of temperature profiles measured in air, water, ethylene glycol, and oil, $0.7 \leq Pr \leq 170$. He also gave curve fits for the entire temperature profile—sublayer, overlap layer, and outer layer—for zero or moderate pressure gradients. Kays and Crawford (1993) report a similar data survey. Other $T^+(y^+)$ models have been proposed. Sucec (1999) uses a power law, $T^+ \approx 6(y^+)^{0.17} + 13.2 Pr - 9.37$. Volino and Simon (1997) avoid the Couette flow approximation, Eq. (6-163), and include a pressure gradient and convection in their near-wall analysis. Their formula for T^+ is complex and not in closed-form but is in good agreement with experiment. Huang and Bradshaw

(1995) analyze pressure gradient effects and find best results with the K–ω two-equation model. Cruz and Silva-Freire (2002) have developed a thermal law of the wall for separating and recirculating flow. These correlations are for smooth, impermeable walls. Wall transpiration, especially blowing, has a very strong effect, as discussed by Faraco-Medeiros and Silva-Freire (1992). Hollingsworth et al. (1992) have analyzed the effect of surface curvature.

6-10.3 The Reynolds Analogy for Stanton Number

Recall that in laminar-boundary-layer flow, Chap. 4, there is a proportionality between skin friction and heat transfer for flat-plate flow:

Pr near 1:
$$C_h \approx \frac{C_f}{2Pr^{2/3}} \quad (4\text{-}80)$$

Similar approximations are possible for turbulent flat-plate flow and also for pipe flow because the velocity and temperature wakes are small. We simply evaluate the two log-laws at the edges of the boundary layer:

$$\frac{T_w - T_e}{T^*} \approx \frac{Pr_t}{\kappa} \ln \frac{\delta_T v^*}{\nu} + A$$

$$\frac{U_e}{v^*} \approx \frac{1}{\kappa} \ln \frac{\delta v^*}{\nu} + B$$

where δ and δ_T are the velocity and thermal boundary-layer thicknesses, respectively. Now subtract these two expressions, assuming with little loss in accuracy that $Pr_t \approx 1$ and $\delta \approx \delta_T$, and rewrite as follows:

$$\frac{q_w}{\rho c_p U_e (T_w - T_e)} \left[\frac{U_e}{v^*} + (A - B) \right] \approx \frac{v^*}{U_e}$$

The coefficient on the left is the Stanton number. The right-hand side equals $(C_f/2)^{1/2}$. Therefore, with A from Eq. (6-166), we can rewrite this expression as an approximate Reynolds analogy for turbulent flat-plate flow:

$Pr \geq 0.7$:
$$C_h \approx \frac{C_f/2}{1 + 13(Pr^{2/3} - 1)(C_f/2)^{1/2}} \quad (6\text{-}168)$$

For Pr near unity, this expression is well approximated by Eq. (4-80). It is valid only for smooth, impermeable, isothermal walls. The skin-friction coefficient for a smooth flat plate should be evaluated by Eq. (6-78).

6-10.3.1 FULLY DEVELOPED TURBULENT PIPE FLOW. In pipe flow, the heat-transfer coefficient is based not on the centerline temperature but rather on the bulk or *cup-mixing* temperature, as in Chap. 3 assuming constant c_p:

$$T_m = \frac{\int \rho \overline{u} \overline{T} \, dA}{\int \rho \overline{u} \, dA} \quad (3\text{-}65)$$

The pipe Stanton number is defined as

$$C_h = \frac{q_w}{\rho c_p u_{av}(T_w - T_m)}$$

u_{av} for fully developed turbulent pipe flow has been evaluated from the log-law in Eq. (6-53). If we now substitute the velocity log-law Eq. (6-38a) and temperature log-law Eq. (6-165) into Eq. (3-65) to evaluate T_m, the result, after substantial manipulation, can be expressed as follows:

$$C_{h,\text{pipe}} \approx \frac{\Lambda/8}{1 + 13(Pr^{2/3} - 1)(\Lambda/8)^{1/2}} \quad 0.5 \leq Pr \leq 10^5 \quad (6\text{-}169)$$

where $\Lambda = 4C_f$ is the pipe-friction factor, to be evaluated from Eq. (6-54). This formula, first proposed by Petukhov (1970), is valid only for smooth walls without blowing or suction.

It is customary to display pipe heat-transfer correlations in the form of a Nusselt number rather than a Stanton number:

$$Nu_D = \frac{q_w D}{k(T_w - T_m)} = C_h Re_D Pr$$

For liquid metals, $Pr < 0.1$, where Eq. (6-169) is not valid, a recommended experimental correlation is given by Sleicher and Rouse (1975):

$Pr < 0.1$: $\qquad Nu_D = 6.3 + 0.00167 Re_D^{0.85} Pr^{0.93} \qquad (6\text{-}170)$

Figure 6-43 shows Nusselt numbers plotted from Eqs. (6-169) and (6-170). These values are in good agreement with fully developed pipe-flow experiments for both

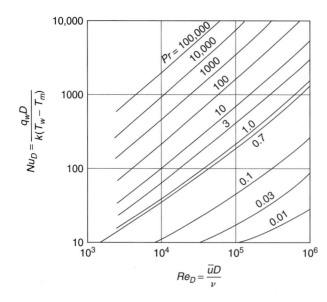

FIGURE 6-43
Nusselt numbers for fully developed turbulent flow in pipes, from Eqs. (6-169) and (6-170).

constant wall temperature and constant wall heat flux. The Nusselt number is typically higher in the entrance of the pipe. Kays and Crawford (1993, Chap. 14) summarize the available data and theories for the entry region.

6-10.4 Turbulent Convection with Suction or Blowing

Wall transpiration has a strong effect on turbulent heat transfer, as reviewed by Moffat and Kays (1984). Figure 6-44 shows some measurements by the Stanford University group of velocity and temperature profiles in flat-plate flow at $Re_x \approx 2E6$. The velocity profiles in Fig. 6-44a rise dramatically with blowing rate, not because of any striking change in turbulent momentum transfer but rather because of change in the shear-stress distribution from Eqs. (6-83). The effect of wall suction ($v_w < 0$) is not so strong, and the measured profile for $v_w/U_e = -0.0024$ agrees well with Stevenson's law of the wall, Eq. (6-86). However, Eq. (6-86) predicts values of u^+ for blowing that are too high, leading Simpson (1968) to propose an alternate law of the wall with transpiration:

$$\frac{2}{v_w^+}\left[(1 + v_w^+ u^+)^{1/2} - (1 + 11 v_w^+)^{1/2}\right] = \frac{1}{\kappa}\ln\left(\frac{y^+}{11}\right) \qquad (6\text{-}171)$$

This relation is formulated so that all profiles start at $(u^+, y^+) = (11, 11)$. Equation (6-171) agrees with the two blowing curves in Fig. 6-44a—that have considerable scatter because of the difficulty of measuring skin friction—but not with other data, where Stevenson's law Eq. (6-86) is better. This controversy—still unresolved—between two competing wall-blowing velocity correlations is discussed in detail by Schetz (1980, pp. 151–155).

Figure 6-44b presents the temperature profiles in wall coordinates, showing much less of an effect of suction and blowing compared to Fig. 6-44a. The same would be true if we added in pressure gradient as a parameter: a dramatic effect on u^+, especially with freestream deceleration, and less of an effect on T^+ [Moffat and Kays (1984)].

Wall transpiration has a strong effect on both skin friction and heat transfer. Figure 6-45 compiles data from the Stanford group [Moffat and Kays (1984)] on the local Stanton number in turbulent flat-plate flow with blowing and suction. We see that suction (or blowing) can cause an order of magnitude increase (or decrease) in wall heat transfer. Kays and Crawford (1980, pp. 180–182), by integrating Eqs. (6-97) and (6-163) across the boundary layer and assuming that eddy viscosity is not affected by transpiration, develop the following correlation for flat-plate flow:

$$\frac{C_f(Re_x, v_w/U_e)}{C_f(Re_x, 0)} \approx \frac{C_h(Re_x, v_w/U_e)}{C_h(Re_x, 0)} \approx \frac{\zeta}{e^\zeta - 1} \qquad (6\text{-}172)$$

where $\zeta = 2(v_w/U_e)/C_f(Re_x, 0)$ or $(v_w/U_e)/C_h(Re_x, 0)$ for wall friction and wall heat transfer, respectively. The reference value $C_f(Re_x, 0)$ is computed from Eq. (6-78), after which $C_h(Re_x, 0)$ is computed for the given Prandtl number from Eq. (6-168). Figure 6-45 shows three representative calculations, for suction ($F = -0.0046$),

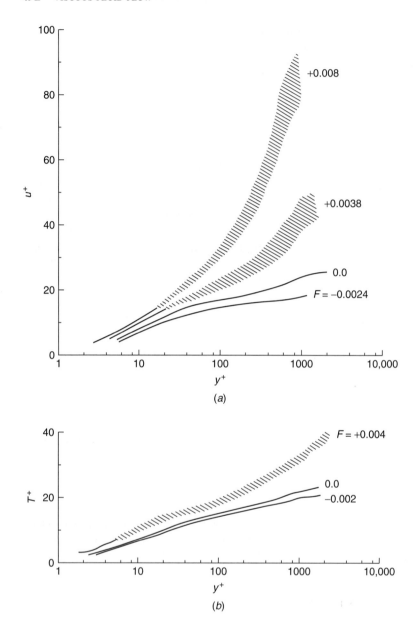

FIGURE 6-44
Experimental (*a*) velocity and (*b*) temperature profiles in flat-plate flow with suction and blowing. The parameter $F = v_w/U_e$. [*After Moffat and Kays (1984)*.]

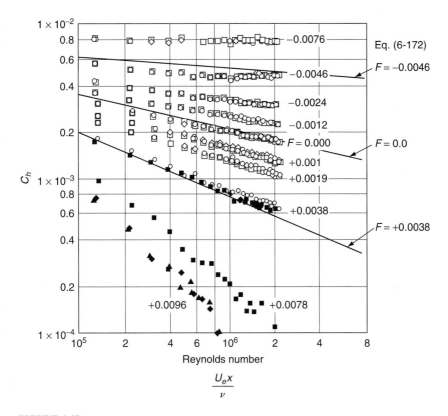

FIGURE 6-45
Measured local Stanton numbers for turbulent flat-plate flow with uniform suction or blowing, where $F = v_w/U_e$. The theory for $F = 0$ uses Eqs. (6-78) and (6-168). [*After Moffat and Kays (1984).*]

impermeable wall ($F = 0$), and blowing ($F = +0.0038$). The agreement is satisfactory for all three cases and indeed fits all the data in the figure. For strong suction, as Re_x increases, the Stanton number asymptotically approaches $C_h = -F$.

6-10.5 Flat-Plate Heat Transfer with Varying Wall Temperature

The previous analyses are for constant temperature difference $T_w - T_e$, whereas in practical applications, the wall temperature is often variable. One approach is to superimpose a series of step changes as was done for laminar flow in Sec. 4-8.4. A comprehensive study of turbulent flat-plate flow with variable $T_w(x)$ was provided in four reports by Reynolds et al. (1958). Suppose that the flow is isothermal, $T_w = T_e$, until a position $x = x_0$, where the wall temperature suddenly changes to $T_w \neq T_e$. Reynolds et al. did an integral analysis, similar to the development of

Eq. (4-28) for laminar flow, assuming that both u and $T - T_w$ vary as $y^{1/7}$. The result is the following formula for local Stanton number aft of x_0:

$$C_h(x, x_0) \approx C_h(x, 0)\left[1 - \left(\frac{x_0}{x}\right)^{9/10}\right]^{-1/9} \tag{6-173}$$

where $C_h(x, 0)$ is the isothermal solution from, say, Eq. (6-168). This result agrees well with their experiments on step change in wall temperature [Reynolds et al. (1958, part 2)]. Now suppose that $(T_w - T_e)$ varies continuously with x. By analogy with the laminar-flow analysis, Eq. (4-169), the total heat transfer over a plate of length x is given by

$$q_w(x) = \int_0^x h(x, x_0)\, d(T_w - T_e)$$

$$h(x, x_0) = \frac{q_w(x, x_0)}{T_w - T_e} \tag{6-174}$$

If the temperature difference contains a number of discontinuities ΔT_i occurring at points $x_0(i)$, Eq. (6-174) separates into a Riemann integral plus a summation of indicial functions, exactly as with Eq. (4-170):

$$q_w(x) = \int_0^x h(x, x_0)\frac{d}{dx_0}(T_w - T_e)\, dx_0 + \sum_{i=1}^{N} h[x, x_0(i)]\Delta T_i \tag{6-175}$$

This development is exactly the same as for laminar flow (Sec. 4-8.4). To provide a numerical example for turbulent flow, suppose that ΔT is a polynomial

$$T_w - T_e = a_0 + a_1 x + a_2 x^2 + a_3 x^3 + \cdots$$

Then the Stanton number referred to the local temperature difference would be

$$\frac{C_{h,\,\text{actual}}}{C_{h,\,\text{plate at }\Delta T}} = \frac{1}{\Delta T(x)}\int_0^x \left[1 - \left(\frac{x_0}{x}\right)^{9/10}\right]^{-1/9}\left(a_0 + \sum_{j=1}^{N} j a_j x_0^{j-1}\right)dx_0$$

Again, as before, the integrals are readily evaluated in terms of gamma functions:

$$\frac{C_{h,\,\text{actual}}}{C_{h,\,\text{plate at }\Delta T}} = \frac{1}{\Delta T(x)}\left[a_0 + \sum_{j=1}^{N} \frac{10}{9} j a_j x^j \frac{\Gamma(10j/9)\Gamma(8/9)}{\Gamma(10j/9 + 8/9)}\right] \tag{6-176}$$

By evaluating $\Gamma(\frac{8}{9}) = 1.07776$, for example, we can write the first five terms

$$\frac{C_{h,\,\text{actual}}}{C_{h,\,\text{plate at }\Delta T}} = \frac{a_0 + 1.1340 a_1 x + 1.2026 a_2 x^2 + 1.2497 a_3 x^3 + 1.2858 a_4 x^4 + \cdots}{a_0 + a_1 x + a_2 x^2 + a_3 x^3 + a_4 x^4 + \cdots} \tag{6-177}$$

Thus the heat transfer changes at a faster rate than one would estimate by simply applying the isothermal-wall formula at the local $\Delta T(x)$. However, the equivalent changes in laminar flow, Eq. (4-173), are much greater. Reynolds et al. (1958, part 3) also study other cases of varying wall temperature, such as a delayed ramp or a suddenly insulated wall. Note that the previous solution is valid only for flat-plate flow with a smooth, impermeable wall. The same superposition scheme would work for other conditions, such as favorable or adverse pressure gradients, but new indicial functions analogous to Eq. (6-173) would have to be found and integrated.

6-10.6 Turbulent Convection with Pressure Gradients

For turbulent heat transfer with both wall temperature *and* freestream velocity varying, several analyses are possible: (1) a remarkably simple quadrature scheme; (2) an inner variable integral method; or (3) various differential methods, including eddy conductivity computations and the K–ϵ two-equation approach. Basically, temperature plays the role of a passive scalar in the energy Eq. (6-21c), which is added to the continuity and momentum relations.

A simple and reliable theory was given by Ambrok (1957), who solved a modeled energy-integral equation so that the local turbulent Stanton number can be calculated by simple quadrature:

$$C_h(x) \approx 0.0295 Pr^{-0.4} \frac{r_0^{0.25}(T_w - T_e)^{0.25}\mu^{0.2}}{\left[\int_0^x r_0^{1.25}(T_w - T_e)^{1.25}\rho U_e \, dx\right]^{0.2}} \quad (6\text{-}178)$$

where $r_0(x)$ is the surface radius of an axisymmetric body (Fig. 4-34). If the body is two-dimensional, one simply drops out the r_0 terms. Note that in the limit of constant $T_w - T_e$ and constant U and r_0, Eq. (6-178) reduces to

$$C_h \approx 0.0295 Pr^{-0.4} Re_x^{-0.2} \approx \frac{C_f}{2 Pr^{0.4}} \quad (6\text{-}179)$$

This is approximately the same as the traditional Reynolds analogy in Sec. 6-10.3.

By extending the momentum method of Das (1988) to the thermal-energy equation, Sucec (1999) developed an inner variable, turbulent-boundary-layer integral method for heat transfer. The result is an added "thermal" first-order differential equation for temperature parameters, similar to Eq. (6-130). His analysis included variable wall blowing and suction, and the results are in good agreement with heat-transfer data for flat plates and for pressure gradients.

Another inner variable integral method that includes both skin friction and wall heat flux was developed by White et al. (1973). By defining and integrating simplified wall functions for velocity and temperature, they gave two coupled first-order ordinary differential equations in the dimensionless variables

$$\lambda_1 = \frac{U_e}{v^*}\left(\frac{T_w}{T_e}\right)^{1/2}$$

$$\lambda_2 = \frac{q_w}{\rho U_e^3}\frac{T_e}{T_w} \quad (6\text{-}180)$$

Assuming known $U_e(x)$, $T_e(x)$, and $T_w(x)$—arbitrary Mach number $Ma_e(x)$ is also allowed—one solves for $\lambda_1(x)$ and $\lambda_2(x)$, whence $v^*(x)$ and $q_w(x)$ can be calculated.

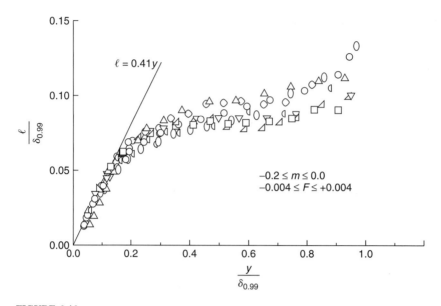

FIGURE 6-46
Measured values of turbulent mixing length for decelerating flows, $U_e = Cx^m$, and for wall blowing and suction, $F = v_w/U_e$. [After Moffat and Kays (1984).]

The differential methods are easily extended to include heat transfer by finite-difference solution of the energy Eq. (6-21c), using the known velocity components from the momentum analysis of Sec. 6-8.3.1 and the assumption of constant Pr_t, so that eddy conductivity $k_t = c_p \mu_t / Pr_t$. The overlap and outer layer eddy-viscosity models of Fig. 6-23 are still valid, even with heat transfer, pressure gradient, and wall transpiration. This is shown from mixing-length measurements by the Stanford group [Moffat and Kays (1984)] in Fig. 6-46. The data are well approximated by $\ell = \kappa y$ in the overlap layer and $\ell = 0.09\delta$ in the outer wake.

Mixing length *does* vary near the wall. The scale of Fig. 6-46 is too large to see the dramatic changes in the sublayer damping constant A from Eq. (6-90). Moffat and Kays (1984) recommend the following empirical algebraic correlation for the van Driest damping constant:

$$A \approx \frac{26}{1 + a[v_w^+ + bp^+/(1 + cv_w^+)]} \quad (6\text{-}181)$$

where p^+ is defined by Eq. (6-100) and where

$a = 7.1 \quad \text{if } v_w^+ \geq 0 \quad \text{and} \quad a = 9.0 \quad \text{if } v_w^+ < 0$

$b = 5.24 \quad \text{if } p^+ \leq 0 \quad \text{and} \quad b = 2.9 \quad \text{if } p^+ > 0$

$c = 10.0 \quad \text{if } p^+ \leq 0 \quad \text{and} \quad c = 0.0 \quad \text{if } p^+ > 0$

Moffat and Kays, also correlate A with the local roughness Reynolds number. Compare Eq. (6-180) with Granville's suggestion, Eq. (6-100). Similar eddy conductivity correlations have been used successfully for heat-transfer calculations by Cebeci and Smith (1974) and By Herring and Mellor (1968). Their formulations include compressibility and axisymmetrical flows. See also Wilcox (1998).

The two-equation methods can also easily accommodate the energy equation for heat-transfer computations. Jones and Launder (1972) show good agreement of their K–ϵ model with four different runs in the experiments of Moretti and Kays (1965) for varying $U_e(x)$ and $T_w(x)$.

Figure 6-47 compares theory and experiment for one run of Moretti and Kays (1965), for which the freestream velocity has a favorable gradient and the

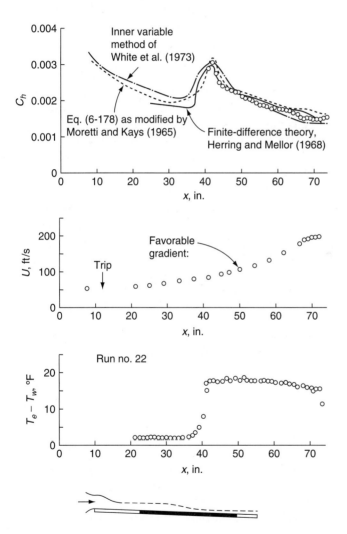

FIGURE 6-47
Comparison of theory and experiment for the variable-velocity and wall temperature experiment of Moretti and Kays (1965).

wall temperature changes abruptly in the middle of the run. Good agreement is shown in Fig. 6-47 with the theories of Ambrok (1957), White et al. (1973), and Herring and Mellor (1968) mentioned before. Jones and Launder (1972) did not report results for this particular run but showed excellent results for four related tests.

SUMMARY

This chapter presents an introduction to the analysis of turbulent time-averaged flows. No detailed treatment is given of the statistical theory of turbulence. The discussion emphasizes engineering properties of turbulent shear layers, such as mean velocity and temperature profiles, wall friction, flow separation, and heat transfer.

After deriving the Reynolds time-averaged equations of turbulent flow, semi-empirical correlations are presented for the inner, outer, and overlap velocity layers in wall-related flows. Simple analyses are then presented for turbulent flow in ducts and past flat plates. For more complex flows, turbulence *modeling* is introduced: eddy viscosity, mixing length, the K–ϵ model, Reynolds stress formulations, and algebraic stress modeling.

Turbulent boundary layers with pressure gradients are studied by both integral and finite-difference methods. Free turbulence—jets, wakes, and mixing layers—are then discussed. The subject of turbulent heat transfer is briefly treated: the temperature law of the wall, the turbulent Prandtl number, eddy conductivity, the Reynolds analogy, and parametric effects such as pressure gradients, wall suction and blowing, and variable surface temperature.

Because of its practical importance and inherent complexity, turbulent-flow research is the most active field in fluid mechanics. Since turbulent fluctuation length scales and frequencies cover such a broad spectrum, present supercomputers—or even future mega-supercomputers—are not able to resolve the full details of high Reynolds number flow. Therefore, for the immediate future, turbulence research will concentrate on detailed experiments and improved semiempirical models of turbulent flow.

PROBLEMS

6-1. By direct substitution of the fluctuation definitions Eqs. (6-6) and the use of averaging rules Eqs. (6-7), develop the three-dimensional time-averaged x-momentum equation and show what reductions occur in a steady two-dimensional turbulent boundary layer.

6-2. Using the Navier–Stokes equations for cylindrical coordinates from App. B and the concepts of Eqs. (6-6) and (6-7), develop a three-dimensional Reynolds stress z-momentum relation for turbulent flow. Simplify this to axial boundary-layer flow along the outside of a cylinder.

6-3. Consider turbulent flow past an isothermal flat plate of width b and length L with constant (ρ, μ, c_p, k). Assume $\delta_u \approx \delta_T$, that is, $Pr \approx 1$. At $x = 0$, the flow has uniform velocity U and temperature T_e. At $x = L$, the mean flow may be approximated by one-seventh power-law profiles:

$$\frac{u}{U} \approx \frac{T - T_w}{T_e - T_w} \approx \left(\frac{y}{\delta}\right)^{1/7}$$

There is no information about the flow structure between the leading and trailing edges. Use a control-volume analysis to estimate, on one side of the plate, (a) the total friction drag and (b) the total heat transfer in terms of the boundary-layer thickness.

6-4. The experiment of Clauser (1954), flow 2200 of Coles and Hirst (1968), used air at 24°C and 1 atm. At the first station, $x = 6.92$ ft, the turbulent-boundary-layer velocity data are as follows:

y, in.	u, ft/s	y, in.	u, ft/s
0.1	16.14	0.8	22.88
0.15	17.02	0.9	23.70
0.2	17.54	1.0	24.38
0.25	18.16	1.25	26.51
0.3	18.69	1.5	28.21
0.4	19.60	2.0	31.22
0.5	20.49	2.5	32.27
0.6	21.24	3.0	32.44
0.7	22.03	3.5	32.50

The boundary-layer thickness was 3.5 in., and the local freestream velocity gradient was $dU/dx \approx -1.06$ s^{-1}. Analyze these data, with suitable plots and formulas, to establish (a) the inner law and wall shear stress, (b) the outer law with Clauser's parameter β and the Coles wake parameter Π, and (c) the logarithmic overlap.

6-5. Analyze the velocity data of Prob. 6-4 to achieve a power-law overlap approximation, as in Eq. (6-50). Find α and C and compare with Eq. (6.50). To save work, use Eq. (6-69) to estimate the momentum thickness. At this station, the wall shear stress is approximately 0.0027 lb$_f$/ft^2. (Hint: Use only the first 10 data points, the rest are well into the wake region.)

6-6. For developed turbulent smooth wall pipe flow, assuming that the log-law analysis of Sec. 6-5.1 is valid with $\kappa = 0.41$, show that the maximum pipe velocity may be computed from

$$\frac{u_{max}}{u_{av}} \approx 1 + 1.29\Lambda^{1/2}$$

where Λ is the Darcy friction factor.

6-7. Water at 20°C flows through a smooth pipe of diameter 3 cm at 30 m^3/h. Assuming developed flow, estimate (a) the wall shear stress (in Pa), (b) the pressure drop (in Pa/m), and (c) the centerline velocity in the pipe. What is the maximum flow rate for which the flow would be laminar? What flow rate would give $\tau_w = 100$ Pa?

6-8. The overlap region of Clauser's velocity profile in Prob. 6-4 may be fit by a power-law estimate $u^+ \approx 7.5(y^+)^{1/6}$. Use this result to estimate the wall shear stress in lb$_f$/ft^2 and the shape factor H.

6-9. Consider fully developed turbulent flow through a duct of square cross section. Taking advantage of the double symmetry, analyze this problem using the log-law, Eq. (6-38a), plus a suitable assumption about variation of shear stress around the cross section. Compare your result for Λ with the hydraulic-radius concept.

6-10. In the overlap layer, turbulent shear is dominant, and the effect of viscosity is small. Suppose that we neglect μ and replace Eq. (6-32) by the approximate gradient relationship

$$\frac{d\overline{u}}{dy} \approx fcn(y, \tau_w, \rho)$$

Show, by dimensional analysis, that this leads directly to the logarithmic overlap law Eq. (6-38a).

6-11. Use the log-law Eq. (6-38a) to analyze turbulent flow through a smooth concentric annulus of inner radius r_i and outer radius r_o. Find expressions for the velocity distribution $u(r)$ and the friction factor Λ. Compare your friction result with the hydraulic-radius approximation. (The maximum velocity does not occur midway in the annular region but could be assumed so as a first approximation.)

6-12. Modify the flat-plate integral analysis of Sec. 6-6.1 by using Eq. (6-71) rather than Eq. (6-69) to combine with Eq. (6-67). (Numerical integration may be necessary.) Compare your results for friction factor and boundary-layer thickness with Eqs. (6-70).

6-13. The flat-plate formulas of Sec. 6-6 assume turbulent flow beginning at the leading edge ($x = 0$). More likely, there is an initial region of laminar flow, as in Fig. P6-13.

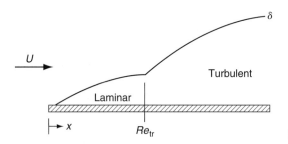

FIGURE P6-13

Devise a scheme to compare $\delta(x)$ and $c_f(x)$ in the turbulent region, $Re > Re_{tr}$, by accounting for the laminar part of the flow.

6-14. Water at 20°C and 1 atm flows at 6 m/s past a smooth flat plate 1 m long and 60 cm wide. Estimate (a) the trailing-edge displacement thickness, (b) the trailing-edge wall shear stress, and (c) the drag of one side of the plate, if $Re_{x,\,tr} = 10^6$.

6-15. Modify Prob. 6-14 to pose the same questions if the plate has an average roughness height of 0.1 mm. Also estimate the value of the log-law shift ΔB at the trailing edge.

6-16. Derive Stevenson's law of the wall with suction or blowing, Eq. (6-86).

6-17. Rewrite Stevenson's relation Eq. (6-86) in the form of a wall-friction law with suction or blowing. Show that the ratio of C_f to the impermeable-wall value C_{f0} is approximately a function only of a "blowing parameter" $B = (2v_w)/(U_e C_f)$. Plot C_f/C_{f0} vs. B in the range $-0.5 < B < 2.0$ and compare with the correlation $[\ln(1 + B)]/B$ recommended by Kays and Crawford (1980, p. 181).

6-18. Water at 20°C flows through a smooth permeable pipe of diameter 8 cm. The volume flow rate is 0.06 m³/s. Estimate the wall shear stress, in pascals, if the wall velocity is (a) 0.01 m/s blowing, (b) 0 m/s, and (c) 0.01 m/s suction. To avoid excessive iteration, assume that the ratio of average to centerline velocity is 0.85.

6-19. Use Stevenson's relation Eq. (6-86) to develop a formula for wall friction in pipe flow with blowing or suction. Apply your result to Prob. 6-7 modified so that the wall blowing rate is 3 cm/s.

6-20. As an alternative to Eq. (6-62), Bergstrom et al. (2002) suggest the following formula for the downshift of the log-law Eq. (6-60) due to uniform surface roughness of height k:

$$\Delta B \approx \frac{1}{\kappa} \ln(k^+) - 3.5 \quad \text{for} \quad k^+ \geq 4.2$$

First compare this correlation, with a sketch or graph, to Eq. (6-62). Then apply this correlation to derive a formula for pipe-friction factor Λ, similar to Eq. (6-64).

6-21. Use numerical quadrature to evaluate and sketch Eq. (6-96) for zero pressure gradient. Compare your results with Eq. (6-41) and Fig. 6-11.

6-22. Use the log-law, Eq. (6-38a), to analyze Couette flow between parallel plates a distance $2h$ apart, with the upper plate moving at velocity U. Show that the turbulent-flow velocity profile is S-shaped, as in Fig. 3-5. Sketch the profile for $Uh/\nu = 10^5$ and compute the ratio $\tau_w h/\mu U$ for this condition.

6-23. Use the log-law Eq. (6-38a) to analyze steady turbulent flow about a long cylinder of radius R rotating at angular rate ω in an infinite fluid. Derive and sketch the variation of skin friction, $C_f = 2\tau_w/\rho\omega^2 R^2$, vs. Reynolds number, $Re = \omega R^2/\nu$.

6-24. Consider the flat-plate flow of Prob. 6-14. Use any integral method from Sec. 6-8 to compute the distribution of $\tau_w(x)$, $\delta^*(x)$, and $H(x)$ along the plate surface. Compare your results with the traditional algebraic flat-plate formulas. For simplicity, neglect the laminar region and begin at $x = 4$ cm with $C_f \approx 0.005$.

6-25. To illustrate the increased resistance of a turbulent boundary layer to separation, compute the Howarth freestream velocity distribution, $U = U_0(1 - x/L)$, using an integral method from Sec. 6-8. Assume turbulent flow from the leading edge ($x = 0$) and find the separation point x_{sep}/L for $U_0 L/\nu =$ (a) 10^6, (b) 10^7, and (c) 10^8.

6-26. Assume potential flow past a cylinder, $U = 2U_0 \sin(x/a)$, with a turbulent boundary layer starting from the stagnation point. Use any turbulent-boundary-layer integral method to compute the value of the separation point ϕ_{sep} for a Reynolds number $U_0 a/\nu$ equal to (a) 10^6, (b) 10^7, and (c) 10^8. Compare with the laminar value $\phi_{sep} \approx 105°$.

6-27. Use one of integral methods of Sec. 6-8 to compute the skin-friction distribution for experiment II of Clauser (1956), flow 2300 of Coles and Hirst (1968). The experimental freestream velocity and skin friction are tabulated. The kinematic viscosity of the fluid (air) was 0.000165 ft²/s. Compare your computed results with the experiment. The average deviation of the seven "good" methods at the Stanford Conference for this experiment was approximately ±15 percent.

x, ft	U, ft/s	C_f (exp)
7.5	26.1	0.00127
9.0	24.8	0.00115
11.0	23.5	0.00110
12.67	22.8	0.00104
16.17	21.3	0.00105
19.17	20.2	0.00105
23.92	18.9	0.00095
26.67	18.1	0.00085

6-28. Use any integral method from Sec. 6-8 to compute $C_f(x)$ for the experiment of Moses, case 5 [flow 4000 of Coles and Hirst (1968)]. The fluid was air with $\nu = 0.000166$ ft^2/s. The experimental freestream velocity and skin-friction data are tabulated. Plot your computer results and compare with the data.

x, ft	U, ft/s	C_f
0.0	82.00	0.00471
0.162	76.26	0.00429
0.323	69.09	0.00347
0.484	61.36	0.00233
0.646	55.49	0.00146
0.865	51.14	0.00090
1.058	49.54	0.00090
1.303	48.93	0.00156
1.516	48.65	0.00215
1.734	48.79	0.00251
1.979	48.86	0.00278
2.198	48.51	0.00280
2.438	48.86	0.00294
2.683	49.13	0.00306
2.928	48.16	0.00305

6-29. Using the similarity concepts of Sec. 6-9, derive power-law expressions, similar to Table 6-3, for the variation of total mass flux with x in the developed region of (*a*) a plane jet, (*b*) a circular jet, and (*c*) a plane mixing layer. Compare these with laminar-flow results.

6-30. At a certain cross section in a developed turbulent plane water jet, the maximum velocity in 3 m/s and the mass flow is 800 kg/s per meter of width. Estimate (*a*) the jet width, (*b*) maximum velocity, and (*c*) total mass flow, at a position 2 m further downstream.

6-31. Air at 20°C and 1 atm issues at 0.001 kg/s from a 4 mm diameter orifice into still air. At a section in the jet 1 m downstream of the orifice, estimate (*a*) the maximum velocity, (*b*) the jet width, and (*c*) the ratio μ_t/μ.

6-32. As part of a low-temperature thermal-power design, a long 5 m diameter vertical cylinder is placed in the ocean. The current across the cylinder is 60 cm/s. At a point

INCOMPRESSIBLE TURBULENT MEAN FLOW 503

1 km downstream of the cylinder, estimate (a) the wake width (in m) and (b) the maximum velocity defect (in cm/s).

6-33. Evaluate the temperature law of the wall Eq. (6-164) numerically, using the van Driest eddy viscosity Eq. (6-90), for $Pr_t = 1.0$ and various values of Pr. Compare your results with Eqs. (6-165) and (6-166).

6-34. Air at 20°C and 1 atm flows at 60 m/s past a smooth flat plate 1 m long and 60 cm wide. The plate surface temperature is 50°C. Estimate the total heat loss (in W) from one side of the plate.

6-35. Modify Prob. 6-34 if the fluid is water flowing at 6 m/s.

6-36. Modify Prob. 6-34 (for airflow) if the plate is permeable, with a uniform blowing velocity of 2 cm/s.

6-37. Water at 20°C and 1 atm flows at 4 kg/s into a smooth tube of diameter 3 cm and length 2 m. If the tube wall temperature is 40°C, estimate the average outlet temperature of the water.

6-38. Reconsider the liquid film flowing down an inclined plane from Prob. 3-15. This time, let the flow be *turbulent* and of constant depth h. Using the log-law for the velocity, (a) find an expression for the bottom wall shear stress as related to ρ, g, μ, θ, h, and the surface velocity U_s. (b) Find a formula for the flow rate Q per unit width and compare to laminar flow, $Q \propto h^3$.

6-39. Use the method of Ambrok, Eq. (6-178), to estimate the heat transfer from an isothermal cylinder at $Re_D = 10^6$ and $Pr = 1.0$. Compare your overall Nusselt number with the experimental value $Nu_D \approx 1400$.

6-40. Investigate the possibility of modifying the explicit or implicit finite-difference boundary-layer methods of Secs. 4-7.1 and 4-7.2 to a two-dimensional *turbulent* boundary layer, using an eddy-viscosity model. Explain any difficulties that arise.

6-41. Assuming eventual success with the development of a turbulent finite-difference technique in Prob. 6-40, apply the method to computation of $C_f(x)$ in flat-plate flow at $Re_L = 10^6$. Take the flow to be turbulent from the leading edge.

6-42. Use the log-law Eq. (6-38a) to analyze turbulent flow near a rotating disk of radius R and angular velocity ω. Let the crossflow velocity w be given by a parabolic hodograph assumption:

$$\overline{w} \approx \overline{u} \tan\beta \left(1 - \frac{y}{\delta}\right)^2 \qquad \beta \approx 12°$$

Recall the hodograph concept of Fig. 4-42. Derive expressions for skin friction C_f and moment coefficient $C_M = 4M/(\rho\omega^2 R^5)$ as a function of disk Reynolds number $Re = \omega R^2/\nu$. Compare your results with the formulas given by Kármán (1921): $C_f \approx 0.053/Re^{1/5}$ and $C_M \approx 0.146/Re^{1/5}$ (see Fig. 3-30).

6-43. Consider a two-dimensional flat-walled diffuser, as in Fig. P6-43. Assume incompressible flow with a one-dimensional freestream velocity $U(x)$ and entrance velocity U_o at $x = 0$. The entrance height is W and the constant depth into the paper is $b = 4W$. Assume turbulent flow at $x = 0$, with momentum thickness $\theta/W = 0.02$, $H(0) = 1.3$, and $U_o W/\nu = 10^5$. Using the method of Head, Sec. 6-8.1.2, numerically estimate the angle ϕ for which separation will occur at $x = 1.5\ W$.

504 VISCOUS FLUID FLOW

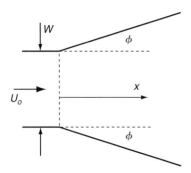

FIGURE P6-43

6-44. Solve Prob. 6-43 instead by the Kármán method of Sec. 6-8.1.1. If time permits, investigate three different Reynolds numbers, $U_o W/\nu = 10^5, 10^6$, and 10^7. Do you expect the separation angle ϕ to increase as the Reynolds number increases?

6-45. Solve Prob. 6-43 again by Head's method if the diffuser is an expanding *cone* with an inlet pipe of diameter W. This time there is no "depth into the paper b." For a given Reynolds number, do you expect a cone to have a smaller separation angle ϕ than a flat-walled diffuser?

6-46. Sink flow, Fig. 4-16, if begun far out with a turbulent boundary layer, may relaminarize as it approaches the origin. Show that the acceleration parameter K_{crit} relates to sink flow and also relates to the Reynolds number of the flow. According to K_{crit} data, what Reynolds number should cause sink flow to relaminarize?

6-47. As an improvement on Stevenson's inner velocity law with suction and blowing, Eq. (6-86), Oljaca and Sucec (1997) offer the following inner/overlap/outer velocity profile formula based upon wall transpiration data compiled by Donald Coles in 1971:

$$u^+ \approx \sigma + \frac{2\Pi}{\kappa} f + \frac{v_w^+}{4}\left(\sigma^2 + \frac{4\Pi}{\kappa}\sigma f + \frac{4\Pi^2}{\kappa^2}f^2\right)$$

where
$$\sigma = \frac{1}{\kappa}\ln(y^+) + A$$

The function $f \approx 3(y/\delta)^2 - 2(y/\delta)^3$ is the wake shape, as in Eq. (6-46). The constant $A \approx 5$ for moderate blowing or suction. Show how this profile differs from the Fig. 6-22 curves for $v_w^+ = \pm 0.02$ and also compare it with Fig. 6-44a for $v_w^+ = 0.0038$.

CHAPTER 7

COMPRESSIBLE-BOUNDARY-LAYER FLOW

7-1 INTRODUCTION: THE COMPRESSIBLE-BOUNDARY-LAYER EQUATIONS

So far the entire text has been devoted to the analysis of incompressible viscous flow. If we now allow for compressibility of a fluid, it follows by definition that we must consider density to be a new variable. In a compressible boundary layer, accelerations are important, and continuity is not trivial. Therefore density is an important variable in this chapter. Further, even in a compressible boundary layer, the normal pressure gradient is usually negligible, and $p \approx p_e(x)$, so that both pressure and density vary and hence the temperature must also vary through a thermodynamic state relation $T = T(p, \rho)$. Since most compressible flows are gases, by far the most common state relation is the perfect-gas relation $p = \rho RT$, and we must be careful to note that T in what follows denotes absolute temperature through the perfect-gas requirement.

In deriving the two-dimensional compressible, laminar, boundary-layer equations, with (x, y) being parallel and normal to the wall, respectively, the same approximations hold: $v \ll u$ and $\partial/\partial x \ll \partial/\partial y$. We need only remember that density is variable and that the variation of k and μ with temperature may be important. Then the two-dimensional boundary-layer equations for an arbitrary compressible fluid are

Continuity:
$$\frac{\partial \rho}{\partial t} + \frac{\partial}{\partial x}(\rho u) + \frac{\partial}{\partial y}(\rho v) = 0 \qquad (7\text{-}1a)$$

x momentum:
$$\rho\left(\frac{\partial u}{\partial t} + u\frac{\partial u}{\partial x} + v\frac{\partial u}{\partial y}\right) \approx -\frac{\partial p_e}{\partial x} + \frac{\partial}{\partial y}\left(\mu\frac{\partial u}{\partial y}\right) \qquad (7\text{-}1b)$$

y momentum:
$$\frac{\partial p}{\partial y} \approx 0 \qquad (7\text{-}1c)$$

Energy:
$$\rho\left(\frac{\partial h}{\partial t} + u\frac{\partial h}{\partial x} + v\frac{\partial h}{\partial y}\right) \approx \frac{\partial p_e}{\partial t} + u\frac{\partial p_e}{\partial x}$$
$$+ \frac{\partial}{\partial y}\left(k\frac{\partial T}{\partial y}\right) + \mu\left(\frac{\partial u}{\partial y}\right)^2 \qquad (7\text{-}1d)$$

where $h = e + p/\rho$ is fluid enthalpy. The y-momentum equation is included simply to show that the pressure variation is governed by the freestream. We have three equations in five unknowns (u, v, ρ, h, T) and therefore need two supplementary relations to correlate the two extra thermodynamic variables

$$T = T(p, \rho) \qquad h = h(p, \rho)$$

Also, the freestream pressure gradient is related to velocity and enthalpy gradients through the Bernoulli equations for inviscid nonconducting flow

$$\frac{\partial p_e}{\partial x} = -\rho_e\left(\frac{\partial U_e}{\partial t} + U_e\frac{\partial U_e}{\partial x}\right) = \frac{\rho_e}{U_e}\frac{\partial h_e}{\partial t} + \rho_e\frac{\partial h_e}{\partial x} - \frac{1}{U_e}\frac{\partial p_e}{\partial t} \qquad (7\text{-}2)$$

Finally, it is assumed that we know how the transport coefficients vary

$$\mu = \mu(T) \qquad k = k(T)$$

through, say, the power-law or Sutherland law formulas discussed in Chap. 1. The boundary conditions are, as usual, no slip ($u = v = 0$) and no temperature jump ($T = T_w$) at the surface and, at the outer edge, a match to the freestream conditions, ($u \to U_e, T \to T_e, h \to h_e$), with no requirement on v at the outer edge.

An alternate relation that will shortly be very useful is obtained by rewriting the energy Eq. (7-1d) in terms of the total enthalpy $H = h + u^2/2$ ($v^2/2$ being entirely negligible):

$$\rho\frac{DH}{Dt} = \frac{\partial p_e}{\partial t} + \frac{\partial}{\partial y}\left(k\frac{\partial T}{\partial y} + \mu u\frac{\partial u}{\partial y}\right) \qquad (7\text{-}3)$$

This is seldom analyzed for unsteady conditions.

This chapter is an overview of viscous compressible flow. For further reading, there are entire monographs on this subject: Park (1990), Smits and Dussage (1996), Oosthuizen and Carscallen (1997), Laney (1998), Anderson (2000, 2002), Chattot (2002), Felcman et al. (2003), and Ockendon and Ockendon (2004).

Somewhat to our disadvantage, the primary emphasis of these texts is usually inviscid flow, not boundary-layer flow. There is a specialized book on supersonic jet flow, Morris et al. (2002), plus comprehensive review articles by Bradshaw (1977), Spina et al. (1994), and Monnoyer (1997). And at least two boundary-layer texts, Schetz (1992) and Schlichting and Gersten (2000), have excellent sections on compressible flow.

7-1.1 Steady Isentropic Flow of a Perfect Gas

We are concerned here with *viscous* flow of gases. But the freestream outside a compressible boundary layer is generally frictionless and adiabatic or even isentropic. Although one can deal computationally with a nonperfect gas, here we assume a perfect gas

$$p = \rho RT \quad R = \text{gas constant} \approx 287 \text{ J/(kg·K)} \quad \text{for air}$$
$$dh = c_p \, dT \quad c_p(T) = c_v(T) + R \quad \gamma(T) = c_p/c_v \tag{7-4}$$

If we further assume constant specific heats, then $h = c_p T$ and $\gamma = \text{constant} \approx 1.40$ for air. The reader is asked to review undergraduate compressible-flow formulas, e.g., Chap. 9 of White (2003). For steady flow with negligible friction and heat transfer, Eq. (7-3) integrates to

$$H = \text{constant} = h + u^2/2 = c_p T + u^2/2$$

or

$$T_o = \text{stagnation temperature} = T + \frac{u^2}{2c_p} = \text{constant}$$

We have assumed *adiabatic* flow, which is generally true only in the freestream, $u = U$. Substituting the perfect-gas relations $c_p = \gamma R/(\gamma - 1)$ and $a = (\gamma RT)^{1/2}$, we obtain

$$\frac{T_o}{T} = \left(\frac{a_o}{a}\right)^2 = 1 + \frac{\gamma - 1}{2} Ma^2 \quad Ma = \frac{U}{a} = \text{Mach number} \tag{7-5}$$

Thus, for *adiabatic* flow, as the Mach number and kinetic energy increase, the freestream temperature and speed of sound decrease, slowly at first, then faster, as shown in Fig. 7.1. If we further assume *isentropic* flow, then pressure and density follow similar relations:

$$\frac{p_o}{p} = \left(\frac{T_o}{T}\right)^{\gamma/(\gamma-1)} = \left(1 + \frac{\gamma - 1}{2} Ma^2\right)^{\gamma/(\gamma-1)}$$
$$\frac{\rho_o}{\rho} = \left(\frac{T_o}{T}\right)^{1/(\gamma-1)} = \left(1 + \frac{\gamma - 1}{2} Ma^2\right)^{1/(\gamma-1)} \tag{7-6}$$

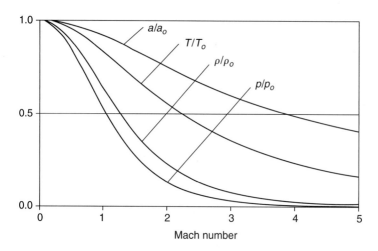

FIGURE 7-1
Adiabatic (T/T_o and a/a_o) and isentropic (p/p_o and ρ/ρ_o) perfect-gas properties versus Mach number for $\gamma = 1.4$.

These two relations are also shown in Fig. 7-1 and are seen to drop off faster with Mach number. We shall compare these adiabatic/isentropic formulas with the viscous boundary-layer behavior.

7-1.2 Steady Viscous Flow: The Crocco–Busemann Relations

Compressible-boundary-layer flow is neither adiabatic nor isentropic. Even if the wall and the freestream are adiabatic, the boundary layer between is not and generates both dissipation and heat transfer. However, certain perfect-gas assumptions lead to interesting energy integrals.

It is not necessary to assume constant specific heats in what follows. If we now consider only steady flow, the boundary-layer equations reduce to

$$\frac{\partial}{\partial x}(\rho u) + \frac{\partial}{\partial y}(\rho v) = 0 \tag{7-7a}$$

$$\rho u \frac{\partial u}{\partial x} + \rho v \frac{\partial u}{\partial y} = -\frac{dp_e}{dx} + \frac{\partial}{\partial y}\left(\mu \frac{\partial u}{\partial y}\right) \tag{7-7b}$$

$$\rho u \frac{\partial h}{\partial x} + \rho v \frac{\partial h}{\partial y} = u \frac{dp_e}{dx} + \frac{\partial}{\partial y}\left(\frac{\mu}{Pr} \frac{\partial h}{\partial y}\right) + \mu \left(\frac{\partial u}{\partial y}\right)^2 \tag{7-7c}$$

where we have introduced the Prandtl number $Pr = \mu c_p/k$ into the energy equation. The total enthalpy relation Eq. (7-3) takes on a particularly interesting form

$$\rho u \frac{\partial H}{\partial x} + \rho v \frac{\partial H}{\partial y} = \frac{\partial}{\partial y}\left(\frac{\mu}{Pr} \frac{\partial H}{\partial y}\right) + \frac{\partial}{\partial y}\left[\left(1 - \frac{1}{Pr}\right)\mu u \frac{\partial u}{\partial y}\right] \tag{7-8}$$

If the Prandtl number is unity (a fair approximation for gases), the last term vanishes, and this relation then admits to a particular solution $H = $ const throughout the boundary layer. However, since $H = h + u^2/2$, $H = $ const implies that $\partial h/\partial y$ vanishes at the wall because u vanishes there (no slip). Thus our particular solution corresponds to zero heat transfer at the wall and is the first of two energy integrals discovered by Busemann (1931) and independently by Crocco (1932):

$$Pr = 1, \text{ adiabatic wall:} \qquad H = h + \frac{u^2}{2} = \text{const} \qquad (7\text{-}9)$$

Thus $Pr = 1$ implies a perfect balance between viscous dissipation and heat conduction so as to keep the stagnation enthalpy constant in an adiabatic boundary layer, just as would be true in outer potential flow. Note that the pressure gradient (if any) is immaterial and there is no requirement for constant μ, k, or c_p. However, the fluid must be a perfect gas or at least satisfy $dh = c_p \, dT$ to good approximation.

A second energy integral is possible for $Pr = 1$ and the further restriction of zero pressure gradient. We may note that the momentum and energy relations, Eqs. (7-7b) and (7-7c), are similar in mathematical character if the pressure gradient is zero, and it almost seems as if u and h could be interchanged except for the dissipation term. In other words, we may test the particular solution $h = h(u)$, noting, for example, that

$$\frac{\partial h}{\partial y} = \frac{dh}{du}\frac{\partial u}{\partial y}$$

in this case. Substituting in the energy relation and assuming $Pr = 1$, we have

$$\frac{dh}{du}\left[\rho u \frac{\partial u}{\partial x} + \rho v \frac{\partial u}{\partial y} - \frac{\partial}{\partial y}\left(\mu \frac{\partial u}{\partial y}\right)\right] = \left(1 + \frac{d^2 h}{du^2}\right)\mu \left(\frac{\partial u}{\partial y}\right)^2 \qquad (7\text{-}10)$$

for $Pr = 1$ and zero pressure gradient. The left-hand side vanishes by virtue of the momentum relation Eq. (7-7b). Therefore the right-hand side must vanish, which can be true only if

$$\frac{d^2 h}{du^2} = -1$$

or $\qquad (7\text{-}11)$

$$h = -\frac{u^2}{2} + C_1 u + C_2$$

By inspection, $C_2 = h_w$ since $u = 0$ there. The constant C_1 may be either related to the wall heat transfer or else to freestream conditions. Choosing the latter, we see that $h = h_e$ at $u = U_e$ or $C_1 = (h_e + U_e^2/2 - h_w)/U_e$. Putting together the final relation, we have the *second* Crocco–Busemann energy integral

For $Pr = 1$, $\dfrac{dp}{dx} = 0$:

$$H = h + \frac{u^2}{2} = h_w + (H_e - h_w)\frac{u}{U_e} \qquad (7\text{-}12)$$

Thus the total enthalpy varies linearly with velocity across the boundary layer in zero pressure gradient with $Pr = 1$. Note that the *first* Crocco–Busemann relation is contained in the second as the special case $H_e = h_w = H = \text{const}$, so that the wall is then adiabatic.

If we further require constant c_p, so that $h = c_p T + \text{const}$, then Eq. (7-12) becomes a relation between temperature and velocity

For $Pr = 1$, $\dfrac{dp}{dx} = 0$, $c_p = \text{const}$:

$$T = T_w + \left(T_e + \frac{U_e^2}{2c_p} - T_w\right)\frac{u}{U_e} - \frac{u^2}{2c_p} \tag{7-13}$$

Equation (7-13) is a temperature–velocity relation which is very useful in formulating approximate theories of the compressible boundary layer when Pr is near unity (gases) and dp/dx is not too large. Noting that $T_e + U_e^2/2c_p$ is the adiabatic wall temperature T_{aw}, we can rewrite the relation as

$$T = T_w + (T_{aw} - T_w)\frac{u}{U_e} - \frac{u^2}{2c_p} \tag{7-14}$$

Differentiating, we can relate the wall heat flux and skin friction

$$q_w = k_w\left(\frac{\partial T}{\partial y}\right)_w = \frac{(T_{aw} - T_w)k_w \tau_w}{U_e \mu_w}$$

or

$$C_h = \frac{q_w}{\rho_e U_e c_p(T_{aw} - T_w)} = \frac{C_f}{2Pr} \tag{7-15}$$

This is a form of the Reynolds analogy. It is strictly valid only for $Pr = 1$ and zero pressure gradient but is a good approximation for gases under both laminar and turbulent arbitrary conditions, particularly if we modify the correction factor Pr to $Pr^{2/3}$.

The Prandtl number also affects the insulated-wall temperature in high-speed flow. When the (inviscid) freestream is adiabatically decelerated to zero velocity, it approaches the "ideal" stagnation temperature T_{0e}, which, for a perfect gas with constant c_p, has the value $[T_e + U_e^2/2c_p]$. However, when a viscous high-speed boundary-layer flow is decelerated to zero velocity at an insulated wall, it approaches an adiabatic wall temperature T_{aw}, which differs from T_{0e} as defined by a dimensionless *recovery factor*, r:

$$r = \frac{T_{aw} - T_e}{T_{0e} - T_e} \approx fcn\,(Pr) \tag{7-16}$$

Accurate evaluation of r is important, since q_w in high-speed flow is proportional to $(T_{aw} - T_w)$, *not* to $(T_w - T_e)$, as we shall see. The Crocco integral and recovery factor concepts have been extended by Van Oudheusden (1997, 2004).

7-2 SIMILARITY SOLUTIONS FOR COMPRESSIBLE LAMINAR FLOW

The compressible-boundary-layer relations Eqs. (7-7) are formidable coupled nonlinear partial differential equations. They can be attacked by finite-difference methods, but analytic techniques give more understanding. Since the classic compressible-flow analyses came in the precomputer era, a great deal of sophisticated effort was applied to simplifying these equations. The early workers developed a number of beautiful transformations relating compressible boundary layers to equivalent incompressible flows. There are also a variety of similarity variables that produce ordinary differential equations for certain freestream and wall conditions. We discuss only one of these transformations due to Illingworth (1950).

7-2.1 The Illingworth Transformation

The stream function $\psi(x, y)$ for compressible flow is defined as

$$\frac{\partial \psi}{\partial y} = \rho u$$
$$\frac{\partial \psi}{\partial x} = -\rho v \tag{7-17}$$

This eliminates the continuity relation identically. It immediately suggests that the variable $\int \rho \, dy$ should replace y in compressible flow. Then the general idea followed by the early workers was to presuppose the existence of *two* similarity variables (ξ, η) and to see whether ψ and u could take the following "split" forms:

$$\psi(\xi, \eta) = \int \rho u \, dy = G(\xi) f(\eta)$$
$$u(\xi, \eta) = U_e(\xi) f'(\eta) \tag{7-18}$$

Then one substitutes in the momentum Eq. (7-7b) and tries to enforce the requirement that the resulting differential equation *be independent of* ξ. Here the analysis could (and did) take several paths and led to differing transformations. The path taken here was discovered by Illingworth (1950), who found that it was convenient to account for viscosity effects in ξ and for density effects in η, with the result

$$\xi = \int_0^x \rho_e(x) U_e(x) \mu_e(x) \, dx = \xi(x) \text{ only}$$
$$\eta = \frac{U_e}{\sqrt{2\xi}} \int_0^y \rho \, dy = \eta(x, y) \tag{7-19}$$

The student should attempt to show that when these variables are substituted in the momentum Eq. (7-7b), the following ordinary (more or less) differential equation results for the function $f(\eta)$:

$$(Cf'')' + ff'' + \frac{2\xi}{U_e}\frac{dU_e}{d\xi}\left(\frac{\rho_e}{\rho} - f'^2\right) = 0. \tag{7-20}$$

where
$$C = \frac{\rho\mu}{\rho_e\mu_e} \approx C(\eta)$$

The primes denote differentiation with respect to η. By examining Eqs. (7-18), we see that the boundary conditions for an impermeable wall are

No slip: $\quad f(0) = f'(0) = 0$

Freestream: $\quad f'(\infty) = 1.0$ $\tag{7-21}$

Two special cases of Eq. (7-20) immediately come to mind. First, on a flat plate at low speed and modest heat transfer, $U_e, \rho,$ and μ are constant; $C \approx 1$ and $dU_e/d\xi = 0$, and we obtain

$$f''' + ff'' = 0 \tag{7-22}$$

which we recognize as the Blasius Eq. (4-45). Second, if ρ and μ are constant but the freestream varies as $U_e = Kx^m$, then $C \approx 1$ and we obtain

$$f''' + ff'' + \frac{2m}{m+1}(1 - f'^2) = 0 \tag{7-23}$$

which is none other than the Falkner–Skan Eq. (4-71). Thus we match smoothly with incompressible analyses.

The boundary-layer energy Eq. (7-7c) also reduces to an ordinary differential equation if we split the enthalpy into a magnitude multiplie by a shape. Let

$$h(x, y) = h_e(\xi) g(\eta) \tag{7-24}$$

By substituting in Eq. (7-7c), we obtain for $g(\eta)$

$$\left(\frac{C}{Pr}g'\right)' + fg' = \left(\frac{\xi}{H_e}\frac{dH_e}{d\xi}\right)f'\left(2g + \frac{U_e^2}{h_e}f'^2\right) - \frac{U_e^2}{h_e}Cf''^2 \tag{7-25}$$

where $H_e = h_e + U_e^2/2$ is the freestream stagnation enthalpy. For a perfect gas with constant specific heats, the Eckert-type parameter U_e^2/h_e is equivalent to a Mach number squared:

$$\frac{U_e^2}{h_e} = \frac{U_e^2}{c_p T_e} = \frac{(\gamma - 1)U_e^2}{\gamma RT_e} = (\gamma - 1)Ma_e^2 \qquad \gamma = \frac{c_p}{c_v} \tag{7-26}$$

In many cases, the entire right-hand side of Eq. (7-25) is negligible. For example, in low-speed flow near the stagnation point, $Ma_e \approx 0$ and $H_e \approx$ const, with $C \approx 1$ and $Pr \approx$ const, so that Eq. (7-25) reduces to

$$g'' + Prfg' \approx 0 \tag{7-27}$$

which we recognize as Eq. (3-168) for incompressible stagnation-flow heat transfer. The boundary conditions on $g(\eta)$ could reflect either adiabatic or heat-transfer conditions:

Adiabatic wall: $\quad\quad g(0) = g_w \quad g'(0) = 0 \quad g(\infty) = 1.0 \quad$ (7-28a)

Heat transfer: $\quad\quad g(0) = g_w \quad g'(0) \neq 0 \quad g(\infty) = 1.0 \quad$ (7-28b)

For adiabatic flow, the problem is to find the correct value of g_w, whereas for heat transfer, g_w is specified, and one must find the correct value of $g'(0)$.

Examining Eqs. (7-20) and (7-26), we see that similarity is not achieved unless certain coefficients are constant or functions only of η:

1. $C = $ const or is related to f and g.
2. $Pr = $ const or is related to f and g.
3. ρ_e/ρ is related to f and g.
4. $\dfrac{2\xi}{U_e}\dfrac{dU_e}{d\xi} = $ const.
5. $\dfrac{U_e^2}{h_e} = $ const or is negligible (or $Pr = 1$).
6. $\dfrac{2\xi}{H_e}\dfrac{dH_e}{d\xi} = $ const.

This is quite an array of conditions, but most are reasonable. Conditions 1 to 3 are satisfied if we make the perfect-gas assumption

$$\frac{\rho_e}{\rho(y)} = \frac{T(y)}{T_e} \approx \frac{h(y)}{h_e} \quad\quad (7\text{-}29)$$

since the pressure is approximately constant across the boundary layer. This means that C and the density can be related approximately to g:

$$\frac{\rho_e}{\rho} \approx g(\eta)$$

$$\frac{h_e}{h}\frac{\mu}{\mu_e} \approx \left(\frac{h}{h_e}\right)^{n-1} = g^{n-1} = C(\eta) \quad\quad (7\text{-}30)$$

where we have made a power-law approximation for the viscosity–temperature relation of gases. Values of n are given in Table 1-2. For air, $n \approx 2/3$, hence $C \approx g^{-1/3}$ to good accuracy. Further, from Fig. 1-26, the Prandtl number is approximately constant for most gases, which takes care of condition 2. Conditions 4 and 5 are satisfied by Falkner–Skan distributions

$$\begin{aligned} U_e &\sim \xi^m \\ H_e &\sim \xi^j \end{aligned} \quad\quad (7\text{-}31)$$

The most common assumptions are constant freestream stagnation temperature ($j = 0$) and flat-plate flow ($m = 0$) or stagnation flow ($m = 1$). Finally, condition 5 is satisfied for flat-plate flow ($U_e^2/h_e \approx$ const) and for stagnation flow ($U_e^2/h_e \approx 0$) but not in general. For Falkner–Skan flows, for example, U_e^2/h_e would vary with ξ unless $j = 2m$, which is not very likely; we would have to agree to neglect this term by making the assumption $Pr \approx 1$ or $U_e^2 \ll h_e$.

We should also note that similarity is violated unless the wall enthalpy $g_w = g(0)$ is taken constant, independent of ξ. Thus, unless otherwise specified, our solutions are valid only for constant H_e and constant T_w.

Putting all these simplifying assumptions together, we can rewrite Eqs. (7-20) and (7-26) in final similarity form for the laminar flow of a perfect gas

Momentum: $$(Cf'')' + ff'' = \beta(f'^2 - g) \tag{7-32a}$$

Energy: $$(Cg')' + Prfg' = -PrC(\gamma - 1)Ma_e^2 f''^2 \tag{7-32b}$$

where $$\beta = \frac{2\xi}{Ma_e} \frac{dMa_e}{d\xi} \qquad C = \frac{\rho\mu}{\rho_e\mu_e} \approx g^{n-1} \approx g^{-1/3} \quad \text{for air}$$

Note that, except for the variable parameter $C(\eta)$, Eqs. (7-32) are similar to the Falkner–Skan incompressible-flow solutions from Secs. 4-3.4 and 4-3.5. Compressible similarity solutions of these—and related—equations have been given by a number of investigators, as detailed in the monograph by Stewartson (1964). Here we discuss, in Sec. 7-4, only the Falkner–Skan-type solutions found by Cohen and Reshotko (1956).

7-3 SOLUTIONS FOR LAMINAR FLAT-PLATE AND STAGNATION-POINT FLOW

The two most relevant applications of laminar, compressible, boundary-layer theory are (1) flat-plate flow (for estimating the friction and heat transfer on slender bodies) and (2) stagnation-point flow (where the heat-transfer rate is usually the highest on blunt bodies). Both cases give insight into the general effect of compressibility on boundary-layer flows.

7-3.1 Approximate Analysis of the Flat-Plate Recovery Factor

Consider flat-plate flow ($\beta = 0$) with constant wall temperature. Further assume that $C = 1$, that is, μ proportional to T, because the dependence of the recovery factor on temperature and Mach number is extremely weak. Then Eqs. (7-32), for an adiabatic wall, reduce to

$$f''' + ff'' \approx 0 \tag{7-33a}$$

$$g'' + Prfg' = -Bf''^2 \tag{7-33b}$$

subject to $\quad f(0) = f'(0) = g'(0) = 0 \quad$ and $\quad f'(\infty) = g(\infty) = 1.$

For brevity, we have defined $B = Pr(\gamma - 1)Ma_e^2$. These are the Blasius relations, and $f(\eta)$ is already known from Fig. 4-6 and Table 4-1. The solution of the (linear, nonhomogeneous) energy Eq. (7-33b) is

$$g = g_w + A \int_0^\eta G \, d\zeta + B \int_0^\eta G \left(\int_0^\zeta \frac{f''^2 \, d\lambda}{G} \right) d\zeta \tag{7-34}$$

where
$$G(\eta, Pr) = \exp\left(-Pr \int_0^\eta f(\zeta) \, d\zeta\right)$$

This solution was first given by Pohlhausen (1921). The constant A must be such that $g(\infty) = 1$. Further, for an *adiabatic* wall, the requirement that $g'(0) = 0$ means that $A = 0$, as can be seen by differentiating Eq. (7-34). For this case, then, $g_w = T_{aw}/T_e$, and rearrangement of Eq. (7-34) gives

$$T_{aw} = T_e + r \frac{U_e^2}{2c_p} \qquad r = 2Pr \int_0^\infty G \left(\int_0^\zeta \frac{f''^2}{G} d\lambda \right) d\zeta \tag{7-35}$$

Thus $r \approx r(Pr)$ to excellent accuracy. For $Pr = 1$, $r = 1$, the plate recovers the freestream stagnation temperature at an adiabatic wall because conduction and dissipation exactly balance in the boundary layer. For $Pr \neq 1$, the integral in Eq. (7-35) may be evaluated by adding three more statements to the computer program for the Blasius solution from Eq. (4-47):

$$F(4) = Y(3)$$
$$F(5) = Y(1) * Y(1) * \text{EXP}(PR * Y(4)) \tag{7-36}$$
$$F(6) = 2 * PR * \text{EXP}(-PR * Y(4)) * Y(5)$$

The output variable $Y(6)$ approaches r, as η becomes large. Some computed recovery factors are listed in Table 7-1. Two curve-fit approximations to these

TABLE 7-1
Flat-plate recovery factors from Eqs. (7-34) and (7-35)

Pr	r(Pr)	Pr	r(Pr)
0.001	0.02946	1.0	1.0
0.003	0.05125	2.0	1.4089
0.006	0.07279	3.0	1.7088
		6.0	2.3571
0.01	0.09434	10.0	2.9622
0.03	0.16534	30.0	4.7238
0.06	0.23610	60.0	6.2467
0.1	0.30731		
0.3	0.54197	100.0	7.6272
0.6	0.77286	300.0	11.5483
Air 0.72	0.84771	600.0	14.881
		1000.0	17.881

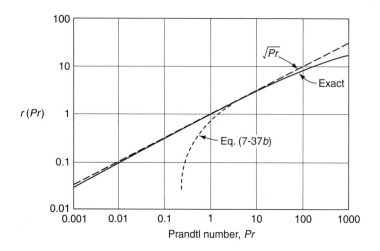

FIGURE 7-2
Flat-plate recovery factors (exact and approximate).

results are

$0.1 < Pr < 3.0$: $\quad\quad\quad\quad r \approx \sqrt{Pr}$ (7-37a)

$3.0 < Pr$: $\quad\quad\quad\quad r \approx 1.905\, Pr^{1/3} - 1.15$ (7-37b)

These are plotted, along with the exact values, in Fig. 7-2. In practice, only formula (7-37a) is important, since most compressible-flow problems involve gases, $Pr \approx 0.7$, $r \approx 0.85$.

7-3.2 Laminar Flat-Plate Friction at High Mach Numbers

The quantity $C = \rho\mu/\rho_e\mu_e$, called the *Chapman–Rubesin parameter* after a pioneering paper by Chapman and Rubesin (1949), has little effect on the recovery factor but does affect friction and heat transfer. The following simplified analysis shows why. For a flat plate, $\beta = 0$ and Eq. (7-32a) becomes

$$(Cf'')' + ff'' = 0$$

The variables may be separated and integrated twice across the boundary layer, with the result

$$\int_0^\infty Cf''\,d\eta = C_w f''(0) \int_0^\infty \exp\left(-\int_0^\eta \frac{f}{C}\,d\zeta\right) d\eta$$

An approximate solution can be found with the near-wall simplifications

$$C \approx C_w \approx \text{const} \quad\quad f \approx f''(0)\frac{\eta^2}{2} + \cdots$$

The integrals may then be evaluated for the approximate result

$$C_w \approx C_w f''(0)\frac{1}{3}\left[\frac{6C_w}{f''(0)}\right]^{1/3} \Gamma\left(\frac{1}{3}\right)$$

or

$$f''(0) \approx 0.48 C_w^{-1/2} \quad (7\text{-}38)$$

In a more accurate analysis, the constant 0.48 would be the Blasius value for incompressible flow, $f''(0)_{\text{Blasius}} = 0.469600$. Since $f''(0)$ is proportional to wall shear stress, we may convert Eq. (7-38) into a skin-friction approximation for compressible flat-plate flow:

$$C_f = \frac{2\tau_w}{\rho_e U_e^2} \approx \frac{0.664 C_w^{1/2}}{Re_x^{1/2}} \quad (7\text{-}39)$$

where $C_w = \rho_w \mu_w / \rho_e \mu_e \approx (T_w/T_e)^{-1/3}$ for gases. Similarly, the wall heat transfer $C_h \approx (C_{h,\,Ma_e=0})(C_w)^{1/2}$. The agreement is good for adiabatic walls but not so good for hot or cold walls.

7-3.3 The Reference Temperature Concept

It was found in the 1950s that the simplified expression Eq. (7-39) could be modified to give good accuracy for all flat-plate compressible gas flows by evaluating the fluid properties (ρ, μ, c_p, k) at a *reference temperature*, T^*, which varies with Mach number and wall temperature. The Chapman–Rubesin parameter would be evaluated as

$$C^* = \frac{\rho^* \mu^*}{\rho_e \mu_e} \approx \left(\frac{T^*}{T_e}\right)^{-1/3} \quad (7\text{-}40)$$

The flat-plate formulas, modified for reference temperature, become

$$\frac{T_{aw}}{T_e} \approx 1 + \sqrt{Pr^*}\,\frac{\gamma - 1}{2} Ma_e^2 \qquad \text{laminar recovery factor} \quad (7\text{-}41a)$$

$$\frac{2\tau_w}{\rho_e U_e^2} \approx \frac{0.664\sqrt{C^*}}{\sqrt{Re_{xe}}} \qquad \text{laminar plate flow} \quad (7\text{-}41b)$$

$$\frac{C_{fe}}{2 C_{he}} \approx Pr^{*2/3} \qquad \text{laminar or turbulent} \quad (7\text{-}41c)$$

The most popular correlation for T^* is due to Eckert (1955):

$$\frac{T^*}{T_e} \approx 0.5 + 0.039 Ma_e^2 + 0.5\frac{T_w}{T_e} \quad (7\text{-}42)$$

Although originally proposed simply as an empirical correlation, the text by Dorrance (1962) shows that the reference temperature is a direct consequence of the similarity relations for compressible flow. Equation (7-42) may also be used to modify *turbulent* flat-plate formulas, but even better approximations are possible in that case.

7-3.4 Exact Laminar Flat-Plate Solutions

We may compare our approximate flat-plate compressible-flow formulas with an array of exact numerical laminar-flow computations by van Driest (1952a). The calculations assume $Pr = 0.75$ and a Sutherland viscosity law for air, Eq. (1-36). Similar formulas for $c_p(T)$ and $k(T)$ were also used. The theory uses a different coordinate transformation from Eqs. (7-19), but the basic idea is quite similar.

Figure 7-3 shows the computed velocity, temperature, and Mach number profiles in the boundary layer for an adiabatic wall. The velocity profiles in Fig. 7-3a are nearly straight lines at high Mach numbers, an approximation often used in hypersonic flow analyses, e.g., Truitt (1960, art. 5-3). The temperature profiles

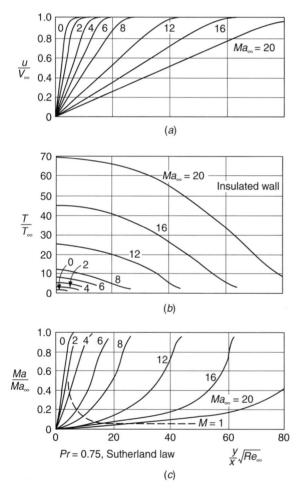

FIGURE 7-3
Calculations of the laminar, compressible, boundary layer on an insulated flat plate by van Driest (1952a): (a) velocity profiles; (b) temperature profiles; (c) Mach number profiles.

show the dramatic heating effect of viscous dissipation in the boundary layer at high Mach numbers. If the velocity is known, the temperature in Fig. 7-3b is well approximated by Eq. (7-13) for $T_w = T_{aw}$, as follows:

$$\frac{T}{T_e} \approx 1 + Pr^{1/2}\left(\frac{\gamma - 1}{2}\right)Ma_e^2\left(1 - \frac{u^2}{U_e^2}\right) \qquad (7\text{-}43)$$

The recovery factor relation $r \approx Pr^{1/2}$ is quite accurate. For example, at $Ma_e = 20$, $r = (0.75)^{1/2}$ gives $T_{aw}/T_e = 70.3$, in excellent agreement with Fig. 7-3b. Finally, the Mach number profiles in Fig. 7-3c show that most of a high-velocity boundary layer is in a supersonic condition.

Figure 7-4 shows flat-plate computations by van Driest (1952a) for a very cold wall, $T_w = T_e/4$. The velocity profiles remain rounded even at Mach number 20, and we recall from Chap. 5 that such profiles are more stable and resistant to transition. If the velocity is known, the temperature profiles are well fit by the Crocco–Busemann relation, Eq. (7-13).

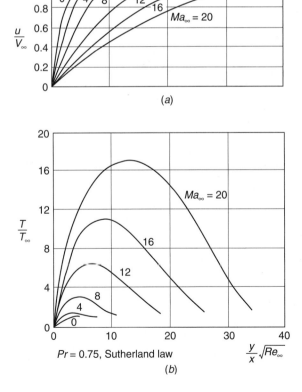

FIGURE 7-4
Calculations by van Driest (1952a) of the compressible, laminar, boundary layer on a cold flat plate ($T_w = T_e/4$): (a) velocity profiles; (b) temperature profiles.

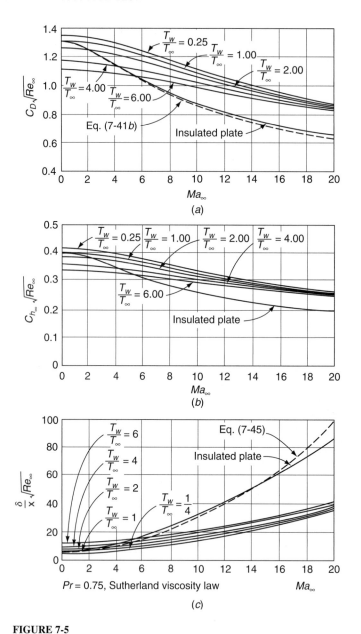

FIGURE 7-5
Calculations by van Driest (1952a) of the drag, heat transfer, and velocity thickness of a laminar, flat-plate, compressible, boundary layer: (a) total drag coefficient $C_D = 2C_f$; (b) local Stanton number; (c) boundary-layer thickness.

Figure 7-5 shows van Driest's calculations for laminar, flat-plate, total drag ($C_D = 2C_f$), local Stanton number C_h, and boundary-layer thickness. Note that the Stanton number for compressible flow is defined as

$$C_h = \frac{q_w}{\rho_e U_e C_{pe}(T_{\text{aw}} - T_w)} \qquad (7\text{-}44)$$

with q_w positive for a cold wall. As the Mach number approaches zero, the temperature difference ($T_{\text{aw}} - T_w$) approaches the value ($T_e - T_w$) used for incompressible flow. Both C_D and C_h are well approximated by the reference temperature formulas Eqs. (7-41). For example, at $Ma_e = 10$ and $T_w/T_e = 4$, we read $C_h(Re_{xe})^{1/2} \approx 0.30$ from Fig 7-5b. For comparison, the reference temperature for this case, from Eq. (7-42), is $T^*/T_e \approx 5.9$, whence $C^* \approx 0.58$ and Eq. (7-41c) predicts $C_h(Re_{xe})^{1/2} \approx 0.306$, or about 2 percent high. This is typical of the accuracy of reference temperature formulas.

The dimensionless boundary-layer thickness in Fig. 7-5c is seen to grow with the Mach number for both adiabatic and nonadiabatic walls. A semiempirical formula to approximate this thickness is

$$\frac{\delta}{x} Re_{xe}^{1/2} \approx C_w^{1/2} \left[5.0 + \left(0.2 + 0.9 \frac{T_w}{T_{\text{aw}}} \right)(\gamma - 1) Ma_e^2 \right] \qquad (7\text{-}45)$$

The accuracy is adequate, if unspectacular, and the prediction of Eq. (7-45) for an adiabatic wall is shown as the dashed curve in Fig. 7-5c.

Van Driest's calculations of the recovery factor, $r \approx Pr^{1/2}$, and the Reynolds analogy factor, $(C_f/2C_h) \approx Pr^{2/3}$, show an insignificant drop-off with Mach number. Note, however, these traditional approximations are for *flat-plate flow*, not pressure gradients. Calculations by Van Oudheusden (1997, 2004) and by this writer show a strong effect of a pressure gradient on the recovery factor. Later, for the Reynolds analogy factor, see Fig. 7-10b. For compressible heat transfer with pressure gradients, an accurate CFD result is preferred to Prandtl number formulas.

7-3.5 High-Speed Plane Stagnation Flow

We can readily extend our previous discussion to stagnation flow. Here, the external temperature T_e is the stagnation, or adiabatic wall temperature, and the difference between T_e and T_w is the correct driving temperature for heat transfer. The similarity equations for $\beta = 1$ and $Ma_e \approx 0$ are

$$(Cf'')' + ff'' + g - f'^2 = 0 \qquad (7\text{-}46a)$$

$$(Cg')' + Prfg' = 0 \qquad (7\text{-}46b)$$

$$U_e = Kx \quad \text{where } K = \text{stagnation velocity gradient} \qquad (7\text{-}46c)$$

The stagnation region is subsonic and, if T_w is not too different from T_e, $C \approx 1$ and also $g \approx 1$, so that the momentum Eq. (7-46a) approximately equals the incompressible Hiemenz solution, Eq. (3-151) or Table 3-4, with $f''(0) \approx 1.2326$.

The energy Eq. (7-46b) can be separated and integrated once, with the result

$$Cg' = C_w g'(0) \exp\left(-Pr \int_0^{\eta} \frac{f}{C} d\zeta\right)$$

Using the same approximations which led to Eq. (7-38), we may integrate again from zero to infinity, with the result

$$C_w[1 - g(0)] \approx g'(0) C_w \Gamma\left(\frac{4}{3}\right) \left[\frac{6C_w}{Pr f''(0)}\right]^{1/3} \qquad f''(0) \approx 1.2326$$

or

$$g'(0) \approx 0.66[1 - g(0)]\left(\frac{Pr}{C_w}\right)^{1/3}$$

This is the correct formulation, but the numbers are slightly off. Extensive computations by Fay and Riddell (1958), using the Sutherland viscosity law, are correlated nicely by slightly adjusted constants:

$$g'(0) \approx 0.570[1 - g(0)]\left(\frac{Pr}{C_w}\right)^{0.4} \qquad \text{[Fay and Riddell (1958)]} \quad (7\text{-}47)$$

We recognize the constant 0.570 from the Hiemenz solution for incompressible plane stagnation flow in Chap. 3. Since $g'(0) \sim q_w$ and $[1 - g(0)] \sim (h_e - h_w)$, Eq. (7-47) is equivalent to specifying a heat-transfer rate for stagnation flow

Plane flow: $\qquad q_w = 0.570 Pr^{-0.6}(\rho_e \mu_e K)^{1/2}\left(\frac{\rho_w \mu_w}{\rho_e \mu_e}\right)^{0.1}(h_e - h_w) \qquad (7\text{-}48a)$

Axisymmetrical flow: $\qquad$ Replace 0.570 by 0.763 $\qquad (7\text{-}48b)$

As noted, since the local Mach number is negligible, the only perceptible difference between plane and axisymmetrical flow is the coefficient (0.570 or 0.763) and the fact that the local velocity gradient K might be different. The heat transfer depends strongly upon K, and no practical computations can be made unless K is known from external considerations. Consider flow past a typical aerodynamic body (cylinder or sphere, say) of finite size D and approach velocity V_∞ (see Fig. 7-6). The local velocity gradient K at the stagnation point of this body depends upon both D and V_∞ and, to a lesser extent, upon the approach Mach number Ma_∞ and the particular body shape. By dimensional analysis, we have

$$\frac{KD}{V_\infty} = f(Ma_\infty, \text{body shape}) \qquad (7\text{-}49)$$

At low speeds (subsonic flow), first-order calculations can be made by the Rayleigh–Janzen procedure [Shapiro (1953, Chap. 12)]. Some results are

Cylinder: $\qquad \dfrac{KD}{V_\infty} = 4\left(1 - 0.416 Ma_\infty^2 - 0.164 Ma_\infty^4 + \cdots\right) \qquad (7\text{-}50a)$

Sphere: $\qquad \dfrac{KD}{V_\infty} = 3\left(1 - 0.252 Ma_\infty^2 - 0.0175 Ma_\infty^4 + \cdots\right) \qquad (7\text{-}50b)$

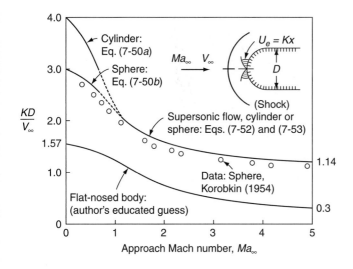

FIGURE 7-6
Theoretical, experimental, and estimated stagnation-point velocity gradients on cylinders, spheres, and flat noses.

Flat-nosed cylinder:

$$\frac{KD}{V_\infty} = \frac{\pi}{2}(1 - \cdots) \qquad (7\text{-}50c)$$

A flat-nosed cylinder has a considerably smaller gradient (this author does not know the series expansion), and one can therefore reduce stagnation-point heat flux by flattening the nose. At supersonic speeds, where these formulas are invalid, one estimate of the gradient comes from newtonian impact theory, Anderson (2000), which predicts the surface-pressure distribution as

$$\frac{p - p_\infty}{\rho_\infty U_\infty^2} \approx \cos^2\phi \qquad (7\text{-}51a)$$

or

$$\left.\frac{dp}{dx}\right|_{x \to 0} = -\rho_e K^2 x = -2\rho_\infty V_\infty^2 \phi \frac{d\phi}{dx} \qquad (7\text{-}51b)$$

where ϕ is the angle between the surface normal and the approaching freestream; for example, $\phi = 0$ at the nose. For a circular nose of diameter D, whether plane or spherical, $\phi = 2x/D$, and we can substitute in Eq. (7-51b) and cancel x to obtain

Round nose at supersonic speeds:

$$\frac{KD}{V_\infty} \approx \sqrt{\frac{8\rho_\infty}{\rho_e}} \qquad (7\text{-}52)$$

Remembering that, in this case, subscript e means stagnation conditions, we can compute ρ_∞/ρ_e from normal shock-wave theory. For a perfect gas, we have

$$\frac{\rho_e}{\rho_\infty} = \frac{(\gamma+1)Ma_\infty^2}{(\gamma-1)Ma_\infty^2 + 2}\left[1 + \frac{\gamma-1}{2}\frac{(\gamma-1)Ma_\infty^2 + 2}{2\gamma Ma_\infty^2 - \gamma + 1}\right]^{1/(\gamma-1)} \quad (7\text{-}53)$$

This formula, though impressive, is slowly varying. For $\gamma = 1.4$, Fig. 7-6 shows that KD/V_∞ from Eq. (7-52) decreases from 2.25 at $Ma = 1$ to about 1.12 at an infinite Mach number. These values merge for subsonic flow with the low-speed formulas Eqs. (7-50a) and (7-50b). Data taken by Korobkin at the Naval Ordinance Laboratory (Report 3841) agree with this trend but fall consistently about 10 percent low.

Equation (7-52) is not valid for flat noses, where it predicts $K = 0$, since D approaches infinity. Would that it were so, a flat nose would eliminate stagnation-point heat transfer! However, there is no question that flat noses *reduce K*, and this author has sketched an estimated flat-nose curve in Fig. 7-6, based on asymptotic values at $Ma = 0$ and $Ma = \infty$.

Equation (7-52) is best applied to a round nose that merges smoothly with the afterbody, such as a hemispheric cylinder. Meanwhile, a flat nose reduces stagnation-point heat transfer by up to 50 percent, since q_w is proportional to $K^{1/2}$. In between these two cases is a hybrid body consisting of a shallow spherical cap attached to a cylinder, resulting in a sharp corner at the shoulder. Figure 7-7 shows hypersonic velocity gradient data for such bodies, as given by Trimmer (Report AEDC-TR-68-99, 1968). For shallow noses with $R/r > 1.42$, the sonic point occurs at the sharp shoulder, and the newtonian (supersonic) theory predicts too low a gradient. For rounder noses, $1 < R/r < 1.42$, the sonic point occurs at $\theta \approx 45°$ and $KD/V_\infty \approx 1.12$, as predicted by Eq. (7-51) for hypersonic flow.

Assuming that K is known, the previous stagnation-point theory for heat flux is in excellent agreement with experiment. Figure 7-8 shows the comparison with

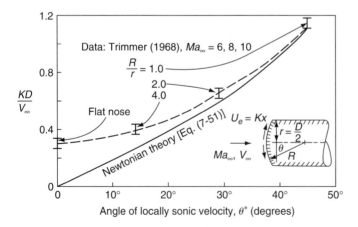

FIGURE 7-7
Comparison of theory and experiment for the stagnation-point velocity gradient K on cylinders with spherical noses.

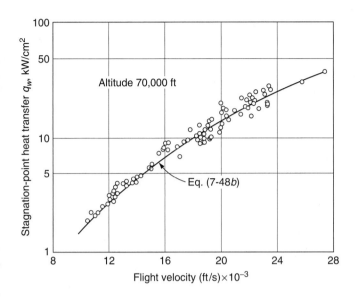

FIGURE 7-8
Theoretical stagnation-point heat flux compared with shock-tube experiments by Rose and Stark (1958) on a hemispheric cylinder.

shock-tube experiments by Rose and Stark (1958) on a hemispheric cylinder at hypersonic speeds approximating reentry conditions. It is thought that discrepancies, if any, are due almost entirely to the difficulty of estimating gas viscosity at the very high temperatures associated with the stagnation point in hypersonic flow.

7-4 COMPRESSIBLE LAMINAR BOUNDARY LAYERS UNDER ARBITRARY CONDITIONS

Although the flat plate and stagnation point are certainly the most important compressible-flow solutions, there are practical problems of variable velocity $U_e(x)$ and temperature $T_e(x)$ or $T_w(x)$ where we need a method of computing, at least approximately, the wall skin-friction and heat-flux distributions. The English translation of the monograph by Walz (1969) is essentially devoted to the development of such approximate computations, mainly for laminar but also for turbulent boundary layers. Also, the monograph by Stewartson (1964) is an excellent source for approximate theories of the compressible, laminar, boundary layer.

At least three possibilities come to mind for the construction of an approximate laminar-flow theory:

1. Modification and extension of the compressible similar solutions Eqs. (7-32)
2. Extension to the integral momentum and energy relations of compressible flow
3. Finite-difference computation of the exact, compressible, laminar equations

Let us have a brief look at each of these approaches.

7-4.1 Modification of Compressible Similar Solutions

A complete array of velocity and stagnation enthalpy distributions has been given by Cohen and Reshotko (1956) for Falkner–Skan-type compressible similar flows. However, they made the simplifying assumptions $C = Pr = 1$, so that variable viscosity effects are only approximate and Prandtl number effects are absent. The similarity Eqs. (7-32) reduce to

Momentum: $\quad\quad\quad\quad f''' + ff'' + \beta(g - f'^2) = 0 \quad\quad\quad$ (7-54a)

Enthalpy: $\quad\quad\quad\quad g'' + fg' = -(\gamma - 1)Ma_e^2 f''^2 \quad\quad\quad$ (7-54b)

Stagnation enthalpy: $\quad S'' + fS' = 0 \quad\quad S = \dfrac{H}{H_e} - 1 \quad\quad$ (7-54c)

Cohen and Reshotko chose to work with the stagnation enthalpy variable S because for Prandtl number unity, S is independent of the Mach number, depending only upon the wall heat transfer S_w and the pressure gradient parameter β. Their computed velocity profiles are shown in Fig. 7-9 for a hot wall ($S_w = +1.0$) and a cold wall ($S_w = -1.0$). The insulated wall would lie in between ($S_w = 0$ or $H = \text{const}$) and would be identical to the incompressible Falkner–Skan profiles (Fig. 4-11a). Note that the similarity coordinate is of the Falkner–Skan type and not the same as η defined here [Eq. (7-19)]. At least two observations about these curves are of interest:

1. The separation point is affected by heat transfer, β_{sep} being -0.326 for the cold wall ($S_w = -1.0$), -0.1988 for the insulated wall ($S_w = 0$), and -0.1305 for the

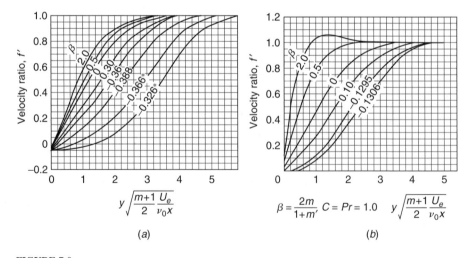

FIGURE 7-9
Similarity solutions for a compressible, laminar, boundary layer: (a) cold wall ($S_w = -1.0$); (b) hot wall ($S_w = +1.0$). [After Cohen and Reshotko (1956).]

hot wall ($S_w = +1.0$). Hot walls suffer earlier separation, and cold walls resist it, as we have seen before.

2. Hot-wall profiles overshoot the freestream velocity for positive β (favorable gradient). This is obvious for $\beta = 2.0$ and barely visible for $\beta = 0.5$.

We should also remark that the coordinate $y\sqrt{(m+1)U_e/2\nu_0 x}$ in Fig. 7-9 contains the quantity ν_0, which denotes the kinematic viscosity evaluated at the freestream stagnation temperature T_0.

An extensive set of tables of compressible similarity solutions are given in the paper (and related report) by Bae and Emanuel (1989).

Figure 7-10 shows the computed shear stress $f''(0)$ and Reynolds analogy factor $C_f/C_h Pr^{2/3}$ as a function of β and S_w. We see that both pressure gradient and wall temperature give strong effects.

For nonsimilar flows, it seems reasonable as a first approximation to compute the local value of $\beta(x)$ and use Fig. 7-10 to estimate the wall heat flux and skin friction. This was suggested by Lees (1956), who also gave a clever approximate analysis for the forward region of a blunt body. As we move away from the stagnation point of a cylinder or sphere, the heat flux drops off, but not as any simple

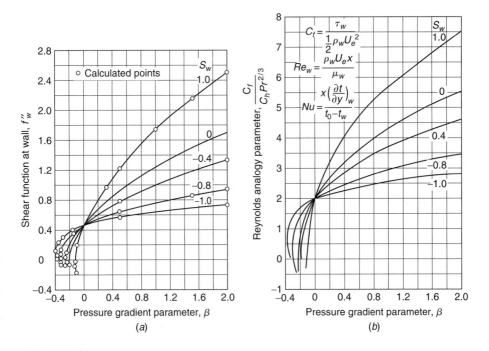

FIGURE 7-10
Computations of skin friction and the Reynolds analogy for laminar compressible similar flows: (a) skin friction $f''(0)$; (b) Reynolds analogy parameter. [After Cohen and Reshotko (1956).]

function of x or ϕ. The local Falkner–Skan parameter is

$$\beta(x) = \frac{2\xi}{U_e}\frac{dU_e}{d\xi}$$

where

$$\xi = \int_0^x \rho_e U_e \mu_e \, dx \tag{7-55}$$

For a perfect gas, we can compute $U_e(x)$ from the relation $U_e = Ma_e\sqrt{\gamma R T_e}$, and the Mach number $Ma_e(x)$ can be estimated from the newtonian approximation Eq. (7-51a) if the approach flow is supersonic:

$$\frac{p_e(x)}{p_0} \approx \frac{p_\infty}{p_0}\left[1 + \gamma Ma_\infty^2 \cos^2\phi(x)\right] \tag{7-56}$$

The ratio $p_0/p_\infty \approx [1 + \gamma Ma_\infty^2]$ from normal-shock theory at high Mach numbers, e.g., White (2003, p. 619). Therefore, since $\phi = 2x/D$ for a round nose, Eq. (7-56) is a formula to predict surface Mach number Ma_e^2, supplemented by the ideal-gas isentropic relations

$$\frac{p_0}{p_e} = \left[1 + \frac{1}{2}(\gamma - 1)Ma_e^2\right]^{\gamma/(\gamma-1)}$$

$$\frac{T_0}{T_e} = 1 + \frac{1}{2}(\gamma - 1)Ma_e^2 \tag{7-57}$$

For a round-nosed blunt body, Lees (1956) combined these approximations with a modified form of the Illingworth transformation Eq. (7-19) and used the concept of "local similarity": that heat transfer $q_w(x)$ is determined by the local gradient $\beta(x)$ from Eq. (7-55). The result is the following formula for local heat flux:

$$\frac{q_w(x)}{q_w(0)} \approx \frac{F(x)}{(2^{j+1}K/U_\infty)^{1/2}} \quad \text{[Lees (1956)]} \tag{7-58}$$

where

$$F(x) = \frac{(C_e U_e/C_0 U_\infty)r_0^j}{\int_0^x (C_e U_e/C_0 U_\infty)r_0^{2j}\, dx}$$

Here K is the stagnation-point velocity gradient (Figs. 7-6 and 7-7), and $r_0(x)$ is the body surface radius, with $j = 1$ for axisymmetric flow and $j = 0$ for plane flow. Also, $C = \rho\mu/\rho_e\mu_e$ is the Chapman–Rubesin parameter; hence $C_e/C_0 = \rho_e\mu_e/\rho_0\mu_0$ where 0 refers to the stagnation point. The formula is quite simple in appearance, but the required quadrature is somewhat tedious. Shortly thereafter, Kemp et al. (1959) slightly refined the analysis and proposed adding a correction factor

$$\frac{q_w(x)}{q_w(0)} \approx \left.\frac{q_w(x)}{q_w(0)}\right|_{\text{Lees}} \left(0.936 + 0.090\sqrt{\beta(x)}\right) \tag{7-59}$$

This amounts to about a 10 percent correction at most. In either formula Eqs. (7-58) or (7-59), one should estimate $q_w(0)$ as accurately as possible, using, say, Eq. (7-48).

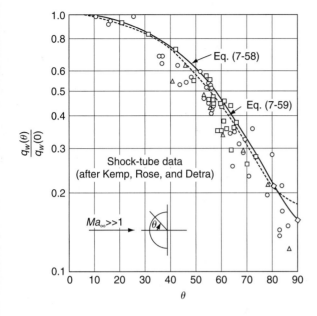

FIGURE 7-11
Comparison of theory and experiment for local heat transfer in hypersonic laminar flow past a hemisphere.

Figure 7-11 compares these two formulas with experimental heat flux measured on a hemisphere in hypersonic flow. Both are in fairly good agreement.

7-4.2 Integral Relations for a Compressible Boundary Layer

A second possible approach to an arbitrary, compressible, boundary layer is through the Kármán-type integral relations, similar to the incompressible-flow theories developed in Sec. 4-6. The approach is nearly the same, but the integral relations must be rederived from the compressible, boundary-layer Eqs. (7-7); the velocity v is eliminated through continuity Eq. (7-7a), and Eqs. (7-7b) and (7-7c) are integrated outward from the wall to infinity. The first result is the momentum-integral relation

$$\frac{d\theta}{dx} + \frac{\theta}{U_e}\frac{dU_e}{dx}\left(2 + H + \frac{U_e}{\rho_e}\frac{d\rho_e}{dU_e}\right) + \frac{1}{\rho_e U_e^2}\frac{d}{dx}\left(p_e\delta - \int_0^\delta p\,dy\right) = \frac{C_f}{2} \quad (7\text{-}60)$$

where
$$C_f = \frac{2\tau_w}{\rho_e U_e^2} \qquad H = \frac{\delta^*}{\theta}$$

$$\delta^* = \int_0^\infty \left(1 - \frac{\rho}{\rho_e}\frac{u}{U_e}\right)dy = \text{compressible displacement thickness}$$

$$\theta = \int_0^\infty \frac{\rho}{\rho_e}\frac{u}{U_e}\left(1 - \frac{u}{U_e}\right)dy = \text{compressible momentum thickness}$$

and, for an adiabatic freestream,

$$\frac{U_e}{\rho_e}\frac{d\rho_e}{dU_e} \equiv -Ma_e^2$$

This relation is valid for either laminar or turbulent flow. For low-speed flow at constant density, Eq. (7-60) reduces to the Kármán integral relation Eq. (4-122). The effect of compressibility has added three complicating factors to the analysis: (1) the density variation must be included in the definition of the integral thicknesses, (2) the freestream Mach number is an important part of the pressure gradient term, and (3) at high speeds, the pressure variation integrated across the boundary layer [last term on the left of Eq. (7-60)] may be too large to neglect.

7-4.3 The Laminar Boundary Layer

For laminar flow, it is customary to neglect the cross-stream pressure variation and consider only adiabatic freestreams, for which Eq. (7-60) reduces to

$$\frac{d\theta}{dx} + \frac{\theta}{U_e}\frac{dU_e}{dx}\left(2 + H - Ma_e^2\right) \approx \frac{C_f}{2} \qquad (7\text{-}61)$$

In computing θ and δ^* from their definitions in Eqs. (7-60), one could avoid use of the energy equation by approximating $\rho_e/\rho = T/T_e$ from the Crocco–Busemann relation, Eq. (7-19).

The text by Walz (1969) discusses several approaches to the solution of Eq. (7-61). Here we present only the method of Gruschwitz (1950), who extended the classic Kármán–Pohlhausen technique by approximating both velocity and density as polynomials in y/δ. The single parameter is the dimensionless pressure gradient, λ or Λ:

$$\lambda = \frac{T_w}{T_e}\frac{\theta^2}{\nu_e}\frac{dU_e}{dx} = \Lambda\left(\frac{\theta}{\Delta}\right)^2 \qquad [\text{Pohlhausen (1921)}] \qquad (7\text{-}62)$$

where $\quad \Delta = \int_0^\delta \frac{\rho}{\rho_e}\,dy \quad$ and $\quad \dfrac{\theta}{\Delta} = \dfrac{37 - \Lambda/3 - 5\Lambda^2/144}{315}$

Note that for low-speed adiabatic flow, $\rho \approx \rho_e$, $\Delta \approx \delta$, and Λ reduces to Pohlhausen's original parameter

$$\frac{\delta^2}{\nu}\frac{dU}{dx}$$

Gruschwitz substituted his polynomial distributions in the momentum-integral and mechanical-energy integral relations and found, after considerable algebra, that the problem reduced to a single first-order differential equation for the momentum thickness $\theta(x)$:

$$\frac{\theta U_e}{\nu_e}\frac{d\theta}{dx} = F_1(\lambda) - \lambda\frac{T_e}{T_w}\left[2 - Ma_e^2 F_2(\lambda)\right] \qquad (7\text{-}63)$$

It is assumed that the variation of U_e, ν_e, Ma_e, and T_e/T_w are known functions of x. Gruschwitz gave polynomial expressions for F_1 and F_2 which include the Prandtl number and specific-heat ratio as parameters:

$$F_1 = 2\frac{\theta}{\Delta}\left(1 - \frac{\Lambda}{15} + \frac{\Lambda^2}{240}\right) \approx 0.235 \exp(-5.4\lambda)$$

$$F_2 = 1 - \frac{\gamma - 1}{2}\frac{\Delta}{\theta}\left[\frac{\delta_3}{\Delta} - \frac{Pr(12 + \Lambda)^2}{2160}\right] \approx 0.78 \exp(1.5\lambda)$$

(7-64)

where

$$\frac{\delta_3}{\Delta} = \frac{798{,}048 - 4656\Lambda - 758\Lambda^2 - 7\Lambda^3}{4{,}324{,}320}$$

The exponential approximations are for the special case $Pr = 0.72$, $\gamma = 1.4$, which approximates air quite closely and avoids unnecessary algebra. Again it is believed that this method of Gruschwitz is approximately equal in accuracy to many far more sophisticated techniques.

Once $\lambda(x)$ has been found from the (numerical) solutions of Eqs. (7-63) and (7-64) and $\Lambda(x)$ computed from Eqs. (7-62), the local skin friction and separation point may be approximated by

Skin friction: $\quad C_f = \dfrac{2\tau_w(x)}{\rho_e U_e^2} \approx \dfrac{4\nu_e}{U_e \Delta}\left(1 + \dfrac{\Lambda}{7}\right)$

(7-65)

Separation: $\quad \lambda \approx -0.1 \quad \Lambda \approx -7$

These equations, though algebraically complicated, are easy to program and solve by a numerical technique such as the Runge–Kutta method of App. C. Further details of the laminar-flow method of Gruschwitz (1950) are discussed in the text by Schreier (1982).

7-4.4 Finite-Difference Methods for Compressible Laminar Flow

Since the laminar-boundary-layer equations are well defined and essentially free from empiricism, they are, in principle, solvable to any accuracy by finite-difference or finite-element methods. Many workers have suggested numerical approaches, and some of these are discussed in the texts by Cebeci and Smith (1974), Schreier (1982), and Anderson (2000). Most of these published methods use coordinate transformations before applying the numerical approximations. Here we will present a finite-difference approach in the raw variables (u, v, ρ, p, T) as an extension of the incompressible-flow methods of Sec. 4-7. The results are easy to understand and program but perhaps are not the most efficient algorithms.

We consider only two-dimensional, steady, compressible flow. We denote the (x, y) location of a mesh point by subscripts (m, n) as in Fig. 7-12, which is a convenient repetition of Fig. 4-25. The variables at each node are denoted by

532 VISCOUS FLUID FLOW

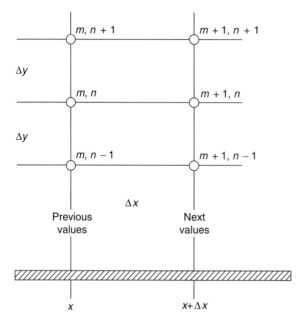

FIGURE 7-12
Finite-difference mesh for a two-dimensional boundary layer.

$u_{m,n}$, $v_{m,n}$, $\rho_{m,n}$, and $T_{m,n}$. The transport properties are also variable because temperature varies: $\mu_{m,n}$, $k_{m,n}$, and $c_{p_{m,n}}$. The freestream properties are assumed known functions of x and hence have only one coordinate subscript: $U_{e,m}$, $T_{e,m}$, $p_{e,m}$, and $\rho_{e,m}$. The Prandtl number is assumed constant.

7-4.4.1 AN EXPLICIT MODEL.
The steady-flow form of the x-momentum Eq. (7-1b) is

$$\rho u \frac{\partial u}{\partial x} + \rho v \frac{\partial u}{\partial y} = \rho_e U_e \frac{dU_e}{dx} + \frac{\partial \mu}{\partial y}\frac{\partial u}{\partial y} + \mu \frac{\partial^2 u}{\partial y^2} \qquad (7\text{-}66)$$

where we have broken up the viscous term into two parts. By exact analogy with the development of Sec. 4-7.1, the finite-difference model of this equation is

$$\rho_{m,n} u_{m,n} \frac{u_{m+1,n} - u_{m,n}}{\Delta x} + \rho_{m,n} v_{m,n} \frac{u_{m,n+1} - u_{m,n-1}}{2\Delta y}$$

$$\approx \frac{\rho_{e,m}(U_{m+1}^2 - U_m^2)}{2\Delta x} + \frac{(\mu_{m,n+1} - \mu_{m,n-1})(u_{m,n+1} - u_{m,n-1})}{(2\Delta y)^2}$$

$$+ \mu_{m,n} \frac{u_{m,n+1} - 2u_{m,n} + u_{m,n-1}}{\Delta y^2}$$

If we clean this up and solve for the (single) downstream velocity, we obtain

$$u_{m+1,n} \approx (\alpha^* - \beta + \zeta)u_{m,n+1} + (1 - 2\alpha^*)u_{m,n} + (\alpha^* + \beta - \zeta)u_{m,n-1}$$
$$+ \frac{\rho_{e,m}\left(U_{m+1}^2 - U_m^2\right)}{2\rho_{m,n}u_{m,n}} \quad (7\text{-}67)$$

where

$$\alpha^* = \frac{\mu_{m,n}\Delta x}{\rho_{m,n}u_{m,n}\Delta y^2} \qquad \beta = \frac{v_{m,n}\Delta x}{2u_{m,n}\Delta y} \qquad \zeta = \frac{(\mu_{m,n+1} - \mu_{m,n-1})\Delta x}{4\rho_{m,n}u_{m,n}\Delta y^2}$$

Equation (7-67) is the explicit compressible analog of formula Eq. (4-146) developed for incompressible flow.

The steady-flow form of the energy Eq. (7-1d) is

$$\rho u c_p \frac{\partial T}{\partial x} + \rho v c_p \frac{\partial T}{\partial y} = u\frac{dp_e}{dx} + \mu\left(\frac{\partial u}{\partial y}\right)^2 + \frac{\partial k}{\partial y}\frac{\partial T}{\partial y} + k\frac{\partial^2 T}{\partial y^2} \quad (7\text{-}68)$$

By exact analogy with the development of Sec. 4-8.2, an explicit finite-difference model of this equation is

$$(\rho u c_p)_{m,n}\frac{T_{m+1,n} - T_{m,n}}{\Delta x} + (\rho v c_p)_{m,n}\frac{T_{m,n+1} - T_{m,n-1}}{2\Delta y}$$
$$\approx -\frac{u_{m,n}\rho_{e,m}\left(U_{m+1}^2 - U_m^2\right)}{2\Delta x} + \mu_{m,n}\left(\frac{u_{m,n} - u_{m,n-1}}{\Delta y}\right)^2$$
$$+ \frac{(k_{m,n+1} - k_{m,n-1})(T_{m,n+1} - T_{m,n-1})}{(2\Delta y)^2}$$
$$+ k_{m,n}\frac{T_{m,n+1} - 2T_{m,n} + T_{m,n-1}}{\Delta y^2}$$

This may be rearranged into the following explicit model:

$$T_{m+1,n} \approx \left(\frac{\alpha^*}{Pr} - \beta + \frac{\zeta}{Pr}\right)T_{m,n+1} + \left(1 - \frac{2\alpha^*}{Pr}\right)T_{m,n} + \left(\frac{\alpha^*}{Pr} + \beta - \frac{\zeta}{Pr}\right)T_{m,n-1}$$
$$- \frac{\rho_{e,m}\left(U_{m+1}^2 - U_m^2\right)}{2(\rho u c_p)_{m,n}} + \frac{\alpha^*}{c_{p,m,n}}(u_{m,n} - u_{m,n-1})^2 \quad (7\text{-}69)$$

This is the compressible analog of Eq. (4-163). Finally, the continuity Eq. (7-1a) is modeled as

$$\frac{(\rho u)_{m+1,n} - (\rho u)_{m,n}}{2\Delta x} + \frac{(\rho u)_{m+1,n-1} - (\rho u)_{m,n-1}}{2\Delta x} + \frac{(\rho v)_{m+1,n} - (\rho v)_{m+1,n-1}}{\Delta y} \approx 0$$

This is rearranged to solve for the normal velocity at the next mesh point:

$$v_{m+1,n} \approx \frac{(\rho v)_{m+1,n-1}}{\rho_{m+1,n}}$$

$$-\frac{\Delta y}{2\rho_{m+1,n}\Delta x}[(\rho u)_{m+1,n} - (\rho u)_{m,n} + (\rho u)_{m+1,n-1} - (\rho u)_{m,n-1}] \quad (7\text{-}70)$$

This is the compressible analog of Eq. (4-148). The values of density at $(m + 1)$ would have already been computed from the ideal-gas law, once the temperatures are known from Eq. (7-69). In general,

$$\frac{\rho_{m,n}}{\rho_e} = \frac{T_e}{T_{m,n}} \quad (7\text{-}71)$$

The transport properties are computed from the known temperature. For air, for example, power-law approximations may be used from Tables 1-2 and 1-3:

$$\frac{\mu_{m,n}}{\mu_e} \approx \left(\frac{T_{m,n}}{T_e}\right)^{0.666}$$

$$\frac{k_{m,n}}{k_e} \approx \left(\frac{T_{m,n}}{T_e}\right)^{0.81} \quad (7\text{-}72)$$

Finally, compute $c_{p_{m,n}} = k_{m,n}Pr/\mu_{m,n}$. Equations (7-67) and (7-69) to (7-72) must be solved together for $(u, T, v, \rho, \mu, k, c_p)$ at the new point $(m + 1, n)$. Since the method is explicit, the following inequalities must hold for numerical stability, assuming $Pr < 1$, that is, a gas:

$$\frac{\alpha^*}{Pr} < 0.5 \qquad \beta < \frac{\alpha^*}{Pr} \qquad \frac{\zeta}{Pr} < \beta \quad (7\text{-}73)$$

These parameters should be monitored during the computation.

7-4.4.2 AN IMPLICIT MODEL. By analogy with Sec. 4-7.2, an implicit finite-difference model is formed if one evaluates the second derivatives of u and T at the next station, $(m + 1)$. We then obtain two tridiagonal relations:

$$-\alpha^* u_{m+1,n+1} + (1 + 2\alpha^*)u_{m+1,n} - \alpha^* u_{m+1,n-1}$$

$$\approx u_{m,n} + \frac{\rho_{e,m}(U_{m+1}^2 - U_m^2)}{2\rho_{m,n}u_{m,n}} - (\beta - \zeta)(u_{m,n+1} - u_{m,n-1}) \quad (7\text{-}74)$$

$$-\frac{\alpha^*}{Pr}T_{m+1,n+1} + \left(1 - \frac{2\alpha^*}{Pr}\right)T_{m+1,n} - \frac{\alpha^*}{Pr}T_{m+1,n-1}$$

$$\approx T_{m,n} + \frac{\alpha^*}{c_{p,m,n}}(u_{m,n} - u_{m,n-1})^2 - \frac{\rho_{e,m}(U_{m+1}^2 - U_m^2)}{2(\rho u c_p)_{m,n}}$$

$$-\left(\beta - \frac{\zeta}{Pr}\right)(T_{m,n+1} - T_{m,n-1}) \quad (7\text{-}75)$$

These are to be solved, in conjunction with Eq. (7-70), for the new velocities and temperatures, using the tridiagonal matrix algorithm described in Sec. 4-7.2.1. Generally this implicit computation, though more complex algebraically, is faster for a given accuracy than the explicit formulation. The density and transport properties are again computed by Eqs. (7-71) and (7-72).

The boundary conditions for either the explicit or implicit method are

Known freestream conditions: U_e, T_e, and ρ_e (7-76a)

No slip: $u_{m,1} = v_{m,1} = 0$ (7-76b)

No temperature jump: $T_{m,1} = T_w$ (7-76c)

Known wall heat flux: $T_{m,2} \approx T_{m,1} + \dfrac{q_w \Delta y}{k_{m,1}}$ (7-76d)

Known initial profiles: $u_{1,n}$, $v_{1,n}$, and $T_{1,n}$ (7-76e)

The implicit method has no stability limitations, but a small mesh size (Δx, Δy) should nevertheless be chosen in order to limit truncation errors.

7-4.4.3 EXAMPLE: A FLAT PLATE WITH ADIABATIC WALL.

The finite-difference marching methods discussed before can be applied, in principle, to any laminar, compressible, boundary layer with known freestream and wall-temperature conditions. As an example, we take a flat-plate flow with $Ma_e = 3.0$, assuming air with a freestream temperature of 300 K and a Prandtl number of 0.71. Viscosity and thermal conductivity were computed by the power-law approximations, Eqs. (7-72). The transverse mesh size Δy was chosen so that the velocity profile would contain about 40 points. The streamwise step size Δx was chosen so that velocities and temperatures would change no more than 5 percent for each step. To avoid instability, the implicit method of Sec. 7-4.4.2 was used.

Figure 7-13 shows the computed results for both an adiabatic and a cold wall ($T_w = T_e$). Compare these with the exact analytical results in Figs. 7-3 and 7-4. The adiabatic wall temperature is slightly low, corresponding to a recovery factor $r \approx 0.81$. The cold wall greatly suppresses the high temperatures generated by viscous dissipation and creates a thinner, more rounded velocity profile that is more stable (resistant to transition). The computed values of $C_D \sqrt{Re_x} \approx 1.18$ and 1.28 for adiabatic and cold walls, respectively, are very close to the exact values in Fig. 7-5a. The numerical approach is judged to be a reasonable success.

7-4.4.4 EXAMPLE: SUPERSONIC LINEARLY DECELERATING FLOW.

As a second example involving an adverse pressure gradient, consider an extension of the Howarth flow from Sec. 4-5.1, where $U = U_0(1 - x/L)$, to a compressible, linearly decreasing, freestream Mach number:

$$Ma_e(x) = Ma_0\left(1 - \frac{x}{L}\right) \quad (7\text{-}77)$$

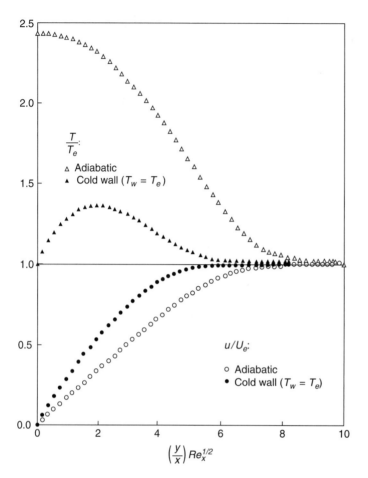

FIGURE 7-13
Digital-computer results for velocity and temperature profiles in flat-plate flow at $Ma = 3.0$, from Eqs. (7-70), (7-74), and (7-75).

For incompressible conditions, $Ma_0 \ll 1$, this laminar flow separated at $x/L \approx 0.12$ (see Figs. 4-20 or 4-26). For higher Mach numbers, let us assume an adiabatic wall and compare an integral method with a finite-difference solution. Again let us take the freestream as air at 300 K with $Pr = 0.71$.

It is quite easy to introduce Eq. (7-77) into the Gruschwitz integral method, Eq. (7-63). One needs only an estimate for the wall temperature, which could be supplied by assuming a "local" recovery factor $r \approx \sqrt{Pr} \approx 0.84$:

$$\frac{T_w}{T_e} \approx 1 + (0.71)^{1/2}\left(\frac{1.4 - 1}{2}\right)Ma_e^2 \qquad (7\text{-}78)$$

Integration of Eq. (7-63) by, say, the Runge–Kutta method of App. C yields $\theta(x)$, after which $\lambda(x)$ and $C_f(x)$ are computed from Eqs. (7-62) and (7-65), respectively.

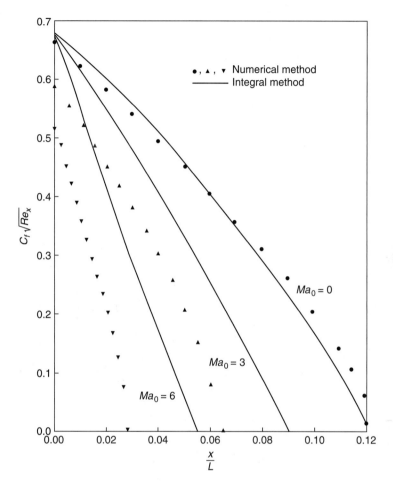

FIGURE 7-14
Boundary layer with linearly decreasing freestream Mach number: comparison of an integral method, Eq. (7-63), with an implicit finite-difference method, Eqs. (7-70), (7-74), and (7-75).

Concurrently, the implicit numerical method of Sec. 7-4.4.2 was also programmed for Eq. (7-77), with an adiabatic wall ($q_w = 0$) approximation, $T_{m,2} = T_{m,1}$. The boundary layer contained at least 40 mesh points Δy at each x station. The local skin friction computed by each method is shown in Fig. 7-14. The case $Ma_0 = 0$ corresponds to our earlier study, Fig. 4-26. We see that compressibility, in this case, causes earlier separation of the laminar boundary layer. A published numerical solution for this problem is not known to this writer, but if we take the numerical results in Fig. 7-14 to be accurate, then the integral method of Gruschwitz is qualitatively correct but not too precise at higher Mach numbers.

An interesting sidelight of this example was that the computed adiabatic wall temperatures were lower than expected, especially near separation. The following table of computed recovery factors near separation illustrates this point:

Ma_0	0	1	2	3	4	5	6
r_{sep}	0.84	0.82	0.75	0.68	0.61	0.55	0.51

The reason for this behavior is that, near separation for a given value of U_e, the near-wall shear stresses decrease, so the viscous dissipation term $[\mu(\partial u/\partial y)^2]$ is smaller and generates less internal energy compared to $(U_e^2/2c_p)$. This lowering of T_{aw}/T_e is difficult to estimate in advance and thus could not reasonably be preprogrammed into the Gruschwitz method of Eq. (7-63).

7-4.4.5 OTHER NUMERICAL SOLUTIONS. The aerospace industry has developed many CFD approaches for compressible viscous flows. Laminar (fluctuation-free) flow is especially successful, even for three-dimensional geometries, since supercomputers can now supply up to 1 billion mesh points. One can now compute the supersonic flow field around a complex aircraft shape, as reviewed by Agarwal (1999). However, the turbulence-modeled portions of aircraft flow are not so accurate.

Compressible flow CFD methods, primarily inviscid, are reviewed in monographs by Laney (1998), Anderson (2002), Chattot (2002), and Felcman et al. (2003). Shock-wave discontinuities are often present and a priori unknown, but excellent shock-capturing and shock-fitting CFD schemes have been developed. Most boundary-layer calculations use finite differences, but the finite-element method is also successful, Hytopoulos et al. (1993). Compressible-boundary-layer integral methods are also available from the Internet Applets of Devenport and Schetz (2002).

Boundary-layer computer methods began in the 1960s with the work of A. M. O. Smith and colleagues at McDonnell-Douglas Corp. Figure 7-15 shows the

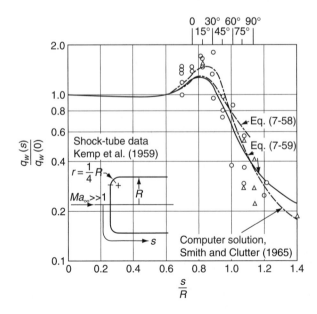

FIGURE 7-15
Comparison of theory and experiment for surface heat flux in laminar hypersonic flow past a flat-nosed cylinder.

computations of Smith and Clutter (1965) for hypersonic flow past a flat-nosed cylinder with a rounded shoulder. The agreement with data from Kemp et al. (1959) is good. The simple correlation formulas Eqs. (7-58) and (7-59) from Lees (1956) are also quite adequate. Such results require a preliminary inviscid calculation to yield accurate freestream distributions $U_e(x)$ and $T_e(x)$.

7-5 SPECIAL TOPICS IN COMPRESSIBLE LAMINAR FLOW

The problem of compressible (supersonic) viscous flow is a broad one, and there are many topics which will not be covered here, such as (1) high-temperature breakdown of air into separate species, notably dissociation of oxygen and nitrogen, with resulting species diffusion (mass transfer); (2) combustion, ionization, and other nonequilibrium thermodynamic processes; (3) radiation from a hot-gas boundary layer; and (4) foreign-gas injection into the boundary layer. Here we briefly discuss two topics. For further reading, see Anderson (2000).

7-5.1 The Laminar Supersonic Cone Rule

In the theory of inviscid supersonic flow past a cone at zero incidence [e.g., Shapiro (1954, Chap. 17)], for $Ma_\infty > 1.2$ and a cone half-angle less than $55°$, the resulting shock wave is attached to the cone vertex, and the flow at the cone surface is at constant velocity, pressure, and temperature. Thus the boundary layer in supersonic cone flow at zero incidence has properties akin to flat-plate flow. The pressure gradient is zero, and let us assume that T_w is constant.

The boundary-layer flow is axisymmetric, and Lees (1956) modified the Illingworth transformation Eqs. (7-18) for such flows as follows:

$$\xi = \int_0^x \rho_e U_e \mu_e r_0^{2j} \, dx$$

and $\hspace{10cm}$ (7-79)

$$\eta = \frac{\rho_e U_e r_0^j}{\sqrt{(2\xi)}} \int_0^y \frac{\rho}{\rho_e} \, dy$$

where $r_0(x)$ is the body surface radius and $j = 1$ for axisymmetric flow and $j = 0$ for plane flow. Note the resemblance to the incompressible Mangler transformation, Eqs. (4-187). When substituted in the boundary-layer momentum and energy equations, similarity relations Eqs. (7-32) result.

For a cone of half-angle ϕ, $r_0 = bx$, where $b = \tan^{-1}\phi$. Then the variables ξ and η for this case become

$$\xi_{cone} = \frac{\rho_e U_e \mu_e b^2 x^3}{3}$$

$$d\eta_{cone} = \left(\frac{3\rho_e U_e}{2x\mu_e}\right)^{1/2} \rho \, dy$$

(7-80)

From the expression for $d\eta/dy$, we can then compute the skin-friction coefficient at the surface

$$C_{f,\text{cone}} = \frac{2\tau_w}{\rho_e U_e^2} = \sqrt{6C_w}\frac{f''(0)}{(\rho_e U_e x/\mu_e)^{1/2}} \qquad (7\text{-}81)$$

where C is the Chapman–Rubesin parameter and $f''(0)$ is computed from the similarity theory Eqs. (7-32) for zero pressure gradient ($\beta = 0$).

If we now make the same analysis for flat-plate flow with constant wall temperature, we obtain

$$\xi_{\text{plate}} = \rho_e U_e \mu_e x$$

$$d\eta_{\text{plate}} = \left(\frac{\rho_e U_e}{2x\mu_e}\right)^{1/2} \rho\, dy \qquad (7\text{-}82)$$

$$C_{f,\text{plate}} = \sqrt{2C_w}\frac{f''(0)}{(\rho_e U_e x/\mu_e)^{1/2}}$$

Now, if T_w/T_e is the same for the cone and the plate, then C_w is the same for both, and if the Mach number and specific-heat ratio are the same for both, then $f''(0)$ is the same, since it arises from exactly the same similarity equations for $\beta = 0$. Then we can compare the two C_f formulas and arrive at a cone rule for laminar compressible flow. We may take at least two points of view:

$$\text{If } Re_{xe,\text{cone}} = Re_{xe,\text{plate}} \quad \text{then} \quad C_{f,\text{cone}} = \sqrt{3}C_{f,\text{plate}} \qquad (7\text{-}83a)$$

$$\text{If } C_{f,\text{cone}} = C_{f,\text{plate}} \quad \text{then} \quad Re_{xe,\text{cone}} = 3Re_{xe,\text{plate}} \qquad (7\text{-}83b)$$

The view in Eq. (7-83a) is the usual one stated in the literature: If a cone and a plate have the same local Reynolds number, Mach number, and wall-temperature ratio, then the cone skin friction is 73.2 percent higher than the plate's local friction. Further, if both have the same Prandtl number, then the cone local heat flux is also 73.2 percent higher than the plate heat flux. This follows from the Reynolds analogy [Eq. (7-41c)]. Remember that this result is for laminar flow only. There is also a turbulent cone rule (Sec. 7-8.4), but in that case the ratio of cone to plate C_f and C_h is only about 1.1 to 1.15, not 1.732. Nor is it an exact square root or anything of the sort, so that from Eq. (7-83b), a Reynolds number ratio is the way a turbulent cone rule is stated.

7-5.2 Hypersonic Flow: Shock-Wave Boundary-Layer Interactions

As supersonic flow increases to *hypersonic* flow, interactions between the boundary layer and the freestream become important. As Anderson (2002) remarks, the definition of hypersonic (Greek, *hyper*: over, above, excessive) is flexible. Some say $Ma > 3$ and some say $Ma > 12$. The key is the interaction. If it is strong, the flow is hypersonic.

Interactions can be classified as (1) pressure induced by the displacement effect of the boundary layer, (2) blast-wave effects due to blunting of a leading

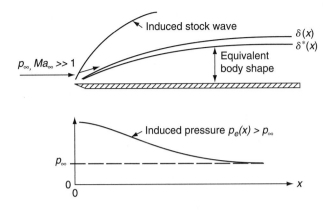

FIGURE 7-16
Flat-plate leading-edge shock wave and pressure distribution induced by boundary-layer displacement.

edge, (3) flow separation in a supersonic compression corner, (4) a shock wave from elsewhere impinging on a boundary layer, and (5) boundary layer changes due to vorticity introduced from a curved upstream shock wave. The interaction problem is intriguing, since it requires the outer inviscid and inner viscous flow equations to be solved together. Reviews of hypersonic flow theory are given in the classic texts by Hayes and Probstein (1959) and Dorrance (1962) and, more recently, by Rasmussen (1994), Smits and Dussauge (1996), and Anderson (2000).

Let us consider the specific problem of supersonic flow past a sharp-edged flat plate, as shown in Fig. 7-16. The boundary layer has a finite but small displacement effect $\delta^*(x)$ that would be negligible at low speeds but causes a significant induced leading-edge shock wave to appear at high speeds. The shock increases pressures $p_e(x)$ over the leading-edge region, as sketched in Fig. 7-16. Associated with this pressure rise are increased skin friction and surface heat transfer.

In a hypersonic boundary layer, near-wall temperatures are high and densities low compared to the freestream. Thus, since ρ/ρ_e is small, the displacement thickness $\delta^* \approx \delta$ from Eq. (7-60). Our correlation Eq. (7-45) then indicates that, if $Ma_e \approx Ma_\infty$ is large,

$$\frac{\delta^*}{x} \approx \frac{\delta}{x} \approx (\text{const}) Ma_\infty^2 \left(\frac{C_w}{Re_x}\right)^{1/2} \qquad (7\text{-}84)$$

The displacement thickness presents a growing "body shape" to the approaching supersonic flow, which must deflect by a local angle $\phi \approx d\delta^*/dx$. If Ma_∞ is large, the oblique-shock formulas [e.g., White (2003, p. 652)] show that the pressure rise across the shock is approximately a function only of the specific-heat ratio and the single parameter $K = Ma_\infty \phi$:

$$\frac{p_e}{p_\infty} \approx 1 + \frac{\gamma}{4}(\gamma + 1)K^2 + \gamma K \left\{ 1 + \left[\frac{(\gamma + 1)K}{4}\right]^2 \right\}^{1/2} \qquad (7\text{-}85)$$

This formula is valid whenever $K = \mathcal{O}(1)$ and ϕ is less than about 20°.

Differentiation of Eq. (7-84) shows that $d\delta^*/dx$ is also proportional to $Ma_\infty^2(C_w/Re_x)^{1/2}$. Then the parameter K may be written as

$$K = Ma_\infty \left(\frac{d\delta^*}{dx}\right) = (\text{const})\chi \qquad \chi = Ma_\infty^3 \left(\frac{C_w}{Re_{x\infty}}\right)^{1/2} \qquad (7\text{-}86)$$

The new grouping, χ, is called the *hypersonic interaction parameter*. The strength of the leading-edge interactions in Fig. 7-16 depends on the size of this parameter. The following order-of-magnitude estimates apply:

$\chi \ll 1$: negligible interaction effects.

$\chi = \mathcal{O}(1)$: weak interaction; can be computed by simply assuming supersonic inviscid flow over the (uncoupled) body shape $\delta^*(x)$.

$\chi \gg 1$: strong interaction; $\delta^*(x)$ and the external supersonic flow are strongly interdependent and must be solved simultaneously.

As it turns out, one can derive an ordinary differential equation valid for all these regimes, and therefore there is no real need to distinguish between weak and strong interactions. An excellent review of this type of theory is given by Bertram and Blackstock (1961).

Since $p_e(x)$ decreases with x, as shown in Fig. 7-16, the pressure gradient is favorable, which tends to decrease the displacement thickness somewhat. Extensive calculations by Bertram and Blackstock (1961) have resulted in the following algebraic correlation for flat plates:

$$\frac{\delta^*}{x} \approx 0.85\left(\frac{T_w}{T_{aw}} + 0.35\right)(\gamma - 1)Ma_\infty^2 \left(\frac{C_w p_\infty}{Re_{x\infty} p_e}\right)^{1/2} \qquad (7\text{-}87)$$

Differentiating with respect to x and evaluating Eq. (7-86), we obtain a more precise estimate for the parameter K:

$$K = \frac{a\chi}{\sqrt{P}}\left(1 + \frac{\chi}{2P}\frac{dP}{d\chi}\right) \qquad P = \frac{p_e}{p_\infty} \qquad a = 0.425\left(\frac{T_w}{T_{aw}} + 0.35\right)(\gamma - 1)$$
$$(7\text{-}88)$$

Equations (7-86) and (7-88) may be solved simultaneously for the shock-induced pressure distribution $P(\chi)$, with the initial condition $P(0) = 1$. A numerical solution is necessary, but the results are nearly linear with χ. Thus a linear approximation will suffice:

$$P \approx 1 + a\chi\left[\tfrac{9}{8}\gamma(\gamma + 1)\right]^{1/2} \qquad (7\text{-}89)$$

To this order of approximation, then, the hypersonic interaction parameter χ successfully correlates the leading-edge induced pressures. For $\gamma = 1.4$ and an adiabatic wall, $a \approx 0.23$ and Eq. (7-89) becomes

Adiabatic wall: $$\frac{p_e}{p_\infty} \approx 1 + 0.45\chi \qquad (7\text{-}90)$$

This simple expression is compared with experimental data for adiabatic flat plates in Fig. 7-17. The theory somewhat underestimates the data for very strong interactions, but the linear behavior is evident. For further details, see Bertram and Blackstock (1961).

The theory has been extended by White (1962b) to a flat surface at an angle of attack. The results are in good agreement with hypersonic force measurements on wedges at angles of attack. Anderson (2000) gives an interesting discussion of viscous interactions on various hypersonic vehicles. Generally speaking, the higher pressures induced by viscous displacement increase the body drag force more than its lift, decreasing its lift–drag ratio.

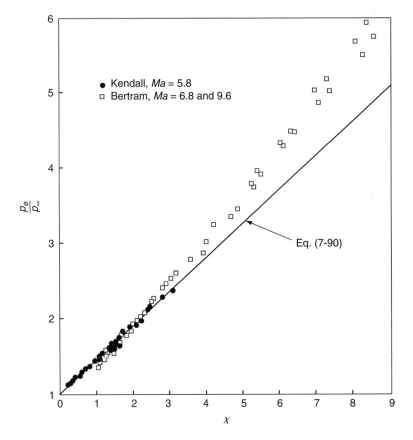

FIGURE 7-17
Correlation of measured shock-induced pressures on an adiabatic flat plate in air at high Mach numbers. [*After Bertram (1958)*.]

7-6 THE COMPRESSIBLE-TURBULENT-BOUNDARY-LAYER EQUATIONS

As in laminar flows, turbulent compressible flows have variable density and thus need supplementing by an equation of state. Turbulent flow also has the additional complication of having to account for density fluctuations

$$\rho = \bar{\rho} + \rho' \tag{7-91}$$

with the usual notation of a bar for average value and a prime for fluctuating value. Since density keeps occurring together with other variables, additional time-averaged quantities are obtained, involving products of fluctuating variables. For example, all three conservation relations contain the product ρu, which becomes, after time averaging,

$$\overline{\rho u} = \bar{\rho}\,\bar{u} + \overline{\rho' u'} \tag{7-92}$$

The quantity $\bar{u}$, when broken out from density, is called the *mass-averaged* velocity, also known as *Favre-averaging*, after the original derivation by Favre (1965). Smits and Dussauge (1996) give the complete Favre-averaged equations of three-dimensional, compressible, turbulent flow. Here we study only turbulent-boundary-layer equations.

Another complication of compressible turbulent flows is that density fluctuations automatically imply temperature fluctuations through the equation of state. Then, if $T = \bar{T} + T'$ and T' is important, there is a possibility that fluctuations in the transport properties, which depend upon T, will also be important:

$$\begin{aligned} \mu(T) = \bar{\mu} + \mu' \qquad & k(T) = \bar{k} + k' \\ c_p(T) = \bar{c}_p + c'_p \qquad & Pr(T) = \overline{Pr} + Pr' \end{aligned} \tag{7-93}$$

This would be very discouraging, although not impossible, but fortunately it appears that transport-property fluctuations can be neglected.

7-6.1 Morkovin's Hypothesis—The Boundary-Layer Equations

In studying supersonic flow, Morkovin (1962) postulated that "the essential dynamics of compressible shear flows will follow the incompressible pattern," that is, fluctuations in density and enthalpy will not modify the turbulence structure as long as fluctuations in Mach number are much less than unity. Subsequent experiments verify this hypothesis for Mach numbers as high as 5. Thus, most of our concepts from incompressible turbulence (Chap. 6) will remain valid.

A visual demonstration of Morkovin's hypothesis is shown in Fig. 7-18, which is a schlieren image of a Mach 3 boundary layer from Garg and Settles (1998). Compare this with the low-speed smoke-flow image in Fig. 6-1*b*. We see that the structures and length scales are similar, indicating that compressible and incompressible flow dynamics are much the same.

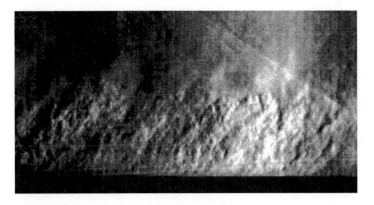

FIGURE 7-18
Visual support for Morkovin's hypothesis: focusing schlieren image of a Mach 3 turbulent boundary layer from Garg and Settles (1998). Compare this to Fig. 6-1b. [*Reprinted by permission of the authors and the journal Experiments in Fluids.*]

The derivation of the compressible two-dimensional turbulent-boundary-layer equations uses the same order-of-magnitude considerations as laminar flow, with the additional assumption that the fluctuations are small:

$$\bar{v} \ll \bar{u}$$

$$\frac{\partial}{\partial x} \ll \frac{\partial}{\partial y} \quad (7\text{-}94)$$

$$Q' \ll \bar{Q} \quad \text{for all } Q$$

These considerations were first applied to the compressible Navier–Stokes equations by van Driest (1951) in a stunning display of expertise in handling time-averaged quantities. The boundary-layer equations thus derived are

Continuity: $\quad \dfrac{\partial}{\partial x}(\bar{\rho}\,\bar{u}) + \dfrac{\partial}{\partial y}(\bar{\rho}\bar{v}) = 0 \quad (7\text{-}95a)$

x momentum: $\quad \bar{\rho}\,\bar{u}\dfrac{\partial \bar{u}}{\partial x} + \bar{\rho}\bar{v}\dfrac{\partial \bar{u}}{\partial y} = \rho_e U_e \dfrac{dU_e}{dx} + \dfrac{\partial \tau}{\partial y} \quad (7\text{-}95b)$

y momentum: $\quad \dfrac{\partial \bar{p}}{\partial y} = -\dfrac{\partial}{\partial y}(\overline{\rho v' v'}) \ll \left|\dfrac{\partial p}{\partial x}\right| \quad (7\text{-}95c)$

Energy: $\quad \bar{\rho}\,\bar{u}\dfrac{\partial \bar{h}}{\partial x} + \bar{\rho}\bar{v}\dfrac{\partial \bar{h}}{\partial y} = \bar{u}\dfrac{d\bar{p}}{dx} + \dfrac{\partial q}{\partial y} + \tau\dfrac{\partial \bar{u}}{\partial y} \quad (7\text{-}95d)$

where $\quad \tau = \bar{\mu}\dfrac{\partial \bar{u}}{\partial y} - \overline{\rho u' v'} \qquad q = \bar{k}\dfrac{\partial \bar{T}}{\partial y} - \overline{\rho v' h'}$

Again, as for incompressible flow, there is a slight pressure gradient in the y direction that can be ignored in the present context (hence $\partial \bar{p}/\partial x \approx -\rho_e U_e dU_e/dx$) but becomes important in a Kármán-type integral method, for example, Eq. (7-60). Note that the turbulent-shear and heat-flux terms are now triple correlations because density is variable, but the semiempirical treatment of these terms is very little affected by this fact.

In analyzing these relations, the typical first step is to eliminate $\overline{\rho v}$ from momentum and energy through use of the continuity relation, leaving only two equations in the three variables $\overline{\rho}, \overline{u}, \overline{T}$. Naturally, another relation is needed, which is of course the equation of state of the fluid. It is sufficient for our purposes to adopt the perfect-gas approximation

$$\overline{p} = \overline{\rho}R\overline{T} \qquad R = \text{gas constant}$$
$$d\overline{h} = c_p\,d\overline{T} \qquad c_p \approx \text{const} \tag{7-96}$$

Since $\overline{p}$ is nearly constant through the boundary layer, the density profile is directly related to the temperature profile by the perfect-gas law

$$\frac{\overline{\rho}(y)}{\rho_e} \approx \frac{T_e}{\overline{T}(y)} \tag{7-97}$$

which simplifies the algebra greatly. The boundary conditions are the usual:
At the wall ($y = 0$):

$$\overline{u} = \overline{\rho v} = 0 \qquad \overline{T} = T_w \qquad \overline{\rho} = \rho_w \tag{7-98a}$$

At the outer edge ($y = \delta$):

$$\overline{u} \to U_e(x) \qquad \overline{T} \to T_e(x) \qquad \overline{\rho} \to \rho_e(x) \tag{7-98b}$$

If the turbulent shear and heat flux are properly correlated, these equations and boundary conditions are well defined and solvable in parabolic marching style, like their laminar counterparts. Eddy viscosity and eddy conductivity assumptions are reasonably effective if the density variation is properly modeled.

7-6.2 The Turbulent Crocco–Busemann Relation

Suppose now we make eddy transport definitions as for incompressible flow. Let

$$-\overline{\rho u'v'} = \mu_t \frac{\partial \overline{u}}{\partial y} \qquad -\overline{\rho v'h'} = k_t \frac{\partial \overline{T}}{\partial y} \tag{7-99}$$

Now rewrite the momentum equation in terms of μ_t and rewrite the energy equation using the variable $H = \overline{h} + \overline{u}^2/2$:

Momentum:

$$\overline{\rho}\,\overline{u}\frac{\partial \overline{u}}{\partial x} + \overline{\rho v}\frac{\partial \overline{u}}{\partial y} = -\frac{d\overline{p}}{dx} + \frac{\partial}{\partial y}\left[(\mu + \mu_t)\frac{\partial \overline{u}}{\partial y}\right]$$

Energy:

$$\overline{\rho}\,\overline{u}\frac{\partial H}{\partial x} + \overline{\rho}\,\overline{v}\frac{\partial H}{\partial y} = \frac{\partial}{\partial y}\left[\left(\frac{\mu}{Pr} + \frac{\mu_t}{Pr_t}\right)\frac{\partial H}{\partial y}\right] \quad (7\text{-}100)$$

$$+ \frac{\partial}{\partial y}\left[\mu\left(1 - \frac{1}{Pr}\right) + \mu_t\left(1 - \frac{1}{Pr_t}\right)\right]\frac{\partial}{\partial y}\left(\frac{\overline{u}^2}{2}\right)$$

where $Pr_t = \dfrac{\mu_t c_p}{k_t}$ is the turbulent Prandtl number

It is immediately obvious that if one makes the reasonable assumption $Pr \approx Pr_t \approx 1$ (approximately true for air) and if the pressure gradient is negligible, the previous two equations are identical in mathematical form and a particular solution is

$$H = C_1 + C_2\overline{u}$$

or $\qquad\qquad\qquad\qquad\qquad\qquad\qquad\qquad\qquad\qquad\qquad\qquad\qquad\qquad (7\text{-}101)$

$$\overline{h} = C_1 + C_2\overline{u} - \frac{\overline{u}^2}{2}$$

This is identical to the laminar relation discussed earlier Eq. (7-13). When the boundary conditions T_w, U_e are introduced, we obtain

Gases: $\qquad\qquad Pr = Pr_t \approx 1 \qquad \dfrac{d\overline{p}}{dx} \approx 0$

$\qquad\qquad\qquad\qquad\qquad\qquad\qquad\qquad\qquad\qquad\qquad\qquad\qquad\qquad (7\text{-}102)$

$$\overline{T} \approx T_w + (T_{\text{aw}} - T_w)\frac{\overline{u}}{U_e} - \frac{r\overline{u}^2}{2c_p}$$

Thus the Crocco–Busemann relation between temperature and velocity is also valid to good accuracy for turbulent flows with a negligible pressure gradient. It is even a good approximation with pressure gradients. Note that we have assumed constant c_p (so that $\overline{h} \approx c_p\overline{T}$) and have slipped in the recovery factor r for greater accuracy. For turbulent flow of gases, $r = Pr^{1/3} \approx 0.89$ for air. In the analyses that follow, we shall make good use of Eq. (7-102).

This integral momentum relation for compressible turbulent flow is identical to the laminar formula Eq. (7-60) when time-averaged velocity, density, and temperature are used. The integral-energy equation is also the same for either laminar or turbulent flow:

$$q_w = \frac{d}{dx}\left[\int_0^\infty \overline{\rho}\,\overline{u}\left(\overline{h} + \tfrac{1}{2}\overline{u}^2 - h_e - \tfrac{1}{2}U_e^2\right)dy\right] \quad (7\text{-}103)$$

One may, in principle, solve these integral relations by having suitable approximations for the (turbulent) velocity, density, and enthalpy profiles.

7-7 WALL AND WAKE LAWS FOR TURBULENT COMPRESSIBLE FLOW

We assume fully turbulent flow. Recall from Figs. 5-15 to 5-19 that (1) laminar compressible flows have more than one mode of instability; (2) as Ma increases, the first mode can be unstable in many different (transverse) directions; and (3) for $Ma > 4$, the second mode is the most unstable. Compressible and incompressible transition processes are similar, as predicted by Morkovin's hypothesis. Direct numerical simulation of flat-plate flow at $Ma = 4.5$, by Adams and Kleiser (1993), predicts second-mode instability that develops into subharmonic staggered Λ-vortices and then into detached high-shear layers. Fully turbulent compressible flow is thus anticipated for all developing compressible, laminar, boundary layers.

In Chap. 6, we were able to develop accurate law-of-the-wall and law-of-the-wake relations for incompressible turbulent mean flow. In keeping with Morkovin's hypothesis that compressible boundary-layer flows "follow the incompressible pattern," we can find modified wall and wake laws up to at least $Ma \approx 5$. The results seem quite reasonable when compared with the gold mine of incompressible- and compressible-flow data compiled by Fernholz and Finley (1980, 1996). Both velocity and temperature are correlated by these formulas.

7-7.1 Compressible Law of the Wall: the van Driest Transformation

This subject is nicely reviewed by Bradshaw (1977) and by Smits and Dussauge (2000). Since density and viscosity vary across the boundary layer, it is appropriate to define inner law dimensionless variables based upon *wall* values of density and viscosity:

$$u^+ = \frac{\bar{u}}{v^*}$$

$$y^+ = \frac{yv^*}{\nu_w} \tag{7-104}$$

$$v^* = \left(\frac{\tau_w}{\rho_w}\right)^{1/2}$$

This also ensures that, in the linear viscous sublayer, $u^+ = y^+$ without any extra coefficients.

Edward R. van Driest (1951) anticipated Morkovin's insight with a near-wall mixing-length theory that accounted for variable density. Neglecting the sublayer, he assumed

$$\tau \approx \tau_t = \bar{\rho}\,\ell^2\left(\frac{d\bar{u}}{dy}\right)^2$$

where
$$\ell \approx \kappa y \tag{7-105}$$

To relate density to velocity, van Driest then assumed a perfect gas and adopted the Crocco–Busemann approximation Eqs. (7-102) for Prandtl number unity,

$$\frac{\rho_w}{\bar{\rho}} = \frac{\bar{T}}{T_w} = 1 + \left(\frac{T_{aw}}{T_w} - 1\right)\frac{\bar{u}}{U_e} - \frac{\gamma-1}{2}Ma_e^2\frac{T_e}{T_w}\left(\frac{\bar{u}}{U_e}\right)^2 \tag{7-106}$$

By substituting Eq. (7-106) in Eq. (7-105) and making Prandtl's original assumption that $\tau \approx \tau_w$ (which is accurate only for a flat plate), van Driest (1951) was able to integrate to obtain $\bar{u}(\bar{y})$ in closed form:

$$u_{eq} = \frac{U_e}{a}\left(\sin^{-1}\frac{2a^2\bar{u}/U_e - b}{Q} + \sin^{-1}\frac{b}{Q}\right) = v^*\left(\frac{1}{\kappa}\ln y^+ + B\right) \quad (7\text{-}107)$$

where

$$a = \left(\frac{\gamma-1}{2}Ma_e^2\frac{T_e}{T_w}\right)^{1/2} \qquad b = \left(\frac{T_{aw}}{T_w} - 1\right) \qquad Q = (b^2 + 4a^2)^{1/2}$$

Here v^* and y^+ are defined exactly as in Eq. (7-104). The right-hand side of Eq. (7-107) is the ordinary incompressible law of the wall. The left-hand side has been given the symbol u_{eq}, the van Driest *effective velocity*.

For zero heat transfer, $b = 0$, $Q = 2a$, and the effective velocity becomes

Adiabatic walls: $\qquad u_{eq} = \dfrac{U_e}{a}\sin^{-1}\left(\dfrac{a\bar{u}}{U_e}\right) \qquad a^2 = 1 - \dfrac{T_e}{T_{aw}} \qquad (7\text{-}108)$

To first order, the van Driest effective velocity correlates supersonic and hypersonic near-wall flows with $\kappa \approx 0.41$ and $B \approx 5.0$, as in incompressible flow. It can also be used to define a compressible "defect law" by analogy with Eq. (6-36). See, for example, Fernholz and Finley (1980).

The success of van Driest's effective velocity is seen in Fig. 7-19, reproduced from Fernholz and Finley (1980). For Mach numbers between 3.5 and 7.2, the raw velocities $\bar{u}$ (solid symbols) are too low for near-adiabatic flow and too high for cold-wall flow. When transformed to u_{eq} from Eq. (7-107), all data cling nicely to

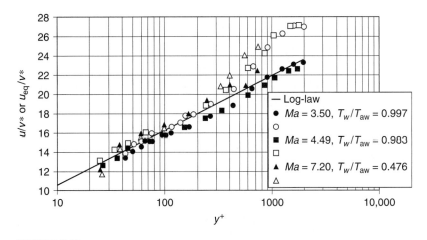

FIGURE 7-19
Verification of the van Driest transformation, Eq. (7-107), for compressible, turbulent, boundary layers: natural velocity (solid symbols) and effective velocity (open symbols). [*After Fernholz and Finley (1980)*.]

the incompressible log-law up to $y^+ \approx 400$, above which they form a law-of-the-wake shape.

So et al. (1994) used the inner–outer matching procedure of Sec. 6-4.1 to show that the effective velocity fits log-laws in both the overlap layer and the outer defect layer. However, the log-law constants (k, B, A) were found to vary with Mach number, heat flux, and $\gamma = c_p/c_v$.

7-7.2 Compressible Law of the Wake

It has also been found that the van Driest transformation makes compressible-flow data fit Coles' incompressible law of the wake, Eq. (6-47), that is, if one defines v^* based on wall properties and replaces $\bar{u}$ by u_{eq}, the wake formula should still be valid:

$$\frac{u_{eq}}{v^*} \approx \frac{1}{\kappa}\ln\left(\frac{yv^*}{\nu_w}\right) + B + \frac{2\Pi}{\kappa}w\left(\frac{y}{\delta}\right), \quad w \approx 3\left(\frac{y}{\delta}\right)^2 - 2\left(\frac{y}{\delta}\right)^3 \quad (7\text{-}109)$$

where $\kappa \approx 0.41$ and $B \approx 5$. For flat-plate data with $1.7 < Ma_e < 10.3$, Fernholz and Finley (1980) show that $\Pi \approx 0.55 \pm 0.05$ for $Re_\theta > 2000$, just as Coles (1956) had shown for incompressible, turbulent flow. Similarly, for $Re_\theta < 2000$, the value of Π drops off slowly to zero, just as Coles showed. For compressibility, it is postulated that Π varies with pressure gradient, freestream turbulence, and wall heat flux, but data are scarce for verification. For further discussion, see Sec. 7.4 of Smits and Dussauge (1996).

The van Driest transformation is equivalent to modifying the velocity by the square root of the integrated density ratio:

$$\bar{u}_{\text{effective}} = \int_0^{\bar{u}} \left(\frac{\bar{\rho}}{\rho_e}\right)^{1/2} d\bar{u} \quad (7\text{-}110)$$

This type of velocity will correlate supersonic data up to Mach 7 and for walls as cold as $T_w/T_{aw} = 0.3$. Meanwhile, the success of the Lees–Illingworth coordinate transformations in laminar compressible flow has prompted many attempts to try something similar for turbulent flows. Bradshaw (1977) reviews these alternate transformations, which have not been widely accepted compared to the van Driest idea above.

7-7.3 A Compressible Law of the Wall with Heat Transfer and a Pressure Gradient

The van Driest transformation, Eq. (7-107), is elegant but complicated algebraically and difficult to implement in a compressible boundary-layer calculation procedure. The present subsection develops an alternate algebraic compressible law of the wall that accounts for three important effects, from White and Christoph (1972). It is intended for preliminary engineering calculations and perhaps, who knows, for problem assignments.

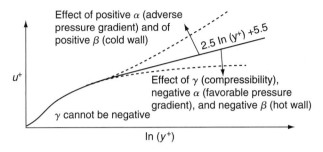

FIGURE 7-20
Sketch illustrating the effect of various parameters on the law of the wall for compressible turbulent flow. [*After White and Christoph (1972).*]

The three effects considered are pressure gradient, compressibility, and wall heat flux. They are nondimensionalized, using wall variables, as follows:

Pressure gradient: $\quad \alpha = \dfrac{\nu_w}{\tau_w v^*} \dfrac{dp_e}{dx}$

Heat flux: $\quad \beta = \dfrac{q_w \nu_w}{T_w k_w v^*}$ \hfill (7-111)

Compressibility: $\quad \gamma = \dfrac{r v^{*2}}{2 c_p T_w}$

where r is the recovery factor and q_w is positive for a cold wall. All three parameters typically take on small (fractional) numerical values. As shown in Fig. 7-20, the effect of positive α and β is to raise the profile above the incompressible log-law, whereas γ and negative α and β tend to lower the profile.

White and Christoph (1972) used the mixing-length approximation, Eq. (7-105), and introduced the near-wall effect of a pressure gradient:

$$\tau \approx \tau_w + \frac{dp_e}{dx} y + \mathcal{O}(y^3) = \bar{\rho} \kappa^2 y^2 \left(\frac{d\bar{u}}{dy}\right)^2 \quad (7\text{-}112)$$

They then rewrote the Crocco–Busemann relation, Eq. (7-102), in terms of wall variables:

$$\frac{\rho_w}{\bar{\rho}} = \frac{\bar{T}}{T_w} \approx 1 + \beta u^+ - \gamma u^{+2} \quad (7\text{-}113)$$

where β and γ are defined by Eqs. (7-111). Combine Eqs. (7-112) and (7-113), solve for the velocity derivative, and rewrite in wall variables:

$$\frac{du^+}{dy^+} = \frac{(1 + \alpha y^+)^{1/2}}{\kappa y^+ (1 + \beta u^+ - \gamma u^{+2})^{1/2}} \quad (7\text{-}114)$$

where α is the pressure gradient parameter from Eq. (7-111). Note that this differential equation contains no overt thickness parameter (δ, δ^*, etc.) and thus may be integrated indefinitely outward from the wall.

Equation (7-114) may be separated and integrated in closed form, starting at an initial point (u_0^+, y_0^+). The result is

$$u^+(y^+, \alpha, \beta, \gamma)$$

$$= \frac{1}{2\gamma}\left(\beta + Q\sin\left\{\phi + \frac{\sqrt{\gamma}}{\kappa}\left[2(S - S_0) + \ln\left(\frac{S-1}{S+1}\frac{S_0+1}{S_0-1}\right)\right]\right\}\right) \quad (7\text{-}115)$$

where

$$\phi = \sin^{-1}\frac{2\gamma u_0^+ - \beta}{Q} \qquad Q = (\beta^2 + 4\gamma)^{1/2} \qquad S = (1 + \alpha y^+)^{1/2}$$

This formula does not account for the viscous sublayer, nor does the van Driest theory from Sec. 7-7.2. The initial conditions are chosen so that the formula reduces to the low-speed logarithmic law $[u^+ = (1/\kappa)\ln(y^+) + B]$ when $\alpha = \beta = \gamma = 0$. We choose the point where the log-law goes to zero:

$$u_0^+ = u^+(y^+, 0, 0, 0) = 0 \quad (7\text{-}116)$$

at

$$y_0^+ = e^{-\kappa B} = 0.1287 \quad \text{for } \kappa = 0.41, B = 5.0$$

Some special cases of Eq. (7-115) for zero pressure gradient, zero wall heat flux, etc., are

$$u^+(y^+, 0, \beta, \gamma) = \frac{1}{2\gamma}\left[\beta + Q\sin\left(\phi + \frac{\sqrt{\gamma}}{\kappa}\ln\frac{y^+}{y_0^+}\right)\right] \quad (7\text{-}117a)$$

$$u^+(y^+, 0, \beta, 0) = \frac{1}{\kappa}\ln\frac{y^+}{y_0^+} + \frac{\beta}{4\kappa^2}\ln^2\frac{y^+}{y_0^+} \quad (7\text{-}117b)$$

$$u^+(y^+, 0, 0, \gamma) = \frac{1}{\sqrt{\gamma}}\sin\left(\frac{\sqrt{\gamma}}{\kappa}\ln\frac{y^+}{y_0^+}\right) \quad (7\text{-}117c)$$

$$u^+(y^+, 0, 0, 0) = \frac{1}{\kappa}\ln\frac{y^+}{y_0^+} = \frac{1}{\kappa}\ln y^+ + B \quad (7\text{-}117d)$$

These will be very useful, for example, in studying flat-plate flow in the next section.

Some of the typical effects contained within Eq. (7-115) are shown in Fig. 7-21. The left chart shows the effect of both compressibility γ and an adverse pressure gradient (positive α) on the law of the wall, compared with the low-speed logarithmic law, which is the straight line labeled (0, 0, 0). Note that, with or without a pressure gradient, compressibility drives the profiles downward, in agreement with experiment. Also note that nothing external is showing, such as Ma_e or δ or T_e. Any point along any one curve could be considered the edge of the boundary layer in a given calculation; the

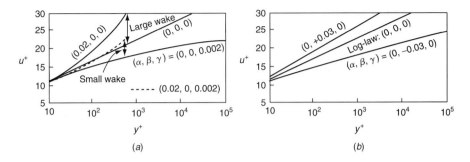

FIGURE 7-21
The law of the wall for compressible flow, from the first-order mixing-length theory, Eq. (7-115): (a) effect of compressibility and adverse pressure gradient; (b) effect of wall heat flux in low-speed flow. Note: β is positive for a cold wall.

resulting kink in the profile would be almost invisible on a linear (rather than logarithmic) plot. For example, as we move out along the lowest curve (0, 0, 0.002), the effective freestream Mach number is continuously increasing, from about 3.8 (for air) at $y^+ = 1000$ to about 19.8 at $y^+ = 10^5$. In other words, this single curve is equivalent to an infinite spectrum of adiabatic compressible flat-plate turbulent flows. To the author, this concept is the beauty of the law of the wall as an analytical tool for turbulent flows.

A second concept in Fig. 7-21a, shown by the arrows, is that compressibility decreases the outer "wake" effect. The dashed-line high-speed wake ($\gamma = 0.002$) is about one-half of the low-speed ($\gamma = 0$) wake for the same value of α. In fact, as γ increases to 0.005 (hypersonic flow), the same wake becomes negligibly small. Thus a wall-related theory of pressure gradient effects in supersonic flow may be more accurate than the (already successful) use of such a theory in incompressible flow.

Figure 7-21b shows the effect of cold and hot walls on the law of the wall in low-speed flow with a zero pressure gradient, as computed from Eq. (7-115).

7.8 COMPRESSIBLE TURBULENT FLOW PAST A FLAT PLATE

A classic problem is compressible turbulent flow past a flat plate at arbitrary Mach number and wall temperature. Practically everyone in the field has had a try at it. Figure 7-22 shows a classic display of the 21 adiabatic wall theories in existence in 1953; 11 years later, Spalding and Chi (1964) reviewed and discussed 32 theories. Now, 40 years later, there are over 50 theories, plus numerous finite-difference computations. The problem is irresistible, like entry flow in a tube.

The ordinate in Fig. 7-22 is the ratio of the skin-friction coefficient at a given Mach number and Reynolds number to the skin friction at the same Reynolds number and $Ma_e = 0$. Clearly the coefficient decreases with Mach number (although

554 VISCOUS FLUID FLOW

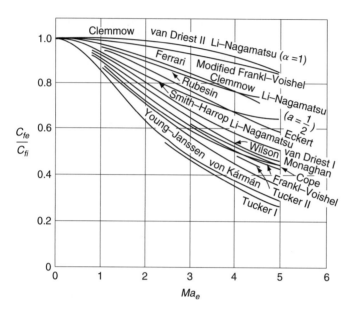

FIGURE 7-22
An array of adiabatic flat-plate theories compiled by Chapman and Kester (1953).

the wall shear itself increases because τ_w is nondimensionalized by U_e^2). The 21 theories are supposed to predict the same thing, namely, adiabatic skin friction. Typical data—that vary quite a bit with Reynolds number—fall along the curve labeled "Frankl–Voishel," which was not a formula but rather some tabulated values. More recent theories, with explicit formulas, will be presented here. Figure 7-22 does not tell the whole story, because as the Reynolds number varies from 10^5 to 10^8, the skin friction wall vary up to 25 percent.

7-8.1 The Flat-Plate Theory of van Driest (1956)

The theory labeled "van Driest II" in Fig. 7-22 looks poor, but in fact it was plotted for an incorrect viscosity–temperature relation: μ proportional to T rather than $T^{0.67}$ for air. Later, van Driest (1956) modified the theory and derived a formula which is still the most popular and, in this writer's opinion, the best of its type ever developed.

These early theories all used the Kármán integral relation Eq. (7-60) that reduces, for zero pressure gradient, to

$$C_f = 2\frac{d\theta}{dx} \qquad \theta = \int_0^\infty \frac{\overline{\rho}}{\rho_e} \frac{\overline{u}}{U_e}\left(1 - \frac{\overline{u}}{U_e}\right) dy \qquad (7\text{-}118)$$

To compute $\theta(x)$, van Driest (1956) used the density and velocity profiles from Eqs. (7-106) and (7-107), respectively. The integration, involving a complicated quotient of polynomial, trigonometric, and logarithmic functions, was approximated in series form and adjusted into the following formula for local skin friction:

$$\frac{\sin^{-1} A + \sin^{-1} B}{\sqrt{C_{fe}(T_{aw}/T_e - 1)}} \approx 4.15 \log\left(Re_{xe} C_{fe} \frac{\mu_e}{\mu_w}\right) + 1.7 \quad (7\text{-}119)$$

where
$$A = \frac{2a^2 - b}{(b^2 + 4a^2)^{1/2}} \quad \text{and} \quad B = \frac{b}{(b^2 + 4a^2)^{1/2}}$$

The parameters a and b were defined earlier in Eq. (7-107). For low-speed flow ($Ma_e \approx 0$), Eq. (7-119) reduces to the Kármán–Schoenherr incompressible relation Schoenherr (1932)

$$\frac{1}{\sqrt{C_f}} \approx 4.15 \log(Re_x C_f) + 1.7 \quad (7\text{-}120)$$

which is still the formula of choice for naval architects estimating the skin friction of ship hulls.

Equation (7-119) can be interpreted as an incompressible formula which has been "stretched" by compressibility and viscosity effects. It can be rewritten in a canonical manner as follows:

$$C_{fe} = \frac{1}{F_c} C_{f_{\text{inc}}}(Re_{xe} F_{Re_x}) = \frac{1}{F_c} c_{f_{\text{inc}}}(Re_{\theta e} F_{Re_x} F_c) \quad (7\text{-}121)$$

where
$$F_c = \frac{T_{aw}/T_e - 1}{(\sin^{-1} A + \sin^{-1} B)^2} \quad \text{and} \quad F_{Re_x} F_c = \frac{\mu_e}{\mu_w}$$

This generic form, common to many flat-plate theories, was suggested by Spalding and Chi (1964) as a *compressibility transformation* for flat-plate friction. One computes an equivalent incompressible Reynolds number $Re_{xe} F_{Re_x}$ and uses one's favorite incompressible skin-friction formula to compute $C_{f_{\text{inc}}}$, dividing the result by F_c to obtain the desired (compressible) C_f. The local Stanton number follows from the Reynolds analogy:

$$C_{he} = \frac{q_w}{\rho_e U_e c_p(T_{aw} - T_w)} \approx \frac{C_{fe}}{2Pr^{2/3}} = 0.62 C_{fe} \quad \text{for air} \quad (7\text{-}122)$$

Figure 7-23 shows computed Stanton numbers on an adiabatic flat plate for both laminar and turbulent flow, using Eqs. (7-41b), (7-119), and (7-122).

Meanwhile, van Driest (1956) also showed that the same stretched variables work for the total drag of a plate:

$$\frac{\sin^{-1} A + \sin^{-1} B}{\sqrt{C_{De}(T_{aw}/T_e - 1)}} \approx 4.13 \log\left(Re_L C_{De} \frac{\mu_e}{\mu_w}\right) \quad (7\text{-}123)$$

This reduces at low speed to the Kármán–Schoenherr incompressible drag formula, $C_D^{-1/2} \approx 4.13 \log(Re_x C_D)$.

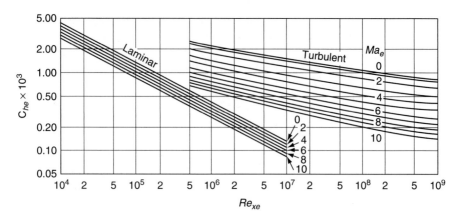

FIGURE 7-23
Theoretical local-wall, heat-flux coefficients on an adiabatic flat plate at high Mach numbers; laminar: Eq. (7-41b); turbulent: Eq. (7-119).

In his review article, Bradshaw (1977) asserts that the van Driest II solution Eq. (7-119) is the most accurate, compared to available data, of all popular formulas for compressible, flat-plate, skin friction.

7-8.2 Compressible Flat-Plate Flow Using Inner Variables

An alternate, and very accurate, method by White and Christoph (1972) uses inner, or law-of-the-wall, variables to extend the analysis of Sec. 6-6.2 to compressible flow. It is assumed that the velocity profile is correlated by the wall law Eq. (7-115) with $\alpha = 0$:

$$u^+ = \frac{\bar{u}(x, y)}{v^*(x)} = f(y^+, \beta, \gamma) \tag{7-124}$$

Now the continuity Eq. (7-95a) is satisfied by the compressible stream function $\psi(x, y)$:

$$\frac{\partial \psi}{\partial y} = +\bar{\rho}\bar{u} \qquad \frac{\partial \psi}{\partial x} = -\bar{\rho}\bar{v} \tag{7-125}$$

When we introduce Crocco's approximation for $\bar{\rho}$ and the law of the wall for $\bar{u}$, we find that the stream function is also a law-of-the-wall variable

$$\psi = \mu_w \int_0^{y^+} \frac{u^+ dy^+}{1 + \beta u^+ - \gamma u^{+2}} = \psi(y^+, \beta, \gamma) \tag{7-126}$$

Introducing these variables into the momentum Eq. (7-95b) for zero pressure gradient gives

$$\rho v^* u^{+} \frac{\partial}{\partial x}(v^* u^{+}) - \frac{\partial \psi}{\partial x} \frac{v^*}{\nu_w} \frac{\partial}{\partial y^{+}}(v^* u^{+}) = \frac{v^*}{\nu_w} \frac{\partial \tau}{\partial y^{+}} \qquad (7\text{-}127)$$

The differentiation with respect to y^{+} is left untouched, but differentiation with respect to x is carried out using the chain rule

$$\frac{\partial}{\partial x} = \frac{\partial y^{+}}{\partial x} \frac{\partial}{\partial y^{+}} + \frac{\partial \beta}{\partial x} \frac{\partial}{\partial \beta} + \frac{\partial \gamma}{\partial x} \frac{\partial}{\partial \gamma}$$

After this, in exact analogy with Sec. 6-6.2, integrate Eq. (7-127) with respect to y^{+} from the wall ($y = 0$, $\tau = \tau_w$) to the outer edge ($y = \delta$, $\tau = 0$). Also introduce the same dimensionless variables as before

$$x^* = \frac{x}{L} \qquad \lambda = \sqrt{\frac{2}{C_f}} \qquad V = \frac{U_e}{U_0} (=1.0 \text{ here}) \qquad (7\text{-}128)$$

Then the differential Eq. (7-127) becomes

$$G \frac{d\lambda}{dx^*} = Re_L = \frac{U_0 L}{\nu_e} \frac{\mu_e}{\mu_w} \left(\frac{T_e}{T_w}\right)^{1/2} \qquad (7\text{-}129)$$

where

$$G(\lambda, \beta, \gamma) = \int_0^{\delta^{+}} \frac{\bar{\rho}}{\rho_w} \left(u^{+2} - \beta u^{+} \frac{\partial u^{+}}{\partial \beta} + 2\gamma u^{+} \frac{\partial u^{+}}{\partial \gamma} \right.$$
$$\left. + \frac{\beta}{\mu_w} \frac{\partial \psi}{\partial \beta} \frac{\partial u^{+}}{\partial y^{+}} - \frac{\gamma}{\mu_w} \frac{\partial \psi}{\partial \gamma} \frac{\partial u^{+}}{\partial y^{+}} \right) dy^{+}$$

This is identical in form to the incompressible relation, Eq. (6-76), except that (1) the nominal Reynolds number Re_L contains additional viscosity and temperature terms (here assumed constant) and (2) the function G contains more terms than its incompressible form. Evaluation of G is merely a matter of dogwork quadrature and curve fitting, and this has been accomplished by White and Christoph (1972):

$$G(\lambda, \beta, \gamma) \approx 8.0 \exp \frac{0.48\lambda}{S} \qquad (7\text{-}130)$$

where
$$S = \frac{(T_{aw}/T_e - 1)^{1/2}}{\sin^{-1} A + \sin^{-1} B}$$

Note that S is the square root of van Driest's parameter F_c from Eq. (7-121). It expresses the combined effect of compressibility and heat transfer without bringing in β or γ explicitly. β and γ are related to $U_e^{+} = \gamma(T_e/T_w)^{1/2}$ by rewriting their definitions from Eq. (7-111)

$$\gamma U_e^{+2} = \frac{T_{aw} - T_e}{T_w}$$
$$\beta U_e^{+} = \frac{T_{aw}}{T_w} - 1 \qquad (7\text{-}131)$$

As in the incompressible analysis of Chap. 6, we see that Eq. (7-129) can be integrated immediately to yield a flat-plate, skin-friction relation

$$Re_{xe} = \frac{U_e x}{\nu_e} = \frac{\mu_w}{\mu_e}\left(\frac{T_w}{T_e}\right)^{1/2} \int_0^\lambda G(\lambda, \beta, \gamma)\, d\lambda$$

Substituting for G from Eq. (7-130), integrating, and rearranging to solve for the skin-friction coefficient $C_{fe} = 2/\lambda^2$, we obtain the following simple formula for turbulent skin friction on a flat plate:

$$C_{fe} \approx \frac{0.455}{S^2 \ln^2\left(\dfrac{0.06}{S} Re_{xe} \dfrac{\mu_e}{\mu_w}\sqrt{\dfrac{T_e}{T_w}}\right)} \quad (7\text{-}132)$$

This formula is very accurate over the entire practical range of turbulent Reynolds numbers, Mach numbers, and wall temperatures. By comparing Eq. (7-132) with the incompressible formula Eq. (6-78), we see that the same compressibility-transformation idea of van Driest's method holds here also:

$$C_{f,\text{comp}} = \frac{1}{F_c} C_{f,\text{incomp}}(Re_x F_{Re_x}) \quad [\text{White and Christoph (1972)}] \quad (7\text{-}133)$$

where
$$F_c = S^2 \quad \text{and} \quad F_{Re_x} = \frac{(\mu_e/\mu_w)(T_e/T_w)^{1/2}}{S}$$

As mentioned, others have attacked the compressible, turbulent, flat-plate problem. Huang et al. (1993) use the van Driest transformed velocity expressed as a wall–wake formula, plus a clever alternate form of the Crocco-law, temperature-velocity relation:

$$T = T_w - \frac{Pr_t q_w \bar{u}}{c_p \tau_w} - \frac{Pr_t \bar{u}^2}{2 c_p} \quad (7\text{-}134)$$

They allow for the variation of Coles' wake parameter Π with Re_θ and outline an iterative numerical integration which yields values of C_f and the Stanton number. No algebraic friction formula results, but the agreement with experiment is excellent for air and helium up to $Ma = 11$. Barnwell and Wahls (1991) combine a wall–wake law with the equilibrium power-law defect stream function of Clauser (1956) to derive skin friction for adiabatic flow with pressure gradients. Their flat-plate formula is in agreement with experiment but, to this writer, is extremely complicated algebraically.

We close this section with Fig. 7-24, required for any textbook, showing the approximate effect of the Mach number and wall temperature on flat-plate skin friction. The chart is strictly valid only for $Re_x = 10^7$ and will vary ± 20 percent over the range $10^6 < Re_x < 10^9$.

7-8.3 The Flat-Plate Turbulent Recovery Factor

In laminar flat-plate flow, the recovery factor is an output of (nearly exact) similarity analysis, Eq. (7-35). In turbulent flow, the theory is not so well developed.

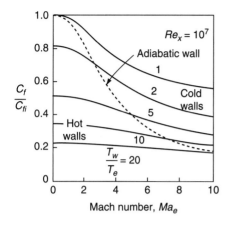

FIGURE 7-24
Ratio of turbulent, flat-plate, skin friction $C_f(M_e, T_w/T_e, Re_{xe})$ to the incompressible value $C_f(0, 1.0, Re_{xe})$ for $Re_x = 10^7$, as computed by the method of White and Christoph (1972) for air, Eq. (7-132).

Dorrance (1962) develops a simple but effective result by breaking the law of the wall into three parts: a linear layer for $y^+ < 5$, a buffer layer for $5 < y^+ < 30$, and a log layer for $y^+ > 30$. By assuming a constant turbulent Prandtl number, Dorrance integrates the velocity and temperature profiles to obtain the following estimate for the turbulent recovery factor:

$$r_{\text{turb}} \approx Pr_t + (Pr - Pr_t)\left(\frac{11.5}{U_e^+}\right)^2 \quad (7\text{-}135)$$

There is no overt effect of the Mach number predicted. From Fig. 7-21a for compressible flow (nonzero γ), a typical value of U_e^+ is about 20. If $Pr_t \approx 1$, then Eq. (7-135) becomes

$$r \approx 1 + (Pr - 1)\left(\frac{11.5}{20}\right)^2 \approx 1 + 0.33(Pr - 1)$$

If Pr is near unity (gases), these are the first two terms in the binomial expansion of the cube root of Pr. Therefore we adopt the customary approximation for the turbulent flat-plate recovery factor:

$$r_{\text{turb}} \approx Pr^{1/3} \quad \text{[Dorrance (1962)]} \quad (7\text{-}136)$$

This formula would be less accurate for flow with strong favorable or adverse pressure gradients, as we have seen for laminar flow in Sec. 7-4.4.4. A formula to account for pressure gradient effects is not known to this writer.

Figure 7-25 shows recovery factors measured in supersonic flow at zero incidence past cones with adiabatic surfaces, after Mack (1954). Transition to turbulence occurs at Reynolds numbers between 2×10^6 and 4×10^6, depending partly upon Mach number and partly upon the particular disturbances in the flow. We see that the simple estimates, $r_{\text{lam}} \approx Pr^{1/2} \approx 0.84$ and $r_{\text{turb}} \approx Pr^{1/3} \approx 0.89$, are in good agreement with these measurements.

7-8.4 The Turbulent Cone Rule

As mentioned before, supersonic flow past a cone at zero incidence generally results in an attached shock wave and a zero surface pressure gradient. For laminar

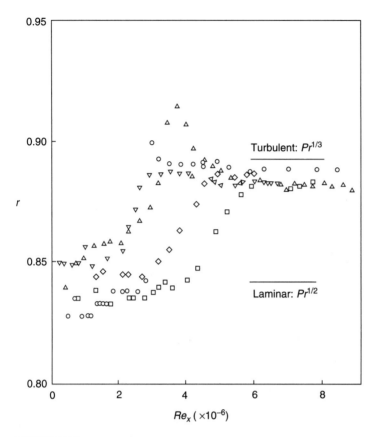

FIGURE 7-25
Measured recovery factors on cones at Mach numbers from 1.2 to 6.0, for air, $Pr = 0.71$. [*Data from Mack (1954).*]

flow, we derived an exact cone rule, Eqs. (7-83). For turbulent flow, van Driest (1952b) derived an approximate cone rule, using the axisymmetric momentum-integral relation with a zero pressure gradient:

$$\frac{d\theta}{dx} + \frac{\theta}{r_0}\frac{dr_0}{dx} = \frac{1}{2}C_f \qquad r_0(x) = \text{surface radius} \qquad (7\text{-}137)$$

For a cone of half-angle ϕ, $r_0 = x \sin \phi$, hence Eq. (7-137) becomes

Cone flow: $$C_f = \frac{2}{x}\frac{d}{dx}(x\theta)$$

If we assume that friction and momentum thickness are related by a power law, $C_f = (\text{const})\theta^{-m}$, where $m = 1$ for laminar flow and $\frac{1}{8} < m < \frac{1}{4}$ for turbulent flow, we may integrate in closed form and compare the cone to the plate. The result is a cone rule in terms of the Reynolds number:

$$\text{If } C_{f,\text{cone}} = C_{f,\text{plate}} \quad \text{then} \quad Re_{xe,\text{cone}} = (2 + m) Re_{xe,\text{plate}} \quad (7\text{-}138)$$

The case $m = 1$ corresponds to the previous laminar result, Eq. (7-38b). The result can also be expressed as a skin-friction ratio. If $Re_{xe,\text{cone}} = Re_{xe,\text{plate}}$, then, for turbulent flow,

$$\frac{C_{f,\text{cone}}}{C_{f,\text{plate}}} \approx (2 + m)^{m/(m+1)} = \begin{cases} 1.087 & m = \frac{1}{8} \\ 1.176 & m = \frac{1}{4} \end{cases} \quad (7\text{-}139)$$

Thus, as mentioned earlier, the effect for turbulent flow is only a 10 to 15 percent increase in friction (and heat transfer), whereas in laminar flow ($m = 1$), the increase is $\sqrt{3}$ or 73 percent.

7-9 COMPRESSIBLE-TURBULENT-BOUNDARY-LAYER CALCULATION WITH A PRESSURE GRADIENT

There are many different approaches to calculation of compressible, turbulent, boundary layers with a pressure gradient. Here we present only a single Kármán-based method, after Walz (1969), and an inner variable approach, after White and Christoph (1972). Other calculation methods are reviewed by Delery and Marvin (1986) and in the texts by Schetz (1992), Cebeci and Cousteix (1998), Schlichting and Gersten (2000), and Cebeci (2003). The Internet Applets of Devenport and Schetz (2002) include compressible, turbulent-flow, integral methods. Compressible two-equation and second-moment models are reviewed by Wilcox (1998). And there is progress on direct numerical simulation (DNS) of shock-wave, boundary-layer interactions, reported by Yoon and Chung (1996).

7-9.1 The Kármán-Based Method of Walz (1969)

As an illustration of a momentum-thickness approach, we select the method developed by Walz (1969) and co-workers. As variables, they choose a momentum-thickness parameter and a shape factor:

$$Z = \theta \left(\frac{\rho_e U_e \theta}{\mu_w} \right)^n$$

$$W = \frac{\delta_3}{\theta} \quad (7\text{-}140)$$

where $n = 0.268$ for turbulent flow and δ_3 is the (compressible) kinetic-energy thickness:

$$\delta_3 = \int_0^\infty \frac{\rho}{\rho_e} \frac{\bar{u}}{U_e} \left(1 - \frac{\bar{u}^2}{U_e^2} \right) dy \quad (7\text{-}141)$$

Substitution in the momentum- and mechanical-energy integral relations gives two coupled first-order differential equations:

Momentum:
$$\frac{dZ}{dx} + \frac{F_1}{U_e}\frac{dU_e}{dx}Z = F_2 \quad (7\text{-}142a)$$

Mechanical-energy:
$$\frac{dW}{dx} + \frac{F_3}{U_e}\frac{dU_e}{dx}W = \frac{F_4}{Z} \quad (7\text{-}142b)$$

The four functions F_i are algebraic functions of the Mach number, Re_θ, shape factor $H = \delta^*/\theta$, and a heat-transfer parameter $\Theta(x)$, assumed known:

$$\Theta(x) = \frac{T_{aw}(x) - T_w(x)}{T_{aw}(x) - T_e(x)} \quad (7\text{-}143)$$

To simplify the formulas, two intermediate parameters, a and b, are defined and used to correlate skin friction, as follows:

$$a \approx 0.0394(W - 1.515)^{0.7} \quad (7\text{-}144)$$

$$b \approx 1 + 0.88\frac{\gamma - 1}{2}Ma_e^2(W - \Theta)(2 - W) \quad \text{turbulent flow}$$

$$C_f(x) \approx 2\frac{a}{b}\left(\frac{\rho_e U_e \theta}{\mu_w}\right)^{-0.268}$$

Then the functions F_1 to F_4 given by Walz (1969) are

$$F_1 = 2.268 + 1.268H - Ma_e^2 \quad (7\text{-}145)$$

$$F_2 = 1.268\frac{a}{b}$$

$$F_3 = 1 - H + 0.88(\gamma - 1)Ma_e^2\left(1 - \frac{\Theta}{W}\right) \quad \text{turbulent flow}$$

$$F_4 = \frac{1}{b}\left[2\beta\left(\frac{\rho_e U_e \theta}{\mu_w}\right)^{0.168} - aW\right]$$

where
$$H \approx 1 + 1.48(2 - W) + 104(2 - W)^{6.7}$$

and
$$\beta = \left[\frac{1 + 0.587(\gamma - 1)Ma_e^2(1 - 0.75\Theta)}{1 + 0.44Ma_e^2(1 - \Theta)(\gamma - 1)}\right]^{n \approx 0.7}$$

The exponent n in the definition of the parameter β is the viscosity–temperature exponent, which is about 0.67 for air.

Separation, if it occurs, is defined by $a = 0$, or $W = 1.515$. Initial values are needed for $Z(0)$ and $W(0)$. Equations (7-142) to (7-145) may easily be integrated by any standard numerical technique—Subroutine RUNGE in App. C will do quite well for this task.

7-9.2 An Inner Variable Approach

The flat-plate theory which led to Eqs. (7-129) and (7-132) was extended by White and Christoph (1972) to pressure gradients by including the parameter $\alpha(x)$ and letting $u^+ = fcn\,(y^+, \alpha, \beta, \gamma)$ from Eq. (7-115). The momentum Eq. (7-127) is extended to include pressure gradients:

$$\rho v^* u^+ \frac{\partial}{\partial x}(v^* u^+) - \frac{\partial \psi}{\partial x} \frac{v^*}{v_w} \frac{\partial}{\partial y^+}(v^* u^+) = \rho_e U_e \frac{dU_e}{dx} + \frac{v^*}{v_w}\frac{\partial \tau}{\partial y^+} \quad (7\text{-}146)$$

Also, the chain rule must be extended to include the effect of $\alpha(x)$:

$$\frac{\partial}{\partial x} = \frac{\partial y^+}{\partial x}\frac{\partial}{\partial y^+} + \frac{\partial \alpha}{\partial x}\frac{\partial}{\partial \alpha} + \frac{\partial \beta}{\partial x}\frac{\partial}{\partial \beta} + \frac{\partial \gamma}{\partial x}\frac{\partial}{\partial \gamma} \quad (7\text{-}147)$$

Now, exactly as in the previous procedure (Sec. 7-8.2), we carry out the differentiation with respect to x and then integrate the whole equation across the boundary layer from $y^+ = 0$ to $y^+ = \delta^+$, writing the final result in terms of the same dimensionless variables defined by Eqs. (7-128). We obtain

$$(G - 3\alpha H)\frac{d\lambda}{dx^*} + \frac{V'}{V}\lambda(\lambda^2 \delta^+ - G) - \lambda^4 \frac{H(1/V)''}{Re_L} = Re_L V \quad (7\text{-}148)$$

where
$$H = H(\lambda, \alpha, \beta, \gamma) = \int_0^{\delta^+} \frac{\rho}{\rho_w}\left(u^+\frac{\partial u^+}{\partial \alpha} - \frac{1}{\mu_w}\frac{\partial \psi}{\partial \alpha}\frac{\partial u^+}{\partial y^+}\right) dy^+$$

The functions Re_L and G are defined exactly as before, Eq. (7-129), except that now u^+ and ψ must also vary with α and hence the final result $G = G(\lambda, \alpha, \beta, \gamma)$. Since the functions F and G are difficult to evaluate under general conditions and no accurate curve-fit correlation could be devised, White and Christoph (1972) chose to rearrange the basic differential Eq. (7-148) into a two-part approximate form, using the van Driest parameter S from Eq. (7-130) and a "stretched" Reynolds number Re^* related to the pressure gradient:

$$Re^* = \frac{Re_L}{(1/V)'} \quad (7\text{-}149)$$

The separation point is defined as a large but finite value of λ, where $d\lambda/dx$ becomes unbounded:

Separation: $\qquad \lambda = \lambda_{max} = 8.7S\,\log(Re^*) \qquad (7\text{-}150)$

Depending upon λ/λ_{max}, the basic differential equation has two different forms:

$\dfrac{\lambda}{\lambda_{max}} > 0.36$:

$$\dfrac{d\lambda}{dx^*} \approx \dfrac{\dfrac{(1/V)'}{1/V}(1 + 9S^{-2}g^*Re^{*0.07}) + \dfrac{(1/V)''}{(1/V)'}(3S^2 g^* Re^{*0.07})}{0.16f^* S^3} \qquad (7\text{-}151a)$$

$\dfrac{\lambda}{\lambda_{max}} < 0.36$ or $Re^* < 0$:

$$\dfrac{d\lambda}{dx^*} \approx \dfrac{1}{8} Re_L V \exp\left(-0.48\dfrac{\lambda}{S}\right) + 5.5 \dfrac{(1/V)'}{1/V} \qquad (7\text{-}151b)$$

The functions f^* and g^* depend only upon λ/λ_{max} and may be curve fit as follows:

$\dfrac{\lambda}{\lambda_{max}} \geqq 0.36$:

$$f^* = (2.434Z + 1.443Z^2)\exp(-44.0\,Z^6) \qquad (7\text{-}152)$$

$$g^* \approx 1 - 2.3Z + 1.76Z^3$$

where $\qquad Z = 1 - \dfrac{\lambda}{\lambda_{max}}$

Equations (7-151) may be solved by any standard numerical method such as Subroutine RUNGE in App. C. It is assumed that $U_e(x)$, $Ma_e(x)$, T_e, and T_w are known. Once $\lambda(x)$ is computed, the local skin friction follows from $C_f(x) = 2/\lambda^2$. To avoid ragged behavior of the solution, the freestream velocity $V(x^*)$ should be fit to a continuous analytic function, so that $(1/V)'$ and $(1/V)''$ are smooth functions. The freestream Mach number $Ma_e(x)$ and the ratio T_w/T_e can be approximated more crudely.

7-9.3 The Experiment of Zwarts

In his thesis, Zwarts (1970) produced a supersonic flow which was sharply decelerated from Mach 4 to Mach 3 in about 8 in. thereafter remaining nearly constant at Mach 3. The measured freestream velocity and Mach number distributions are shown in Fig. 7-26. With $U_0 = 2204$ ft/s, White and Christoph (1972) proposed two different curve fits to the velocity distribution:

$$V \approx 1.0013 - 0.00051x^* - 0.003885x^{*2}$$
$$+ 0.0004466x^{*3} - 0.0000139x^{*4} \qquad (7\text{-}153a)$$

$$V \approx 0.9165 + 0.0835\exp(-0.03x^* - 0.05x^{*2}) \qquad (7\text{-}153b)$$

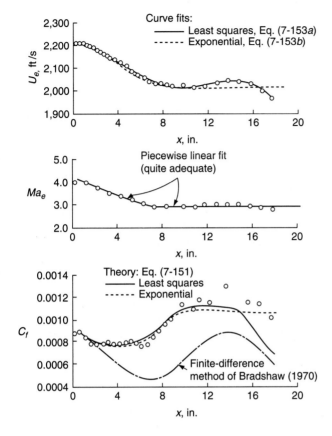

FIGURE 7-26
Comparison of theory with the supersonic relaxing-flow experiment of Zwarts (1970). [*After White and Christoph (1972).*]

where x^* is in inches. The polynomial is seen to be an excellent fit, whereas the exponential form does not reproduce the final curvature at the end of the run. The Mach number $Ma_e(x^*)$ was fitted by two piecewise-straight lines, as shown in Fig. 7-26. The wall temperature was approximately adiabatic, hence $T_w/T_e \approx 1 + r(\gamma - 1)Ma_e^2/2$ was used. These formulas were applied to Eq. (7-151), and the computed skin friction is also shown in Fig. 7-26. Both curve fits are seen to be in good agreement with the data.

Figure 7-26 shows that the skin-friction predictions of the method of White and Christoph (1972) are somewhat sensitive to the quality of the curve fit of $U_e(x)$, as was the incompressible inner variable method of Das (1988) from Sec. 6-8.2. The method avoids the use of momentum thickness or shape factor, so no predictions of these parameters were made. Meanwhile, the method of Walz from Sec. 7-9.1 could not be programmed for this flow, since Zwarts (1970) did not report data on kinetic-energy thickness $\delta_3(x)$. Other theories have been applied to Zwarts' experiment, notably the turbulent-energy differential method of Bradshaw and Ferriss (1971). Later, at the Stanford Conference [Kline et al. (1982, pp. 1263–1264)], four competitors—all with multi-equation differential models—attempted to predict this flow, called case 8411. Of the four predictions, two were excellent for skin friction and two were poor, all four were excellent for predicting shape factor, and none could reproduce the measured momentum-thickness distribution.

SUMMARY

This chapter gives an introduction to the analysis of steady compressible boundary layers for both laminar and turbulent flows. Such flows can be handled either by approximate (integral or correlative) techniques or by "exact" (finite-difference) computations, and some of the best known methods are discussed. While the main emphasis is upon flat-plate and stagnation-point flows, the effects of pressure gradient, heat transfer, and shock-wave interaction are also briefly discussed.

It is not possible in one short chapter to treat many of the viscous-flow phenomena that have been studied under high-speed flow conditions. Some topics omitted are mass transfer at the wall, species diffusion due to dissociation or foreign-gas injection, chemical reactions, three-dimensional flows, shock-wave impingement and leading-edge bluntness effects, wall roughness, and supersonic jets and wakes. The reader is referred to the extensive aerospace literature on such flows and also to the monographs by Dorrance (1962), Schreier (1982), Schetz (1984), Smits and Dussauge (1996), Anderson (2000), and Settles (2004). It is hoped that the present chapter whets the reader's appetite for further study of this interesting field of compressible viscous flow.

PROBLEMS

7.1. Consider compressible-laminar-boundary-layer flow past a flat plate with a cold wall. If $Pr \approx 1$, find the value of T_w, in terms of (T_e, c_p, U_e), for which the maximum temperature in the boundary layer occurs at the point where $u = 0.5U_e$. Sketch the profile $T(y)$.

7.2. Consider the flow of an ideal gas, $\gamma = 1.4$, past a flat plate ($\beta = 0$) at $Ma_e = 3.0$. Solve the similarity relations, Eqs. (7-32), numerically. Plot the resulting velocity and temperature profiles and compare them with (a) the finite-difference results in Fig. 7-13 and (b) the Crocco–Busemann approximation Eq. (7-14). Solve for either an adiabatic wall or a cold wall, $T_w = T_e$.

7.3. Air at 0°C and 13 kPa is in laminar flow past a flat plate at 800 m/s. Estimate (a) the adiabatic wall temperature in °C and (b) the heat-transfer rate in W/m^2 at $x = 15$ cm if the wall temperature is 20°C.

7-4. Evaluate the flat-plate recovery factor numerically from Eq. (7-35) for a value of Pr not listed in Table 7-1. Compare your results with the approximations Eqs. (7-37).

7-5. Air at 20°C and 20 kPa flows at 1200 m/s past an insulated flat plate. At $x = 10$ cm, assuming laminar flow with no shock-wave interactions, estimate (a) the wall shear stress (in Pa) and (b) the wall heat-transfer rate (in W/m^2).

7-6. Helium at 20°C and 14 kPa flows axially at 1600 m/s toward a 25 cm diameter cylinder with a hemispheric nose. Using normal-shock, perfect-gas theory, estimate the velocity gradient K in s^{-1} at the nose of the hemisphere.

7-7. Air at 20°C and 40 kPa flows at $Ma = 3.5$ toward a sphere of diameter 10 cm whose surface is at 400°C. Assuming an ideal gas, estimate the heat-transfer rate (in W/m²) at the stagnation point.

7-8. In hypersonic flow past an insulated surface, the density is small near the wall and causes the shape factor $H = \delta^*/\theta$ from Eqs. (7-60) to be quite large. For laminar flow, assuming a linear velocity profile $u/U_e \approx y/\delta$ (see Fig. 7-3a) and the Crocco–Busemann temperature correlation Eq. (7-13), show that

$$\frac{\delta^*}{\delta} \approx 1 - \frac{1}{2a}\ln(1+a) \qquad a = \frac{\gamma-1}{2}Ma_e^2$$

Compute the value of the shape factor if $\gamma = 1.4$ and $Ma_e = 8.0$.

7-9. Consider laminar flow at constant U_e and Ma_e past an adiabatic flat plate. Use the method of Gruschwitz from Sec. 7-4.3 to solve for the variation of momentum thickness and skin-friction coefficient along the wall. Compare with other flat-plate formulas. Hint: Closed form integration is possible.

7-10. Program the finite-difference method of Sec. 7-4.4, with either an explicit or implicit model, for laminar flat-plate flow with $Ma_e = 4.0$ and $\gamma = 1.40$. Assume an adiabatic wall. Compute enough downstream sections to establish the validity of the similarity variable $(y/x)Re_x^{1/2}$. Compare the computed velocities, temperatures, and skin-friction coefficients with the van Driest results in Figs. 7-2 to 7-4.

7-11. Modify Prob. 7-10 for a cold wall, $T_w = T_e/4$.

7-12. Air at 20°C and 8000 Pa is in laminar flow past a flat plate at 4200 m/s. Using Fig. 7-17, estimate the shock-induced pressure p_e, in pascals, at the position $x = 8$ cm along the plate.

7-13. As shown in Fig. P7-13, air at 20°C and 30 kPa flows at $Ma = 3.5$ toward a symmetric two-dimensional wedge of half-angle 15°. If the wedge surface is insulated, estimate its temperature (in °C) for an assumed laminar-flow condition.

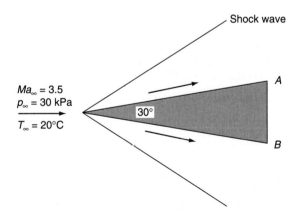

FIGURE P7-13

7-14. Modify Prob. 7-13 to find the total heat transfer per unit width on the upper surface of the wedge if the base AB has a length of 10 cm. Let $T_w = 300$ K.

568 VISCOUS FLUID FLOW

7-15. Modify Prob. 7-13 for an assumed turbulent-flow condition.

7-16. Modify Prob. 7-14 for an assumed turbulent-flow condition.

7-17. Figure P7-17 is a shadowgraph of airflow at $Ma = 2.85$ striking a 20° ramp, from Settles et al. (1979). The white line at left shows the thickness of the approaching turbulent boundary layer. (*a*) Determine if the shock-wave angle (the thick black line) is in agreement with oblique-shock theory. (*b*) Discuss how the details of the shadowgraph can be explained by boundary-layer concepts.

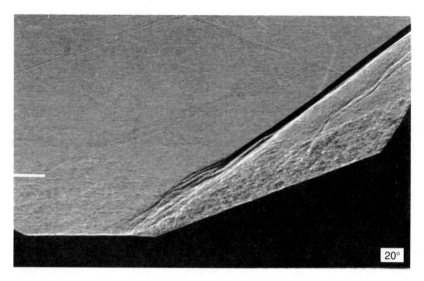

FIGURE P7-17

7-18. Solve the hypersonic-interaction Eq. (7-88) numerically for $\gamma = 1.4$ and $T_w = T_{aw}$ in the range $0 < \chi < 10$. Compare your results with Fig. 7-17 and with the simple linear approximation, Eq. (7-90).

7-19. Air at 20°C and 25 kPa flows at 2200 m/s past an insulated flat plate. At the point $x = 2$ cm, use hypersonic-interaction theory to estimate the actual surface pressure (in Pa).

7-20. Consider the turbulent-boundary-layer energy Eq. (7-95*d*) with negligible viscous shear, heat transfer, and pressure gradient. Near the wall, $\tau \approx \tau_w$, and convection is negligible. Integrate the energy equation to obtain $q = q_w + \bar{u}\tau_w$. Then, introducing the eddy viscosity and turbulent Prandtl number, integrate to find the temperature in the form of Eq. (7-134), as found by Huang et al. (1993).

7-21. At a certain position in a turbulent flat-plate boundary layer, the air freestream conditions are $U_e = 900$ m/s and $T_e = 20°C$. The wall temperature is 300°C and the wall heat transfer is 30 kW/m² into the wall. The boundary-layer thickness is 6 mm. Using the compressible law of the wall, estimate the local wall shear stress (in Pa). Compare your estimate with other flat-plate formulas. Let $p_e = 1$ atm.

7-22. Modify Prob. 7-5 to study the position $x = 1$ m, where the boundary-layer flow is turbulent.

7-23. In the experiment of Coles (1954), for smooth adiabatic-wall flat-plate airflow at three Mach numbers and three x stations, some of the reported data may be listed as follows:

Ma_e	$U_e x/\nu_e$	C_f
2.6	4.84E6	0.00181
2.6	8.32E6	0.00166
3.7	3.98E6	0.00162
3.7	7.25E6	0.00138
4.5	3.37E6	0.00155
4.5	3.52E6	0.00148
4.5	6.91E6	0.00123

Compare these data with predictions by the turbulent flat-plate methods of van Driest and White and Christoph in Sec. 7-8. Also show comparative predictions for $Ma_e = 0$.

7-24. The data from the experiment of Zwarts (1970), used in preparing Fig. 7-26, may be listed as follows:

x, in.	U_e, ft/s	Ma_e	C_f (exp)
0.75	2204	4.016	0.000850
2.25	2174	3.727	0.000767
3.75	2132	3.464	0.000778
5.25	2082	3.266	0.000810
6.75	2037	2.987	0.000774
8.25	2023	2.958	0.000950
9.75	2025	2.974	0.001145
11.75	2024	2.950	0.001175
13.75	2047	3.083	0.001295
15.75	2020	2.947	0.001151
17.75	1967	2.751	0.000998

The freestream kinematic viscosity was $1.06E-4$ ft^2/s. Make your own curve fit to the velocity and Mach number data, and, starting at the first position, compute a theoretical skin-friction distribution for comparison, using the method of Walz, Eqs. (7-144) and (7-145). (Numerical integration is required.)

7-25. Repeat Prob. 7-24 but use the inner variable boundary-layer method of White and Christoph, Eqs. (7-151).

7-26. The answer to Prob. 7-13 is $T_{aw} \approx 919$ K. (*a*) Again assuming laminar flow, find the wedge half-angle for which $T_{aw} \approx 950$ K. (*b*) Find the adiabatic-wall temperature of a flat plate (wedge angle = 0) with a stream Mach number of 3.5. (*c*) One of your answers (*a*, *b*) will be less than the other. Give a thermodynamic explanation for this result.

7-27. In a wind tunnel experiment, air at 20°C and 8000 Pa flows past an adiabatic flat plate at $Ma \approx 4.5$. Turbulent-boundary-layer velocities are measured as follows:

y, mm	5.4	8.6	14.0	21.6
$\bar{u}$, m/s	1100	1160	1240	1310

Using the van Driest transformation, estimate the wall shear stress in pascals.

APPENDIX A

TRANSPORT PROPERTIES OF VARIOUS NEWTONIAN FLUIDS

THE VISCOSITY AND THERMAL CONDUCTIVITY OF WATER

Kestin and Wakeham (1988) recommend the following formula for the viscosity of water at atmospheric pressure:

$$\log_{10}\left(\frac{\mu}{1.002}\right) \approx \frac{20 - T}{T + 96}[1.2378 - 1.303\text{E} - 3(20 - T)$$
$$+ 3.06\text{E} - 6(20 - T)^2 + 2.55\text{E} - 8(20 - T)^3]$$

with μ in (mPa·s) and T in (°C). Numerical values are tabulated below. They further state that $k(T)$ is not known very accurately for any liquid. Estimated values of $k(T)$ for water are given below.

Temperature T, °C	Viscosity μ, mPa·s	Thermal conductivity k, W/m·K	Temperature T, °C	Viscosity μ, mPa·s	Thermal conductivity k, W/m·K
0	1.792	0.564	55	0.504	0.648
5	1.519	0.574	60	0.467	0.653
10	1.307	0.584	65	0.434	0.657
15	1.138	0.593	70	0.405	0.662
20	1.002	0.602	75	0.379	0.665
25	0.890	0.610	80	0.355	0.669
30	0.797	0.617	85	0.334	0.673
35	0.719	0.624	90	0.315	0.676
40	0.653	0.631	95	0.298	0.678
45	0.596	0.637	100	0.282	0.681
50	0.547	0.643	Uncertainty:	±0.4%	±2.5%

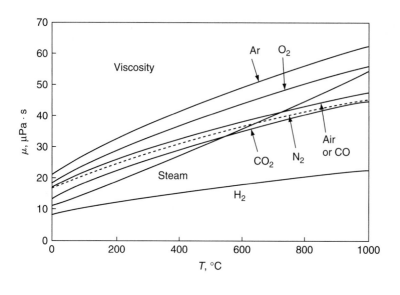
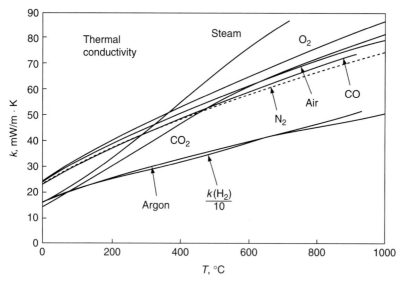

FIGURE A-1
Viscosity and thermal conductivity of common gases at low pressures.

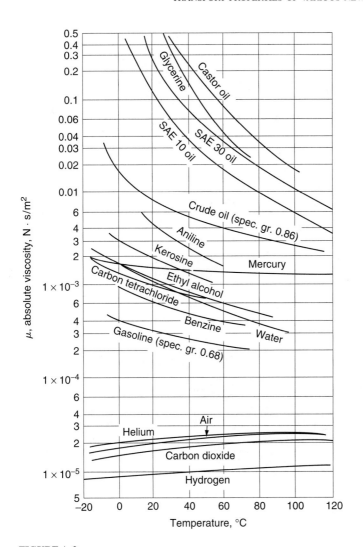

FIGURE A-2
Absolute viscosity of common fluids at atmospheric pressure.

574 VISCOUS FLUID FLOW

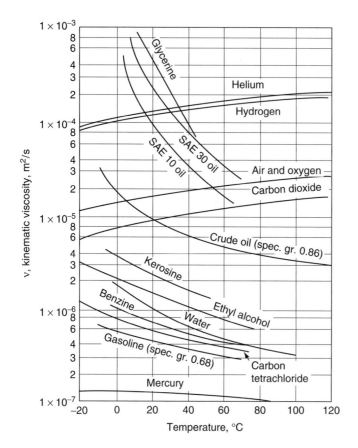

FIGURE A-3
Kinematic viscosity of common fluids at atmospheric pressure.

TABLE A-1
Properties of saturated water at atmospheric pressure

Temperature T, °C	Density ρ, kg/m³	Viscosity μ, mPa·s	Surface tension [‡] $\mathcal{T}$, N/m	Vapor pressure p_v, kPa	Bulk modulus K, MPa
0	1000	1.792	0.0757	0.61	2062
20	998	1.002	0.0727	2.34	2230
40	992	0.653	0.0696	7.38	2304
60	983	0.467	0.0662	19.92	2301
80	972	0.355	0.0627	47.35	2235
100	958	0.282	0.0589	101.3	2120
150	915	0.182	0.0488	461	1692
200	863	0.136	0.0377	1580	1190
250	797	0.107	0.0261	3970	716
300	707	0.086	0.0144	8560	342
350	487	0.068	0.0038	16,500	82
374 [†]	315	0.019	0.0	22,100	0

[†] Critical point.
[‡] In contact with air.

TABLE A-2
Properties of air at atmospheric pressure

T, °C	ρ, kg/m³	μ, μPa·s	k, mW/m·K	c_p, J/kg·K	Pr
0	1.294	17.2	24.2	1003	0.71
50	1.093	19.5	27.6	1006	0.71
100	0.947	21.7	31.0	1010	0.71
150	0.835	23.8	34.4	1016	0.70
200	0.747	25.7	37.6	1024	0.70
250	0.675	27.6	40.8	1034	0.70
300	0.616	29.3	43.9	1045	0.70
400	0.525	32.5	49.7	1069	0.70
500	0.457	35.5	55.3	1093	0.70
600	0.405	38.3	60.9	1114	0.70
700	0.363	40.9	65.9	1135	0.70
800	0.329	43.4	70.3	1153	0.71
900	0.301	45.7	74.7	1170	0.72
1000	0.277	47.9	78.6	1184	0.72

Source: White (1988).

TABLE A-3
Properties of common liquids at 1 atm and 20°C (68°F)

Liquid	ρ, kg/m^3	μ, kg/(m·s)	$\mathcal{T}$, N/m*	p_v, N/m^2	Bulk modulus, K, N/m^2
Ammonia	608	2.20E−4	2.13E−2	9.10E+5	—
Benzene	881	6.51E−4	2.88E−2	1.01E+4	1.4E+9
Carbon tetrachloride	1590	9.67E−4	2.70E−2	1.20E+4	9.65E+8
Ethanol	789	1.20E−3	2.28E−2	5.7E+3	9.0E+8
Ethylene glycol	1117	2.14E−2	4.84E−2	1.2E+1	—
Freon 12	1327	2.62E−4	—	—	—
Gasoline	680	2.92E−4	2.16E−2	5.51E+4	9.58E+8
Glycerin	1260	1.49	6.33E−2	1.4E−2	4.34E+9
Kerosene	804	1.92E−3	2.8E−2	3.11E+3	1.6E+9
Mercury	13,550	1.56E−3	4.84E−1	1.1E−3	2.55E+10
Methanol	791	5.98E−4	2.25E−2	1.34E+4	8.3E+8
SAE 10W oil	870	1.04E−1‡	3.6E−2	—	1.31E+9
SAE 10W30 oil	876	1.7E−1‡	—	—	—
SAE 30W oil	891	2.9E−1‡	3.5E−2	—	1.38E+9
SAE 50W oil	902	8.6E−1‡	—	—	—
Water	998	1.00E−3	7.28E−2	2.34E+3	2.19E+9
Seawater (30‰)	1025	1.07E−3	7.28E−2	2.34E+3	2.33E+9

*In contact with air.
‡Representative values. The SAE oil classifications allow a viscosity variation of up to ±50 percent.

TABLE A-4
Properties of common gases at 1 atm and 20°C (68°F)

Gas	Molecular weight	R, m^2/(s^2·K)	ρg, N/m^3	μ, N·s/m^2	Specific-heat ratio
H$_2$	2.016	4124	0.822	9.05E−6	1.41
He	4.003	2077	1.63	1.97E−5	1.66
H$_2$O	18.02	461	7.35	1.02E−5	1.33
Ar	39.944	208	16.3	2.24E−5	1.67
Dry air	28.97	287	11.8	1.80E−5	1.40
CO$_2$	44.01	189	17.9	1.48E−5	1.30
CO	28.01	297	11.4	1.82E−5	1.40
N$_2$	28.02	297	11.4	1.76E−5	1.40
O$_2$	32.00	260	13.1	2.00E−5	1.40
NO	30.01	277	12.1	1.90E−5	1.40
N$_2$O	44.02	189	17.9	1.45E−5	1.31
Cl$_2$	70.91	117	28.9	1.03E−5	1.34
CH$_4$	16.04	518	6.54	1.34E−5	1.32

TRANSPORT PROPERTIES OF VARIOUS NEWTONIAN FLUIDS **577**

TABLE A-5
Critical-point constants for common fluids

Substance	Molecular weight	T_c, °R	p_c, atm	μ_c, μPa·s	k_c, mW/(m·K)
H_2	2.016	60.0	12.8	3.47	90.0
He	4.003	9.47	2.26	2.54	20.8
Ar	39.944	272	48.0	26.4	29.8
Air	28.97[†]	238[†]	36.4[†]	19.3[†]	38.1[†]
CO_2	44.01	548	72.9	34.3	51.1
CO	28.01	239	34.5	19.0	36.2
N_2	28.02	227	33.5	18.0	36.3
O_2	32.00	278	49.7	25.0	44.1
NO	30.01	324	64	25.8	49.5
N_2O	44.02	557	71.7	33.2	54.9
Cl_2	70.91	751	76.1	42.0	40.7
CH_4	16.04	343	45.8	15.9	66.1

[†] Values for air are pseudocritical properties computed for the average composition of sea-level dry air.

TABLE A-6
Sutherland and power-law constants for gas viscosity

Gas	T_0, °R	μ_0, slugs/ft·h	S, °R	n
Ammonia	491.6	0.000722	679	0.981
Chlorine	491.6	0.000970	585	1.00
Deuterium	545.7	0.000974		0.699
Ethylene	491.6	0.000722	407	
Helium	491.6	0.001406	143	0.666
Hydrogen chloride	491.6	0.001030	643	1.03
Hydrogen sulfide	491.6	0.000880	596	
Krypton	491.6	0.001744	338	
Methane	491.6	0.000902	356	0.873
Methyl chloride	491.6	0.000743	817	
Neon	491.6	0.002233	110	0.657
Nitric oxide	491.6	0.001346	230	0.78
Nitrous oxide	491.6	0.001015	493	0.89
Sulfur dioxide	491.6	0.000880	749	
Xenon	491.6	0.001579	454	

Source: Chapman and Cowling (1970).

578 VISCOUS FLUID FLOW

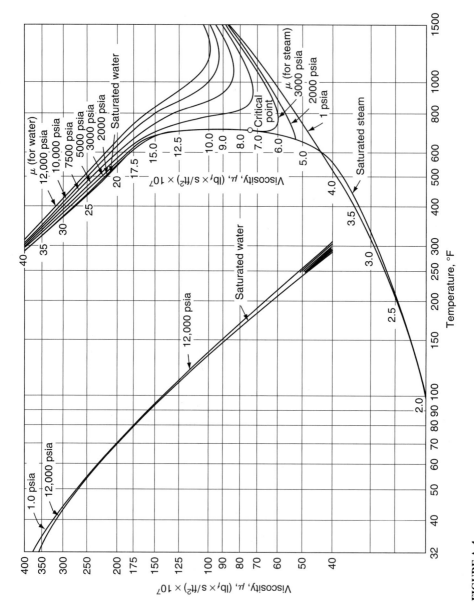

FIGURE A-4
Viscosity of steam and water. (*From the 1967 ASME Steam Tables.*)

TRANSPORT PROPERTIES OF VARIOUS NEWTONIAN FLUIDS 579

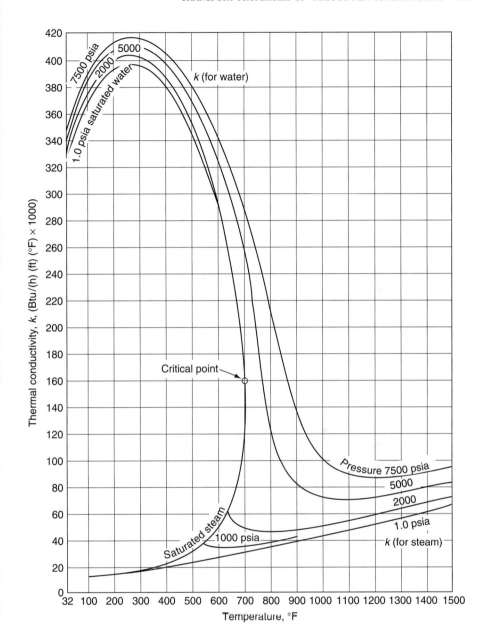

FIGURE A-5
Thermal conductivity of steam and water. (*From the 1967 ASME Steam Tables.*)

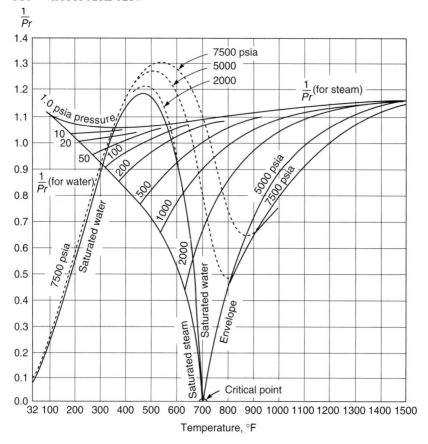

FIGURE A-6
Reciprocal Prandtl number $1/Pr$ for steam and water. (*From the 1967 ASME Steam Tables.*)

TABLE A-7
Molecular parameters for dilute-gas transport properties, Eqs. (1-33) and (1-41)

Gas	σ, Å	T_e, K
H_2	2.827	59.7
He	2.551	10.22
Ar	3.542	93.3
Air	3.711	78.6
CO_2	3.941	195.2
CO	3.690	91.7
O_2	3.467	106.7
N_2	3.798	71.4
NO	3.492	116.7
N_2O	3.828	232.4
Cl_2	4.217	316.0
CH_4	3.758	148.6

Source: R.A. Svehla, *NASA Tech. Rep. R-132*, 1962.

APPENDIX B

EQUATIONS OF MOTION OF INCOMPRESSIBLE NEWTONIAN FLUIDS IN CYLINDRICAL AND SPHERICAL POLAR COORDINATES

The general equations of motion for incompressible flow with constant transport properties are

Continuity: $\quad\quad\quad \nabla \cdot \mathbf{V} = 0 \quad\quad\quad$ (B-1a)

Navier–Stokes: $\quad\quad\quad \dfrac{D\mathbf{V}}{Dt} = -\dfrac{1}{\rho}\nabla p + \mathbf{g} + \dfrac{1}{\rho}\nabla \cdot \tau_{ij} \quad\quad\quad$ (B-1b)

Energy: $\quad\quad\quad \rho c_p \dfrac{DT}{Dt} = k\nabla^2 T + \Phi \quad\quad\quad$ (B-1c)

where $\quad\quad\quad \Phi = \mu\left(\dfrac{\partial u_i}{\partial x_j} + \dfrac{\partial u_j}{\partial x_i}\right)\dfrac{\partial u_i}{\partial x_j}$

CYLINDRICAL COORDINATES

These coordinates (r, θ, z) are related to the Cartesian (x, y, z) by

$$x = r\cos\theta \quad y = r\sin\theta \quad z = z \tag{B-2}$$

The velocity components are (v_r, v_θ, v_z). The equations of motion become
Continuity:

$$\frac{1}{r}\frac{\partial}{\partial r}(rv_r) + \frac{1}{r}\frac{\partial}{\partial \theta}(v_\theta) + \frac{\partial}{\partial z}(v_z) = 0 \tag{B-3}$$

Convective time derivative:

$$\mathbf{V}\cdot\nabla = v_r\frac{\partial}{\partial r} + \frac{1}{r}v_\theta\frac{\partial}{\partial \theta} + v_z\frac{\partial}{\partial z} \tag{B-4}$$

Laplacian operator:

$$\nabla^2 = \frac{1}{r}\frac{\partial}{\partial r}\left(r\frac{\partial}{\partial r}\right) + \frac{1}{r^2}\frac{\partial^2}{\partial \theta^2} + \frac{\partial^2}{\partial z^2} \tag{B-5}$$

r momentum:

$$\frac{\partial v_r}{\partial t} + (\mathbf{V}\cdot\nabla)v_r - \frac{1}{r}v_\theta^2 = -\frac{1}{\rho}\frac{\partial p}{\partial r} + g_r + \nu\left(\nabla^2 v_r - \frac{v_r}{r^2} - \frac{2}{r^2}\frac{\partial v_\theta}{\partial \theta}\right) \tag{B-6}$$

θ momentum:

$$\frac{\partial v_\theta}{\partial t} + (\mathbf{V}\cdot\nabla)v_\theta + \frac{v_r v_\theta}{r} = -\frac{1}{\rho r}\frac{\partial p}{\partial \theta} + g_\theta + \nu\left(\nabla^2 v_\theta + \frac{2}{r^2}\frac{\partial v_r}{\partial \theta} - \frac{v_\theta}{r^2}\right) \tag{B-7}$$

z momentum:

$$\frac{\partial v_z}{\partial t} + (\mathbf{V}\cdot\nabla)v_z = -\frac{1}{\rho}\frac{\partial p}{\partial z} + g_z + \nu\nabla^2 v_z \tag{B-8}$$

Energy:

$$\rho c_p\left[\frac{\partial T}{\partial t} + (\mathbf{V}\cdot\nabla)T\right] = k\nabla^2 T + \mu\left[2\left(\epsilon_{rr}^2 + \epsilon_{\theta\theta}^2 + \epsilon_{zz}^2\right) + \epsilon_{\theta z}^2 + \epsilon_{rz}^2 + \epsilon_{r\theta}^2\right] \tag{B-9}$$

where

$$\epsilon_{rr} = \frac{\partial v_r}{\partial r} \quad \epsilon_{\theta\theta} = \frac{1}{r}\frac{\partial v_\theta}{\partial \theta} + \frac{v_r}{r} \quad \epsilon_{zz} = \frac{\partial v_z}{\partial z}$$

$$\epsilon_{\theta z} = \frac{1}{r}\frac{\partial v_z}{\partial \theta} + \frac{\partial v_\theta}{\partial z} \quad \epsilon_{rz} = \frac{\partial v_r}{\partial z} + \frac{\partial v_z}{\partial r} \quad \epsilon_{r\theta} = \frac{1}{r}\frac{\partial v_r}{\partial \theta} + \frac{\partial v_\theta}{\partial r} - \frac{v_\theta}{r}$$

Vorticity components:

$$\omega_r = \frac{1}{r}\frac{\partial v_z}{\partial \theta} - \frac{\partial v_\theta}{\partial z} \qquad \omega_\theta = \frac{\partial v_r}{\partial z} - \frac{\partial v_z}{\partial r} \qquad \omega_z = \frac{1}{r}\frac{\partial}{\partial r}(rv_\theta) - \frac{1}{r}\frac{\partial v_r}{\partial \theta}$$

SPHERICAL POLAR COORDINATES

These coordinates (r, θ, ϕ) are related to (x, y, z) by the transformations

$$x = r \sin\theta \cos\phi \qquad y = r \sin\theta \sin\phi \qquad z = r \cos\theta \qquad \text{(B-10)}$$

The equations of motion then become, for velocity components (v_r, v_θ, v_ϕ):

Continuity:

$$\frac{1}{r^2}\frac{\partial}{\partial r}(r^2 v_r) + \frac{1}{r \sin\theta}\frac{\partial}{\partial \theta}(v_\theta \sin\theta) + \frac{1}{r \sin\theta}\frac{\partial}{\partial \phi}(v_\phi) = 0 \qquad \text{(B-11)}$$

Time derivative following the particle:

$$\frac{D}{Dt} = \frac{\partial}{\partial t} + v_r \frac{\partial}{\partial r} + \frac{v_\theta}{r}\frac{\partial}{\partial \theta} + \frac{v_\phi}{r \sin\theta}\frac{\partial}{\partial \phi} \qquad \text{(B-12)}$$

Laplacian operator:

$$\nabla^2 = \frac{1}{r^2}\frac{\partial}{\partial r}\left(r^2 \frac{\partial}{\partial r}\right) + \frac{1}{r^2 \sin\theta}\frac{\partial}{\partial \theta}\left(\sin\theta \frac{\partial}{\partial \theta}\right) + \frac{1}{r^2 \sin^2\theta}\frac{\partial^2}{\partial \phi^2} \qquad \text{(B-13)}$$

r momentum:

$$\frac{Dv_r}{Dt} - \frac{1}{r}(v_\theta^2 + v_\phi^2) = -\frac{1}{\rho}\frac{\partial p}{\partial r} + g_r$$
$$+ \nu\left(\nabla^2 v_r - \frac{2v_r}{r^2} - \frac{2}{r^2}\frac{\partial v_\theta}{\partial \theta} - \frac{2v_\theta \cot\theta}{r^2} - \frac{2}{r^2 \sin\theta}\frac{\partial v_\phi}{\partial \phi}\right)$$
$$\text{(B-14)}$$

θ momentum:

$$\frac{Dv_\theta}{Dt} + \frac{1}{r}(v_r v_\theta - v_\phi^2 \cot\theta) = -\frac{1}{\rho r}\frac{\partial p}{\partial \theta} + g_\theta$$
$$+ \nu\left(\nabla^2 v_\theta + \frac{2}{r^2}\frac{\partial v_r}{\partial \theta} - \frac{v_\theta}{r^2 \sin^2\theta} - \frac{2\cos\theta}{r^2 \sin^2\theta}\frac{\partial v_\phi}{\partial \phi}\right)$$
$$\text{(B-15)}$$

ϕ momentum:

$$\frac{Dv_\phi}{Dt} + \frac{1}{r}(v_r v_\phi + v_\theta v_\phi \cot\theta)$$
$$= -\frac{1}{\rho r \sin\theta}\frac{\partial p}{\partial \phi} + g_\phi$$
$$+ \nu\left(\nabla^2 v_\phi - \frac{v_\phi}{r^2 \sin^2\theta} + \frac{2}{r^2 \sin^2\theta}\frac{\partial v_r}{\partial \phi} + \frac{2\cos\theta}{r^2 \sin^2\theta}\frac{\partial v_\phi}{\partial \phi}\right) \qquad \text{(B-16)}$$

Energy:
$$\rho c_p \frac{DT}{Dt} = k \nabla^2 T + \mu \left[2\left(\epsilon_{rr}^2 + \epsilon_{\theta\theta}^2 + \epsilon_{\phi\phi}^2\right) + \epsilon_{\theta\phi}^2 + \epsilon_{\theta r}^2 + \epsilon_{r\phi}^2 \right] \quad \text{(B-17)}$$

where

$$\epsilon_{rr} = \frac{\partial v_r}{\partial r} \qquad \epsilon_{\theta\theta} = \frac{1}{r}\frac{\partial v_\theta}{\partial \theta} + \frac{v_r}{r} \qquad \epsilon_{\phi\phi} = \frac{1}{r\sin\theta}\frac{\partial v_\phi}{\partial \phi} + \frac{v_r}{r} + \frac{v_\theta \cot\theta}{r}$$

$$\epsilon_{\theta\phi} = \frac{\sin\theta}{r}\frac{\partial}{\partial \theta}\left(\frac{v_\phi}{\sin\theta}\right) + \frac{1}{r\sin\theta}\frac{\partial v_\theta}{\partial \phi} \qquad \epsilon_{\theta r} = r\frac{\partial}{\partial r}\left(\frac{v_\theta}{r}\right) + \frac{1}{r}\frac{\partial v_r}{\partial \theta}$$

$$\epsilon_{r\phi} = \frac{1}{r\sin\theta}\frac{\partial v_r}{\partial \phi} + r\frac{\partial}{\partial r}\left(\frac{v_\phi}{r}\right)$$

APPENDIX C

A RUNGE–KUTTA SUBROUTINE FOR N SIMULTANEOUS DIFFERENTIAL EQUATIONS

Amateur programmers who use digital computers on a relatively casual basis often express a desire for a simple subroutine to perform numerical quadrature of ordinary differential equations. Because it is highly accurate and does not need starting solutions, the Runge–Kutta method would serve this purpose if its logic were not rather complex.

This appendix presents a Runge–Kutta subroutine in the FORTRAN and BASIC languages for use with an arbitrary number of ordinary differential equations. The program is adapted from a similar routine developed for the MAD language at the University of Michigan. This author has found it accurate for all of the many types of differential equations he has encountered over the past 40 years. It requires the programmer only to write expressions for the derivatives needed in the particular equation. The programmer also selects the step size H, which may vary during the run. The subroutine is called by the FORTRAN statement

$$\text{CALL RUNGE (N, Y, F, X, H, M, K)}$$

The arguments N to K may be described in order of appearance:

>N = number of differential equations to be solved (set by the programmer).
>Y = array of N dependent variables (with initial values set by the programmer).
>F = array of the N derivatives of the variables Y [the programmer must provide an expression in the main program for the calculation of each F(I)].
>X = independent variable (initialized by the programmer).
>H = step size ΔX (set by the programmer).
>M = index used in the subroutine which must be set equal to zero by the programmer before the first CALL.
>K = integer from the subroutine which is used as the argument of a computed GO TO statement in the main program, such as GO TO (10,20), K. Statement 10 below calculates the derivatives F(I) and statement 20 prints the answers, X and Y(I).

For example, a simple FORTRAN program to use this CALL is as follows:

```
   DIMENSION Y(10),F(10)
   READ X,N,XLIM,H,M,(Y(I),I = 1,N)
 8 IF(X-XLIM)6,6,7
 6 CALL RUNGE(N,Y,F,X,H,M,K)
   GO TO (10,20),K
10 F(1) = the derivative of Y(1)
   F(2) = the derivative of Y(2)
   . . . . . . . . . . . . . . . . . . . . .
   F(N) = the derivative of Y(N)
   GO TO 6
20 WRITE X,(Y(I),I = 1,N)
   GO TO 8
 7 STOP
   END
```

It is only necessary that one place the proper expressions for the derivatives in the statements for F(I) = ···.

Note that it is not necessary to increment X in the main program. This is automatically done correctly in the subroutine. One could change H at any time by inserting a logic expression for changing H underneath the WRITE statement (statement 20).

The array size for Y(I) and F(I) is arbitrary and could be increased or decreased at will. The subroutine contains an additional array Q(I) which should be of like size. The subroutine FORTRAN listing is as follows:

```
      SUBROUTINE RUNGE (N, Y, F, X, H, M, K)
C     THIS ROUTINE PERFORMS RUNGE-KUTTA CALCULA-
C     TION BY GILLS METHOD
      DIMENSION Y(10), F(10), Q(10)
      M = M + 1
      GO TO (1, 4, 5, 3, 7), M
    1 DO 2 I = 1, N
    2 Q(I) = 0
      A = .5
      GO TO 9
    3 A = 1.707107
C     IF YOU NEED MORE ACCURACY, USE
C     A = 1.7071067811865475244
    4 X = X + .5*H
    5 DO 6 I = 1, N
      Y(I) = Y(I) + A*(F(I)*H − Q(I))
    6 Q(I) = 2.*A*H*F(I) + (1. − 3.*A)*Q(I)
      A = .2928932
C     IF YOU NEED MORE ACCURACY, SET
C     A = .2928932188134524756
      GO TO 9
    7 DO 8 I = 1, N
    8 Y(I) = Y(I) + H*F(I)/6. − Q(I)/3.
      M = 0
      K = 2
      GO TO 10
    9 K = 1
   10 RETURN
      END
```

The same algorithm can be written in BASIC for a personal computer:

```
DIM y(10),f(10),q(10)
DATA 0,1,2,2,0,1
READ x,n,xlim,h,m,y(1)
48 IF x > xlim THEN STOP
GOTO 500
50 IF k = 1 THEN GOTO 60
IF k = 2 THEN GOTO 90
60 f(1) = (THE GIVEN 1st DERIVATIVE)
GOTO 500
```

```
90 PRINT x,y(1)
GOTO 48
500 REM The Routine for Runge:
m = m + 1
IF m = 1 THEN GOTO 501
IF m = 2 THEN GOTO 504
IF m = 3 THEN GOTO 505
IF m = 4 THEN GOTO 503
IF m = 5 THEN GOTO 507
501 FOR i = 1 TO n
q(i) = 0
NEXT i
a = .5
GOTO 509
503 a = 1.707107
504 x = x + .5*h
505 FOR i = 1 TO n
y(i) = y(i) + a*(f(i)*h − q(i))
q(i) = 2*a*h*f(i) + (1 − 3*a)*q(i)
NEXT i
a = .2928932
509 k = 1
GOTO 50
507 FOR i = 1 TO n
y(i) = y(i) + h*f(i)/6 − q(i)/3
NEXT i
m = 0
k = 2
GOTO 50
END
```

Higher order equations are simply broken into a sequence of first-order differential equations. For example, $y'' = f(x, y, y')$ can be written as two equations: $z' = f(x, y, z)$ and $y' = z$. Thus, if the variables are defined in reverse order, e.g., $Y(1) = y'$ and $Y(2) = y$, the first statement, $F(1) = \cdots$, spells out the basic differential equation and subsequent statements $F(I) = \cdots$ for $I > 1$, are merely integrations. Three examples of the use of the program are as follows:

Example 1. A second-order equation

$$y'' = x^2 - y^2 - 3y'$$

Program: $F(1) = X * X - Y(2) * Y(2) - 3. * Y(1)$

$F(2) = Y(1)$

Example 2. A third-order equation:
$$y''' = y'^2 - 4y$$

Program:
$$F(1) = Y(2) * Y(2) - 4. * Y(3)$$
$$F(2) = Y(1)$$
$$F(3) = Y(2)$$

Example 3. Four integrations:
$$z = \int\int\int\int f(x)\, dx: \text{ equivalent to } z'''' = f(x)$$

Program:
$$F(1) = f(x)$$
$$F(2) = Y(1)$$
$$F(3) = Y(2)$$
$$F(4) = Y(3)$$

BIBLIOGRAPHY

Abramovich, G. N. (1963), *The Theory of Turbulent Jets*, M.I.T. Press, Cambridge, Mass.
Adams, N. A., and L. Kleiser (1993), "Numerical Simulation of Transition in a Compressible Flat Plate Boundary Layer," in Kral and Zang, pp. 101–110.
Agarwal, R. (1999), "Computational Fluid Dynamics of Whole-Body Aircraft," *Annu. Rev. Fluid Mech.*, vol. 31, pp. 125–169.
Agullo, O., and A. D. Verga (1997), "Exact Two Vortices Solution of Navier–Stokes Equations," *Phys. Rev. Lett.*, vol. 78, no. 12, p. 2361.
Ambrok, G. S. (1957), "Approximate Solutions of Equations for the Thermal Boundary Layer with Variations in the Boundary Layer Structure," *Soviet Phys.-Tech. Phys.*, vol. 2, no. 9, pp. 1979–1986.
Ames, W. F. (1965), *Nonlinear Partial Differential Equations in Engineering*, Academic, New York.
Andereck, C. D., S. Liu, and H. Swinney (1986), "Flow Regimes in a Circular Couette System with Independently Rotating Cylinders," *J. Fluid Mech.*, vol. 164, pp. 155–183.
Anderson, J. D., Jr. (1995), *Computational Fluid Dynamics: The Basics with Applications*, McGraw-Hill, New York.
—— (2000), *Hypersonic and High Temperature Gas Dynamics*, Amer. Inst. of Aero. and Astro., Reston, Va.
—— (2001), *Fundamentals of Aerodynamics*, 3rd ed., McGraw-Hill, New York.
—— (2002), *Modern Compressible Flow with Historical Perspective*, 3rd ed., McGraw-Hill, New York.
Antonia, R. A., and R. W. Bilger (1973), "An Experimental Investigation of an Axisymmetric Jet in a Co-Flowing Air Stream," *J. Fluid Mech.*, vol. 61, pp. 805–822.
Apelt, C. J. (1961), "The Steady Flow of a Viscous Fluid Past a Circular Cylinder at Reynolds Numbers 40 and 44," *Aero. Res. Council London*, R & M 3175.
Aref, H. (2004), *Introduction to Computational Fluid Dynamics*, Cambridge Univ. Press, New York.
Aris, R. (1990), *Vectors, Tensors, and the Basic Equations of Fluid Mechanics*, Dover, New York.
Arnal, D. (1984), "Description and Prediction of Transition in Two-Dimensional Incompressible Flow," pp. 2-1–2-71 of AGARD, 1984.
Arpaci, V. S. (1984), *Convective Heat Transfer*, Prentice-Hall, Upper Saddle River, NJ.
Au-Yang, M. K. (2001), *Flow-Induced Vibration of Power and Process Plant Components: A Practical Workbook*, ASME Press, New York.
Bae, Y. Y., and G. Emanuel (1989), "Boundary-Layer Tables for Similar Compressible Flow," *AIAA J.*, vol. 27, no. 9, pp. 1163–1164.

Bake, S.D., G. W. Meyer, and U. Rist (2002), "Turbulence Mechanism in Klebanoff Transition: A Quantitative Comparison of Experiment and Direct Numerical Simulation," *J. Fluid Mech.*, vol. 459, pp. 217–243.
Baker, A. J. (1983), *Finite Element Computational Fluid Mechanics*, Hemisphere, New York.
Baker, C. J. (1979), "The Laminar Horseshoe Vortex," *J. Fluid Mech.*, vol. 95, pt. 2, pp. 347–368.
Baker, R. C. (2003), *Introductory Guide to Flow Measurement*, ASME Press, New York.
Baldwin, B. S., and H. Lomax (1978), "Thin Layer Approximation and Algebraic Model for Separated Turbulent Flow," AIAA Paper 78-257.
Baranblatt, G. I., A. J. Chorin, O. H. Hald, and V. M. Prostokishin (1997), "Scaling Laws for Fully Developed Turbulent Flow in Pipes," *Appl. Mech. Rev.*, vol. 50, pp. 413–429.
Baritaud, T. (1996), *Direct Numerical Simulation for Turbulent Reacting Flows*, Editions Technip, Paris.
Barnwell, R. W., and R. A. Wahls (1991), "Skin Friction Law for Compressible Turbulent Flow," *AIAA J.*, vol. 29, no. 3, pp. 380–386.
Barrett, M. J., and D.K. Hollingsworth (2003), "Heat Transfer in Turbulent Boundary Layers Subjected to Freestream Turbulence: Part I, Experimental Results, Part II, Analysis and Correlation," *J. Turbomachinery*, vol. 125, April, pp. 232–251.
Bear, J. (2000), *Dynamics of Fluids in Porous Media*, 2nd ed., Elsevier, New York.
Becker, A. A. (2004), *Introductory Guide to Finite Element Analysis*, ASME Press, New York.
Beer, G. (2001), *Programming the Boundary Element Method: An Introduction for Engineers*, Wiley, New York.
Bejan, A. (1994), *Convection Heat Transfer*, 2nd ed., Wiley, New York.
Benjamin, T. B. (1960), "Effects of a Flexible Boundary on Hydrodynamic Stability," *J. Fluid Mech.*, vol. 9, pt. 4, pp. 513–532.
Berger, S. A. (1971). *Laminar Wakes*, Elsevier, New York.
Bergstrom, D. J., N. A. Kotey, and M. F. Tachie (2002), "The Effects of Surface Roughness on the Mean Velocity Profile in a Turbulent Boundary Layer," *J. Fluids Eng.*, vol. 124, September, pp. 664–670.
Berker, A. R. (1963), "Intégration des équations du mouvement d'un fluide visqueux incompressible," in S. Flügge (ed.), *Encyclopedia of Physics*, vol. 8, pt. 2, pp. 1–384, Springer, Berlin.
Berman, A. S. (1953), "Laminar Flow in Channels with Porous Walls," *J. Appl. Phys.*, vol. 24, no. 9, pp. 1232–1235.
Bernard, P. S., and J. M. Wallace, (2002), *Turbulent Flow: Analysis, Measurement, and Prediction*, Wiley, New York.
Bertram, M. H. (1958), "Boundary Layer Displacement Effects in Air at Mach Numbers of 6.8 at 9.6," NACA Technical Note 4133, February.
Bertram, M. H., and T. A. Blackstock (1961), "Some Simple Solutions to the Problem of Predicting Boundary Layer Self-Induced Pressures," National Aeronautics and Space Administration Technical Note D-798, April.
Betchov, R., and A. B. Szewczyk (1963), "Stability of a Shear Layer Between Parallel Streams," *Phys. Fluids*, vol. 6, pp. 1391–1396.
Betchov, R., and W. O. Criminale (1967), *Stability of Parallel Flows*, Academic, New York.
Bird, G. A. (1994), *Molecular Gas Dynamics and the Direct Simulation of Gas Flows*, Clarendon Press, Oxford.
Bird, R. B., R. C. Armstrong, and O. Hassager (1977), *Dynamics of Polymeric Liquids*, vols. 1–2, Wiley, New York.
Bird, R. B., W. E. Stewart, and E. N. Lightfoot (2001), *Transport Phenomena*, 2nd ed., Wiley, New York.
Blackwell, B. F. (1973), "The Turbulent Boundary Layer on a Porous Plate: An Experimental Study of the Heat Transfer Behavior with Adverse Pressure Gradients," Ph.D. thesis, Stanford Univ., Stanford, Calif.
Blasius, H. (1908), "Grenzschichten in Flüssikeiten mit kleiner Reibung," *Z. Angew. Math. Phys.*, vol. 56, pp. 1–37 [English translation in NACA Technical Memo. 1256].
—— (1913), "Das Ähnlichkeitsgesetz bei Reibungsvorgängen in Flüssikeiten," *Forsch. Arb. Ing.-Wes.*, no. 134, Berlin.

Blazek, J. (2001), *Computational Fluid Dynamics: Principles and Applications*, Elsevier, New York.
Blevins, R. D. (1977), *Flow-Induced Vibrations*, Van Nostrand, New York.
Bluman, G. W., and J. D. Cole (1974), *Similarity Methods for Differential Equations*, Springer, New York.
Blyth, M. G., P. Hall, and D. T. Papageorgiou (2003), "Chaotic Flows in Pulsating Cylindrical Tubes: A Class of Exact Navier–Stokes Solutions," *J. Fluid Mech.*, vol. 481, pp. 187–213.
Bödewadt, U. T. (1940), "Die Drehströmmung über festem Grund," *Z. Angew. Math. Mech.*, vol. 20, pp. 241–253.
Boris, J. P. (1989), "New Directions in Computational Fluid Dynamics," *Ann. Rev. Fluid Mech.*, vol. 21, pp. 345–385.
Bourchtein, A. (2002), "Exact Solutions of the Generalized Navier–Stokes Equations for Benchmarking," *Int. J. Numerical Methods Fluids*, vol. 39, no. 11, pp. 1053–1071.
Bradshaw, P. (1977), "Compressible Turbulent Shear Layers," *Annu. Rev. Fluid Mech.*, vol. 9, pp. 33–54.
––––– (1987), "Turbulent Secondary Flows," *Annu. Rev. Fluid Mech.*, vol. 19, pp. 53–74.
Bradshaw, P., and D. H. Ferriss (1971), "Calculation of Boundary Layer Development Using the Turbulent Energy Equation: Compressible Flow on Adiabatic Walks," *J. Fluid Mech.*, vol. 46, pp. 86–110.
Bramley, J. S., and D.M. Sloan (1987), "Numerical Solution for Two-Dimensional Flow in a Branching Channel Using Boundary-Fitted Coordinates," *Comput. Fluids*, vol. 15, no. 3, pp. 297–311.
Branover, H., et al. (Eds.) (1999), *Turbulence and Structures: Chaos, Fluctuations, and Helical Self-Organization in Nature and the Laboratory*, Academic, New York.
Brauer, H., and D. Sucker (1976), "Umströmung von Platten, Zylinder und Kugeln," *Chem.-Ing.-Tech.*, vol. 48, pp. 665–671 [English translation in *Int. Chem. Eng.*, vol. 18, pp. 367–374, 1978].
Brennen, C. E. (1995), *Cavitation and Bubble Dynamics*, Oxford Univ. Press, New York.
Breuer, M. (1998), "Large Eddy Simulation of the Subcritical Flow Past a Circular Cylinder: Numerical and Modeling," *Int. J. Numerical Methods Fluids*, vol. 28, no. 9, pp. 1281–1302.
Brinich, P. F. (1954), "Boundary Layer Transition at Mach 3.12 with and Without Single Roughness Element," NACA Technical Note 3267.
Brown, G. L., and A. Roshko (1974), "On Density Effects and Large Structure in Turbulent Mixing Layers," *J. Fluid Mech.*, vol. 64, pt. 4, pp. 775–816.
Brown, W. B. (1962), "Exact Numerical Solutions of the Complete Linearized Equations for the Stability of Compressible Boundary Layers," Norair Report NOR-62-15, Northrop Aircraft Inc., Hawthorne, Calif.
Brush, S. G. (1972), *Kinetic Theory*, vols. 1–3, Pergamon, New York.
Brutin, D., and Tadrist, L. (2003), "Experimental Friction Factor of a Liquid Flow in Microtubes," *Phys. Fluids*, vol. 15, no. 3, pp. 653–661.
Bruun, H. H. (1995), *Hot-Wire Anemometry: Principles and Signal Analysis*, Oxford Univ. Press, New York.
Buddenberg, J. W., and C. R. Wilke (1949), "Calculation of Gas-Mixture Viscosities," *Ind. Eng. Chem.*, vol. 41, pp. 1345–1347.
Burgers, J. M. (1924), "The Motion of a Fluid in the Boundary Layer along a Plane Smooth Surface," *Proc. First Int. Cong. Appl. Mech.*, p. 113, Delft.
Burmeister, L. C. (1993), *Convective Heat Transfer*, 2nd ed., Wiley, New York.
Buschmann, M. H., and M. Gad-el-Hak (2003), "Debate Concerning the Mean Velocity Profile of a Turbulent Boundary Layer," *AIAA J.*, vol. 41, no. 4, pp. 565–572.
Busemann, A. (1931), *Handbuch der Physik*, vol. 4, pt. 1, p. 366, Geest and Portig, Leipzig.
Cantwell, B., D. Coles, and P. Dimotakis (1978), "Structure and Entrainment in the Plane of Symmetry of a Turbulent Spot," *J. Fluid Mech.*, vol. 87, pp. 641–672.
Carey, G. F. (1997), *Computational Grids: Generations, Adaptation, and Solution Strategies*, Taylor & Francis, Washington, D.C.
Carlson, D. R., S. E. Widnall, and M. F. Peeters (1982), "A Flow-Visualization Study of Transition in Plane Poiseuille Flow," *J. Fluid Mech.*, vol. 121, pp. 487–505.

Carr, B., and V. E. Young (1996), "Videotapes and Movies on Fluid Dynamics and Fluid Machines," in *Handbook of Fluid Dynamics and Fluid Machinery*, vol. II, J. A. Schetz and A. E. Fuhs (Eds.), Wiley, New York.
Carslaw, H. S., and J. C. Jaeger (1959), *Conduction of Heat in Solids*, 2nd ed., Clarendon, Oxford.
Cassel, K. W., A. I. Rubin, and J. D. A. Walker (1996), "Influence of Wall Cooling on Hypersonic Boundary Layer Separation and Stability," *J. Fluid Mech.*, vol. 321, pp. 189–216.
Caughey, D. A., and M. M. Hafez (Eds.) (2004), *Frontiers of CFD 2002*, World Scientific, River Edge, N.J.
Cebeci, T. (1984), "Problems and Opportunities with Three-Dimensional Boundary Layers," AGARD Report 719, Paper 6.
—— (1988), *Physical and Computational Aspects of Convective Heat Transfer*, Springer, New York.
—— (2003), *Modeling and Computation of Turbulent Flows*, Elsevier, New York.
Cebeci, T. (Ed.) (1982), *Numerical and Physical Aspects of Aerodynamic Flows*, Springer-Verlag, New York.
Cebeci, T., and J. Cousteix (1998), *Modeling and Computation of Boundary-Layer Flows*, Horizons, Long Beach, Calif.
Cebeci, T., and A. M. O. Smith (1974), *Analysis of Turbulent Boundary Layers*, Academic, New York.
CECAM (2004), "Dynamics of Fluids at Interfaces," European Centre for Atomic and Molecular Computations, CECAM, Lyon, France, July 26–29.
Cercignani, C. (2000), *Rarefied Gas Dynamics*, Cambridge Univ. Press, New York.
Chandrasekhar, S. (1961), *Hydrodynamic and Hydromagnetic Stability*, Oxford Univ. Press, London.
Chang, M. W., and B. A. Findlayson (1987), "Heat Transfer in Flow Past Cylinders at Re<150: Part 1, Calculations for Constant Fluid Properties," *Numerical Heat Transfer*, vol. 12, pp. 179–195.
Chapman, D. R., and M. W. Rubesin (1949), "Temperature and Velocity Profiles in the Compressible Laminar Boundary Layer with Arbitrary Distribution of Surface Temperature," *J. Aeronaut. Sci.*, vol. 16, pp. 547–565.
Chapman, D. R., and R. H. Kester (1953), "Measurements of Turbulent Skin Friction on Cylinders in Axial Flow at Subsonic and Supersonic Velocities," *J. Aeronaut. Sci.*, vol. 20, pp. 441–448.
Chapman, S., and T. G. Cowling (1970), *The Mathematical Theory of Nonuniform Gases*, 3rd ed., Cambridge Univ. Press, London.
Charbeneau, R. J. (1999), *Groundwater Hydraulics and Pollutant Transport*, Prentice-Hall, Upper Saddle River, NJ.
Chattot, J. J. (2002), *Computational Aerodynamics and Fluid Dynamics*, Springer, New York.
Chen, C-J., and S-Y. Jaw (1997), *Fundamentals of Turbulence Modeling*, Taylor & Francis, Washington, D.C.
Chesnakas, C.J., and R.L. Simpson (1997), "Detailed Investigation of the Three-Dimensional Separation about a 6:1 Prolate Spheroid," *AIAA J.*, vol. 35, no. 6, pp. 990–999.
Chevray, R. (1989), "Chaos and the Onset of Turbulence," in W. K. George and R. Arndt (Eds.), *Advances in Turbulence*, pp. 127–158, Hemisphere, New York.
Choi, C-H., K. J. A. Westin, and K. S. Breuer (2003), "Apparent Slip Flows in Hydrophilic and Hydrophobic Microchannels," *Phys. Fluids*, vol. 15, no. 10, pp. 2897–2902.
Chung, T. J. (2002), *Computational Fluid Dynamics*, Cambridge Univ. Press, New York.
Churchill, S. W. (1988), *Viscous Flows—The Practical Use of Theory*, Butterworth, Stoneham, Mass.
Churchill, S. W., and H. Ozoe (1973)," A Correlation for Laminar Free Convection from a Vertical Plate," *J. Heat Transfer*, vol. 95, pp. 540–541.
Churchill, S. W., and H. S. Chu (1975), "Correlating Equations for Laminar and Turbulent Free Convection from a Vertical Plate," *Int. J. Heat Mass Transfer*, vol. 18, pp. 1323–1329.
Churchill, S. W., and R. Usagi (1972), "A General Expression for the Correlation of Rates of Transfer and Other Phenomena," *AIChE J.*, vol. 18, pp. 1121–1128.
Clauser, F. H. (1954), "Turbulent Boundary Layers in Adverse Pressure Gradients," *J. Aeronaut. Sci.*, vol. 21, pp. 91–108.
—— (1956), "The Turbulent Boundary Layer," *Adv. in Appl. Mech.*, vol. 4, pp. 1–15, Academic, New York.
Clift, R., J. R. Grace, and M. E. Weber (1978), *Bubbles, Drops, and Particles*, Academic, New York.

Cochran, W. G. (1934), "The Flow Due to a Rotating Disk," *Proc. Cambridge Philos. Soc.*, vol. 30, pp. 365–375.
Cohen, C. B., and E. Reshotko (1956), "Similarity Solutions for the Compressible Laminar Boundary Layer with Heat Transfer and Arbitrary Pressure Gradient," NACA Report 1293 [see also NACA Report 1294].
Coles, D. E. (1954), "Measurements of Turbulent Friction on a Smooth Flat Plate in Supersonic Flow," *J. Aeronaut. Sci.*, vol. 21, pp. 433–448.
—— (1956), "The Law of the Wake in the Turbulent Boundary Layer," *J. Fluid Mech.*, vol. 1, pp. 191–226.
—— (1965), "Transition in Circular Couette Flow," *J. Fluid Mech.*, vol. 21, pp. 385–425.
Coles, D. E., and E. A. Hirst (1968), "Computation of Turbulent Boundary Layers—1968 AFOSR-IFP Stanford Conference," *Proc. 1968 Conf.*, vol. 2, Stanford Univ., Stanford, Calif.
Comini, G., and B. Sunder (Eds.) (2000), *Computational Analysis of Convective Heat Transfer*, WIT Press/Computational Mechanics, Boston.
Constantin, P., and C. Foias (1988), *Navier–Stokes Equations*, Univ. Chicago Press, Chicago.
Constantinescu, G. S., et al. (2002), "Numerical Investigation of Flow Past a Prolate Spheroid," *J. Fluids Eng.*, vol. 124, pp. 904–910.
Constantinescu, V. N. (1995), *Laminar Viscous Flow*, Spring-Verlag, New York.
Corrsin, S., and A. L. Kistler (1955), "Freestream Boundaries of Turbulent Flows," NACA Report 1244.
Cottet, G-H., and P. D. Koumoutsakos (1999), *Vortex Methods: Theory and Practice*, Cambridge Univ. Press, New York.
Couette, M. (1890), "Etudes sur le frottement des liquides," *Ann. Chim. Phys.*, ser. 6, vol. 21, pp. 433–510.
Cousteix, J. (1986), "Three-Dimensional and Unsteady Boundary-Layer Computations," *Annu. Rev. Fluid Mech.*, vol. 18, pp. 173–196.
Craft, T. J., A. V. Gerasimov, H. Iacovides, and B. E. Launder (2002), "Progress in the Generalization of Wall-Function Treatments," *Int. J. Heat Fluid Flow*, vol. 23, pp. 148–160.
Craik, A. D. D. (1971), "Nonlinear Resonant Instability in Boundary Layers," *J. Fluid Mech.*, vol. 50, pp. 393–413.
Crank, J., and P. Nicolson (1947), "A Practical Method for Numerical Evaluation Solutions of Partial Differential Equations of the Heat-Conduction Type," *Proc. Cambridge Philos. Soc.*, vol. 43, pp. 50–67.
Crocco, L. (1932), "Sulla trasmissione del calore da una lamina piana a un fluido scorrente ad alta velocita," *L'Aerotecnica*, vol. 12, pp. 181–197 [translated as NACA Technical Memo. 690].
Crolet, J. M. (Ed.) (2000), *Computational Methods for Flow and Transport in Porous Media*, Kluwer Academic, New York.
Cruz, D. O. A., and A. P. Silva-Freire (2002), "Note on a Thermal Law of the Wall for Separating and Recirculating Flows," *Int. J. Heat Mass Transfer*, vol. 45, no. 7, pp. 1459–1465.
Cunningham, W. J. (1963), "The Concept of Stability," *Am. Sci.*, vol. 51, pp. 425–436.
Curle, N. (1958), "Accurate Solutions of the Laminar Boundary Layer Equations, for Flows Having a Stagnation Point and Separation," *Aero. Res. Council London*, R & M 3164.
Currie, I. G. (1993), *Fundamental Mechanics of Fluids*, 2nd ed., McGraw-Hill, New York.
Darcy, H. P. G. (1856), *Les fontaines publiques de la ville de Dijon*, Victor Dalmont, Paris.
Das, D. K. (1987), "A Numerical Study of Turbulent Separated Flows," *Am. Soc. Mech. Eng.* Forum on Turbulent Flows, FED vol. 51, pp. 85–90.
—— (1988), "A Simple Theory for Calculating Turbulent Boundary Layers under Arbitrary Pressure Gradients," *Int. J. Eng. Fluid Mech.*, vol. 1, no. 1, pp. 83–99.
Das, D. K., and F. M. White (1986), "Integral Skin Friction Prediction for Turbulent Separated Flows," *J. Fluids Eng.*, vol. 108, pp. 476–482.
Dauenhauer, E. C., and J. Majdalani (2003), "Exact Self-Similar Solution of the Navier–Stokes Equations for a Porous Channel with Orthogonally Moving Walls," *Phys. Fluids*, vol. 15, no. 6, pp. 1485–1495.
Davis, S. H. (1976), "The Stability of Time-Periodic Flows," *Annu. Rev. Fluid Mech.*, vol. 8, pp. 57–74.
—— (1987), "Thermocapillary Instabilities," *Annu. Rev. Fluid Mech.*, vol. 19, pp. 403–435.

Debnath, L., and D. N. Riahi (Eds.) (2000), *Nonlinear Instability, Chaos, and Turbulence*, vol. 2 of *Advances in Fluid Mechanics*, WIT Press/Computational Mechanics, Boston.
Defant, A. (1961), *Physical Oceanography*, vols. 1–2, Pergamon, New York.
Delery, J. M., and J. G. Marvin (1986), "Shock-Wave Boundary Layer Interaction," AGARD-AG-280.
Denn, M. M. (2001), "Extrusion Instabilities and Wall Slip," *Annu. Rev. Fluid Mech.*, vol. 33, pp. 265–287.
Dennis, S. R. C., and J. Dunwoody (1966), "The Steady Flow of a Viscous Fluid Past a Flat Plate," *J. Fluid Mech.*, vol. 24, pp. 577–595.
Devenport, W. J., and J. A. Schetz (2002), Boundary Layer Applets, URL: http://www.engapplets.vt.edu/fluids/
Dhawan, S., and R. Narasimha (1958), "Some Properties of Boundary Layer Flow During Transition from Laminar to Turbulent Motion," *J. Fluid Mech.*, vol. 3, pp. 418–436.
Dimas, A. A., J. P. Collins and P. S. Bernard (1998), "Fast Parallel Vortex Method for Turbulent Flow Simulation," Paper FEDSM98-5000, ASME 1998 Fluids Eng. Conf., Washington, D.C.
Dorrance, W. H. (1962), *Viscous Hypersonic Flow*, McGraw-Hill, New York.
Drazin, P. G. (2002), *Introduction to Hydrodynamic Stability*, Cambridge Univ. Press, New York.
Drazin, P. G., and W. H. Reid (1981), *Hydrodynamic Stability*, Cambridge Univ. Press, London.
Dresner, L. (1983), *Similarity Solutions of Nonlinear Partial Differential Equations*, Pitman Advanced Pub. Program, Boston, Mass.
Dryden, H. L. (1934), "Boundary Layer Flow Near Flat Plates," *Proc. Fourth Int. Congr. Appl. Mech.*, p. 175, Cambridge Univ. Press, London.
—— (1953), "Review of Published Data on the Effect of Roughness on Transition from Laminar to Turbulent Flow," *J. Aeronaut. Sci.*, vol. 20, pp. 477–482.
Dunham, J. (1972), "Prediction of Boundary Layer Transition on Turbomachinery Blades," AGARD Meeting on Boundary Layers in Turbomachines.
Dunn, D. W., and C. C. Lin (1955), "On the Stability of the Laminar Boundary Layer in a Compressible Fluid," *J. Aeronaut. Sci.*, vol. 22, pp. 455–477.
Durbin, O. A., and R. B. A. Pettersson (2001), *Statistical Theory and Modeling for Turbulent Flows*, Wiley, New York.
Durbin, P. A. (1995), "Separated Flow Computations with the k-e-v^2 Model," *AIAA J.*, vol. 33, pp. 659–664.
Durst, F. (1992), *Principles and Practice of Laser-Doppler Anemometry*, 2nd ed., Academic, New York.
Dwoyer, D. L., and M. Y. Hussaini (Eds.) (1987), *Stability of Time Dependent and Spatially Varying Flows*, Springer, New York.
Dwyer, H. A. (1981), "Some Aspects of Three-Dimensional Laminar Boundary Layers," *Annu. Rev. Fluid Mech.*, vol. 13, pp. 217–229.
Eckert, E. R. G. (1942), "Die Berechnung des Wärmeübergangs in der laminaren Grenzschicht," *Forsch. Ver. Deutsch. Ing.*, vol. 416, pp. 1–23.
—— (1955), "Engineering Relations for Friction and Heat Transfer to Surfaces in High Velocity Flow," *J. Aeronaut. Sci.*, vol. 22, pp. 585–587.
Eckert, E. R. G., and R. M. Drake, Jr. (1972), *Analysis of Heat and Mass Transfer*, McGraw-Hill, New York.
Eckert, E. R. G., and T. F. Irvine, Jr. (1957), "Incompressible Friction Factor, Transition and Hydrodynamic Entrance Length Studies of Ducts with Triangular and Rectangular Cross Sections," *Proc. Fifth Midwestern Conf. Fluid Mech.*, pp. 122–145, Univ. Michigan Press, Ann Arbor, Mich.
Eckert, E. R. G., P. L. Donoughe, and B. J. Moore (1957), "Velocity and Friction Characteristics of Laminar Viscous Boundary Layer and Channel Flow over Surfaces with Ejection and Suction," NACA Technical Note 4102.
Ehlers, W., and J. Bluhm (Eds.) (2002), *Porous Media*, Springer-Verlag, New York.
Eichelbrenner, E. A. (1973), "Three-Dimensional Boundary Layers" *Annu. Rev. Fluid Mech.*, vol. 5, pp. 339–360.
Ekman, V. W. (1905), "On the Influence of the Earth Rotation on Ocean Currents," *Arch. Mat. Astron. Fys.*, bd. 2, no. 11, pp. 1–53.

Emmons, H. W. (1951), "The Laminar-Turbulent Transition in a Boundary Layer: Part I," *J. Aeronaut. Sci.*, vol. 18, pp. 490–498.

—— (1970), "Critique of Numerical Modeling of Fluid-Mechanics Phenomena," *Annu. Rev. Fluid Mech.*, vol. 2, pp. 15–36.

Engquist, B., and A. Rizzi (2004), *Computational Fluid Dynamics*, Cambridge Univ. Press, New York.

Eucken, A. (1913), *Phys. Z. Leipz.*, vol. 14, pp. 324–332.

Evans, H. (1968), *Laminar Boundary Layers*, Addison-Wesley, Reading, Mass.

Fage, A. (1936), "Experiments on a Sphere at Critical Reynolds Numbers," *Aero. Res. Council London*, R & M 1766.

Falkner, V. M., and S. W. Skan (1931), "Some Approximate Solutions of the Boundary Layer Equations," *Philos. Mag.*, vol. 12, no. 7, pp. 865–896 [see also *Aero. Res. Council London*, R & M 1314, 1930].

Faraco-Medeiros, M.A., and A. P. Silva-Freire (1992), "Transfer of Heat in Turbulent Boundary Layers with Injection or Suction: Universal Laws and Stanton Number Equations," *Int. J. Heat Mass Transfer*, vol. 35, no. 4, pp. 991–995.

Favre, A. (1965), "Equations des Gaz Turbulents Compressibles," *Journal de Mecanique*, vol. 4, pp. 361–421.

—— (1995), *Chaos and Determinism: Turbulence as a Paradigm for Complex Systems Converging Toward Final States*, Johns Hopkins Univ. Press, Baltimore, Md.

Fay, J. A., and F. R. Riddell (1958), "Theory of Stagnation Point Heat Transfer in Dissociated Air," *J. Aeronaut. Sci.*, vol. 25, pp. 73–85.

Felcman, J., I. Straskraba, and M. Feistauer (2003), *Mathematical and Computational Methods for Compressible Flow*, Oxford Univ. Press, New York.

Feng, Z-G., and E. E. Michaelides (2001), "Drag Coefficients of Viscous Spheres at Intermediate and High Reynolds Numbers," *J. Fluids Eng.*, vol. 123, pp. 841–849.

Fernholz, H. H., and E. Krause (Eds.) (1982), *Three-Dimensional Turbulent Boundary Layers*, Springer, Berlin.

Fernholz, H. H., and P. J. Finley (1980), "A Critical Commentary on Mean Flow Data for Two-Dimensional Compressible Turbulent Boundary Layers," AGARDograph 253 [see also AGARDographs 223 and 263].

—— (1996), "Incompressible Zero-Pressure-Gradient Turbulent Boundary Layer: An Assessment of the Data," *Prog. Aerosp. Sci.*, vol. 32, no. 4, pp. 245–311.

Ferziger, J. H. (1998), *Numerical Methods for Engineering Application*, 2nd ed., Wiley, New York.

Ferziger, J. H., A. A. Lyrio, and J. C. Bardina (1982), "New Skin Friction and Entrainment Correlations for Turbulent Boundary Layers," *J. Fluids Eng.*, vol. 104, pp. 537–540.

Ferziger, J. H., and M. Peric (2001), *Computational Methods for Fluid Dynamics*, Springer, New York.

Fjørtoft, R. (1950), "Application of Integral Theorems in Deriving Criteria of Stability for Laminar Flows and for the Baroclinic Circular Vortex," *Geofys. Pub. Oslo*, vol. 17, no. 6, pp. 1–52.

Flachsbart, O. (1932), "Winddruck auf Gasbehälter," AVA Reports, Göttingen, series IV, pp. 134–138.

Frankl, F. I. (1934), *Trans. Central Aero-Hydrodyn. Inst. Moscow*, no. 176.

Freitas, C. J. (1995), "Perspective: Selected Benchmarks from Commercial CFD Codes," *J. Fluids Eng.*, vol. 117, June, pp. 208–218.

Friedrichs, K. O. (1942), "Theory of Viscous Fluid," in *Fluid Dynamics*, ch. 4, Brown Univ. Press, Providence, R. I.

Fulford, G. D. (1964), "The Flow of Liquids in Thin Films," *Adv. in Chem. Eng.*, vol. 5, pp. 151–236.

Gad-el-Hak, M. (1996), "Modern Developments in Flow Control," *Appl. Mech. Rev.*, vol. 49, July, pp. 365–380.

Gad-el-Hak, M. (Ed.) (2001), *The MEMS Handbook (Mechanical Engineering)*, CRC Press, Boca Raton, Fla.

Galbraith, R. A. McD., S. A. Sjolander, and M. R. Head (1977), "Mixing Length in the Wall Region of Turbulent Boundary Layers," *Aeronaut. Q.*, vol. 27, pt. 2, pp. 97–110.

Garg, S., and G. S. Settles (1998), "Measurements of a Supersonic Turbulent Boundary Layer by Focusing Schlieren Deflectometry," *Exp. Fluids*, vol. 25, no. 3, pp. 254–264.

Garg, V. K. (Ed.) (1998), *Applied Computational Fluid Dynamics*, Marcel Dekker, New York.
Gaster, M. (1962), "A Note on the Relation Between Temporally-Increasing and Spatially-Increasing Disturbances in Hydrodynamics Stability," *J. Fluid Mech.*, vol. 14, pp. 222–224.
Gazley, C., Jr., and A. R. Wazzan (1984), "Control of Water Boundary Layer Stability and Transition by Surface Temperature Distributions," in V. V. Kozlov (Ed.), *Laminar-Turbulent Transition*, Springer, New York.
Gebhart, B. (1969), "External Natural Convection Flow," *Appl. Mech. Rev.*, vol. 22, pp. 691–701.
Gebhart, B., and R. L. Majahan (1982), "Instability and Transition in Buoyancy-Induced Flows," *Adv. Appl. Mech.*, vol. 22, pp. 231–315.
Gebhart, B., Y. Jaluria, R. L. Mahajan, and B. Sammakia (1988), *Buoyancy-Induced Flows and Transport*, Hemisphere, New York.
George, W. K. (1989), "The Self-Preservation of Turbulent Flows and Its Relation to Initial Conditions and Coherent Structures," in W. K. George and R. Arndt (Eds.), *Advances in Turbulence*, pp. 39–73, Hemisphere, New York.
George, W. K., and L. Castillo (1997), "Zero Pressure Gradient Turbulent Boundary Layer," *Appl. Mech. Rev.*, vol. 50, no. 12, pp. 689–729.
Gersten, K. (1959), "Corner Interference Effects," AGARD Report 299.
Geurts, B. (2003), *Elements of Direct and Large Eddy Simulation*, R. T. Edwards, Flourtown, Penn.
Gibbings, J. C. (1959), "On Boundary Layer Transition Wires," Aeronautical Research Council, CP-462.
Giedt, W. H. (1949), "Investigation of Variation of Point Unit Heat Transfer Coefficient Around a Cylinder Normal to an Air Stream," *ASME Trans.*, vol. 71, pp. 375–381.
Glauert, M. B. (1956), "The Wall Jet," *J. Fluid Mech.*, vol. 1, pp. 625–643.
Glauert, M. B., and M. J. Lighthill (1955), "The Axisymmetric Boundary Layer on a Long Thin Cylinder," *Proc. R. Soc. London Ser. A*, vol. 230, pp. 188–203.
Godrèche, C., and P. Manneville (Eds.) (1998), *Hydrodynamics and Nonlinear Instabilities*, Oxford Univ. Press, New York.
Goldberg, P. (1958), "A Digital Computer Solution for Laminar Flow Heat Transfer in Circular Tubes," M. S. thesis, Mech. Engineering Dept., M.I.T., Cambridge, Mass.
Goldstein, R. J. (Ed.) (1996), *Fluid Mechanics Measurements*, 2nd ed., Taylor & Francis, Washington, DC.
Goldstein, S. (1930), "Concerning Some Solutions of the Boundary Layer Equations in Hydrodynamics," *Proc. Cambridge Philos. Soc.*, vol. 26, pp. 1–30.
—— (1938), *Modern Developments in Fluid Dynamics*, vols. 1–2, Oxford Univ. Press, London.
—— (1948), "On Laminar Boundary Layer Flow Near a Position of Separation," *Q. J. Mech. Appl. Math.*, vol. 1, pp. 43–69.
Görtler, H. (1942), "Berechnung von Aufgaben der freien Turbulenz auf Grund eines neuen Näherungsansatzes," *Z. Angew. Math. Mech.*, vol. 22, pp. 244–254.
—— (1955), "Dreidimensionales zur Stabilitätstheorie Laminarer Grenzschichten," *Z. Angew. Math. Mech.*, vol. 35, pp. 362–363.
—— (1957), "A New Series for the Calculation of Steady Laminar Boundary Layer Flows," *J. Math. Mech.*, vol. 6, pp. 1–66.
Graetz, L. (1883), "Über die Wärmeleitungsfähigkeit von Flüssigkeiten," *Ann. Phys. Chem.*, pt. 1, vol. 18, pp. 79–94; pt. 2, vol. 25, pp. 337–357.
Granville, P. S. (1953), "The Calculation of the Viscous Drag of Bodies of Revolution," David Taylor Model Basin Report 849.
—— (1987), "Baldwin–Lomax Factors for Turbulent Boundary Layers in Pressure Gradients," *AIAA J.*, vol. 25, no. 12, pp. 1624–1627.
—— (1989), "A Modified Van Driest Formula for the Mixing Length of Turbulent Boundary Layers in Pressure Gradients," *J. Fluids Eng.*, vol. 111, pp. 94–97.
Greenwood, D. T. (1988), *Principles of Dynamics*, 2nd ed., Prentice-Hall, Englewood Cliffs, N.J.
Grégoire, G., M. Favre-Marinet, and F. Julien Saint Amand (2003), "Modeling of Turbulent Fluid Flow over a Rough Wall With or Without Suction," *J. Fluids Eng.*, vol. 125, pp. 636–642.

Gregory, N., J. T. Stuart, and W. S. Walker (1955), "On the Stability of Three-Dimensional Boundary Layers with Application to the Flow Due to a Rotating Disc," *Philos. Trans. R. Soc. London Ser. A*, vol. 248, pp. 155–199.

Gross, A., and H. Burchard (2002), *Applied Turbulence in Marine Waters*, Springer-Verlag, New York.

Grove, A. S., F. H. Shair, E. E. Petersen, and A. Acrivos (1964), "An Experimental Investigation of the Steady Separated Flow Past a Circular Cylinder," *J. Fluid Mech.*, vol. 19, pp. 60–80.

Gruschwitz, E. (1950), "Calcul approché de la couche limite laminaire en écoulement compressible sur une paroi nonconductrice de la chaleur," ONERA Pub. 47, Paris.

Guo, B-Y, and P-Y Kuo (1998), *Spectral Methods and Their Applications*, World Scientific, River Edge, N.J.

Haaland, S. E. (1983), "Simple and Explicit Formulas for the Friction Factor in Turbulent Pipe Flow," *J. Fluids Eng.*, vol. 105, pp. 89–90.

Hadamard, J. (1911), "Movement permanent lent d'une sphere liquide visqueuse dans un liquid visqueux," *C. R. Acad. Sci. Paris Sér. A–B*, vol. 152, pp. 1735–1739.

Hadjiconstantinou, N. G. (2003), "Comment on Cercignani's Second-Order Slip Coefficient," *Phys. Fluids*, vol. 15, no. 8, pp. 2352–2354.

Hagen, G. (1839), "Über die Bewegung des Wassers in engen zylinderschen Rohren," *Poggendorff's Ann. Phys. Chem.*, vol. 46, pp.423–442.

Hamdan, M. H. (1998), "Alternative Approach to Exact Solutions of a Special Class of Navier–Stokes Flows," *Appl. Math. Computation (New York)*, vol. 93, no. 1, pp. 83–90.

Hamel, G. (1917), "Spiralförmige Bewegung zäher Flüssigkeiten," *Jahresber. Deutsch. Math. Ver.*, vol. 25, pp. 34–60.

Hanjalic, K. (1994), "Advanced Turbulence Closure Models: A View of Current Status and Future Prospects," *Int. J. Heat Fluid Flow*, vol. 15, pp. 178–203.

Hanjalic, K., and B. E. Launder (1972), "A Reynolds Stress Model of Turbulence and Its Application to Thin Shear Flows," *J. Fluid Mech.*, vol. 52, pp. 609–638.

—— (1976), "Contribution Towards a Reynolds-Stress Closure for Low-Reynolds-Number Turbulence," *J. Fluid Mech.*, vol. 74, pt. 4, pp. 593–610.

Hansen, A. G. (1964), *Similarity Analyses of Boundary Value Problems in Engineering*, Prentice-Hall, Englewood Cliffs, N. J.

—— (1967), *Fluid Mechanics*, Wiley, New York.

Hansen, C. F. (1958), "Approximations for Thermodynamic and Transport Properties of High Temperature Air," NACA Technical Note 4150.

Happel, J., and H. Brenner (1983), *Low Reynolds Number Hydrodynamics*, 2nd ed., Prentice-Hall, Englewood Cliffs, N.J.

—— (1983), *Low Reynolds Number Hydrodynamics: With Special Applications to Particulate Media*, Martinus Nijhoff, Boston.

Harlow, F. H., and J. E. Welch (1965), "Numerical Calculation of Time-Dependent Viscous Incompressible Flow of Fluid with Free Surface," *Phys. Fluids*, vol. 8, pp. 2182–2189.

Hartnett, J. P., and E. R. G. Eckert (1957), "Mass Transfer Cooling in a Laminar Boundary Layer with Constant Properties," *ASME Trans.*, vol. 79, pp. 247–254.

Hartree, D. R. (1937), "On an Equation Occurring in Falkner and Skan's Approximate Treatment of the Equations of the Boundary Layer," *Proc. Cambridge Philos. Soc.*, vol. 33, pp. 223–239.

—— (1939), "The Solution of the Equations of the Laminar Boundary Layer for Schubauer's Observed Pressure Distribution for an Elliptic Cylinder," *Aero. Res. Council London*, R & M 2427.

Hausen, H. (1943), "Darstellung des Wärmeuberganges in Rohren durch verallgemeinerte Potenzbeziehungen," *Z. VDI Beih. Verfahrenstech*, no. 4, p. 91.

Hayes, W. D., and R. F. Probstein (1959), *Hypersonic Flow Theory*, Academic, New York.

Head, M. R. (1958), "Entrainment in the Turbulent Boundary Layer," *Aero. Res. Council London*, R & M 3152.

Heinz, S. (2003), *Statistical Mechanics of Turbulent Flows*, Springer-Verlag, New York.

Hele-Shaw, H. S. (1898), "Investigation of the Nature of Surface Resistance of Water and of Stream Motion under Certain Experimental Conditions," *Trans. Inst. Naval Arch.*, vol. 11, p. 25.

Herbert, T. (1983), "Secondary Instability of Plane Channel Flow to Subharmonic Three-Dimensional Disturbances," *Phys. Fluids*, vol. 26, pp. 871–874.

Hermann, R. (1936), "Wärmeübertragung bei freier Strömung am waagerechten Zylinder in zwei-atomigen Gasen," VDI Forschungsheft 379 [see also NACA Technical Memo. 1366, 1954].

Herring, H. J., and G. L. Mellor (1968), "A Method of Calculating Compressible Turbulent Boundary Layers," National Aeronautics and Space Administration CR-1144, September [see also NASA SP-216, pp. 27–132].

Herring, H. J., and J. F. Norbury (1967), "Some Experiments on Equilibrium Turbulent Boundary Layers in Favorable Pressure Gradients," *J. Fluid Mech.*, vol. 27, pp. 541–549.

Hiemenz, K. (1911), "Die Grenzschicht an einem in den gleichförmigen Flüssigkeitsstrom eingetauchten geraden Kreiszylinder," *Dingler's Polytech. J.*, vol. 326, p. 321.

Hill, R., and G. Power (1956), "Extremum Principles for Slow Viscous Flow and the Approximate Calculation of Drag," *Q. J. Mech. Appl. Math.*, vol. 9, pp. 313–319.

Hilpert, R. (1933), "Wärmeabgabe von geheizten Drähten und Rohren im Luftstrom," *Forsch. Arb. Ing.-Wes.*, vol. 4, pp. 215–224.

Hilsenrath, J., et al. (1955), *Tables of Thermodynamic and Transport Properties*, Nat. Bur. Std. Circular 564; reprinted (1960) by Pergamon.

Hinze, J. O. (1975), *Turbulence*, 2nd ed., McGraw-Hill, New York.

Hirata, M., and N. Kasagi (Eds.) (1988), *Transport Phenomena in Turbulent Flows*, Hemisphere, New York.

Hirsch, C. (1988), *Numerical Computation of Internal and External Flows*, Wiley, New York.

Hirschfelder, J. O., C. F. Curtiss, and R. B. Bird (1954), *Molecular Theory of Gases and Liquids*. Wiley, New York.

Ho, C.-M., and P. Huerre (1984), "Perturbed Free Shear Layers," *Annu. Rev. Fluid Mech.*, vol. 16, pp. 365–424.

Hoffmann, K. A., and S. T. Chiang (1998), 2 vols., *Computational Fluid Dynamics*, Engineering Education Systems, Austin, Tex.

Hollingsworth, D. K., R. J. Moffat, and W. M. Kays (1992), "Effect of Concave Surface Curvature on the Turbulent Prandtl Number and the Thermal Law of the Wall," *Exp. Thermal Fluid Sci.*, vol. 5, no. 3, pp. 299–306.

Holmes, P., et al. (1998), *Turbulence, Coherent Structures, Dynamical Systems, and Symmetry*, Cambridge Univ. Press, New York.

Holstein, H., and T. Bohlen (1940), "Ein einfaches Verfahren zur Berechnung laminarer Reibungsschichten, die dem Näherungsverfahren von K. Pohlhausen genügen," *Ber. Lilienthal Ges. Luftfahrtforsch.*, S-10, pp. 5–16.

Homann, F. (1936), "Der Einfluss grosser Zähigkeit bei der Strömung um den Zylinder und um die Kugel," *Z. Angew. Math. Mech.*, vol. 16, pp. 153–164.

Howarth, L. (1938), "On the Solution of the Laminar Boundary Layer Equations," *Proc. R. Soc. London Ser. A*, vol. 164, pp. 547–579.

Huang, P. G., and P. Bradshaw (1995), "Law of the Wall for Turbulent Flows in Pressure Gradients," *AIAA J.*, vol. 33, no. 4, pp. 624–632.

Huang, P. G., P. Bradshaw, and T. J. Coakley (1993), "Skin Friction and Velocity Profile Family for Compressible Turbulent Boundary Layers," *AIAA J.*, vol. 31, no. 9, pp. 1600–1604.

Huebner, K. H., et al. (2001), *The Finite Element Method for Engineers*, 3rd ed., Wiley, New York.

Hughes, T. H., and W. H. Reid (1965), "On the Stability of the Asymptotic Suction Boundary Layer Profile," *J. Fluid Mech.*, vol. 23, pp. 715–735.

Hussain, A. K. M. F. (1986), "Coherent Structures and Turbulence," *J. Fluid Mech.*, vol. 173, pp. 303–356.

Hutton, J. F., et al. (1989), *An Introduction to Rheology*, Elsevier Health Sciences, Philadelphia.

Hytopoulos, E., J. A. Schetz, and M. Gunzburger (1993), "Numerical Solution of the Compressible Boundary Layer Equations Using the Finite Element Method," *AIAA J.*, vol. 31, no. 1, pp. 6–7.

Iaccarino, G. (2001), "Predictions of a Turbulent Separated Flow Using Commercial CFD Codes," *J. Fluids Eng.*, vol. 123, Dec., pp. 819–828.

Iglisch, R. (1944), "Exakte Berechnung der laminaren Reibungsschicht an der längsangeströmten ebenen Platte mit homogener Absaugung," *Schr. Deutsch. Akad. Luftfahrtforschung*, ser. B, vol. 8, pp. 1–51 (translated as NACA Technical Memo. 1205).

Iida, O., and Y. Nagano (1998), "Relaminarization Mechanisms of Turbulent Channel Flow at Low Reynolds Numbers," *Turbulence and Combustion*, vol. 60, no. 2, pp. 193–213.

Illingworth, C. R. (1950), "Some Solutions of the Equations of Flow of a Viscous Compressible Fluid," *Proc. Cambridge Philos. Soc.*, vol. 46, pp. 469–478.

Imai, I. (1957), "Second Aproximation to the Laminar Boundary Layer Flow over a Flat Plate," *J. Aeronaut. Sci.*, vol. 24, pp. 155–156.

Ingham, D. B., and I. Pop (2002), *Transport Phenomena in Porous Media*, 2 vols., Pergamon, New York.

Jacobson, S. A., and W. C. Reynolds (1998), "Active Control of Streamwise Vortices and Streaks in Boundary Layers," *J. Fluid Mech.* vol. 360, pp. 179–211.

Jaffe, N. A., T. T. Okamura, and A. M. O. Smith (1970), "Determination of Spatial Amplification Factors and Their Application to Predicting Transition," *AIAA J.*, vol. 8, no. 2, pp. 301–308.

Jaluria, Y. (1980), *Natural Convection: Heat and Mass Transfer*, Pergamon, New York.

Jaluria, Y., and K. E. Torrance (1986), *Computational Heat Transfer*, Hemisphere, New York.

Jeffery, G. B. (1915), "The Two-Dimensional Steady Motion of a Viscous Fluid," *Philos. Mag.*, vol. 29, pp. 455–465.

Johnston, J. P. (1960), "On Three-Dimensional Turbulent Boundary Layers Generated by Secondary Flow," *J. Basic Eng.*, vol. 82, pp. 233–248.

—— (1998), "Review: Diffuser Design and Performance Analysis by a Unified Integral Method," *J. Fluids Eng.*, vol. 120, no. 1, pp. 6–18.

Jones, O. C. (1976), "An Improvement in the Calculation of Turbulent Friction in Rectangular Ducts," *J. Fluids Eng.*, vol. 98, pp. 173–181.

Jones, O. C., and J. C. M. Leung (1981), "An Improvement in the Calculation of Turbulent Friction in Smooth Concentric Annuli," *J. Fluids Eng.*, vol. 103, pp. 615–623.

Jones, W. P., and B. E. Launder (1972), "The Prediction of Laminarization with a Two-Equation Model of Turbulence," *Int. J. Heat Mass Transfer*, vol. 15, pp. 301–314.

—— (1973), "The Calculation of Low Reynolds Number Phenomena with a Two-Equation Model of Turbulence," *Int. J. Heat Mass Transfer*, vol. 16, pp. 1119–1130.

Jordinson, R. (1970), "The Flat Plate Boundary Layer, Part 1, Numerical Integration of the Orr–Sommerfeld Equation," *J. Fluid Mech.*, vol. 43, pp. 801–811.

Joseph, D. D. (1976), *Stability of Fluid Motions*, Springer-Verlag, Berlin.

Joslin, R. D., G. Erlebacher, and M. Y. Hussaini (1996), "Active Control of Instabilities in Laminar Boundary Layers: Overview and Concept Validation," *J. Fluids Eng.*, vol. 118, pp. 494–497.

Kader, B. A. (1981), "Temperature and Concentration Profiles in Fully Turbulent Boundary Layers," *Int. J. Heat Mass Transfer*, vol. 24, no. 9, pp. 1541–1544.

Kakaç, S., and Y. Yener (1994), *Convective Heat Transfer*, 2nd ed., CRC Press, Boca Raton, Fla.

Kakaç, S., W. Aung, and R. Viskanta (1985), *Natural Convection: Fundamentals and Applications*, Taylor & Francis, Washington, D.C.

Kamemoto, K., and M. Tsutahara (Eds.) (2000), *Vortex Methods*, World Scientific, River Edge, N.J.

Kaplan, R. E. (1964), "The Stability of Laminar Incompressible Boundary Layers in the Presence of Compliant Boundaries," Aeroelastic and Structures Research Lab. Report. ASRL-TR 166-1, M.I.T., Cambridge, Mass.

Karim, S. M., and L. Rosenhead (1952), "The Second Coefficient of Viscosity of Liquids and Gases," *Rev. Modern Phys.*, vol. 24, pp. 108–116.

Kármán, T. von (1911), "Über den Mechanismus des Widerstandes, den ein bewegter Körper in einer Flüssigkeit erzeugt," *Nachr. Ges. Wiss. Göttingen Math.-Phys. Kl. II*, pp. 509–517, 547–556.

—— (1921) "Über laminare und turbulente Reibung," *Z. Angew. Math. Mech.*, vol. 1, pp. 233–252 (English translation in NACA Technical Memo. 1092).

—— (1939), "The Analogy between Fluid Friction and Heat Transfer," *ASME Trans.*, vol. 61, pp. 705–710.

—— (1964), *The Wind and Beyond*, Little Brown, Boston, Mass.

Karniadakis, G. E., and A. Bestok (2001), *Micro Flows*, Springer-Verlag, New York.
Karniadakis, G. E., and S. J. Sherwin (1999), *Spectral/Hp Element Methods for CFD*, Oxford Univ. Press, New York.
Kaviany, M. (2001), *Principles of Convective Heat Transfer*, 2nd ed., Springer-Verlag, New York.
Kays, W. M., and M. E. Crawford (1993), *Convective Heat and Mass Transfer*, 3rd ed., McGraw-Hill, New York.
Kee, R. J., M. C. Elliott, and P. Glarborg (2003), *Chemically Reacting Flow: Theory and Practice*, Wiley, New York.
Keenan, J. H. (1941), *Thermodynamics*, Wiley, New York.
Kegelman, J. T., and T. J. Mueller (1986), "Experimental Studies of Spontaneous and Forced Transition on an Axisymmetric Body," *AIAA J.*, vol. 24, no. 3, pp. 397–403.
Keh, H. J., and P. Y. Chen (2001), "Slow Motion of a Droplet Between Two Parallel Plane Walls," *Chem. Eng. Sci.*, vol. 56, no. 24, pp. 6863–6871.
Keller, H. B., and H. Takami (1966), "Steady Two-Dimensional Viscous Flow of an Incompressible Fluid Past a Circular Cylinder," in *Proc. Symp. Numerical Solutions of Nonlinear Differential Equations*, pp. 115–140 [see also *Phys. Fluids Supp.*, vol. 12, pp. II-51–II-56 (1969)].
Kemp, N. H., R. H. Rose, and R. W. Detra (1959), "Laminar Heat Transfer Around Blunt Bodies in Dissociated Air," *J. Aerosp. Sci.*, vol. 26, pp. 421–430.
Kennard, E. H. (1938), *Kinetic Theory of Gases*, McGraw-Hill, New York.
Kestin, J., and L. N. Persen (1962), "The Transfer of Heat Across a Turbulent Boundary Layer at Very High Prandtl Numbers," *Int. J. Heat Mass Transfer*, vol. 5, pp. 355–371.
Kestin, J., and W. A. Wakeham (1988), *Transport Properties of Fluids: Thermal Conductivity, Viscosity, and Diffusion Coefficient*, Center for Information and Numerical Data Analysis and Synthesis, Data Series, vol. I-1, Hemisphere, New York.
Kestin, J., P. F. Maeder, and H. H. Sogin (1961), "The Influence of Turbulence on the Transfer of Heat to Cylinders Near the Stagnation Point," *Z. Angew. Math. Phys.*, vol. 12, pp. 115–132.
Khonsari, M. M., and E. R. Booser (2001), *Applied Tribology: Bearing Design and Lubrication*, Interscience, New York.
Kim, K., H. J. Sung, and M. K. Chung (2002), "Assessment of Local Blowing and Suction in a Turbulent Boundary Layer," *AIAA J.*, vol. 40, no. 1, pp. 175–177.
Klebanoff, P. S. (1955), "Characteristics of Turbulence in a Boundary Layer with Zero Pressure Gradient," NACA Report 1247.
Klebanoff, P. S., and K. D. Tidstrom (1972), "Mechanism by Which a Two-Dimensional Roughness Element Induces Boundary Layer Transition," *Phys. Fluids*, vol. 15, no. 7, pp. 1173–1188.
Klebanoff, P. S., K. D. Tidstrom, and L. M. Sargent (1962), "The Three-Dimensional Nature of Boundary Layer Instability," *J. Fluid Mech.*, vol. 12, pp. 1–24.
Kleiser, L., and T. A. Zang (1991), "Numerical Simulation of Transition in Wall-Bounded Shear Flows," *Annu. Rev. Fluid Mech.*, vol. 23, pp. 495–537.
Kline, S. J., B. J. Cantwell, and G. M. Lilley (1982), "The 1980-81 AFOSR-HTTM Stanford Conference on Complex Turbulent Flows: Comparison of Computation and Experiment," vols. 1–3, Mech. Engineering Dept., Stanford Univ., Stanford, Calif.
Kline, S. J., D. E. Abbott, and R. W. Fox (1959), "Optimum Design of Straight-Walled Diffusers," *J. Basic Eng.*, ser. D, vol. 81, pp. 305–320.
Kline, S. J., M. V. Morkovin, G. Sovran, and D. J. Cockrell (1968), "Computation of Turbulent Boundary Layers—1968 AFOSR-IFP Stanford Conference," *Proc. 1968 Conf.*, vol. 1, Stanford Univ., Stanford, Calif.
Knauss, J. A. (1978), *Introduction to Physical Oceanography*, Prentice-Hall, Englewood Cliffs, N.J.
Knowles, C. P., and B. Gebhart (1968), "Stability of the Laminar Natural Convection Boundary Layer," *J. Fluid Mech.*, vol. 34, pp. 657–686.
Kobayashi, R., Y. Kohama, and Ch. Takamadate (1980), "Spiral Vortices in Boundary Layer Transition Regime on a Rotating Disk," *Acta Mech.*, vol. 35, pp. 71–82.
Kohr, M., and I. Pop (Eds.) (2004), *Viscous Incompressible Flow: For Low Reynolds Numbers*, WIT Press/Computational Mechanics, Boston.

Korkegi, R. H. (1956), "Transition Studies and Skin Friction Measurements on an Insulated Flat Plate at a Mach Number of 5.8," *J. Aerosp. Sci.*, vol. 23, pp. 97–102.
Korobkin, I. (1954), Naval Ordnance Lab., Report No. 3841.
Koschmieder, E. L. (1993), *Bénard Cells and Taylor Vortices*, Cambridge Univ. Press, New York.
Koumoutsakos, P., and A. Leonard (1995), "High-Resolution Simulations of the Flow Around an Impulsively Started Cylinder Using Vortex Methods," *J. Fluid Mech.*, vol. 296, pp. 1–38.
Kovaznay, L. S. G., V. Kibens, and R. F. Blackwelder (1970), "Large Scale Motion in the Intermittent Region of a Turbulent Boundary Layer," *J. Fluid Mech.*, vol. 41, pp. 283–325.
Kral, L. D., and T. A. Zang (Eds.) (1993), *Transitional and Turbulent Compressible Flows—1993*, ASME Fluids Engineering Division, vol. 151, New York.
Kramer, M. O. (1957), "Boundary Layer Stabilization by Distributed Damping," *J. Aeronaut. Sci.*, vol. 24, pp. 459–460.
Kramers, H. (1946), "Heat Transfer from Spheres to Flowing Media," *Physica*, vol. 12, pp. 61–80.
Kreiss, H.-O., and J. Lorenz (1989), *Initial-Boundary Value Problems and the Navier–Stokes Equations*, Academic, New York.
Kreyszig, E. (1999), *Advanced Engineering Mathematics*, 8th ed., Wiley, New York.
Kuehn, T. H., and R. J. Goldstein (1980), "Numerical Solution to the Navier–Stokes Equations for Laminar Natural Convection about a Horizontal Isothermal Circular Cylinder," *Int. J. Heat Mass Transfer*, vol. 23, pp. 971–980.
Lachmann, G. V. (1961), *Boundary Layers and Flow Control*, vols. 1–2, Pergamon, London.
Lachowicz, J. T., N. Chokani, and S. P. Wilson (1996), "Boundary Layer Stability Measurements in a Hypersonic Quiet Tunnel," *AIAA J.*, vol. 34, pp. 2496–2500.
Ladyzhenkaya, O. A. (1969), *The Mathematical Theory of Viscous Incompressible Flow*, 2nd ed., Gordon and Breach, New York.
Lai, W. M., E. Krempl, and D. Rubin (1995), *Introduction to Continuum Mechanics*, 3rd ed., Butterworth-Heinemann, Woburn, Mass.
Lamb, H. (1932), *Hydrodynamics*, 6th ed., Cambridge Univ. Press, London (reprinted by Dover in 1945).
Lamb, P. (Ed.) (1987), *Proc. Tenth U.S. Natl. Congr. Appl. Mech.*, Amer. Soc. Mech. Engineers, New York.
Landahl, M. T., and E. Mollo-Christensen (1992), *Turbulence and Random Processes in Fluid Mechanics*, 2nd ed., Cambridge Univ. Press, New York.
Landau, L. D., and E. M. Lifschitz (1959), *Fluid Mechanics*, Pergamon, London.
Laney, C. B. (1998), *Computational Gasdynamics*, Cambridge Univ. Press, New York.
Langlois, W. E. (1964), *Slow Viscous Flow*, Macmillan, New York.
Lapidus, L., and G. F. Pinder (1999), *Numerical Solution of Partial Differential Equations in Science and Engineering*, Wiley Interscience, New York.
Lauchle, G. C., and G. B. Gurney (1984), "Laminar Boundary Layer Transition on a Heated Underwater Body," *J. Fluid Mech.*, vol. 144, pp. 79–101.
Launder, B. E., B. J. Reece, and W. Rodi (1975), "Progress in the Development of a Reynolds Stress Turbulence Closure," *J. Fluid Mech.*, vol. 68, pp. 537–566.
Launder, B., and N. Sandham, (Eds.) (2001), *Closure Strategies for Turbulent and Transitional Flows*, Cambridge Univ. Press, New York.
Lee, C. (2002), "Large Eddy Simulation of Rough-Wall Turbulent Boundary Layers," *AIAA J.*, vol. 40, no. 10, pp. 2127–2130.
Lee, J. H. W., and V. Chu (2003), *Turbulent Jets and Plumes: A Lagrangian Approach*, Kluwer Academic, New York.
Lees, L. (1956), "Laminar Heat Transfer over Blunt-Nosed Bodies at Hypersonic Flight Speeds," *Jet Propul.*, vol. 26, pp. 259–269, 274.
Lees, L., and C. C. Lin (1946), "Investigation of the Stability of the Laminar Boundary Layer in a Compressible Fluid," NACA Technical Note 1115.
Lefebvre, P. J., and F. M. White (1989), "Experiments on Transition to Turbulence in a Constant-Acceleration Pipe Flow," *J. Fluids Eng.*, vol. 111, pp. 428–432.
Lefebvre, P. J., and F. M. White (1991), "Further Experiments on Transition to Turbulence in Constant-Acceleration Pipe Flow," *J. Fluids Eng.*, vol. 113, pp. 223–227.

Legendre, R. (1965), "Lignes de courant d'un écoulement continu," *Rech. Aérospat.*, vol. 105, pp. 3–9.
Leibenson, L. S. (1935), "The Energy Form of the Integral Condition in the Theory of the Boundary Layer," *Dokl. Akad. Nauk SSSR*, vol. 2, pp. 22–24.
Lesieur, M. and O. Métais (1996), "New Trends in Large Eddy Simulations of Turbulence," *Annu. Rev. Fluid Mech.*, vol. 28, pp. 45–82.
Lewis, J. A., and G. F. Carrier (1949), "Some Remarks on the Flat Plate Boundary Layer," *Q. Appl. Math.*, vol. 8, pp. 228–234.
Libby, P. A. (1996), *An Introduction to Turbulence*, Taylor & Francis, New York.
Libby, P. A., and T. M. Liu (1967), "Further Solutions of the Falkner–Skan Equation," *AIAA J.*, vol. 5, pp. 1040–1042.
Liepmann, H. W. (1943), "Investigations on Laminar Boundary-Layer Stability and Transition on Curved Boundaries," NACA Wartime Report W107 (ACR3H30) [see also NACA Technical Memo. 1196 (1947) and NACA Report 890 (1947)].
Lim, F. J., and Schowalter, W. R. (1989), "Wall Slip of Narrow Molecular Weight Distribution Polybutadienes," *J. Rheol.*, vol. 33, no. 8, pp. 1359–1382.
Lin, C. C. (1955), *The Theory of Hydrodynamic Stability*, Cambridge Univ. Press, London.
Lin, C.-S., and Q.-S. Han (1991), "Boundary Element Analysis of Low Reynolds Number Viscous Fluid Flow," in *Computational Mechanics*, A. A. Balkema, Hong Kong, pp. 1481–1486.
Lindgren, E. R. (1965), "Experimental Study on Turbulent Pipe Flows of Distilled Water," Oklahoma State Univ., Civil Engineering Dept., Report 1AD621071.
Lingwood, R. J. (1995), "Absolute Instability of the Boundary Layer on a Rotating Disk," *J. Fluid Mech.*, vol. 299, pp. 17–23.
—— (1996), "An Experimental Study of Absolute Instability of a Rotating-Disk Boundary Layer," *J. Fluid Mech.*, vol. 314, pp. 373–405.
List, E. J. (1982), "Turbulent Jets and Plumes," *Annu. Rev. Fluid Mech.*, vol. 14, pp. 189–212.
Little, W. J. (1963), *Arnold Engr. Dev. Ctr.*, Report AEDC-TDR-63-190, Tullahoma, Tenn.
Lock, R. C. (1951), "The Velocity Distribution in the Laminar Boundary Layer Between Parallel Streams," *Q. J. Mech. Appl. Math.*, vol. 4, pp. 42–63.
Löhner, R. (2001), *Applied CFD Techniques: An Introduction Based on Finite Element Methods*, Wiley, New York.
Lomax, H., T. H. Pulliam, and D. Zingg (2001), *Fundamentals of Computational Fluid Dynamics*, Springer-Verlag, New York.
Loos, H. G. (1955), "A Simple Laminar Boundary Layer with Secondary Flow," *J. Aeronaut. Sci.*, vol. 22, pp. 35–40.
Lucquin, B., and O. Pironneau (1998), *Introduction to Scientific Computing*, Wiley, New York.
Ludwieg, H., and W. Tillmann (1949), "Untersuchungen über die Wandschubspannung in Turbulenten Reibungsschichten," *Ing.-Arch.*, vol. 17, pp. 288–299 (translated as NACA Technical Memo. 1285).
Lumley, J. L. (1989), "The State of Turbulence Research," in W. K. George and R. Arndt (Eds.), *Advances in Turbulence*, pp. 1–10, Hemisphere, New York.
Mack, L. M. (1954), "An Experimental Investigation of the Temperature Recovery Factor," Calif. Inst. Tech. Report 20-80, Pasadena, Calif.
—— (1969), "Boundary Layer Stability Theory," Document 900-277, Rev. A, Jet Propulsion Lab., Pasadena, Calif.
—— (1977), "Transition Prediction and Linear Stability Theory," pp. 1-1–1-22 of AGARD (1977).
—— (1977a), "Transition and Laminar Instability," Publication 77-15, Jet Propulsion Lab., Pasadena, Calif.
—— (1984), "Boundary Layer Stability Theory," pp. 3-1–3-81 of AGARD (1984).
—— (1987), "Review of Linear Compressible Stability Theory," in D. L. Dwoyer and M. Y. Hussaini (Eds.), *Stability of Time Dependent and Spatially Varying Flows*, pp. 164–187, Springer, New York.
Malik, M. R. (1986), "The Neutral Curve for Stationary Disturbances in Rotating Disk Flow," *J. Fluid Mech.*, vol. 164, pp. 275–287.

Malvern, L. E. (1997), *Introduction to Mechanics of a Continuous Medium*, Prentice-Hall, Upper Saddle River, N.J.
Mangler, W. (1945), "Boundary Layers on Bodies of Revolution in Symmetrical Flow," *Ber. Aerodyn. Versuchsanst. Goett.*, Report 45/A/17.
—— (1948), "Zusammenhang zwischen ebenen und rotationsymmetrischen Grenzschichten in kompressiblen Flüssigkeiten," *Z. Angew. Math. Mech.*, vol. 28, p. 97–103.
Mathieu, J., and J. Scott (2000), *An Introduction to Turbulent Flow*, Cambridge Univ. Press, New York.
Mattingly, G. E., and W. O. Criminale (1972), "The Stability of an Incompressible Two-Dimensional Wake," *J. Fluid Mech.*, vol. 51, pp. 233–272.
Maxwell, C. (1860), *Scientific Papers of Clerk Maxwell*, vol. 2, p. 1, Dover, New York.
Mayle, R. E. (1991), "The Role of Laminar-Turbulent Transition in Gas Turbine Engines," *J. Turbomachinery*, vol. 113, October, pp. 509–537.
Meier, H. U., and P. Bradshaw (Eds.) (1987), *Perspectives in Turbulence Studies*, Springer, Berlin.
Meksyn, D. (1961), *New Methods in Laminar Boundary Layer Theory*, Pergamon, London.
Mellor, G. L., and D. M. Gibson (1966), "Equilibrium Turbulent Boundary Layers," *J. Fluid Mech.*, vol. 24, pp. 225–253.
Menna, J. D., and F. J. Pierce (1988), "The Mean Flow Structure Around and Within a Turbulent Junction or Horseshoe Vortex—Part I," *J. Fluids Eng.*, vol. 110, pp. 406–414 [see also Part II, pp. 415–423].
Merzkirch, W. (1974), *Flow Visualization*, Academic, New York.
—— (1987), *Flow Visualization*, 2nd ed., Elsevier, New York.
Meseguer, A. (2003), "Streak Breakdown Instability in Pipe Poiseuille Flow," *Phys. Fluids*, vol. 15, no. 5, pp. 1203–1213.
Messiter, A. F. (1970), "Boundary Layer Flow near the Trailing Edge of a Flat Plate," *SIAM J. Appl. Math.*, vol. 18, pp. 241–257.
Michaelides, E. E. (2003), "Hydrodynamic Force and Heat/Mass Transfer from Particles, Bubbles, and Drops—The Freeman Scholar Lecture," *J. Fluids Eng.*, vol. 125, March, pp. 209–238.
Michel, R. (1952), "Etude de la transition sur les profils d'aile-establissment d'un point de transition et calcul de la trainée de profil en incompressible," ONERA Report No. 1/1578A.
Middleman, S. (1995), *Modeling Axisymmetric Flows: Dynamics of Films, Jets, and Drops*, Elsevier Science & Technology, New York.
Miles, J. P., and M. Farrashkhalvat (2003), *Basic Structural Grid Generation*, Butterworth-Heinemann, Woburn, Mass.
Mills, R. D. (1968), "Numerical Solutions of Viscous Flow through a Pipe Orifice at Low Reynolds Number," *J. Mech. Eng. Sci.*, vol. 10, no. 2, pp. 133–140.
Millsaps, K., and K. Pohlhausen (1953), "Thermal Distribution in Jeffery–Hamel Flows between Nonparallel Walls," *J. Aeronaut. Sci.*, vol. 20, pp. 187–196.
Milne, W. E. (1953), *Numerical Solution of Differential Equations*, Wiley, New York.
Milne-Thomson, L. M. (1968), *Theoretical Hydrodynamics*, 5th ed., Macmillan, New York.
Minkowycz, W. J., E. M. Sparrow, G. E. Schneider, and R. H. Pletcher (1988), *Handbook of Numerical Heat Transfer*, Wiley, New York.
Miyake, Y., K. Tsujimoto, and M. Nakaji (2001), "Direct Numerical Simulation of Rough-Wall Heat Transfer in a Turbulent Channel Flow," *Int. J. Heat Fluid Flow*, vol. 22, no. 3, pp. 237–244.
Moffat, R. J., and W. M. Kays (1984), "A Review of Turbulent Boundary Layer Heat Transfer Research at Stanford, 1958–1983," *Adv. Heat Transfer*, vol. 16, pp. 241–365.
Moin, P. (2001), *Fundamentals of Engineering Numerical Analysis*, Cambridge Univ. Press, New York.
Moin, P., and K. Mahesh (1998), "Direct Numerical Simulations: A Tool in Turbulence," *Annu. Rev. Fluid Mech.*, vol. 30, pp. 539–578.
Monin, A. S., and A. M. Yaglom (1972), *Statistical Fluid Mechanics*, 2 vols., M.I.T. Press, Cambridge, Mass.

Monnoyer, F. (1997), "Hypersonic Boundary Layer Flows," Chap. 12 of *Flows at Large Reynolds Numbers*, H. Schmitt (Ed.), WIT Press, Computational Mechanics, Boston.
Moody, L. F. (1944), "Friction Factors for Pipe Flow," *ASME Trans.*, vol. 66, pp. 671–684.
Moore, F. K. (1958), "On the Separation of the Unsteady Laminar Boundary Layer," in *Boundary Layer Research*, H. G. Görtler (Ed.), Springer, Berlin.
Moretti, P. M., and W. M. Kays (1965), "Heat Transfer to a Turbulent Boundary Layer with Varying Freestream Velocity and Varying Surface Temperature: An Experimental Study," *Int. J. Heat Mass Transfer*, vol. 8, pp. 1187–1202.
Morgan, K. (2004), *Introduction to Computational Fluid Dynamics*, Cambridge Univ. Press, New York.
Morkovin, M. V. (1962), "Effects of Compressibility on Turbulent Flows," in A. Favre (Ed.), *Mécanique de la Turbulence*, CNRS, Paris, pp. 367–380.
—— (1969), "Critical Evaluation of Transition from Laminar to Turbulent Shear Layers with Emphasis on Hypersonically Traveling Bodies," AFFDL-TR-68-149, Wright-Patterson AFB, Ohio [see also C. S. Wells (1969), pp. 1–31].
—— (1969a), "The Many Faces of Transition," in *Viscous Drag*, C. S. Wells (Ed.), Plenum Press, New York.
Morris, P. J., G. Raman, and D. McLaughlin (Eds.) (2002), *High Speed Jet Flows: Fundamentals and Applications*, Taylor & Francis, New York.
Morrison, G. L. (2003), "Euler Number Based Orifice Discharge Coefficient Relationship," *J. Fluids Eng.*, vol. 125, January, pp. 189–191.
Moses, H. L., R. R. Jones III, and J. F. Sparks (1979), "An Integral Method for the Turbulent Boundary Layer with Separated Flows," in *Turbulent Boundary Layers*, ASME FED vol. G00145, pp. 69–73.
Mueller, T. (Ed.) (2004), *Flow Visualization XI: Proceedings 11th Int. Symp.*, CD-ROM, Univ. of Notre Dame, South Bend, Ind.
Muggli, F. A. (1997), "Flow Analysis in a Pump Diffuser–Part 2: Validation and Limitations of CFD for Diffuser Flows," *J. Fluids Eng.*, vol. 119, no. 4, pp. 978–984.
Nachtsheim, P. R. (1963), "Stability of Free Convection Boundary Layer Flow Investigation by Integration of Disturbance Differential Equations," National Aeronautics and Space Administration Technical Note D-2089.
—— (1964), "An Initial Value Method for the Numerical Treatment of the Orr–Sommerfeld Equation for the Case of Plane Poiseuille Flow," National Aeronautics and Space Administration Technical Note D-2414.
Nakayama, A. (1995), *PC-Aided Numerical Heat Transfer and Convective Flow*, CRC Press, Boca Raton, Fla.
Nakayama, A., and H. R. Rahai (1984), "Measurement of Turbulent Flow Behind a Flat Plate Mounted Normal to the Wall," *AIAA J.*, vol. 22, pp. 1817–1819.
Nakayama, Y. (ed.) (1988), *Visualized Flow*, Pergamon, Oxford.
Nakayama, Y., and Y. Tanida (Eds.) (1996), *Atlas of Flow Visualization, Vol. II*, CRC Press, Boca Raton, Fla.
Nansen, F. (1902), "The Oceanography of the North Polar Basin: The Norwegian North Polar Expedition," *Sci. Res. (Christianaia)*, vol. 3, p. 357.
Narasimha, R., and K. R. Sreenivasan (1979), "Relaminarization of Fluid Flows," *Adv. Appl. Mech.*, vol. 19, pp. 221–309.
Navier, C. L. M. H. (1823), "Mémoire sur les lois du mouvement des fluides," *Mem. Acad. R. Sci. Paris*, vol. 6, pp. 389–416.
Nayfeh, A. H. (2000), *Perturbation Methods*, Wiley, New York.
Newman, B. G. (1951), "Some Contributions to the Study of the Turbulent Boundary Layer Near Separation," Aust. Dept. Supply, Report ACA-53.
Nikuradse, J. (1933), "Strömungsgesetze in rauhen Rohren," *Forsch. Arb. Ing.-Wes.*, no. 361.
Obi, S., K. Aoki, and S. Masuda (1993), "Experimental and Computational Study of Turbulent Separating Flow in an Asymmetric Plane Diffuser," *Proc. 9th Symp. Turbulent Shear Flows*, pp. 305–312.

Obot, N. T. (1988), "Determination of Incompressible Flow Friction in Smooth Circular and Noncircular Passages: A Generalized Approach Including Validation of the Nearly Century-Old Hydraulic Diameter Concept," *J. Fluids Eng.*, vol. 110, pp. 431–440.
Obremski, H. J., and A. A. Fejer (1967), "Transition in Oscillating Boundary Layer Flow," *J. Fluid Mech.*, vol. 29, pp. 93–111.
Obremski, H. J., and M. V. Morkovin (1969), "Application of a Quasi-Steady Stability Model to Periodic Boundary Layer Flows," *AIAA J.*, vol. 7, no. 7, pp. 1298–1301.
Obremski, H. J., et al. (1969), "A Portfolio of Stability Characteristics of Incompressible Boundary Layers," AGARDograph No. 134, NATO, Paris.
Ockendon, H., and J. R. Ockendon (1995), *Viscous Flow*, Cambridge Univ. Press, New York.
—— (2004), *Waves and Compressible Flow*, Springer-Verlag, New York.
Oleinik, O. A., and V. N. Samokhin (1999), *Mathematical Models in Boundary Layer Theory*, CRC Press, Boca Raton, Fla.
Oljaca, M., and J. Sucec (1997), "Prediction of Transpired Turbulent Boundary Layers with Arbitrary Pressure Gradients," *J. Fluids Eng.*, vol. 119, Sept., pp. 526–532.
Oosthuizen, P. H. (1999), *Introduction to Convective Heat Transfer*, McGraw-Hill, New York.
Oosthuizen, P. H., and W. E. Carscallen (1997), *Compressible Fluid Flow*, McGraw-Hill, New York.
Oran, E. S., C. K. Oh, and B. Z. Cybyk (1998), "Direct Simulation Monte Carlo: Recent Advances and Applications," *Annu. Rev. Fluid Mech.*, vol. 30, pp. 403–441.
Orr, W. M'F. (1907), "The Stability or Instability of the Steady Motions of a Perfect Liquid and of a Viscous Liquid," *Proc. R. Irish Acad. Sect. A*, vol. 27, pp. 9–68; 69–138.
Orszag, S. A., R. Glowinski, and C. A. J. Fletcher (1991), *Computational Techniques for Fluid Dynamics: Fundamental and General Techniques*, Springer-Verlag, New York.
Oseen, C. W. (1910), "Über die Stokes'sche Formel und über eine verwandte Aufgabe in der Hydrodynamik," *Ark. f. Math. Astron. och. Fys.*, vol. 6, no. 29.
—— (1927), "Neure Methoden und Ergebnisse in der Hydrodynamik," Akademische Verlag, Geest & Portig, Leipzig.
Osterlund, J. M. (1999), "Experimental Studies of Zero-Pressure-Gradient Turbulent Boundary Layer Flow," Ph.D. Dissertation, Dept. of Mechanics, Royal Institute of Technology, Stockholm, Sweden.
Osterlund, J. M., A. V. Johanssen, and H. M. Nagib (2000), "Comment on 'A Note on the Intermediate Region in Turbulent Boundary Layers," *Phys. Fluids*, vol. 12, no. 9, pp. 2360–2363.
Ostrach, S. (1953), "An Analysis of Laminar Free-Convection Flow and Heat Transfer about a Flat Plate Parallel to the Direction of the Generating Body Force," NACA Report 1111.
—— (1982), "Low Gravity Fluid Flows," *Annu. Rev. Fluid Mech.*, vol. 14, pp. 313–345.
Owens, E. J., and G. Thodos (1957), "Thermal Conductivity Reduced-State Correlation for the Inert Gases," *AIChE J.*, vol. 3, pp. 454–461.
Owens, R. G., and T. N. Phillips (2002), *Computational Rheology*, Imperial College Press, London.
Pai, S. I. (1954), *Fluid Dynamics of Jets*, Van Nostrand, New York.
Paneras, A. G. (1997), "On the Calculation of Turbulent Incompressible Flow about a 6:1 Prolate Spheroid at High Incidence," in *Advances in Fluid Mechanics*, vol. 11, pp. 323–360, Computational Mechanics, Boston.
Panton, R. L. (1996), *Incompressible Flow*, 2nd ed., Wiley, New York.
Papanastasiou, T., A. N. Alexandrou, and G. Georgiou (1999), *Viscous Fluid Flow*, CRC Press, Boca Raton, Fla.
Park, C. (1990), *Nonequilbrium Hypersonic Aerothermodynamics*, Wiley, New York.
Parry, W. T., J. C. Bellows, J. S. Gallagher, and A. H. Harvey (2000), *ASME International Steam Tables for Industrial Use*, ASME Press, New York (software also available).
Patankar, S. V. (1980), *Numerical Heat Transfer and Fluid Flow*, Hemisphere, New York.
Patel, V. C. (1998), "Perspective: Flow at High Reynolds Numbers and Over Rough Surfaces—Achilles Heel of CFD," *J. Fluids Eng.*, vol. 120, pp. 434–444.
Patel, V. C., W. Rodi, and G. Scheuerer (1985), "Turbulence Models for Near-Wall and Low Reynolds Number Flows: A Review," *AIAA J.*, vol. 23, no. 9, pp. 1308–1319.
Peaceman, D. W., and H. H. Rachford (1955), "The Numerical Solution of Parabolic and Elliptic Differential Equations," *J. Soc. Ind. Appl. Math.*, vol. 3, pp. 28–41.

Petukhov, B. S. (1970), "Heat Transfer and Friction in Turbulent Pipe Flow with Variable Physical Properties," *Adv. Heat Trans.*, vol. 6, pp. 504–564.
Peyret, R. (2002), *Spectral Methods for Incompressible Viscous Flow*, Springer-Verlag, New York.
Peyret, R., and T. D. Taylor (1983), *Computational Methods for Fluid Flow*, Springer, New York.
Pfenninger, W., and V. D. Reed (1966), "Laminar Flow Research and Experiments," *Astronaut. Aeronaut.*, vol. 4, no. 7, pp. 44–50.
Phillips, O. M. (1978), *Dynamics of the Upper Ocean*, 2nd ed., Cambridge Univ. Press, New York.
Piercy, N. A. V., M. S. Hooper, and H. F. Winny (1933), "Viscous Flow through Pipes with Cores," *Philos. Mag.*, vol. 15, no. 7, pp. 647–676.
Pipes, L. A. (1958), *Applied Mathematics for Engineers and Physicists*, McGraw-Hill, New York.
Piquet, J., J. A. Richards, and X. Jia (2001), *Turbulent Flows: Models and Physics*, Springer-Verlag, New York.
Pironneau, O. (1989), *Finite Element Methods for Fluids*, Masson, Paris.
Pirro, D. M., J. G. Wills, and A. Wessol (2001), *Lubrication Fundamentals*, Marcel Dekker, New York.
Plapp. J. E. (1957), "The Analytic Study of Laminar Boundary Layer Stability in Free Convection," *J. Aeronaut. Sci.*, vol. 24, pp. 318–319.
Plesniak, M. W., and J. P. Johnston (1988), "The Effects of Stabilizing and Destabilizing Curvature on a Plane Mixing Layer," in M. Hirata, and N. Kasagi (Eds.), *Transport Phenomena in Turbulent Flows*, pp. 377–390, Hemisphere, New York.
Pletcher, R. H. (1978), "Prediction of Incompressible Turbulent Separating Flows," *J. Fluids Eng.*, vol. 100, pp. 427–433.
—— (1988), "Progress in Turbulent Forced Convection," *J. Heat Transfer*, 50th Anniv. Issue, pp. 1129–1144.
Pohlhausen, E. (1921), "Der Wärmeaustausch zwischen festern Körpen und Flüssigkeiten mit kleiner Reibung und kleiner Wärmeleitung," *Z. Angew. Math. Mech.*, vol. 1, pp. 115–121.
Poiseuille, J. L. M. (1840), "Recherches expérimentelles sur le mouvement des liquids dans les tubes de trés petits diamétres," *Comptes Rendus*, vol. 11, pp. 961–967; 1041–1048.
Polderman, H. G., G. Velraeds, and W. Knol (1986), "Turbulent Lubrication Flow in an Annular Channel," *J. Fluids Eng.*, vol. 108, pp. 185–192.
Polymeropoulos, C. E., and B. Gebhart (1967), "Incipient Instability in Free Convection Boundary Layers," *J. Fluid Mech.*, vol. 30, pp. 225–239 [see also *AIAA J.*, vol. 4, p. 2066–2068 (1966)].
Pope, S. B. (2000), *Turbulent Flows*, Cambridge Univ. Press, New York.
Prandtl, L. (1904), "Über Flüssigkeitsbewegung bei sehr kleiner Reibung," *Proc. Third Int. Math. Congr. Heidelberg* [English translation in NACA Technical Memo. 452].
—— (1921), "Bemerkund über die Entstehung der Turbulenz," *Z. Angew. Math. Mech.*, vol. 1, pp. 431–436.
—— (1925), "Über die ausgebildete Turbulenz," *Z. Angew. Math. Mech.*, vol. 5, pp. 136–139.
—— (1945a), "Über ein neues Formelsystem für die ausgebildete Turbulenz," *Nachr. Akad. Wiss. Göttingen Math.–Phys. Kl. II*, p. 6.
—— (1945b), "Über Reibungsschicht bei dreidimensionalen Strömunden," *Betz-Festschrift 1945*, pp. 134–141.
—— (1961), *Collected Works*, vols. 1–3, Springer, Berlin.
Present, R. D. (1958), *Kinetic Theory of Gases*, McGraw-Hill, New York.
Profilo, G., G. Soliani, and C. Tebaldi (1998), "Some Exact Solutions of the Two-Dimensional Navier–Stokes Equations," *Int. J. Eng. Sci.*, vol. 36, no. 4, pp. 459–471.
Proudman, I., and J. R. A. Pearson (1957), "Expansions at Small Reynolds Numbers for the Flow Past a Sphere and a Circular Cylinder," *J. Fluid Mech.*, vol. 2, pp. 237–262.
Raffel, M., C. E. Willert, and J. Kompenhaus (1998), *Particle Image Velocimetry: A Practical Guide*, Springer-Verlag, New York.
Ragab, S. A., and A. H. Nayfeh (1982), "A Comparison of the Second Order Triple-Deck Theory with Interacting Boundary Layers," in T. Cebeci (Ed.), *Numerical and Physical Aspects of Aerodynamic Flows*, p. 237, Springer, New York [see also AIAA Paper 80–0072, 1980].
Rasmussen, M. (1994), *Hypersonic Flow*, Interscience, New York.
Raupach, M. R., R. A. Antonia, and S. Rajagopalan (1997), "Rough Wall Turbulent Boundary Layers," *Appl. Mech. Rev.*, vol. 44, pp. 1–25.

Rayleigh, Lord (1880), *Scientific Papers*, vol. 1, pp. 474–487, Dover, New York, 1964.
—— (1916), "On the Dynamics of Revolving Fluids," *Proc. R. Soc. London Ser. A*, vol. 93, pp. 148–154.
Reichardt, H. (1956), "Über die Geschwindigkeitsverteilung in einer geradlinigen turbulenten Couette-Strömmung," *Z. Angew. Math. Mech.*, Sonderheft, vol. 36, pp. 26–29.
Reid, R. C., J. M. Pravsnitz, and T. K. Sherwood, (1987), *The Properties of Gases and Liquids*, 4th ed., McGraw-Hill, New York.
Reiner, M. (1969), *Deformation, Strain and Flow: An Elementary Introduction to Rheology*, 3rd ed., H. K. Lewis, London.
Reshotko, E. (1976), "Boundary Layer Stability and Transition," *Annu. Rev. Fluid Mech.*, vol. 8, pp. 311–349.
—— (1979), "Drag Reduction by Cooling in Hydrogen-Fueled Aircraft," *J. Aircraft*, vol. 16, no. 9, pp. 584–590.
—— (1985), "Control of Boundary Layer Transition," AIAA Paper 85-0562, March 12–14, Boulder, Colo.
—— (1987), "Stability and Transition, How Much Do We Know?," in Lamb (Ed.), *Proc. Tenth U.S. Natl. Congr. Appl. Mech.*, pp. 421–434, Amer. Soc. Mech. Engineers, New York.
—— (1988), "Stability and Transition of Boundary Layers," in H. Branover, M. Mond, and Y. Unger (Eds.), *Progress in Astronautics and Aeronautics*, vol. 112, pp. 278–311, AIAA, Washington, D.C.
Reynolds, O. (1874), "On the Extent and Action of the Heating Surface for Steam Boilers," *Manchester Lit. Philos. Soc.*, vol. 14, pp. 7–12.
—— (1883), "On the Experimental Investigation of the Circumstances Which Determine Whether the Motion of Water Shall be Direct or Sinuous, and the Law of Resistance in Parallel Channels," *Philos. Trans. R. Soc. London Ser. A*, vol. 174, pp. 935–982.
—— (1886), "On the Theory of Lubrication and Its Application to Mr. Beauchamp Tower's Experiments Including an Experimental Determination of the Viscosity of Olive Oil," *Philos. Trans. R. Soc. London Ser. A*, vol. 177, pp. 157–234.
—— (1895), "On the Dynamical Theory of Incompressible Viscous Fluids and the Determination of the Criterion," *Philos. Trans. R. Soc. London Ser. A*, vol. 186, pp. 123–164.
Reynolds, W. C., W. M. Kays, and S. J. Kline (1958), "Heat Transfer in the Turbulent Incompressible Boundary Layer: Part 1, Constant Wall Temperature; Part 2, Step-Wall Temperature Distribution; Part 3, Arbitrary Wall Temperature and Heat Flux; Part 4, Effect of Location of Transition and Prediction of Heat Transfer in a Known Transition Region," NACA Memos. No. 12-1-58W–12-4-58W.
Riahi, D. N. (2000), *Flow Instability*, WIT Press/Computational Mechanics, Boston.
Richardson, E. G., and E. Tyler (1929), "The Transverse Velocity Gradients near the Mouth of a Pipe in Which an Alternating or Continuous Flow of Air is Established," *Proc. Phys. Soc. London*, vol. 42, pp. 1–15.
Richardson, J., and H. Power (1996), "Boundary Element Analysis of Creeping Flow Past Two Porous Bodies of Arbitrary Shape," *Eng. Anal. Boundary Elements*, vol. 17, no. 3, pp. 193–204.
Rist, U., and H. Fasel (1995), "Direct Numerical Simulation of Controlled Transition in a Flat-Plate Boundary Layer," *J. Fluid Mech.*, vol. 298, pp. 211–248.
Rivkin, S. L. (1988), *Thermodynamic Properties of Gases*, 4th ed., Hemisphere, New York.
Roache, P. J. (1976), *Computational Fluid Dynamics*, revised ed., Hermosa Press, Albuquerque.
Robertson, J. M. (1965), *Hydrodynamics in Theory and Application*, Prentice-Hall, Englewood Cliffs, N.J.
Rodi, W. (1976), "A New Algebraic Relation for Calculating Reynolds Stress," *Z. Angew. Math. Mech.*, vol. 56, pp. 331–340.
—— (1984), *Turbulence Models and Their Application in Hydraulics*, Brookfield, Brookfield, Vt.
Rogers, D. F. (1992), *Laminar Flow Analysis*, Cambridge Univ. Press, New York.
Rogers, M. G., and G. N. Lance (1960), "The Rotationally Symmetric Flow of a Viscous Fluid in the Presence of an Infinite Rotating Disk," *J. Fluid Mech.*, vol. 7, pp. 617–631.
Roll, H. U. (1965), *Physics of the Marine Atmosphere*, International Geophysics Series, vol. 7, Academic, New York.
Rose, P. H., and W. I. Stark (1958), "Stagnation Point Heat Transfer Measurements in Dissociated Air," *J. Aeronaut. Sci.*, vol. 25, pp. 86–97.

Rosenhead, L. (1940), "The Steady Two-Dimensional Radial Flow of Viscous Fluid Between Two Inclined Planes," *Proc. R. Soc. London Ser. A*, vol. 175, pp. 436–467.
Rosenhead, L. (Ed.) (1963), *Laminar Boundary Layers*, Oxford Univ. Press, London.
Ross, J. A., and F. H. Barnes (1970), "The Flat Plate Boundary Layer, Part 3, Comparison of Theory with Experiment," *J. Fluid Mech.*, vol. 43, pp. 819–832.
Rott, N. (1964), "Theory of Time-Dependent Laminar Flows," *in High Speed Aerodynamics and Jet Propulsion*, Princeton Univ. Press, vol. IV, pp. 395–438.
Rott, N., and L. F. Crabtree (1952), "Simplified Laminar Boundary Layer Calculation for Bodies of Revolution and for Yawed Wings," *J. Aeronaut. Sci.*, vol. 19, pp. 553–565.
Rotta, J. C. (1956), "Experimenteller Beitrag zur Entstehung turbulenter Strömung im Rohr," *Ing.-Arch.*, vol. 24, pp. 258–281.
—— (1986), "Experience of Second Order Turbulent Closure Models," *Zeit. Flugwiss. Weltraumforsch.*, vol. 10, pp. 401–407.
Roumeliotis, J., and G. R. Fulford (2000), "Droplet Interactions in Creeping Flow," *Comput. Fluids*, vol. 29, no. 4, pp. 435–450.
Rouse, H., and S. Ince (1957), *History of Hydraulics*, State Univ. of Iowa, Institute of Hydraulic Research, Iowa City, Ia.
Rybczynski, W. (1911), "Über die fortschreitende Bewegung einer flüssigen Kugel in einem zähen Medium," *Bull. Int. Acad. Sci. Cracov.*, vol. 1911A, pp. 40–46.
Sachdev, P. L. (2000), *Self-Similarity and Beyond: Exact Solutions of Nonlinear Problems*, CRC Press, Boca Raton, Fla.
Sagaut, P., and M. Germano (2002), *Large Eddy Simulation for Incompressible Flows: An Introduction*, Springer-Verlag, New York.
Sandham, N. D., and W. C. Reynolds (1987), "Some Inlet Plane Effects on the Numerically Simulated Spatially Developing Two-Dimensional Mixing Layer," *Proc. Sixth Symp. Turbulent Shear Flow*, pp. 22-4-1–22-4-6, Toulouse, France.
Saric, W. S., and A. S. Nayfeh (1975), "Nonparallel Stability of Boundary Layer Flows," *Phys. Fluids*, vol. 18, pp. 945–950.
Saric, W. S., and A. S. W. Thomas (1984), "Experiments on the Subharmonic Route to Turbulence in Boundary Layers," in T. Tatsumi (Ed.), *Turbulence and Chaotic Phenomena in Fluids*, pp. 117–122, Elsevier, New York.
Saric, W. S., and H. L. Reed (1987), "Three-Dimensional Stability of Boundary Layers," in H. U. Meier, and P. Bradshaw (Eds.), *Perspectives in Turbulence Studies*, pp. 71–92, Springer, Berlin.
Saric, W. S., H. L. Reed, and E. J. Kershen (2002), "Boundary-Layer Receptivity to Freestream Disturbances," *Annu. Rev. Fluid Mech.*, vol. 34, pp. 291–319.
Sasmal, G. P., and J. I. Hochstein (1994), "Marangoni Convection with a Curved and Deforming Surface in a Cavity," *J. Fluids Eng.*, vol. 116, Sept., pp. 577–582.
Schetz, J. A. (1980), *Injection and Mixing in Turbulent Flow*, AIAA, vol. 69, New York.
—— (1984), *Foundations of Boundary Layer Theory for Momentum, Heat, and Mass Transfer*, Prentice-Hall, Englewood Cliffs, N.J.
—— (1992), *Boundary Layer Analysis*, Prentice-Hall, Upper Saddle River, N.J.
Schlichting, H. (1932), "Berechnung ebener periodischer Grenzschichtströmungen," *Phys. Z.*, vol. 33, pp. 327–335.
—— (1933a), "Laminare Strahlenausbreitung," *Z. Angew. Math. Mech.*, vol. 13, pp. 260–263.
—— (1933b), "Zur Entstehung der Turbulenz bei der Plattenstromung," *Nachr. Ges. Wiss. Göttingen. Math.-Phys. Kl. II*, pp. 182–208.
—— (1979), *Boundary Layer Theory*, 7th ed., McGraw-Hill, New York.
Schlichting, H., and K. Bussmann (1943), "Exakte Losungen für die laminare Grenzschicht mit Absaugung und Ausblasen," *Schr. Deutsch. Akad. Luftfahrtforschung*, ser. B, vol. 7, no. 2.
Schlichting, H., and K. Gersten (2000), *Boundary Layer Theory*, 8th ed., Springer, New York.
Schmid, P. J., and D. S. Henningson (2001), *Stability and Transition in Shear Flows*, Springer, New York.
Schmidt, E., and K. Wenner (1941), "Wärmeabgabe über den Umfang eines angeblasenen geheizten Zylinders," *Forsch. Arb. Ing.-Wes.*, vol. 12, pp. 65–73.

Schmidt, E., and W. Beckmann, with E. Pohlhausen (1930), "Das Temperatur- und Geschwindigkeitsfeld von einer Wärme abgebenden, senkrechten Platte bei natürlicher Konvektion," *Forsch. Arb. Ing.-Wes.*, vol. 1, pp. 391–404.
Schmidt, F. W., and B. Zeldin (1969), "Laminar Flow in the Inlet Section of Tubes and Ducts," *AIChE J.*, vol. 15, pp. 612–614.
Schmitt, H. (Ed.) (1997), *Flow at Large Reynolds Numbers*, vol. 11 of *Advances in Fluid Mechanics*, Computational Mechanics, Boston.
Schoenherr, K. E. (1932). "Resistance of Plates," *Trans. Soc. Nav. Architects Mar. Eng.*, vol. 40.
Schreier, S. (1982), *Compressible Flow*, Willey, New York.
Schubauer, G. B. (1935), "Airflow in a Separating Laminar Boundary Layer," NACA Report 527.
Schubauer, G. B., and H. Skramstad (1947), "Laminar Boundary Layer Oscillations and Transition on a Flat Plate," *J. Res. Nat. Bur. Stand.*, vol. 38, pp. 251–292 [see also NACA Report 909 (1948)].
Schubauer, G. B., and P. S. Klebanoff (1955), "Contributions on the Mechanics of Boundary Layer Transition," NACA Tech. Note 3489 [see also NACA Report 1289].
Scott, J. L., J. A. Brennan, and D. M. Blakeslee (1994), "GRI/NIST Orifice Meter Discharge Coefficient Database, Vol. 1," NIST Standard Reference Database 45, U. S. Dept. of Commerce, Natl. Inst. of Standards, Gaithersburg, Md.
Sears, W. R. (1948), "Boundary Layer of Yawed Cylinders," *J. Aeronaut. Sci.*, vol. 15, pp. 49–52 [see also *Appl. Mech. Rev.*, vol. 7, pp. 281–285 (1954)].
Sedov, L. I. (1959), *Similarity and Dimensional Methods in Mechanics*, Academic, New York.
Sellars, J. R., M. Tribus, and J. S. Klein (1956), "Heat Transfer to Laminar Flow in a Round Tube or Flat Conduit: The Graetz Problem Extended," *ASME Trans.*, vol. 78, pp. 441–448.
Senecal, V. E., and R. R. Rothfus (1953), "Transition Flow of Fluids in Smooth Tubes," *Chem. Eng. Prog.*, vol. 49, pp. 533–538.
Settles, G. S. (2004), *Schlieren and Shadowgraph Techniques; Visualizing Phenomena in Transparent Media*, Springer-Verlag, New York.
Settles, G. S., T. J. Fitzpatrick, and S. M. Bogdonoff (1979), "Detailed Study of Attached and Separated Compression Corner Flowfields in High Reynolds Number Supersonic Flow," *AIAA J.*, vol. 17, no. 6, pp. 579–585.
Sexl, T. (1930), "Über den von E. G. Richardson entdeckten 'Annulareffekt'," *Z. Phys.*, vol. 61, pp. 349–362.
Shah, R. K. (1978), "A Correlation for Laminar-Hydrodynamic Entry Length Solutions for Circular and Noncircular Ducts," *J. Fluids Eng.*, vol. 100, pp. 177–179.
Shah, R. K., and A. L. London (1978), *Laminar Flow Forced Convection in Ducts*, Academic, New York.
Shank, M. E. (1954), "Brittle Failure of Nonship Steel-plate Structures," *Mech. Eng.*, vol. 76, pp. 23–28.
Shapiro, A. H. (1953, 1954), *The Dynamics and Thermodynamics of Compressible Fluid Flow*, vols. 1 and 2, Ronald Press, New York.
Sharatchandra, M. C., M. Sen, and M. Gad-el-Hak (1997), "Navier–Stokes Simulations of a Novel Viscous Pump," *J. Fluids Eng.*, vol. 119, June, pp. 372–382.
Sharipov, F., and D. Kalempa (2003) "Velocity Slip and Temperature Jump Coefficients for Gaseous Mixtures. 1. Viscous Slip Coefficient," *Phys. Fluids*, vol. 15, no. 6, pp. 1800–1806.
Sharipov, F., and V. Seleznev (1998), "Data on Internal Rarefied Gas Flows", *J. Phys. Chem. Ref. Data*, vol. 27, p. 657.
Shen, S. F. (1954), "Calculated Amplified Oscillations in Plane Poiseuille and Blasius Flows," *J. Aeronaut. Sci.*, vol. 21, pp. 62–64.
Sherman, F. S. (1990), *Viscous Flow*, McGraw-Hill, New York.
Shima, N. (1993), "Prediction of Turbulent Boundary Layers with a Second Moment Closure: Part 1. Effects of Periodic Pressure Gradient, Wall Transpiration, and Freestream Turbulence," *J. Fluids Eng.*, vol. 115, no. 1, pp. 56–63.
Sibulkin, M. (1952), "Heat Transfer near the Forward Stagnation Point of a Body of Revolution," *J. Aeronaut. Sci.*, vol. 19, pp. 570–571.
Simpson, R. L. (1968), "The Turbulent Boundary Layer on a Porous Wall," Ph.D. thesis, Stanford Univ., Stanford, Calif.

—— (1989), "Turbulent Boundary Layer Separation," *Annu. Rev. Fluid Mech.*, vol. 21, pp. 205–234.
Simpson, R. L., D. C. Whitten, and R. J. Moffat (1970), "An Experimental Study of the Turbulent Prandtl Number of Air with Injection and Suction," *Int. J. Heat Mass Transfer*, vol. 13, pp. 125–143.
Sirignano, W. A. (1999), *Fluid Dynamics and Transport of Droplets and Sprays*, Butterworth-Heinemann, Woburn, Mass.
Slattery, J. C., and R. B. Bird (1958), "Calculation of the Diffusion Coefficient of Dilute Gases and of the Self-Diffusion Coefficient of Dense Gases," *AIChE J.*, vol. 4, 137–142.
Sleicher, C. A., and M. W. Rouse (1975), "A Convenient Correlation for Heat Transfer to Constant and Variable-Property Fluids in Turbulent Pipe Flow," *Int. J. Heat Mass Transfer*, vol. 18, pp. 677–683.
Smith, A. G., and D. B. Spalding (1958), "Heat Transfer in a Laminar Boundary Layer with Constant Fluid Properties and Constant Wall Temperature," *J. R. Aero. Soc.*, vol. 62, pp. 60–64.
Smith, A. M. O., and D. W. Clutter (1963), "Solution of the Incompressible Boundary Layer Equations," *AIAA J.*, vol. 1, pp. 2062–2071.
—— (1965), "Machine Calculations of Compressible Laminar Boundary Layers," *AIAA J.*, vol. 3, pp. 639–647.
Smith, A. M. O., and N. Gamberoni (1956), "Transition, Pressure Gradient, and Stability Theory," Douglas Aircraft Report ES-26388 [see also *Proc. Ninth Int. Cong. Appl. Mech.*, vol. 4, pp. 234–244 (1957)].
Smith, F. T. (1977), "The Laminar Separation of an Incompressible Fluid Streaming Past a Smooth Surface," *Proc. R. Soc. London A*, vol. 30, pp. 143–156.
Smith, S. H. (1994), "Exact Solution of the Unsteady Navier–Stokes Equations Resulting from a Stretching Surface," *J. Appl. Mech.*, vol. 61, no. 3, pp. 629–633.
Smits, A. J., and J.-P. Dussauge (1996), *Turbulent Shear Layers in Supersonic Flow*, Springer-Verlag, New York.
Smits, A. J., and T. T. Lim (Eds.) (2000), *Flow Visualization: Techniques and Examples*, Imperial College Press, London.
Smoluchowski, M. von (1898), *Sitzungsber. Akad. Wiss. Wien*, vol. 107, Abt. 2a, pp. 304–329; vol. 108, Abt. 2a, pp. 5–23.
So, R. M. C., et al. (1994), "Logarithmic Laws for Compressible Turbulent Boundary Layers," *AIAA J.*, vol. 32, no. 11, pp. 2162–2168.
Sod, G. A. (1985), *Numerical Methods in Fluid Dynamics*, Cambridge Univ. Press, New York.
Sofialidis, D., and P. Primos (1996), "Wall Suction Effects on the Structure of Fully Developed Pipe Flow," *J. Fluids Eng.*, vol. 118, no. 1, pp. 33–39.
—— (1997), "Fluid Flow and Heat Transfer in a Pipe with Wall Suction," *Int. J. Heat Mass Transfer*, vol. 40, no. 15, pp. 3627–3640.
Sokolnikoff, I. S. (1946), *Mathematical Theory of Elasticity*, McGraw-Hill, New York.
Sommerfeld, A. (1908), "Ein Beitrag zur hydrodynamischen Erklaerung der turbulenten Flussigkeitsbewegungen," *Proc. Fourth Int. Cong. Math.*, Rome, vol. III, pp. 116–124.
Sowerby, L. (1954), "Secondary Flow in a Boundary Layer," *Aero. Res. Council London*, Report 16832.
Spain, B. (2003), *Tensor Calculus: A Concise Approach*, Dover, Mineola, N.Y.
Spalart, P. R. (1988), "Direct Simulation of a Turbulent Boundary Layer up to $Re_\theta = 1400$," *J. Fluid Mech.*, vol. 187, pp. 61–98.
Spalding, D. B. (1961), "A Single Formula for the Law of the Wall," *J. Appl. Mech.*, vol. 28, pp. 455–457.
Spalding, D. B., and S. W. Chi (1964), "The Drag of a Compressible Turbulent Boundary Layer on a Smooth Flat Plate with and without Heat Transfer," *J. Fluid Mech.*, vol. 18, pp. 117–143.
Spalding, D.B., and W.M. Pun (1962), "A Review of Methods for Predicting Heat Transfer Coefficients for Laminar Uniform-Property Boundary Layer Flows," *Int. J. Heat Mass Transfer*, vol. 5, pp. 239–250.
Spangler, I. G., and C. S. Wells, Jr. (1968), "Effect of Freestream Disturbances on Boundary Layer Transition," *AIAA J.*, vol. 6, pp. 534–545.

Sparrow, E. M. (1955), "Analysis of Laminar Flow Convection Heat Transfer in the Entrance Region of Flat Rectangular Ducts," NACA Technical Note 3331.
Sparrow, E. M., and J. L. Gregg (1956), "Laminar Free-Convection Heat Transfer from the Outer Surface of a Vertical Circular Cylinder," *ASME Trans.*, vol. 78, pp. 1823–1829 [see also vol. 96, pp. 178–183].
Spells, K. E. (1952), "A Study of Circulation Patterns within Liquid Drops Moving Through a Liquid," *Proc. Phys. Soc.*, ser. B, vol. 65, pp. 541–546.
Speziale, C. G. (1991b), "Analytical Methods for the Development of Reynolds-Stress Closures in Turbulence," *Annu. Rev. Fluid Mech.*, vol. 23, pp. 107–157.
Speziale, C. G., S. Sarkar, and T. B. Gatski (1991a), "Modeling the Pressure-Strain Correlation of Turbulence: An Invariant Dynamical Systems Approach," *J. Fluid Mech.*, vol. 227, pp. 245–272.
Spina, E. F., A. J. Smits, and S. K. Robinson (1994), "The Physics of Supersonic Turbulent Boundary Layers," *Annu. Rev. Fluid Mech.*, vol. 26, pp. 287–319.
Squire, H. B. (1933), "On the Stability of Three-Dimensional Distribution of Viscous Fluid Between Parallel Walls," *Proc. R. Soc.*, ser. A, vol. 142, pp. 621–628.
—— (1951), "The Round Laminar Jet," *Q. J. Mech. Appl. Math.*, vol. 4, pp. 321–329.
Sreenivasan, K. R. (1982), "Laminarescent, Relaminarizing, and Retransitional Flows," *Acta Mech.*, vol. 44, pp. 1–48.
Sreenivasan, K. R., and R. Narasimha (1982), "Equilibrium Parameters for Two-Dimensional Turbulent Wakes," *J. Fluids Eng.*, vol. 104, pp. 167–170.
Steinheuer, J. (1968), "Similar Solutions for the Laminar Wall Jet in a Decelerating Outer Flow," *AIAA J.*, vol. 6, pp. 2198–2200.
Stetson, K. F. (1988), "On Nonlinear Aspects of Hypersonic Boundary Layer Stability," *AIAA J.*, July, pp. 883–885.
Stevenson, T. N. (1963), "A Law of the Wall for Turbulent Boundary Layers with Suction or Injection," Cranfield College, Aero. Report 166.
Stewartson, K. (1954), "Further Solutions of the Falkner–Skan Equation," *Proc. Cambridge Philos. Soc.*, vol. 50, pp. 454–465.
—— (1964), *The Theory of Laminar Boundary Layers in Compressible Fluids*, Oxford Univ. Press. London.
—— (1969), "On the Flow near the Trailing Edge of a Flat Plate II," *Mathematika*, vol. 16, pp. 106–121.
—— (1974), "Multistructured Boundary Layers on Flat Plates and Related Bodies," *Adv. Appl. Mech.*, vol. 14, pp. 145–239.
Stokes, G. G. (1845), "On the Theories of Internal Friction of Fluids in Motion," *Trans. Cambridge Philos. Soc.*, vol. 8, pp. 287–305.
—— (1851), "On the Effect of the Internal Friction of Fluids on the Motion of Pendulums," *Cambridge Philos. Trans.*, vol. 9, pp. 8–106.
Stratford, B. S. (1954), "Flow in the Laminar Boundary Layer Near Separation," *Aero. Res. Council London*, R & M 3002.
—— (1959), "Prediction of Separation of the Turbulent Boundary Layer," *J. Fluid Mech.*, vol. 5, pp. 1–16 [see also pp. 17–35].
Strawn, R. C., and S. J. Kline (1983), "A Stall Margin Design Method for Planar and Axisymmetric Diffusers," *J. Fluids Eng.*, vol. 105, pp. 28–33.
Stuart, J. T. (1987), "Instability, Three-Dimensional Effects, and Transition in Shear Flows," in H. U. Meier, and P. Bradshaw (Eds.), *Perspectives in Turbulence Studies*, pp. 1–25, Springer, Berlin.
Sucec, J. (1999), "Prediction of Heat Transfer in Turbulent Transpired Boundary Layers," *J. Heat Transfer*, vol. 121, February, pp. 186–190.
Sucker, D., and H. Brauer (1975), "Fluiddynamik bei der angeströmten Zylindern," *Wärme und Stoffübertragung*, vol. 8, pp. 149–158.
Sumitani, Y., and N. Kasagi (1995), "Direct Numerical Simulation of Turbulent Transport with Uniform Wall Injection and Suction", *AIAA J.*, vol. 33, no. 7, pp. 1220–1228.
Sutherland, W. (1893), "The Viscosity of Gases and Molecular Force," *Philos. Mag.*, vol. 5, pp. 507–531.
Swinney, H. L., and J. P. Gollub (1981), *Hydrodynamic Instabilities and the Transition to Turbulence*, Springer, New York.

Sychev, V. V. (1972), "Laminar Separation," *Fluid Dynamics*, vol. 7, pp. 407–417.
Szeri, A. Z. (1998), *Fluid Film Lubrication: Theory and Design*, Cambridge Univ. Press, New York.
Szymanski, F. (1932), "Quelques solutions exactes des équations de l'hydrodynamiquede fluide visqueux dans le cas d'un tube cylindrique." *J. Math. Pures Appl.*, (9), vol. 11, pp. 67–107.
Talpaert, Y. R. (2003), *Tensor Analysis and Continuum Mechanics*, Dover, Mineola, N.Y.
Taneda, S. (1956), "Experimental Investigation of the Wakes Behind Cylinders and Plates at Low Reynolds Numbers," *J. Phys. Soc. Jpn.*, vol. 11, pp. 302–307.
―――― (1979), "Visualization of Separating Stokes Flows," *J. Phys. Soc. Jpn.*, vol. 46, pp. 1935–1942.
Tani, I. (1949), "On the Solution of the Laminar Boundary Layer Equations," *J. Phys. Soc. Jpn.*, vol. 4, pp. 149–154.
―――― (1969), "Boundary Layer Transition," *Annu. Rev. Fluid Mech.*, vol. 1, pp. 169–196.
Tannehill, J. C., D. A. Anderson and R. H. Pletcher (1997), *Computational Fluid Mechanics and Heat Transfer*, Taylor & Francis, Washington, D.C.
Tanner, R. I. (1993), "Stokes Paradox for Power-Law Flow Around a Cylinder," *J. Non-Newtonian Fluid Mech.*, vol. 50, no. 2–3, pp. 217–224.
Tatsumi, T. (1952), "Stability of the Laminar Inlet-Flow Prior to the Formation of Poiseuille Regime," *J. Phys. Soc. Jpn.*, vol. 7, pp. 489–495.
Tatsumi, T. (Ed.) (1984), *Turbulence and Chaotic Phenomena in Fluids*, Elsevier, New York.
Taulbee, D. B. (1989), "Engineering Turbulence Models," in W. K. George and R. Arndt (Eds.), *Advances in Turbulence*, pp. 75–125, Hemisphere, New York.
Taylor, G. I. (1923), "Stability of a Viscous Liquid Contained Between Two Rotating Cylinders," *Philos. Trans. R. Soc. London Ser. A*, vol. 223, pp. 289–343.
Taylor, R. P., H. W. Coleman, and B. K. Hodge (1985), "Prediction of Turbulent Rough-Wall Skin Friction Using a Discrete Element Approach," *J. Fluids Eng.*, vol. 107, pp. 251–257.
Taylor, R. P., W. F. Scaggs, and H. W. Coleman (1988), "Measurement and Prediction of the Effects of Nonuniform Surface Roughness on Turbulent Flow Friction Coefficients," *J. Fluids Eng.*, vol. 110, pp. 380–384.
Telionis, D. P. (1981), *Unsteady Viscous Flows*, Springer-Verlag, New York.
Tennekes, H., and J. L. Lumley (1972), *A First Course in Turbulence*, M.I.T. Press, Cambridge, Mass.
Terrill, R. M. (1960), "Laminar Boundary Layer Flow with Separation with and Without Suction," *Philos. Trans. Ser. A*, vol. 253, pp. 55–100.
Theodorsen, T., and A. Regier (1944), "Experiments on Drag of Revolving Discs, Cylinders, and Streamline Rods at High Speeds," NACA Report 793.
Thilmany, J. (2003), "How Does Your Fluid Flow?," *Mech. Eng.*, Dec. 2003, pp. 35–37.
Thom, A. (1933), "Flow Past Circular Cylinders at Low Speeds," *Proc. R. Soc. London Ser. A*, vol. 141, pp. 651–669.
Thomas, L. C., and H. M. Kadry (1990), "A One Parameter Integral Method for Turbulent Boundary Layer Flow," *J. Fluids Eng.*, vol. 112, pp. 433–436.
Thomas, L. C., and S. M. F. Hasani (1989), "Supplementary Boundary-Layer Approximations for Turbulent Flow," *J. Fluids Eng.*, vol. 111, pp. 420–427.
Thompson, J. F., et al. (Eds.) (1998), *Handbook of Grid Generation*, CRC Press, Boca Raton, Fla.
Thwaites, B. (1949), "Approximate Calculation of the Laminar Boundary Layer," *Aeronaut. Q.*, vol. 1, pp. 245–280.
Timoshenko, S., and J. N. Goodier (1970), *Theory of Elasticity*, 3rd ed., McGraw-Hill, New York.
Tobak, M., and D. J. Peake (1982), "Topology of Three-Dimensional Separated Flows," *Annu. Rev. Fluid Mech.*, vol. 14, pp. 61–85.
Tollmien, W. (1929), "Über die Entstehung der Turbulenz," *Nachr. Ges. Wiss. Göttingen Math.-Phys. Kl. II*, pp. 21–44 [translated in NACA Technical Memo. 609].
―――― (1931), "Grenzschichttheorie," *Handbuch der experimentalischen Physik*, vol. IV, pt. 1, pp. 241–287, Leipzig.
Tomotika, S., and T. Aoi (1951), "An Expansion Formula for the Drag on a Circular Cylinder Moving through a Viscous Fluid at Small Reynolds Numbers," *Q. J. Mech. Appl. Math.*, vol. 4, pp. 401–406.
Tomotika, S., T. Aoi, and H. Yosinabu (1953), "On the Forces Acting on a Circular Cylinder Set Obliquely in a Uniform Stream at Lower Values of Reynolds Number," *Proc. R. Soc. London Ser. A*, vol. 219, pp. 233–244.

Townsend, A. A. (1984), "Axisymmetric Couette Flow at Large Taylor Numbers," *J. Fluid Mech.*, vol. 144, pp. 329–362.
Trefethen, A. E., L. N. Trefethen, and P. J. Schmid (1999), "Spectra and Pseudospectra for Pipe Poiseuille Flow," *Comput. Methods Appl. Mech. Eng.*, vol. 175, pp. 413–420.
Trefethen, L. N. (2001), *Spectral Methods in MATLAB*, Society for Industrial & Appl. Math., Philadelphia.
Trefethen, L. N., A. E. Trefethen, S. C. Reddy, and T. A. Driscoll (1993), "Hydrodynamic Stability Without Eigenvalues," *Science*, vol. 261, pp. 578–584.
Tretheway, D. C., and C. D. Meinhart (2002), "Apparent Fluid Slip at Hydrophobic Microchannel Walls," *Phys. Fluids*, vol 14, no. 3, pp. L9–L12.
Trimmer, L. L. (1968), Arnold Engineering Development Center, Tenn., Report AEDC-TR-68–99.
Tritton, D. J. (1959), "Experiments on the Flow Past a Circular Cylinder at Low Reynolds Numbers," *J. Fluid Mech.*, vol. 6, pp. 547–567.
Trogdon, S. A., and M. T. Farmer (1991), "Unsteady Axisymmetric Creeping Flow from an Orifice," *Acta Mech.*, vol. 88, no. 1–2, pp. 61–75.
Truesdell, C. (1954), "The Present Status of the Controversy Regarding the Bulk Viscosity of Liquids," *Proc. R. Soc. London Ser. A*, vol. 226, pp. 1–69.
Truitt, R. W. (1960), *Aerodynamic Heating*, Ronald Press, New York.
Tsangaris, S., and N. W. Vlachakis (2003), "Exact Solution of the Navier–Stokes Equations for the Fully Developed, Pulsating Flow in a Rectangular Duct with a Constant Cross-Sectional Velocity," *J. Fluids Eng.*, vol. 125, March, pp. 382–385.
Tsou, F. K., and E. M. Sparrow (1970), "Hydrodynamic Stability of Boundary Layers with Surface Mass Transfer," *Appl. Sci. Res.*, vol. 22, pp. 273–286.
Tzabiras, G. (Ed.) (2003), *Calculation for Complex Turbulent Flows*, WIT Press/Computational Mechanics, Boston.
Uchida, S. (1956), "The Pulsating Viscous Flow Superposed on the Steady Laminar Motion of Incompressible Fluid in a Circular Pipe," *Z. Angew. Math. Phys.*, vol. 7, pp. 403–422.
Uyehara, O. A., and K. M. Watson (1944), "High Pressure Vapor-Liquid Equilibria," *Natl. Pet. News*, vol. 36, pp. R623-R635 [see also O. A. Hougen, and K. M. Watson, *Chemical Process Principles Charts*, Wiley, New York, 1960].
Vainshtein, P., P. Shapiro, and C. Gutfinger (2002), "Creeping Flow Past and Within a Permeable Spheroid," *Int. J. Multiphase Flow*, vol. 28, no. 12, pp. 1945–1963.
Van Dommelen, L. L., and S. F. Shen (1981), "The Spontaneous Generation of the Singularity in a Separating Laminar Boundary Layer," *J. Comp. Phys.*, vol. 38, pp. 125–140.
—— (1982), "The Genesis of Separation," Chap. 17 of Cebeci (1982).
Van Driest, E. R. (1951), "Turbulent Boundary Layer in Compressible Fluids," *J. Aeronaut. Sci.*, vol. 18, pp. 145–160.
—— (1952a), "Investigation of Laminar Boundary Layer Compressible Fluids Using the Crocco Method," NACA Technical Note 2597.
—— (1952b), "Turbulent Boundary Layer on a Cone in a Supersonic Flow at Zero Angle of Attack," *J. Aeronaut. Sci.*, vol. 19, pp. 55–57, 72.
—— (1956), "The Problem of Aerodynamic Heating," *Aero. Eng. Rev.*, vol. 15, no. 10, pp. 26–41 [see also *J. Aeronaut. Sci.*, vol. 23, pp. 1007–1011, 1036, 1956].
—— (1956a), "On Turbulent Flow Near a Wall," *J. Aeronaut. Sci.*, vol. 23, pp. 1007–1011.
—— (1959), C. C. Lin (Ed.), *Turbulent Flows and Heat Transfer*, pp. 339–427, Princeton Univ. Press, Princeton, N.J.
Van Driest, E. R., and C. B. Blumer (1963), "Boundary Layer Transition, Freestream Turbulence, and Pressure Gradient Effects," *AIAA J.*, vol. 1, pp. 1303–1306.
Van Dyke, M. (1964), *Perturbation Methods in Fluid Mechanics*, Academic, New York [reprinted in 1975 by Parabolic Press, Stanford, Calif.].
—— (1969), "Higher Order Boundary Layer Theory," *Annu. Rev. Fluid Mech.*, vol. 1, pp. 265–292.
—— (1982), *An Album of Fluid Motion*, Parabolic Press, Stanford, Calif.
Van Ingen, J. L. (1956), "A Suggested Semi-Empirical Method for the Calculation of the Boundary Layer Transition Region," Inst. of Tech., Dept. of Aeronautics and Eng., Report VTH-74, Delft, Holland.

Van Oudheusden, B. W. (1997), "Complete Crocco Integral for Two-Dimensional Laminar Boundary Layer Flow over an Adiabatic Wall for Prandtl Numbers Near Unity," *J. Fluid Mech.*, vol. 353, Dec. 25, pp. 313–330.
—— (2004), "Compressibility Effects on the Extended Crocco Relation and the Thermal Recovery Factor in Laminar Boundary Layer Flow," *J. Fluids Eng.*, vol. 126, January, pp. 32–41.
Varley, E., and B. R. Seymour (1994), "Applications of Exact Solutions to the Navier–Stokes Equations: Free Shear Layers," *J. Fluid Mech.*, vol. 274, pp. 267–291.
Versteeg, H. K., and W. Malalasekera (1996), *An Introduction to Computational Fluid Dynamics: The Finite Volume Method*, Addison-Wesley, Reading, Mass.
Volino, R. J., and T. W. Simon (1997), "Velocity and Temperature Profiles in Turbulent Boundary Layer Flows Experiencing Streamwise Pressure Gradients," *J. Heat Transfer*, vol. 119, no. 3, pp. 433–439.
Volker, J. (2003), *Large Eddy Simulation of Turbulent Incompressible Flows*, Springer-Verlag, New York.
Wagner, R. D., and M. C. Fischer (1984), "Fresh Attack on Laminar Flow," *Aerosp. Am.*, vol. 22, no. 3, pp. 72–76.
Walker, J. E., G. A. Whan, and R. R. Rothfus (1957), "Fluid Friction in Noncircular Ducts," *AIChE J.*, vol. 3, pp. 484–489.
Walz, A. (1969), *Boundary Layers of Flow and Temperature*, M.I.T. Press, Cambridge, Mass.
Wang, C. Y. (1990), "Exact Solutions of the Navier–Stokes Equations: The Generalized Beltrami Flows, Review and Extension," *Acta Mech.*, vol. 81, no. 1–2, pp. 69–74.
Wang, K. C. (1997), "Features of Three-Dimensional Separation and Separated Flow Structure," Chap. 1 of *Flows at Large Reynolds Numbers*, H. Schmitt (Ed.), WIT Press/Computational Mechanics, Boston.
Warnack, D., and H. H. Fernholz (1998), "Effects of a Favorable Pressure Gradient and of the Reynolds Number on an Incompressible Axisymmetric Turbulent Boundary Layer. Part 2. The Boundary Layer with Relaminarization," *J. Fluid Mech.*, vol. 359, March 25, pp. 357–381.
Watson, J. (1958), "A Solution of the Navier–Stokes Equations, Illustrating the Response of a Laminar Boundary Layer to a Given Change in the External Stream Velocity," *Q. J. Mech.*, vol. 11, pp. 302–325.
Wazzan, A. R. (1975), "Spatial Stability of Tollmien–Schlichting Waves," *Prog. Aerosp. Sci.*, vol. 16, no. 2, pp. 99–127.
Wazzan, A.R., C. Gazley, Jr., and A.M.O. Smith (1979), "Tollmien–Schlichting Waves and Transition," *Prog. Aerosp. Sci.*, vol. 18, no. 2, pp. 351–392.
—— (1981), "H-R_x Method for Predicting Transition," *AIAA J.*, vol. 19, no. 6, pp. 810–812.
Wazzan, A. R., H. Taghavi, and G. Keltner (1984), The Effect of Mach Number on the Spatial Stability of Adiabatic Flat Plate Flow to Oblique Disturbances," *Phys. Fluids*, vol. 27, no. 2, pp. 331–341.
Wazzan, A. R., T. T. Okamura, and A. M. O. Smith (1968a), "The Stability of Water Flow over Heated and Cooled Flat Plates," *J. Heat Transfer*, vol. 90, pp. 109–114.
—— (1968b), "Spatial and Temporal Stability Charts for the Falkner–Skan Boundary Layer Profiles," Douglas Aircraft Co. Report DAC-67086.
Weber, H. C. (1939), *Thermodynamics for Chemical Engineers*, Wiley, New York.
Weissberg, H. L. (1959), "Laminar Flow in the Entrance Region of a Porous Pipe," *Phys. Fluids*, vol. 2, pp. 510–516.
Wells, C. S. (1967), "Effects of Freestream Turbulence on Boundary Layer Transition," *AIAA J.*, vol. 5, no. 1, pp. 172–174.
Wells, C. S. (Ed.) (1969), *Viscous Drag Reduction*, Plenum Press, New York.
Wesseling, P. (2001), *Principles of Computational Fluid Dynamics*, Springer, New York.
White, F. M. (1959), "Laminar Flow in Porous Ducts," Ph.D. Thesis, Georgia Institute of Technology.
—— (1962a), "Laminar Flow in a Uniformly Porous Tube," *J. Appl. Mech.*, vol. 29, pp. 201–204.
—— (1962b), "Hypersonic Laminar Viscous Interactions on Inclined Flat Plates," *J. Am. Rocket Soc.*, vol. 32, pp. 780–781.
—— (1969), "A New Integral Method for Analyzing the Turbulent Boundary Layer with Arbitrary Pressure Gradient," *J. Basic Eng.*, vol. 91, pp. 371–378.

―― (1988), *Heat and Mass Transfer*, Addison-Wesley, Reading, Mass.
―― (2003), *Fluid Mechanics*, 5th ed., McGraw-Hill, New York.
White, F. M., B. F. Barfield, and M. J. Goglia (1958), "Laminar Flow in a Uniformly Porous Channel," *J. Appl. Mech.*, vol. 25, pp. 613–617.
White, F. M., and G. H. Christoph (1972), "A Simple Theory for the Two-Dimensional Compressible Turbulent Boundary Layer," *J. Basic Eng.*, vol. 94, pp. 636–642.
White, F. M., R. C. Lessmann, and G. H. Christoph (1973), "Calculations of Turbulent Heat Transfer and Skin Friction," *AIAA J.*, vol. 11, pp. 1046–1048.
Whitfield, J. D., and F. A. Iannuzzi (1969), "Experiments on Roughness Effects on Cone Boundary Layer Transition up to Mach 16," *AIAA J.*, vol. 7, pp. 465–470.
Wieghardt, K., and W. Tillmann (1951), "On the Turbulent Friction Layer for Rising Pressure," NACA Technical Memo. 1314.
Wieselberger, C. (1914), "Der Luftwiderstand von Kugeln," *Z. Flugtech. Motor*, vol. 5, pp. 140–144 [see also *Z. Phys.*, vol. 22, pp. 321–328; vol. 23, pp. 219–224 (1921)].
Wilcox, D. C. (1998), *Turbulence Modeling for CFD*, 2nd ed., DCW Industries, La Cañada, Calif.
Wilcox, D. C., and M. W. Rubesin (1980), "Progress in Turbulence Modeling for Complex Flow Fields Including Effects of Compressibility," NASA TP–1517.
Wild, J. M. (1949), "The Boundary Layer of Yawed Infinite Wings," *J. Aeronaut. Sci.*, vol. 16, pp. 41–45.
Wilke, C. R. (1950), "A Viscosity Equation for Gas Mixtures," *J. Chem. Phys.*, vol. 18, pp. 517–519 [see also *Ind. Eng. Chem.*, vol. 41, pp. 1345–1347 (1949)].
Wrobel, L. C. (2002), *The Boundary Element Method*, vol. 1, Wiley, New York.
Wu, J. C. (1961), "On the Finite Difference Solution of Laminar Boundary Layer Problems," *Proc. 1961 Heat Transfer Fluid Mech. Inst.*, pp. 55–69, Stanford Univ. Press, Stanford, Calif.
Wygnanski, I., F. Champagne, and B. Marasli (1986), "On the Large Scale Structures in Two-Dimensional, Small-Deficit Turbulent Wakes," *J. Fluid Mech.*, vol. 168, pp. 31–71.
Wygnanski, I., and H. Fiedler (1969), "Some Measurements in the Self-Preserving Jet," *J. Fluid Mech.*, vol. 38, pt. 3, pp. 577–612.
Wygnanski, I., M. Sokolov, and D. Friedman (1976), "On a Turbulent 'Spot' in a Laminar Boundary Layer," *J. Fluid Mech.*, vol. 78, pt. 4, pp. 785–819.
Yakhot, V., and L. M. Smith (1992), "The Renormalization Group, The ε-Expansion, and Derivation of Turbulence Models," *J. Sci. Computing*, no. 7, pp. 35–61.
Yang, W.-J. (1989), *Handbook of Flow Visualization*, Hemisphere, New York.
―― (1994), *Computer-Assisted Flow Visualization*, Begell House, New York.
Yoon, K. T., and T. J. Chung (1996), "Three-Dimensional Mixed Explicit-Implicit Generalized Galerkin Spectral Element Methods for High-Speed Turbulent Compressible Flows," *Comput. Methods Appl. Mech. Eng.*, vol. 135, Sept, 1, pp. 343–367.
Yuan, S. W., and A. B. Finkelstein (1956), "Laminar Pipe Flow with Injection and Suction through a Porous Wall," *ASME Trans.*, vol. 78, pp. 719–724.
Zagarola, M. V., and A. J. Smits (1998), "Mean Flow Scaling in Turbulent Pipe Flow," *J. Fluid Mech.*, vol. 373, pp. 33–79.
Zukauskas, A., and J. Ziukzda (1985), *Heat Transfer on a Cylinder in Crossflow*, Hemisphere, New York.
Zwarts, F. J. (1970), "The Compressible Turbulent Boundary Layer in a Pressure Gradient," thesis, McGill Univ., Montreal.

INDEX

Acceleration of a fluid, 17
  relaminarization, 472–473
Accommodation coefficient, 48
Acoustic streaming, 320–321
ADI method, 189
Adiabatic wall temperature, 515, 535
Adverse pressure gradient, 229, 240, 357, 465–466
  in equilibrium flow, 421
  prediction of, 465–466, 561–566
Airfoil, flow past, 5–8
Air-water interface, 51, 252–253
Algebraic stress model, 454
Angle of attack, 6
Angular momentum, 95
Angular velocity of a fluid, 20
Annulus flow, 114–115
Applets for boundary layers, 272, 538, 561
Asymptotic expansions, 300–307
  composite functions, 304
  inner and outer solutions, 304–306
  matching principle, 303–305
  triple-deck theory, 306–307
Axisymmetric boundary layers, 291–300
  cone flow, 291–293, 539–540, 559–561
  along a cylinder, 295–297
  round jet, 212, 297–299
  round wake, 299–300
  sphere flow, 294–295
  Thwaites-type method, 293–295

Baldwin–Lomax model, 447
BASIC Runge–Kutta subroutine, 587–588

Bernoulli equation, 69, 85, 226, 412, 506
Bessel function roots, 126
Biharmonic equation, 88, 166
Bingham plastic, 24
Blasius equation, 231, 512
  compressible laminar flow, 512
  flat-plate formulas, 221, 233, 235
  linearized stability, 352–353
  similarity solution, 231–235
Blasius pipe friction, 110, 426
Blowing (*see* Suction flows)
Blowing parameter, 249, 492, 501
Body forces, 62–63
Boltzmann constant, 39
Boundary conditions, 45–54, 74–75
  in boundary-layer theory, 227, 291, 412, 506, 546
  for finite-difference methods, 191, 192, 535
  in free convection, 323
  at a free liquid surface, 49–51, 74
  at an inlet or exit, 53–54, 74–75
  at a liquid-vapor interface, 51–52, 74
  for the Orr–Sommerfeld equation, 347
  at a permeable wall, 49
  at a solid surface, 45–46, 74
Boundary-element methods, 177–178
Boundary-fitted coordinates, 202–204
Boundary layer
  concept of, 3, 149, 215–216, 221, 226–227
  on a curved wall, 228–229
  equations of motion, laminar, 225–228, 506
  turbulent, 411–413, 544–546

618 VISCOUS FLUID FLOW

Boundary layer—*Cont.*
  Prandtl's simplified model, 94
  stability, 344–357
  thickness, definition, 216
    on a flat plate, 220, 233, 434, 520
    on an oscillating plate, 133
    on a rotating disk, 158
    on a suddenly moved plate, 131
    in stagnation flow, 149, 152
    with turbulent pressure gradients, 454–461, 561–566
Boussinesq viscosity model, 441, 485
Brinkman number, 100, 104
Buffer layer, 419–420
Bulk modulus, 43, 44
  of various liquids, 576
  of water, 44
Bulk viscosity coefficient, 66
Buoyancy of various fluids, 322
Bypass transition, 378, 387

Canonical equation of state, 37
Cartesian coordinates, 76
Cavitation number, 84
Centrifugal instability, 368–370
Chaos and turbulence, 392–394
Chapman–Rubesin parameter, 516, 517, 528
Chemical reactions, 16
Circular pipe flow, laminar, 13, 108–110
  roughness values, 432
  turbulent, 425–427, 429–433
Classification, of equations, 59–60
  of solutions to Navier–Stokes, 96–98
Clauser parameter, 330, 420, 456, 459
Clauser eddy viscosity model, 211, 443, 476
Clauser's experiments, 422, 465–466, 499, 501
Coherent structures, 400–401
Colebrook friction formula, 432
Coles' law (*see* Law of the wake)
Collision diameter and integrals, 27, 33
Complementary error function, 130
Complex turbulent flows, 469–473
Compliant boundaries, 366–367
Compressibility factor, 40–41, 551

Compressibility transformation
  for laminar flow, 511–514
  for turbulent flow, 548–549, 555, 558
Compressible viscous flow
  basic equations, 506
  laminar boundary layers, 505–507
  Morkovin's hypothesis, 544–545, 548
  stability, 362–366
  turbulent boundary layers, 552–566
Computational fluid dynamics, 183–204
  of laminar boundary layers, 271–278
    in axisymmetric flow, 294
    for compressible flow, 531–539
    for temperatures, 281–283
  of the Navier–Stokes equations, 190–196
    illustrative examples, 196–204, 538–539
    pressure correction, 195–196
    primitive variables, 192–194
    stream function-vorticity approach, 191–192
  for one-dimensional unsteady flow, 187–189
  for turbulent boundary layers, 531–538
Concentration gradient, 31
Conditioned sampling, 404
Conduction thickness, 279
Cone rule, supersonic
  laminar, 539–540
  turbulent, 559–561
Conformal mapping for laminar duct flow, 114
Conservation laws, 59–60
  of energy, 69–72, 91
    integral forms, 90–91, 413–414, 529–530
  of mass, 60–61, 90
  of momentum, 62–64, 90
Continuity equation, 61
  for turbulent flow, 407
Control of transition, 391–392
Control-volume formulations, 89–92
  flat-plate boundary-layer analysis, 216–225
  wall-temperature change, 222–224
Convective acceleration, 17
Convective heat transfer
  laminar, 278–287
  turbulent, 485–498

Coriolis effects, 89, 142
Corner flow, 315–316
Corresponding states, principle of, 25
Couette flows, 98–106
  combined with Poiseuille flow, 110–112
  between parallel plates, 98–101, 133–135
  between rotating cylinders, 103–105, 209
  between sliding cylinders, 101–103
  stability of, 105–106, 357–359, 368–370
  unsteady, 133–135
Crank–Nicolson method, 332
Creeping motion, 88, 165–183
  about a cylinder, 175–176
  about a disk, 171
  about a droplet or bubble, 177
  in a duct entrance, 290
  about a flat plate, 175
  heat transfer in, 178–180
  about immersed bodies, 167–174
  in Jeffrey–Hamel flow, 211
  lubrication theory, 180–183
  through an orifice, 201
  Oseen, improvement of, 174–176
  about a sphere, 168–171
  about a spheroid, 171–172
Critical point correlations
  for the equation of state, 40
  for gas compressibility, 40
  for self-diffusion, 33
  for thermal conductivity, 29
  for viscosity, 25–26
Critical point data for various fluids, 577
Critical Reynolds number concept, 337
Crocco–Busemann relation
  laminar flow, 309, 508–510
  turbulent flow, 546–547
Cup-mixing temperature, 119, 489
Curved-wall boundary layers, 228
Curvilinear coordinates, 75–77, 228, 308–309
  for rotationally symmetric flow, 290
Cylinder flow
  normal to the freestream, 8–11, 269–271
    drag, 175–176
    finite-difference solution, 196–198, 277
    heat transfer, 179, 283–284
    integral solution, 269–271, 283
    parallel to the freestream, 295–297
    rotating, 104
Cylindrical polar coordinates, 76, 92, 582–583

d'Alembert paradox, 2, 9
Damping factor, van Driest, 443, 445, 496
Darcy friction factor, 109, 426
Decay of a vortex, 207, 209
Defect thickness, 420
  velocity profiles in outer layer, 416–421
Deformation of a fluid element, 18–19, 65–66
  newtonian description, 65
Density of air and water, 575
  of various fluids, 576
Diffuser flow, 12–13, 503–504
Diffusion, 31–34, 78–80
  velocity, 32
Diffusivity, 32, 79
Digital computer solutions (*see* Computational fluid dynamics)
Dilatant fluid, 24
Dilatation, 20, 61
Dimensionless parameters, 81–84
  in the boundary conditions, 83–84
  in the equations of motion, 81–82
  in free convection, 82–83, 322
  in a turbulent boundary layer, 415–418
  variables defined, 81
Direct numerical simulation
  of transition, 382–383
  of turbulence, 401, 441, 453
Discharge coefficient of an orifice, 201–202
Discriminant function, 77
Displacement thickness
  definition, 149, 217, 529
  in Falkner–Skan flows, 243
  in flat-plate flow, 220, 233
  in laminar hypersonic flow, 541, 567
  in stagnation flow, 149
  in turbulent flow, 413, 423
Dissipation, 72, 100, 410, 509
  function, 72
  integral, 264, 414

Dissipation—*Cont.*
  thickness, 264
  turbulent, 410
Dissociation of gases, 41
Disturbance equation, 339, 341, 345
Drag coefficient, 170, 219, 235, 436
Drag crisis, 11
Drag force
  in axial cylinder flow, 295–297
  of a cylinder in cross flow, 11, 176
  of a disk, 171
  of a droplet or bubble, 177, 199–200
  of a flat-plate
    laminar, 218, 235, 237, 306, 520
    turbulent, 436–437, 555
  of a sphere, 170, 176, 199–200
Duct flows
  laminar, steady, 106–125
    inlet effects, 287–290
    noncircular, 112–118
    porous, 137–140
    temperature distribution, 118–125
    thermal entrance problem, 120–125
  turbulent, 425–428
    noncircular, 428–429
  unsteady, 125–129
    oscillatory, 127–129
    starting flow, 125–127
Dynamic pressure, 109

Eccentric annulus, 114–115
Eckert number, 82, 226
Eckert reference temperature, 517
Eddy coefficients
  of conductivity, 485–487, 546
  of viscosity, 211, 441–445, 546
Effective diameter concept, 429
Efflux principle, 2
Eigenvalues in the Graetz problem, 122
  in stability theory, 342, 348
Einstein summation convention, 409
Ekman spiral flow, 142–144, 211
Elliptic
  cylinder flow, 261, 315
  differential equation, 77–78
  duct flow, 113
  integral, 163
Emmons spots, 371, 374–377

Energy
  definition, 36, 69
  equation of, 69–72, 91, 225, 506
    for turbulence, 409, 485, 506, 545
    integral relation, 222–223, 263–264, 278, 413
Enthalpy of a fluid, 36, 72
Enthalpy thickness, 223
Entrainment, integral, 414, 457
  in jet flow, 255, 298, 475
  shape factor, 457
Entropy, 36–37
Entry flow in ducts, 106, 120, 287–290
  circular pipe, 12, 106, 288
  friction factor, 288
  thermal-entrance length, 124
Equation of state, 36–37, 507
  of a perfect gas, 39–40, 507, 546
  of water, 55
Equations of motion, 73, 97
  in cylindrical coordinates, 582–583
  in general orthogonal coordinates, 75–77, 228, 308–309
  in spherical polar coordinates, 583–584
  for turbulent flow, 406–411, 545
Equilibrium turbulent flow, 419, 422, 465–467
  Clauser parameter, 330, 420, 456, 459
Error function, 130
Euler's equation, 69, 309
Eulerian coordinates, 16
Evaporation condition, 52
Explicit finite-difference method
  for the diffusion equation, 188
  for laminar boundary layers, 272–276, 281–283, 532–534

Falkner–Skan equation, 240, 514
  application to heat transfer, 279–281
  for compressible flow, 512, 526–528
  nonuniqueness, 244
  Nusselt numbers, 245–246
  relation to cone flows, 291–292
  solutions, 243, 526–528
  stability of, 353–354
  transition, 383, 386
  wedge flow application, 239–246
Fanning friction factor, 109

Favorable pressure gradient, 149, 229, 240, 417, 551
   effect on heat transfer, 247
   effect on the law of the wall, 551
   in Jeffrey–Hamel flow, 165
   in stagnation flow, 149, 155
Favre turbulence averaging, 544
Fick's law, 32
Film flow, 207, 208
Finite-difference methods (see Computational fluid dynamics)
Finite-element methods, 184, 190, 203
First law of thermodynamics, 36, 69, 91, 409
Flat-plate flow
   Blasius solution, 231, 512
     asymptotic expansion, 305–306
   compressible, laminar, 514–521, 535–536
   turbulent, 553–559
   differential analysis, 231–239
   laminar flow, integral analysis, 216–225
     with suction or blowing, 248–249
     developing, 259–261
     three-dimensional, 311–314
   turbulent flow, 433–438, 554–559
     fluctuations, 400
     roughness effects, 437–438
     visualization, 399
Fluid, definition of, 15
FORTRAN solutions
   Falkner–Skan wedge flows, 242
   flat-plate flow, 233, 238
   recovery factor, 515
   rotating disk, 157
   Runge–Kutta subroutines, 583–589
   stagnation flow, 147, 153
Fourier law of conduction, 28–29, 70
Fractals in turbulence, 393, 404
Free convection, 43, 82, 94, 321–327
   basic equations, 323
   on a horizontal cylinder, 327
   on inclined plates, 326
   stability of, 367–368
   on a vertical cylinder, 327
   on a vertical plate, 94, 323–326
Free surface, 49–51, 74
Free turbulence (see Free-shear flows)
Free-shear flows
   laminar, 251–257

    mixing layer, 251–253
    plane jet, 253–256
    plane wake, 256–257
    round jet, 212, 297–299
    round wake, 299–300
    stability of, 360–361
   turbulent, 473–485
    mixing layer, 401, 479–481, 482
    plane jet, 476–477
    round jet, 477–479
    wake, 481–485
Freestream disturbances, 384
Freestream turbulence, 180, 284, 385
   effect on transition, 385–387
Friction factor, definition, 100, 109, 219, 426
   applications (see Skin friction)
Friedrichs, model equation of, 301–303, 334
Froude number, 84, 226
Fully developed duct flow, 12, 107
Fully rough flow, 430, 431, 438

Gamma functions, 285, 494
Gas constant, 39, 507
   of various fluids, 576
Gaussian velocity distribution, 257, 299, 483
   intermittency profiles, 405
Geophysical flows, 88–89, 141–144
Goldstein singularity, 258–259
Görtler vortices, 180, 370
Graetz duct-temperature problem, 120–125
Graetz number, 124
Grashof number, 83, 321–322
   modified for heat flux, 326
Grid selection for CFD, 191
Gruschwitz, method of, 530–531

Hagen–Poiseuille flow, 108
Heat, definition, 36, 69
Heat flux, turbulent inertia term, 409
Heat transfer, 28
   on a circular cylinder, 179, 283–284
   compressible flow parameter, 551
   in Falkner–Skan wedge flow, 527
   on a flat plate, 238–239, 517, 520, 555
   in incompressible turbulent flow, 485–498
   in stagnation flow, 153–154, 522, 525

Heat transfer—*Cont.*
  with suction and blowing, 248–249, 491–493
  with variable wall temperature, 284–287, 493–495
Heat-conduction equation, 78, 129, 257
Heat-transfer coefficient, definition, 101, 284
  applications (*see* Stanton number)
Helmholtz vorticity equation, 86
Historical outline, 1–4
Hodograph, laminar flow, 311–312
  for a rotating disk, 334
  for turbulent flow, 503
Horseshoe vortex, 317
Hot-wire anemometer, 180
Howarth, linearly retarded flow of, 258–259, 275–276
  compressible form, 535–537
  by the method of Thwaites, 268–269
Hydraulic diameter concept, 116
  laminar, 115–118
  turbulent, 428
Hydraulically smooth wall, 430
Hydrostatics, 64–65
  pressure, 65
Hyperbolic differential equation, 77–78
Hypersonic flow
  inviscid theory, 524, 541
  leading-edge interaction, 540–543
  viscous interaction parameter, 542

Ideal gas (*see* Perfect gas)
Illingworth transformation, 511–514
  modification by Lees, 528
Implicit finite-difference method
  for the diffusion equation, 188–189
  for laminar boundary layers, 276–278, 282–283, 534–535
Incompressible flow, 68, 73, 84
Independence principle for swept wings, 313, 315
Inner layer, turbulent, 418–419
Inner-variable analysis, 457–460
  for compressible turbulent flow, 556–558, 563–564
  for flat-plate turbulent flow, 434–436, 556–558

  for pressure gradient flow, 458–460, 563–564
Integral methods
  compressible flow, 529–531, 561–564
  inner-variable type, 457–460, 563–564
  inverse type, 467–469
  laminar, momentum, 264–271, 529–530
    temperature, 278–281
  turbulent, compressible, 561–564
    incompressible, 454–461
Integral relations for boundary layers
  energy, 223, 263–264, 413
  entrainment, 457–458
  laminar, compressible, 529–531
    incompressible, 261–268
  mechanical energy, 264, 414
  momentum, 218, 262–263, 413, 529
  turbulent, compressible, 547, 554, 561
    incompressible, 413–414
Intermittency factor, 372, 402, 404, 405, 444, 479
Invariants of a tensor, 21
Inverse integral method, 467–469
Inviscid flow (*see* Potential flow)
Irrotational flow, 20, 85, 93
Isentropic compressible flow, 507–508, 528
Isotropic fluid, 29, 65
Isotropic turbulence, 405

Jeffery–Hamel flow, 161–165
  separation criterion, 165
Jet flow
  laminar, two-dimensional, 253–256, 336
    axisymmetric, 212, 297–299
  stability of, 360–361
  turbulent, 475–479
Junction vortex, 317

K–ϵ modeling, 449–452, 473
Kármán integral relation (*see* Momentum integral relation)
Kármán viscous pump, 158, 211
Kármán vortex street, 9–10
Kármán–Pohlhausen method, 530
Kármán–Schoenherr drag formula, 555
Kelvin, theorem of, 86

Kelvin–Helmholtz instability, 340–344, 400
  breaking waves, 344
Kinematic properties, 16–22
Kinematic viscosity, 32, 79
  of various fluids, 574
Kinetic energy, 69
  thickness, 561
  turbulent mean-flow equation, 410
Kinetic Reynolds number, 127
Kinetic theory of gases
  for binary diffusion, 33
  for perfect gases, 39
  for slip velocity, 47, 213
  for temperature jump, 48
  for thermal conductivity, 30
  for viscosity, 26–28
King's law, 180
Knudsen number, 46, 83
Kolmogorov length scale, 400
Kronecker delta function, 65

Lag, in boundary layers, 460–461
Lagrange, theorem of, 86
Lagrangian coordinates, 16
Laminar boundary-layer equations, 226–227
  compressible flow, 506
  in free convection, 323
  three-dimensional, 308–309
  unsteady flow, 227
Laminar flow theory
  for compressible boundary layers, 505–531
  exact solutions, 97–98
  for incompressible boundary layers, 215–328
Laplace's equation, 78, 79, 166
Laplacian, finite-difference approximation, 187
Large-eddy simulation, 13–14, 441
Law of the wake, 421–424, 455
  application to skin friction, 424
  for compressible flow, 550
  integral theory application, 455, 456, 458
Law of the wall, 418–419
  for compressible flow, 548–549, 551–552

  with suction or blowing, 439–440, 491–493
  for temperature profiles, 487–489
Lewis number, 32, 79
Lift coefficient of an airfoil, 6
Liquid film draining, 207, 208
Literature explosion, 4, 184–185
Logarithmic law, turbulent, 416, 549
  applications, 425–433
  roughness effect, 430
  with suction or blowing, 439–440, 491–493
Logistic equation, 393
Log-mean temperature difference, 123
Long-cylinder parameter, 296
Low-density limit for transport properties, 26–27, 30, 33
Lubrication theory, 180–183
Ludwieg–Tillmann shear stress correlation, 455

Mach number, 47, 82, 363, 507
Magnetohydrodynamics, 60
Mangler transformation, 293, 539
Marangoni convection, 51
Marching methods, 184, 227, 272, 546
Mass-diffusivity coefficient, 32
Matched asymptotic expansions, 304–305
Mean free path, 26, 46
Mechanical energy integral relation, 264
  turbulent flow, 414
Mechanical pressure, 67
Metric stretching factors, 75, 308
Microflows, 110, 202
Micropump, 202–203
Mixing layer (*see* Free-shear flows)
Mixing length, turbulent, 442–443, 496
Models of turbulence (*see* Turbulence modeling)
Molecular theory, Stockmayer potential, 26–27
  transport parameters for gases, 27, 580
Molecular weight, 39
  of mixtures of gases, 39–40
  of various gases, 576
Mollier chart, 37
Moment coefficient for rotating disk, 160

624 VISCOUS FLUID FLOW

Momentum equation for turbulence, 408
Momentum integral relation
   for laminar flows, 218, 262–263, 529
   for turbulent flows, 455, 554, 560–563
Momentum thickness
   compressible form, 529
   definition, 218, 529
   for Falkner–Skan flows, 243
   on a flat plate, 220, 235, 433
   transition correlation, 379, 381, 385, 386
   in a turbulent boundary layer, 413, 433
Moody chart, 432
Morkovin's hypothesis, 544–545, 548
Moving-boundary flows, 129–134
   oscillating plate, 131–133
   suddenly accelerated plate, 129–131
   between two plates, 133–135
MRS separation criterion, 319

Natural convection (*see* Free convection)
Navier–Stokes equations, 67–68, 86
   dimensionless form, 81
Neutral curves, 351, 356
Newtonian fluids, 2, 23, 65
Newtonian hypersonic impact theory, 523–524
Newton's second law, 62
Noncircular ducts, 112–118, 428–429
Noninertial coordinates, 88–89
Nonnewtonian fluid, 24–25
Non-uniqueness problems, 140, 163, 244
Normal modes, 342
Normal pressure gradient effect, 229
No-slip condition, 8, 23, 45, 54, 74
   for turbulent flow, 47, 412
No-temperature-jump condition, 45, 74, 412
Numerical stability, 188, 274, 282, 534
Nusselt number, definition, 83, 101
   in creeping motion, 178–180
   for cylinders and spheres, 179
   for duct flows, 118, 119, 124
   in Falkner–Skan flow, 245–246
   on a flat plate, 224, 239
   in free convection, 324–326
   in general laminar boundary layers, 281
   in stagnation flow, 154

One-dimensional flow approximation, 91
One-equation model, 448–449
One-seventh power law, 433
Open-channel flow, 51
Orifice flow, 200–202
Orr–Sommerfeld equation, 347, 350
   numerical solution, 351
   parametric effects, 357–368
Orthogonal coordinate systems, 75–77, 228, 308–309
   cylindrical polar, 76, 582
   spherical polar, 76–77, 583
Oscillatory motion, 127–129, 131–133
   transition in, 389–391
Oseen, theory of, 174–176
Oseenlet, 213
Outer layer, turbulent, 418, 419–421
Overlap layer, turbulent, 416
   power-law approximation, 424, 433
   roughness effect, 430

Parabolic differential equation, 77–78, 227, 272, 546
Parabolic flow on a flat plate, 311–314
Parallel stream interaction, 251–253
Peclet number, 82, 120
Perfect fluid theory (*see* Potential flow)
Perfect gas law, 39–40, 507, 546
   applications, 507, 513, 546
Pipe flow (*see* Circular pipe flow)
Pohlhausen–Gruschwitz polynomials, 530–531
Point-sink flow, 249–250
Poiseuille flow, in a pipe, 108–110
   between parallel plates, 110–112
   stability analysis, 357–358
   with wall slip, 110, 213
Poiseuille number, 100, 110
   for various ducts, 117
Poisson equation, 107, 192, 347
Porous ducts (*see* Suction flows)
Porous media flows, 49
Potential energy, 69, 71, 91
Potential flow analysis, 6, 8, 241
   for a cylinder in cross flow, 8–9, 269
   for Falkner–Skan wedge flow, 241
   for a sphere, 169
Potential vortex, 104

Power-law approximation
  for cone flow, 291, 560–561
  for equilibrium flow, 420, 465
  for free-shear flows, 485
  for nonnewtonian fluids, 24–25, 205
  for overlap velocity profiles, 424–425
  for thermal conductivity, 30
  for turbulent flat-plate flow, 433
  for viscosity, 27–28, 513, 534, 577
  for wedge flows, 241
Prandtl number, 30, 79, 82, 508
  of gases, 42
  of steam and water, 580
  turbulent definition, 547
  of various fluids, 80
Pressure drop in ducts, laminar, 287–289
  turbulent, 425–433
Pressure jump at an interface, 50, 56
Pressure-gradient parameter, 416, 418, 446, 551
Principal axes for strains and stresses, 21, 66
Prolate spheroid flow, 13–15, 171–172
Properties of a fluid, 15–44
  kinematic, 16–22
  surface tension, 50–51
  thermodynamic, 36–44
  transport, 22–35
Pseudoplastic fluid, 24
Pseudoresonance, 359

Radius of curvature, 50, 228
Rayleigh number, 325
Rayleigh stability theorems, 348–349
Reattachment, 470–471
Receptivity of boundary layers, 378, 391
Recovery factor, laminar, 510, 515–516, 521, 538
  turbulent, 547, 559–560
Recovery temperature, 510, 515
Reduced properties, 25–26, 29, 33
Reference-temperature concept, 517
Reflection, coefficient of, 46
Relaminarization, 472–473
Relaxing flows, 460
Reversibility of creeping flows, 174, 182
Reynolds analogy, 48, 154, 247, 486, 521, 540, 555
  compressible flow, 510, 555

  effect of pressure gradient, 247
  laminar flat-plate flow, 247
  turbulent flat-plate flow, 489, 555
Reynolds dye experiment, 371–372
Reynolds equations of turbulence, 407–409
Reynolds lubrication equation, 182, 183
Reynolds number, 5, 82, 100, 225
Reynolds rules of averaging, 407
Reynolds stresses, 408, 410
  basic equation, 410
  modeling, 452–454, 462
Reynolds transport theorem, 90
Rheology, 15
Richardson's annular effect, 128
Rossby number, 89
Rotating cylinders, 103–105, 368–370
  disk, 155–161
Rotationally symmetric flow, 290–300
Roughness effects
  of commercial pipes, 432
  at high Mach numbers, 389
  on transition, 387–389
  in turbulent flow, 429–433, 437–438
Round jet, laminar, 212, 297–299
  turbulent, 477–479
Rules of time-averaging, 402, 407
Runge–Kutta subroutines, 585–589

Saddle-point separation, 317
Sand-grain roughness, 431
Schmidt number, 32, 79
Second coefficient of viscosity, 23, 66
Second law of thermodynamics, 36
Secondary flow, 160–161, 310–311
Second-moment closure model, 441, 452
Self-diffusion, 31, 33
Self-preserving free-shear flows, 474–475
Separation, concept of, 6, 9, 111, 229–230
  in corner flow, 173
  in Couette–Poiseuille flow, 111
  in creeping flow, 173
  on a cylinder in cross flow, 9–10, 197–198, 270
  in a diffuser, 12–13
  in Falkner–Skan flow, 242

626 VISCOUS FLUID FLOW

Separation, concept of—*Cont.*
  in Jeffery–Hamel flow, 165
  in linearly retarded flow, 258, 269, 275
  prediction criteria
    laminar flow, 111, 165, 267, 293, 531
    turbulent flow, 455, 458, 563
    reattachment, 470–471
    singular behavior of, 258–259
    on a sphere, 199, 294–295
    three-dimensional, 6–8, 315–318
Series solution for laminar boundary layers, 258
Shape factor, 219, 244, 265, 359, 455
  of Clauser, 420
  effect on stability, 359–360, 381
  effect on transition, 381
  for entrainment, 413, 457
  in laminar flow, 220, 529
  Thwaites' correlation, 267–268
  in turbulent flow, 413, 455, 561, 562
Shear layer (*see* Free-shear flows)
Shear stress in a boundary layer, 230, 309
  Thwaites' correlation, 267
  turbulent, 446
Shock-wave boundary-layer interaction, 540–543
Similarity solutions
  basic concept, 18, 144, 230, 239
  for compressible flow, 513–514, 525–528
  for Falkner–Skan flows, 239–247, 330
  for free convection, 324
  for laminar boundary layers, 231, 240, 250, 254, 312
  of the Navier–Stokes equations, 145–165
    Jeffery–Hamel flow, 161–165
    rotating disk, 155–161
    stagnation flow, 145–155
  point-sink flow, 250
  for Stokes' moving-wall problems, 130, 133
  for turbulent jets, 476
Similarity transformation, 511
SIMPLE algorithm, 196
Singular perturbation problems, 301
Skewed boundary layers, 310–311

Skin-friction coefficient, 47, 100
  definition, 47, 219
  in duct flow, laminar, 109, 116, 288
    turbulent, 426
  on a flat plate, laminar, 220, 235, 517
    turbulent, 433–434, 438, 555, 558
  on a long cylinder, 297
  in stagnation flow, 150
  in supersonic flow, 517
  in a turbulent boundary layer, 424, 455–456
Slip flow in tubes, 110
Slip length, 45
Slip velocity, 46–47
Small-disturbance stability concept, 339
Specific heats, definition, 38
  ratio, 38, 42, 83
  of various gases, 38, 576
Spectral numerical methods, 184
Speed of sound, 39, 42, 507
Sphere flow, 168–171, 177, 178, 523
  finite-difference solution, 198–200
  for a laminar boundary layer, 294–295
  stagnation point velocity gradient, 146, 522–524
Spherical polar coordinates, 583–584
Squire's theorem, 346, 365
Stability of laminar flows, 337–370
  basic concept, 337–339
  of Couette flows, 105–106, 357–359, 368–370
  inviscid theory, 348–350, 368, 395
  of numerical methods, 188, 274, 282, 534
  of parallel flows, 344–357
  of Poiseuille flows, 357–359
  of rotating-disk flow, 160–161, 370
  temporal versus spatial, 348, 350
  of viscous flows
    basic equations, 350–351
    Blasius flow, 352–353
    centrifugal effects, 368–370
    compliant boundaries, 366–367
    compressibility effects, 362–366
    experiments, 355–357
    of Falkner–Skan flow, 353–354
    of free-convection flows, 367–368

of free-shear flows, 360–361
  heating and cooling effects, 361–362
  Orr–Sommerfeld equation, 347,
    350, 362
  parametric effects, 357–370
Stagnation enthalpy, 222, 263, 309,
    413, 506
Stagnation temperature, 507
Stagnation-point flow, 145–155, 521–525
  axisymmetric, 150–151, 522
  compressible form, 521–525
  freestream velocity gradient, 146,
    522–524
  plane, 145–150, 522
  temperature distributions, 151–155
Stall of an airfoil, 5–8
Stanford conference, in 1968, 414, 443,
    448, 456
  in 1980–81, 414, 452, 462, 565
Stanton number, 48, 155, 223, 489, 490
  for compressible flow, 521, 555
  with suction and blowing, 248, 491
Start-up, of pipe flow, 125–127
  of cylinder flow, 318–319
  transition of, 389–391
Stationary turbulence, 403
Steady-flow energy equation, 91–92
Stokes' first and second problems, 130–133
Stokes flow (see Creeping flow)
Stokes' hypothesis, 67, 72
Stokes' paradox, 167–168
Stokes' stream function, 168, 211
Strain-rate tensor, 21
  in curvilinear coordinates, 582, 584
Streaklines, 9–10
Stream function, 61–62, 86–88, 138, 145,
    168, 211
  axisymmetric, 93
  for Blasius flow, 231
  compressible form, 62, 556
  of Stokes, 168, 211
Streamlines, 8, 62
Stress tensor, 63
  turbulent inertia form, 408
Strouhal number, 10
  for a circular cylinder, 11
Sublayer, viscous, 419, 548
Substantial derivative, 17

Suction flows
  on a flat plate, 135–136, 248–249,
    439–440
  developing, 259–261
  between parallel plates, 136–140
  in a porous pipe, 140, 210
  stability of, 359–360
  in turbulent flow, 439–440
Superlayer, 404, 405
Surface forces, 63
Surface tension, 50–51
  of various fluids in air, 576
  of water in air, 51, 575
Sutherland law approximation
  for thermal conductivity, 30
  for viscosity, 28, 577
Swept wings, 6–8, 314–315
Swirl flow, 206, 290

Taylor number, 106, 369
  vortices, 209, 369–370
TDMA, 188, 276–278, 282, 535
Temperature gradient, 29
  jump, 48
  law of the wall, 487–489
Tensor invariants, 21
Thermal boundary layer
  definition, 152
  laminar flows, 237–239, 278–287
  thickness, 152, 238
  turbulent flows, 489
Thermal conductivity, 28–31
  of gases, 29–31, 572
  of liquids, 573
  of mixtures of gases, 34
  of steam and water, 571, 579
Thermal decoupling, 68, 97, 411
Thermal diffusivity, 32, 79
Thermal entrance in a duct, 120–125
Thermal expansion coefficient, 43, 83, 322
  of steam and water, 44
Thermal-energy integral relation, 91,
    263–264, 413
  compressible form, 547
  for turbulent flow, 413
Thermodynamic equilibrium, 16
Thermodynamic pressure, 67

Thermodynamic properties, 36–44
Three-dimensional boundary layers
  laminar, 307–318
    basic equations, 308–309
    corner flow, 315–316
    flat-plate flow, 311–314
    separation concepts, 7–8, 315–318
    yawed infinite cylinder, 314–315
  turbulent, 469–470
Thwaites, integral method of, 265–271
  application to transition, 379, 381–382
  for axisymmetric flow, 293–294
Time-averaging, 401–403, 406–408
Timelines, 9–10
Tollmien–Schlichting waves, 345, 353, 371–372, 400
Total enthalpy (*see* Stagnation enthalpy)
Transition to turbulence, 105, 110, 370–378
  control of, 391–392
  on a flat plate or cone, 376–377, 379, 559
  freestream turbulence effects, 383–387
  prediction methods, 378–392
  pressure-gradient effects, 379–383, 386
  Reynolds dye experiment, 371–372
  roughness effects, 387–389
  shape factor effects, 381
  spanwise vorticity, 372–374
  turbulent spots, 371, 374–377
  unsteady flow, 389–391
Transpiration, 49, 439–440
Transport properties, 22–35, 60
Transverse-curvature effect, 296
Traveling-wave disturbance, 342, 345
Tridiagonal matrix algorithm, 276–278, 282, 535
Trip wire, 388
Triple-deck theory, 306–307
Turbulence kinetic energy, 409
  basic equation, 410, 412
  modeling, 448–452
Turbulence modeling, 440–454
  for boundary layers, 445
  direct numerical simulation, 401, 441, 464–465, 482
  eddy conductivity, 485–487, 546
  eddy viscosity, 441–447, 449

higher order models, 447–454
K–ϵ model, 449–452
  at low Reynolds numbers, 451–452
  mixing length, 442–443, 496
Turbulent boundary layers, 433–440, 454–469
  with pressure gradient, 454–461
Turbulent dissipation, 409, 410, 412, 449
Turbulent flows
  basic equations, 407–411
    boundary layer, 411, 433, 544
  channel flow, 427–428
  complex, 469–472
  compressible, 544–546, 553–566
  convective heat transfer, 485–498, 555
  on a flat plate, 399, 404, 433–438, 445, 553–558
  integral relations, 413–414, 433
  jets and wakes, 473–485
  mixing layer, 479–482
  pipe flow, 425–427
  roughness effects, 429–432, 437–438
  three-dimensional, 469–470
  unsteady, 407
Turbulent fluctuations, 400, 402, 404, 478, 480, 484
  compressible, 544
  frequency spectra, 405–406
  hot-wire measurements, 404
  statistical theory, 403
  time-averaging, 401–403, 406–408
Turbulent heat flux, 409
Turbulent inertia (*see* Reynolds stresses)
Turbulent Prandtl number, 449, 486–487, 547
  for kinetic energy and dissipation, 449
Turbulent shear stress, 403, 408, 411, 413
Turbulent-energy methods, 448–452
Two-equation models, 449–452, 462

Uncoupling of temperature, 68, 97, 411
Unfavorable gradient (*see* Adverse pressure gradient)
Unheated starting length, 224, 493–494
Unit Reynolds number effect, 371
Unsteady boundary layers, 318–321
  transition of, 389–391

Unsteady Couette flows, 129–135
Upwind differencing, 193
U-tube, flow in, 206

Van der Pol equation, 395
Van Driest, damping factor, 443, 447, 496
  effective velocity, 549
  flat-plate theory, 554–556
  transformation, 548
Vapor pressure of water, 44, 575
  of various fluids, 576
Variable-wall-temperature solutions, 284–287, 493–495
Velocity defect, 416, 418, 420
Velocity potential, 85, 93
Velocity profile, definition, 99, 216
  for turbulent flow, 414
Velocity vector, 16
Viscometry, 57, 105, 114
Viscosity, 22–28
  of air and water, 571, 575
  coefficient definition, 23, 66
  of gases, 26, 572–573, 576
  of liquids, 26, 34–35, 574, 576
  of mixtures of gases, 34
  second coefficient of, 23, 66
  of steam and water, 571–573, 578
Viscous sublayer, length scale, 419–420
Visualization of flow, 398–400
Vortex breakdown, 374, 376
Vortex, decay, 207, 209
  methods, 75, 184
  shedding, 10–11
  stretching, 86
Vortex sheet, 340
Vortex street, 9–10

Vorticity, 20, 79, 85
  basic equation, 79, 86
  in cylindrical coordinates, 583

Wake flow
  Coles parameter, 421, 456
  laminar, two-dimensional, 256–257
    axisymmetric, 299–300
  stability of, 360–361
  turbulent, 481–485
Wall functions, 451–452
Wall jet, laminar, 244, 336
Wall Reynolds number, 139
Wall transpiration, 439–440
Wall-conduction temperature, 487
Wall-friction velocity, 416, 435
Wall-wake law, 421, 423, 434, 458, 548–559
Walz, method of, 561–563
Wave cancellation control, 392
Wave equation, 78
Weber number, 84
Wedge flows, 241–247
Wind stress on the ocean, 141
Wind-driven flows, 141–144
  Ekman spiral flow, 142–144
  penetration depth, 143, 211
  starting flow, 141–142
Work, definition, 36, 69, 91

Yawed cylinder flow, 314–315
Yielding fluid, 24

Zwarts, supersonic experiment of, 564–566